EXPLORING THE UNKNOWN

ISBN 0-16-049546-6

NASA SP-4407

EXPLORING THE UNKNOWN

Selected Documents in the History of the
U.S. Civilian Space Program
Volume III: Using Space

John M. Logsdon, Editor
with Roger D. Launius, David H. Onkst
and Stephen J. Garber

The NASA History Series

National Aeronautics and Space Administration
NASA History Division
Office of Policy and Plans
Washington, D.C.   1998

Library of Congress Cataloguing-in-Publication Data

Exploring the Unknown: Selected Documents in the History of the U.S. Civil Space Program/John M. Logsdon, editor ... [et al.]
p. cm.—(The NASA history series) (NASA SP: 4407)

Includes bibliographical references and indexes.
Contents: v. 1. Organizing for exploration
1. Astronautics—United States—History.   I. Logsdon, John M., 1937–
                          II. Series.   IV. Series: NASA SP: 4407.
TL789.8.U5E87   1998   96-9066
387.8'0973–dc20   CIP

For sale by the U.S. Government Printing Office
Superintendent of Documents, Mail Stop: SSOP, Washington, DC 20402-9328
ISBN 0-16-049546-6

Dedicated to
Arthur C. Clarke,
William Pecora,
and Leonard Jaffe

Pioneers in the Use of Space for Benefits on Earth

Contents

Acknowledgments xix

Introduction xxi

Biographies of Volume III Essay Authors and Editors xxv

Glossary xxvii

Chapter One

Essay: "The History of Satellite Communications," by Joseph N. Pelton 1

Documents

I-1 and I-2 Arthur C. Clarke, "The Space-Station: Its Radio Applications,"
May 25, 1945; and Arthur C. Clarke, "Extra-Terrestrial Relays:
Can Rocket Stations Give World-Wide Radio Coverage?,"
Wireless World, October 1945, pp. 305–308 11

I-3 John R. Pierce, "Exotic Radio Communications," *Bell Laboratories
Records*, September 1959, pp. 323–329 22

I-4 Memorandum from S. G. Lutz to A.V. Haeff, "Commercial Satellite
Communication Project; Preliminary Report of Study Task Force,"
October 22, 1959 31

I-5 H.A. Rosen and D.D. Williams, *Commercial Communications Satellite*,
Report RDL/B-1, Engineering Division, Hughes Aircraft Company,
January 1960 35

I-6 "Memorandum for Conference on Communications Satellite
Development," December 7, 1960 39

I-7 White House Press Secretary, "Statement by the President,"
December 30, 1960 41

I-8 Federal Communications Commission, "FCC Relation to Space
Communication," Public Notice-G, 1627, March 14, 1961 42

I-9 and I-10 F.R. Kappel, President, American Telephone and Telegraph
Company, to the Honorable James E. Webb, Administrator,
NASA, April 5, 1961 (with several attachments); and
James E. Webb, Administrator, to F.R. Kappel, President,
American Telephone and Telegraph Company, April 8, 1961 45

I-11   President John F. Kennedy to Honorable Newton Minow, Chairman,
             Federal Communications Commission, May 15, 1961 60

I-12   Ben F. Waple, Acting Secretary, Federal Communications
             Commission, "An Inquiry Into the Administrative and Regulatory
             Problems Relating to the Authorization of Commercially Operable
             Space Communications Systems: First Report," FCC Report 61-676,
             4774, Docket No. 14024, May 24, 1961 61

I-13   National Aeronautics and Space Council, "Communication
             Satellites," July 14, 1961 65

I-14   Emanuel Celler, Chairman, Committee on the Judiciary, House of
             Representatives, et al., to the President, August 24, 1961 67

I-15   Frederick G. Dutton, Assistant to the President, Memorandum for
             the President, November 13, 1961 71

I-16–I-18   Senator Robert S. Kerr, "Amendment to the National Aeronautics
             and Space Act of 1958, Space Communications," November 28, 1961;
             E.C. Welsh, Executive Secretary, National Aeronautics and Space
             Council, Executive Office of the President, Memorandum to the
             President, April 11, 1962; and "Communications Satellite Act of 1962,"
             Public Law 87-624, 76 Stat. 419, signed by the President on
             August 31, 1962. 72

I-19   Edward A. Bolster, Department of State, to Mr. Johnson,
             Memorandum, "Space Communication," May 3, 1962; with attached:
             "Role of the Department of State in Space Communication
             Development" 85

I-20   Project Telstar, "Preliminary Report, Telstar I, July–September 1962,"
             Bell Telephone Laboratories, Inc., 1962. 89

I-21   Memorandum from J.D. O'Connell, Special Assistant to the President
             for Telecommunications and Director of the Office of Telecommuni-
             cations Management, to the Secretary of State, Secretary of Defense,
             Secretary of Commerce, Administrator, National Aeronautics and
             Space Administration, and Chairman, Federal Communications
             Commission, "Policy Concerning U.S. Assistance in the Development
             of Foreign Communications Satellite Capabilities," September 17,
             1965, with attached: National Security Action Memorandum 338 91

I-22   National Security Action Memorandum No. 342, "U.S. Assistance in
             the Early Establishment of Communications Satellite Service for
             Less-Developed Nations," March 4, 1966 95

I-23   David Bruce, U.S. Ambassador to the United Kingdom, to the
            Secretary of State, "Transfer of U.S. Communications Satellite
            Technology," Telegraphic Message, November 9, 1966 96

I-24   Memorandum from J.D. O'Connell for the President,
            February 8, 1967, with attached: "A Global System of Satellite
            Communications: The Hazards Ahead," February 8, 1967. 99

I-25   Leonard H. Marks, Ambassador, Chairman, "Report of the United
            States Delegation to the Plenipotentiary Conference on Definitive
            Arrangements for the International Telecommunications Satellite
            Consortium (First Session), Washington, D.C., February 24–
            March 21, 1969," April 10, 1969 108

I-26   "Second Report and Order in the Matter of Establishment of
            Domestic Communications-Satellite Facilities by Non-Governmental
            Entities," Docket No. 16495, June 16, 1972 120

1-27   George M. Low, Deputy Administrator, NASA, "Personal Notes,"
            December 23, 1972 132

I-28   Committee on Satellite Communications, Space Applications Board,
            Assembly of Engineering, National Research Council, "Federal
            Research and Development for Satellite Communications," 1977 135

I-29   John J. Madison, Legislative Affairs Specialist, NASA, Memorandum
            for the Record, "Advanced Communications Technology Satellite
            (ACTS) program meeting, October 13, 1983" 145

I-30   William Schneider, Under Secretary of State for Security Assistance,
            Science, and Technology, and David J. Markey, Assistant Secretary of
            Commerce for Communications and Information, "A White Paper
            on New International Satellite Systems," Senior Interagency Group
            on International Communication and Information Policy,
            February 1985 147

Chapter Two

**Essay: "Observing the Earth From Space," by Pamela E. Mack
and Ray A. Williamson** 155

Documents

II-1   Dr. Harry Wexler, "Observing the Weather from a Satellite Vehicle,"
            Journal of the British Interplanetary Society 7 (September 1954): 269–276. . 177

II-2   S.M. Greenfield and W.W. Kellog, "Inquiry into the Feasibility of
            Weather Reconnaissance from a Satellite Vehicle," The RAND
            Corporation, R-365, August 1960, pp. v–vi, 1–23, 31 183

II-3   Hugh L. Dryden, for T. Keith Glennan, NASA, and Roy W. Johnson,
            Department of Defense, "Agreement Between the Department of
            Defense and the National Aeronautics and Space Administration
            Regarding the TIROS Meteorological Satellite Project," April 13, 1959. 203

II-4   U.S. Department of Commerce, Weather Bureau, "National Plan for
            a Common System of Meteorological Satellites," Technical Planning
            Study No. 3, Preliminary Draft, October 1960, pp. 1–3 204

II-5   Hugh L. Dryden, Deputy Administrator, for James E. Webb,
            Administrator, NASA, and Luther H. Hodges, Secretary of
            Commerce, "Basic Agreement Between U.S. Department of
            Commerce and the National Aeronautics and Space Administration
            Concerning Operational Meteorological Satellite Systems,"
            January 30, 1964 206

II-6   Robert M. White, Administrator, Environmental Science Services
            Administration, National Environmental Satellite Center,
            U.S. Department of Commerce, to Dr. Homer E. Newell, Associate
            Administrator for Space Science and Applications, NASA,
            August 15, 1966 211

II-7 and II-8   George E. Brown, Jr., Chairman, Committee on Science, Space, and
            Technology, U.S. House of Representatives, to D. James Baker,
            Acting Under Secretary for Oceans and Atmosphere, U.S. Department
            of Commerce, February 22, 1993; and Jim Exon, Chairman,
            Subcommittee on Nuclear Deterrence, Arms Control and Defense
            Intelligence, U.S. Senate, to Ron Brown, Secretary of Commerce,
            June 2, 1993. 213

II-9   National Performance Review, Department of Commerce,
            "Establish a Single Civilian Operational Environmental Polar
            Satellite Program," September 30, 1993 216

II-10   Presidential Decision Directive/NSTC-2, The White House,
            "Convergence of U.S. Polar-orbiting Operational Environmental
            Satellite Systems," May 5, 1994 221

II-11 and   D. James Baker, Under Secretary for Oceans and Atmosphere,
II-12   U.S. Department of Commerce, to John Morgan, Director,
            EUMETSAT, May 6, 1994; and D. James Baker, Under Secretary
            for Oceans and Atmosphere, U.S. Department of Commerce, to
            Jean-Marie Luton, Director, European Space Agency, May 6, 1994 . . . 224

II-13   Peter C. Badgley, Program Chief, Natural Resources, NASA,
             "Current Status of NASA's Natural Resources Program," *Proceedings*
             of the Fourth Symposium on Remote Sensing of Environment held 12, 13,
             14, April 1966 (Ann Arbor, MI: University of Michigan, 1966),
             pp. 547–558. 226

II-14   "Prepared by Jaffe and Badgley at Seamans' Request:
             NASA Natural Resources Program," May 13, 1966 237

II-15   Leonard Jaffe, Director, Space Applications Programs, OSSA, to
             Deputy Administrator, thru Homer S. Newell, Associate
             Administrator for Space Science and Applications, "Meeting at
             the U.S. Geological Survey (USGS), August 25, 1966, regarding
             Remote Sensing and South America," August 31, 1966, with
             attached: Robert G. Reeves, For the Record, "Meeting at the U.S.
             Geological Survey (USGS), 10 a.m., August 25, 1966,"
             August 31, 1966. 240

II-16   Office of the Secretary, U.S. Department of the Interior,
             "Earth's Resources to be Studied from Space," News Release,
             September 21, 1966. 244

II-17   Charles F. Luce, Under Secretary, U.S. Department of the Interior,
             to Dr. Robert C. Seamans, Jr., Deputy Administrator, NASA,
             October 21, 1966, with attached: "Operational requirements for
             global resource surveys by earth-orbital satellites: EROS Program" . . . 246

II-18   Irwin P. Halpern, Director, Policy Staff, NASA, Memorandum for
             General Smart, "Earth Resources Survey Program,"
             September 5, 1967. 248

II-19   Jacob E. Smart, Assistant Administrator for Policy, NASA,
             Memorandum for Dr. Mueller, *et al.,* "Earth Resources Survey
             Program," October 3, 1967, with attached: Draft Memorandum
             for Mr. Webb, Dr. Seamans, Dr. Newell, "Issues Re: The Earth
             Resources Survey Program" 250

II-20   Edgar M. Cortright for George E. Mueller, Associate Administrator
             for Manned Space Flight, Memorandum to Assistant Administrator
             for Policy, "Earth Resources Study Program," November 17, 1967. . . . 253

II-21   Interior Department, "Appeal of 1971 Budget Allowance: EROS,"
             November 25, 1969 256

II-22   Robert P. Mayo, Director, Bureau of the Budget, to
             Honorable Walter J. Hickel, Secretary of the Interior,
             April 14, 1970, with attached: "Statement for Senator Mundt" 257

II-23 Arnold W. Frutkin, Memorandum to Dr. Fletcher, Administrator,
NASA, *et al.,* "Some Recent International Reactions to ERTS-1,"
December 22, 1972 259

II-24 James V. Zimmerman for Arnold W. Frutkin, Assistant Administrator
for International Affairs, to Dr. John V.N. Granger, Acting Director,
Bureau of International Scientific and Technological Affairs,
Department of State, September 12, 1974, with attached:
"Foreign Policy Issues Regarding Earth Resource Surveying
by Satellite: A Report of the Secretary's Advisory Committee
on Science and Foreign Affairs," July 24, 1974 262

II-25 Clinton P. Anderson, Chairman, Senate Committee on
Aeronautical and Space Sciences, U.S. Senate, to Dr. James C.
Fletcher, Administrator, NASA, October 14, 1972 269

II-26 Walter C. Shupe, Chief, GAO Liaison Activities, NASA,
Memorandum to Distribution, "GAO Report to Congress 'Crop
Forecasting by Satellite: Progress and Problems,' B-183184,
April 7, 1978," April 21, 1978 272

II-27 Charles J. Robinove, Director, EROS Program Office, Geological
Survey, U.S. Department of the Interior, Memorandum to Staff
of the EROS Program, "Optimism vs. pessimism *or* where do
we go from here? (some personal views)," December 10, 1975 275

II-28 James C. Fletcher, Administrator, NASA, to Mr. John C. Sawhill,
Associate Director, Office of Management and Budget,
October 19, 1973. 277

II-29 Christopher C. Kraft, Jr., Director, Johnson Space Center, to
Associate Administrator for Applications, NASA Headquarters,
"Private Sector Operation of Landsat Satellites," March 12, 1976 281

II-30 Bruno Augenstein, Willis H. Shapley, and Eugene B. Skolnikoff,
"Earth Information From Space by Remote Sensing," report
prepared for Dr. Frank Press, Director, Office of Science and
Technology Policy, June 2, 1978, pp. ii–iv, 1–14 282

II-31 Zbigniew Brzezinski, The White House, Presidential Directive/
NSC-54, "Civil Operational Remote Sensing," November 16, 1979 . . . 294

II-32 David S. Johnson, Chairman, Satellite Task Force, *Planning for a
Civil Operational Land Remote Sensing Satellite System: A Discussion of
Issues and Options* (Rockville, MD: U.S. Department of Commerce,
National Oceanic and Atmospheric Administration, June 20, 1980),
pp. 1–16. 296

II-33   Ed Harper, Office of Management and Budget, Memorandum to
             Craig Fuller/Martin Anderson, "Resolution of Issues Related to
             Private Sector Transfer of Civil Land Observing Satellite Activities,"
             July 13, 1981 306

II-34   Government Technical Review Panel, "Report of the Government
             Technical Review Panel on Industry Responses on
             Commercialization of the Civil Remote Sensing Systems,"
             November 10, 1982, pp. 1–25 309

II-35   "Transfer of Civil Meteorological Satellites," House Concurrent
             Resolution 168, November 14, 1983 321

II-36   "Land Remote-Sensing Commercialization Act of 1984,"
             Public Law 98–365, 365, 98 Stat. 451, July 17, 1984 329

II-37   Office of the Press Secretary, The White House, "Statement by the
             Press Secretary," June 1, 1989 344

II-38   Office of the Press Secretary, The Vice President's Office,
             "Vice President Announces Landsat Policy," February 13, 1992,
             with attached: "Landsat Remote Sensing Policy" 345

II-39   Department of Defense and National Aeronautics and Space
             Administration, "Management Plan for the Landsat Program,"
             March 10, 1992 347

II-40   "Land Remote-Sensing Policy Act of 1992," Public Law 102–555,
             106 Stat. 4163, October 28, 1992 352

II-41–II-44 George E. Brown, Jr., Chairman, Committee on Science,
             Space, and Technology, U.S. House of Representatives, to John H.
             Gibbons, Assistant to the President, Office of Science and Technology
             Policy, August 9, 1993; John Deutch, Under Secretary of Defense, to
             George E. Brown, Jr., Chairman, Committee on Science, Space, and
             Technology, U.S. House of Representatives, December 9, 1993;
             John H. Gibbons, Director, Office of Science and Technology Policy,
             to George E. Brown, Jr., Chairman, Committee on Science, Space,
             and Technology, U.S. House of Representatives, December 10, 1993;
             and George E. Brown, Jr., Chairman, Committee on Science,
             Space, and Technology, U.S. House of Representatives, to John H.
             Gibbons, Assistant to the President, Office of Science and
             Technology Policy, December 14, 1993. 368

II-45   The White House, Presidential Decision Directive/NSTC-3, "Landsat
             Remote Sensing Strategy," May 5, 1994. 372

II-46–II-48   Gregory W. Withee, Acting Assistant Administrator for Satellite
               and Information Services, NOAA, to Walter S. Scott, President
               and Chief Executive Officer, World View Imaging Corporation,
               January 4, 1993; Duane P. Andrews, Assistant Secretary of Defense,
               to Gregory W. Withee, Acting Assistant Administrator for Satellite
               and Information Services, NOAA, December 24, 1992; and Ralph
               Braibanti, Deputy Director, Office of Advanced Technology,
               Bureau of Oceans and International Environmental and Scientific
               Affairs, U.S. Department of State, to Michael Mignogno, Chief,
               Landsat Commercialization Division, NOAA, October 19, 1992 375

II-49   Office of the Press Secretary, The White House, "U.S. Policy on
               Licensing and Operation of Private Remote Sensing Systems,"
               March 10, 1994 379

II-50   Robert S. Winokur, Assistant Administrator for Satellite and
               Information Services, NOAA, to Albert E. Smith, Vice President,
               Advanced Government and Commercial Systems, Lockheed
               Missile and Space Company, Inc., April 22, 1994 381

Chapter Three

Essay: "Space as an Investment in Economic Growth," by Henry R. Hertzfeld 385

Documents

III-1   Jack G. Faucett, President, Jack Faucett Associates, Inc., to
               Willis H. Shapley, Associate Deputy Administrator, NASA,
               November 22, 1965, with attachment omitted 401

III-2   Roger W. Hough, "Some Major Impacts of the National Space
               Program," Stanford Research Institute, Contract NASW-1722,
               June, 1968, pp. 1–2, 19–22, 36 402

III-3   "Economic Impact of Stimulated Technological Activity,"
               Final Report, Midwest Research Institute, Contract
               NASW-2030, October 15, 1971, pp. 1–11 408

III-4   Michael K. Evans, "The Economic Impact of NASA R&D Spending,"
               Executive Summary, Chase Econometric Associates, Inc.,
               Bala Cynwyd, Pennsylvania, Contract NASW-2741, April 1976,
               pp. i–iii, 1–18. 414

III-5   Robert D. Shriner, Director of Washington Operations,
               Chase Econometrics, to Henry Hertzfeld, NASA, April 15, 1980 426

III-6   "Economic Impact and Technological Progress of NASA
Research and Development Expenditures," Executive Summary,
Midwest Research Institute, Kansas City, Missouri, for the National
Academy of Public Administration, September 20, 1988, pp. 1–4 427

III-7   "NASA Report May Overstate the Economic Benefits of Research
and Development Spending," Report of the Comptroller General
of the United States, PAD-78-18, October 18, 1977, pp. i–iii 430

III-8   Martin D. Robbins, John A. Kelley, and Linda Elliott, "Mission-
Oriented R&D and the Advancement of Technology: The Impact
of NASA Contributions," Final Report, Industrial Economics
Division, Denver Research Institute, University of Denver,
Contract NSR 06-004-063, May 1972, pp. iii–iv, 25–39, 59 432

III-9   "Quantifying the Benefits to the National Economy from
Secondary Applications of NASA Technology—Executive Summary,"
NASA CR-2674, Mathematica, Inc., March 1976. 445

III-10   "Economic Effects of a Space Station: Preliminary Results,"
NASA, June 16, 1983, pp. 1–2, 20–21 450

III-11   "The Economic Impact of the Space Program: A Macro and
Industrial Perspective," prepared for Rockwell International
by The WEFA Group, Bala Cynwyd, Pennsylvania, May 1994,
pp. 1–4. 451

III-12   Office of the Press Secretary, The White House, "The President's
Space Policy and Commercial Space Initiative to Begin the Next
Century," Fact Sheet, February 11, 1988. 455

III-13   National Space Policy Directive 3, "U.S. Commercial Space Policy
Guidelines," The White House, February 12, 1991 460

III-14   "Fact Sheet, National Space Policy," The White House, National
Science and Technology Council, September 19, 1996 463

III-15   "Commercial Space Industry in the Year 2000: A Market
Forecast," The Center for Space Policy (CSP), Inc., Cambridge,
Massachusetts, June 1985 473

III-16   William M. Brown and Herman Kahn, "Long-Term Prospects for
Developments in Space (A Scenario Approach)," Hudson
Institute, Inc., Croton-on-Hudson, New York, Contract
NASW-2924, October 30, 1977, pp. 257–274 480

III-17   Robert Dunn, "NASA Policy to Enhance Commercial Investment
in Space," internal NASA document, September 13, 1983. 488

III-18   "Space Commercialization Meeting," memo with agenda,
                participants, and outline of policy issues, The White House,
                August 3, 1983. 498

III-19   Craig L. Fuller, The White House, Memorandum for the
                Cabinet Council on Commerce and Trade, "Commercial Space
                Initiatives," April 10, 1984, with attached: "Private Enterprise in
                Space—An Industry View," pp. iv–v 501

III-20   "Feasibility Study of Commercial Space Manufacturing, Phase II
                Final Report," Volume I: Executive Summary, MDC E1625,
                McDonnell Douglas Astronautics Company, East St. Louis, Missouri,
                January 15, 1977, pp. 1–2, 8–20 504

III-21   "Space Industrialization: Final Report," Volume 1. Executive
                Summary, SD 78-AP-0055-1, Rockwell International Space
                Division, Contract NAS8-32198, April 14, 1978, pp. 1–8 517

III-22   "Space Industrialization: An Overview," Final Report, Volume 1,
                SAI-79-662-HU, Science Applications, Inc., April 15, 1978,
                pp. 1–5, 10–12, 15–17 527

III-23   "Feasibility of Commercial Space Manufacturing: Production
                of Pharmaceuticals," Final Report, Volume I, Executive Summary,
                MDC E2104, McDonnell Douglas Astronautics Company,
                St. Louis Division, Contract NAS8-31353, November 9, 1978,
                pp. 1–3, 26–30. 534

III-24   James Beggs, Administrator, NASA, to William Clark, Assistant
                to the President for National Security Affairs, August 26, 1983,
                with attached: John F. Yardley, President, McDonnell Douglas
                Astronautics Company, to James Beggs, Administrator, NASA,
                August 23, 1983. 539

III-25   L. Smith, McDonnell Douglas Corporation, "Electrophoresis
                Operations in Space," briefing charts, September 1983,
                pp. 6–7, 30. 541

III-26   U.S. General Accounting Office, "Commercial Use of Space:
                Many Grantees Making Progress, but NASA Oversight Could
                be Improved," Executive Summary, GAO/NSIAD-91-142,
                May 1991, pp. 2–5 543

III-27   Leo S. Packer, Special Assistant to Associate Administrator,
                Office of Advanced Research and Technology, NASA, "Proposal
                for Enhancing NASA Technology Transfer to Civil Systems,"
                September 26, 1969, pp. 1–9. 546

III-28   F. Douglas Johnson, Panayes Gastseos, and Emily Miller, with
          assistance from Charles F. Mourning, Thomas Basinger, Nancy
          Gundersen, and Martin Kokus, "NASA Tech Brief Program: A Cost
          Benefit Evaluation," Executive Summary, University of Denver
          Research Institute, Contract NASW-2892, May 1977, pp. i–iii 553

III-29   Robert J. Anderson, Jr., William N. Lanen, and Carson E. Agnew,
          with Faye Duchin and E. Patrick Marfisi, "A Cost-Benefit Analysis of
          Selected Technology Utilization Office Programs," Executive
          Summary, MathTech, Contract NASW-2731,
          November 7, 1977, pp. 1–6 555

III-30   Richard L. Chapman, Loretta C. Lohman, and Marilyn J.
          Chapman, "An Exploration of Benefits from NASA 'Spinoff,'"
          Chapman Research Group, Contract 88-01 with NERAC, Inc.,
          June 1989, pp. 1–5, 23–28 559

III-31   H.R. Hertzfeld, "Technology Transfer White Paper," internal
          NASA document, June 23, 1978 563

III-32   "NASA Technology Transfer: Report of the Technology
          Transfer Team," December 21, 1992. 574

Biographical Appendix 579

Index 591

The NASA History Series 605

Acknowledgments

This volume is the third in a series that had its origins almost a decade ago. The individuals involved in initiating the series and producing the first two volumes have been acknowledged in those volumes [Volume I—*Organizing for Space* (1995); Volume II—*External Relationships* (1996)]; those acknowledgments will not be repeated here. An exception must be made for NASA Chief Historian Roger D. Launius, who has become not only a strong supporter of this series but also an essential collaborator in its implementation.

We owe thanks to the individuals and organizations that have searched their files for potentially useful materials, and for the staffs at various archives and collections who have helped us locate documents. Without question, first among them is Lee D. Saegesser of the History Office at NASA Headquarters, who has helped compile the NASA Historical Reference Collection that contains many of the documents selected for inclusion in this work. All those in the future who will write on the history of the U.S. space program will owe a debt of thanks to Lee; those who have already worked in this area realize his tireless contributions.

At the Space Policy Institute, research associate David H. Onkst made so many contributions to the organization of material for this volume that he deservedly has been listed as co-editor. Graduate students Erin Hatch, Becky Dodder, Garth Henning, and David Vaughn also helped in the preparation of the volume, and research Dwayne A. Day has continued his involvement with the series while concentrating on his own research. The overview essays for the satellite communications and remote-sensing sections were written several years ago, before the decision to expand the series beyond the originally planned two volumes. (The total is now up to six.) Ray A. Williamson helped update and expand Pamela E. Mack's discussion of remote sensing, and he has been added as the second author of the essay. I made far fewer modifications to Joseph N. Pelton's original essay on satellite communications. When it became clear that a focus on economic issues would be valuable, Henry Hertzfeld graciously agreed to oversee the collection of documents for that section and to write the overview essay. Trish Mastrobuono and Julie Hudson of the Institute staff have supported the effort throughout and, with graduate student Irena Slage, helped create the document-scanning capability used in the final stages of the project.

My thanks go to all those mentioned above, and again to those who helped get this effort started almost a decade ago.

John M. Logsdon, George Washington University

****

There are numerous people at NASA associated with historical study, technical information, and the mechanics of publishing who helped in myriad ways in the preparation of this documentary history. Stephen J. Garber prepared the biographical appendix, helped in the final proofing, and prepared the index of the work and deservedly is listed as a co-editor. Nadine J. Andreassen of the NASA History Office performed editorial and proof-reading work on the project; and the staffs of the NASA Headquarters Library, the Scientific and Technical Information Program, and the NASA Document Services Center provided assistance in locating and preparing for publication the documentary materials in this work. The NASA Headquarters Printing and Design Office developed the layout and handled printing. Specifically, we wish to acknowledge the work of Jane E. Penn, Patricia M. Talbert, Jonathan L. Friedman, and Kathleen Gasparin for their design and editorial work. In addition, Michael Crnkovic and Stanley Artis saw the book through the publication process. Thanks are due to all of them.

Roger D. Launius, NASA

Introduction

One of the most important developments of the twentieth century has been the movement of humanity into space with machines and people. The underpinnings of that movement—why it took the shape it did; which individuals and organizations were involved; what factors drove a particular choice of scientific objectives and technologies to be used; and the political, economic, managerial, and international contexts in which the events of the space age unfolded—are all important ingredients of this epoch transition from an Earthbound to a spacefaring people. This desire to understand the development of spaceflight in the United States sparked this documentary history series.

The extension of human activity into outer space has been accompanied by a high degree of self-awareness of its historical significance. Few large-scale activities have been as extensively chronicled so closely to the time they actually occurred. Many of those who were directly involved were quite conscious that they were making history, and they kept full records of their activities. Because most of the activity in outer space was carried out under government sponsorship, it was accompanied by the documentary record required of public institutions, and there has been a spate of official and privately written histories of most major aspects of space achievement to date. When top leaders considered what course of action to pursue in space, their deliberations and decisions often were carefully put on the record. There is, accordingly, no lack of material for those who aspire to understand the origins and evolution of U.S. space policies and programs.

This reality forms the rationale for this series. Precisely because there is so much historical material available on space matters, the National Aeronautics and Space Administration (NASA) decided in 1988 that it would be extremely useful to have a selective collection of many of the seminal documents related to the evolution of the U.S. civilian space program that was easily available to scholars and the interested public. While recognizing that much space activity has taken place under the sponsorship of the Department of Defense and other national security organizations, the U.S. private sector, and other countries around the world, NASA felt that there would be lasting value in a collection of documentary material primarily focused on the evolution of the U.S. government's civilian space program, most of which has been carried out under the agency's auspices since 1958. As a result, the NASA History Office contracted with the Space Policy Institute of George Washington University's Elliott School of International Affairs to prepare such a collection. This is the third volume in the documentary history series; three additional ones detailing programmatic developments with respect to space transportation, space science, and human spaceflight will follow.

The documents collected during this research project were assembled from a diverse number of both public and private sources. A major repository of primary source materials relative to the history of the civil space program is the NASA Historical Reference Collection of the NASA History Office, located at the agency's headquarters in Washington, D.C. Project assistants combed this collection for the "cream" of the wealth of material housed there. Indeed, one purpose of this series from the start was to capture some of the highlights of the holdings at headquarters. Historical materials housed at the other NASA installations, institutions of higher learning, and presidential libraries were other sources of documents considered for inclusion, as were papers in the archives of individuals and firms involved in opening up space for exploration.

Copies of more than 2,500 documents in their original form collected during this project (not just the documents selected for inclusion), as well as a database that provides a guide to their contents, will be deposited in the NASA Historical Reference Collection. Another complete set of project materials is located at the Space Policy Institute at George Washington University. These materials in their original form are available for use by researchers seeking additional information about the evolution of the U.S. civilian space program or wishing to consult the documents reprinted herein in their original form.

The documents selected for inclusion in this volume are presented in three major chapters, each covering a particular aspect of the utilization of space capabilities and the unique characteristics of the space environment. These chapters address: (1) communicating via satellite; (2) observing the Earth from space for practical purposes (Earth science will be covered in a later volume); and (3) the various ways in which space activities have had economic impacts. Volume I in this series covered the antecedents to the U.S. space program, as well as the origins and evolution of U.S. space policy and of NASA as an organizational institution. Volume II addressed the relationship between the civilian space program of the United States and the space activities of other countries, the relationship between the U.S. civilian and national security space and military efforts, and NASA's relationship with industry and academic institutions. As mentioned above, future volumes will cover space transportation, space science, and human spaceflight.

Each section in this volume is introduced by an overview essay, prepared by individuals particularly well-qualified to write on the topic. In the main, these essays are intended to introduce and complement the documents in the section and to place them in a chronological and/or substantive context. Each essay contains references to the documents in the section it introduces, and many also contain references to documents in other sections of the collection. These introductory essays were the responsibility of their individual authors, and the views and conclusions contained therein do not necessarily represent the opinions of either George Washington University or NASA.

The documents included in each section were chosen by the project team in concert with the essay writer from those assembled by the research staff for the overall project. The contents of this volume emphasize primary documents or long-out-of-print essays or articles and material from the private recollections of important actors in shaping space affairs. Key legislation and policy statements are also included. The contents of this volume thus do not comprise in themselves a comprehensive historical account; they must be supplemented by other sources, those both already available and to become available in the future. Indeed, a few of the documents included in this collection are not complete; some portions of them were still subject to security classification as this volume went to print.

Each document is assigned its own number in terms of the chapter in which it is placed. As a result, the first document in the third section of this volume is designated "Document III-1." Each document is accompanied by a headnote setting out its context and providing a background narrative. These headnotes also provide specific information about people and events discussed. We have avoided the inclusion of explanatory notes in the documents themselves and have confined such material to the headnotes.

The editorial method we adopted for publishing these documents seeks to preserve spelling, grammar, paragraphing, and use of language as in the original. We have sometimes changed punctuation where it enhances readability. We have used ellipses ("...") to note where sections of a document have not been included in this publication, and we have avoided including words and phrases that had been deleted in the original document unless they contribute to an understanding of what was going on in the mind of the writer in making the record. Marginal notations on the original documents are inserted into the text of the documents in brackets, each clearly marked as a marginal comment. When deletions to the original document have been made in the process of declassification, we have noted this with a parenthetical statement in brackets. Except insofar as illustrations and figures are necessary to understanding the text, those items have been omitted from this printed version. Page numbers in the original document are noted in brackets internal to the document text. Copies of all documents in their original form, however, are available for research by anyone interested at the NASA History Office or the Space Policy Institute of George Washington University.

We recognize that there are certain to be quite significant documents left out of this compilation. No two individuals would totally agree on all documents to be included from the more than 2,500 that we collected, and surely we have not been totally successful in locating all relevant records. As a result, this documentary history can raise an immediate question from its users: Why were some documents included while others of seemingly equal importance were omitted? There can never be a fully satisfactory answer to this question. Our own criteria for choosing particular documents and omitting others rested on three interrelated factors:

- Is the document the best available, most expressive, most representative reflection of a particular event or development important to the evolution of the space program?

- Is the document not easily accessible except in one or a few locations, or is it included (for example, in published compilations of presidential statements) in reference sources that are widely available and thus not a candidate for inclusion in this collection?

- Is the document protected by copyright, security classification, or some other form of proprietary right and thus unavailable for publication?

As general editor of this volume, I was ultimately responsible for the decisions about which documents to include and for the accuracy of the headnotes accompanying them. It has been an occasionally frustrating but consistently exciting experience to be involved with this undertaking. My associates and I hope that those who consult it in the future find our efforts worthwhile.

John M. Logsdon
Director
Space Policy Institute
Elliott School of International Affairs
George Washington University

Biographies of Volume III Essay Authors and Editors

Stephen J. Garber is a policy analyst in the NASA History Office, Washington, D.C. He is the author of numerous articles on aerospace history and space policy.

Henry R. Hertzfeld, Senior Research Scientist at the Space Policy Institute, George Washington University, is an expert in the economic, legal, and policy issues of space and advanced technological development. He has served as a Senior Economist and Policy Analyst at both NASA and the National Science Foundation, and he has been a consultant to many agencies and organizations. He is the co-editor of *Space Economics* (AIAA, 1992), as well as many articles on space economic issues. Dr. Hertzfeld holds a Ph.D. in economics from Temple University and a J.D. from George Washington University.

Roger D. Launius is Chief Historian of the National Aeronautics and Space Administration, Washington, D.C. He has produced several books and articles on aerospace history, including *Frontiers of Space Exploration* (Greenwood Press, 1998); *Organizing for the Use of Space: Historical Perspectives on a Persistent Issue* (Univelt, Inc., AAS History Series, Volume 18, 1995), editor; *NASA: A History of the U.S. Civil Space Program* (Krieger Publishing Co., 1994); *History of Rocketry and Astronautics: Proceedings of the Fifteenth and Sixteenth History Symposia of the International Academy of Astronautics* (Univelt, Inc., AAS History Series, Volume 11, 1994), editor; *Apollo: A Retrospective Analysis* (Monographs in Aerospace History, Volume 3, 1994); and *Apollo 11 at Twenty-Five,* an electronic picture book issued on computer disk (Space Telescope Science Institute, Baltimore, MD, 1994).

John M. Logsdon is Director of the Space Policy Institute of George Washington University's Elliott School of International Affairs, where he is also a professor of political science and international affairs and Director of the Center for International Science and Technology Policy. He holds a B.S. in physics from Xavier University and a Ph.D. in political science from New York University. He has been at George Washington University since 1970, and he previously taught at The Catholic University of America. He is also a faculty member of the International Space University and Director of the District of Columbia Space Grant Consortium. Dr. Logsdon is an elected member of the International Academy of Astronautics and the Board of Trustees of the International Space University and Chair of the Advisory Council of the Planetary Society. He has lectured and spoken to a wide variety of audiences at professional meetings, colleges and universities, international conferences, and other settings, and he has testified before Congress on numerous occasions. He is frequently consulted by the electronic and print media for his views on various space issues. Dr. Logsdon has been a Fellow at the Woodrow Wilson International Center for Scholars and was the first holder of the Chair in Space History of the National Air and Space Museum. He is a Fellow of the American Association for the Advancement of Science and an Associate Fellow of the American Institute of Aeronautics and Astronautics. In addition, he is North American editor for the journal *Space Policy*.

David H. Onkst is a Research Associate at the George Washington University Space Policy Institute and a member of the History Department at American University, both located in Washington, D.C. He has won numerous scholarly awards, including the 1998 Robert H. Goddard Historical Essay Award, a George Meany Memorial Archives Fellowship, and several teaching fellowships at American University. Onkst has written articles on subjects ranging from the possibility of life on Mars and Europa to a history of southern African-American World War II veterans and their use of the G.I. Bill of Rights for such periodicals as *Space Times* and the *Journal of Social History*. He also served as a behind-the-scenes editor for *Eye in the Sky: The Story of the CORONA Spy Satellites* (Smithsonian Institution Press, 1998). Onkst is a specialist in U.S. social and cultural history since 1941.

Joseph N. Pelton is a professor of telecommunications at the University of Colorado at Boulder, having returned from the post of Vice President of Academic Programs and Dean of the International Space University (ISU). Prior to going to ISU, Dr. Pelton held a dual appointment as Director of the Interdisciplinary Telecommunications Program at the University of Colorado, as well as Director of the Center for Advanced Research in Telecommunications and Training. He has worked in a variety of positions at INTELSAT for the past two decades, and he has been involved with satellite applications since 1965, in positions with Rockwell International, NASA, Communications Satellite Corporation (COMSAT), and George Washington University. He also taught at American University in Washington, D.C. After receiving an undergraduate degree in physics, Dr. Pelton went on to receive a master's degree from New York University and a Ph.D. from Georgetown University.

Ray A. Williamson is a Senior Research Scientist in the Space Policy Institute, focusing on the history, programs, and policy of Earth observations, space transportation, and space commercialization. He joined the Institute in 1995. Previously, he was a Senior Associate and Project Director in the Office of Technology Assessment (OTA) of the U.S. Congress. He joined OTA in 1979. While at OTA, Dr. Williamson was project director for more than a dozen reports on space policy, including: *Russian Cooperation in Space* (1995); *Civilian Satellite Remote Sensing: A Strategic Approach* (1994); *Remotely Sensed Data: Technology, Management, and Markets* (1994); *Global Change Research and NASA's Earth Observing System* (1994); and *The Future of Remote Sensing from Space: Civilian Satellite Systems and Applications* (1993). Dr. Williamson has written extensively about the U.S. space program. He holds a B.A. in physics from Johns Hopkins University and a Ph.D. in astronomy from the University of Maryland. He spent two years on the faculty of the University of Hawaii studying diffuse emission nebulae and ten years on the faculty of St. John's College in Annapolis, Maryland. He is a member of the faculty of the International Space University and is a member of the editorial board for the journal *Space Policy*.

Glossary

ABMA . . . . .Army Ballistic Missile Agency
ACDA . . . . .Arms Control and Disarmament Agency
AFB . . . . .Air Force Base
AID . . . . .Agency for International Development (State Department)
ARPA . . . . .Advanced Research Projects Agency
ASTP . . . . .Apollo-Soyuz Test Project
ASVT . . . . .Application System Verification and Test
ATS . . . . .Applications Technology Satellite
AVHRR . . . . .Advanced Very High Resolution Radiometer
BSS . . . . .Broadcast Satellite Service
CBD . . . .*Commerce Business Daily*
CFR/C.F.R. . . . . .Code of Federal Regulations
CIA . . . . .Central Intelligence Agency
COSC . . . . .Committee on Satellite Communications
COSMIC . . . . .Computer Software Management Information Center
CSP . . . . .Center for Space Policy
CSTI . . . . .Civil Space Technology Initiative
db/dB . . . . .Decibel
DBS . . . . .Direct Broadcast Satellite
DCI . . . . .Director of Central Intelligence
DCS . . . . .Data collection system
DDR&E . . . . .Director, Defense Research and Engineering
DDT&E . . . . .Design, Development, Test and Engineering
DMSP . . . . .Defense Meteorological Satellite Program
DOC/DoC . . . . .Department of Commerce
DOD/DoD . . . . .Department of Defense
DOE/DoE . . . . .Department of Energy
DOI/DoI . . . . .Department of the Interior
ELV . . . . .Expendable Launch Vehicle
EOS . . . . .Earth Observing System *or* Electrophoresis Operations in Space
                            (program)
EOSAT . . . . .Earth Observation Satellite Company
EPA . . . . .Environmental Protection Agency
EROS . . . . .Earth Resources Observational Satellites (later Systems)
ERS . . . . .Earth Resources Survey (program)
ERTS . . . . .Earth Resources Technology Satellite (later known as Landsat)
ESA . . . . .European Space Agency
ESSA . . . . .Environmental Science Services Administration
ETM . . . . .Enhanced Thematic Mapper
EUMETSAT . . . . .European Organisation for the Exploitation of Meteorological
                            Satellites
EVA . . . . .Extravehicular activity
FAO . . . . .Food and Agriculture Organization (U.N.)
FCC . . . . .Federal Communications Commission
FM . . . . .Frequency-modulation
FEDD . . . . .For Early Domestic Distribution
FSS . . . . .Fixed Satellite Service
FTE . . . . .Full-Time Equivalent
FY . . . . .Fiscal year

GAO . . . . .General Accounting Office
GDP . . . . .Gross Domestic Product
GE . . . . .General Electric
GHz . . . . .Gigahertz
GIS . . . . .Geographic Information System
GNP . . . . .Gross National Product
GOES . . . . .Geostationary Operational Environmental Satellite
GPO . . . . .Government Printing Office
GSP . . . . .Gross Space Product
GWP . . . . .Gross Worldwide Product
HF . . . . .High frequency
HIRS . . . . .High-Resolution Infrared Sounder
HRMSI . . . . .High Resolution Multispectral Stereo Imager
ICUS . . . . .Interim Communication Satellite Committee
IGI . . . . .Industrial Guest Investigator (agreement)
IPO . . . . .Integrated Program Office
IPOMS . . . . .International Polar Orbiting Meteorological Satellite Group
IR . . . . .Infrared
IRAC . . . . .Interdepartment Radio Advisory Committee
ITOS . . . . .Improved TIROS Operational Satellite
ITU . . . . .International Telecommunications Union
JEA . . . . .Joint Endeavor Agreement
JSC . . . . .Johnson Space Center
KC . . . . .Kilocycle
KW/kW . . . . .Kilowatt
LACIE . . . . .Large Area Crop Inventory Experiment
LCG . . . . .Landsat Coordinating Group
LDC . . . . .Less Developed Country
LEO . . . . .Low Earth Orbit
LSI . . . . .Large Scale Integrated (circuit)
MDAC . . . . .McDonnell Douglas Astronautics Company
MIT . . . . .Massachusetts Institute of Technology
MOD . . . . .Ministry of Defense
MPS . . . . .Materials Processing in Space
MRI . . . . .Midwest Research Institute
MSC . . . . .Manned Space Center (later known as Johnson Space Center)
MSS . . . . .Multispectral Scanner *or* Mobile Satellite Service
MTCR . . . . .Missile Technology Control Regime
MTF . . . . .Man-Tended Facility
MTT . . . . .Message toll telephone
NACA . . . . .National Advisory Committee for Aeronautics
NAPA . . . . .National Academy of Public Administration
NASA . . . . .National Aeronautics and Space Administration
NESDIS . . . . .National Environmental Satellite, Data, and Information
                            Service
NIH . . . . .National Institutes of Health
NMI . . . . .NASA Management Instruction
NOAA . . . . .National Oceanic and Atmospheric Administration
NOMSS . . . . .National Operational Meteorological Satellite System
NPOESS . . . . .National Polar-Orbiting Environmental Satellite System
NRC . . . . .National Research Council
NRO . . . . .National Reconnaissance Office
NSAM . . . . .National Security Action Memorandum

NSC . . . . .National Security Council
NSDD . . . . .National Security Decision Directive
NSF . . . . .National Science Foundation
NSTC . . . . .National Science and Technology Council
OCDM . . . . .Office of Civil and Defense Mobilization
OLS . . . . .Operational Linescan System
OMB . . . . .Office of Management and Budget (formerly the Bureau of the
             Budget)
OMV . . . . .Orbital Maneuvering Vehicle
OPEC . . . . .Organization of the Petroleum Exporting Countries
OSSA . . . . .Office of Space Science and Applications (NASA)
OSTP . . . . .Office of Science and Technology Policy (White House)
PAM . . . . .Payload Assist Module
PDD . . . . .Presidential Decision Directive
POES . . . . .Polar-orbiting Operational Environmental Satellite
R&D . . . . .Research and development
RCA . . . . .Radio Corporation of America
RDSS . . . . .Radio Determination Satellite Service
RFI . . . . .Request for information
RFP . . . . .Request for proposals
ROI . . . . .Return on Investment
SBIR . . . . .Small Business Innovation Research (program)
SI . . . . .Space Industrialization
SIG . . . . .Senior Interagency Group
SLAR . . . . .Side Looking Airborne Radar
SOCC . . . . .Satellite Operations Control Center
SPOT . . . . .Satellite Pour l'Observation de la Terre
SSI . . . . .Space Services, Inc.
SSM/I . . . . .Special Sensor Microwave/Imager
START . . . . .Strategic Arms Reduction Treaty
STS . . . . .Space Transportation System
TCC . . . . .Telecommunication Coordinating Committee
TDRSS . . . . .Tracking Data and Relay Satellite System
TEA . . . . .Technical Exchange Agreement
TIROS . . . . .Television and Infrared Operational Satellite
TM . . . . .Thematic Mapper
TOS . . . . .Transfer Orbit Stage *or* TIROS Operational Satellite
TOVS . . . . .TIROS Operational Vertical Sounder
TSP . . . . .Technical Support Package
TUO . . . . .Technology Utilization Office
UHF . . . . .Ultrahigh frequency
U.N./UN . . . . .United Nations
UNGA . . . . .U.N. General Assembly
U.S./US . . . . .United States
USC/U.S.C. . . . . .United States Code
USDA . . . . .U.S. Department of Agriculture
USGS . . . . .U.S. Geological Survey
USG . . . . .U.S. Government
UV . . . . .Ultraviolet
VHF . . . . .Very high frequency

Chapter One

The History of Satellite Communications

by Joseph N. Pelton

Although the idea of using artificial Earth satellites to relay messages from one point on the Earth to another had been discussed in several places prior to 1945,[1] most accounts of the development of satellite communications begin by discussing Arthur Clarke's landmark works on the topic during that year. In two 1945 papers—one privately circulated and one published in *Wireless World*—Clarke discussed the special characteristics of geosynchronous orbit that would enable three satellites in that orbit to provide global communications.[2] [I-1, I-2] Clarke noted that in an orbit of 22,300 miles above the Earth, the velocity of a satellite exactly matched the velocity of the Earth's surface as the planet rotated about its axis; thus from the Earth, a satellite would appear to remain in a fixed position in the sky. In such an orbit, a satellite could "see" 40 percent of the equatorial plane. Clarke noted the benefits of such an orbital perspective, especially for telecommunications, because the curvature of the Earth's surface and atmospheric interference placed limits on ground-based transmissions. In addition, the use of satellites in geosynchronous orbit would make the design of a ground antenna simpler in terms of tracking and pointing mechanisms.

For these insights, Arthur C. Clarke is frequently called the "Father of Satellite Communications," and there have been ongoing efforts to officially designate the geosynchronous orbit as the "Clarke Orbit." Ironically, however, while a visionary in many respects, Clarke did not foresee how quickly communications satellites would become a reality. This is because he did not anticipate the invention of the transistor, which greatly reduced the necessary weight of a communications satellite and dramatically increased its reliability and lifetime. From the pre-transistor perspective of 1945, Clarke envisioned that communicating via satellite would in effect require a space station—an orbital platform weighing many tons with an on-board crew to replace burned-out vacuum tubes.[3] And while Clarke may not have been totally prescient, he can be credited with identifying a line of technological development that bore fruit in less than twenty years.

---

1. Delbert D. Smith, *Communication via Satellite: A Vision in Retrospect* (Boston: A.W. Sijthoff, 1976), pp. 15–19.

2. The more frequently cited of Clarke's semi-annual papers is Arthur C. Clarke, "Extra-Terrestrial Relays: Can Rocket Stations Give World-Wide Radio Coverage?," *Wireless World* 51 (October 1945): 305–08. A May 25, 1945, typed paper about geosynchronous satellites, "The Space Station: Its Radio Applications," was sent to members of the British Interplanetary Society and other addressees some five months before the more famous *Wireless World* article. This earlier paper was finally published in *Spaceflight* 10 (March 3, 1968): 85–86.

3. Personal interview by the author with Arthur C. Clarke, Sri Lanka, May 1984.

The Early Years of Concept and Experimentation (1945–1963)

In the decade that followed Clarke's article, increasingly powerful rockets were developed, largely in the context of the Cold War competition between the United States and the Soviet Union. The launch of Sputnik 1 by the Soviets in October 1957 triggered a number of U.S. space initiatives. The creation of the National Aeronautics and Space Administration (NASA) from the National Advisory Committee for Aeronautics (NACA) and the surge of funding for U.S. rocket programs, such as Vanguard, Thor, Atlas, and Titan, were immediate results. In the 1960 U.S. presidential election, the "Missile Gap" debates between John F. Kennedy and Richard M. Nixon set the stage for a strong U.S. space program for the 1960s, perhaps regardless of the election's outcome.[4] One issue under discussion at the time was whether space development would be almost exclusively a result of government activities or whether private enterprise would play a significant role.

The creation of communications satellite research and development programs within NASA and the Department of Defense in the late 1950s and early 1960s proved to be a highly effective means of establishing U.S. capability in this field; however, these government efforts were paralleled by private-sector communications satellite research and development activities. These communications satellite initiatives came at a key time in terms of the overall development of international communications. American Telephone and Telegraph (AT&T) and others kept private-enterprise interests alive with parallel research and development efforts of their own.

In the late 1950s and early 1960s, a number of new submarine telephone cables were being laid across the Atlantic and Pacific Oceans. These new cables replaced outmoded telegraph cables and stimulated the rapid growth of international telecommunications. The leaders of NASA, aerospace manufacturers, and telecommunications organizations all recognized that high-capacity communications satellites could also support the rapid growth of global communications. Only several years later did the idea emerge that communications satellites could also support regional or domestic telecommunications needs. [5]

Because of the U.S. lead in micro-electronics, the U.S. Signal Corps was able to launch the world's first communications satellite on an Atlas rocket soon after Sputnik. The first U.S. communications project was known as SCORE (Signal Communication by Orbital Relay Equipment), a broadcast-only satellite launched on December 18, 1958. SCORE lasted only twelve days and could only send to Earth a pre-recorded message from President Dwight D. Eisenhower: "Peace on earth, good will toward men."[6]

The first artificial satellite that actually relayed a real-time voice message from the Earth to orbit and back was Echo 1, launched on August 12, 1960. The Echo program was a successor to an International Geophysical Year effort to measure the density of the upper atmosphere by observing the orbit of a twelve-foot-diameter balloon-like satellite. In 1959 John Pierce, an AT&T scientist and one of the pioneers of the communications

---

4.   See John M. Logsdon, *The Decision to Go to the Moon: Project Apollo and the National Interest* (Cambridge, MA: MIT Press, 1970), and Walter A. McDougall, . . . *The Heavens and the Earth: A Political History of the Space Age* (New York: Basic Books, 1985), for discussions of early U.S. space policy.

5.   Joseph N. Pelton, *Global Communications Satellite Policy: Intelsat, Politics and Functionalism* (Mt. Airy, WA: Lomond Systems, 1974), pp. 44–102. The Soviet Union with its vast size and northern latitudes had to use a highly elliptical orbit for its communications satellites. This orbit required less rocket power to reach from Soviet launch sites than did geosynchronous orbit. The combination of internal need and ease of orbital access led the Soviet Union to initiate domestic communications satellite service much sooner than other countries.

6.   *Ibid.*, pp. 46–48.

satellite field, suggested using the larger 100-foot-diameter Echo satellite to test transatlantic radio communications. [I-3] NASA accepted Pierce's suggestion, and the orbiting satellite was successfully used as a passive reflector (that is, there were no electronic systems to amplify the signal aboard the satellite) of an August 18 message from New Jersey to France. (Similar experiments had been conducted earlier using the Earth's Moon as a passive reflector.)

Although this and many other experiments using the passive Echo 1 and Echo 2 satellites (Echo 2 was launched in January 1964) were successful, in the late 1950s and early 1960s, industry and government attention increasingly focused on active communications satellites carrying on-board electronics that received a signal from the Earth, amplified it, and sent it back to the Earth. Such satellites had more predictable orbits than passive satellites; fewer were required to create a communications network; and signals relayed through active satellites required less expensive ground stations and had much higher capacity.[7]

The first artificial communications satellite that foreshadowed today's active satellite technology, Courier 1B, was designed and launched by the U.S. military in October 1960. It featured solar and battery power, an active antenna for transmission and reception, and electronic repeaters capable of frequency conversion from uplink to downlink signals. Despite the technological gains that Courier 1B represented, it still had a capability of only sixteen teletype channels. In short, it was little more than an experimental device. Submarine cables still had 100 times this capacity.

NASA, AT&T, and the Department of Defense, however, were by this time moving ahead with tremendous energy and quickly achieved impressive results. On July 10, 1962, only a year and a half after Courier 1B, a quantum leap in capability was achieved with the launch of Telstar, an AT&T-designed and -built experimental satellite with sufficient capacity to relay a television signal. Telstar was launched into a medium orbit with a 570-mile perigee; at this orbit, a number of satellites (about twelve to fifteen) would be required for an operational system. The success of the Telstar experiment immediately changed the world's view of the potential of this new technology. [I-20][8] Recognition grew that communications satellites could have three to four times the capacity of then-current submarine cables. Almost overnight, their commercial viability was advanced from remote to highly likely. Quickly thereafter, on December 14, 1962, the NASA-funded Relay 1 of the Radio Corporation of America (RCA) was launched into an orbit with a 660-mile perigee, demonstrating many of the same features as Telstar but with a longer lifetime in orbit. The technical feasibility of active communications satellites was thus clearly demonstrated by the second half of 1962.[9] The development of these early communications satellites represented the first steps toward the practical use of space and began a debate about whether such an enterprise should be public or private in nature.

Then, a satellite built by Hughes Aircraft, developed with both company and NASA funding and launched on December 14, 1963, demonstrated the final technical feature required for communications satellites to become commercially viable—stable and continuous operation in geosynchronous orbit. Beginning in 1959, Hughes had been work-

---

7.   John R. Pierce, *The Beginnings of Space Communications* (San Francisco, CA: San Francisco Press, 1968), p. 103; Arthur C. Clarke, *The Promise of Space* (New York: Harper and Row, 1968), pp. 100–101; Smith, *Communications via Satellite*, pp. 51–55.

8.   It should be noted here that because the document about Project Telstar was produced in the second half of 1962, it appears in chronological order after this essay but is referenced out of order within the essay. Also note that the references to Documents I-14, I-22, and I-26 are out of order.

9.   Leonard Jaffe, *Communications in Space* (New York: Holt, Reinhart and Winston, 1966), p. 86; Orrin E. Dunlap, Jr., *Communications in Space* (New York: Harper and Row, 1960), p. 151.

ing on such a satellite, first with the company's own funds and since 1961 under contract to NASA. [I-4, I-5] This satellite, Syncom 2 (Syncom 1 suffered a system failure), was followed by a Syncom 3 mission that was even more successful in demonstrating the feasibility of high-capacity telecommunications operations in geosynchronous orbit.[10]

This lightning-like development was paralleled by progress with military satellites, particularly the Lincoln Experimental Satellite (LES) series from Lincoln Laboratory, which tested secure transponders for strategic communications from the National Command Authority. Together, these accomplishments set the stage for the practical exploitation of this exciting new technology.

The Creation of Comsat and INTELSAT (1962–1965)

The international civilian communications program soon evolved toward a global network of "stabilized" satellites in geosynchronous orbits a tenth of the way to the Moon. The civilian system began with an initial satellite over the Atlantic Ocean (1965), then the Pacific Ocean obtained service (1967), and finally global coverage was completed with Indian Ocean service (1969), just as Arthur C. Clarke had envisioned it twenty-four years earlier.

Although most critical technical choices had been made by 1965, the issue of how to institutionalize the civilian communications satellite system was far from clear-cut or easily decided. During 1961 and 1962, there was intense debate in the United States about public versus private ownership and operations. Political control and financing were also items of disagreement. Not surprisingly, these issues led to a major political debate in the United States.

The Eisenhower administration supported the development of satellite communications, but only if that development was based on private-sector initiatives.[11] [I-6, I-7] When John F. Kennedy took office in early 1961, however, he expressed a strong support for a leading government role in communications satellite development.[12] Achieving his objective, however, meant sorting out within the Kennedy administration the appropriate role of the government in communications satellite research and development, regulation, and ownership and operation. [I-8, I-9, I-10, I-11, I-12, I-13, I-15]

Once the administration had developed its position, it had to gain the assent of Congress. This was not a straightforward task; many in Congress had views on the issue that differed from the proposed White House policy. [I-14] The net result was that three

---

10.   Dunlap, *Communications in Space*, pp. 152–55.
11.   On December 30, 1960, in one of his last speeches in office, President Eisenhower stated: "This nation has traditionally followed a policy of conducting international telephone, telegraph and other communication services through private enterprise subject to governmental control, licensing, and regulation. We have achieved communications facilities second to none among nations of the world. Accordingly, the government should aggressively encourage private enterprise in the establishment and operation of satellite relays for revenue producing services." *Public Papers of the Presidents of the United States: Dwight D. Eisenhower, 1960* (Washington, DC: U.S. Government Printing Office, 1979), p. 888.
12.   The now-famous Kennedy speech of May 25, 1961, that established the goal of sending humans to the Moon and returning them to Earth also called for the establishment of a global satellite system for communications that would benefit all countries, promote world peace, and allow nondiscriminating access for all countries of the world. It called for a "constructive role for the U.N. in international space communications." *Public Papers of the Presidents of the United States, John F. Kennedy, 1961* (Washington, DC: U.S. Government Printing Office, 1962), pp. 529–31. Kennedy's position on communications satellites thus set the stage for the United Nations to act on this subject as well. In September 1961, the United Nations General Assembly adopted Resolution 1721, Section P, concerning the establishment of a global communications satellite system. Section P stated that "communications by means of satellite should be available to the millions of the world as soon as possible on a global and non-discriminating basis."

different versions of national legislation to create a framework for satellite communications emerged during 1961—and especially during 1962—within Congress. The bill of Senator Robert S. Kerr (D–OK) would have made space communications entirely private. From the opposite perspective, the bill of Senator Estes Kefauver (D–TN) would have made such communications entirely a governmental enterprise. Finally, the Kennedy administration's bill sought a compromise between private and public ownership and among various policy objectives. [I-16, I-17]

After months of debate and a filibuster led by liberal Democrats, complete with a cloture vote, the Communications Satellite Act of 1962 finally emerged. [I-18] This law called for the creation of a new entity to be known as the Communications Satellite Corporation (Comsat), with ownership divided fifty-fifty between the general public and telecommunications corporations, such as AT&T, International Telephone and Telegraph (ITT), RCA, and Western Union International.[13] Comsat's Board of Directors consisted of six representatives from the public stockholders, six representatives of the telecommunications industry, and three presidential appointees. Comsat was designated as the official representative of the United States for global satellite communications. Two years later, the corporation became the manager of the emerging global system known as the International Telecommunications Satellite Consortium (INTELSAT), which was formed on August 20, 1964.[14] [I-19]

Between the creation of Comsat as a new corporation on August 31, 1962, and the creation of INTELSAT, Comsat contracted with the Hughes Aircraft Company (the designer of Syncom 1, 2, and 3) to build an upgraded version of the Syncom satellite. This satellite was initially designated HS 303; it later became known officially as INTELSAT I. The world, however, came to know it by its popular name, "Early Bird." The satellite, which was the first operational geosynchronous communications satellite, weighed eighty-five pounds and was launched in April 1965. It had a lifetime of eighteen months and a capacity of 240 voice circuits or, alternatively, a black-and-white television channel. This transatlantic satellite, with three times the capacity of the largest submarine cable then available and the ability to provide real-time television transmission, captured the world's attention. Early Bird ushered in a new age of international television communications.[15] Also in 1965, the U.S. Department of Defense deployed a low-Earth-orbit satellite system known as the Initial Defense Satellite Communication System (IDSCS), while the Soviet Union deployed its first highly elliptical satellite system known as Molniya ("Lightning").

---

13.   It is interesting to note that John A. Johnson, General Counsel of NASA, was temporarily detailed to Senator Kerr's office to write draft legislation for the Communications Satellite Act, then later requested by the Kennedy administration to draft the version that actually became law (personal interview by the author with John A. Johnson, February 1984, INTELSAT Archives). See also J.O. Pastore, *The Story of Communications* (New York: MacFadden-Bertell, 1964), pp. 67–92.

14.   Over the years, the Communications Satellite Act of 1962 was amended to allow telecommunications organizations to sell off their Comsat holdings, to restructure the Comsat Board of Directors, and to allow Comsat to be the official U.S. participant in the International Maritime Satellite Organization (INMARSAT), another consortium formed over a decade later. In most respects, Comsat's legislatively defined role has remained the same. Over time, however, through actions of the Federal Communications Commission (FCC), the U.S. executive branch, and the courts, Comsat has given up its ownership of Earth stations, entered the U.S. satellite communications market on a competitive basis with other corporations, and found its monopoly role in INTELSAT and INMARSAT questioned. These changes were largely the result of a changing regulatory environment within the U.S. government. During the Nixon, Carter, and Reagan administrations, there have been increasing efforts to move toward a deregulatory and competitive approach to most telecommunications activities that had traditionally been carried out by monopolies. Despite the erosion of the legislative and regulatory framework within which Comsat operated with respect to INTELSAT, which occurred between 1965 and 1990 (especially the loss of technical management of the INTELSAT global system between 1973 and 1979), Comsat remains one of the few monopolies left in the United States.

15.   ICSC Document, ICSC-7-4E (April 1965), pp. 8–9.

Twenty-Five Years of Communications Satellite Developments (1965–1990)

As the engineering and design of the world's first operational communications satellites proceeded, the nature of the institutional arrangements for global satellite communications became a topic of international discussion and dispute.[16] The United States initially thought in terms of a series of bilateral agreements between the United States and major users of international telecommunications. However, those on the other side of the Atlantic, developing a common position through the Committee on European Posts and Telecommunications (the association of European communications entities, most of which were government-owned monopolies), made it clear in February 1964 that a multilateral agreement was the only acceptable approach. The Europeans also were reluctant to hand all technical, operational, and policy control over to the United States, while the United States wished to preserve, as long as possible, its technological advantages in this new commercial sector. [I-21, I-23, I-24] The United States also wanted to ensure that the benefits of communicating via satellites were available to all countries, whatever their stage of economic development. [I-22]

The United States quickly dropped its insistence on bilateral arrangements and worked toward an acceptable multinational framework. Negotiations with Australia, Canada, Japan, and Europe lasted two years; their outcome hinged on several key issues. France preferred three separate regional satellite systems—one for Europe, one for the Americas, and one for Asia—but finally agreed to a single global system with a capital ceiling for a space segment investment of $500 million. This was a much higher initial investment than several other European countries were willing to accept. Japan and Australia played important roles in promoting compromise. They also promoted the use of geosynchronous satellites, because medium-altitude systems would have created gaps in coverage, with the largest gaps being in the Pacific Ocean region.[17] Perhaps the most important compromise between the United States and other countries was U.S. acceptance of the position that the initial agreement was only valid on an interim basis—after five years of experience with a U.S.-dominated organization, the agreement would be reopened in 1969 for review and potential revision. [I-25]

One of the key issues debated and reviewed in the 1962–1964 negotiations was what type of services INTELSAT should provide. Would it furnish all forms of public telecommunications services for both domestic and international service? The Interim Arrangements of 1964 and the Definitive Arrangements of 1971 that followed both specified that INTELSAT, with special approval, could provide a wide range of "specialized services" that included but were not limited to radio navigation services, broadcasting

---

16.   Smith, *Communication via Satellite*, pp. 135–41.

17.   It is a commonly held belief that the successful launch and operation of Early Bird in geosynchronous orbit effectively ended the debate about the "right orbit." In fact, ITT was selected to undertake a detailed study of medium-altitude satellites after the launch of Early Bird. Over the years, the issue of the "best" orbit for communications satellites has arisen again and again. Because of its northern latitude, not easily covered by satellites in geosynchronous orbit over the equator, the Soviet Union used highly elliptical orbits for its Molniya satellites. Most recently, the idea of creating a low-Earth-orbit grid of satellites interconnected by intersatellite links has been proposed by at least one potential land mobile communications satellite system operator. Despite the early nongeosynchronous systems, such as the Soviet Union's Molniya system and the Department of Defense's IDSCS, and despite the proposed new low- and medium-altitude communications satellite systems, use of the geosynchronous orbit is still predominant. Well over 95 percent of all communications satellites for domestic and international fixed satellite services, as well as for domestic, regional, and international fixed satellite services, plus those for military, mobile, and direct-broadcast satellite services, have been launched into geosynchronous orbit.

satellite services, space research services, meteorological services, and Earth resource services. Throughout its history, however, INTELSAT has confined its activities to fixed satellite service public telecommunications for several reasons:

- These public telecommunications services were the most well-established, prevalent, and "desirable" services for its constituent members in terms of revenues.
- The special financial conditions and agreements needed to embark on "specialized" services posed a barrier to moving into these new areas.
- These other "new" services were largely unproven in terms of market viability.
- Other national, regional, or global ventures and institutional entities providing such services as maritime communications, regional communications, direct-broadcasting to home antennas, and remote sensing grew up over the years. Thus it was not easy for INTELSAT to expand as these organizations developed more specialized markets.

In the quarter of a century that followed the first commercial satellite operations, a remarkable array of technical developments has ensued. Key innovations have included: multidestination services among and between very small aperture antennas; the use of more frequency bands; three-axis stabilization, rather than satellites rotating about their vertical axis; and large, high-performance antennas on board the satellites themselves.

The last three decades of satellite technology development can be best shown perhaps by the evolution of the satellites of the INTELSAT system. During this period, many major technological advances occurred, including the seven shown in Table I–1. The cumulative result of these technological gains has been to produce fixed satellite designs that are overall approximately 1,000 times more cost-effective than the Early Bird satellite. In total, the last three decades have produced satellites that are at least eighty times more effective in terms of power, are 100 times more frequency efficient, and have more than ten times greater lifetimes. It is perhaps because fiber optic cables have achieved parallel developments in cost-efficiency on the Earth that the remarkable and sustained technological breakthroughs in satellite telecommunications are not more widely recognized or celebrated.

Most advances in communications satellite technology have originated within the United States; leading developers have been the scientists and engineers of such aerospace manufacturers as Ball Aerospace, Fairchild, Hughes Aircraft, Lockheed-Martin (now including General Electric, or GE, and RCA), TRW, and Ford Aerospace (now Space Systems/Loral). There have been many other contributors, such as NASA, the Department of Defense, the National Science Foundation, universities, and research laboratories such as Lincoln Laboratory, Johns Hopkins Advanced Physics Laboratories, Comsat Laboratories, and the Jet Propulsion Laboratory (JPL) and other NASA centers.[18] In the last decade in particular, the spread of satellite technology has become truly global. Major capabilities exist in Europe, Russia, Canada, and Japan, and more are emerging in India, China, Korea, Brazil, Israel, and Australia.

---

18.   See John H. McElroy, ed., *Space Science and Applications* (New York: IEEE Press, 1986), pp. 183–284. Although it is difficult to single out precisely and without omission all of the individuals who played the most important roles in communications satellite development over this period, some of the most important players were: Wernher von Braun, Richard Marsten, Robert Lovell, and Leonard Jaffe of NASA; Harold Rosen and Albert "Bud" Wheelon of Hughes Aircraft; Adolph Thiel of TRW; Jack Harrington of Lincoln Laboratory; Sigfried Rieger, Ernst Dietrich, Martin Votaw, and John Johnson of Comsat and INTELSAT; Joseph Campanella, Wilbur Pritchard, and Burton Edelson of Comsat Laboratories; Kenneth Rose of Ford Aerospace; Jack Kiegler of GE/RCA; William Pickering of JPL; J.O. Pastore of the U.S. Senate; Edward Welsh of the National Space Council; and John Pierce of AT&T's Bell Laboratories.

Table I–1
Technological Advances

Area of Development   Measured Advance   Key New Technologies Developed

Area of Development	Measured Advance	Key New Technologies Developed
1. Satellite Power	Power increased 80 times	• Sun-oriented solar cell array • High-efficiency solar cells • High-performance N_2H_2 batteries
2. Effective Use of Radio Frequencies	Radio frequencies available at 100 times greater	• Use of hybrid frequency bands • Frequency reuse by spatial separation • Spot-beam antennas • Frequency reuse by polarization discrimination
3. Satellite Lifetime	Increase in lifetime from 1.5 to 15 years	• Longer life batteries • Higher performance thrusters and propellants • Solid-state electronics • Enhanced satellite control techniques (including option for inclined orbit operation)
4. Digital Communication Techniques and Digital Circuit Multiplication and Compression Techniques	Up to 1,000-percent increase through use of TDMA and CDMA plus digital compression techniques	• Development of 155.5-megabyte TDMA • Digital speech interpolation • 32- and 16-kilobyte-per-second voice bringing two- to fourfold gain, respectively
5. On-Board Satellite Switching	Exact gain not easily measured; expanded use of spot-beam antennas—and thus expanded frequency reuse—optimized by on-board interconnection of beams and cross strapping	• Satellite-switched TDMA • Hybrid frequency connection between uplink and downlink • On-board switching fault detection and diagnostics
6. Earth Station Antennas	Decrease in costs of Earth stations four- to tenfold while decrease in size of antennas by factor of ten	• Use of solid state electronics • Elimination of most cryogenics • Enhanced low-cost construction materials and improved construction
7. Launch Vehicle Technology	Launch reliability increased to nearly 90 percent and lift capability by several thousand percent; cost-efficiency of launching, however, on a pre-pound basis, not changed significantly	• Enhanced rocket motor design with greater thrust • Enhanced guidance systems

This unique development of new space communications technologies in the 1970s and 1980s did not simply spring up spontaneously. Within NASA, a series of experimental satellites under the Applications Technology Satellite (ATS) program were developed and tested during the 1967–1976 period. These satellites expanded the technological reach of satellite communications in terms of higher frequency bands, new antenna size and performance, and satellite power and stabilization. The ATS program also helped demonstrate the new technology required to boost overall communications satellite performance. These experimental satellites not only demonstrated new technology, but provided valuable educational services to the Caribbean, the South Pacific, Brazil, and India, as well as to rural and remote parts of the United States. NASA joined with Canada in the Communications Technology Satellite (CTS) program; after its launch in 1976, CTS demonstrated new techniques of space telecommunications that could operate with very small terminals.

As the communications satellite industry matured into the only major successful commercial application of space, controversy arose during the 1970s over continuing a government-funded research and development program in support of that industry. Some in the Nixon administration argued that such a program constituted a subsidy to a particular segment of the private sector—a role the government should not play. NASA, faced with this argument and the need to adjust to a rapidly declining budget in the post-Apollo period, decided in 1972 to terminate its support of communications-related research and development. [I-27]

This decision remained controversial for a number of years; by the late 1970s, NASA was being urged to reenter the area. [I-28] The program that NASA proposed, the Advanced Communications Technology Satellite (ACTS), had a difficult time getting White House approval for most of the 1980s; congressional and some mixed industrial pressure finally led to the program's going forward. [I-29] After its launch in 1993, however, ACTS went on to demonstrate a variety of new techniques for enhancing the performance of communications satellites, especially with regard to operating in the new Ka band (thirty to 200 hertz) and on-board processing of signals so as to interconnect a very large number of spot beams (narrow, very-high-power beams), which boosted frequency reuse and increased satellite throughout capacity. However, it is hard to measure the significant impact. By the end of September 1996, the Federal Communications Commission (FCC) filing deadline, about fifteen new Ka-band satellite systems with a combined estimated value of approximately $50 billion have been proposed to provide high-data-rate multimedia video sources to North America and/or the world. If completely displayed, this would mean more than 1,200 new satellites in geosynchronous, medium, and low-Earth orbits.[19]

Furthermore, the development of the Tracking and Data Relay Satellite System (TDRSS) has led the way in such areas as intersatellite links and communications between low and geostationary orbits, satellite-switched time division multiple access techniques that allow multiple users to employ the same transponder, and combined fixed and mobile satellite communications. Today, the Orion Satellite System is operating intersatellite links, and many of the proposed new multimedia satellites will offer intersatellite link capabilities.

The experimental programs funded by the Department of Defense have also been contributors to technology development. In addition to Lincoln Laboratory's LES-1 to LES-9 experimental satellites, there have been numerous missions designed by the

---

19. *Space 30: Thirty Year Overview of Space Applications and Exploration* (Washington, DC: Society of Satellite Professionals, 1989).

Aerospace Corporation and various tactical communication and Defense Satellite Communications System (DSCS) spacecraft launched over the years. Because many of these military satellites were built by contractor firms that also constructed commercial satellites, there was often effective technology transfer within those firms.

Domestic Communications Satellite Systems

The first domestic satellite systems were deployed in the Soviet Union (Molniya in 1965) and Canada (Anik in 1971). A June 1972 decision by the FCC opened the way to the use of satellites for domestic communications within the United States, thus opening up a large new market for satellite telecommunications. [I-26] The ability of such systems to provide service to rural and remote areas and to relay television and other broadcast services to very small aperture antennas was quickly proven. Over the last fifteen years, these early successes have resulted in seventeen countries developing their own operational or experimental domestic satellite systems and placing satellites in the geosynchronous orbit.[20] In addition, approximately fifty countries are obtaining domestic communications satellite service through their participation (typically through the lease of transponders) in international and regional satellite systems.

The beginning of the evolution toward regional systems can be attributed to EUTELSAT, the European Telecommunications Satellite Organization. This organization started in its provisional form on June 30, 1977. This was followed by ARABSAT, which became operational in 1985, even though the idea was developed eight years earlier. After ARABSAT, the trend shifted away from public consortia closely modeled on a scaled-down version of INTELSAT. Newer systems used a privately owned—and more competitive—approach. ASTRA was established in the 1980s to provide low-power direct-broadcast service in Europe. PanAmSat began providing private transatlantic services in 1988. In Asia, Palapa in 1980 and later ASIASAT in 1990 began to offer certain forms of both regional and domestic service, with APSTAR following suit four years later.

Conclusion

In the last quarter century, tremendous progress has been made in space communications. New and expanded frequency bands have been operationally proven, and various uses of communications satellites have been successfully demonstrated. Consistent gains have been made in frequency reuse, spacecraft power, reliability, and lifetime. Improvements in Earth station delay and performance, digital modulation, and digital coding help complete a picture of total performance gains of more than 1,000 times in the last three decades. The range of space communications services has evolved from international and domestic fixed satellite services to mobile satellite services, broadcast satellite services, and even intersatellite links.

In the past, the distinction was quite clear between the satellite servers known as Fixed Satellite Service (FSS), Mobile Satellite Service (MSS), Broadcast Satellite Service (BSS), and Radio Determination Satellite Service (RDSS). These designations, as developed by the International Telecommunications Union (ITU), were used to allocate frequencies. Ironically, as the ITU has gone to more and more precise definitions of frequency allocations, such as aeronautical mobile, maritime mobile, and land mobile satellite services,

---

20. The seventeen countries that have launched one or more domestic communications satellite systems are: Australia, Brazil, Canada, China, France, Germany, India, Indonesia, Italy, Japan, Luxembourg, Mexico, Spain, Sweden, the United States, the United Kingdom, and the Soviet Union.

the technology has been moving in the opposite direction. The satellites' characteristics have been moving together in terms of satellite power, antenna beams, and on-board processing. Many of the latest Ka-band satellites, such as General Electric's GE Star. can actually offer fixed, mobile broadcast, and navigational services.

Equally significant is that these new satellite systems, because they can work to microterminals (fifty to sixty-five centimeters in diameter) and to handheld transceivers, can "bypass" conventional terrestrial networks. Thus, it can be said that satellite communications systems are now becoming a truly large, mass consumer business that are starting to rival terrestrial telecommunications systems.

Innovation has not been limited to the technological arena. Beginning with a single global telecommunications satellite entity, INTELSAT, there has been a proliferation of organizational forms for bringing the promise of communications satellites into reality. The 1984 decision in the United States to modify the traditional U.S. position that INTELSAT was the only authorized provider of global communications satellite services was a key to this development. [I-30] Both public and private forms of institutionalizing communications satellite services have emerged, as have several creative hybrid public-private organizations. Clearly, today, new private and competitive forms of satellite communications are becoming predominant as both INTELSAT and INMARSAT are spinning off new commercial entities to provide new forms of satellite services.

The ability to communicate words, images, and data instantaneously around the globe has fundamentally changed the character of international and intercultural relations. Through this application of space technology, a "global village" has truly come into being.

Document I-1

Document title: Arthur C. Clarke, "The Space-Station: Its Radio Applications," May 25, 1945.

Source: National Air and Space Museum Archives, Smithsonian Institution, Washington, D.C.

Document I-2

Document title: Arthur C. Clarke, "Extra-Terrestrial Relays: Can Rocket Stations Give World-Wide Radio Coverage?," *Wireless World*, October 1945, pp. 305–308.

The Russian theorist Konstantin Tsiolkovsky was the first to note that a satellite orbiting 22,300 miles above the Earth's surface would travel at a speed that would make it appear to be stationary from Earth because its orbital velocity would be the same as the speed at which the Earth was rotating. In 1928, Herman Potôcnik, an Austrian Imperial Army officer, writing under the pseudonym Noordung, proposed a crewed space station in such a "geosynchronous" orbit, to be used for meteorology, reconnaissance, and Earth mapping. However, it was Arthur C. Clarke that first called widespread attention to the utility of the geosynchronous orbit for communications. In May 1945, Clarke, a physicist and at that time the secretary of the British Interplanetary Society, circulated six copies of his paper "The Space-Station: Its Radio Applications" to his society colleagues. (The paper was not actually published until 1968, when it appeared in the society's Spaceflight *magazine.) A second paper, written in June 1945, appeared in the October 1945 issue of* Wireless World.

Document I-1

[no page number]

The Space-Station: Its Radio Applications

Arthur C. Clarke
25 May 1945

[1] 1.   The Space-station was originally conceived as a refueling depot for ships leaving the Earth. As such it may fill an important though transient role in the conquest of space, during the period when chemical fuels are employed. Other uses, some of them rather fantastic, have been suggested for the space-station, notably by Hermann Noordung.[1] However, there is at least one purpose for which the station is ideally suited and indeed has no practical alternative. This is the provision of world-wide ultra-high-frequency radio services, including television.

2.   In the following discussion the word "television" will be used exclusively but it must be understood to cover all services using the u.h.f. spectrum and higher. It is probable that television may be among the least important of these as technical developments occur. Other examples are frequency modulation, facsimile (capable of transmitting 100,000 pages an hour[2]), specialized scientific and business services, and navigational aids.

3.   Owing to bandwidth considerations television is restricted to the frequency range above 50-Mc/sec [megacycles per second], and there is no doubt that very much higher frequencies will be used in the immediate future. The American Telephone and Telegraph Company are [sic] already building an experimental network using frequencies up to 12,000 megacycles.[3] Waves of such frequencies are transmitted along quasi-optical paths and accordingly receiver and transmitter must lie not far from the line of sight. Although refraction increases the range, it is fair to say that the service radius for a television station is under 50 miles. (The range of the London service was rather less than this.) *As long as radio continues to be used for communication, this limitation will remain, as it is a fundamental and not a technical restriction.*

4.   Wide-band frequency-modulation, one of the most important of radio developments, comes in the same category. FM can give much better quality and freedom from interference than normal amplitude-modulation, and many hundreds of stations are being planned for the post-war years in America alone. The technical requirements of FM make it essential that only the direct signal be used, and ionospheric reflexions cannot be employed. The range of the service is thus limited by the curvature of the Earth, precisely as for television.

5.   To provide services over a large area it is necessary to build numerous stations on high ground or with radiators on towers several hundred feet high. These stations have to be linked by landline or subsidiary radio circuits. Such a system is practicable in a small country such as Britain, but even here the expense will be enormous. It is quite prohibitive in the case of a large continent and it therefore seems likely that only highly populated communities will be able to have television services.

6.   An even more serious problem arises when an attempt is made to link television systems in different parts of the globe. Theoretical studies[2] indicate that using a radio relay system, repeater stations will be necessary at intervals of less than fifty miles. These will take the form of towers several hundred feet high, carrying receivers, amplifiers and transmitters. To link regions several thousand miles apart will thus cost many millions of pounds, and the problem of trans-oceanic services remains insoluble.

7.   In the near future, the large airliners which will fly great circle routes over oceans and uninhabited regions of the world will require television and allied services and there is no known [2] manner in which these can be provided.

8.   All these problems can be solved by the use of a chain of space-stations with an orbital period of 24 hours, which would require them to be at a distance of 42,000 Km [kilometers] from the centre of the earth. (Fig 1.) There are a number of possible arrangements for such a chain but that shown is the simplest. The stations would lie in the earth's equatorial plane and would thus always remain fixed in the same spots in the sky, from the point of view of terrestrial observers. Unlike all other heavenly bodies they would never rise nor set. This would greatly simplify the use of directive receivers installed on the earth.

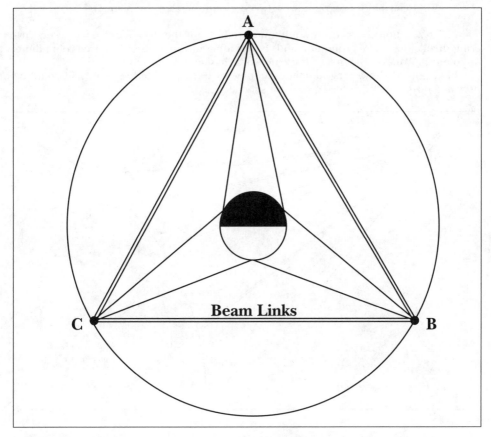

Figure 1.

9.   The following longitudes are provisionally suggested for the stations to provide the best service to the inhabited portions of the globe, though all parts of the planet will be covered.

30 E—Africa and Europe.

150 E—China and Oceana.

90 W—The Americas.

10. Each station would broadcast programmes over about a third of the planet.

Assuming the use of a frequency of 3,000 megacycles, a [3] reflector only a few feet across would give a beam so directive that almost all the power would be concentrated on the earth. Arrays a metre or so in diameter could be used to illuminate single countries if a more restricted service was required.

11. The stations would be connected with each other by very-narrow-beam, low-power links, probably working in the optical spectrum or near it, so that beams less than a degree wide could be produced.

12. The system would provide the following services which cannot be realized in any other manner:

    a) Simultaneous television broadcasts to the entire globe, including services to aircraft.

    b) Relaying of programmes between distant parts of the planet.

13. In addition the stations would make redundant the network of relay towers covering the main areas of civilisation and representing investments of hundreds of millions of pounds. (Work on the first of these networks has already started.)

14. Figure II shows diagrammatically some of the specialised services that could be provided by the use of differing radiator systems.

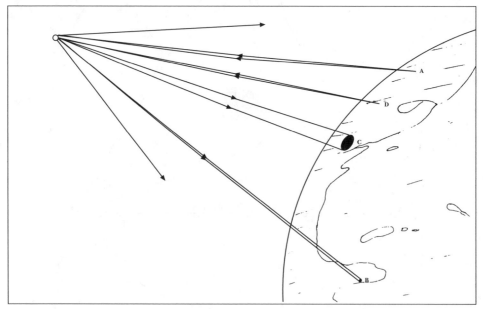

Figure II.

    Programme from A being relayed to point B and area C.
    Programme from D being relayed to whole hemisphere.

[4] 15. The numerous technical problems involved in this communication system cannot be discussed here but it can be stated that none of them present any difficulties even at the present time, thanks to the development of hyperfrequency engineering. It is hoped to discuss them in a later paper when security conditions permit.

16. The receiving equipment at the earth end would consist of small parabolas perhaps a foot in diameter with dipole pickup. These would be sufficiently directive to prevent interference in the three doubly-illuminated zones. They would be aimed towards the station with the least zenithal distance and once adjusted need never be touched again. Mobile equipment would require automatic following which presents slight mechanical complications (a few valves and a servo motor) but no technical difficulties.

17. The efficiency of the system would be nearly 100%, since almost all the power would fall on the service area. A preliminary investigation shows that the world broadcast would require about ten kilowatts, while the beam relay services would require only fractions of a kilowatt. These powers are very small compared with present-day broadcasting stations, some of which radiate hundreds of kilowatts. All the power required for a large number of simultaneous services could be obtained from solar generators with mirrors about ten metres in radius, assuming an efficiency of about 40%. In addition, the conditions of vacuum make it easy to use large and fully demountable valves.

18. No communication development which can be imagined will render the chain of stations obsolete and since it fills what will eventually be an urgent need, its economic value will be enormous.

19. For completeness, other major uses of the station are listed below:—
    a)   Research.—Astrophysical, Physical, Electronic.

These applications are obvious. The space-station would be justified on these grounds alone, as there are many experiments which can only be conducted above the atmosphere.
    b)   Meteorological.

The station would be absolutely invaluable for weather forecasting as the movement of fronts, etc. would be visible from space.
    c)   Traffic.

This is looking a good deal further ahead, but ultimately the chain will be used extensively for controlling and checking, possibly by radar, the movement of ships approaching or leaving the earth. It will also play an extremely important role as the first link in the solar communication system.

References

1. Noordung, Hermann. "Das Problem der Befahrung des Weltraums."
2. Hansell, C. W. "Radio-Relay-Systems Development." (*Proceedings of the Institute of Radio Engineers,* March 1945, pp 156 - 168.)
3. Guy, Raymond F. Address to I.R.E., Philadelphia, December 7th, 1944.

Document I-2

October 1945
Wireless World

[305]

Extra-Terrestrial Relays:
Can Rocket Stations Give World-Wide Radio Coverage?

By Arthur C. Clarke

[original set in three columns of newspaper style text per page]

Although it is possible, by a suitable choice of frequencies and routes, to provide telephony circuits between any two points or regions of the earth for a large part of the time, long-distance communication is greatly hampered by the peculiarities of the ionosphere, and there are even occasions when it may be impossible. A true broadcast service, giving constant field strength at all times over the whole globe would be invaluable, not to say indispensable, in a world society.

Unsatisfactory though the telephony and telegraph position is, that of television is far worse, since ionospheric transmission cannot be employed at all. The service area of a television station, even on a very good site, is only about a hundred miles across. To cover a small country such as Great Britain would require a network of transmitters, connected by coaxial lines, waveguides or VHF relay links. A recent theoretical study[1] has shown that such a system would require repeaters at intervals of fifty miles or less. A system of this kind could provide television coverage, at a very considerable cost, over the whole of a small country. It would be out of the question to provide a large continent with such a service, and only the main centres of population could be included in the network.

The problem is equally serious when an attempt is made to link television services in different parts of the globe. A relay chain several thousand miles long would cost millions, and transoceanic services would still be impossible. Similar considerations apply to the provision of wide-band frequency modulation and other services, such as high-speed facsimile[,] which are by their nature restricted to the ultra-high-frequencies.

Many may consider the solution proposed in this discussion too far-fetched to be taken very seriously. Such an altitude is unreasonable, as everything envisaged here is a logical extension of developments in the last ten years—in particular the perfection of the long-range rocket of which V2 was the prototype. While this article was being written, it was announced that the Germans were considering a similar project, which they believed possible within fifty to a hundred years.

Before proceeding further, it is necessary to discuss briefly certain fundamental laws of rocket propulsion and "astronautics." A rocket which achieved a sufficiently great speed in flight outside the earth's atmosphere would never return. This "orbital" velocity is 8 km per sec. (5 miles per sec.), and a rocket which attained it would become an artificial satellite, circling the world for ever with no expenditure of power—a second moon, in fact. The German transatlantic rocket A10 would have reached more than half this velocity.

It will be possible in a few more years to build radio controlled rockets which can be steered into such orbits beyond the limits of the atmosphere and left to broadcast scientific information back to the earth. A little later, manned rockets will be able to make similar flights with sufficient excess power to break the orbit and return to earth.

There are an infinite number of possible stable orbits, circular and elliptical, in which a rocket would remain if the initial conditions were correct. The velocity of 8 km/sec.

applies only to the closest possible orbit, one just outside the atmosphere, and the period of revolution would be about 90 minutes. As the radius of the orbit increases the velocity decreases, since gravity is diminishing and less centrifugal force is needed to balance it. Fig. 1 shows this graphically. The moon, of course, is a particular case and would lie on the curves of Fig. 1 if they were produced. The proposed German space-stations [Figure 1] would have a period of about four and a half hours.

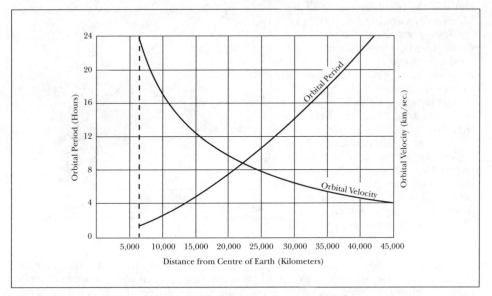

Figure 1. Variation of orbital period and velocity with distance from the center of the earth.

It will be observed that one orbit, with a radius of 42,000 km, has a period of exactly 24 hours. A body in such an orbit, if its plane coincide with that of the [306] earth's equator, would revolve with the earth and would thus be stationary above the same spot on the planet. It would remain fixed in the sky of a whole hemisphere and unlike all other heavenly bodies would neither rise nor set. A body in a smaller orbit would revolve more quickly than the earth and so would rise in the west, as indeed happens with the inner moon of Mars.

Using material ferried up by rockets, it would be possible to construct a "space-station" in such an orbit. The station could be provided with living quarters, laboratories and everything needed for the comfort of its crew, who would be relieved and provisioned by a regular rocket service. This project might be undertaken for purely scientific reasons as it would contribute enormously to our knowledge of astronomy, physics and meteorology. A good deal of literature has already been written on the subject.[2]

Although such an undertaking may seem fantastic, it requires for its fulfillment rockets only twice as fast as those already in the design stage. Since the gravitational stresses involved in the structure are negligible, only the very lightest materials would be necessary and the station could be as large as required.

Let us now suppose that such a station were built in this orbit. It could be provided with receiving and transmitting equipment (the problem of power will be discussed later) and could act as a repeater to relay transmissions between any two points on the hemisphere beneath, using any frequency which will penetrate the ionosphere. If directive arrays were used, the power requirements would be very small, as direct line of sight transmission would be used. There is the further important point that arrays on the earth, once set up, could remain fixed indefinitely.

Moreover, a transmission received from any point on the hemisphere could be broadcast to the whole of the visible face of the globe, and thus the requirements of all possible services would be met (Fig. 2).

It may be argued that we have as yet no direct evidence of radio waves passing between the surface [Figure 2] of the earth and outer space; all we can say with certainty is that the shorter wavelengths are not reflected back to earth. Direct evidence of field strength above the earth's atmosphere could be obtained by V2 rocket technique, and it is to be hoped that someone will do something about this soon as there must be quite a surplus stock somewhere! Alternatively, given sufficient transmitting power, we might obtain the necessary evidence by exploring for echoes from the moon. In the mean-

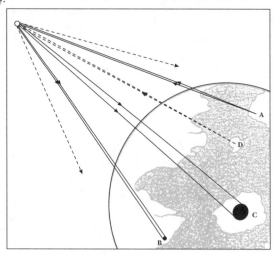

Figure 2. Typical extra-terrestrial relay services. Transmission from A being relayed to points B and area C; transmission from D being relayed to whole hemisphere.

time we have visual evidence that frequencies at the optical end of the spectrum pass through with little absorption except at certain frequencies at which resonance effects occur. Medium high frequencies go through the E layer twice to be reflected from the F [Figure 3] layer and echoes have been received from meteors in or above the F layer. It seems fairly certain that frequencies from, say, 50 Mc/s to 100,000 Mc/s could be used without undue absorption in the atmosphere or the ionosphere.

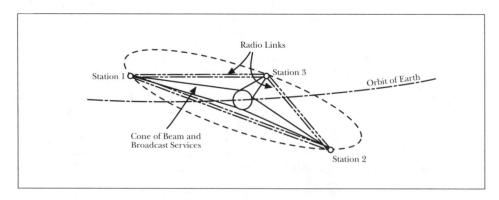

Figure 3. Three satellite stations would ensure complete coverage of the globe.

A single station could only provide coverage to half the globe, and for a world service three would be required, though more could be readily utilized. Fig. 3 shows the simplest arrangement. The stations would be arranged approximately equidistantly around the earth, and the following longitudes appear to be suitable:—

30 E—Africa and Europe.

150 E—China and Oceana.

90 W—The Americas.

The stations in the chain would be linked by radio or optical beams, and thus any conceivable beam or broadcast service could be provided.

The technical problems involved in the design of such stations are extremely interesting,[3] but only a few can be gone into here. Batteries of parabolic reflectors would be provided, of apertures depending on the frequencies employed. Assuming the use of 3,000 Mc/s waves, mirrors about a metre across would beam almost all the power on to the earth. Larger reflectors could be used to illuminate single countries or regions for the more restricted services, with [307] consequent economy of power. On the higher frequencies it is not difficult to produce beams less than a degree in width, and, as mentioned before, there would be no physical limitations on the size of the mirrors. (From the space station, the disc of the earth would be a little over 17 degrees across). The same mirrors could be used for many different transmissions if precautions were taken to avoid cross modulation.

It is clear from the nature of the system that the power needed will be much less than that required for any other arrangement, since all the energy radiated can be uniformly distributed over the service area, and none is wasted. An approximate estimate of the power required for the broadcast service from a single station can be made as follows:—

The field strength in the equatorial plane of a $\lambda/2$ dipole in free space at a distance of d metres is

$$e = 6.85 \frac{\sqrt{P}}{d} \text{ volts/metre, where P is the power radiated in watts.}$$

Taking d as 42,000 km (effectively it would be less), we have $P = 37.6\ e^2$ watts. (e now in μV/metre.)

If we assume e to be 50 microvolts/metre, which is the F.C.C. standard for frequency modulation, P will be 94 kW [kilowatts]. This is the power required for a single dipole, and not an array which would concentrate all the power on the earth. Such an array would have a gain over a simple dipole of about 80. The power required for the broadcast service would thus be about 1.2 kW.[4]

Ridiculously small though it is, this figure is probably much too generous. Small parabolas about a foot in diameter would be used for receiving at the earth end and would give a very good signal/noise ratio. There would be very little interference, partly because of the frequency used and partly because the mirrors would be pointing towards the sky which could contain no other source of signal. A field strength of 10 microvolts/metre might well be ample, and this would require a transmitter output of only 50 watts.

When it is remembered that these figure relate to the broadcast service, the efficiency of the system will be realised. The point-to-point beam transmissions might need powers of only 10 watts or so. These figures, of course, would need correction for ionospheric and atmospheric absorption, but that would be quite small over most of the band. The slight falling off in field strength due to this cause towards the edge of the service area could be readily corrected by a non-uniform radiator.

The efficiency of the system is strikingly revealed when we consider that the London Television service required about 3 kW average power for an area less than fifty miles in radius.[5]

A second fundamental problem is the provision of electrical energy to run the large number of transmitters required for the different services. In space beyond the atmosphere, a square metre normal to the solar radiation intercepts 1.35 kW of energy.[6] Solar engines have already been devised for terrestrial use and are an economic proposition in tropical countries. They employ mirrors to concentrate sunlight on the boiler of a low-pressure steam engine. Although this arrangement is not very efficient it could be made much more so in space where the operating components are in a vacuum, the radiation is intense and continuous, and the low-temperature end of the cycle could be not far from absolute zero. Thermo-electric and photo-electric developments may make it possible to utilize the solar energy more directly.

Though there is no limit to the size of the mirrors that could be built, one fifty metres in radius would intercept over 10,000 kW and at least a quarter of this energy should be available for use.

The station would be in continuous sunlight except for some weeks around the equinoxes, when it would enter the earth's shadow for a few minutes every day. Fig. 4 shows the state of affairs during the eclipse period. For [308] this calculation, it is legitimate to consider the earth as fixed and the sun as moving round it. The station would gaze the earth's shadow at A, on the last day in February. Every day, as it made its diurnal revolution, it would cut more deeply into the shadow, undergoing its period of maximum [Figure 4] eclipse on March 21st. On that day it would only be in darkness for 1 hour 9 minutes. From then onwards the period of eclipse would shorten, and after April 11th (B) the station would be in continuous sunlight again until the same thing happened six months later at the autumn equinox, between September 12th and October 14th. The total period of darkness would be about two days per year, and as the longest period of eclipse would be little more than an hour there should be no difficulty in storing enough power for an uninterrupted service.[7]

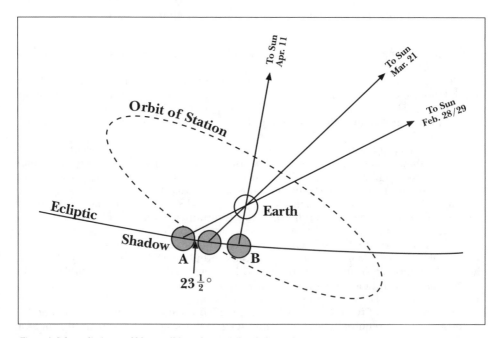

Figure 4. Solar radiation would be cut off for a short period each day at the equinoxes.

Conclusion

Briefly summarized, the advantages of the space station are as follows:—

(1) It is the only way in which true world coverage can be achieved for all possible types of service.

(2) It permits unrestricted use of a band at least 100,000 Mc/s wide, and with the use of beams an almost unlimited number of channels would be available.

(3) The power requirements are extremely small since the efficiency of "illumination" will be almost 100 percent. Moreover, the cost of the power would be very low.

(4) However great the initial expense, it would only be a fraction of that required for the world networks replaced, and the running costs would be incomparably less.

Appendix—Rocket Design

The development of rockets sufficiently powerful to reach "orbital" and even "escape" velocity is not only a matter of years. The following figures may be of interest in this connection.

The rocket has to acquire a final velocity of 8 km/sec. Allowing 2 km/sec. for navigational corrections and air resistance loss (this is legitimate as all space-rockets will be launched from very high country) gives a total velocity needed of 10 km/sec. The fundamental equation of rocket motion is[2]

$$V = v \, log_e R$$

where V is the final velocity of the rocket, v the exhaust velocity and R the ratio of initial mass to final mass (payload plus structure). So far v has been about 2–2.5 km/sec for liquid fuel rockets but new designs and fuels will permit of considerably higher figures. (Oxy-hydrogen fuel has a theoretical exhaust velocity of 5.2 km/sec and more powerful combinations are known.) If we assume v to be 3.3 km/sec, R will be 20 to 1. However, owing to its finite acceleration, the rocket loses velocity as a result of gravitational retardation. If its acceleration (assumed constant) is a metres/sec.[2], then the necessary ratio R_g is increased to

$$R_g = R^{\frac{\alpha + g}{\alpha}}$$

For an automatically controlled rocket α would be about $5g$ and so the necessary R would be 37 to 1. Such ratios cannot be realised with a single rocket but can be attained by "step-rockets,"[2] while very much higher ratios (up to 1,000 to 1) can be achieved by the principle of "cellular construction."[3]

Epilogue—Atomic Power

The advent of atomic power has at one bound brought space travel half a century nearer. It seems unlikely that we will have to wait as much as twenty years before atomic-powered rockets are developed, and such rockets could reach even the remoter planets with a fantastically small fuel/mass ratio—only a few percent. The equations developed in the appendix still hold, but v will be increased by a factor of about a thousand.

In view of these facts, it appears hardly worth while to expend much effort on the building of long-distance relay chains. Even the local networks which will soon be under construction may have a working life of only 20–30 years.

References

1. "Radio-Relay Systems," C. W. Hansell, *Proc. I.R.E.*, Vol 33, March, 1945.
2. "Rockets," Willy Ley. (Viking Press, N.Y.)
3. "Das Problem der Befahrung des Weltraums," Hermann Noordung.
4. "Frequency Modulation," A. Hund. (McGraw Hill.)
5. "London Television Service," MacNamara and Birkinshaw. *J.I.E.E.*, Dec., 1938.
6. "The Sun," C. G. Abbott. (Appleton-Century Co.)
7. *Journal of the British Interplanetary Society*, Jan., 1939.

Document I-3

Document title: John R. Pierce, "Exotic Radio Communications," *Bell Laboratories Records*, September 1959, pp. 323–329 (reprinted with permission).

Source: AT&T Archives, Warren, New Jersey.

In the 1950s, John R. Pierce and his research team at AT&T's Bell Laboratories began exploring the possibility of communicating via satellites. This review article, his second major one on the subject, examines the potential for using satellites in either geosynchronous orbit or in lower orbits to receive signals from the Earth and retransmit them to another location. Similar to his April 1955 article for Jet Propulsion, *which discussed in more technical terms some of the initial concepts for different types of communications satellites and orbital radio relays, this article describes some of the "state of the art" experiments that he and his team proposed to carry out using the large Echo 1 satellite to be orbited by NASA as a passive reflector of signals originating from the Earth. By 1960, Bell Labs had decided that the technical obstacles to a medium-altitude active repeater communications satellite could be solved, and AT&T consequently proposed the experimental Telstar program to the government.*

---

[323]   J. R. Pierce

Exotic Radio Communications

[original set in two columns of newspaper style text per page]

Pioneering work often seems exotic in its inception. Only a very few years ago, the idea of launching an artificial satellite seemed exotic, if not scatterbrained. But satellites have become almost commonplace. Today's exoticism is space flight by human beings, and we do not know what tomorrow's might be.

In the early part of this century, it would have taken an incorrigible visionary to foresee the present Bell System direct distance dialing network, undersea telephone cables, coaxial cable systems, and transcontinental microwave radio-relay routes. These all grew out of work which in its inception seemed far from any practical reality. It is an important part of Bell Laboratories activities to look far ahead—to study possible future communications services and thus build a fund of knowledge to draw upon if these services should become economically attractive.

In this article I shall deal primarily with some of the pioneering work at Bell Laboratories which may someday be important to the Bell System in providing broadband transoceanic radio communication. And to introduce this subject, I shall first briefly review some of our past accomplishments. My purpose is not merely to present a list of important radio research projects. Rather, I hope to illustrate the importance of good scientific and engineering work and to show the value of the Bell System pattern of careful study, measurement and design.

Radio itself seemed exotic in an earlier day. Before the founding of Bell Laboratories in 1925, the A.T.&T. Co. and Western Electric contributed heavily to the technology of radio broadcasting. Even earlier, in 1915, A.T.&T. and Western made use of newly developed power vacuum tubes in demonstrating radio communication between Arlington, Virginia, and both Hawaii and Paris. This showed a potentiality for transoceanic communication which could not be overlooked. One result was the first use of radio for commercial telephone service, from the mainland in California to Catalina Island in 1920. The work also led directly to experiments in transatlantic telephony as early as 1923, and to the inauguration of commercial transatlantic telephone service in 1927.

[324]

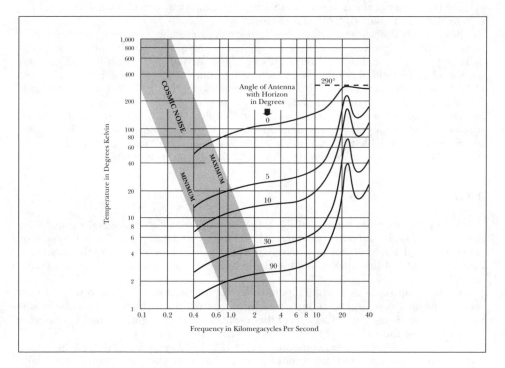

Theoretical atmospheric noise (in degrees K) versus frequency as an ideal antenna points at various angles to the horizon. The color region indicates the expected range of cosmic noise.

In this early long-wave work, accurate measurements of field strengths were made. A form of modulation was adopted—in this case the first use on radio of the single-sideband technique—which was best suited to the nature of the medium and the needs of the system.

The value of this type of approach was again illustrated with short-wave radio, put into commercial service in 1928. A tricky sort of communication, short-wave propagation shows both long-term variations of signal strength and rapid fading. Careful measurements by Bell Laboratories workers showed that such fading has a multipath nature—that is, radio waves in bouncing different numbers of times between the earth and the ionosphere alternately add and subtract in the radio receiver. These measurements also showed a rapid variation in the direction from which signal components arrive at a receiving antenna and especially in the vertical angle of arrival.

Such extensive and accurate measurements made it clear that operating frequencies should be changed from time to time to suit the condition of the ionosphere. As a replacement for the early narrow-band antenna arrays, the simple rhombic antenna invented at Bell Laboratories permitted effective operation over the required wider band of frequencies. In following the more rapid variations in angle of arrival, the MUSA system—an array of rhombic antennas interconnected with phase-changing networks—made it possible for a receiver to track the observed changes in the vertical angle of arrival of the radio signal.

As a part of the careful studies of short-wave phenomena at Bell Laboratories from 1929 through 1931, K. G. Jansky investigated noise in the short-wave bands at the Holmdel Laboratory. In the course of these studies, he detected radio noise of extra-terrestrial origin—work which laid the basis of radio astronomy. In recognition of Mr. Jansky's discovery, the laboratory at the new National Radio Observatory at Green Bank, West Virginia, is to be named the Karl G. Jansky Laboratory.

Besides this short-wave work, higher frequencies were also explored, and much fundamental knowledge was gained. This was applied in providing a number of over-water circuits and in mobile radio. However, the next large-scale Bell System application of radio was found in the field of microwaves, which have frequencies of thousands of megacycles. G. C. Southworth started his microwave work as early as 1932, long before any use for such frequencies could be assured. H. T. Friis and his associates took up this work in 1938.

Here again we see how early scientific and exploratory work led to extensive measurements and studies, and to the development of a sound technical art. The knowledge so gained was invaluable to radar during World War II, and later made possible the experimental New York-Boston System in 1947 and the Transcontinental TD-2 Radio-Relay System in 1951.

Reliable Microwave Service

As in previous cases, there was a lot to learn. Studies of microwave paths proved the value of using highly directive antennas. These studies set a pattern in the Bell System of using very good, narrow-beam antennas that allow the use of low power and that minimize interference. All of this work showed that microwaves could provide very reliable service indeed.

We now approach more contemporary developments, and it is time to remind ourselves again that it is largely an illusion to think of such past achievements as commonplace. They were, and certain aspects of them still are, as challenging as anything we have in mind for the future.

Current thinking in radio communications [325] still emphasizes the use of higher and higher frequencies, but direction of propagation is another important factor. Many intriguing problems and possibilities arise when we direct antennas toward the troposphere and the ionosphere, and toward satellites. Some of these problems were foreshadowed at Bell Laboratories as early as 1934 when A. M. Skellett and W. M. Goodall, in their studies of the ionosphere, looked for reflections from ionized meteor trails. The frequencies of 2 to 6 mc [megacycles] used in these studies were too low to give a strong signal, but statistical analysis of the data seemed to yield evidence for such reflections.

In other studies of the ionosphere, workers outside the Laboratories proposed in 1951 to use the turbulence of the ionosphere to achieve beyond-the-horizon scatter propagation. At a frequency of 50 mc, a 776-mile circuit was established between Cedar Rapids, Iowa, and Sterling, Virginia. Bell Laboratories monitored these signals, and with carefully designed antennas was able to receive teletypewriter messages during 1951 to 1954. In the course of this work, very high signal strengths were detected for very short periods. These observations indicated strong reflections from meteor trails, a verification of Skellett's and Goodall's early ideas.

However, ionospheric scattering proved disappointing for long-distance telephone circuits. For both turbulence scatter and meteor-trail scatter, the bandwidth is too narrow and transmission is too erratic. The more important region for the scatter technique proved to be the lower-altitude troposphere.

Kenneth Bullington did the pioneering work in this field. During 1950–1951, he collected data on and tested what we now know to be tropospheric scatter propagation, over paths 200 to 300 miles long. He pointed out the possibilities of this mode of transmission in historic papers published in the *Proceedings of the I.R.E.* in 1950 and 1953.

Beginning in 1955, further studies of scatter propagation were carried out over a path between a 60-foot scanning antenna at the Holmdel Laboratory and a transmitter on a farm in Pharsalia, New York, 171 miles away. The effects of antenna size, signal strength, depth and speed of fading, and angles of arrival were investigated. These data were compared with the predictions of a theory worked out by H. T. Friis, A. B. Crawford and D. C. Hogg. This theory supposes that the scattering is caused by a large number of randomly positioned but nearly horizontal discontinuities in dielectric constant in the first few miles above the earth's surface. The measurements fit the theoretical predictions very well in many respects. The knowledge acquired in these studies of tropospheric scatter is now widely used in designing scatter circuits.

Scatter Circuits in Operation

Scatter circuits designed by Bell Laboratories and installed by Western Electric are currently in operation over the DEW [Defense Early Warning] line in the far north and over the "White Alice" system in Alaska. In addition, a broad-band scatter system for commercial telephone and television service was established between Florida City and Havana in 1957. This Florida-Cuba circuit handles 36 telephone channels and has the capability of handling 120 or more.

If we now turn our attention to the problem of future broad-band radio transmission between North America and Europe, scatter circuits are an obvious suggestion. It might be possible, for example, to set up a series of relay stations via [326] Greenland, various North Atlantic islands, and Scotland. Our studies indicate, however, that this type of communications would be very expensive. Large antennas and high-power transmitters would have to be built and maintained at remote arctic locations. Further, the multipath nature of scatter transmission would probably result in a poor broad-band circuit by television standards, although several dozen telephone channels might be provided.

What, then, is another type of possible intercontinental radio communications? As early as 1954, we considered the use of artificial earth satellites as relays, and I published a technical paper on the subject. At that time, however, problems of launching such satellites were unexplored.

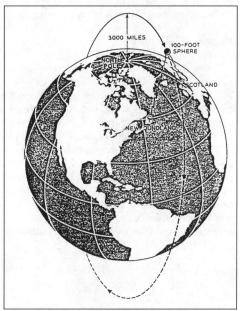

The geometry of a passive-reflector satellite in a polar orbit at an altitude of 3,000 miles, with terminals located in Newfoundland and Scotland.

The first Sputnik in October 1957 changed the picture radically, and we began to look at these possibilities much more seriously.

If satellite communication ever becomes a reality, it will be no exception to past Bell System experience. That is, it will necessarily be preceded by the established pattern of meticulous study and experimental work. At the moment, we do not have enough knowledge or experience to describe in detail a practicable system or to state exactly how it might be used. We can do little more than speculate on the various possibilities.

One proposal is to place satellites 22,400 miles above the equator. At this height, a satellite would rotate in step with the earth and seem always to hang in the same position in the sky. Such satellites would be "active" relay stations—that is, they would be equipped with receivers and transmitters, and probably with accurately pointed directive antennas. This is an apparently attractive proposal, but for the present it raises at least two serious questions: the problems of accurate rocketry to launch and orient such satellites are indeed formidable, and the problems of equipment life in such relay stations are, to put it mildly, severe.

A second proposal is to place active satellites in orbit only a few thousand miles above the earth. These would not be stationary in relation to the earth, but with a sufficient number of them, signals could be relayed from each whenever it is in a usable section of its orbit. With this second proposal, rocket accuracy is somewhat eased, but equipment life in a low-altitude relay station is as serious a question as in the case of the 22,400 mile satellite.

With low-altitude satellites, however, a transmitter and receiver on the satellite are not essential. Instead, one may put in orbit a group of passive reflectors. Large, high-power transmitters would then transmit to a satellite reflector, and signals would bounce from it and thus reach a distant receiver. The satellites would be aluminized plastic spheres—"balloons" perhaps 100 feet in diameter—with a high reflectivity to microwaves.

As an exercise to explore possibilities and problems, we have studied in some detail a theoretical system using passive satellites for transatlantic broad-band transmission. As shown in the illustration [of the passive-reflector satellite], the terminals were considered to be in Newfoundland and on an island in the Hebrides off Scotland. The satellites are to be imagined as traveling in polar orbits.

Regions of Visibility

On the polar projection shown below, closed contour lines are drawn for various heights of orbit. These define areas in which a single satellite would be simultaneously visible to both the Newfoundland and Scotland terminals. At a height of 2,000-miles, for example, a satellite would be visible to both terminals anywhere within the 2,000-mile contour, even along the outer edges of the area.

The first of the accompanying tables (*see below*) shows calculations relating to the orbit [327] heights. The shortest and longest visibility times in the second and third horizontal rows are of particular interest. For the 1,000-mile height, as an example, the zero for shortest visibility time indicates that for some passes, a satellite would not be visible at all at both of the two terminals. A 3,000-mile satellite, however, would be visible at least 31.4 minutes, and as long as 55.4 minutes, for every revolution around the earth. On the average it would be visible 22 percent of the time. Thus, even with only one satellite, one might get quite long stretches of broadband communication.

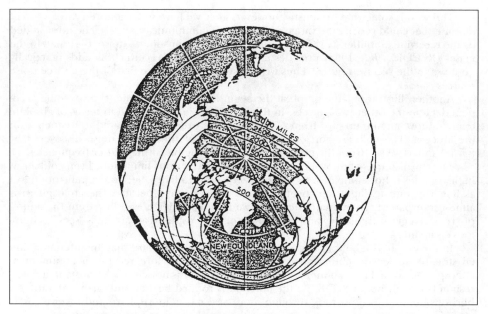

Oval-shaped areas define regions in which polar-orbit satellites at various heights would be simultaneously visible to both terminals at Newfoundland and Scotland. The usable areas would be somewhat smaller, however, since transmission is difficult when [the] satellite is near the horizon.

Visibility Times at Various Heights of Orbit

Height of Satellite Above Surface of Earth in Miles

	500	1000	1500	2000	2500	3000
Time of one revolution (minutes)	100.4	118.0	136.6	155.0	175.2	195.2
Shortest visibility (minutes)	0	0	8.0	12.5	23.8	31.4
Longest visibility (minutes)	14.7	20.0	29.6	36.6	46.2	55.4
Average visibility (percent)	3.5	6.9	12.9	17.7	19.6	22.0

Assumptions: Terminals in Newfoundland and Hebrides; polar orbits; refraction effects ignored; and visibility from horizon to horizon.

This view is somewhat optimistic, however, since we have so far ignored three sources of noise that could restrict the range of this type of communications: (1) The noise added by the receiving amplifier, (2) Cosmic noise, and (3) Atmospheric noise. Fortunately, the maser (RECORD, *July*, 1958) provides us with a microwave receiver that adds practically no noise to the received signal. Thus we can largely neglect the first of these three noise sources.

Another illustration [the graph at the beginning of this document] gives some pertinent data on the other two sources. In this graph, noise is described in terms of absolute temperature, ranging upward from 1° Kelvin on the ordinate, as related to frequency on the abscissa. The color region describes the range of the cosmic noise discovered by Jansky, and as we see from the graph, it becomes negligible at the higher frequencies.

The third source—atmospheric noise—is a more serious limitation. Even cold air at high altitudes is hot compared to absolute zero, so it radiates electromagnetic noise just as hot iron radiates light and heat. The radiation is small because the atmosphere is almost transparent. To evaluate the noise, we must consider how transparent the atmosphere is at a given frequency and also how much atmosphere an antenna "sees" as it follows a satellite.

In the graph, the bottom curve labeled 90 degrees illustrates that an antenna pointed straight up sees a minimum of atmosphere and therefore receives a minimum of atmospheric noise. From about 2 to 10 kilomegacycles this noise is fairly constant and corresponds to only about 2.5° K. As the antenna is rotated farther and farther toward the horizon, however, it must look through more and more atmosphere and receive correspondingly more noise. The zero curve at the top is the case where the antenna points horizontally; here it sees a very long atmospheric path, and the noise actually approaches the assumed atmospheric temperature (290° K) at very short wavelengths for which air is not very transparent to microwaves.

[328] Note, however, that the curves for the various angles are displaced downward toward the lower noise values as the angle above the horizon is increased. Even as close as 10 degrees above the horizon, an antenna will see only about 13° K of atmospheric noise. These curves make us feel that we can realize the advantages of the maser if we use signals from a satellite reflector only when it is 7 degrees or more above the horizon. This limitation in effect contracts the contour areas in the polar-projection map, which were drawn with the assumption that signals would be received right down to the horizon.

How serious is this limitation? Suppose we consider satellites 3,000 miles high and use them only when they are at least 7 degrees above the horizon. Average visibility per rotation will thus be less than the 22 percent listed in the first table, but if we put more and more satellites up, the result is an increase in the percentage of time that at least one satellite is visible. For 24 satellites, at least one satellite would be available to both Newfoundland and Scotland for 99 percent of the time. The interruptions would occur at predictable intervals, and would therefore be less serious than if they were random in time. The second table (*see below*) lists some other possibilities for different minimum angles and percentages of service interruptions.

**Number of Randomly Spaced Satellites Needed for Various
Minimum Elevation Angles and Percentages of Interruption**

Minimum Elevation Angle in Degrees	Percentage of Interruption		
	10%	5%	1%
0	9	12	19
3.25	11	14	21
7.25	12	15	24
12.60	17	22	33

Assumptions: Terminals in Newfoundland and Hebrides; Polar Orbits at 3000-mile height

The next obvious question is whether transmitters and antennas are available for such communication. Assuming an operating frequency of 2 kilomegacycles, a 40 db [decibel] signal-to-noise ratio, and 100-foot spherical reflectors at 3,000 miles, we have calculated that we would need antennas 150 feet in diameter and transmitter power of 100 kilowatts. Antennas of this size have been used, and the required power could be obtained by paralleling ten commercially available tubes. At present, however, we cannot be sure that the required type of satellite reflector would withstand the conditions of space and maintain its shape in orbit.

Need for Knowledge

At this stage the reader may feel that this is exoticism with a vengeance. We have perhaps raised more questions than those we have tried to answer. Aside from the problem of costs, which can hardly be handled definitively at this early date, the technical problems are extensive. But herein lies the point of this discussion—we need more fundamental knowledge of the possibilities before we can begin to think realistically of actual systems. And the only way we know of to get this knowledge is to continue our traditions of careful search, study of the problems, and measurement.

On March 19, T. Keith Glennan, Administrator of the National Aeronautics and Space Administration (NASA), announced plans to launch several satellite spheres next year. These experimental spheres, fabricated from an aluminized plastic, are to be 100 feet in diameter. The announcement also mentioned the plan to establish communications between an 85-foot tracking antenna at Goldstone, California, and communications facilities on the East Coast, including [329] Bell Laboratories equipment at Holmdel. The Goldstone antenna is operated by the Jet Propulsion Laboratory of Pasadena, California, which is owned by NASA and operated under contract to the NASA by the California Institute of Technology.

Plastic sphere, 100 feet in diameter, of [the] type to be used in satellite transmission experiments. In orbit, thin plastic with aluminized surface will reflect microwaves.

In connection with this type of work, then, what are some of the specific problems on which we have worked, and what are some of the problems concerning which we need additional knowledge?

While we have a very good maser in the 6,000 mc range, there is still some room for improvement, and we need masers for other frequencies. A related problem is that some types of antennas tend to pick up noise from all directions, so we are adapting the horn-reflector type antenna, which does not have this defect. With such equipment, we have already made measurements of sky temperatures which check the theoretical curves shown on page 324. We believe that these antennas may also have many uses in radio astronomy.

We have made some studies of the effect of ultraviolet light on the properties of aluminized Mylar, and for this material we have investigated absorption at various wavelengths to tell us the temperature a satellite might attain in space.

Other obvious fields for additional work are those of propagation measurements, guidance, and the many components besides masers and antennas that must go into any experimental system. A very important need is highly reliable components for experiments with active satellite repeaters. We have inaugurated work on such components.

This need for further study should emphasize that today we have no proven answers to the problem of overseas broad-band radio communications. If we ever turn on our television sets and view a European event beamed by radio across the Atlantic, it may come to us over a system no one has even thought of yet. But in the meantime, to assure the possibility we must continue to pursue a vigorous and effective research program.

Document I-4

Document Title: Memorandum from S.G. Lutz to A.V. Haeff, "Commercial Satellite Communication Project; Preliminary Report of Study Task Force," October 22, 1959.

Source: Hughes Space and Communications Company, Los Angeles, California (used with permission).

By 1959, work on communications satellite research and development was going on in several industrial firms besides Bell Laboratories. The Department of Defense had taken the lead in sponsoring research on active repeater satellites, while NASA concentrated its initial efforts on passive reflectors. In particular, the Department of Defense was supporting research on a complex satellite project called Advent, which intended to develop a satellite for use in geosynchronous orbit. An engineering team at Hughes Aircraft in mid-1959, led by Harold Rosen, devised a proposal for a much simpler geosynchronous satellite and asked the company to support its development. This memorandum reports to Hughes vice president for research, A.V. Haeff, the conclusions of an internal task force set up to assess the proposal of Rosen and his team, which also included Donald Williams and Thomas Hudspeth. (The appendices referred to in this memo are not included.) Over the following months, Hughes managers debated whether to provide support for the proposal from company funds or to seek government support for the project. Enough corporate funds were made available to keep the project going, but it was not until NASA contracted with Hughes to develop and demonstrate what became known as Syncom that the project became the foundation for the many geosynchronous satellites to follow.

[1]

Hughes Aircraft Company
Interdepartmental Correspondence

To: A. V. Haeff   **cc:** See Distribution   **Date:** 22 October 1959

Subject: Commercial Satellite Communication Project;   **From:** S. G. Lutz
        Preliminary Report of Study Task Force

1. It is the unanimous opinion of the Task Force working members* that the satellite communication system proposed by Dr. H. A. Rosen is technically feasible, is possible of realization within close to the estimated price and schedule, has great potential economic attractiveness and should not encounter too serious legal or political obstacles.

2. The Task Force has, of necessity, concentrated on technical aspects of the program and has not been able to make an adequate market survey. The phraseology, "great potential economic attractiveness[,]" is justified by the following:

    a. A rapidly increasing demand for new long-distance communication facilities is being created by: (1) Population increase, (2) Shrinkage of travel time via commercial jet aircraft, (3) Increasing foreign industrialization and international commerce, (4) Increasing military communication loads, and (5) Forthcoming decrease in HF [high-frequency] communication capability because of the declining sunspot cycle. Rather than being able to open more HF radio circuits to carry the increasing traffic, new circuits (cable, scatter or satellite) will be needed to pick up perhaps a third of the traffic now carried by HF circuits.

---

*   Task Force working members are: E. D. Felkel, S. G. Lutz, D. E. Miller, H. A. Rosen and J. H. Striebel.

b. The Bell System, which formerly depended on radio for intercontinental phone circuits, has been investing heavily <u>and</u> profitably in long submarine cables; four in the past few years. The first trans-Atlantic phone cable provided thirty-six circuits (about 140 kc [kilocycle] bandwidth), cost about $30,000,000.00, and reportedly paid out in its first two years. A second trans-Atlantic cable soon will be placed in service at a reported cost of $40,000,000.00, presumably for a similar number of circuits. Tropospheric scatter radio chains are comparable in cost and are geographically constrained.

c. Comparing the proposed satellite system ($5,000,000.00 for 4500 kc bandwidth) with submarine cable, it could carry up to thirty times as much traffic at one-sixth the investment!

[2] 3. Converting "potential" into "actual" economic attractiveness will depend on acquiring communication traffic, most probably via cooperative agreement with one or more communication common carriers. General Telephone may be the best prospect (certainly a better one than the complacent Bell System) because it is trying to gain stature despite Bell's long-distance monopoly. The proposed satellite system could bypass Bell land-lines in linking General's east-coast and west-coast systems, in addition to giving it non-Bell circuits to Europe and other continents. General Telephone also could negotiate more efficiently with the communication services of other countries and even other domestic companies (Western Union, etc.) than [Hughes Aircraft] could; not being a common carrier. This and related market survey problems seemed too sensitive to be explored adequately by the engineers of this task force, even if time and suitable contracts had been available. General Telephone need not be the only potential partner, of course, for even a smaller common carrier might supply enough traffic to get started. As few as six circuits (30 kc out of the available 4500 kc) to Europe should justify a five-million-dollar investment in proportion to submarine telephone cables.

4. . . . (15 October [Interdepartmental Correspondence] from Lutz to Haeff, Jerrems) lists three questions which define the scope of the market survey believed to be desirable. To this list should be added a study of the relative costs and outage times for splicing a broken cable vs replacing a dead satellite repeater. As a preliminary estimate, keeping a launching in readiness on Jarvis Island should be less expensive than keeping a cable ship in readiness and a new satellite could be put up in hours, instead of the weeks required to locate and repair a cable-break.

5. Technical aspects of the proposed program have been evaluated in more detail, and with higher confidence in the conclusions, than was possible with the preceeding [sic] economic aspects. The crux of the technical attractiveness of this program (and an important economic consideration as well) lies in quick-reaction capability at low cost. By being able to keep the weight of a simple broad-band repeater payload below 25 lbs, it can be put in stationary orbit by an inexpensive (one-third million dollars) solid-fuel Scout booster. Everyone else (NASA, RCA, Space Electronics, Signal Corps) has viewed a stationary orbit repeater as a more sophisticated, hence heavier device, with attitude control to use high gain antenna *beams* on the satellite. More payload weight requires a larger *liquid*-fueled rocket and severe logistic problems in transporting or making liquid oxygen for an equatorial launch. The alternative of launching from the U.S. and "dog-legging" into an equatorial orbit increases guidance problems and requires Saturn thrusts. Thus, NASA and others consider the stationary orbit communications repeater as a high-cost program for 1965-70. This Task Force has convinced itself of the feasibility of puttin [sic] 25 lbs, or possibly 30 lbs, into a useful quasi-stationary orbit with a Scout booster, of achieving a 4500 kc bandwidth repeater within this weight and of doing this within a year of the date that full funding is provided.

6. How can Hughes expect to do so much better than others? The answer does not lie in any startling but questionable innovations, inventions or break-throughs. Rather, the answer lies chiefly in application of the Hughes brand of System Engineering, plus exploiting Hughes competence in low-noise reception and traveling-wave-tube development. The starting point was to assume a quasi-stationary orbit (satellites held within about 5° angular limits of desired point on the stationary orbit), to be put there by a Scout booster. The limited payload weight to 30 lbs on the basis of Chance-Vought performance predictions, or to 25 lbs on derating the predicted velocity by 800 fps. This obviously limits the satellite transmitting power, energized from solar cells, to a watt or so. [3] Transmission at or near 2 kmc (the accepted optimum frequency for space communication) favors high antenna gain and use of traveling-wave-tubes. The nearest to a break-through was the assurance by Dr. J. T. Mandel of the feasibility of developing a 2.5 watt periodic PM focused 2 kmc high efficiency traveling-wave-tube of one pound, including its INDOX VI focusing magnets. The low satellite power is handled at the earth terminals by low noise (cooled maser or parametric) reception and very high antenn [sic] gain (58 db). In achieving the latter at reasonable costs, the quasi-stationary position of the satellite avoids the need for full azimuth and elevation control which has been made even 80 ft steerable parabolas so expensive. At similar cost, the beam from a 150 ft truncated parabola can be steered through a ±5° range. Thus, the burden is put on the earth-terminals, where it belongs. The satellite antenna design is a compromise between using an omni-directional antenna for maximum simplicity and using a 17° beam for maximum gain. While either of these extremes could be fatal, the compromise of a spin-stabilized doughnut pattern provides 6 to 9 db gain, with simplicity. Finally, with adequate design for a 14 db S/N ratio, the addition of frequency modulation raises the S/N ration to a commercial 32 db.

7. Because of the importance of assessing feasibility of staying within the weight capability of the Scout booster, Ed Felkel was named to the Task Force to analyze the weight of the payload package. His report (attached in Appendix B) shows confidence of keeping it safely within weight.

8. Putting the satellite in orbit and keeping it in position entails a sequence of individually-practicable operations within today's state of the art. Cumulatively, however, the multiplicity of stages plus operations of velocity adjustment, de-spinning, re-spinning and incremental orbit adjustment present a currently-indeterminable hazard to the success of any one firing. It is believed that a combination of (a) careful and conservative engineering with step-by-step pre-testing, (b) adequate training on analog simulators, (c) study of any troubles in earlier NASA Scout firings, and (d) adequate determination of the cause of any initial Hughes failure, will result in adequate probability of success within the programmed three tries. Admittedly, there can never be certainty of success in only three attempts. However, a fourth or subsequent firings should not increase the program cost proportionately.

9. As might be expected, the Task Force study has resulted in significant system improvements, by Dr. Rosen as well as by Task Force members and others. For example, the payload configuration has been broadened to improve spin-stability and has been stiffened by a central column. More important, perhaps, has been the swing away from design primarily for television relaying, with additional narrower i-f channels for other communication services, toward the simpler and more flexible and potentially rewarding approach of coordinated use of a broad-band single-channel repeater simultaneously by several earth-terminals. This mode of operation requires that earth-terminals equalize their transmitting powers by monitoring the spectrum from the satellite, rather than depending on AGC of separate i-f channels in the satellite to prevent a too-strong earth-signal from weakening other retransmissions. Also, this mode of operation provides flexi-

bility of bandwidth reapportionment between earth-terminals in accordance with shifting relative traffic loads. In short, this approach overcomes the "two at a time" limitation of most prior proposals and thus approached more closely the eventual many-user "exchange in orbit" concept. Furthermore, it accomplishes this without sacrificing television capability, requiring only that other traffic be limited during a television program and be kept out of the television band.

[4] 10. Determination and resolution of possible legal and political problems and governmental restrictions obviously is beyond the scope of this Task Force. A few of the possible problems will be mentioned. The usual difficulties with the Federal Communications Commission can be expected in obtaining a license for a new type radio service for frequencies have not yet been allocated. Similar, or worse, difficulties can be expected with the corresponding regulatory bodies of other nations where earth-terminals are located. Characteristically, the FCC makes no precedent-setting decisions without holding industry-wide hearings and these could be competitively detrimental. Furthermore, the State Department might become involved because of the international nature of this venture. Next, some governmental agency probably has control of Jarvis Island and would insist on approving its use. Finally, NASA probably would have to sanction the commercial sale and use of Scout boosters and could impose other controls on the program, such as requiring provision for removing dead repeaters from orbit, or provision for disabling their electronics in event that the project is abandoned with repeaters still in orbit. As a ray of sunshine, NASA's mission is non-military space technology. They have expressed encouragement toward commercial projects which would not require NASA funds. If NASA becomes "sold" on the proposed project, they might provide inestimable assistance in surmounting the other governmental obstacles. One recognizes that exploration by a Hughes representative of the above governmental restrictions could readily "leak" to competitors, or even to the press, and be highly detrimental. This danger can be avoided, it is believed, by retaining a consultant to make this preliminary investigation without disclosing his client or the details of the project.

11. The impact of the proposed program on the military services could be both good and bad. It would be conclusive proof of Hughes' competence to execute a major space program and in Hughes' confidence and initiative in undertaking it without governmental funds. Thus, it should put us in better competitive position for managing future governmental space projects. It could have a bad impact, however, in "showing up" the inefficiency of military satellite programs.

12. It is known that Bell, RCA and probably other large companies recognize the potential attractions of satellite communication and probably have program plans. It is reasonable to assume that Bell would plan to invest several times the cost of the trans-Atlantic cable in a big stationary orbit project, timed to the availability of big boosters, five or ten years hence. Pressure for additional international circuits may lead them to re-examine the feasibility of moving faster by using a smaller booster and lighter payload, much as we propose. Certainly they could be expected to do this if they learned that their chief competitor, General Telephone, planned such a program in cooperation with Hughes. Most of the prestige value and a portion of the economic value would be sacrificed if our communication satellite were not the first. This indicates the need for a quick decision and a fast program under tight security.

* * * * * *

RECOMMENDATIONS

I.   If another company gets into orbit first, much of the publicity and prestige value will be lost and we would have to compete for traffic. Furthermore, this must be a low-cost program and delays increase costs. Consequently, the program should be planned to start development now. The expensive commitments (for rockets, ground installations, etc.) can be deferred for a few months without delaying the launching date.

[5] II.   Fund the traveling-wave-tube development separately as a commercial product. A one-pound tube of this capability should find application in Signal Corps portable microwave relay repeaters, possibly in field television transmitters, as well as in other programs. A quarter-million for its development seems a normally good product development risk. This tube is the heart of the proposed satellite electronic system and will be its longest lead-time component.

III.   Fund the remainder of the payload development and earth-terminal (antenna and low-noise receiver), in an amount of about $850,000.00. Also, take an option three Scout boosters, plus necessary real-estate, etc. If this is too large a commitment in advance of completion of the comprehensive market survey and negotiations with potential customers, fund a sufficient fraction to carry the development program this long. Delaying the start of development would delay completion of the program correspondingly.

IV.   Explore with General Telephone Company, at top management level, their interest in a non-Bell long-distance and overseas capability and their willingness to cooperate as the common carrier in the proposed program. Avoid disclosing details which might permit General's electronic subsidiary, Sylvania, to attempt to replace us. Reach a working agreement which will permit prompt working-level discussions of General's cooperation in the program. If negotiations with General fail, try the next best company.

V.   A task force, or project team, consisting of key personnel loaned as required from several organizations—Communications Division, Research Laboratories, Systems Development Laboratories—should be set up to carry out the program.

Document I-5

Document title: H.A. Rosen and D.D. Williams, *Commercial Communications Satellite*, Report RDL/B-1, Engineering Division, Hughes Aircraft Company, January 1960.

Source: Hughes Space and Communications Company, Los Angeles, California (used with permission).

By the end of 1959, Harold A. Rosen and his team had reworked their initial mid-1959 design for a geosynchronous communications satellite into a form that was very close to what was actually first launched as Syncom 1 in 1963. This report describes that design; only relevant excerpts appear here. The report anticipated a NASA program for communications satellite research and development that might provide a source of funding to Hughes for developing the satellite; however, NASA first chose to support a lower orbit satellite proposed by RCA called Relay. During discussions with NASA in 1960, the space agency suggested to Hughes the use of a larger Thor Delta booster rather than the Scout booster specified in this report. This would allow the satellite to be launched from Cape Canaveral in Florida rather than the Jarvis Island launch site discussed in the report.

[i]
Commercial Communications Satellite

[1]   ## 1. INTRODUCTION - PROGRAM OBJECTIVES

This document describes an inexpensive communication satellite system for inter-continental transmission of television, telephonic, and teletype messages on a commercial basis.

The system proposed uses an active repeater in a satellite having a circular orbit in the plane of the earth's equator with a period of 24 hours. Such a satellite is generally recognized as the ultimate communication satellite because it remains stationary to the earth.

The NASA has a program which is expected to lead eventually to such a satellite. The schedule for the program is not firm, but NASA testimony to Congress* indicates that the goal is four to five years away. This conclusion is reached from the technical specifications that NASA has until now believed are necessary, involving heavy (800 to 3000 pounds), complex payloads with two to three years' life as an objective. As a more immediate program, NASA will put a number of 100-foot diameter passive balloon reflectors into orbit during this year. These balloons will be tracked by several organizations, and will provide valuable scientific information. However, such reflectors are not of any real commercial value because [of] large amounts of power per unit bandwidth and immense tracking antennas required to give even the intermittent coverage afforded by low-altitude orbits.

There are several military communication satellite programs now under way. None of these conflict with the commercial program proposed here; the military programs use high power active repeaters in low-altitude orbits in order to avoid any requirements for large antennas at the terminals.

The presently proposed commercial system can be put into operation within one year. This radical improvement in schedule is achieved primarily through the design of a very light (25 pound) satellite repeater, a design based on realistic objectives for satisfactory commercial [2] application. The light payload required with the present concept permits use of an inexpensive solid-propellant booster, the Scout. This results in a program cost of 5 million dollars.

The advantages of such a program would be severalfold. Financially, it is believed that the initial development, terminal installations, and launching costs could be recovered in a fraction of the first year's operation. It is expected that the useful life of a repeater will be about one year, and that the cost of replacing the repeater in orbit will be about 0.5 million dollars.

The suggested communication system is capable of large growth. The first repeater will cover most of the continental United States, all of Europe, all of South America, and much of Africa. An additional repeater would cover Hawaii, Australia, Japan, and other parts of the Orient. In addition to extension of geographic coverage, the existence of the communication link will result in an increase in foreign business which in turn will result in greater use of the facility. Extrapolation of recent trends in overseas messages shows that the present cable capacity between the U.S. and Europe will be exceeded within the next two years. Since the proposed facility is much less expensive than a cable, it is logical to expect this overflow to be handled by the proposed facility.

In addition to its commercial value, the proposed communication satellite should contribute greatly to national prestige and friendly foreign relations.

----

* "Hearings before the NASA Authorization Subcommittee of the Committee on Aeronautical and Space Sciences, United States Senate," U.S. Government Printing Office, 1959.

[3]   2. SUMMARY OF TECHNICAL FEATURES

The proposed communication system consists of a satellite repeater in a synchronous, equatorial orbit operating in conjunction with two or more ground terminals, each of which is linked by land lines or microwave relays to the appropriate domestic communication systems.

The repeater consists of a transistorized UHF receiver and an L-band (2 KMC) transmitter having a power output of 2.5 watts. Since the electrical power is supplied by solar cells, the useful life of the repeater is expected to be limited only by the life of the transmitting tube to about one year. Besides serving as the communication repeater, the receiver-transmitter is also used as a guidance signal repeater, and the receiver additionally acts as a command receiver.

The payload also contains a compressed nitrogen attitude and vernier velocity control system, which provides for proper illumination of the solar cells, correct aiming of the directional antenna, and precise adjustments of the orbit.

The ground terminals consist of a large aperture antenna shared by the 25-KW [kilowatt] UHF transmitter and the low noise L-band receiver. The antenna reflector will be fixed, and the small departures of the payload from an exactly stationary orbit will be followed by moving the antenna feed.

The satellite is launched using the NASA Scout, and two additional solid-propellant rockets are used to establish the desired orbit. The launching site will be Jarvis Island, an equatorial island approximately 1300 nautical miles south of Hawaii. The use of this suitably located equatorial site results in a large decrease in required propulsion system performance and guidance complexity.

Further technical detail is furnished by the following sections of this proposal. . . .

[25]   6. PROGRAM COST

An estimate of the development cost of the communication system is given in Table 6-1, and an estimate of the cost of the entire program is given in Table 6-2.

The amount of confidence which can be placed in these figures is worth some discussion. The costs of the Scout rocket, attitude guidance, launcher, and ground support equipment were obtained from the Vought Astronautics brochure, "Space Research Vehicle Systems Developed from NASA Scout," published in August, 1959. The UHF TV transmitter is a production item and its cost is firm. The cost of the ground antenna was estimated by an experienced supplier of such devices. Island construction costs were estimated by an overseas construction company which has had considerable experience with [Atomic Energy Commission] projects in the Marshall Islands.

The development cost estimates were obtained from the individuals who would be responsible for the various items. Although some variation in cost of particular items is to be expected, the chances that the total will remain under the 1.2 million dollar figure seems quite good, because of the strong appeal of the project to creative engineers and the subsequent high degree of enthusiasm with which the job will be performed.

[26]

Payload
TWT	$0.15 M
Electronics	0.15 M
Structure	0.05 M
5th and 6th Stages	0.25 M
Environmental Testing	0.10 M
Total	$0.70 M

Terminal
    Antenna Design   0.05 M
    Transmitter Modifications   0.05 M
    Low Noise Receiver Design   0.10 M
    Total   0.20 M

Guidance - Perigee
    Transmitter Design   0.05 M
    Receiver Design   0.09 M
    Antennas   0.02 M
    Computer   0.04 M
    Total   0.20 M

Guidance - Apogee
    Auxiliary Antennas   0.05 M
    Computer   0.05 M
    Total   0.10 M

Total Development Cost $1.20 M

Table 6-1
Development Cost

[27]
Development Cost

Terminal Cost

Antenna	$0.30 M		
Transmitter	0.12 M		
Receiver	0.03 M		
Building and Land	0.10 M		
	$0.55 M	x 2	$1.10 M (2)

Jarvis Island Installation

Construction of Buildings	$0.25 M		
Construction of Airstrips	0.25 M		
Launcher	0.25 M		
Ground Support Equipment	0.70 M		
Transportation	0.25 M		
	$1.70 M	x 2	$3.40 M (2)

Launchings

Scout with Attitude Guidance	$0.361 M		
Payload	0.072 M		
	$0.433 M	x 3	$1.30 M (3)

Miscellaneous

Salaries of Field Personnel	$0.200 M		
Reserve	0.200 M		
	$0.400 M		$0.40 M
Total Program Cost			$5.00 M

Table 6-2
Program Cost

[28]   7. CONCLUSIONS AND RECOMMENDATIONS

It is concluded that it is technically feasible, within the present state of the art of rocket and electronic technology, to establish a commercial 24-hour communication satellite using the Scout rocket vehicle. It is recommended that NASA encourage such a program and recognize it as an important new application of the Scout. This program can be accomplished by the Hughes Aircraft Company within a year at a cost of 5 million dollars.

Document I-6

Document title: "Memorandum for Conference on Communications Satellite Development," December 7, 1960.

Source: NASA Historical Reference Collection, History Office, NASA Headquarters, Washington, D.C.

In the closing months of the Eisenhower administration, NASA Administrator T. Keith Glennan and his associates paid much attention to the appropriate relationship between government and the private sector in the development of communications satellites. AT&T's active interest in trying to establish a leading position in this new technology stimulated Glennan and his colleagues to focus on this issue. This memorandum, with no credited author, but almost certainly prepared by Robert Nunn, Glennan's special assistant for communications satellites, summarizes the situation as of December 1960.

[1]

Memorandum for Conference on Communications Satellite Development

December 7, 1960

1. Basic Mutual Recognition
    a. AT&T uniquely has the greatest and most obvious private business interest in satellites because of its overseas telephone business.
    b. Other companies have expressed varying degrees of interest in and shown varying degrees of initiative with respect to participation in the development of elements of a communication satellite system, although none has given evidence of desiring to spend company funds in substantial amounts as has AT&T.
    c. NASA alone has the statutory responsibility to the nation for developing space technology and facilitating its civil application to communications.
    d. It follows that all these respective interests must be harmonized on a common ground.
[handwritten notes in margin: "1) May - June 1961 - firm up our coop. utilization of A's," "2) Obtain assurance of A-participation in comm.," and "3) NASA must preserve bid n (and) competition."]
2. Common Ground
    a. Neither AT&T nor NASA should pre-empt the other's central area of responsibility and competence.

      b.   Both AT&T and NASA should avoid duplication and waste, from the total national point of view, including manpower, time, and money.

      c.   Both AT&T and NASA favor private ownership and operation of communication satellites.

      d.   Both AT&T and NASA must consider the national interest, which includes all of the various competitive interests, in communication satellite system operation.

[2]   e.   Neither AT&T nor NASA can afford to let the development of the utilization of the commercial satellite systems become a "political" issue.

      f.   It follows that a common objective and a disciplined common approach to it must be mutually understood.

   3.   <u>AT&T's Position As Presently Understood</u>

      a.   AT&T's approach may give others the impression that AT&T is seeking to pre-empt responsibility for the nation's non-military, satellite-based, communications program.

      b.   This can not be allowed to occur either in reality or in appearance for the following reasons:

         1.   It would seem to constitute an abandonment by NASA of its responsibility under the law.

         2.   It would seem to constitute a "give away" by NASA, impliedly [sic] sanctioning AT&T as the chosen owner and operator, which is beyond NASA's legal authority.

         3.   It would contribute fuel to the fire of the arguments by those who favor Government ownership and operation of all satellites.

         4.   It would stimulate debate instead of action and thereby engender delay and diversion from the main technological task ahead of us.

         5.   The nation can be assured of continued development only by Government activity and control of R&D [research and development] programs, since AT&T does not and probably can not provide assurance of continuity of effort should unforeseen obstacles arise.

[3] 4.   <u>NASA's Approach</u>

      a.   AT&T must fit its activities, even when using its own funds, into the communications program of NASA as a part of it. Present here is the implication that the interests and technical approaches of AT&T and NASA will be substantially the same.

      b.   This does not involve NASA's pre-empting AT&T because NASA is in the forefront of the proponents of private ownership and operation.

      c.   This is otherwise essential for the following reasons:

         1.   It assures the avoidance of wasteful duplication.

         2.   It supports NASA in its affirmation of private ownership and its belief in the feasibility of an industry-Government "partnership" for developmental purposes.

         3.   It assures the Congress of the Government control of R&D and therefore improves the possibility of getting on with the job with the minimum of legislative delay which might be caused if the Congress gets the impression that the Executive Branch is not giving adequate direction to the total national effort.

         4.   It takes account of the fact that vehicles and launch facilities are scarce national assets and must be utilized under Governmental control.

         5.   The only way the Government can enter into a "partnership" is to reserve unto itself the authority to identify and the authority to protect valid public interests.

    d.   NASA's approach includes competitive hardware acquisition for at least two Thor-based flights at Government expense and competitive launchings (including satellite, vehicle and launch costs) at [4] private or Government expense using at least two Atlas-based vehicles.

        1.   The key to this approach is "competition" because it is the only way in which NASA can assure the Congress that its approach is not preferential.

        2.   Competition may also result in a better system.

5.   Special Problems

    a.   <u>Reimbursement</u> depends upon a ruling by the Comptroller General. If favorable, then NASA can develop a relationship with industry which is not dependent upon the budget-authorization-appropriation cycle. If not favorable, then NASA would have to seek legislation to authorize it to credit such reimbursements to its own appropriations.

    b.   <u>Patents</u> depend upon the application of the present law, unless NASA is successful in getting its statute amended. This means that AT&T's case for waiver under the present law is one way the ownership of inventions might be determined and the application of the law to a "cooperative agreement" is a less preferred way of determining ownership. NASA intends to seek a legislative amendment of its statute in the same terms as were proposed to the last Congress.

    c.   <u>Follow-on R&D and prototype launchings</u> must be planned since it is hardly likely that a four-shot program will be adequate to develop an operational prototype satellite and an operational system. Back-up vehicles should be available for repetitive shots within three months of any failure.

    d.   <u>Participation by AT&T</u> in all satellite communications experiments under NASA programs in a manner similar to that employed in the Echo program seems desirable from the standpoint of all concerned.

[5]   e.   <u>Publicity</u> by AT&T should avoid "predictions" involving launchings and avoid the impression that AT&T can "go it alone" in the R&D phase. There should be closer cooperation between [public information] offices in AT&T and NASA so that AT&T releases are available to NASA in a timely manner.

Document I-7

Document title: White House Press Secretary, "Statement by the President," December 30, 1960.

Source: NASA Historical Reference Collection, NASA History Office, NASA Headquarters, Washington, D.C.

Shortly before he left office, President Dwight D. Eisenhower issued a number of policy statements in an attempt to set the future agenda on various issues, including communications satellite policy. In a statement released to the press on December 30, 1960, Eisenhower reiterated his position that private industry should establish and operate communications satellite systems.

Statement by the President

The commercial application of communication satellites, hopefully within the next several years, will bring all the nations of the world closer together in peaceful relationships as a product of this nation's program of space exploration.

The world's requirements for communication facilities will increase several fold during the next decade and communication satellites promise the most economical and effective means of satisfying these requirements.

Increased facilities for overseas telephone, international telegraph, and other forms of long-distance person-to-person communications, as well as new facilities for transoceanic television broadcasts, through the use of man-made satellites, will constitute a very real benefit to all the peoples of the world.

This nation has traditionally followed a policy of conducting international telephone, telegraph and other communications services through private enterprise subject to Governmental licensing and regulation. We have achieved communications facilities second to none among the nations of the world. Accordingly, the government should aggressively encourage private enterprise in the establishment and operation of satellite relays for revenue-producing purposes.

To achieve the early establishment of a communication satellite system which can be used on a commercial basis is a national objective which will require the concerted capabilities and funds of both Government and private enterprise and the cooperative participation of communications organizations in foreign countries.

Various agencies of Government, including the Department of State, the Department of Defense and the Office of Civil and Defense Mobilization, have important interests and responsibilities in the field of communications.

With regard to communication satellites, I have directed the National Aeronautics and Space Administration to take the lead within the Executive Branch both to advance the needed research and development and to encourage private industry to apply its resources toward the earliest practicable utilization of space technology for commercial civil communications requirements. In carrying out this task NASA will cooperate closely with the Federal Communications Commission to make certain that the high standards of this nation for communications services will be maintained in the utilization of communication satellites.

[handwritten note—"Drafted Dec 23, 1960"]
[handwritten note—"Released Dec 30, 1960"]

Document I-8

Document title: Federal Communications Commission, "FCC Relation to Space Communication," Public Notice-G, 1627, March 14, 1961.

Source: NASA Historical Reference Collection, History Office, NASA Headquarters, Washington, D.C.

The question of authority over determining space communications policy was unclear at first. NASA and the Department of Defense had both been assigned responsibility for technology development, but the Federal Communications Commission (FCC) had a say in the allocation of frequencies for satellite communications. In addition, several other committees and organizations, both international and within the U.S. government, were responsible for different aspects of the subject.

[1]
PUBLIC-G, 1627   March 14, 1961

FCC Relation to Space Communication

GENERAL

The Federal Communications Commission activities in connection with space communication have increased greatly because of the many new and unique problems posed
by rapid technological and scientific developments in this field.

Although the Commission is not responsible for any over-all space program or any
particular space vehicle launching project, the mounting activity in space communication
has an impact on its regulation of non-Government radio users. This stems from its obligations under the Communications Act which, among other things, requires the FCC to
"study new uses for radio, provide for experimental uses of frequencies, and generally
encourage the larger and more effective use of radio in the public interest" as well as to
"make available, so far as possible, to all the people of the United States a rapid, efficient,
Nation-wide, and world-wide wire and radio communication service."

This involves the allocation and assignment of frequencies for space communication
and the authorization of privately conducted research and experimentation looking
toward the use of natural or man-made satellites to provide civil communication services
on a regular basis. Radio signals "bounced" or relayed from such satellites would permit
the transmission of large amounts of telephone, telegraph and other traffic, including
television, over great distances. Such developments present a new and complex array of
technical problems. Not the least of these is finding suitable and sufficient frequencies
and insuring compatibility between space communication systems and surface systems so
that the public interest will best be served. Many regulatory problems will flow from
adding space communication to radio's already manifold uses.

COORDINATION AND COOPERATION

The achievement of these purposes involves both national and international considerations. Consequently, the Commission is working closely with the interests involved.

This coordination and cooperation requires particularly close relationship by the FCC
with the National Aeronautics and Space Administration (NASA), which directs the
Nation's non-military space effort. On February 28, 1961, the FCC and NASA announced
a joint "memorandum of understanding" for delineating and coordinating their respective responsibilities in civil communication space activities.

[2] Other interagency activities include FCC participation in the following:

The Telecommunication Coordinating Committee (TCC) [all underlining is
handwritten] of the Department of State, which has an ad hoc working group
under the chairmanship of FCC Commissioner T.A.M. Craven to draft foreign
policy recommendations on space communication systems;

The Telecommunication Planning Committee (TPC) which advises the
Office of Civil and Defense Mobilization (OCDM), with FCC representation on
space study panels;

The FCC and the OCDM have joint responsibility for national frequency allo-
cations, with staff work through joint meetings of FCC representatives with the
Interdepartment Radio Advisory Committee (IRAC) and its Subcommittee on
Frequency Allocations (SFA);

The U.S. Committee for Study Groups IV and VIII of the International Radio Consultative Committee (CCIR) of the International Telecommunication Union (ITU), with which the FCC participates through subgroups;

The Space Science Board (SSB), of which the FCC's Chief Engineer is a member of the International Relations Committee concerned with international basic space research activities, working internationally through the Committee on Space Research (COSPAR);

The International Radio Scientific Union (URSI), which has FCC participation and, in turn, is a member of the International Council of Scientific Unions (ICSU); and

The National Bureau of Standards Central Radio Propagation Laboratory, with which the FCC maintains liaison through membership on the Interdepartment Council on Radio Propagation and Standards.

INTERNATIONAL CONSIDERATIONS

The International Administrative Radio Conference, held at Geneva in 1959 under the auspices of the ITU, adopted an international table of frequency allocations which, for the first time, provided bands of frequencies for space and earth-space services. These bands, however, are for research purposes only and are useful principally for tracking, control and telemetry functions. Although no bands were allocated internationally for space satellite relay communication, a special ITU Administrative Radio Conference was scheduled tentatively for late 1963 to deal specifically with space problems on the basis of developments as of that time. At the request of the Department of State, preparatory work toward formulating the United States position at that conference has been initiated jointly by the FCC and the IRAC.

[3] Domestically, steps have been taken by the FCC to implement the 1959 Geneva Radio Regulations nationally pending ratification of that treaty by the President upon the advice and consent of the Senate.

FCC PROCEEDINGS

As a result of developments in space communication during 1960, the Commission reopened its proceeding in the general inquiry relative to the allocation of frequencies above 890 Megacycles (Docket 11866) to determine, in the light of evidence then available, whether the frequency requirements for communication via space satellites would require modification of the Commission's decision to permit some additional classes of users to establish communications systems on frequencies between 1,000 and 10,000 Mc. After a careful analysis of all the evidence then on hand, the Commission concluded that its earlier decision need not be modified at that time.

However, in view of rapid developments in space communication, the Commission instituted an inquiry (Docket 13522) as to space frequency needs on a longer-range basis. This information will assist the Commission in its preparatory work leading to a United States position for [a] future international conference on space needs and usage. The inquiry was augmented to consider conditions for sharing space bands with other radio services and whether protected areas might be established and held in reserve for future earth terminals for civil communication systems using space relays.

EXPERIMENTATION

The Commission is encouraging experimentation in this new field in the hope that private industry can develop considerable additional technical information which will serve to further the country's over-all space program.

In this regard, an experimental authorization was granted in January of this year to the ITT Laboratories, Nutley, N.J., to bounce signals off the moon and passive (non-radio-equipped) earth satellites for basic research and study.

Also in January of this year, an experimental authorization was granted to the American Telephone and Telegraph Co. to permit it to go forward with plans to develop an experimental program wherein earth terminal facilities at Holmdel, N.J., would transmit to and receive from active (radio-equipped) earth satellites which also are undergoing development by AT&T.

MONITORING

Another FCC activity is the continued monitoring of channels being used for space communication. This started with its long range direction finding work in tracing Sputnik I, before the Government established special installations to track space objects. Commission monitoring is to prevent unauthorized use by other stations of channels employed for space communication, and to identify and locate sources of interference on those channels. At a number of FCC monitoring stations, special equipment includes sensitive receivers, high gain directional antennas and automatic frequency scanning devices.

[4]   RADIO ASTRONOMY

Related to space communication is the use of radio in astronomy. The Geneva 1959 conference, for the first time in history, provided for protecting specific frequencies utilized in radio astronomy. The FCC has completed the groundwork for putting these provisions into effect domestically when the Geneva agreement is ratified by the United States.

Meanwhile, the Commission has adopted rules to minimize interference to frequencies used for radio astronomy observations in this country, particularly at the National Radio Astronomy Observatory at Green Bank and the Naval Radio Research Observatory at Sugar Grove, both in West Virginia.

Document I-9

Document title: F.R. Kappel, President, American Telephone and Telegraph Company, to the Honorable James E. Webb, Administrator, NASA, April 5, 1961 (with several attachments).

Source: NASA Historical Reference Collection, History Office, NASA Headquarters, Washington, D.C.

Document I-10

Document title: James E. Webb, Administrator, to F.R. Kappel, President, American Telephone and Telegraph Company, April 8, 1961.

Source: NASA Historical Reference Collection, History Office, NASA Headquarters, Washington, D.C.

The new Democratic administration of President John F. Kennedy was less sympathetic than its Republican predecessor to AT&T's plans to establish the leading position in the development of communications satellites. This exchange of letters reflects the position taken by new NASA Administrator James E. Webb—that it was desirable to re-examine the government role in communications satellite development before deciding that the government should take a secondary position in that development to AT&T and possibly other U.S. communications carriers. The position of AT&T President Fred R. Kappel to that stance, as reflected in his letter to Webb, is supported by a series of attachments indicating AT&T's plans as they had developed in the preceding months.

Document I-9

[1]

April 5, 1961

THE HONORABLE JAMES E. WEBB, Administrator
National Aeronautics and Space Administration
1520 H Street Northwest
Washington 25, D.C.

Dear Mr. Webb:

It has come to my attention that the *Wall Street Journal* on March 29, 1961 carried an article stating that invitations were issued last year to companies such as American Telephone and Telegraph Company to come forward with partnership proposals with the Government (on communications satellites), but that NASA has yet to receive a firm proposal from any company.

In view of events which have taken place during the past few months, this statement, which we understand grew out of a press conference which you held with respect to NASA's budget, is of deep concern to me. The specific events to which I refer are as follows.

On September 15, 1960, Mr. G. L. Best of this Company, wrote to Dr. Glennan (Attachment No. 1) saying that we had under way the development of an active communications satellite and associated ground radio facilities and would hope that NASA would be willing to launch trial satellites for us at our expense if this proved to be the most practicable arrangement.

This letter was followed by several informal discussions after which Dr. E. I. Green of the Bell Telephone Laboratories wrote Dr. Glennan on October 20, 1960 (Attachment No. 2) and enclosed a statement of the objectives and principal features of the experiment which the Bell System proposed to make.

[2] Subsequently, there were several discussions during which Dr. Glennan and his people outlined some of the problems which NASA felt were involved in accepting our original proposal. During these conversations, various possibilities of a joint NASA-Bell System project were discussed, and on December 14, 1960 I wrote to Dr. Glennan outlining in some detail several specific proposals as to how a joint undertaking might be accomplished (Attachment No. 3).

Shortly after, NASA decided to ask for bids covering the construction of an active satellite of its own design and to seek the cooperation of private industry here and of the telephone administrations of Great Britain and France in trials using such a satellite. We were offered an opportunity to bid on such a project and did so, making substantial allowance in our bids for the value which we thought the telephone industry might get from such an experiment. A copy of the transmittal letter which accompanied these bids is also enclosed (Attachment No. 4).

During discussions which preceded our bid, we pointed out that whether such a project would take the place of one step in the Bell System's proposed program would depend on the type of satellite which was placed in orbit. At that time we expressed the hope that if our requirements were not met by NASA's project, NASA would launch a Bell System satellite later at our expense.

In our studies of satellite communications, while paying primary attention to the needs of international common carriers, we have also considered certain specific Defense Department needs—for example, mobility and the provision by the Defense Department of a few reliable voice channels to remote locations. The early experiments and tests of satellite relays can be made to serve both civilian and defense objectives and we have discussed such matters of common interest with agencies of the Department of Defense. I believe Dr. Fisk and Mr. Dingman of this Company discussed this briefly with you and Dr. Dryden a few weeks ago.

Summing up these events, I think it is clear that we have made every effort to find a way of getting this very vital experimental work done promptly and that, contrary to our not having made any specific proposals, we have actually made three specific proposals to NASA. Mr. Best has gone over this matter by telephone with Dr. Dryden but I thought that you might like to have this statement of the situation from me.

[3] We are extremely anxious to avoid any further delay in getting trials under way for a number of reasons, the most important of which can be summarized as follows:

1. There is a need for point-to-point space communications systems—to help meet the growing demand for overseas communications.
2. Such systems would be a natural means of augmenting existing connecting links between the common carrier networks of this country and those of foreign countries and would also provide alternate routes for reliability.
3. Our estimates of costs and traffic volumes lead us to the conclusion that a satellite system such as we propose is economically feasible.
4. Trials of active communications satellites are needed to determine the basic facts upon which a commercial communications satellite system may be designed.
5. We are prepared to move ahead rapidly if permitted to do so.
6. If this country does not maintain the leading position in space communications for peaceful purposes, which is now within its grasp, others will take the lead.
7. The severity and frequency with which sunspot disturbances are occurring threaten to disrupt existing forms of overseas radio communication seriously during the next several years. This is of great importance to military as well as to civilian communications.

There need be no fear that this Company is seeking a monopoly in international communications through the use of satellites. Our only interest in satellites is their use as another means of connecting the Bell System's communications network in this country with similar networks in foreign countries.

We have stated both to the Government and publicly that any satellite system which we sponsor will be available to all United States international communications common [4] carriers—either through lease or ownership arrangements—for any services authorized by the Federal Communications Commission. We have also stated that rockets and launching facilities will be provided by private suppliers under appropriate arrangements with the Government. (See Mr. Dingman's letter of March 21, 1961 to the Federal Communications Commission—Attachment No. 5.) Furthermore, the creation of the satellite system we propose, to do our public service job, will not preclude in any way the development of other space communications systems for other purposes.

In view of the urgency of this whole matter, I should like very much to drop in and [handwritten underlining] discuss some of its aspects with you in more detail. I am sure there can be no important differences between us as to ultimate objectives, and perhaps we can by discussion at this time advance the attainment of those objectives. Will you please call me at your convenience.

> Sincerely yours,
> [hand-signed: "F.R. Kappel"]
> President

Attachments

****

[1]

Attachment No. 1

September 15, 1960

Dr. T. Keith Glennan, Administrator
National Aeronautics and Space Administration
1520 H Street, N.W.
Washington 25, D.C.

Dear Dr. Glennan:

During the discussion which Messrs. Botkin, Pierce and I had with you and your people several weeks ago, we expressed the view that the commercial satellite communication systems of the future should be owned and operated in this country by communications carriers. In other words, we believe that existing national policy and practice in the communications field should be extended to embrace the new medium.

By so doing, it will be possible to achieve efficient integration of the planning, construction, and operation of overseas cable systems, satellite systems, and other radio facilities, and also assure the integration of domestic and overseas operations that is so necessary to the orderly planning and development of telephone service in particular. Communication with other countries by any medium, of course, requires the cooperation of the organizations or administrations responsible for furnishing external communication services in those countries.

It is assumed that there will be some form of Government supervision of the launching and orbiting of satellites, as well as, of course, regulation of the communication services rendered and the radio frequencies used. It also seems to us that any international action which our Government may feel desirable in order to adequately coordinate with other governments the use of satellites may be taken without Governmental ownership or management of the facilities required to furnish commercial communications.

During our conversation there was also considerable discussion as to what part of the work of developing a practical satellite communication system might be undertaken by commercial communications carriers, and what part of the expense of such work should be borne by them. At the close of this discussion you asked that we set down our thoughts on [2] these matters, or, more specifically, state what the Bell System companies' plans are for the future and what part they were looking to the National Aeronautics and Space Administration to do. Our present views on this subject are outlined below.

The Bell System now has under way the development of an active satellite and associated ground radio facilities and would like to proceed with an experimental trial of these facilities in intercontinental communications as soon as possible. The telephone administrations of England, France, and Germany have all indicated a desire to participate in such a project. We hope to be ready for a transatlantic trial in eighteen to twenty months, or less. The experimental satellite, or satellites, would be placed at an orbital altitude of perhaps 2,200 miles and would carry a repeater designed to make initial use of a 5-megacycle radio-frequency bandwidth. We are willing to assume the cost involved in this experiment, except that we would expect the participating foreign administrations to pay at least the cost of their own ground stations.

Our present thinking is that we would design and construct the trial satellites for our own use, making sure that the mechanical design would be compatible with the design and capability of whatever launching vehicle was used. While it is probably too early to know just what facilities for launching could be made available, we would hope that the National Aeronautics and Space Administration would be willing to launch these trial satellites for us, at our expense, if this proved to be the most practicable arrangement.

Although our primary interest is in proceeding with a trial of active satellites, we shall be glad to cooperate with your organization in any further tests of passive satellites that you may wish to conduct, using not only the ground equipment now available but also the equipment that would be developed for active satellite trials.

I am sure you understand that these thoughts may be subject to some modification as the program develops, but I believe they will hold basically. We would, of course, seek the advice of your organization in all phases of the work and keep you informed of our progress.

We would welcome any comments you may care to make about any part of this proposed program.

Sincerely yours,

G. L. BEST
Vice President

****

[no pagination]

Attachment No. 2

October 20, 1960

DR. T. KEITH GLENNAN, Administrator
National Aeronautics and Space Administration
1520 H Street, N.W.
Washington, DC

Dear Dr. Glennan:

Dr. Fisk agreed to provide you with a brief statement of the objectives and principal features of the experiment the Bell System proposes to conduct on long distance communication via an active satellite.

As indicated in the statement, the experiment is an important part of a continuing Bell System development program directed toward large scale application of radio satellites for broad-band communications.

Since Dr. Fisk is out of town, I am enclosing the statement, which has his agreement.

Sincerely yours,

E. I. GREEN
Executive Vice President

Enc.

***

[1]

PROPOSED BELL SYSTEM EXPERIMENT ON
ACTIVE SATELLITE COMMUNICATION

OBJECTIVE

To carry out an experiment in transoceanic communications with a satellite carrying an active repeater suitable for multichannel telephony and for television. The experiment is an important part of a continuing Bell System development program directed toward large scale application of radio satellites for broad-band communications. The program includes extensive laboratory research and development work leading to long life and reliable operation of such a system.

SYSTEM CHARACTERISTICS

Proposed Operating Frequencies: 6775-6875 mc [megacycle] ground to satellite.
6425-6525 mc satellite to ground.

Baseband Width: 2 mc.

Modulation: FM with ± 10 mc swing.

Transmission Path: Europe to U.S.A. and reverse.

Satellite: Microwave receiver; 2-watt microwave transmitter; circularly polarized reception and radiation; solar cells for primary power; nickel-cadmium storage battery. Separate beacon transmitter, 150 milliwatts at about 136 mc for tracking. Satellite essentially spherical, about 4' diameter, weight 175 lbs. or less. The initial satellite will not be engineered primarily for the long life needed for commercial operation. Orbit of satellite should be as nearly polar as possible, at an altitude of 2,000 to 5,000 miles. (Note: This experiment is directed particularly toward normal telephone communication in which one way transmission delay of 1/4 second, such as would be encountered with a 24-hour satellite, would be intolerable.)

Ground Receiving Station at Holmdel: Existing 20' x 20' horn reflector antenna, used in Echo I. Improved antenna control system. (Construction of a larger 60' x 60' horn reflector antenna is to be started immediately, [2] but this may not be in operation until a few months after the first active satellite experiment.) Maser for operation at 6475 mc. Wideband FM feedback receiver.

Ground Transmitting Station at Holmdel: Existing 60' dish, used in Echo I, modified for 6875 mc operation, or a new similar dish. Improved antenna control system. 1 kw transmitter, using either a commercially available klystron or a traveling wave tube under development by Bell Laboratories, choice to be determined by engineering considerations.

Ground Station or Stations in Europe: Same as at Holmdel, possibly with variations in detail required to satisfy foreign partners.

Antenna Pointing: Use Minitrack data. Have Goddard Space Flight Center compute orbit parameters to be transmitted to BTL. Track computation and antenna control orders by Bell Laboratories.

Estimated Performance: With satellite 7.5° above horizon, and assuming total noise temperature of 34° K, 2 db [decibels] satellite antenna gain and achievable accuracy in antenna pointing, system would provide peak-to-peak signal-to-rms noise ratio of 47 db. Such performance will provide a path for about 450 one-way, high-grade telephone channels, or in the order of 100 two-way channels. Alternatively, the experimental system will provide one-way-at-a-time transmission of a black-and-white television picture of a quality only slightly inferior to American commercial standards, and not noticeable on the average home receiver.

Schedule: The system is expected to be operational in 12 months, assuming no undue delays are encountered in: (1) assignment of the necessary experimental radio frequencies, (2) availability of a suitable satellite launching vehicle, and (3) agreements with the foreign partner(s) and execution of their agreed-upon technical tasks.

October 20, 1960

***

[no pagination]

28 September 1960

Mr. G. L. Best, Vice President
American Telephone and Telegraph Company
195 Broadway
New York 7, New York

Dear Mr. Best:

Thank you for your letter dated September 15, 1960, which you handed me on that date. It was a pleasure to see you and Dr. Baker again.

The proposal of the Bell System to integrate satellite systems into its commercial operations is of considerable interest to NASA.

From a broad point of view, as you no doubt appreciate, your request that NASA launch trial satellites for the Bell System raises issues of national policy which we are currently studying. Accordingly, there is no simple or ready-made response that I can give you at this time.

It is helpful to me to know the position of your company and its particular plans for the future. If you are able to be more specific about any facet of your program at any time in the future, please be assured that I would appreciate being informed.

Sincerely yours,

T. Keith Glennan
Administrator

****

[1]

Attachment No. 3

December 14, 1960

DR. T. KEITH GLENNAN, Administrator
National Aeronautics and Space Administration
1520 H Street, N.W.
Washington 25, D.C.

Dear Dr. Glennan:

Our discussion of December 7 regarding satellites for communication purposes raised a number of questions relative to the respective efforts of NASA and A.T.&T. Co. in this area of scientific exploration. I believe there was no disagreement on the need for advancing scientific knowledge as fast as possible to the point where commercial communications by satellites can be undertaken by the common carrier communications companies and their international counterparts, each in their respective areas of service.

In our conversation, we indicated we are proceeding at our own expense with the development and construction of experimental active satellites which will be ready for launching within a year, this to be an initial step toward the establishment in a few years of a commercial satellite system. You in turn advised that NASA is also planning experimentation in the field of active communications satellites, with the view to developing further scientific information in this area.

It was agreed that our objectives, stemming as they do from our respective areas of responsibility and competence, have much in common. Moreover, I believe we were in accord that the national interest could be best served if our efforts were combined in this field so as to avoid wasteful duplication and delay in the development of a final system.

[2] It is recognized that NASA has broad responsibilities to advance "the role of the United States as a leader in aeronautical and space technology and in the application thereof to the conduct of peaceful activities within and outside the atmosphere." This charges your organization with a broad obligation to assure that the activities necessary to achieve this result are being carried out. It does not, however, seem to require that NASA itself duplicate work now being done or planned by private industry but only that it encourage industry in its efforts and be prepared to move in should industry be unable or unwilling to take full advantage of its opportunities.

The common carriers of the United States have the responsibility to the American public to furnish the best possible communications service not only within this country but also internationally. In the course of discharging A.T.&T.'s responsibilities, we have developed overseas radiotelephone service and overseas telephone cables and have established telephone communications with nearly every country in the world. The technical problems which were solved in bringing these facilities into service are comparable to those which are faced in satellite communication systems.

We are also constantly seeking to improve the communications art and find better and more economical means of doing the job to which we are dedicated. In so doing we maintain the most extensive communications research and development laboratories in the world, Bell Telephone Laboratories, whose responsibility is to explore every possible way of improving communication. The exploration of the use of satellites as a means of radio communication is a natural part of the Bell Laboratories overall program and it has been devoting substantial effort to activities in this area for a number of years. In this connection, our Bell System technology has already been the source of the essential components which enable a satellite to act as a communications relay. These include transistors, diodes, solar cells, and reliable traveling wave tubes. Out of the work which the Bell Laboratories has done in the Echo experiment, and from its long and extensive participation in radar have come a series of developments directly applicable to satellite communications, such as high-quality antennas of the horn and Casegranian types, microwave masers as low noise receivers, FM circuits with feedback, also for low noise, and various techniques for precision tracking.

[3] The achievement of a communications satellite system will depend not only on effective satellite relays and expert communications systems engineering. These must be joined in a unique fashion with space technology where NASA's primary responsibility lies. Moreover, space relaying of signals, like other isolated communications techniques, can provide useful communications service only when combined with existing land facilities. It is this joining of the communications and space arts and facilities which indicates the desirability for the joint efforts of our organizations.

It is our thought that such joint efforts would have as their objectives both demonstrating transatlantic TV (which we understand to be one of NASA's primary objectives) and other experiments which will represent the first step in an orderly developmental program for an operating communications system.* Concurrently we would have extensive effort on ground stations for transmitting and receiving, as well as tracking facilities for controlling the antennas. The problems of the entire communication system, including economic problems as well as such important matters as optimum bandwidth, operating margins, systems balance and reliability of components would receive prime attention.

This experiment in its public communications aspects would, we believe, provide information and an opportunity for experimentation not only to us but also to other interested common carrier communication companies. This can be accomplished by inviting other international common carriers to use the satellite circuits experimentally for their own forms of communication.

With this as background, we would like to offer the following specific proposals:

a. That NASA and we join in the setting of performance specifications for the first experimental active satellite.

b. That we develop and build the first satellite taking advantage of research already done and developments well under way. We are prepared to pay for this work in its entirety, or for such part of the expense as would reflect our respective interests in the project.

[4] c. That NASA launch the first satellite and provide tracking data from its Minitrack stations. In this connection, we are willing to bear the whole cost of launching and tracking or to share these costs with NASA in any way you feel will properly reflect our respective interests in the project.

d. That the existing ground station at Holmdel be made available and modified at our expense for the purpose of making the necessary communications tests. (This station is, of course, compatible with the communications network of this Company.)

e. That, taking advantage of our long established working relations with overseas communications operating agencies, arrangements be made with at least one of them for one or more overseas ground stations.

f. That other common carriers be invited to use the satellite circuits experimentally when such circuits are operational.

g. That full information on satellite performance be made available to NASA.

---

*   A program for the Development of an Active Satellite Communication System has been prepared by Bell Telephone Laboratories and is available for detailed discussion.

As is common practice, we would expect that much of the work on the communications system would be contracted to other private companies as was done in the Echo experiment, for example, on the transmitting antenna. NASA may also wish to contract with others for many of the other items involved, such as mechanisms for multiple launching, satellite orientation arrangements, etc.

If, as we are confident, this experiment is successful, it is our plan to move as promptly as possible toward the establishment of a commercial satellite communications system which will be integrated with existing common carrier communications facilities, both here and abroad. When this system is operational circuits will be made available to other communications common carriers for use in their business, just as circuits are now available in overseas telephone cables. The proposals that we are making should [5] be of substantial value to the military and other government departments as well as to the other users of our services and those of other communications carriers.

I hope that this outline will offer a useful basis for approaching the problems which we discussed last week.

Sincerely,

F. R. KAPPEL

****

[1]   Attachment No. 4

Western Electric Company
INCORPORATED
Defense Activities Division
120 Broadway, New York 5, N.Y.
Area Code 212 571-5761

C. R. SMITH
  VICE PRESIDENT

March 20, 1961

The National Aeronautics and Space Administration
Goddard Space Flight Center
Greenbelt, Maryland

Attention: Procurement and Supply Division
             JDC:241:mj

Gentlemen:

The attached proposal is submitted in response to your Request for Proposal No. GS-1861, Low Altitude Active Communications Satellite, dated January 4, 1961 and the Telegraphic Amendment thereto dated February 24, 1961.

This response contains three separate and complete proposals. Proposal 1 is based on the use of frequencies in the 400-500 and 2,200-2,300 mc bands. Proposals 2 and 3 are based on the use of frequencies in the 5,925-6,425 and 3,800-4,200 mc bands, as requested in the Telegraphic Amendment of February 24, 1961. Proposal 3 differs from Proposal 2 in that it includes a contractor-furnished radiation experiment package.

Bell Telephone Laboratories has been actively pursuing a program of research and development in satellite communications. This work has been undertaken because of the Bell System's position as a major U.S. international communications carrier and the obligation this imposes upon it to develop and provide any new means of communications that hold promise of improving its services to the general public and the government. The Bell System's program, for developing a satellite communications system and for placing such a system in commercial service in collaboration with the telecommunications agencies of other countries, is expected to parallel in many respects the System's achievements in the development and establishment of transoceanic submarine telephone cable systems. As was done with the cables, the Bell System will under- [2] take to work out mutually satisfactory arrangements with the other United States international carriers whereby they can obtain facilities for the services furnished by them.

We believe that NASA and the Bell System have a common interest in pointing experimental work in the field of satellite communications toward the realization of a commercial system as quickly as possible with a minimum of cost and without duplication of effort. For this reason, we strongly favor Proposal 2 or Proposal 3, since the 6 kmc and 4 kmc frequencies are already being used in common carrier communications systems in both the United States and Europe. The Bell Telephone Laboratories' program, which is based on the use of these frequencies, is well under way and maximum progress toward our mutual goal will, we believe, be achieved with the experimental satellite contemplated under Proposal 2 or 3. Not only will this permit the testing, in the experimental satellite, of components of the kind that will be used in later prototypes of commercial satellites, but valuable information will he obtained on the problems of sharing the proposed frequencies by terrestrial and satellite common carrier systems.

In response to a specific NASA request, an offer to undertake this program on a cost-plus-fixed-fee basis is associated with each of the proposals. In view of the Bell System interest expressed above, however, each proposal also contains an offer to undertake the program on a cost-sharing basis. These offers involve billing NASA an amount equal to about one-fourth of the expense associated with Proposal 1 or a considerably smaller part of the expense associated with Proposals 2 or 3, since the work to be undertaken under these latter proposals will make a larger contribution to our own research and development program than the work under Proposal 1. All of these cost-sharing offers are on a cost reimbursement basis. Each offer, however, includes a maximum dollar limit of cost to be billed to NASA.

The A. T. & T. Company has offered to provide ground station equipment and operation in the United States and to undertake to arrange for related ground station equipment and operation overseas. NASA has been assured that the United States ground station will be operational in time to meet the planned launching schedule and that these facilities will be made available to NASA for this experiment.

[3] Every attempt has been made to include in the three parts of our response, General Evaluation Information, Scientific and Technical Proposal, and Cost Proposal, all of the information requested in connection with this procurement. The representation relating to small business is attached as a separate item.

This quotation, in response to NASA Request for Proposal GS-1861, is firm for a period of ninety days from the date of this letter. Questions in connection with this quotation should be directed to Mr. R. P. Wilson of this office on Extension 5735.

Sincerely

"C. R. Smith" [hand-signed]

Att.
Proposal
Exhibit "A"

****

[1]

Attachment No. 5

March 21, 1961

Mr. Ben F. Waple, Acting Secretary
Federal Communications Commission
Washington 25, D.C.

Re: Docket No. 13522

Dear Sir:

The comments filed by A. T. & T. Co. in Docket No. 13522 were directed to the specific questions posed in the Notice of Inquiry and Supplement. It is our understanding that the Commission was seeking technical information concerning frequency requirements for space communications without discussion at this time of legal or economic question. However, in view of the comments of this character in some of the other responses and the publicity they have been given, we believe a brief statement should be made to forestall any misunderstanding of the Bell System position and bring the comments into a better perspective.

Our interest in satellite communications is simply stated. There is a need for point-to-point space communications system—to help meet the growing demand for international communications of all kinds, and to provide alternate routes from a reliability standpoint. Such space communications systems are a natural supplement to and extension of existing common carrier networks.

The traditional communications policy in this country has been to have common carriers serve both domestic and international needs for public communications. This policy was recently restated in FCC Public Notice G1271 dated February 28, 1961 that ". . . overseas public communications are provided by private enterprise, subject to Government regulation. . . ." This notice also included the following:

   "(1) The earliest practicable realization of a commercially operable communication satellite system is a national objective.
[2] "(2) The attainment of this urgent national objective in the field of communications may be accomplished through concerted action by existing agencies of Government and private enterprise.
   "(3) In accordance with the traditional policy of conducting international communications service through private enterprise subject to Government regulation, private enterprise should be encouraged to undertake development and utilization of satellite systems for public communication services."

We do not seek a monopoly in satellite communications. We do not wish to exclude other international carriers either from establishing such systems or from sharing the use of the system we propose. We seek only the opportunity to employ private initiative, management and capital in the public interest and under public regulation in a manner wholly consistent with traditional public policy with respect to international communications.

Our estimates of costs and traffic volumes lead us to the conclusion that a satellite system such as we propose is economically feasible. We are prepared to move ahead as rapidly as possible and it is important that we be permitted to do so, not only to meet the service requirements for the near future but also to make sure that this country will lead the way in international space communications for peaceful purposes.

Ownership of the facilities involved could be handled in the traditional way. The foreign terminals would be owned by the foreign telecommunication agencies. We have had many years of mutually satisfactory operating experience with these agencies all over the world and are completely confident that we can come to an equitable arrangement with them concerning the ownership and use of the satellites.

Use of the United States portion of the satellite system would be made available, of course, to all international communications carriers serving the United States for any services they now are, or may in the future be, authorized to provide by the FCC under the Communications Act. Here, too, the facilities would be made available on an equitable basis either by ownership participation through pro rata payment of capital investment and operating expenses or by lease arrangements. These arrangements would preserve competition in the international communications field to the extent that it is determined by the FCC to be in the public interest.

We believe the low-orbit system proposed by AT&T is the preferred system at this time. The technology is well advanced for the low-orbit satellite. On the other hand, there are drawbacks [3] to the synchronous high-altitude satellite. To begin with, there is a .6 second round-trip delay which would be a very serious degradation of telephone service. Further, there are the very difficult problems of placing the high-altitude satellite in proper orbit, maintaining it on the station, stabilizing and accurately pointing its directional antenna. The solution to these latter problems is at a minimum several years away and it's imperative to get on with the job now—not years hence.

The producers of electronic gear and other products and services would benefit from the introduction of this new mode of communications which will broaden their markets. A substantial part of the ground station equipment and many of the components of the satellites themselves will be obtained on a competitive basis from industrial suppliers. Rockets and launching facilities will be provided by private suppliers under appropriate arrangements with the Government.

As stated at the outset, we believe that the questions to which the Commission is seeking the answers in this proceeding are essentially technical in character, and they must be answered promptly if the United States is to maintain its leadership in the communications field. The purpose of this letter is to provide information which may be helpful to the FCC as it considers policy decision vital to the vigorous advancement of the nation's space communications program.

Very truly yours,

"J.E. Dingman" [hand-signed]

Document I-10

[1]

April 8, 1961

Mr. F. R. Kappel
President
American Telephone and Telegraph Company
195 Broadway
New York 7, New York

Dear Mr. Kappel:

Thank you for your letter of April 5th. I will be happy to see you whenever you plan to be in Washington. I am appearing before the House Committee on Science and Astronautics on Monday morning, April 10th, and am engaged all day on Tuesday, the 11th, with the President's Committee on Equal Employment Opportunity, which is having its first meeting. Otherwise, I will be glad to reschedule my appointments so we can meet at your convenience.

In order that you may understand, perhaps more fully than a report in the *Wall Street Journal*, the questions and answers at my press conference on the budget, I am enclosing the release. You will note, on page 3, my statement is as follows:

"In order to take full advantage of the potentialities of the communications satellite for both industry and governmental uses, industry financing of research and development costs is postponed and full governmental financing is provided. Ten million dollars is added for this purpose."

[2] On page 8, you will note my statement that:

"The basic change is simply to postpone, until we know more than we know today, the real decision as to how this new result of space sciences and technology can be most usefully applied."

Again, on page 9, you will note Dr. Dryden's statement that:

". . . the program is the same, John (Finney), the program of four flights that you have heard outlined in great detail. This is merely an estimate as to whether the Treasury would recover money. It seems to be such an uncertain thing at this time that we prefer to have the money in hand, to carry it forward to the test program."

Further, in answer to the question as to whether the addition of the ten million dollars to the budget represented any modification of policy, I stated:

"It represents a policy decision to have a good hard look at this before making commitments."

Since you have referred to the discussion in my office with Dr. Fisk and Mr. Dingman of your company, I suggest you ask them if the above does not represent what I told them was going through my mind as the only sensible way to approach a decision of such magnitude and significance far beyond the communications industry, as well as long-range implications and importance to many segments of the communication industry over and above its great significance to your own company.

Last night I took your letter home and read the attachments. Your letter of December 14th to Dr. Glennan does state, on page 3, that you would

". . . like to offer the following specific proposals:

[3] "a. That NASA and we join in the setting of performance specifications for the first experimental active satellite.

"b. That we develop and build the first satellite taking advantage of research already done and developments well under way. We are prepared to pay for this work in its entirety, or for such part of the expense as would reflect our respective interests in the project.

"c. That NASA launch its first satellite and provide tracking data from its Minitrack stations. In this connection, we are willing to bear the whole cost of launching and tracking or to share these costs with NASA in any way you feel will properly reflect our respective interests in the project.

"d. That the existing ground station at Holmdel be made available and modified at our expense for the purpose of making the necessary communications tests. (This station is, of course, compatible with the communications network of this Company.)

"e. That, taking advantage of our long established working relations with overseas communications operating agencies, arrangements be made with at least one of them for more overseas ground stations.

"f. That other common carriers be invited to use the satellite circuits experimentally when such circuits are operational.

"g. That full information on satellite performance be made available to NASA.

[4] I am told that your letter of December 14th was delivered by a number of your associates, that an extended conference ensued, <u>and that it was made clear that NASA would not permit your company, or any other, to pre-empt the program of the United States in this area.</u> [handwritten highlighting in margin] Later, in a letter dated January 17th, 1961, your proposal (e) as amplified in your telegram of January 12th, to undertake negotiations for overseas land stations on behalf of NASA was not accepted, but instead negotiations were initiated and completed by NASA, with the technical advice of your company.

On January 4, 1961, as indicated in your attachment No. 4, March 26, 1961, the letter from your Mr. C. R. Smith to our Procurement and Supply Division, we requested proposals in accordance with our own performance specifications for an experimental low-altitude, active communication satellite. With the letter of Mr. Smith, you submitted a proposal to meet our performance specifications.

I believe you will agree that our request for proposals was not an acceptance of your proposal of December 14th, but was instead the first step toward a policy of permitting all companies interested in this project to furnish competitive proposals rather than limiting the development of the satellite to arrangements that would be made only with your company.

You will recognize that all of the above either took place before or was underway at the time I took the oath of office on February 14th. It is background for the position I have taken publicly, and mentioned above, "to have a good hard look at this before making commitments." I assume part or all of this falls into the category you have called "events which have taken place during the past few months," and needs to be considered in addition to the "specific events" to which you refer in your letter.

With further reference to the record of my press conference, you will note on page 12 that the question which Dr. Dryden answered related to a presentation by your company before the Federal Communications Commission in connection with, as the questioner put it, your "being interested in the $170 million program to put up their (your) own satellites." The direct question was whether I "had any indication that AT&T has taken a new look at the desirability of this."

[5] Although the conversation Dr. Dryden and I had with Dr. Fisk and Mr. Dingman was of quite a general and exploratory nature and was in no way a negotiation or even delineation of official positions, I did get the impression that your company was making a very thorough examination, doing some real soul-searching, and I so stated in my remarks at the bottom of page 12. If this is not correct, I will appreciate your advice.

I agree completely that we should sit down and straighten out any misunderstandings that may have arisen. If you believe our public statements do not fairly represent the position of your company, I will be more than happy to take any steps necessary to make the real facts clear.

Sincerely yours,

James E. Webb
Administrator

Enclosure

A:Webb:ns
N
cc: Dr. Dryden
     Mr. Nunn
     Mr. Phillips
     BAC

Document I-11

Document title: John F. Kennedy to Honorable Newton Minow, Chairman, Federal Communications Commission, May 15, 1961.

Source: NASA Historical Reference Collection, History Office, NASA Headquarters, Washington, D.C.

During his first State of the Union address on January 30, 1961, President Kennedy had called for an international effort to develop communications satellites. Four months later, he reiterated this position while considering a sweeping acceleration of the U.S. space program.

May 15, 1961

Dear Mr. Chairman:

I am most interested in having facilitated early development of communication satellites and will appreciate prompt determination by the Federal Communications Commission, the National Aeronautics and Space Administration, and other appropriate agencies of the conditions and safeguards under which that can go forward. Subject to establishing the necessary precautions, I am hopeful that the public and private resources of our free society can be brought to bear for significant and early research progress in this field, and, as quickly as possible, for actual operation of satellite telephones, television, and other communication systems that will bring the world closer together.

I will appreciate your keeping me informed of the steps being taken toward that goal and of tangible progress that is made.

Sincerely,

[signed] John F. Kennedy

Honorable Newton Minow
Chairman
Federal Communications Commission
Washington, D.C.

Document I-12

Document title: Ben F. Waple, Acting Secretary, Federal Communications Commission, "An Inquiry Into the Administrative and Regulatory Problems Relating to the Authorization of Commercially Operable Space Communications Systems: First Report," FCC Report 61-676, 4774, Docket No. 14024, May 24, 1961.

Source: NASA Historical Reference Collection, History Office, NASA Headquarters, Washington, D.C.

On May 24, 1961, the Federal Communications Commission (FCC), tasked with outlining the initial policy to determine how the communications satellite system would operate, issued its "First Report." The FCC limited participation in the system to international communications carriers— AT&T, ITT, RCA, and Western Union. This policy excluded aerospace and communications equipment manufacturers and consequently provoked numerous complaints. Forced to respond to the aerospace and communications equipment manufacturers' objections, the FCC stated that such companies' participation in the establishment of the system would be neither "necessary nor beneficial." This issue would later play a major role in the controversy over the Communications Satellite Act.

[475/1]   Before the   FCC 61-676
            FEDERAL COMMUNICATIONS COMMISSION   4774
                       Washington 25, D.C.

In the Matter of   )
)   Docket No. 14024
An Inquiry Into the Administrative and   )
Regulatory Problems Relating to the   )
Authorization of Commercially Operable   )
Space Communications Systems   )

FIRST REPORT

By the Commission:

1. On March 29, 1961, the Commission adopted a Notice of Inquiry (released on April 3, 1961) designed to facilitate an early solution to the administrative and regulatory

problems relating to the future authorization of commercially operable space communication systems. It was stated in the Notice that it may not be feasible to have more than one or a limited number of commercial satellite communication systems due to the substantial capital investment required and limitation of radio spectrum space; and that this raises a problem as to the manner in which such a system or limited number of systems could be accommodated within the Commission's policy of fostering beneficial competition in the international communication field and within the anti-trust laws. Accordingly, the Notice solicited views from all interested parties as to the best plan of [e]nsuring that international communications common carriers, and others, participate on an equitable and non-discriminatory basis in a single or limited number of satellite systems. Views were also solicited as to the legality of the suggested plan; the Commission's authority to prescribe such plan; and the extent to which participants in the plan would be subject to the Commission's jurisdiction. The Notice directed that responses thereto be filed on or before May 1961 and that replies to such responses be filed on or before May 15, 1961.

2.   Responses have been filed by twelve parties, viz., American Rocket Society; American Securities Corporation (for the future Western Union International, Inc.); American Telephone and Telegraph Company; General Electric Company; General Telephone & Electronics Corporation; Hawaiian Telephone Company; International Telephone & Telegraph Corporation (and American Cable & Radio Corporation); Lockheed Aircraft Corporation; Press Wireless, Inc.; Radio Corporation of America (and RCA Communications, Inc.); The Western Union Telegraph Company; and the Department of Justice (commenting only on anti-trust matters).

3.   Replies to such responses were filed by American Telephone and Telegraph Company, General Electric Company, and Lockheed Aircraft [476/2] Corporation.

4.   In general, the respondents were in agreement that for economic and other reasons a single satellite communications system or a limited number of systems, financed and owned by private enterprise, would best serve the public interest. To the extent that the respondents addressed themselves to a specific type of plan, they generally favor a joint venture for the ownership and operation of a system. The principal difference among respondents in this respect related to the composition of such a joint venture. Thus, American Telephone and Telegraph Company and International Telephone and Telegraph Corporation favor ownership in such a system being limited to international communications common carriers, such entities participating in ownership to a degree consistent with their use of the system; General Telephone & Electronics Corporation would limit the ownership to both domestic and international communications common carriers; while Lockheed Aircraft Corporation,[1] General Electric Company, and The Western Union Telegraph Company favor ownership by common carriers, the manufacturing companies, and possibly the public.

5.   Upon consideration of the responses and the replies filed herein the Commission has arrived at certain conclusions, the application of which will serve to foster and accelerate the ultimate establishment of a commercially operable space satellite communication system in the public interest.

6.   We have concluded that the recommendations made herein with respect to the formation or arrangement of a joint venture (or joint undertaking) composed only of existing common carriers engaged in international telephone and telegraph communication is deserving of consideration and exploration as an effective means of promoting the orderly development and effectuation of such a system. We believe that, under

---

1.   Lockheed in its reply comments withdrew its proposal that ownership in a satellite system include private interests other than the international carriers.

Commission regulatory jurisdiction and subject to the conditions and safeguards here-inafter set forth, some form of joint venture by the international common carriers is clear-ly indicated as best serving the public interest for the following reasons:

(a) It appears to be generally accepted that because of considerations of practi-cal economics and technical limitations, it will not be feasible for some time to come to accommodate more than one commercial satellite system.

(b) Communication via satellite will be a supplement to, rather than a substitute for, existing communication systems operated by the international common carriers, thereby becoming an integral part of the total communication system of each such car-rier.

[477/3] (c) The responses filed by the international carriers express a willingness and indicate a capability to marshal their respective resources for the purpose of develop-ing a satellite communication facility.

(d) By reason of their experience in and responsibility for furnishing interna-tional communications service, the international carriers themselves are logically the ones best qualified to determine the nature and extent of the facilities best suited to their needs and those of their foreign correspondents, with whom they have long[-]standing and effective commercial relationships and who necessarily will have a sub-stantial interest in the operations of any satellite system.

(e) Under the Communications Act, the international carriers are obligated to furnish the public with adequate, efficient service at reasonable charges, and this obligation can best be discharged by those carriers maintaining, as far as possible, the greatest degree of direct control and responsibility over the facilities employed in this service.[2]

7.   These considerations, in our opinion, demonstrate the desirability of exploring at this time the means whereby the international common carriers may, collectively, but sub-ject to appropriate regulation and safeguards, take such steps as are necessary to plan and effect the ultimate integration of satellite communication techniques into the fabric of international common carrier service. At the same time these considerations would appear to militate against the suggestions which have been made by certain of the respon-dents that any joint venture with respect to the ownership of satellite communication sys-tems should include participation by the public or by companies in the aerospace and communications equipment manufacturing industries.

8.   We are not unmindful of the substantial interests that these industries have made in the field of space science and the important contributions they have to make to this field. Nor are we unmindful of the potential market that satellite systems represent for the sale of communications and related equipment. However, it appears that the adaptation and integration of satellite communication techniques to international common carrier operations is within the economic means of the existing carriers, although [478/4] requiring cooperative arrangements among them. We fail to see why ownership partici-pation by the aerospace and communications equipment industries will be beneficial or necessary to the establishment of a satellite communication system to be used by the com-mon carrier industry. On the other hand, such participation may well result in encum-bering the system with complicated and costly corporate relationships, disrupting operational patterns that have been established in the international common carrier industry, and impeding effective regulation of the rates and services of the industry.

---

2.   It is recognized that this new technology of communication may present numerous, unique and dif-ficult problems which may involve several approaches and solutions of a type and nature different from those which have been used heretofore in the field of international communications. However, we are satisfied that any such new problems can best be resolved by working within the existing framework of our international com-mon carrier industry.

9. Insofar as the proposal for such participation may have been motivated by concern that without participation the manufacturers of communications equipment will be excluded from this market by the manufacturing companies affiliated with the participating common carriers, the Commission is well aware of this danger. Accordingly, it is the Commission's intention to require that any joint venture that may evolve shall make adequate and effective provision, such as competitive bidding, to [e]nsure that there will be no favoritism in the procurement of communications equipment required for the construction, operation and maintenance of the satellite system. We want to stress that we shall also take all necessary measures and establish regularized procedures to [e]nsure that such a policy is faithfully and conscientiously administered. In this connection, and also to promote the maximum degree of standardization, the Commission will also require that its approval be obtained with respect to the specifications for all equipment used by the common carriers in the satellite system, including the ground terminals. At the same time, before approving any specifications, we shall examine closely into the relevant patent situation to [e]nsure that an undesirable or dominant patent position will not hamper or frustrate the Commission's objectives in this regard.

10. It is neither possible nor feasible for the Commission here to indicate all the specific features which it believes should be incorporated in any joint venture of international common carriers. These matters will, of course, require careful, extended study and formulation by the interested carriers acting under the aegis of the Commission and in accordance with the procedures and policies hereafter to be provided for. However, regardless of organization or type of entity that may subsequently evolve, it must contain clear and definite provisions which will [e]nsure that existing and future international common carriers, whether or not any such carrier participates through ownership in the joint venture, shall have equitable access to, and non-discriminatory use of, the satellite system, under fair and reasonable terms, so as to obtain communication facilities in the system to serve overseas points with the types of services for which they are licensed or authorized by this Commission. The Commission, in issuing licenses or authorizations that may be required to effectuate such joint venture, will take all appropriate measures to implement this policy and to effect such other safeguards as may be required in the public interest.

11. We are making no determination at this time as to the desirability or need for participation in any such joint venture by domestic common carriers.

[479/5] 12. In view of the foregoing, the Commission hereby announces that it will invite all United States international common carriers and certain United States government agencies to attend a conference with the Commission at an early date to explore plans and procedures whereunder consideration of the matters dealt with herein may go forward. A further order will be issued upon conclusion of such consideration.

FEDERAL COMMUNICATIONS COMMISSION

Ben F. Waple
Acting Secretary

Adopted: May 24, 1961

Released: May 24, 1961

Document I-13

Document title: National Aeronautics and Space Council, "Communication Satellites," July 14, 1961.

Source: NASA Historical Reference Collection, History Office, NASA Headquarters, Washington, D.C.

This policy statement outlines the Kennedy administration's approach to the development of communications satellites. It places a much greater emphasis than the Eisenhower administration on the government's role in ensuring that the public and national interests would be served as this new technological capability was brought into being. It also emphasizes the need to develop a truly global system for satellite communications.

[1]   NATIONAL AERONAUTICS AND SPACE COUNCIL

Policy Document Approved—July 14, 1961

Communication Satellites

National Purpose

Science and technology have progressed to such degree that communication through use of space satellites has become possible.

The President has recognized this potentiality and has requested that it be translated into an actuality. In his Message on the State of the Union, the President invited all nations to join with us in a new communication satellite program. On May 25, the President asked the Congress for $50 million of additional funds to accelerate "the use of space satellites for world-wide communications." Again, on June 15, the President requested the Space Council "to make the necessary studies and government-wide policy recommendations for bringing into optimum use at the earliest practicable time, operational communications satellites."

Hence, the national purpose and intent have been made clear.

Program Status

Research and development in the communications satellite field have been conducted over the past few years. This activity has been [2] under government auspices and guidance and has employed primarily the competence and facilities of private industry, through the use of public funds. From these efforts have come prospects for several different types of communication systems, employing passive and active satellites, in either high or low orbit. Much more scientific and technical work needs to be done before an initial system can be selected for commercial operation.

Agencies of the government have been developing a U. S, position with respect to the international allocation of frequencies, in anticipation of an International Telecommunication Union space conference in 1963.

There is a widespread private industry interest in communication satellites, with the anticipation that they can be utilized to meet increased demands for service and for commercial benefit. Also, foreign countries have indicated their interest in communication satellites.

The FCC has instituted proceedings in which problems concerning communication satellite systems are being examined.

The present status of the communication satellite programs, both military and civil, is that of research and development. Neither the arrangements between government and industry for research and development [3] nor the government participation as to preparation of a plan or plans for ownership and operation of a commercial system have contained any commitments as to the operational system.

A communications system using satellites is made up of a number of interconnected parts, of which the satellites are but one part. The full system includes message origination facilities, ground sending stations, ground receiving stations, and message delivery facilities—in addition to the satellites used for continuous receipt and relay of messages. We already have an elaborate communications system between the United States and some parts of the world. Communication satellites must be integrated into the existing system. Adding communication satellites to this system would permit substantially increasing the coverage, increasing the capacity for communication, and enabling television and high speed data, as well as voice and record, to be transmitted and received over great distances.

Problems

As a matter of progressive action, the central question is how to move from a research and development status to an operational status in which the newly emerging technology may be utilized in the public interest.

[4] There are two principal problem areas: one having to do with continuing to advance the state of the art on an accelerated basis and the technical selection of the specifications of an initial operational system; the other having to do with organization and the mode of operation best suited to accommodate the wide range of public interests involved.

Policy

Following are major objectives and policy guidelines for the proper handling of those problems:

1. <u>Time:</u> Operational satellites should become a part of the means of long distance communication at the earliest practicable time and this should be achieved through the leadership of the United States. This means acceleration of effort in research and development, in plans for operation and management, and in cooperative negotiations with other countries.

2. <u>Ownership:</u> There is a wide variety of types, methods, and procedures for the ownership of the U.S. portion of the system. The type of ownership should be that which gives the greatest assurance that the public interest will be best served. Any ownership plan which promises less would be contrary to policy. [5] The type and nature of ownership should not be decided, however, until recommendations submitted by private enterprise have been evaluated by the appropriate agencies of the government to determine whether they meet the policy requirements. If these policy requirements are met, the government will encourage private enterprise to establish and operate a system. This should be decided as soon as practicable in order to maximize the level of national effort.

In addition to the other policy statements in this document, the following criteria and principles should be employed in evaluating recommendations for private ownership of the U.S. portion of the system:

    a.   non-discriminatory use of and equitable access to the system by present and future communications carriers;

    b.   effective competition, such as competitive bidding, in furnishing equipment purchased, leased, or otherwise acquired from non-U.S. government sources;

      c.   full compliance with antitrust legislation and with the regulation of rates, licenses, frequencies, etc.[,] by the appropriate government agencies.

[6] 3. <u>Government Responsibilities:</u> In addition to its regulatory responsibilities, the government should:

      a.   conduct or maintain supervision over international agreements and negotiations;

      b.   conduct and encourage research and development to facilitate accomplishment of these policy objectives and to give maximum assurance of rapid and continuous and technological progress;

      c.   control all launching of U.S. spacecraft;

      d.   make use of the commercial system and avoid competition with it;

      e.   establish separate communication satellite systems, when required to meet unique government needs which cannot, in the national interest, be met by the commercial system;

      f.   assure the effective use of the radio frequency spectrum;

      g.   assure that provision exists for the discontinuance of satellite transmissions when required in the interest of communication efficiency and effectiveness;

      h.   provide technical assistance to newly developing countries in order to attain an effective global system as soon as practicable.

[7] 4. <u>New Uses and Reduced Rates:</u> It is an objective that satellites make available for general use new and expanded international communications services. Transmission of records, voice, and television over great distances should facilitate the exchange of information and ideas throughout the world. These new and expanded uses should, at the earliest possible time, be made available through an economical system, the lower costs of which will be reflected in overseas communication rates. Anticipated greater use and lower costs per channel in a communication satellite system may make lower rates practicable.

     5. <u>Global Coverage:</u> A system of communications designed for "global" coverage is to be contrasted with a system limited to connecting heavy traffic markets and subject to expansion only in response to added demands of sufficient volume as to be profitable per se. Rather, a "global" system is one with the potential and the objective to provide efficient communication service throughout the whole world as soon as technically feasible, including service where individual portions of the coverage are not profitable or even have no expectation of future profit. It is a national objective to have such a global system operable as soon as possible within the limits of technology.

[8] 6. <u>Foreign Participation:</u> It is axiomatic that there be foreign participation in any international commercial communication system. In addition to participation through use, there would be foreign ownership or control of ground facilities outside the United States; international agreements as to frequencies and operating practices; arrangements for connections with other systems; and opportunities through foreign ownership or otherwise in the satellites in the system. The U.S. hopes that practical measures for such foreign participation can be developed.

     7. <u>Relationship with United Nations:</u> The U.S. should examine with other countries the development of the most constructive role for the United Nations, including the [International Telecommunication Union], in international space communications.

Document I-14

Document title: Emanuel Celler, Chairman, Committee on the Judiciary, House of Representatives, *et al.*, to the President, August 24, 1961.

Source: NASA Historical Reference Collection, History Office, NASA Headquarters, Washington, D.C.

The form of ownership of an operational communications satellite system became a controversial issue in 1961 and 1962, as it became clear that such a system would be established within a few years. The Kennedy administration's policy statement on communications satellites (Document I-13 in this volume) was released on July 24, 1961 (approved on July 14), and provoked this response from thirty-five members of Congress who feared AT&T's dominance of an operational system. They urged President Kennedy to wait until any system was fully operational before he made a final decision on the form of ownership. Kennedy did not follow this suggestion; several of the signers of this letter were among those leading the push for public ownership of a communications satellite system during the 1962 congressional debate on the issue.

[1]   CONGRESS OF THE UNITED STATES
                              House of Representatives
                                Washington, D.C.

Emanuel Celler
11th District New York

Chairman
Committee on the Judiciary

                                                                    August 24, 1961

The President
The White House
Washington, D.C.

My dear Mr. President:

Early development of a space satellite communications system is of fundamental national importance. Such a system gives promise of revolutionizing international communications and communications within the United States. It has potentiality for an unprecedented increase in worldwide telephone and telegraph communications and for providing transoceanic television and radio transmission.

We undersigned members of Congress therefore believe that it is crucial that the United States be the first to develop the system. We further believe that the Federal Government should by contract carry out extensive research, experimentation, and development of a satellite communications system. Not a minute should be wasted. After such a system has become fully operational, but not until then, we believe, can decisions be intelligently made as to whether such a system should be publicly or privately owned and under what circumstances.

As you have pointed out, "the present status of the communications satellite programs, both civil and military, is that of research and development. To date, no arrangements between the government and private industry contain any commitments as to an operational system." We believe this is as it should be. Present commitments of any kind as to the control of this system may hinder its rapid development and prejudge vital questions of public interest and international relations.

The course of research and development for this new system have [sic] demonstrated one overwhelming fact: We do not at present know which system can be put into use first, nor which system will be most efficient once in orbit. Given this technological uncertainty, the complicated question of ownership and control of this system must necessarily be covered with an even greater haze of uncertainty. In order to [e]nsure that the rapid development of this new system is not [2] impeded by a premature decision as to ownership, we are of the opinion that prudence requires a further investigation of the broadest aspects of the ownership question. Specifically, we believe that the debate over ownership should be separated from the developmental question until the entire system becomes fully operational. During this period development should proceed with all possible speed while careful study is given to the decision as to the control of these unripened fruits of science.

While we believe that the final question of ownership should not be decided at a time when we have insufficient knowledge, we wish to make it clear that should private concerns be authorized to own and operate the system, the government agencies entrusted with responsibility must, consistent with the antitrust laws, prevent any concern from attaining a monopolistic, dominant, or preferential advantage. Otherwise the national interest would be frustrated for generations to come in a historic achievement which, according to a responsible prediction, may well constitute a multi-billion dollar a year business in ten to fifteen years.

On July 24 you issued a policy statement that essential conditions to private ownership of a space satellite communications system are a "structure of ownership and control which will assure maximum possible competition" and "full compliance with antitrust legislation." We are in complete agreement with these conditions and it is for this reason that we are deeply concerned about orders issued by the Federal Communications Commission on May 24 and July 25, 1961 which clearly contemplate limiting ownership to a specified group of so-called "international carriers" which does not even include all these carriers. These orders are contrary to the policy established by you; they are contrary to the principles of the antitrust laws.

The FCC orders appear for all practical purposes to determine that the satellite communications system is to be owned and operated by this group of ten "international carriers." This would mean that only four concerns would participate in the system's ownership since the other six companies in this group have professed no interest whatsoever in space communications. More important, it would mean that one of these four companies, AT&T, would have a dominant and very probably a monopoly position in ownership of the space communications system. In effect, AT&T would be the chosen instrument of the United States Government to own and control civilian space communications.

This would be intolerable from the standpoint of the public interest. As the Department of Justice has stated, "the continuing opportunity (for AT&T) to favor its own facilities would always be present and would inevitably result in discrimination or suspicion of discrimination no matter how strict might be the policy of (AT&T) to provide [3] equal service to its competitors." Furthermore, "the opportunity to favor the purchase of equipment produced by (AT&T's subsidiary, Western Electric Co.) would be irresistible."

The head of the Justice Department's Antitrust Division has testified that "the degree of concentration in this field may very well be one of the reasons why America is not further advanced in the field today than it is . . . . Our system has not produced as it should, and the public interest has suffered because there has been undue concentration in this field."

We believe that to safeguard the public interest it is essential that any plan permitting private ownership if, indeed, such is preferred to public, of the space satellite system must:

      (1) afford all interested United States communications common carriers, domestic as well as international, opportunity to participate in ownership of the system; and

      (2) afford all interested communications and aerospace manufacturers opportunity to participate in ownership of the system.

We have seen from past experience how the American Telephone & Telegraph Company has been able to expand its monopoly position and strengthen its hold on the American economy by combining, under the aegis of one holding company, its equipment manufacturing concern, the Western Electric Company, and the operating divisions of the Bell Telephone System. Only by insisting upon the widest participant by all interested communications and aerospace manufacturers and operators can there by any hope that such a monopoly can be forestalled in this new and vital field.

The antitrust laws prohibit monopolization of any part of the domestic or foreign commerce of the United States. They also prohibit the acquisition, ownership, control, or operation by an interstate or foreign wire carrier of any station or any system for radio communications or signals between any place in any state in the United States and any place in any foreign country, if the purpose is, and/or the effect thereof may be, to substantially lessen competition or to restrain commerce between any place in any state in the United States and any place in any foreign country, or unlawfully to create monopoly in any line of commerce. In these circumstances, any plan which does not meet *both* the conditions we have specified would, in our considered judgment, be in direct violation of the antitrust laws and would require special legislation by the Congress. No executive order or decree of any agency can override the antitrust laws.

[4] Nor is there any logical or rational basis for excluding U.S. domestic communications common carriers from ownership in the system while granting companies which have no interest and virtually no investment in international communications service opportunity to participate in the system's ownership, particularly since the space satellite could provide domestic as well as international communications services.

Furthermore, it is clear that the space satellite communications system will be vastly different from the conventional common carrier type of operation. Thus there is no justification for excluding communications and aerospace manufacturers, particularly when the record clearly demonstrates that a number of these organizations have a far greater contribution to make in expert technology than any of the ten "communications carriers."

The question of monopoly is only one of many complicated questions involved in the decision as to what kind of an ownership system will best meet the public interest. The ramifications of this remarkable system are likely to be truly revolutionary. And, as with all revolutions, it is clear that our understanding of the implications of a new technology is likely to lag behind developments themselves. Because we believe time and study are essential to wise decision-making, and because we do not want to prejudice the ultimate question of control and ownership during the period of study, we urge that:

1. No decisions concerning ultimate control be made until the entire system becomes fully operational.
2. No contracts, decisions, or acts which may prejudice the ultimate decision as to ownership be agreed to until the entire system becomes fully operational.
3. During this period, the Congress be consulted upon the question of ultimate control and ownership and allowed to exercise its constitutional responsibility to supervise activities of Federal agencies regulating foreign and domestic commerce.
4. During this period, all other interested parties be consulted fully upon the question of ownership and control.
5. During this period, all possible questions of international agreement, cooperation, control and ownership related to other nations and the [United Nations] be thoroughly explored.

The United States can demonstrate to the world what a democratic system can accomplish in developing a space communications satellite system. But if decisions are taken in haste and allowed to cramp and prejudice [5] the rational development of the new gift of science, it is likely that we may not only prejudice a question of vital national concern, but we may hinder the rapid development of the system itself.

Your statement of July 24, 1961 makes it clear that if private ownership is to be favored, the ownership and control system must meet eight stringent conditions. We would like to emphasize that the conditions laid down are a very difficult set of tests for any system of ownership to meet. We believe that if careful thought is given to how or, indeed, whether, these tests can be met, not only will the public interest be served, but the rapid development of this new gift of science will mo[v]e ahead unhindered by a premature struggle over its fruits.

Sincerely yours,

Hubert H. Humphrey	Leonard Farbstein	Joseph M. Montoya
Estes Kefauver	Kenneth J. Gray	John E. Moss
Wayne Morse	Chet Holifield	Abraham J. Multer
	Elmer J. Holland	M. Blaine Peterson
Joseph P. Addabbo	Lester Holtzman	Henry S. Reuss
Thomas L. Ashley	Robert W. Kastenmeier	Ralph J. Rivers
Edward P. Boland	Eugene J. Keogh	James Roosevelt
James A. Burke	Frank Kowalski	William Fitts Ryan
James A. Byrne	Thomas J. Lane	John F. Shelley
Emanuel Celler	Richard E. Lankford	B. F. Sisk
Merwin Coad	Roland V. Libonati	Herman Toll
Jeffery Cohelan	Clem Miller	Al Ullman

Document I-15

Document title: Frederick G. Dutton, Assistant to the President, Memorandum for the President, November 13, 1961.

Source: NASA Historical Reference Collection, History Office, NASA Headquarters, Washington, D.C.

In October 1961, the FCC Ad Hoc Carrier (or Industry) Committee Report proposed that a nonprofit corporation be established to develop and operate the communications satellite system. This corporation would lease circuits to authorized carriers, which would own the satellites as well as their ground stations. The corporation would be run by a board of directors, including representatives of AT&T, ITT, RCA, and Western Union and three public directors appointed by the president. The committee's report resulted in immediate controversy, as noted in Frederick Dutton's memorandum to President Kennedy. ITT, RCA, and Western Union all expressed concern that AT&T would dominate such a corporation, while representatives of other aerospace and electronic manufacturers were unhappy that they would be excluded from participating in such a revolutionary field. Some members of Congress expressed concern that such a corporation involving all of the international carriers would constitute a monopoly. The issue was not settled until the passage of the Communications Satellite Act on August 31, 1962.

November 13, 1961

Memorandum for the President

As a matter of information, you should be aware that the proper kind of entity to own and operate communication satellites is becoming an increasing source of controversy. I have brought together a Task Force of representatives from the interested Federal agencies with the Chairman, Ed Welsh, as Executive Director of the Space Council, to prepare recommendations consistent with your policy statement in this field. The Executive agencies are dissatisfied with the report of the FCC Ad Hoc Industry Committee; and substantial Congressional and press concern continues over AT&T's potential stranglehold over communication satellites. The Task Force will have recommendations ready in December. Senator Kerr is preparing his own legislative recommendations, so the entire matter will undoubtedly come to a head during the coming Congressional session.

Frederick G. Dutton

Document I-16

Document title: Senator Robert S. Kerr, "Amendment to the National Aeronautics and Space Act of 1958, Space Communications," November 28, 1961.

Source: NASA Historical Reference Collection, History Office, NASA Headquarters, Washington, D.C.

Document I-17

Document title: E.C. Welsh, Executive Secretary, National Aeronautics and Space Council, Executive Office of the President, Memorandum to the President, April 11, 1962.

Source: NASA Historical Reference Collection, History Office, NASA Headquarters, Washington, D.C.

Document I-18

Document title: "Communications Satellite Act of 1962," Public Law 87-624, 76 Stat. 419, signed by the President on August 31, 1962.

Source: NASA Historical Reference Collection, History Office, NASA Headquarters, Washington, D.C.

By the spring of 1962, the Kennedy administration had decided its preference regarding what kind of communications satellite organization should be developed. There were some in Congress, however, who wanted public ownership of any such organization, while others argued for totally private control. While the congressional debate was going on, Telstar was launched and successfully operated by AT&T. This further strengthened the case for private-sector operation of international satellite communications. In addition to the administration bill mentioned in the memorandum to President Kennedy from Edward C. Welsh, Executive Secretary for the National Aeronautics and Space Council, there were fifteen other legislative proposals concerning the same subject. These advocated alternatives included government ownership, limited private ownership similar to the administration bill, and open ownership not limited to the international carriers.

John Johnson, NASA's general counsel, was first asked to draft Senator Robert Kerr's bill for a communications satellite corporation and then asked to draft the Kennedy administration's bill for the same proposal. Kerr's bill provided for an entirely privately run corporation, regulated by the government. It represented the more conservative side of the argument over ownership and control. Kerr chaired the Senate Committee on Aeronautical and Space Sciences and served primarily as an advocate for the administration's bill rather than his own. Estes Kefauver represented the other side of the debate and introduced a bill calling for total government ownership. He led the opposition to the administration bill on the Senate floor, first through the addition of a number of amendments and then by a filibuster. The Senate finally moved for cloture for the first time since 1927 to end the debate. In the end, the administration bill was not significantly altered and eventually became the basis for the Communications Satellite Act of 1962.

Document I-16

[1] (AUTOMATIC RELEASE IN A. M.'S OF TUESDAY, NOVEMBER 28, 1961)

FROM THE OFFICE OF SENATOR ROBERT S. KERR

DRAFT

Be it enacted by the Senate and House of Representatives of the United States of America in Congress assembled,

That the National Aeronautics and Space Act of 1958, as amended (42 USC 2451-2476), is amended by adding thereto a Title IV, to read as follows:

TITLE IV—SPACE COMMUNICATIONS

DECLARATION OF POLICY

Sec. 401. The Congress hereby declares that it is the policy of the United States to provide leadership in the establishment of a world-wide communications system involving the use of space satellites at the earliest practicable time, to provide for ownership of the United States portion of the system, and to invite all nations to participate in the system in the interest of world peace and closer brotherhood among peoples throughout the world.

CREATION OF SATELLITE COMMUNICATIONS CORPORATION

Sec. 402. (a) The provisions of this Section shall take effect as provided in Sec. 406 of this Act.

(b) There is hereby created a corporation, to be known as the Satellite Communications Corporation (hereafter referred to as the "Corporation"), whose object and purposes shall be to develop, construct, operate, manage, and promote the use of a communications satellite system in the public interest, and to foster research and development in the field of space telecommunications.

(c) Organization and Operation. The Corporation shall be organized and operated as a communications common carriers' carrier, and shall own the United States portion of the communications satellite system, consisting of the satellites, the earth terminals, and associated ground control and tracking facilities. The Corporation. shall make the facilities of the system available on a nondiscriminatory and equitable basis, at rates to be established by the Federal Communications Commission (hereafter called the "Commission"), to all United States carriers authorized by the Commission to provide communications services via satellite.

(d) Foreign Participation. The Corporation shall also provide opportunities for foreign participation in the communications satellite system, through ownership or otherwise, on an equitable basis and on reasonable terms

(e) Ownership of Corporation. Ownership interests in the Corporation shall be limited to United States communications common carriers who are determined by the Commission to be eligible to participate in such ownership.

[2] (f) <u>Officers of Corporation.</u> There shall be a Board of Directors consisting of two Directors appointed by each carrier or affiliated group of carriers having an ownership interest in the Corporation and two additional Directors designated jointly by United States carriers who are authorized by the Commission to provide communications services via satellite but who do not acquire an ownership interest in the Corporation. The Board of Directors shall choose, or elect by majority vote, the principal officers of the Corporation. Each Director shall have one vote in all matters determined by the Board of Directors.

(g) <u>Financing of Corporation.</u>

(1) The total authorized capital stock of the Corporation shall be 500 million dollars, consisting of five thousand shares of the value of $100,000.00 each. Such stock shall be of one class, shall be nonassessable, and shall be issued only for cash fully paid. Stock held by owning carriers shall not be transferable, except with the approval of the Commission, and then only to other communications common carriers determined by the Commission to be eligible to participate in the ownership of the Corporation.

(2) The minimum amount of stock which shall be held by an owning carrier shall be five shares.

POWERS OF CORPORATION

Sec. 403. The Corporation shall have perpetual succession, and shall have power to do any and all things necessary and proper to carry out the object and purposes of the Corporation, including, without limitation thereto, the following—

(a) to acquire, hold, own, mortgage, lease, and dispose of real and personal property, of every class and description and without limitation as to place;

(b) to lease channels to authorized users of the communications satellite system;

(c) to conduct research and development;

(d) to enter into, make and perform contracts and agreements of every kind and description with any person, firm, association, corporation, municipality, county, state, body politic, or government or colony or dependency thereof.

(e) to sue and be sued;

(f) to accept unconditional gifts of services, money or property, and legacies and devises;

(g) to adopt and alter a corporate seal and, subject to prior approval of the Commission, by-laws not inconsistent with the laws of the United States or of any State;

(h) to establish and maintain offices and facilities for the conduct of the affairs of the Corporation in the District of Columbia and in the several states and territories of the United States, and in foreign countries; and

(i) to purchase, hold, sell, and transfer the shares of its own capital stock.

[3]   RELATIONSHIP BETWEEN THE CORPORATION AND
        THE NATIONAL AERONAUTICS AND SPACE ADMINISTRATION

Sec. 404. (a) The Administration shall be responsible for—

(1) furnishing launch vehicles required for the communications satellite system on a schedule which will facilitate the economical and efficient development and operation of the system;

(2) launching the satellites, and furnishing launch-crew and associated services;

(3) consulting with the Corporation on the technical specifications for satellites and earth terminal station, and on the number and location of such stations; and

(4) to the greatest extent practicable, coordinating its research and development program in the field of space telecommunications with that of the Corporation so as to give maximum assurance of rapid and continuous scientific and technological progress.

(b) The costs of the launch vehicles and launching and related services furnished by the Administration under subparagraphs (a) (1) and (a) (2) above shall be reimbursed by the Corporation as a credit to current appropriations of the Administration.

(c) The Administration is also authorized to furnish other services, on a reimbursable basis, upon the request of the Corporation and as required for the successful development and operation of the communications satellite system.

(d) The Corporation shall consult with the Administration, and coordinate its research program, as provided in subparagraphs (a) (3) and (a) (4) above.

RELATIONSHIP BETWEEN THE CORPORATION AND THE FEDERAL COMMUNICATIONS COMMISSION

Sec. 405. (a) In regulating the Corporation as a communications common carrier under the Communications Act of 1934, as amended, the Commission shall [e]nsure—

(1) that the communications satellite system established by the Corporation is technically compatible with and operationally interconnected with existing communications facilities;

(2) that the rate structure established for the communications services offered by the Corporation will provide a fair return on the capital invested in the Corporation; and

(3) that there will be nondiscriminatory use of and equitable access to the system by present and future authorized communications carriers.

[4] (b) The Commission is also authorized to—

(1) determine which United States communications common carriers shall be eligible to participate in the ownership of the Corporation;

(2) approve the by-laws of the Corporation, or alterations thereto;

(3) require the Corporation to employ competitive bidding in the acquisition of equipment used in the system, to the maximum extent feasible, so as to preserve effective competition;

(4) require the Corporation to provide communications services in areas of the world where it may be uneconomical, so as to make the system global in coverage as soon as technically feasible; and

(5) require the Corporation to provide opportunities for foreign participation, through ownership or otherwise, in the system, on an equitable basis, and on reasonable terms.

COMPLETION OF ORGANIZATION OF THE CORPORATION

Sec. 406. (a) The President is authorized to take all steps necessary to organize the Corporation described in Sec. 402 hereof, including but not limited to the following:

(1) obtaining a determination by the Commission as to which communications common carriers shall be eligible to participate in the ownership of the Corporation;

(2) obtaining commitments from such eligible common carriers as to the amounts they will invest in the Corporation; and

(3) receiving nominations to the Board of Directors of the Corporation, as provided for in Sec. 402 (f) hereof.

(b) The Senate Committee on Aeronautical and Space Sciences and the House Committee on Science and Astronautics shall be notified when all steps necessary to the organization of the Corporation have been completed, including approval of the proposed Corporation by-laws by the Commission. Sec. 402 will take effect thirty days after the date of such notification.

(c) If the organization of the Corporation has not been completed within three months after the date of enactment of this Act, the President shall make an interim report to the Senate Committee on Aeronautical and Space Sciences and the House Committee on Science and Astronautics on the status of such organization.

Document I-17

[1]
Executive Secretary   April 11, 1962

Memorandum to the President

It is understood that Chairman Oren Harris of the House Interstate and Foreign Commerce Committee is scheduled to see you tomorrow. If so, he may want to discuss the Communications Satellite legislation, which is before his committee. He has completed his hearings and is hoping to mark up the bill very soon.

Your bill was introduced in the Senate as S. 2814. It was introduced by Congressman Harris in the House as H.R. 10115. Harris also introduced your bill as modified by the Senate Space Committee. That bill is H.R. 11040, and is currently the one primarily being considered by Chairman Harris.

Harris has been most cooperative in this matter, as illustrated by his invitation to Nick Katzenbach and me to help iron out various questions with him and his staff yesterday. He indicated that he wanted to make the minimum number of changes in the bill and thought it important to act quickly so that efforts by others to obtain Government ownership of the system would not have time to block action this session. This same view is, I believe, held by Senator Kerr and his Space Committee, and Senator Pastore and the Interstate and Foreign Commerce Committee in the Senate.

Our meeting with Chairman Harris was also attended by Newt Minow, who stated that the majority of the FCC, including himself, believed they could "live with and make work" the bill as it now stands. Some clarifying language of various minor points is being worked out by the staff of the Commerce Committee and staff of the Justice Department.

It is suggested that emphasis might well be made on the following:
1. That broad-based ownership is the important principle, and that the provision in H.R. 11040 (50% of stock ownership by the public, and 50% by authorized carriers) is satisfactory.
2. That the new corporation should be authorized to own ground stations, without preventing individual carriers from also [2] owning such terminals, and that it would be best if the legislation left the decision as to ground station ownership up to a finding of public interest, convenience, and necessity on the part of the FCC, without any language in the bill which would prejudice or influence the FCC's decision in any individual case.
3. That legislation should not be delayed any longer than absolutely necessary.
4. That it makes no difference whether the new legislation becomes a separate statute (which you had proposed) or takes the form of amendment to the Communications Act or to the Space Act, and that this is a matter for the Congress to determine.

Attached for possible reference is my recent testimony on the major changes made in your bill and some of the major aspects of the proposed legislation.

[hand-signed] E.C. Welsh

Attachment

Document I-18

[1]   Public Law 87-624
87th Congress, H.R. 11040
August 31, 1962

An Act

76 STAT. 419

To provide for the establishment, ownership, operation, and regulation of a commercial communications satellite system, and for other purposes.

Be it enacted by the Senate and House of Representatives of the United States of America in Congress assembled,

TITLE I—SHORT TITLE, DECLARATION OF POLICY AND DEFINITIONS

SHORT TITLE

SEC. 101. This act may be cited as the "Communications Satellite Communications Act of 1962" [citation in margin: "Communication Satellite Act of 1962"].

DECLARATION OF POLICY AND PURPOSE

SEC. 102. (a) The Congress hereby declares that it is the policy of the United States to establish, in conjunction and in cooperation with other countries, as expeditiously practicable a commercial communications satellite system, as part of an improved global communications network, which will be responsive to public needs and national objectives, which will serve the communication needs of the United States and other countries, and which will contribute to world peace and understanding.

(b) The new and expanded telecommunication services are to be made available as promptly as possible and are to be extended to provide global coverage at the earliest practicable date. In effectuating this program, care and attention will be directed toward providing such services to economically less developed countries and areas as well as those more highly developed, toward efficient and economical use of the electromagnetic frequency spectrum, and toward the reflection of the benefits of this new technology in both quality of services and charges for such services.

(c) In order to facilitate this development and to provide for the widest possible participation by private enterprises United States participation in the global system shall be in the form of a private corporation, subject to appropriate governmental regulation. It is the intent of Congress that all authorized users shall have nondiscriminatory access to the system; that maximum competition be maintained in the provision of equipment and services utilized by the system; that the corporation created under this Act be so organized and operated as to maintain and strengthen competition in the provision of communications services to the public; and that the activities of the corporation created under this Act and or the persons or companies participating in the ownership of the corporation shall be consistent with the Federal antitrust laws.

(d) It is not the intent of Congress by this Act to preclude the use of the communications satellite system for domestic communication services where consistent with the provisions of this Act nor to preclude the creation of additional communications satellite systems, if required to meet unique governmental needs or if otherwise required in the national interest.

DEFINITIONS

SEC. 103. As used in this Act, and unless the context otherwise requires-

    (1) the term "communications satellite system" refers to a system of communications satellites in space whose purpose is to relay telecommunication information between satellite terminal stations, [2] together with such associated equipment and facilities for tracking, guidance, control, and command functions as are not part of the generalized launching, tracking, control, and command facilities for all space purposes;

    (2) the term "satellite terminal station" refers to a complex of communication equipment located on the earth's surface, operationally connected with one or more terrestrial communication systems, and capable of transmitting telecommunications to or receiving telecommunications from a communications satellite system.

    (3) the term "communications satellite" means an earth satellite which is intentionally used to relay telecommunication information;

    (4) the term "associated equipment and facilities" refers to facilities other than satellite terminal stations and communications satellites, to be constructed and operated for the primary purpose of a communications satellite system, whether for administration and management, for research and development, or for direct support of space operations;

    (5) the term "research and development" refers to the conception, design, and first creation of experimental or prototype operational devices for the operation of a communications satellite system, including the assembly of separate components into a working whole, as distinguished from the term "production," which relates to the construction of such devices to fixed specifications compatible with repetitive duplication for operational applications;

    (6) the term "telecommunication" means any transmission emission or reception of signs, signals, writings, images, and sounds or intelligence of any nature by wire, radio, optical, or other electromagnetic systems;

    (7) the term "communications common carrier" has the same meaning as the term "common carrier" has when used in the Communications Act of 1934, as amended, and in addition includes, but only for purposes of sections 303 and 304 [citation in margin: "48 Stat. 1064; 47 USC 609"], any individual, partnership, association, joint-stock company, trust, corporation, or other entity which owns or controls, directly or indirectly, or is under direct or indirect common control with, any such carrier; and the term "authorized carrier," except otherwise provided for purposes of section 304 by section 304 (b) (1), means a communications common carrier which has been authorized by

the Federal Communications Commission under the Communications Act of 1934, as amended, to provide services by means of communications satellites;

(8) the term "corporation" means the corporation authorized by title III of this Act.

(9) the term "Administration" means the National Aeronautics and Space Administration; and

(10) the term "Commission" means the Federal Communications Commission.

[3]   TITLE II—FEDERAL COORDINATION, PLANNING, AND REGULATION

IMPLEMENTATION OF POLICY

SEC. 201. In order to achieve the objectives and to carry out the purposes of this Act—

(a) the President shall—

(1) aid in the planning and development and foster the execution of a national program for the establishment and operation, as expeditiously as possible, of a commercial communications satellite system;

(2) provide for continuous review of all phases of the development and operation of such a system, including the activities of a communications satellite corporation authorized under title III of this Act;

(3) coordinate the activities of governmental agencies with responsibilities in the field of telecommunication, so as to [e]nsure that there is full and effective compliance at all times with the policies set forth in this Act;

(4) exercise such supervision over relationships of the corporation with foreign governments or entities or with international bodies as may be appropriate to assure that such relationships shall be consistent with the national interest and foreign policy of the United States;

(5) [e]nsure that timely arrangements are made under which there can be foreign participation in the establishment and use of a communications satellite system;

(6) take all necessary steps to [e]nsure the availability and appropriate utilization of the communications satellite system for general governmental purposes except where a separate communications satellite system is required to meet unique governmental needs, or is otherwise required in the national interest; and

(7) so exercise his authority as to help attain coordinated and efficient use of the electromagnetic spectrum and the technical compatibility of the system with existing communications facilities both in the United States and abroad.

(b) the National Aeronautics and Space Administration shall—

(1) advise the Commission on technical characteristics of the communications satellite system;

(2) cooperate with the corporation in research and development to the extent deemed appropriate by the Administration in the public interest;

(3) assist the corporation in the conduct of its research and development program by furnishing to the corporation, when requested, on a reimbursable basis, such satellite launching and associated services as the Administration deems necessary for the most expeditious and economical development of the communications satellite system;

(4) consult with the corporation with respect to the technical characteristics of the communications satellite system;

(5) furnish to the corporation, on request and on a reimbursable basis, satellite launching and associated services required for the establishment operation,

and maintenance of the communications satellite system approved by the Commission; and

[4]   (6) to the extent feasible, furnish other services, on a reimbursable basis, to the corporation in connection with the establishment and operation of the system.

   (c) the Federal Communications Commission, in its administration of the provisions of the Communications Act of 1934, as amended, and as supplemented by this Act, shall [citation in margin: "48 Stat. 1064; 47 USC 609"]—

(1) [e]nsure effective competition, including the use of competitive bidding where appropriate, in the procurement by the corporation and communications common carriers of apparatus, equipment, and services required for the establishment and operation of the communications satellite system and satellite terminal stations; and the Commission shall consult with the Small Business Administration and solicit its recommendations on measures and procedures which will [e]nsure that small business concerns are given an equitable opportunity to share in the procurement program of the corporation for property and services, including but not limited to research, development, construction, maintenance, and repair;

(2) [e]nsure that all present and future authorized carriers shall have nondiscriminatory use of, and equitable access to, the communications satellite system and satellite terminal stations under just and reasonable charges, classifications, practices, regulations, and other terms and conditions and regulate the manner in which available facilities of the system and stations are allocated among such users thereof;

(3) in any case where the Secretary of State, after obtaining the advice of the Administration as to technical feasibility, has advised that commercial communication to a particular foreign point by means of the communications satellite system and satellite terminal stations should be established in the national interest, institute forthwith appropriate proceedings under section 214 (d) of the Communications Act of 1934, as amended [citation in margin: "57 Stat. 12; 47 USC 214"], to require the establishment of such communication by the corporation and the appropriate common carrier or carriers;

(4) [e]nsure that facilities of the communications satellite system and satellite terminal stations are technically compatible and interconnected operationally with each other and with existing communications facilities;

(5) prescribe such accounting regulations and systems and engage in such ratemaking procedures as will [e]nsure that any economies made possible by a communications satellite system are appropriately reflected in rates for public communication services;

(6) approve technical characteristics of the operational communications satellite system to be employed by the corporation and of the satellite terminal stations;

(7) grant appropriate authorizations for the construction and operation of each satellite terminal station, either to the corporation or to one or more authorized carriers or to the corporation and one or more such carriers jointly, as will best serve the public interest, convenience, and necessity. In determining the public interest, convenience, and necessity the Commission shall authorize the construction and operation of such stations by communications common carriers or the corporation, without preference to either;

(8) authorize the corporation to issue any shares of capital stock, except the initial issue of capital stock referred to in section 304 (a), or to borrow any moneys, or to assume any [5] obligation in respect of the securities of any other person, upon a finding that such issuance, borrowing, or assumption is com-

patible with the public interest, convenience, and necessity and is necessary or appropriate for or consistent with carrying out the purposes and objectives of this Act by the corporation;

(9) [e]nsure that no substantial additions are made by the corporation or carriers with respect to facilities of the system or satellite terminal stations unless such additions are required by the public interest, convenience, and necessity;

(10) require, in accordance with the procedural requirements of section 214 of the Communications Act of 1934, as amended [citation in margin: "57 Stat. 11; 47 USC 214"], that additions be made by the corporation or carriers with respect to facilities of the system or satellite terminal stations where such additions would serve the public interest, convenience, and necessity; and

(11) make rules and regulations to carry out the provisions of this Act.

TITLE III—CREATION OF A COMMUNICATIONS SATELLITE CORPORATION

CREATION OF CORPORATION

SEC. 301. There is hereby authorized to be created a communications satellite corporation for profit which will not be an agency or establishment of the United States Government. The corporation shall be subject to the provisions of this Act and, to the extent consistent with this Act, to the District of Columbia Business Corporation Act. The right to repeal, alter, or amend this Act at any time is expressly reserved [citation in margin: "68 Stat. 177; D.C. Code 29-901"].

PROCESS OF ORGANIZATION

SEC 302. The President of the United States shall appoint incorporators, by and with the advice and consent of the Senate, who shall serve as the initial board of directors until the first annual meeting of stockholders or until their successors are elected and qualified. Such incorporators shall arrange for an initial stock offering and take whatever other actions are necessary to establish the corporation, including the filing of articles of incorporation, as approved by the President.

DIRECTORS AND OFFICERS

SEC. 303. (a) The corporation shall have a board of directors consisting of individuals who are citizens of the United States, of whom one shall be elected annually by the board to serve as chairman. Three members of the board shall be appointed by the President of the United States, by and with the advice and consent of the Senate, effective the date on which the other members are elected, and for terms of three years or until their successors have been appointed and qualified, except that the first three members of the board so appointed shall continue in office for terms of one, two, and three years, respectively, and any member so appointed to fill a vacancy shall be appointed only for the unexpired term of the director whom he succeeds. Six members of the board shall be elected annually by those stockholders who are communications common carriers and six shall be elected annually by the other stockholders of the corporation. No stockholder who is a communications common carrier and no trustee for such a stockholder shall vote, either directly or indirectly, through the votes of subsidiaries or affiliated companies, nominees, or any persons subject to [6] his direction or control, for more than three candidates for membership on the board. Subject to such limitation, the articles of incorporation to be filed by the incorporators designated under section 302 shall provide for cumulative voting under section 27 (d) of the District of Columbia Business Corporation Act (D.C. Code, sec. 29-911 (d)) [citation in margin: "68 Stat. 191"].

(b) The corporation shall have a president, and such other officers as may be named and appointed by the board, at rates of compensation fixed by the board, and serving at the pleasure of the board. No individual other than a citizen of the United States may be an officer of the corporation. No officer of the corporation shall receive any salary from any source other than the corporation during the period of his employment by the corporation.

FINANCING OF THE CORPORATION

SEC 304. (a) The corporation is authorized to issue and have outstanding, in such amounts as it shall determine, shares of capital stock, without par value, which shall carry voting rights and be eligible for dividends. The shares of such stock initially offered shall be sold at a price not in excess of $100 for each share and in a manner to encourage the widest distribution to the American public. Subject to the provisions of subsections (b) and (d) of this section, shares of stock offered under this subsection may be issued to and held by any person.

(b) (1) For the purposes of this section the term "authorized carrier" [note in margin: "Authorized carrier"] shall mean a communications common carrier which is specifically authorized or which is a member of a class of carriers authorized by the Commission to own shares of stock in the corporation upon a finding that such ownership will be consistent with the public interest, convenience, and necessity.

(2) Only those communications common carriers which are authorized carriers shall own shares of stock in the corporation at any time, and no other communications common carrier shall own shares either directly or indirectly through subsidiaries or affiliated companies, nominees, or any persons subject to its direction or control. Fifty per centum of the shares of stock authorized for issuance at any time by the corporation shall be reserved for purchase by authorized carriers and such carriers shall in the aggregate be entitled to make purchases of the reserved shares in a total number not exceeding the total number of the nonreserved shares of any issue purchased by other persons. At no time after the initial issue is completed shall the aggregate of the shares of voting stock of the corporation owned by authorized carriers directly or indirectly through subsidiaries or affiliated companies, nominees, or any persons subject to their direction or control exceed 50 per centum of such shares issued and outstanding.

(3) At no time shall any stockholder who is not an authorized carrier, or any syndicate or affiliated group of such stockholders, own more than 10 per centum of the shares of voting stock of the corporation issued and outstanding.

(c) The corporation is authorized to issue, in addition to the stock authorized by subsection (a) of this section, nonvoting securities, bonds, debentures, and other certificates of indebtedness as it may determine. Such nonvoting securities, bonds, debentures, or other certificates of indebtedness of the corporation as a communications common carrier may own shall be eligible for inclusion in the rate base of the carrier to the extent allowed by the Commission. The voting [7] stock of the corporation shall not be eligible for inclusion in the rate base of the carrier.

(d) Not more than an aggregate of 20 per centum of the shares of stock of the corporation authorized by subsection (a) of this section which are held by holders other than authorized carriers may be held by persons of the classes described in paragraphs (1), (2),

(3), (4), and (5) of section 310 (a) of the Communications Act of 1934, as amended (47 U.S.C. 310) [citation in margin: "48 Stat. 108"].

(e) The requirement of section 45 (b) of the District of Columbia Business Corporation Act (D.C. Code, sec. 29-920 (b)) as to the percentage of stock which a stockholder must hold in order to have the rights of inspection and copying set forth in that subsection shall not be applicable in the case of holders of the stock of the corporation, and they may exercise such rights without regard to the percentage of stock they hold.

(f) Upon application to the Commission by any authorized carrier and after notice and hearing, the Commission may compel any other authorized carrier which owns shares of stock in the corporation to transfer to the applicant, for a fair and reasonable consideration, a number of such shares as the Commission determines will advance the public interest and the purposes of this Act. In its determination with respect to ownership of shares of stock in the corporation, the Commission, whenever consistent with the public interest, shall promote the widest possible distribution of stock among the authorized carriers.

PURPOSES AND POWERS OF THE CORPORATION

SEC. 305. (a) In order to achieve the objectives and to carry out the purposes of this Act, the corporation is authorized to—

(1) plan, initiate, construct, own, manage, and operate itself or in conjunction with foreign governments or business entities a commercial communications satellite system;

(2) furnish, for hire, channels of communication to United States communications common carriers and to other authorized entities, foreign and domestic; and

(3) own and operate satellite terminal stations when licensed by the Commission under section 201 (c) (7).

(b) Included in the activities authorized to the corporation for accomplishment of the purposes indicated in subsection (a) of this section, are, among others not specifically named—

(1) to conduct or contract for research and development related to its mission;

(2) to acquire the physical facilities, equipment and devices necessary to its operations, including communications satellites and associated equipment and facilities, whether by construction, purchase, or gift;

(3) to purchase satellite launching and related services from the United States Government;

(4) to contract with authorized users, including the United States Government, for the services of the communications satellite system; and

(5) to develop plans for the technical specifications of all elements of the communications satellite system.

(c) To carry out the foregoing purposes, the corporation shall have the usual powers conferred upon a stock corporation by the District of Columbia Business Corporation Act [citation in margin: "68 Stat. 17; D.C. Code 29-901"].

[8] TITLE IV—MISCELLANEOUS

APPLICABILITY OF COMMUNICATIONS ACT OF 1934

SEC. 401. The corporation shall be deemed to be a common carrier within the meaning of section 3 (h) of the Communications Act of 1934, as amended, and as such shall be fully subject to the provisions of title II and title III of that Act [citation in margin: "48 Stat. 1066; 47 USC 153; 48 Stat. 1070; Ante, p. 64; 47 USC 201-222, 301-397"]. The provision of satellite terminal station facilities by one communication common carrier to one or more other communications common carriers shall be deemed to be a common carrier activity fully subject to the Communications Act. Whenever the application of the provisions of this Act shall be inconsistent with the application of the provisions of the Communications Act, the provisions of this Act shall govern.

NOTICE OF FOREIGN BUSINESS NEGOTIATIONS

SEC. 402. Whenever the corporation shall enter into business negotiations with respect to facilities, operations, or services authorized by this Act with any international or foreign entity, it shall notify the Department of State of the negotiations, and the Department of State shall advise the corporation of relevant foreign policy considerations. Throughout such negotiations the corporation shall keep the Department of State informed with respect to such considerations. The corporation may request the Department of State to assist in the negotiations, and that Department shall render such assistance as may be appropriate.

SANCTIONS

SEC. 403. (a) If the corporation created pursuant to this Act shall engage in or adhere to any action, practices, or policies inconsistent with the policy and purposes declared in section 102 of this Act, or if the corporation or any other person shall violate any provision of this Act, or shall obstruct or interfere with any activities authorized by this Act, or shall refuse, fail, or neglect to discharge his duties and responsibilities under this Act, or shall threaten any such violation, obstruction, interference, refusal, failure, or neglect, the district court of the United States for any district in which such corporation or other person resides, or may be found shall have jurisdiction, except as otherwise prohibited by law, upon petition of the Attorney General of the United States, to grant such equitable relief as may be necessary or appropriate to prevent or terminate such conduct or threat.

(b) Nothing contained in this section shall be construed as relieving any person of any punishment, liability, or sanction which may be imposed otherwise than under this Act.

(c) It shall be the duty of the corporation and all communications common carriers to comply, insofar as applicable, with all provisions of this Act and all rules and regulations promulgated thereunder.

REPORTS TO THE CONGRESS

SEC. 404. (a) The President shall transmit to the Congress in January of each year a report which shall include a comprehensive description of the activities and accomplishments during the preceding calendar year under the national program referred to in section 201 (a) (1), together with an evaluation of such activities and accomplishments in terms of the attainment of the objectives of this Act and any recommendations for additional legislative or other action which the President may consider necessary or desirable for the attainment of such objectives.

[9] (b) The corporation shall transmit to the President and the Congress, annually and at such other times as it deems desirable, a comprehensive and detailed report of its operations, activities, and accomplishments under this Act.

(c) The Commission shall transmit to the Congress, annually and at such other times as it deems desirable, (i) a report of its activities and actions on anticompetitive practices as they apply to the communications satellite programs; (ii) an evaluation of such activities and actions taken by it within the scope of its authority with a view to recommending such additional legislation which the Commission may consider necessary in the public interest; and (iii) an evaluation of the capital structure of the corporation so as to assure the Congress that such structure is consistent with the most efficient and economical operation of the corporation.

Approved August 31, 1962, 9:51 a.m.

Document I-19

Document title: Edward A. Bolster, Department of State, to Mr. Johnson, Memorandum, "Space Communication," May 3, 1962, with attached: "Role of the Department of State in Space Communication Development."

Source: Record Group 59, General Records of the Department of State, Archives II, National Archives and Records Administration, College Park, Maryland.

As the debate over the organization and ownership of the U.S. communications satellite system heated up in mid-1962, the question of the relationship of that system to the rest of the world's communications entities was also beginning to be addressed. This internal State Department memorandum summarizes, for an official of the Department's Economic Bureau (organizational code "E"), the state of affairs in May 1962.

Memorandum

TO: E - Mr. Johnson   DATE: May 3, 1962

FROM: TRC - Edward A. Bolster

SUBJECT: Space Communication

The attached material summarizes the Department's interest and responsibilities in the space communications field. As you know, Phil Farley's office (Special Assistant for Atomic Energy and Outer Space) is to be abolished soon; responsibility within the Department for communication satellite matters will probably be transferred to E. I suggest that you obtain Mr. Farley's comments on this subject within the next few days, since he will be leaving in the near future for his assignment in Paris.

Dr. Irvin Stewart, Director of Telecommunications Management, wishes to meet with you as soon as possible to discuss your mutual interest in certain aspects of communications policy. I suggest that you arrange such a meeting soon and that we brief you orally in advance.

You will recall that Senator Pastore told Under Secretary McGhee that he would like to meet with you at your convenience to discuss the Department's handling of communications policy matters.

****

[attachment, page 1]

Role of the Department of State in Space Communication Development

One of the earliest instances of the Department's participation in space communication problems was in the preparation for and participation in the 1958 Los Angeles Assembly of the International Radio Consultative Committee (CCIR) at which significant technical recommendations were formulated, particularly with regard to problems of radio frequency selection and use by such systems. During this time and immediately following that meeting, those concerned with regulatory problems also assisted the Department in formulating U.S. proposals for the 1959 Geneva International Radio Conference. These proposals were substantially non-controversial, although they did elicit lively discussions at the Conference and required some special meetings in Washington during the Conference in order to resolve certain differences involving radio astronomy. These were of such concern that the Science Advisor to the President became involved at one point. The issue related to the recognition of space research as compared with radio astronomy in the international table of frequency allocations. A reasonably satisfactory solution was developed in which the Department played a major role in bringing the opposing views together.

Following the Geneva Conference the Department was immediately involved in policy development plans with various agencies. It worked with NASA in clearing frequencies with foreign countries in connection with tracking satellites, with the FCC and Interdepartment Radio Advisory Committee (IRAC) in space frequency use planning, and with the Senate Committee on Aeronautical and Space Sciences in the issuance on March 19, 1960 of a study of "Radio Frequency Control in Space Telecommunications" by Dr. Edward Wenk, Jr. The private companies, especially AT&T, were becoming interested in the potentialities of relay by satellite. [handwritten underlining] In fact, the pace became so rapid that it was necessary on September 30, 1960 for the Telecommunication Coordinating Committee (TCC), advisory to the Department, to request those companies known to be active not to "make any commitments to foreign entities pending further advice as to the promulgation of relevant policies which will guide and govern such activities." It [handwritten underlining] then set up a committee to develop such policies under the chairmanship of FCC Commissioner T.A.M. Craven. Membership on this committee included those involved in other groups also developing various aspects of space communication policy nor necessarily directly related to international affairs. For example, the IRAC/FCC began developing coordinated views on the radio frequencies needed for operational (as distinct from research) use of space. On December 4, 1960, the [page 2, handwritten underlining] Senate Space Committee issued a second report on "Policy Planning for Space Telecommunications" largely based on replies from Executive Branch Agencies to inquiries from the Committee based on questions posed in the March report.

Meanwhile, the AT&T had replied on October 21, 1960 to the FCC for authority to operate a communications satellite system.

On December 31, 1960, President Eisenhower issued a communication satellite policy statement urging that the Government aggressively encourage private enterprise in the establishment and operation of a revenue producing system and directing the National Aeronautics and Space Administration to cooperate closely with the FCC in facilitating this objective. In January 1961 the FCC licensed ITT and AT&T for communication satellite experiments.

The National Aeronautics and Space Council was reactivated as a result of an amendment on April 25, 1961 to the 1958 Space Act which, among other things, made the Vice President Chairman of the Council. On July 24, 1961, President Kennedy issued a new communication satellite policy prepared by the Council. It is the current basic policy in the field and will be found on pages 45 and 46 of . . . "Communication Satellites: Technical, Economic and International Developments."* It confirmed the policy of ownership and operations to be in private hands and Government responsibility to encourage research developments, provide launching facilities and exercise regulatory functions including the obligation to "examine with other countries the most constructive role for the United Nations, including the ITU (International Telecommunications Union) in international space communications. Foreign participation in the system would be provided "through ownership or otherwise." The Space Council has the responsibility for policy coordination and for making recommendations to the President concerning actions needed "to achieve full and prompt compliance with the policy."

The TCC Space Communication Subcommittee had, in April, formulated an initial draft statement of policy and on June 1, 1961, this was circulated on an Official Use Only basis to the TCC Members noting that, because related studies were being conducted in other areas of the Executive Branch, the Department did not propose further consideration of the matter at the time. It is, however, desirable to summarize its conclusions:

[page 3] 1. The development of communication satellites should be a national objective and immediate action was required.

2. It should be accomplished by joint efforts of Government and private enterprise.

3. The new system should supplement, not replace, existing systems.

4. It should be privately owned and Government regulated.

5. Existing and future common carriers should have non-discriminatory access to the system.

6. Other nations should participate in the civil system.

7. The civil system should meet all Government needs normally provided by privately owned communication systems.

8. They should not be subject to special space law.

9. The International Telecommunication Union should serve as the principal international organization in this field.

10. Interim installation of key message centers is desirable pending establishment of direct circuits.

11. The possibility of global TV and radio relay via communication satellites should be emphasized.

Also on June 5, 1961 the FCC met with the commercial communications carriers in furtherance of its Docket 14024 regarding the form of ownership which might be appropriate to an international communication satellite system. An Ad Hoc Committee was formed which reported in October . . . in favor of a joint venture non-profit corporation with ownership limited to the carriers, access by all carriers needing the service, and participation by foreign carriers "by ownership or otherwise" as provided in the President's policy of July 24.

On January 11, 1962 Senator Robert Kerr, Chairman of the Senate Space Committee, introduced a bill, S.2650 . . . in the Senate providing for private ownership and leaving to the FCC (which had stated its preference for ownership limited to the carriers) the approval of the owners.

---

*   It may be of interest to note that this document was prepared under the direction of Mr. MacQuivery of TD who was detailed from the Department to the Senate Space Committee [rest of the sentence is illegible].

[page 4] Considerable opposition had been expressed by the Department of Justice and some non-carriers to the restriction in ownership on the basis that this would form an undesirable monopoly and this view prevailed in the Executive Branch during the fall of 1961. The Space Council coordinated a draft bill which was forwarded by the President to the Congress and introduced as S.2814 on February 7, 1962. . . . It would have provided for two classes of stock, one of which could be purchased by the general public and the other by the carriers. A third point of view was expressed by a rather small group of Congressmen in favor of Government ownership of the system. Congressman William Fitts Ryan represented this group and introduced H.R. 9907 in support of this view.

Although specific legislation was not introduced until January 1962, hearings had been held before various Congressional committees beginning with that before the House Committee on Science and Astronautics in May 1961. The Department testified before this Committee and since then has testified before the Senate Commerce Subcommittee on Communications, the Senate Small Business Subcommittee on Monopoly, the Senate Committee on Aeronautical and Space Sciences, the Senate Judiciary Subcommittee on Anti-trust and Monopoly, the Senate Commerce Committee and the Subcommittee on Communications of the House Committee on Interstate and Foreign Commerce. The principal witnesses for the Department were Under Secretary George McGhee and Philip Farley, Special Assistant to the Secretary for Atomic Energy and Outer Space (S/AE). Primary responsibility in the Department for space communications had been delegated to S/AE although representatives of E usually accompanied the witnesses from the Department when they testified, and S/AE consulted TRC when ITU matters were involved. It is understood, however, that with the planned dissolution of S/AE, primary responsibility will be transferred to E.

The foregoing discussion has related primarily to domestic developments. They are basic to, and influence greatly, international developments, however. In addition to preparation of positions for the CCIR Los Angeles and ITU Geneva Conferences, the Department is actively concerned now with preparation for the projected ITU Extraordinary Administrative Radio Conference on Space Radiocommunications expected to be held in Geneva in the fall of 1963. Frequency proposals have tentatively been agreed in the FCC/IRAC and circulated by the Department to missions in all ITU Member countries as "Preliminary Views of the U.S. on the Allocation of Radio Frequencies for Space Radiocommunication." . . . Response to this effect is just beginning to be significant.

[page 5] The U.S. was host to an international meeting of Study Group IV (on Space) of the CCIR in Washington in March 1962. In review and amplification of the Los Angeles actions, further scientific and technical recommendations were adopted on various space communication problems.

On March 7, 1962, the President sent a letter to Mr. Khrushchev to which the latter replied on March 20 concerning cooperation in outer space. Related talks were initiated in New York on March 27 between Deputy Director Hugh Dryden of NASA and Mr. Blagonravov of the USSR. Communication satellite cooperation was specifically considered and it is expected that this will lead to further contact between the two administrations in this particular area.

Perhaps the most significant and current international problem in this area is the relation of space communications to other space questions before the United Nations, particularly involving the U.N. Committee on the Peaceful Uses of Outer Space. After nearly two years of dormancy, at least partially due to the refusal of the U.S.S.R. to participate, the Committee was reactivated and given a new lease on life in November 1961. On December 20, 1961 the General Assembly adopted unanimously U.N. Res. 1721 (XVI) dealing with space. . . . Section D of that resolution involves communication satellites and designates the ITU as the responsible agency in this area. After noting plans for the 1963 extraordinary conference it recommends "that the ITU consider at this conference those aspects of space communication in which international cooperation will be required. . . ." [T]his implied invitation to expand the agenda of the EARC beyond consideration of radio frequency allocations and is currently a controversial issue in the U.S. The Bureau of International Organization Affairs (IO) has had the primary responsibility for prepa-ration and coordination of positions for the U.N. discussions and is currently very inter-ested in implementation of U.N. Res. 1721 as regards the ITU. Deputy Assistant Secretary Gardner of IO plans to participate in current meetings of the Administrative Council of the ITU at Geneva at which Mr. Francis Colt de Wolf of TRC represents the U.S.

In summary, the principal issues involving the Department as regards space commu-nications at present are (1) development of the U.S. position as regards the appropriate role of the ITU in this area in relation to our obligations under the U.N. Resolution; (2) development of the U.S. positions to be taken at the ITU Extraordinary Administrative Radio Conference planned for 1963, in Geneva; [page 6] (3) determination of the man-ner in which to approach other interested administrations in the development of the international communication satellite system (as distinct from the U.S. position itself) and (4) how US/USSR joint efforts in the development of communication satellites could be moved forward in accordance with the Kennedy-Khrushchev exchange of letters.

In addition to the communication satellite report, attached also is a statement of the Department's position presented by Under Secretary McGhee before the Subcommittee on Anti-trust and Monopoly of the Senate Committee on the Judiciary on March 30, 1962.

Document I-20

Document title: Project Telstar, "Preliminary Report, Telstar I, July–September 1962," Bell Telephone Laboratories, Inc., 1962.

Source: AT&T Archives, Warren, New Jersey (used with permission).

Telstar I was built and paid for by American Telephone and Telegraph Company, which wanted to investigate the viability of a communications system that used a constellation of medium-altitude satel-lites. When Telstar I was launched on July 10, 1962, it became the first real-time active communica-tions satellite and broadcast live television between the United States and England and France. This was a considerable advance over previous experiments, which had involved only voice and data com-munications using store and forward satellites such as Signal Communication by Orbiting Relay Equipment (SCORE) and Courier. It demonstrated the attractiveness of satellites compared to undersea cables for transatlantic communications. But it also demonstrated the limitations of a medium-altitude satellite, which only stayed in view of a ground station for a brief amount of time. Telstar I had a four-month life span. The following are excerpts of two of the sections of Telstar I's preliminary report.

PROJECT TELSTAR

Preliminary Report
Telstar I

JULY – SEPTEMBER 1962

PREPARED BY
BELL TELEPHONE LABORATORIES, INC.

[3]   **SECTION 1—OBJECTIVES AND EQUIPMENT DESIGN**

INTRODUCTION

The feasibility of communicating by satellites was demonstrated by NASA's Project Echo in 1960 and also by Projects Courier and Score. In March 1959, Messrs. J. R. Pierce and R. Kompfner of Bell Telephone Laboratories, Incorporated wrote a paper showing the feasibility of an active broadband satellite communications repeater. This started Project Telstar.

Objectives

In general terms, the Telstar Communications Experiment is intended to advance the entire area of communications by satellite. Specifically, the experiment is intended to do five things.
1.   The first objective is to test an actual broadband communications satellite. The Telstar satellite was originally intended for experiments in telephony, data transmission, and single-channel television. While not primarily designed for two-way telephony, the system could provide 60 simultaneous conversations.
2.   A second objective is to test the reliability of electronics equipment in the satellite under the stress of launch and the environment of space.
3.   A third objective is to provide measurements of radiation levels in space; this function is completely separate from the communications experiments.
4.   A fourth objective is to provide additional knowledge about the best technique for tracking accurately a moving satellite.
5.   A fifth objective is to provide a real-life test for the ground-station equipment. . . .

[117]   **SECTION 5—CONCLUSIONS**

FUTURE TELSTAR PROGRAM

The results which have been discussed represent the current picture of what we have learned in the field of satellite communications. A considerable amount of data has been gathered on transmission phenomena and propagation. In this area it has been most gratifying to find that transmission at 4 and 6 kmc is exactly according to theory, and there has been no fading or multipath effects that have been observed. The transmission of a variety of signals has indicated that the performance of the link can be completely specified by the standard transmission parameters.

The area where further information would be helpful is that of environment. Here, it is important that we be able to characterize the levels and types of radiation, the incidence and distribution of micrometeoroids and the behavior of the earth's magnetic field. The reliability of components in a space environment will have an important bearing on the economics of satellite communications.

On the ground station, continued effort is being expended to simplify and minimize the equipment required. This is especially true in the area of satellite tracking. The tracking at Andover has been excellent—with no loss of signal level being attributable to tracking error. However, in future systems, it would appear that greater advantage can be derived from autotrack and that programmed tracking can be considerably simplified.

To summarize: The future Telstar Program will consist of:
1.   Further transmission tests to confirm and refine the data already gathered.
2.   Continued observation of radiation effects, temperature and spin axis orientation.
3.   Evaluation of satellite performance to obtain a measure of component reliability.

Document I-21

Document title: Memorandum from J. D. O'Connell, Special Assistant to the President for Telecommunications and Director of the Office of Telecommunications Management, to the Secretary of State, Secretary of Defense, Secretary of Commerce, Administrator, National Aeronautics and Space Administration, and Chairman, Federal Communications Commission, "Policy Concerning U.S. Assistance in the Development of Foreign Communications Satellite Capabilities," September 17, 1965, with attached: National Security Action Memorandum 338.

Source: Record Group 273, Records of the National Security Council, Archives II, National Archives and Record Administration, College Park, Maryland.

An interim International Telecommunications Satellite Organization (INTELSAT) was created in 1964 with the understanding that after five years there would be negotiations to create a more permanent organizational structure for international telecommunications via satellite. Comsat was the manager of the interim INTELSAT system, and its structure institutionalized U.S. dominance of the organization's operations and hardware procurement. The United States hoped to maintain that dominant position for as long as possible. The White House appointed a Special Assistant to the President, General James O'Connell, to further that objective. This national security directive reflects O'Connell's efforts to restrict U.S. assistance to other countries that desired to develop their own communications satellite capability.

[no pagination]

September 17, 1965

MEMORANDUM TO: Secretary of State
                        Secretary of Defense
                        Secretary of Commerce
                        Administrator, National Aeronautics and Space Administration
                        Chairman, Federal Communications Commission

SUBJECT: Policy Concerning U.S. Assistance in the Development of
                        Foreign Communications Satellite Capabilities

    The attached policy statement concerning U.S. assistance in the development of foreign communications satellite capabilities is promulgated in accordance with the approval of the President, as noted in [handwritten underlining] <u>National Security Action Memorandum 338</u>, dated September 15, 1965. This statement was transmitted to the President by my memorandum dated August 25, 1965.

    As noted in NSAM 338, my office will keep the subject policy under constant review. The cooperation and suggestions of the departments and agencies concerned are invited.

[hand-signed "J. D. O'Connell"]
Special Assistant to the President
for Telecommunications and
Director of Telecommunications Management

Information copies:
    Director, Bureau of the Budget
    Executive Secretary, National Aeronautics and Space Council
    Special Assistant to the President for Science and Technology
    President, Communications Satellite Corporation

****

[attachment, no page number]   August 25, 1965

POLICY CONCERNING U.S. ASSISTANCE IN THE
DEVELOPMENT OF FOREIGN COMMUNICATIONS
SATELLITE CAPABILITIES

GENERAL:

It is the policy of the United States to support the development of a [handwritten underlining] single global commercial communications satellite system to provide common carrier and public service communications. The intent of the United States to exploit space technology for the service of all mankind, and to promote its use in support of peace, understanding and world order has been stated clearly in legislation and in Administration speeches and official releases. The U.S. Government is committed to use global commercial communications facilities for general governmental communications purposes wherever commercial circuits of the type and quality needed to meet government requirements can be made available on a timely basis and in accordance with applicable tariff or, in the absence of Federal Communications Commission jurisdiction, at reasonable cost. Separate satellite communications facilities including surface terminals may be established and maintained by the U.S. Government to meet those unique and vital national security needs which cannot be met by commercial facilities. The capacity of these separate facilities shall at all times be limited to that essential to meet such unique needs. These policies underlie the spirit and the letter of the Communications Satellite Act of 1962, its legislative history and the position of the United States in the negotiations leading to the signing of agreements establishing interim arrangements for a global commercial communications satellite system.

Provisions for the establishment of the global commercial communications satellite system and a U.S. national defense communications satellite system consistent with these policies have now advanced to the point where it is desirable to amplify and interpret these policies [page 2] in order to guide United States relations with other countries in the development of communications satellite capabilities, particularly with respect to providing technology and assistance therefor.

No Foreign Dissemination

DISCUSSION:

Most major countries of the world other than the United States provide international public communications services through governmental agencies or chartered chosen instrument corporations partially or wholly owned by the government. Assistance to any of these foreign governments in the development of communications satellite systems can potentially develop competitors seeking to divert traffic from the single global system being developed by the international consortium established as a result of U.S. actions initiated by the Communications Satellite Act of 1962 and now joined by forty-six nations.

The communications satellite activities of U.S. Government agencies, including the Department of Defense and the National Aeronautics and Space Administration, have an important bearing on the U.S. support of the objectives of the Communications Satellite Act of 1962. These activities may contribute to the dissemination of scientific and technical knowledge of the subject to foreign countries which might be used to the detriment of U.S. policy in this field.

A policy to guide government agencies in the dissemination of satellite technology and in the provision of assistance which is consistent with the overall policies enunciated above is necessary. Such policy should be sufficiently comprehensive to give due regard to the specific requirements of national security.

For the purposes of this policy statement it is intended that restrictions upon transfer of technology and provision of assistance [handwritten underlining] refer to detailed engineering drawings, production techniques and equipment, and manufacturing or fabrication process pertaining to complete communications satellites or a significant portion thereof, and to provision of launching services or launch vehicles for communications satellites. It is not intended that this policy statement apply to surface terminals or limit dissemination of information concerning [page 3] systems concepts, description of spacecraft and normal scientific and technical publications of a professional character. Furthermore, it is not intended that this statement shall limit the dissemination of information required to be disclosed under the provisions of the Special Agreement of August 20, 1964, pertaining to the establishment of a global commercial communications satellite system.

Specific principles to guide United States arrangements for assistance to other countries in the development of communications satellite capabilities are:

1.   The United States should conform fully with the 1964 Agreements Establishing Interim Arrangements for a Global Commercial Communications Satellite System.

2.   The United States *should refrain* from providing assistance to other countries which would significantly promote, stimulate or encourage proliferation of communications satellite systems.

3.   The United States *should not* consider requests for launch services or other assistance in the development of communications satellites for commercial purposes except for [handwritten underlining] use in connection with the single global system established under the 1964 Agreements.

4.   The United States should recognize the vital national security needs of other allied nations which can be met by satellite communications and which cannot be met by the commercial system. For example, the United Kingdom has indicated its need for highly reliable satellite communications from England to Australia and to other Far East terminals.

5.   The United States aim is to encourage selected allied nations to use the U.S. national defense communications satellite system rather than to develop independent systems and to accommodate allied needs within the U.S. system (with additional costs normally to be borne by the participants). Recognized needs should be restricted to those, similar to ours, which are vital to the national security of the selected allied nations and which cannot be met by commercial facilities. To accommodate the needs within the U.S. national defense system it may prove necessary to include one or more satellites, synchronous or otherwise, [page 4] whether of the same or different design. In this case, such satellite(s) should be designed to be electronically interoperable with the satellites of the basic U.S. national defense communications satellite system in order to permit mutual usage.

6. Agreements for direct assistance to allies which may significantly promote their communications satellite capability should require satisfactory assurance that the assistance furnished will be used only within the framework of agreements and arrangements to which the United States is a participant and will not be transmitted or transferred to a third nation without prior U.S. authorization. No agreement [handwritten highlighting in the margin through the end of this paragraph] should be concluded with any nation until information has been made known to other allied nations concerning the U.S. willingness to cooperate in meeting other nations' national security needs which are similar to ours.

7. U.S. firms are required to comply with Munitions Control licensing procedure prior to communication satellite or related technology, transferring equipment or components as embraced by the United States Munitions List, [handwritten underlining] including booster technology and launch services, to foreign nations or firms.

8. U.S. firms are also required to comply with the Department of Commerce's export licensing requirements prior to communicating or transferring to foreign nations or firms certain other relevant technology, equipment or components, not covered by the U.S. Munitions List.

9. All transactions approved under paragraphs 7 and 8 involving technology and assistance pertaining to complete communications satellites or a significant portion thereof, and to provision of launching services or launch vehicles for communications satellites should be conditioned upon express (written) assurances to this government by the foreign nation(s). The assurances should be that technology and assistance obtained will be used only within the framework of the existing international consortium agreements for a single global system or the framework of such special agreements as are referred to in paragraph 6 above and will not be transmitted or transferred to a third nation without prior U.S. authorization.

[page 5] 10. The principles and policy set forth in this document should be reviewed and updated as communications satellite system developments progress and definitive requirements are determined and after the global commercial communications satellite system has been established and is in substantial use.

POLICY:

Therefore, in keeping with the above, it is the United States policy to:

1. Promote the prompt establishment and successful operation of a single global common carrier and public service communications satellite system in cooperation with other nations as part of an improved global communications network which will provide expanded telecommunications services and which will contribute to world peace and understanding.

2. Avoid measures which would adversely affect either the continued expansion of participation in the existing international agreement for a single global commercial communications satellite system or acceptability of the basic premises of the present agreements on a permanent basis.

3. Make use of commercial communications facilities for general governmental purposes wherever commercial circuits of the type and quality needed to meet government requirements can be made available on a timely basis and in accordance with applicable tariff or, in the absence of Federal Communications Commission jurisdiction, at reasonable cost. Establish and maintain separate satellite communications facilities including ground terminals with capacity limited to that necessary to meet those unique and vital national security needs which cannot be met by commercial facilities. The capacity of these separate facilities shall at all times be limited to that essential to meet such unique needs.

[page 6] 4. Encourage selected allied nations to use the U.S. national defense communications satellite system rather than to develop independent systems and accommodate their needs within the U.S. system (with additional costs normally to be borne by the participants). Recognized needs should be restricted to those, similar to ours, which are vital to the national security of selected allied nations and which cannot be met by commercial facilities.

5. Withhold provision of assistance to any foreign nation in the field of communications satellites which could significantly promote, stimulate or encourage proliferation of communications satellite systems.

6. Provide technology and assistance in the field of communications satellites to foreign nations: (a) only if such nations are to participate in the U.S. national defense communications satellite system and then only to the extent required for that participant to be effective; or (b) only for use in connection with the single global commercial communications satellite system in accordance with the provisions of the Interim Agreement and Special Agreement of August 20, 1964; and only if there exist appropriate assurances that such technology or assistance will not be transmitted or transferred to a third nation without prior U.S. authorization.

The policies expressed above will be kept under review by the Special Assistant to the President for Telecommunications/Director of Telecommunications Management and the agencies and departments concerned.

Document I-22

Document title: National Security Action Memorandum No. 342, "U.S. Assistance in the Early Establishment of Communications Satellite Service for Less-Developed Nations," March 4, 1966.

Source: Record Group 273, Records of the National Security Council, Archives II, National Archives and Record Administration, College Park, Maryland.

One of President Kennedy's objectives from the start of his involvement with communications satellites was to make sure that any system developed was truly global, served poorer countries, and linked centers of economic activity. Lyndon Johnson's administration continued this policy and issued this directive to emphasize its importance.

---

[1]   March 4, 1966

NATIONAL SECURITY ACTION MEMORANDUM NO. 342

TO: The Secretary of State
      The Secretary of Defense
      The Secretary of Commerce
      The Secretary of Health, Education and Welfare
      The Administrator, National Aeronautics and Space Administration
      The Chairman, Federal Communications Commission
      The Administrator, Agency for International Development
      The Director, United States Information Agency
      The Special Assistant to the President for Telecommunications and Director
         of Telecommunications Management

SUBJECT: U.S. Assistance in the Early Establishment of Communications Satellite Service for Less-Developed Nations

In carrying out his responsibilities under the Communications Satellite Act of 1962, the President has directed that the United States Government take active steps to encourage the construction of earth-station links to the worldwide communications satellite system in selected less-developed countries. Emphasis in this effort is to be on encouraging the selected countries to construct these stations out of their own resources, stressing the many benefits of direct access to the global communications satellites.

The Special Assistant to the President for Telecommunications/Director of Telecommunications Management has been designated by the President as the agent for coordinating this project.

The State Department and AID are to determine (a) the countries to be included in this program and (b) U.S. Government actions, if any, for encouraging the accelerated construction of earth stations and related facilities in these countries. In cases involving possible U.S. technical or financial assistance, the President has directed that no special funds should be requested. All funding of such projects is to be handled out of current AID FY 1966 appropriations or out of the regular FY 1967 funds.

[2] The Department of State is to report its findings to the President, through the Special Assistant to the President for Telecommunications/Director of Telecommunications Management, by July 1, 1966.

The President has directed that the Executive Agent and Manager of the National Communications System [NCS] and U.S. Government agencies operating facilities outside the NCS utilize the global communications satellite system in handling traffic whenever possible and where national security requirements will not be compromised, consistent with sound cost-efficiency and other management considerations.

A Working Group is to be established, in accordance with the President's instruction, to study the possibilities of using the communications satellite system to advance information exchange and educational purposes, in line with his desire that the United States play a greater role in international education efforts, particularly in less-developed countries.

[hand-signed: "Bromley Smith"]

Document I-23

Document title: David Bruce, U.S. Ambassador to the United Kingdom, to the Secretary of State, "Transfer of U.S. Communications Satellite Technology," Telegraphic Message, November 9, 1966.

Source: Record Group 59, General Records of the Department of State, Archives II, National Archives and Record Administration, College Park, Maryland.

European governments and industry knew that the United States was following a restrictive policy regarding the transfer of technology; this became a source of irritation as the United States attempted to increase the intensity of its space cooperation with Europe and as the 1969 negotiations for definitive INTELSAT arrangements approached. This diplomatic cable from the U.S. embassy in London reflects a foreign policy perspective—that the restrictive policy outlined in NSAM 338 (Document I-21) was not in the best overall interest of the nation. Others in Washington and overseas, concerned with international space policy, shared this perspective and urged that the 1965 policy directive be revised. Their arguments were partially successful, and a slightly less restrictive version of NSAM 338 was issued in mid-1967.

[1]   INCOMING TELEGRAM *Department of State*

[rubber stamped: "1966 NOV 9 AM 11 28"]

Action
     R 0913182 NOV 66
     FM AMEMBASSY LONDON
     TO RUEHC/SECSTATE WASHDC

Info   INFO RUFIVC/AMEMBASSY BERN
     RUFHOL/AMEMBASSY BONN
     RUFHBS/AMEMBASSY BRUSSELS
     RUFHCR/AMEMBASSY PARIS
     RUFHRO/AMEMBASSY ROME
     RUFHOL/AMEMBASSY STOCKHOLM
     RUFHOL/AMEMBASSY THE HAGUE
     RUALOT/AMEMBASSY TOKYO
     STATE GRNC
     BT

[Abbreviations in margin: "E, SS, G, SP, SC, L H, EUR, EA, P, USIA, NSC, INR, CIA, NSA, DOD, ACDA, SCI STR, MC, GDP, OC, COM, DTM, FCC, NSF, OST, RSR"]

CONFIDENTIAL LONDON 3872

Transfer of U.S. Communications Satellite Technology

REF: STATE 76929, LONDON'S A-1084 OF NOV. 4, 1966
NSA

1. US policy on the dissemination of information on communications satellite technology has an impact not only on US objectives regarding a permanent single global communications satellite system but also on technology as it relates to European well being. It is the Embassy's premise that an economically strong, technologically advanced and politically cohesive Europe is in the US national interest. Economic strength and technological competence go hand in hand. This is not to say that technological parity in every field is necessary for strong economies but reasonable competence in most advanced sectors appears to be a sine qua non for long term competitiveness even though comparative advantage may lie with one country or another from time to time. This is particularly true in an environment of reduced trade barriers which exposes the industrial sector to keen international competition.

2. In this context it is clear that communications satellite technology encompasses a very narrow slice of technology. The acquisition of greater competence in this field is likely to have only a marginal impact, in practical terms, on narrowing the over-all technological gap. By the same token US initiatives in other single sectors of technology, treated individually, will have minimal effect on the over-all position. However, concentrated cooperative efforts by the US across the board in all possible areas might lead to a significant improvement in Europe's position vis-à-vis the US (and the USSR). But, given the extent of us investment in R&D in the advanced sectors, it is unlikely that Europe under any circumstances could [2] eliminate the gap in the foreseeable future. Any substantial

narrowing of the gap will require a major increase in European investment in R&D which under current circumstances is improbable.

3.   Therefore, on a cost benefit analysis the costs to the US of sharing technology with Europe in terms of Western leadership, markets, etc., are not likely to be great. On the other hand, the benefits could be considerable. The principal gain would be psychological. US initiatives would be regarded as an act of good will—a cooperative gesture from a friend and ally—which would further strengthen Atlantic bonds. It would encourage US/Europe scientific and industrial cooperation and neutralize any tendency to turn to the USSR in frustration. It would tend to still those voices which charge that US policies are directed to establishing and maintaining absolute domination in all advanced areas of technology.

4.   To a lesser degree US technological cooperation, to the extent that it would assist Europe to maintain a reasonably competitive position, will increase specialization and trade for the benegit [sic] of both sides.

5.   Turning now to the specific case of communication satellite technology, much of the above reasoning applies. The advantages to the US of relaxing its objectives is for a substantial US share in a single global communications satellite system. Such psychological factors could well be important. The ambitions of European industry in obtaining a larger share of the INTELSAT contracts are well known. Both industry and government regard communications satellites as the one sector of space investment which promises an early commercial return. Informed Europeans are aware that current US policies prevent European industry from obtaining the know-how which they feel would permit them to compete for and obtain such contracts. Much of the recent publicity in Europe released by industry supporting the concept of regional communications satellite systems, allegedly complementary with the global system, is believed primarily designed to provide Europe a greater share of the market for satellites and other elements of the commercial system. The European Conference on Satellite Communications (CETS) is meeting in The Hague later this month with the specific objective of improving European capabilities in satellite technology. Certainly regional systems will be examined as one means of achieving this objective. While France may favor regional systems as an end in themselves, it is believed most countries are basically concerned with industrial aspects and are [3] quite content with [a] single system concept so long as they obtain [a] fair slice of the equipment cake. Thus, if Europeans feel that the US policies and predominance in a single global system will continue to frustrate what they feel to be their quite legitimate aspirations on production, they may well seek to negotiate an agreement at the 1969 Conference which would permit the establishment of regional systems.

6.   It is recognized that there is fear that the relaxation of restrictions on transfer of satellite technology will give Europeans the tools to establish separate systems. In our view this fear is exaggerated. First of all, as mentioned above, the general European objective is to achieve adequate competence to bid for INTELSAT and possibly IDCSP [Initial Defense Communications Satellite Program] contracts, and not to establish independent systems per se. This is particularly true of the British. Second, the possibility appears remote that the Europeans could launch an independent system by 1969. Aside from the question of satellite development, Europe will not have launch capability until well into the 1970's.

7.   To conclude on this point, an offer to share communications technology with Europe would gain a measure of good will and serve to alleviate European suspicions of US intentions. This should improve the US negotiating posture in 1969. On the other hand the danger to US objectives in such a move would, in practical terms, be negligible since the Europeans would be unable to use such technology over the short term to launch an independent system.

8. Answers to the specific questions posed in the final paragraph of reference telegram are as follows:

(A) Policy directive has not hindered desirable scientific and technological cooperation in any practical way except in the area of communications satellite technology. However, the existence of this policy has undoubtedly colored European views as to the disinterested nature of US offers of cooperation.

(B) Scientific or industry dissatisfaction has not been reflected as irritant on political lines since 1964 negotiations. Political interest may reappear prior to 1969 negotiations.

(C) The divergencies of view among British interests on communications satellite technology were analyzed in Embassy's A-1084 of Nov. 4. To summarize briefly, the GPO is fundamentally concerned with efficient economic communications and less involved in political and industrial considerations. The GPO strongly supports the concept of a single global system. The foreign office is anxious to abtain [sic] political cohesion in Europe and will seek European consensus even [4] at some compromise of domestic ambitions. Industry primary interest is to secure larger share of INTELSAT procurement and will support vigorously any proposal, either for a single system or an independent system, which will improve its competitive position, vis-à-vis US industry. The Ministry of Defense (MOD) is not taking a more active role in the communications satellite question due to participation in IDCSP and possibly ADCSP. Any MOD support for independent European initiative will depend largely on its experience with joint US/UK military projects.

(D) If British are assured of US commitment to genuine cooperative effort in communications satellite technology, we believe they would be prepared to give the required assurances. Clarification on the role of regional and national systems (e.g. ABC proposal for US national TV relay system) would be required. Also Europeans may wish to launch experimental satellites using US launchers. Embassy judgement here is not based on specific comments from industry or government but from the interpretation of expressions of opinion by the Foreign Office, GPO and industry contacts over the past year or so. Bruce

BT

Document I-24

Document title: Memorandum from J.D. O'Connell for the President, February 8, 1967, with attached: "A Global System of Satellite Communications: The Hazards Ahead," February 8, 1967.

Source: Lyndon B. Johnson Presidential Library, Austin, Texas (used with permission).

General James D. O'Connell, the individual responsible for promoting the U.S. policy objective of creating a single global system of satellite communications based on INTELSAT, saw many hazards ahead. This memorandum sketches his perceptions of the challenges to achieving this policy objective.

[1]   MEMORANDUM

The White House
Washington

February 8, 1967

MEMORANDUM FOR THE PRESIDENT

I submit a proposed draft of the President's 1966 report to the Congress as required by the Communications Satellite Act of 1962. This report emphasizes positive accomplishments. It does not describe the hazards which INTELSAT and ComSat face. Some of these hazards are:

a. Actions of certain international record carriers indicate that they consider it to be in their corporate interest to emasculate INTELSAT, the single global system, and ComSat.

b. Certain major aerospace manufacturers both here and abroad deprecate the value of INTELSAT and the single global system. They favor many proliferating domestic and regional systems. Obviously these would provide a larger market for their products.

c. [bolded passages were highlighted with a marker in the original] **France has been promoting within Europe a regional communications satellite system which will compete with INTELSAT,** and **will probably join the Soviet Molnya** [sic: Molniya] **system.**

d. ComSat's studies conclude that there is more business and earnings in the domestic communications field than can be derived from international traffic. This conflict of interest has been demonstrated in recent FCC filings where ComSat has failed to take a clear-cut position as the servant of the international INTELSAT joint venture.

e. Certain members of INTELSAT who derive a favorable balance of payments under present arrangements are not supporting the U.S. policy of actively encouraging the establishment of satellite communications facilities for the developing nations. This has resulted in inadequate progress toward the design of low cost earth terminals and satellite systems—concepts which are needed to promote early effective and economical use in the developing nations.

f. [bolded passages were highlighted with a marker in the original] **Major continental European nations are critical of the "excessively dominant"** position of the United States in the decisions of the International Consortium. **Actions to reduce U.S. dominance and to obtain a manager other than ComSat are expected during the 1969 negotiations to extend the existing Interim Agreement or consummate a more permanent one.**

[2] g. Action by the United States to embark upon separate domestic or regional enterprises prior to 1969 will have a serious negative impact on the single global system, the International Consortium, the 1969 renegotiations, and ComSat's future as Manager for INTELSAT.

h. The recent FCC action to adopt a 50-50 shared ground station ownership formula between ComSat and the communications common carriers has not reduced conflict as had been hoped. ComSat's investment capital potential has been cut in half but the record carriers still want more. A merger of ComSat with the six other U.S. international carriers is becoming increasingly vital.

i. The general disorder of U.S. international telecommunications has been and is a serious obstacle to progress in commercial communication satellites and is a threat to their future. It is also creating increasing pressure to reverse the trend toward greater Government use of the international common carriers and causing serious consideration of programs to step up the capacity of the Government's own communication satellite systems. Diversion of Government traffic from the carriers will further jeopardize the future viability of ComSat and the global system.

j.   [bolded passages were highlighted with a marker in the original] **Communication satellites are in such an early stage** of their technological and systems development that present systems should soon be made obsolete by the new developments. **But research and development efforts by ComSat and NASA are inadequate to push progress fast enough.** I am increasing the efforts of my office to push for faster progress.

The national policy established by the President and Congress is to give first priority to the successful achievement of a single international global system at the earliest time. It is a sound policy which makes paramount the objectives of world peace and understanding. The importance of the single global system to achieve these objectives cannot be overemphasized. Executive Branch departments are working diligently to reduce the hazards and obtain the objectives sought by the Communications Satellite Act, but success is far from certain yet. The trend appears to be toward progressively more serious obstacles. Further discussion of these obstacles is contained in the attachment.

In a subsequent report I will set forth the steps being taken by my office and other government agencies to cope with these hazards. Some of my proposals for Government actions are included in the attached summary.

J. D. O'Connell

****

[i]

A Global System of Satellite Communications
— The Hazards Ahead —

SUMMARY

The hazards to the future success of the International Consortium (INTELSAT) and ComSat appear to be increasing and becoming more serious. Knowledgeable students of the situation are privately expressing the thought that [bolded passages were highlighted with a marker in the original] **it is entirely possible that INTELSAT may fall apart in favor of a series of regional systems.**
If this were to occur it would mean:
- A massive setback in future growth and easy access in international telecommunications.
- The loss of the soundest, simplest, lowest cost system of international telecommunication which can make the largest contribution to world peace and understanding.
- A reversion to reactionary concepts of rich nation domination of zones of communication influence, increased length and lower quality of transmission paths, and higher consumer costs.
- A very serious prestige loss to the United States.
- Financial loss to the shareholders of ComSat.

The most serious threats to INTELSAT and ComSat which are described in the following pages have not yet reached critical or unmanageable stage. But over optimism, lack of vigorous action, or actions which aggravate these trends can cause these problems to rapidly get beyond control.

ACTIONS UNDER WAY

I am bending every effort to clarify this situation and achieve unified government action to overcome the growing obstacles. Among my proposals are the following:

    a. Give *first priority* to development of INTELSAT and the international system. This is in conformance with U.S. policy and statute.

    b. Emphasize the great advantages of the single global system. [bolded passages were highlighted with a marker in the original] **(There is *no* regional need which the single global system cannot meet with better service at lower cost, with better spectrum conservation.)**

[ii] c. Provide greater U.S. aid to developing nations in getting earth stations.

    d. Develop as rapidly as possible the practical use of the higher (and less used) frequency bands for exclusive use of large domestic satellite systems. Service to be available in 4–5 years.

    e. Use the INTELSAT system for early U.S. domestic service growth and ETV [educational television] experiments.

    f. Avoid FCC or Congressional action to constitute a separate U.S. domestic satellite system for the immediate future.

    g. Accelerate the development of low cost earth stations.

    h. Accelerate the development of more efficient multiple access systems to reduce the cost of communication to both rich and poor nations.

[1]   A GLOBAL SYSTEM OF SATELLITE COMMUNICATIONS
— THE HAZARDS AHEAD —

THE U.S. COMMITMENT TO A SINGLE GLOBAL SYSTEM

BACKGROUND

The foundation of our communications satellite policy has been embodied in the concept of a single global system to which all nations could have equal access, and through which international communications could flow free of artificial constraints held over from the colonial traditions of past centuries. The concept stems both from the policy objectives established by the Congress and from our international agreements.

THE COMMUNICATIONS SATELLITE ACT OF 1962

Declaration of Policy and Purpose

Sec. 102. (a) The Congress hereby declares that it is the policy of the United States to establish, in conjunction and in cooperation with other countries, as expeditiously as practicable a commercial communications satellite system, as part of an improved global communications network, which will be responsive to public needs and national objectives, which will serve the communication needs of the United States and other countries, and which will contribute to world peace and understanding.

(b) The new and expanded telecommunications services are to be made available as promptly as possible and are to be extended to provide global coverage at the earliest practicable date. In effectuating this program, care and attention will be directed toward providing such services to economically less developed countries and areas as well as those more highly developed, toward efficient and economical use of the electromagnetic frequency spectrum, and toward the reflection of the benefits of this new technology in both quality of services and charges for such services.

THE INTERNATIONAL AGREEMENT OF AUGUST 20, 1964—55 NATIONS

Desiring to establish a single global commercial communication satellite system as part of an improved global communications network which will provide expanded telecommunications services to all areas of the world and which will contribute to world peace and understanding;

[2] Determined, to this end, to provide, through the most advanced technology available, for the benefit of all nations of the world, the most efficient and economical service possible consistent with the best and most equitable use of the radio spectrum.

BASIS

The single global system is truly a revolutionary concept. It is also intrinsically sound from the viewpoint of supporting the policy objectives established by the Congress and confirmed in our international agreements.

USE OF COMMUNICATIONS SATELLITE TECHNOLOGY TO CONTRIBUTE TO WORLD PEACE AND UNDERSTANDING, IMPROVED WORLD TRADE, AND COMMERCE

The integrity of the global system is vital to our primary goal of using satellite technology to promote world peace and understanding, and to our corollary goals of improved world trade, commerce, and better understanding between nations. We must nurture this global system concept, [bolded passages were highlighted with a marker in the original] **for if we allow it to deteriorate into a series of isolated regional networks we may forever lose the golden opportunity which satellite technology provides for creating a world community in which communications flow freely between nations.**

SPECIAL ATTENTION TO THE COMMUNICATIONS NEEDS OF LESS DEVELOPED COUNTRIES

The single commercial communications satellite system provides the broadly based structure to meeting the demands of the smaller nations. Further, it provides a framework in which the United States can work effectively to promote communications satellite technology designed to aid the developing nations.

EFFECTIVE USE OF THE FREQUENCY SPECTRUM

The demands for communications satellite service already promise to overtax the capability of the frequency spectrum. Only through the economies of scale and engineering efficiency of a truly global system will all of the nations of the world be able to gain equal benefits from communications satellite technology.

[3] THE HAZARDS AHEAD

These are fundamental problems in the field of satellite communications which have national importance and which can profoundly affect the economic, social, and political objectives of this Nation. These problems arise from many sources but may be generally categorized as follows:

1. Interests that conflict with the global system;
2. The impact of U.S. domestic communications issues;
3. The "limited objectives" syndrome;
4. Fear of U.S. domination;
5. The general disorder of U.S. international communications.

INTERESTS THAT CONFLICT WITH THE GLOBAL SYSTEM

A. [bolded passages were highlighted with a marker in the original] **International Carriers**
These organizations, including both the United States and foreign carriers, **view the INTELSAT system as a direct competitor to the established cable and high frequency radio routes.** Long established spheres of influence and methods of operation are also threatened. With a single global communication satellite system, opportunity for national control of international communications routing, ability to charge transit fees and apply other restrictive practices will be lost. Many of the foreign carriers feel that their earnings would be greater and their control maintained if traffic is transmitted over cable and high frequency radio systems where their ownership may be as high as 50 percent in contrast to their 1 to 5 percent ownership which is typical in INTELSAT. Some Administrations in Europe frankly admit using the profits of their international telecommunications traffic to subsidize their domestic costs for both telecommunications and postal services. Some Administrations also feel that where cable and high frequency radio are no longer viable, their interests would be better served through the creation of regionally oriented satellite systems where their ownership share could be increased.
Another arrangement preferred by U.S. record carriers and some foreign Administrations over the present COMSAT-INTELSAT arrangement would be a completely non-profit space segment structure which would be [4] supported by the various nations on the basis of use. Administrations would derive income from earth station charges which could be handled much like cable charges without the added complication of providing income and profits to the space segment owners. The only major impediment in the way of this arrangement is COMSAT. Other nations are not faced with this dilemma. All sophisticated Administrations can be assumed to be thoroughly aware that the U.S. poses the only impediment to this kind of structuring.
Some of the international record carriers are also concerned that the INTELSAT organization will spawn greater sophistication in the less developed nations and encourage them to assume a more prominent role in their internal communications systems as well as the satellite earth terminals used as gateways for international traffic.
In the past, international carriers have often dominated completely communications within a developing nation through control of international communication facilities.

B. Domestic Common Carriers
While our domestic common carriers recognize the need for single ownership and management of international communications systems and for compatibility between international and domestic systems, they are in conflict with certain of COMSAT's proposals for early domestic communications satellite service. A key question affecting domestic service stems from the present international agreements which provide for shared frequencies between terrestrial microwave systems and space services.
In the frequency bands assigned to satellite communications, we also operate a profusion of domestic terrestrial microwave relay systems which involve a capital investment over $2 billion. This is over twice the total capital invested in international telecommunications by all the nations of the world. These terrestrial microwave systems are continuing

to expand rapidly. It is clear that the installation of numbers of domestic satellite earth stations is certain to impede the growth of these terrestrial microwave services. It is also clear that the unrestricted installation of earth stations for domestic purposes will preempt space required to expand the international system. Despite technological advances there is an ultimate limitation to the amount of communication which can be carried in this frequency band.

[5] The solution to this problem lies in the development of new exclusive frequency bands for satellite communications. The course of action is feasible but will require additional research and development and a few years of time—but it will make it possible to exploit satellite communications to its fullest potential.

C. U.S. Space Systems Manufacturers

At present the aerospace industries of the U.S. enjoy a significant technological lead over all foreign competitors in the field of communications satellites. The concept of a single global system prevents rapid, albeit wasteful, proliferation of the space segment hardware and restricts the market to supplying the INTELSAT organization. [bolded passages were highlighted with a marker in the original] **A greater market potential could be created more rapidly through the development of independent national systems or proliferation of regional systems. The aerospace industry thus seeks a proliferation of systems.**

While this approach may produce a short run profit for the aerospace industry, it will produce a long run harvest of international telecommunications chaos and ill will. We cannot afford to be cast in the role of sponsoring these misguided attempts to implement inefficient and unnecessary communications satellite systems that do not have the traffic base to make their operations viable or to achieve the economies of scale possible in a single global system. In this case, the short term commercial interests of our aerospace manufacturers are in conflict with the national and international objectives of creating an effective global system to introduce a new era in world telecommunications and better serve all mankind.

D. National Ambitions of Foreign Governments

Several foreign nations, [bolded passages were highlighted with a marker in the original] **notably France,** feel that they must develop their own communications satellite capability as rapidly as possible to reinforce national prestige.

An important motive for individual nationalistic control of communications satellites stems from a desire [bolded passages were highlighted with a marker in the original, note in margin: "#1"] **to continue to exercise cultural and political leadership in traditional areas of influence without intervention by an international body such as INTELSAT.** [note in margin: "#2"] **Equally important is their desire to create a viable option which can be used as a negotiating point in the 1969 discussions.** [note in margin: "Symphonie"]

[6] THE IMPACT OF U.S. DOMESTIC COMMUNICATIONS ISSUES

A. The Ownership of International Earth Stations

The recent FCC decision to share ground station ownership 50-50 between COMSAT and the communications common carriers has not yet reduced conflict or speeded progress as had been hoped. COMSAT's investment capital potential has been cut in half but the record carriers still want more.

As more and more international traffic shifts communications to satellites, the rate base position of the international carriers will become progressively worse. Further, dividing ownership among COMSAT and the five international carriers seriously jeopardizes the profit position of COMSAT as well. So long as we have the present irrational arrangement of international common carriers, the situation is certain to continue to get worse. A merger of the five U.S. international carriers is becoming increasingly vital. If it does not take place soon some of the wiser heads in the industry see no way of preventing Government ownership as a way of bringing order out of chaos.

B. Patent Problems

The INTELSAT organization's patent provisions are already a problem that promises to become increasingly acute. Some U.S. aerospace firms are balking at a requirement that INTELSAT be given patent rights to all patents used on INTELSAT contracts.

C. TV and Educational TV Interests

The proposals of the Ford Foundation, American Broadcasting Company, and others to create a domestic TV distribution system via communications satellites have confronted the FCC with problems that are fraught with economic, policy, legal, and technical issues which overlap and impact upon our international agreements.

These domestic issues are raising serious apprehensions among our foreign partners that the U.S. intends to place domestic interests and pressures ahead of better world communications. And, of course, the Soviet Union has repeatedly criticized INTELSAT as a rich man's club being run for the primary benefit of the United States.

[7] THE "LIMITED OBJECTIVES" SYNDROME

A. Inadequate Research and Development

Communications satellites are still in a very early stage of development from the viewpoint of both technology and utilization. In order to achieve the goals established by the Communications Satellite Act of 1962, a substantial acceleration of research and development program is needed.

A primary task is to produce advanced multiple access capabilities to serve low traffic density terminals of the developing nations. We also need more stable space platforms, better pointing accuracy in our antennas, and improved primary power sources for the space vehicles.

[bolded passages were highlighted with a marker in the original] **Although COMSAT is seeking to construct a major research and development facility, they are encountering strong opposition from our European partners.** Such a research center would tend to be competitive to foreign manufacturing interests. Certain U.S. aerospace firms have opposed the COMSAT research and development center for the same reason. COMSAT is certain to run into serious trouble unless they adopt a carefully planned philosophy acceptable to their foreign partners and U.S. industry.

B. Conflict as to the Proper Pace of System Implementation

COMSAT has pushed ahead faster than desired by some of the European partners in order to create an operating system with worldwide coverage prior to 1969. A successful system will be the best possible insurance of our continuing the present favorable pattern of international agreements. Within INTELSAT, however, there is an element which feels that the global system should proceed more slowly so that the gap in aerospace technology between the U.S. and other countries can be closed prior to major deployment of the system. There is also a reluctance among those nations who are large owners of telephone cable systems to divert traffic to satellites so long as channels are available in the cables.

FEAR OF U.S. DOMINATION

A.   Technology Gap

Many foreign nations feel that they were forced to accept the 1964 communications satellite agreements on U.S. terms as a result of deficiencies in their own space research and development capabilities.

[8] These nations are working as hard as [bolded passages were highlighted with a marker in the original] **possible to improve their technical position to give them options to strengthen their position and insure more equitable participation in the INTELSAT organization after 1969.** [note in margin, arrow pointing to, and emphasizing, the following sentence] The manner in which the U.S. shares its technology impacts directly on this issue. We are eager to share technology with those nations firmly committed to INTELSAT. **We do not, however, find it in the U.S. interest to provide the tools with which a foreign nation can circumvent the INTELSAT agreements by contributing to the establishment of competing regional systems.**

B.   Leadership in INTELSAT Administration and Management

At present the U.S., through the Communications Satellite Corporation, serves as Chairman of the Interim Communications Satellite Committee, provides the Manager for all technical operations of INTELSAT, and has [bolded passages were highlighted with a marker in the original] **a controlling voting interest of 54 percent** in most decisions of the Consortium. Many foreign nations feel that this is an unacceptable domination of INTELSAT by the U.S. We are already experiencing pressure within the Consortium to reduce the influence of the U.S. and to strip COMSAT of its administrative and technical control. This will undoubtedly be an important point in the 1969 renegotiations.

A GENERAL DISORDER OF U.S. INTERNATIONAL COMMUNICATIONS

A.   Inadequate Responsiveness to Communications Needs of the U.S. Government

The serious conflicts of interest that have existed in recent years among U.S. international communications common carriers have been greatly increased by the advent of communications satellites. This situation has resulted in increased controversy and delay in meeting new Government requirements. A recent Department of Defense requirement for communication service to Japan, Thailand, and the Philippines to provide support for operations in Vietnam has resulted in many conflicting filings with the Federal Communications Commission. This conflict and others like it have raised serious doubts within Government agencies concerning the ability of the U.S. international common carriers, placed as they are in a constant conflict of interest, to provide assured and rapid response to new or emergency communications requirements of the Government.

Such uncertainties create a growing incentive for the Government to reverse previous trends toward increased use of international common carriers and to turn instead to greater dependency upon and use of [9] Government owned communications satellite systems. Since the U.S. Government is by far the largest single user of international commercial communications channels, loss of any portion of this business by the carriers could have a serious adverse effect on all the international carriers, but particularly on the viability of international satellite communications.

B.   Delays in U.S. Earth Station Construction

The continual controversy and divided system responsibility in U.S. international communications have also adversely affected and delayed construction of an adequate U.S. earth terminal complex to keep pace with the growing communications capacity of the space segment. On the East Coast of the U.S. the system needs two full scale operating terminals. One terminal is in operation but the authorization to construct the second terminal has now been delayed for over a year in an attempt to compromise the conflicting interests between the international telegraph carriers, the Communications Satellite Corporation, and our domestic common carriers. By 1968 a second communications satellite terminal will be needed in Hawaii and another one on the West Coast. Both of these stations should be under construction now. They are not. A major factor in the delay has been the general disorder in the U.S. international communications structure and the continuing pattern of conflict between the carriers.

C. Confusion and Conflict Resulting from the Foreign Interests of U.S. International Common Carriers
    The need for communications satellite earth stations in foreign countries where U.S. international common carriers have business interests has created dissension among these carriers and had a very serious impact on the U.S. image overseas. In the Philippines, controversy between U.S. common carriers concerning responsibility for assisting in the organization of communications satellite activities and the construction of an earth terminal resulted in a long period of stalemate and confusion. The ultimate decision on the part of the Philippine Government was to sharply reduce the activity of U.S. carriers within the Philippines.
    A similar situation has developed in Central and South America where certain of our international carriers are fighting a rear guard action against loss of their operating franchises. These conflicts and controversies are delaying the construction of satellite earth terminals in direct opposition to the announced United States policy to promote a rapid growth of satellite communications capability so as to strengthen bonds within this hemisphere.

Document I-25

Document title: Leonard H. Marks, Ambassador, Chairman, "Report of the United States Delegation to the Plenipotentiary Conference on Definitive Arrangements for the International Telecommunications Satellite Consortium (First Session), Washington, D.C., February 24–March 21, 1969," April 10, 1969.

Source: NASA Historical Reference Collection, History Office, NASA Headquarters, Washington, D.C.

The 1964 Interim Agreement that created INTELSAT specified that after five years a conference would be called to develop a definitive agreement for the organization. Accordingly, a Plenipotentiary Conference on Definitive Arrangements for the International Telecommunications Satellite Consortium convened in Washington, D.C., on February 24, 1969. The report of the U.S. delegation on the first session of that conference detailed the many areas of disagreement that would have to be resolved before a definitive agreement was possible. It took several years of difficult negotiations before that objective was achieved. The definitive agreements for INTELSAT went into effect on May 21, 1971.

[1]

Report of the United States Delegation to the Plenipotentiary Conference on Definitive Arrangements for the International Telecommunications Satellite Consortium

(First Session)
Washington, D.C.
February 24–March 21, 1969

Submitted to the Secretary of State:

Leonard H. Marks, Ambassador
Chairman, United States Delegation
April 10, 1969

Summary

The purpose of the Conference is to establish definitive arrangements for the International Telecommunications Satellite Consortium (INTELSAT). During the period February 24 to March 21, 1969, 96 interested member and non-member countries exchanged views on general aspects and specific details of proposed definitive arrangements. No final decisions were taken by the Conference, which stands recessed until November 18, 1969, at which time it is expected that draft definitive arrangements will be considered by the reconvened Conference. In the interim, an intersessional Preparatory Committee is being formed to develop the draft definitive arrangements.

The Conference provided the first occasion since the interim arrangements were signed in 1964 for governments in their sovereign capacity to examine the organization collectively. Hence it is not surprising that political factors influenced the positions taken by many delegations. However, the avoidance of the extraneous political issues often raised at international conferences is notable. There have been no challenges to credentials and no polemics on issues outside the business of the Conference.

The Conference worked in its first session through four working committees and in plenary session. The various Committee reports produced have been referred to the intersessional Preparatory Committee and will provide a basis for that Committee's work.

I. Background
    Following a series of successful communication satellite experiments by Government and industry in the United States in the period 1958–1963, the United States undertook, pursuant to the terms and mandates of the Communications Satellite Act of 1962, to establish, in conjunction and in cooperation with other countries, a single global commercial communication satellite system. After several months of international bilateral and multilateral negotiation [2] in 1963–64, the Government of the United States convened a Plenipotentiary Conference at Washington, D.C. in July 1964, at which texts of two agreements were initialed by 19 participating Governments. The first is the intergovernmental Agreement Establishing Interim Arrangements for a Global Commercial Communications Satellite System, in which the parties agreed to establish a global communication satellite system in cooperation with one another. A related Special Agreement, signed by Governments or telecommunications entities designated by member Governments, contains details relating to operation, financial aspects, procurement,

control and maintenance of the global satellite system. The U.S. Government is a party to the first Agreement and the Communications Satellite Corporation (ComSat), the telecommunications entity designated by the U.S. Government, is a signatory of the second (Special) Agreement. Pursuant to these two Agreements, opened for signature on August 20, 1964, the cooperating member countries brought into existence the International Telecommunications Satellite Consortium (INTELSAT), which now has 68 member countries.

Pursuant to Article IX of the Interim Agreement, the governing body of INTELSAT— the Interim Communications Satellite Committee (ICUS)—produced a report containing the Committee's recommendations, and other shades of opinion, on definitive arrangements for the organization. This report was issued to all INTELSAT member Governments on December 31, 1968.

Under other terms of Article IX the United States was obligated to convene a plenipotentiary conference to consider the ICUS Report within 90 days of the date it was issued. In compliance with this obligation, the United States convened the Plenipotentiary Conference on Definitive Arrangements for the International Telecommunications Satellite Consortium in Washington on February 24, 1969. After four weeks of work, the Conference recessed its first session on March 21, 1969. This report covers that session.

II. Agenda of the Conference
As adopted, the agenda provided for:

1.   Election of the Chairman
2.   Adoption of the Agenda
[3] 3.   Adoption of Conference Rules of Procedure
4.   Election of other Officers
5.   Organization of the Conference
     A.   Credentials Committee
     B.   Editorial Committee
     C.   Working Committees
6.   Report of the Credentials Committee
7.   Consideration of the report and recommendations of the Interim Communications Satellite Committee and of definitive arrangements for the International Telecommunications Satellite Consortium
8.   Signing of definitive arrangements

The Conference followed this agenda, completing items 1–6, including adoption unanimously of the rules of procedure, established its work program, and proceeded in working committees to consider the wide variety of proposals contained in the ICUS report, as well as other proposals introduced during the Conference. It did not complete consideration of item 7, the substantive business of the Conference, and, of course, did not reach item 8, signing of agreements.

III. Participation
Sixty-seven of the sixty-eight members of INTELSAT registered at the Conference. They were Algeria, Argentina, Australia, Austria, Belgium, Brazil, Canada, Ceylon, Chile, China, Colombia, Denmark, Ethiopia, France, Federal Republic of Germany, Greece, Guatemala, India, Indonesia, Iran, Ireland, Israel, Italy, Jamaica, Japan, Jordan, Kenya, Korea, Kuwait, Lebanon, Libya, Liechtenstein, Luxembourg, Malaysia, Mexico, Monaco, Morocco, Netherlands, New Zealand, Nicaragua, Nigeria, Norway, Pakistan, Panama, Peru, Philippines, Portugal, Saudi Arabia, Singapore, Republic of South Africa, Spain, Sudan, Sweden, Switzerland, Syrian Arab Republic, United Republic of Tanzania,

Thailand, Tunisia, Turkey, Uganda, United Arab Republic, United Kingdom, United States, Vatican City State, Venezuela, Republic of Viet Nam, Yemen Arab Republic. Only Iraq was not represented.

Observers were sent from the following twenty-nine non-member countries: Afghanistan, Barbados, Bolivia, Bulgaria, Cambodia, Cameroon, Congo (K), Costa Rica, [4] Czechoslovakia, Ecuador, Finland, Ghana, Hungary, Ivory Coast, Liberia, Maldive Islands, Mauritania, Mauritius, Mongolia, Paraguay, Poland, Romania, Senegal, Somalia Republic, Southern Yemen, Union of Soviet Socialist Republics, Uruguay, Yugoslavia, Zambia. In addition, observers attended from the United Nations and the International Telecommunication[s] Union. Thus a total of ninety[-]eight delegations attended the first session of the Conference.

A complete list of participants is attached as Annex A, including a list of the United States Delegation.

Among the non-member observer countries, a substantial number spoke during various meetings of the Conference. Particularly the USSR, Poland, Romania and some of the African observer delegations presented substantive comments and views on questions being considered by the Working Committees of the Conference.

IV. Organization of the Conference

The United States, as host Government, provided the secretariat and physical facilities for the Conference. The conference facilities in the State Department Building were used. In addition, the main auditorium of the Pan American Health Organization Building was used for several committee meetings. Administrative and secretariat support from the Office of International Conferences was outstanding throughout the Conference and won deserved praise from many delegations at the final plenary session. In addition, the Federal Communications Commission, the Office of the Director of Telecommunications Management, and the Communications Satellite Corporation made available secretarial and administrative support staff and equipment which contributed further to the efficiency of the overall operation.

The Conference was formally opened by the Acting Secretary of State, Elliot L. Richardson. The names of the elected Conference officers and Committee Chairmen and Vice Chairmen are set forth in Annex B. All Conference officers and Committee Chairmen and Vice Chairmen were unanimously elected and there were no objections to the [5] composition of any of the Conference Committees. The established Working Committees of the Conference and their subject matter were the following:

Committee I	Structure and Functions of INTELSAT Consortium, with particular regard to questions of membership, scope of services, organizational structure including structure of major organs, their functions and voting.
Committee II	Legal and Procedural Questions, including definitions, legal status, entry into force, duration, amendment, withdrawal, settlement of disputes.
Committee III	Financial Arrangements.
Committee IV	Other Operational Arrangements, including procurement policy, inventions and data, technical and operational matters.

All four of these Committees were constituted as committees of the whole, i.e. open to participation by all member country delegations, and all committee sessions, though not the sessions of working groups, were open to observers. Most of the work of the Conference was done in the Committees, with plenary meetings held only at the beginning and near the end of the session.

V.   Work of the Committees

Committee I—Structure and Functions

Committee I's work program, including nine specific topics, is set forth in Annex C. To facilitate its work, the Committee formed three working groups, each of which prepared a report, which was reviewed and accepted by the Committee and forwarded to a plenary meeting for consideration.

Working Group A—The report of Working Group A (Com. I/84, Rev. 1) deals with the purposes and objectives of INTELSAT and the scope of INTELSAT's activities. The Working Group developed a draft Preamble for the definitive arrangements, a draft [6] article on "Objectives and Purposes" and a draft article on "Scope of Activities." These draft articles were adopted unanimously by the ten-country Working Group, subject to notes and reservations set forth in the report.

The Preamble is substantially similar to the Preamble of the 1964 intergovernmental Agreement and represents, in substance, the points which the U.S. Delegation sought to have included. France, supported by Syria, Switzerland, Belgium, and Sweden, reserved on the question of the use of the word "single" in the phrase "single global system," arguing that the term is ambiguous, ignores the presence of other communication satellite systems in the world, and, therefore, should be eliminated.

The objectives and purposes of the organization as proposed in the draft article include the creation of a global organization to establish a single global commercial communication satellite system (France reserving again on "single") "intended primarily to provide international public telecommunication services on a commercial basis of high quality and reliability, and sufficient to provide such services to all areas of the world." This statement of objectives and purposes is consistent with U.S. views.

The Working Group submitted a proposed article on the organization's authorized scope of activities. The draft article states that INTELSAT: shall provide the space segment for international public telecommunications services; shall make its global satellite facilities available for domestic public telecommunications services on a non-discriminatory basis if this would not affect adversely the provision of facilities for international public services; may provide facilities in the global space segment for specialized service, presumably domestic or international in scope, if this would be both technically and economically acceptable and does not affect adversely the provision of international public services; may provide separate satellites for domestic public telecommunications services; and may provide separate satellites for specialized telecommunications services, presumably both domestic and international in scope, if this would [7] be both technically and economically acceptable and does not affect adversely the provision of international public services.

Although the Working Group agreed on this text, there are still several areas of less than complete agreement. The status and relative priority of domestic traffic is a matter of particular concern to Denmark, Pakistan, Portugal, the United Kingdom and the United States, all of which have geographically separated areas between which communication satellite traffic is or may be contemplated. (The U.S. concern, however, relates to all domestic traffic.) In connection with "specialized" telecommunication services, France, among others, expressed concern that INTELSAT may be entering into areas or types of service better left to other organizations or to national governments to provide. As drafted, the article was acceptable to the United States.

Working Group B—The report of Working Group B (Com. I/111) deals with the structure of the organization. The Working Group discussed and reported on the nature, composition and functions of the major organs of the organization including an assembly of members, a governing body and a manager. No draft articles were prepared although a number of proposed drafts were submitted to the group and are reflected in the report. The eighteen-country group attempted to identify and record alternative views on structure of the organization but did not attempt to negotiate or reconcile inconsistent or incompatible proposals.

There was unanimous support for creating an assembly, but the question arose as to its composition. Some thought the assembly should be exclusively a governmental body (i.e. participants would be under direct government control), and others suggested it be an assembly of telecommunication operating agencies or entities which are the signatories to the operating or special agreement. The United States, India and the United Kingdom, among others, proposed that the designation of delegations to the assembly be reserved as a matter of discretion of the individual member countries. An alternative solution proposed was to divide the assembly into [8] two assemblies, (1) an assembly of governments, meeting less regularly and concerning itself only with review of programs and progress and possible amendment of the intergovernmental agreement, and (2) an assembly of telecommunication operating entities, meeting more regularly, perhaps, annually, to oversee and consider the management and progress of the system.

There was considerable discussion on the powers of the assembly, and proposals range, in substance, from treating the assembly as the equivalent of a stockholders' meeting to giving the assembly direct responsibility and decision-making powers relating to operation of the system. The broadest support probably is for the relatively less operationally responsible assembly, and proponents of increased assembly powers are fewer as the powers assigned increase. There are two schools of thought on voting in the assembly. Nearly all of those speaking on the point favored one nation-one vote, in many cases, however, subject to the assumption that the assembly will have relatively limited powers. The United States position was to combine one nation-one vote with a weighted vote reflecting relative levels of investment of the members.

There was unanimous opinion in the Committee favoring establishment of a governing body equivalent to the present Interim Communications Satellite Committee (ICUS) of INTELSAT. It was unanimously agreed that representatives to this body should be from the telecommunication operating entities involved. There was a consensus that the size of the body should be limited in the interest of efficiency and effectiveness, although all agreed that equitable arrangements should be made for representation from smaller member countries and all geographical areas. There was no consensus on how to achieve these goals. There also was no specific agreement on functions of the governing body, but it appeared to be intended by most delegations that its functions would be similar in nature and scope to those now performed by the ICUS. Voting in the governing body was discussed without [9] conclusions. There is general agreement that voting should be weighted to reflect relative investment in or use of the system, but there appears to be substantial support for the view that no single country or small group of countries (2 or 3) should be able to impose or block (veto) a decision of the governing body. (The U.S. currently has an effective veto power under the interim arrangements.)

The Working Group reported three principal views on the management arrangements for the future organization and the proponents and supporters of each view proclaimed their desire to ensure efficient, competent management.

The proponents of the first view maintained that the definitive arrangements should establish a firm goal of full internationalization of the management, under a director general, within a specific period of time. This view was supported by Belgium, France, India, Switzerland, the United Kingdom, Canada and Germany, among others.

The proponents of the second view, while not excluding the possibility of partial or complete internationalization of the Manager under a director general, felt that fixing a rigid time period in which this goal must be realized might interfere with the necessity of [e]nsuring efficient and effective management. Australia, Chile, Nigeria and Venezuela favored this approach. Some members of the Working Group who supported this view wanted it made clear that staff should be recruited on the basis of competence rather than on the basis of investment or geographical representation of countries.

A third view, expressed by the United States, specifically rejected the view that internationalization of the Manager should, in itself, be a primary goal or common aim. Efficient management should be the only goal of the organization regarding the structure of its management body. Internationalization of the organization should be addressed in the assembly and the governing body. Subsequently, the U.S. Delegation indicated that we could consider dividing the management function and creating an international administrative [10] management body to handle administrative, financial, and legal functions, with the operational manager continuing to handle other functions. The United States made it clear, however, that the concept of a director general interpositioned between the manager and the governing body was unacceptable.

Working Group C—The report of Working Group C (Com. I/94) deals with eligibility for INTELSAT membership and the question of relationships with non-member countries. The Working Group produced two draft articles, one on membership, which stipulates that only International Telecommunication[s] Union members are eligible for INTELSAT membership; and one setting forth principles of access to the system, which would make direct access to the space segment of the global system available to all signatories and other states, countries or areas not members of the organization. The Group unanimously supported the draft articles, with Tunisia, however, recording the view that the possibility of admitting non-ITU members to membership in INTELSAT should not be excluded. A few other members and several observers also spoke in favor of this view in meetings of the parent Committee, though a considerably larger number of members spoke in favor of ITU membership as a condition.

Committee I discussed the topics considered by its Working Groups before and after the Groups met. In addition, the topics of the rights and obligations of members and INTELSAT relationship with the ITU were discussed in the Committee, but these discussions did not go beyond a few expressions of views and these topics were not assigned to a Working Group. The Secretariat prepared and distributed a summary of the main points touched upon in these discussions (Com. I/107, Rev. 1).

Committee II—Legal and Procedural Questions

Committee II established three Working Groups. Its agenda is set forth in Annex D.

[11] Working Group on Legal Status—Legal status was examined at some length in both the Working Group and the full Committee. All of the other members of the Working Group (Brazil, Chile, Germany, the Philippines, Sweden, Switzerland, and the U.K.) opposed the U.S. position that INTELSAT should continue as a joint venture without legal personality. Instead they favored establishing INTELSAT as a legal entity distinct from the participants. The joint venture was described by the U.S. as a viable means of carrying out the activities and purposes of INTELSAT. The majority view urges that INTELSAT will be better able to contract, own property, sue or be sued, obtain privileges and immunities, and incur and dispose of liabilities appropriately if it is a separate legal personality. The United States position is that all these functions have been performed and can continue to be performed through a joint venture.

The report of the Working Group contains separate statements of the majority and U.S. positions and, with the summary record of the Committee's discussion of the report, was transmitted to Committee I for consideration. However, the matter was not discussed in Committee I.

Com. II/9 (and 11) contains the report of this Working Group.

Working Group on Accession, Supersession and Buy-Out—

Entry Into Force—The U.S. position was that the definitive arrangements should enter into force upon final adherence by two-thirds of the present members who hold, or whose signatories to the Special Agreement hold, a substantial proportion (80% was suggested) of the investment quota under the Special Agreement. This position was adopted as the majority position by the Committee, but the exact percentage was left to be decided by the Plenary. The major point of contention concerned the financial criterion. Some delegates (Chile and Switzerland) strongly objected to a requirement which would enable the largest "shareholder" to block the entry into force of the new agreements. Only a small minority of delegations (Sweden, [12] France and Mexico) asserted that there must, as a matter of international law, be unanimous adherence by all prior members. The remainder appeared to accept adherence by a substantial majority as legally sufficient.

Transfer of Rights and Obligations—The Working Group produced two alternative draft articles. The first of these would transfer the rights and obligations of signatories to the interim Special Agreement to the signatories of the comparable part of the definitive arrangements, including transfer of ownership in undivided shares. The second approach transfers all rights and obligations of interim signatories to INTELSAT under the definitive arrangements. The two approaches are contrasted as appropriate, respectively, for a continued joint venture or the establishment of an organization with a legal personality.

Buy-Out—Committee II in its report formulated the following general legal principles for the buy-out of non-continuing members, leaving the mechanics to Committee III: (1) fair compensation with reasonable expedition; (2) for patents and data, either fair compensation or continued enjoyment; and (3) the amount of compensation to be settled by negotiation between the non-continuing member and INTELSAT. Failing an agreement, the non-continuing member could challenge any determination by the governing body before a neutral arbitral tribunal. Although there appeared to be no opposition to the principle of equitable compensation for non-continuing members, Chile and Sweden argued that a non-continuing member's share cannot be bought out without its consent.

Com. II/10 contains this Group's report.

Working Group on Other Matters—Com. II/15 and 16 are the reports of this Group.

Privileges and Immunities—This item was considered at some length in the full Committee as well as in the Working Group. The report of Committee II on this subject noted that a majority of the delegates favored including in the intergovernmental agreement two general provisions: the first would commit the host [13] state to conclude a headquarters agreement providing for appropriate privileges and immunities within the jurisdiction where the headquarters were located; the second would authorize the board of governors to negotiate with member states on an ad hoc basis those privileges and immunities appropriate for the proper functioning of INTELSAT. It was generally recognized that INTELSAT would obtain tax immunities where necessary, although provision for such immunities should not be specifically made in the intergovernmental agreement. Some delegates expressed the view that INTELSAT would have to have legal personality in order to be granted privileges and immunities under their domestic laws.

Settlement of Disputes—There were three major points of controversy: the proper parties to arbitration; the selection of the panel from which the third member (the president) of an arbitral tribunal is chosen; and, the scope of arbitrable disputes. The U.S. position was that the signatories to the operating agreement, the board of governors, and the assembly should be the only competent parties to an arbitration proceeding. The majority of the delegations wanted to include in the intergovernmental agreement some mechanism for settling disputes among the governments['] parties to that agreement. With respect to the selection of the panel members, the Working Group's report, which was adopted by Committee II, stated that a majority favored selection of the panel by the Assembly without weighted voting. This was inconsistent with the U.S. position that the selection should be made by the board of governors. As for the scope of arbitrable disputes, the U.S. position that the scope should be confined to legal disputes received majority support, although some delegations favored a broader scope. The majority favored limiting the scope to legal disputes but using a different formulation from that proposed by the U.S.

Amendment Processes—The U.S. position that a proposed amendment to the operating agreement be approved by the board of governors was opposed by several delegates. Although there was general agreement that no amendment to the operating agreement should be made [14] without the consent of the parties to the intergovernmental agreement, differing views were expressed as to the manner in which that consent should be manifested.

Liability of the Signatories—This matter was discussed only in the full Committee. A significant issue was whether, if INTELSAT is given separate legal status, the signatories would or should enjoy limited liability for INTELSAT obligations. There was significant opinion that limited liability automatically followed from establishing INTELSAT with legal personality. The U.S., along with Australia and Sweden, felt that it would not. However, Sweden and a number of other delegations expressed the view that limited liability is an advantage which should be afforded the signatories. A second issue was whether exemption of the signatories to the operating agreement from inter se liability should extend beyond consequential damages arising from a breakdown in service. There was general agreement that the definitive arrangements should not impair member states' responsibilities under the Treaty on Outer Space.

Withdrawal, Reservations, Definitions and Number of Agreements—Withdrawal was discussed only briefly in Committee II. No general agreement was discernible.

The matters of reservations and definitions were deferred until the final text of the agreement has been generally established. There appeared to be no opposition at this time to the U.S. position prohibiting reservations.

An overwhelming majority supported the U.S. position for two agreements, with some delegates reserving until more is known of the final text.

Committee III—Financial Arrangements

Committee III's work program is shown in Annex E. Its report is Doc. 16.

[15] There was general agreement in the Committee that investment in the system should be related to use, as the U.S. had proposed, though some countries, principally the Arab group, favored applying the investment/use system only after allocating a base share of investment to each member. There was near agreement on a minimum share for each member, regardless of its use of the system, most delegations favoring 0.05%. However, many delegations thought members with lower use should not be required to accept this minimum.

The Committee divided three ways on the question of what types of use of the system should be counted in determining investment shares, about one-third of those countries that expressed views (including the U.S.) favoring all use of INTELSAT-financed facilities, one-third favoring international traffic only, and another group proposing to count international traffic and that domestic traffic that crosses international boundaries (or would do so if projected on the surface of the earth). The ideas of several delegations on this question clearly were related to the question of voting strength in the organization. It was agreed that shares should be adjusted periodically, but left open whether adjustments would be based on past (U.S.) or projected use or some combination of the two. There was also a division of views on the question of circumstances under which a signatory could decline to accept an increased or decreased share.

It was agreed by all but a few members that there should be a utilization charge, partly as a means of compensating owners whose share is larger than their use prior to adjustment of shares. Working Group 1, which handled most of the Committee's work, recommended the cost of money plus about 2% as the basis of compensation for use of capital in determining utilization charges.

Committee III's second Working Group considered financial provisions relating to transition from the present agreements and possible withdrawals from [16] the organization. While the Working Group did not, and did not attempt to, reach final agreement, there was general agreement on the principles that should be applicable. It should not be too difficult to reach agreement in this area.

Working Group 3 agreed, with respect to financial access to the system by non-members, that the organization, in establishing space segment utilization charges for non-members[,] should take account of the fact that non-members have not borne any of the risks and obligations of membership. This would mean that a rate charged to non-members should take account of both the cost of the members' capital and the risk they have taken in investing in the system.

Committee IV—Other Operational Arrangements

Committee IV, which had two Working Groups, considered only procurement policy (Working Group A) and patent and data policy (Working Group B).

On procurement policies there developed three alternative approaches: (1) a relatively simple provision calling for international tenders and contractor selection on the basis of the best combination of price, quality and timely delivery; (2) retention of the existing interim arrangement provisions, which have allowed international spreading of contracts; and (3) the amplification of (1) above by addition of specific language encouraging international spreading of contracts with distribution roughly proportionate to relative investment percentages of members. There is, in fact, little actual difference in the probable practical effect of alternatives (2) and (3). The United States, and apparently a majority of those countries expressing themselves on this issue, favored alternative (1). The United Kingdom and Japan supported alternative (2) and France supported alternative (3). The proposed specific wording is set forth in the Committee's report (Doc. 12).

[17] With regard to patent and data policy, two alternatives emerged. The United States, the United Kingdom and others favored a provision which would leave to the discretion of the Governing Body the particular patent policies to be applied in each contract negotiation. Canada, Germany, India, France and others proposed a patent provision which would establish in the definitive arrangements a fixed non-exclusive license policy pursuant to which any INTELSAT contractor would get title to inventions and data developed under INTELSAT contracts, and the organization would take a non-exclusive license to use the information only in connection with space segments and would thus forego control over the contractor's use of the information. There was virtually unanimous agreement that to the extent INTELSAT obtains rights in inventions and data, they should be made available on a royalty-free basis for use in the INTELSAT space segment and on a reasonable royalty basis for other uses. The alternative patent and data policies are set forth in Com. IV/10 and 11.

The Credentials Committee

The Credentials Committee was nominated by the Conference Chairman and approved during the initial plenary meeting. Its members were Ireland, Norway, Panama, Philippines and Turkey. The Committee elected the representative of Turkey Chairman. The Committee found all credentials to be in order and its report was accepted without discussion by the Conference.

The Editorial Committee

The Editorial Committee was nominated by the Conference Chairman and approved by the plenary. Its members are Belgium, Canada, Colombia, France, Jamaica, Mexico, Spain, the United Kingdom, and the United States. There being no work to be done on final texts, the Editorial Committee did not meet or function at the first session of the Conference.

[18] The Steering Committee

Composition of the Steering Committee was provided for in the Conference Rules of Procedure as follows: Chairman—Conference Chairman (USA); members—the four regional Vice Chairmen (Netherlands, Venezuela, India and Algeria; but the Representative of The Netherlands left and was replaced by the Swiss Representative in the fourth week) and the Chairmen of the four Working Committees (Argentina, Japan, Australia and Italy). The Steering Committee met regularly throughout the first session, coordinating the program of the Conference. This Committee prepared the proposal adopted by the Conference to establish an intersessional Preparatory Committee (Annex G) which is discussed in detail in Section VII below.

VI. Plenary Session of the Conference

The Conference met in five plenary sessions. At the initial plenary the agenda was adopted, committees were formed, and conference officers were elected. There were four plenary sessions during the final week to receive and consider the Working Committee reports.

The opening session proceeded thorough the agenda as planned without any difficulties. Prior to adoption of the Conference Rules of Procedure, representatives from Italy, Nigeria and India sought assurances, which were given by the Conference Chairman, that the rules with regard to statements of observers would be interpreted liberally to ensure the fullest possible exchange of views. The United Kingdom suggested and was assured that maximum opportunity would be made available to achieve consensus on all matters.

Sweden introduced a comprehensive and novel set of draft definitive arrangements (Doc. 8). This draft was not given much direct attention by the Conference. The only other comprehensive draft agreements submitted were tabled by the United States at the end of the first week of the Conference (Doc. 10). These drafts were not discussed as such, but various articles were considered by the Conference Committees.

[19] At the four plenary sessions held during the final week, reports were received from the Working Committees and were discussed. Because of the number of unresolved issues and the general complexity of definitive arrangements, it became obvious that substantially more time would be required to develop final texts. It was decided, on the recommendation of the Steering Committee, to recess the Conference March 21, 1969 and to refer the Committee reports and all other relevant Conference documents to an intersessional Preparatory Committee for study and work. A proposal to provide for interim work was discussed at some length during the fourth plenary session and the Steering Committee was requested to revise the proposal to reflect the views expressed. A revised paper was submitted to the fifth and final plenary and was adopted unanimously without discussion (Annex G).

VII. Future Meetings

The Conference provided for two types of future meetings. First, member and observer countries are to notify the Conference Secretariat if they intend to participate in or observe meetings of the intersessional Preparatory Committee, which is to convene in Washington as soon as possible after May 20, on a date to be notified by the Conference Secretariat. The Committee is then to meet again thereafter at such times as it may decide as necessary to complete its work.

A second plenipotentiary session of the Conference is scheduled to convene in Washington on November 18, 1969.

The Preparatory Committee is intended to be broadly representative of all areas and attitudes. It is encouraged to resolve in an objective manner differences of views presented during the Conference, although it is not empowered to negotiate definitive arrangements. No country will be bound by the views and positions of the Committee's report, whether or not it is represented on the Committee.

The Committee is instructed to elect its own Chairman and to establish its own procedures and methods of work. Its report is to be circulated through the Secretary General of the Conference at least sixty days prior to the reconvening of the Conference. The report will be in [20] the form of draft agreements, with such alternate drafts of specific articles as may be necessary to reflect differences of significant views. However, degrees of support are not to be reflected in the report.

If the Committee should be unable to complete its work in time, it may postpone the reconvening of the Conference and request the host Government (U.S.) to reconvene the Conference at the earliest convenient date.

VIII. Conclusions

From the outset it was known that the task of establishing definitive arrangements for a global commercial communication satellite system would be demanding, complex and time consuming. During the four weeks of this session, the Conference collected, considered and condensed a great many views. This was an essential and desirable first step.

Machinery has now been established for further significant steps, i.e. preparation of drafts and, hopefully, resolving of differences of views. This work is to be done in the intersessional Preparatory Committee. The Committee's work product will be draft agreements, including alternative provisions where appropriate, which all interested countries can consider prior to and at a reconvened session of the Conference, now scheduled for November 1969.

One overriding value of the first session was the educational benefit it offered to all participating countries. Many countries, particularly those not represented on the ICUS, expressed their views at this session for the first time. Many of the earlier published positions of other countries were explained, elaborated and documented. In the increased international understanding it produced, and in the high level of international cooperation it evidenced, this first session was undoubtedly successful. The individual and joint efforts of all the participating representatives and observers, particularly those of the Conference officers and Committee and Working Group Chairmen, have advanced the prospects of successful conclusion of these considerations immeasurably.

The first session appropriately concluded on a note of constructive optimism and cooperation. The provisions for further work and deliberations should make possible [21] a continued valuable exchange of views and timely conclusion of definitive arrangements for the presently operating global commercial communications satellite system established by INTELSAT.

The Chairman of the Delegation wishes to thank the other members of the Delegation for their very able assistance and sound advice.

[hand-signed: "Leonard H. Marks"]
Leonard H. Marks, Ambassador
Chairman, United States Delegation

Document I-26

Document title: "Second Report and Order in the Matter of Establishment of Domestic
Communications-Satellite Facilities by Non-Governmental Entities," Docket No. 16495,
June 16, 1972.

Source: *Federal Communications Commission Reports: Decisions and Reports of the Federal
Communications Commission of the United States, June 9, 1972 to August 4, 1972*, Volume 35,
Second Series (Washington, DC: U.S. Government Printing Office, 1974), pp. 844–851,
860–867.

*When the original institutional arrangements for communicating via satellite were discussed during
the 1962–1964 period, the White House and Congress focused on setting up a system for international communications as quickly as possible. Little attention was given to issues related to using communications satellites for domestic services within the United States. Beginning in 1966, however,
various companies began to apply to the Federal Communications Commission for authorization to
develop and operate such domestic satellite systems. This order and accompanying report represent the
Federal Communications Commission's decision on the matter, opening the U.S. domestic market to
interested multiple providers of various communications services via satellite.*

---

[844]

Federal Communications Commission Report

F.C.C. 72-531

BEFORE THE
FEDERAL COMMUNICATIONS COMMISSION

WASHINGTON, D.C. 20554

In the Matter of
ESTABLISHMENT OF DOMESTIC COMMUNICATIONS-   }
    SATELLITE FACILITIES BY NON-GOVERNMENTAL   } Docket No. 16495
    ENTITIES   }

SECOND REPORT AND ORDER

(Adopted June 16, 1972; Released June 16, 1972)

BY THE COMMISSION: CHAIRMAN BURCH DISSENTING AND ISSUING A STATE
MENT IN WHICH COMMISSIONERS REID AND WILEY JOIN; COMMISSIONER
JOHNSON CONCURRING AND ISSUING A STATEMENT.

I. PROCEEDINGS BEFORE THE COMMISSION

1. This proceeding was instituted by the Commission on March 2, 1966 (Notice of
Inquiry, 31 F.R. 3507; Supplemental Notice of Inquiry, October 20, 1966, 31 F.R. 13763)

to explore various legal, technical and policy questions associated with the possible authorization of domestic communications satellite facilities to nongovernmental entities. On March 24, 1970, the Commission issued a first *Report and Order* (1970 Report) inviting the submission of applications to assist our determinations (22 FCC 2d 86, 35 F.R. 5356), and consolidated a concurrently issued Notice of Proposal Rule Making (22 FCC 2d 810). In response to the 1970 Report, system applications were filed by the following:

The Western Union Telegraph Company (Western Union)

Hughes Aircraft Company and various telephone operating companies of GTE Service Corporation (Hughes/GTE)

Western Tele-Communications, Inc. (WTCI)

RCA Global Communications Inc. and RCA Alaska Communications, Inc. (RCA Globcom/RCA Alascom or "the RCA applicants")

Communications Satellite Corporation and American Telephone and Telegraph Company (Comsat/AT&T)

Comsat

MCI Lockheed Satellite Corporation (MCI Lockheed)

Fairchild Industries, Inc. (Fairchild)

In addition, applications for earth stations only were filed by:

Hawaiian Telephone Company

Twin County Trans-Video, Inc.

TelePrompTer Corporation

LVO Cable, Inc., and United Video, Inc.

Phoenix Satellite Corporation

[845] 2. Comments and reply comments on the applications and rule making issues were received from the applicants and other interested parties. By a Memorandum Opinion and Order issued on March 17, 1972 (34 FCC 2d 1), the Commission afforded the parties an opportunity to file written comments and to be heard orally on a proposed Second Report and Order (34 FCC 2d 9) recommended by the Chief of the Common Carrier Bureau (staff recommendation). Written comments were received and oral argument before the Commission en banc was held on May 1–2, 1972.[1]

3.   Upon consideration of the entire record, we are of the view that the staff recommendation adequately describes the background of this proceeding, the general nature of the pending applications, and the previously filed comments and reply comments of the parties on the applications and rule making issues. Accordingly, we will adopt the descriptive portions of the staff recommendation without reiterating such material here. However, as stated in the Memorandum Opinion and Order of March 17, 1972, our action in designating the staff recommendation for written and oral comment was taken "before reaching any determinations in this matter" and "therefore does not reflect any predisposition by the Commission with respect to the resolution of the issues involved" (34 FCC 2d at 2). The Commission's determinations, which are set forth below, incorporate the staff's reasoning and conclusions on the issues only as expressly indicated herein or to the extent that they are clearly consistent with our statements of policy and conclusions.

---

1.   The two entities who had not previously participated in this proceeding were granted leave to be heard orally: the Department of Defense and the Network Project (FCC 72-314).The motions of various parties to correct the transcript of oral argument are hereby granted. Some applicants have submitted statements, without leave from the Commission, purportedly in further response to questions from individual Commissioners at the oral argument. While such statements have been placed in the record, we do not rely on them.

II. INTRODUCTORY POLICY STATEMENT

4.   As the Commission recognized in the 1970 Report (22 FCC 2d at 88, 95–96), and as confirmed by the applications and responses filed pursuant to that Report, the satellite technology has the potential of making significant contributions to the nation's domestic communications structure by providing a better means of serving certain of the existing markets and developing new markets not now being served. There are concrete proposals before us for the use of communications satellites to augment the long-haul terrestrial facilities of existing carriers for point-to-point switched transmissions services, and to connect off-shore distant domestic points (i.e., Alaska, Hawaii, Puerto Rico) to the contiguous states. There are also proposals for the use of satellites as a means of providing point-to-multipoint services, such as program transmission, although plans for such use are now most tentative and uncertain. Other proposals reflect the view that the most important value of domestic satellites at the present time lies in their potential for developing new markets and for expanding existing markets for specialized communications services.

5.   Notwithstanding the specific proposals that have been submitted, the true extent and nature of the public benefit that satellites [846] may produce in the domestic field remains [sic] to be demonstrated. The United States has a well-developed and rapidly expanding complex of terrestrial facilities, and advances in terrestrial technology and operations can be expected to continue the present trend toward reduced transmission costs and more efficient services. Although pointing to some increased operational flexibility in the routing of its traffic, the predominant terrestrial carrier, AT&T, disclaims that the satellite technology presently offers any cost savings or other marked advantages over terrestrial facilities in the provision of the switched services that constitute the bulk of its traffic, message toll telephone (MTT) and wide area telephone service (WATS). At the same time, there is an uncertainty, that can only be resolved by actual operating experience, as to whether the time delay inherent in voice communications via synchronous satellites will provide an acceptable quality of service to the general public when domestic telephone traffic is routed indiscriminately and on a large scale basis via satellite and terrestrial facilities.

6.   Although the satellite technology appears to have great promise of immediate public benefit in the specialized communications market, here too there are uncertainties as to how effectively and readily satellite services can develop or penetrate that market. Thus, in the area of point-to-multipoint transmission, the commercial broadcast networks are as yet undecided as the whether to use this technology in whole or in part. We do have a concrete proposal for a CATV network from Hughes, expressions of interest by public broadcasting and other educational entities, and the possibility of interest by independent supplies of program material to CATV and broadcast outlets. Moreover, several system applicants, in addition to seeking to attract program transmission business, have premised their proposals on the sale of other specialized services—in part as a complement to existing or proposed terrestrial offerings, but in the main with the expectation of expanding existing special service markets and developing new markets. To be sure, the applications generally do not identify specific services that are new or innovative. However, in our judgment, the uncertain ties as to the nature and scope of the special markets and innovative services that might be stimulated will only be resolved by the experience with operational facilities.

7.   Under the circumstances, we will be guided by the following objectives in formulating the policies to govern our licensing and regulation of the construction and use of satellite systems for domestic communications purposes, namely:

    (a) to maximize the opportunities for the early acquisition of technical, operational, and marketing data and experience in the use of this technology as a new communications resource for all types of services;

(b) to afford a reasonable opportunity for multiple entities to demonstrate how any operational and economic characteristics peculiar to the satellite technology can be used to provide existing and new specialized services more economically and efficiently than can be done by terrestrial facilities;

[847] (c) to facilitate the efficient development of this new resource by removing or neutralizing existing institutional restraints or inhibitions; and

(d) to retain leeway and flexibility in our policy making with respect to the use of satellite technology for domestic communications so as to make such adjustments therein as future experience and circumstances may dictate.

8. We are further of the view that multiple entry is most likely to produce a fruitful demonstration of the extent to which the satellite technology may be used to provide existing and new specialized services more economically and efficiently than can be done by terrestrial facilities. Though specialized services constitute a relatively small percentage of AT&T's total traffic, it is presently the predominant terrestrial supplier of specialized services. There is some existing and potential competition from Western Union and any new specialized carriers authorized pursuant to the Commission's decision in *Specialized Common Carrier Services* (29 FCC 2d 870). But the capacity of their terrestrial facilities is small compared to those of AT&T or the high capacity facilities proposed by the satellite system applicants.[2] The presence of competitive sources of supply of specialized services, both among satellite system licensees and between satellite and terrestrial systems, should encourage service and technical innovation and provide an impetus for efforts to minimize costs and charges to the public.

9. Of course, the incentive for competitive entry by financially responsible satellite system entrepreneurs to develop specialized markets must be meaningful and not just token. This requires that we take appropriate measures toward the end that a reasonable opportunity for effective entry is not defeated or weakened by AT&T, either directly or through its existing or future relationships with Comsat. In this regard, we cannot ignore the effects upon achievement of our objectives that might result from AT&T's existing economic strength and dominance stemming from its permeating presence and influence in all domestic communications markets. Nor can we ignore the ability of AT&T—an ability not possessed by other applicants—to load a high capacity satellite system with MTT and WATS traffic and thereby control the cost of specialized services furnished via that system. Other applicants, lacking a similar initial traffic nucleus, would be operating—at least initially—with lightly loaded, costly facilities until such time as they might succeed in reducing their unit costs by a substantial specialized traffic fill.

10. In addition, where AT&T combines its monopoly and competitive services on the same facilities, it is difficult to identify AT&T relevant costs associated with specialized services to insure that revenues from the monopoly services are not being used to subsidize any part of its competitive services. Thus, if AT&T were permitted unrestricted use of satellites for both monopoly and specialized services, this might obscure any meaningful comparison of operating costs between satellite and terrestrial facilities for the provision of specialized services as [848] well as curtail any realistic opportunity for entry by others to serve the specialized markets via satellite.

11. We recognize that the problem of cross-subsidy now exists with respect to the establishment of rates and identification of relevant costs for specialized services furnished by AT&T terrestrially. However, this longstanding problem would be exacerbated by permitting the troublesome monopoly and competitive service combinations to

---

2.   The Commission has also authorized terrestrial facilities to various miscellaneous carriers providing program transmission service to CATV systems and broadcasters.

be carried over into this new arena. Moreover, the cross-subsidy aspect is only part of the deterrent to a reasonable opportunity for competitive satellite entry in the specialized field and, even if resolved, would not overcome AT&T's unique advantage of being able to control satellite circuit costs by the extent to which it chooses to load the high capacity satellite facilities with telephone traffic while the specialized field is being developed.[2a]

12. All of the foregoing factors and concerns with respect to AT&T, in our judgment, might well result in discouraging or deterring others from attempting to penetrate the markets for specialized services. As a further consequence, AT&T's dominance in the communications field would be extended rather than lessened in the domestic area. This would derogate from our policy of seeking to promote an environment in which new suppliers of communications services would have a bona fide opportunity for competitive entry. This policy was the basis for our decision in the *Specialized Common Carrier Services* proceeding (29 FCC 2d 870). While this policy explicitly accommodates an opportunity for AT&T and other existing carriers to compete "fully and fairly" with new entrants, it does not preclude the Commission from taking reasonable measures to assure that competitive entry would be a meaningful reality in the high capacity satellite field. Paragraph 104 of the *Specialized Carrier* decision states: "We further stress that our policy determination as to new specialized carrier entry terrestrially, does not afford any measure of protection against domestic communications satellite entry or otherwise prejudge our determination in Docket No. 16495 as to what course would best serve the public interest in the domestic satellite field" (29 FCC 2d at 920).

13. The same considerations lead us to conclude that the achievement of our objectives would be prejudiced by authorizing the Comsat/AT&T proposal based on their contractual arrangement. First, since AT&T is a principal source of the domestic service revenue that Comsat would seek to obtain, it is not realistic to expect Comsat to compete vigorously in the provision of specialized services on an end-to-end or "retail" basis and thereby challenge AT&T's terrestrial domination in this field. Secondly, if Comsat should proceed in the dual capacities proposed in its two pending system applications, the revenues that would be guaranteed to Comsat from the AT&T contractual arrangement would give it an extraordinary advantage and head start over all other potential domestic satellite entrants seeking to develop specialized services in competition with Comsat as well as with AT&T's [849] terrestrial services. If Comsat were given the option of serving AT&T soley [sic] and accepted it, such a course would unnecessarily deprive others of the benefit of Comsat's expertise in the communications satellite field. If Comsat were to elect to serve only entities other than AT&T, its expertise and facilities would be available to the public and carriers other than AT&T. But if Comsat is to be authorized to provide satellite services to AT&T, it should operate exclusively as a carrier's carrier—not engaged in retailing communications services to the public—and provide such service under a tariff offering which would afford an opportunity for other carriers to have non-discriminatory access to the same system.

14. Finally, our consideration of the conditions under which AT&T and Comsat should be permitted to enter the domestic satellite field is necessarily affected by AT&T's ownership of 29 percent of Comsat's stock and its ability to elect three of the 15 Comsat directors. Such ownership was contemplated and encouraged by the Congress in enacting the Communications Satellite Act of 1962 (see Section 394(b)(2)). Thus, this is not a matter over which Comsat has any control. However, that Act, which was formulated to meet

---

2a.   We recognize that AT&T, in its offerings of specialized services, may not, for rate purposes, distinguish between specialized services provided via satellite on the one hand, and terrestrial facilities on the other hand, and thus somewhat alleviate the competitive problem. However, we believe that it will from a regulatory standpoint complicate a definitive comparison between the relative cost and other advantages of satellite and terrestrial facilities in serving the competitive market for specialized services.

the nation's policies and objectives with respect to the earliest possible establishment of a global communications satellite system, does not preclude authorized carriers from voluntarily disposing of their shares of Comsat stock.[3] All of the major carriers who originally owned Comsat stock, except AT&T, have since divested their interests. While the participation of experienced carriers had a useful function when Comsat was newly organized and gaining communications experience, this relationship warrants reassessment in light of current conditions.

15. Aside from the foregoing basic considerations of fairness and equity we reaffirm the staff recommendation in favor of multiple entry. In this connection it is important also to take cognizance of the fact that the initial implementation of domestic satellites does not confront us with a normal or routine situation. Some departures from conventional standards may be required if the public is to realize the potential benefits of this high capacity technology and we are to pursue our objective of competitive entry. This is true not only in the case of AT&T, but also for other applicants because of different factors. For example, as the staff points out, the capacity proposed by most system applicants substantially exceeds the traffic under their control or firm customer commitments. They are relying primarily on speculative business which they hope will materialize after the facilities become operational. We must, of course, make the requisite statutory findings as to an applicant's financial qualification and ability to implement its proposal, and we can require a reasonable showing that there will be no adverse impact on rates or services to customers of carrier applicants now engaged in providing essential communications services to the public. But if we adhere too strictly to conventional standards in this unconventional situation, such as requiring a persuasive showing by new entrants that competition is reasonably feasible and that the anticipated market can economically support its proposed [850] facilities, most such new applicants may in effect be denied any opportunity to demonstrate the merits of their proposals at their own risk and without potential dangers to existing services—thereby depriving the public of the potential benefits to be derived from diverse approaches by multiple entrants. It is our judgment that the potential benefits to the public warrant the application of rules and policies which will afford a reasonable opportunity for domestic satellite facilities to be established initially on a competitive basis. It is also necessary to retain flexibility to alter our initial determination in the light of evolving circumstances.

III. DETERMINATION ON THE ISSUES

A. Number of systems to be authorized initially

16. In light of the foregoing policy objectives, we have concluded that the public interest would be best served at this initial stage by affording a reasonable opportunity for entry by qualified applicants both pending and new, subject to the showings and conditions described below which we believe to be necessary to implement our objectives and to protect the public. We have reached this decision after consideration of the various alternatives discussed in the staff recommendation (paragraphs 45–78) and the views expressed by the parties.

---

3. Indeed, in 1969 Congress amended the 1962 Act to provide for fewer common carrier elected directors in proportion to their decrease in stock ownership in Comsat (47 U.S.C. 733). This schedule contemplates that the percentage of common carrier stock ownership may fall below eight percent, in which event there would be no directors elected by common carriers.

17. Like the staff and most parties, we think it unwise to attempt to select or prescribe one system (either a consortium of all the applicants or selection of one applicant) or to choose one or more systems through comparative hearings. In addition to the reasons given by the staff (staff recommendation, paragraphs 50–61), which we adopt, such a course would not promote our policy objectives discussed above. However, we are not accepting the alternative recommended by the staff (paragraphs 71–78) or requiring of encouraging consolidations of applicants along guidelines prescribed by the Commission. While we recognize that there may well be advantages to and need for voluntary consolidations or sharing arrangements (such as "launch risk pools") undertaken at the applicants' initiative as a matter of prudent business judgment, we do not deem it advisable to structure the architecture of any joint space segment operations. Rather, we will permit and encourage such arrangements so long as they are consistent with the policy conditions set forth herein. Accordingly, we will accord the system applicants a 30-day period within which to apprise the Commission as to whether they intend to pursue their pending applications, as modified to achieve compliance with this *Second Report and Order*, or whether they desire further time to reframe their proposals.

18. Our decision in favor of multiple entry does not mean that we have opted for a policy of "unlimited or unrestricted open entry." Our aim, as outlined above, is to afford qualified applicants a reasonable opportunity to demonstrate the public advantages in use of the satellite technology as a means of communications. But such entry cannot be "open" in the sense that it is without any restrictions or limitations. Pursuant to statute we must require showings of financial, technical and other qualification and make the requisite finding that a grant of the particular proposal will serve the public interest, con-[851] venience and necessity. Although, as discussed in paragraph 15 above, it is our intention to make such determinations with due regard for the unique circumstances involved here, each applicant must make a sufficient showing of potential public benefit to justify the assignment of orbital locations and frequencies. Moreover, we believe it necessary to impose certain conditions to protect the public from possible detriment and to further the implementation of our policy objectives. In addition to the conditions discussed below, we will require a reasonable showing by any common carrier applicant now engaged in providing essential communications services that revenue requirements related to the proposed domestic satellite venture will not be a burden or detriment to customers for such essential services. . . .

IV. ORDER

[860] 44a. Authority for the policies and conditions adopted herein is contained in Sections 1, 2, 3, 4(i) and (j), 201, 202, 212, 213, 214, 218, 219, 220, 301, 303, 307–309, 310(b), 319, 396, 403 and 605 of the Communications Act of 1934 and Section 102 and 201(c)(8) of the Communications Satellite Act of 1962.

45. Accordingly, IT IS ORDERED, That:

a. The policies and conditions set forth herein, and such portions of the staff recommendation (34 FCC 2d 9) as are expressly approved or clearly consistent with the policies and conditions herein, ARE ADOPTED, effective July 25, 1972.

b. Each of the applicants for domestic communications satellite systems named in paragraph 1 above, SHALL APPRISE THE COMMISSION on or before July 25, 1972, as to whether it intends to pursue its pending system applications, in whole or in part, with such modifications as are required to achieve compliance with the policies and conditions specified in this Second Report and Order; or whether it desires additional time for the purpose of reframing its proposal consistently with such policies and conditions.[11]

   c.   The Commission retains full jurisdiction over all aspects of this proceeding.

FEDERAL COMMUNICATIONS COMMISSION
Ben F. Waple, *Secretary*

DISSENTING STATEMENT BY CHAIRMAN BURCH

In this proceeding, the Commission is dealing with matters of extraordinary complexity and even subtlety. We are called on to establish ground rules for an industrial technology that does not yet exist, to serve some present markets and some that are at best speculative—and most difficult of all, the interrelationships between the two. The policy decisions thus arrived at are not in the usual sense definitive: rather, they represent "signals" to the applicants that will cause them to reformulate their proposals, and these in turn will almost surely not be the same as those with which the Commission is here ostensibly dealing. Our objective is to engraft a new and untested technology onto an existing domestic communications complex, whose characteristic problems are essentially independent of satellite technology *per se*.

In approaching such a maze of unpredictables and potential pitfalls, the Commission would have been well advised to adopt a posture of "least is best" (thus making only those decisions necessary to elicit the applicants' genuine intentions), to build from the base of irreducible marketplace realities (namely, AT&T traffic), to discipline itself [861] against the temptation to piggyback on this already complex policy finding its favorite regulatory schemes and hangups (for example, the desire to "get a handle on AT&T"), and to offer all applicants a maximum of options (which might well lead to the evolution of a competitive marketplace in which the consumer will benefit). As a general proposition, I believe the Commission has violated every one of these counsels of caution.

And to whose real benefit? That is most difficult to say. For, although the thread runs through the majority document that its key findings have been made in the interest of "competition," somewhere along the line the overriding purpose of the competitive marketplace seems to have gotten lost: namely, benefit to the consumer in the form of better and/or cheaper goods and services than would otherwise be available. Instead, the Commission has gone off in pursuit of a peculiar and novel form of competition—measured, so far as one can tell, by how many satellite systems go aloft in how many "space segments" (a benchmark that I strongly suspect would strike the typical consumer as irrelevant even if he could grasp its meaning). "Space segment" competition may, of course, translate into consumer benefit one day. Then again it may not. It all depends—and it is here that the majority document leaves pragmatic reality behind and takes off into the blue sky of academic abstraction. For example:

(a) There is repeated reference (see in particular par. 10 and fn. 2a) to "meaningful" and "definitive comparison" between the relative costs "and other advantages" of satellite technology as against terrestrial facilities in providing communications services to the public—most of which services are not unique to satellite technology anyway. This is used as a principal rationale for imposing inhibitions on AT&T, for example. I agree that such "basing point" comparisons are desirable. But this proceeding is not mere academic exercise. Tens of millions of investment dollars are involved, and so are services to the consuming public—present and near-term as well as future. In my judgment, there is an

---

11.   Upon considerations of such responses, the Commission will issue a public notice concerning the procedures we will follow in processing applications.

excessive trade-off of present and near-term benefits for mostly speculative long[-]range developments that, in any case, may be a wash from the consumers' perspective.

(b) Other inhibitions and restrictions are rationalized (see in particular pars. 9 and 11) on the ground that AT&T's "unique advantage" of being able to fill satellite capacity with existing and predictable future traffic will inevitably produce "unfair" competition and somehow disserve the public. I find this an ironic twist indeed—that "success" is to be penalized rather than rewarded and that economies of scale must be foresworn as inconsistent with a theoretical model of pure competition (for traffic that is mostly a gleam in some speculators' eyes). The Commission would have been better advised, in my view, to take existing traffic as a "given" and then attempt to build from there—with safeguards, as specified in the earlier Specialized Common Carrier decision, against undue dominance of these specialized markets by existing carriers. This might have redounded to the immediate benefit of the consuming public, available alike to AT&T's customers and to its competitors', in the form of lowest unit costs.

(c) The Commission majority, by contrast, stands the usual norms of [862] competition on their head. In its attempt to "structure" the marketplace rather than permit full and fair competition between new and existing carriers, the Commission in effect ignores its sound commitment in the Specialized Common Carrier decision not to create any "protective umbrella" for new entrants or "any artificial bolstering of operations that cannot succeed on their own merits." Thus, AT&T is precluded from providing point-to-point private-line services via satellite—even though, as the majority acknowledges, "other applicants, lacking a[n] initial traffic nucleus, would be operating with lightly loaded, costly facilities." All of which presumably means that the consumer will have to pay artificially inflated rates for specialized services during an initial three-year developmental period (unless by terrestrial facilities alone, wholly in line with the "full and fair" competitive entry formula of the earlier decision, AT&T is able to undersell its competitors anyway). And further, because the majority document is open-ended (see par. 21), this initial period could be extended *ad infinitum* at the Commission's sole discretion. Again, there is the question "who benefits"—except possibly the stockholders of a few specialized carriers operating in a protected marketplace, and all in the much abused name of "competition"!

My overriding concern is not so much that this decision will lead to irrational results as that it may lead to no results at all that will be of substantial public benefit. It is doubly ironic, in view of the majority's determination to inhibit AT&T and that company's own downbeat projections as to the cost/benefits of satellite technology, that AT&T may in the end simply apply for a satellite system of its own. And because its monopoly services—MTT, WATS, AUTOVON—constitute the vast preponderance of present traffic, an AT&T system is the only one that could conceivably achieve an immediate fill and thus conclusively demonstrate its economic viability.

The big loser seems to be the one applicant with genuine experience in space-segment management—namely, Comsat. By rejecting the AT&T/Comsat contractual arrangement out of hand, *rather than attaching conditions that might encourage the evolution of real competition,* the Commission majority has reduced Comsat's effective choice to one: that is, electing to become an end-to-end retail carrier. But even here, the option is more apparent than real. Because of a seemingly innocuous sentence at the end of par. 26 ("In the event that Comsat elects to proceed other than as a carrier's carrier, it will be prohibited from owning or operating domestic satellite facilities at any overseas point served by INTELSAT facilities (staff recommendation, paragraph 114)."), Comsat would be barred from serving any noncontiguous state or territory, would lose its present traffic to these points (almost all of which is traffic to the mainland), and would be left with virtually unutilized "white elephant" earth stations in Alaska, Hawaii, and Puerto Rico. Some option.

The other option—becoming a carrier's carrier and leasing transponders on tariff to all comers, including AT&T—is in the end AT&T's choice and not Comsat's at all. And my own strong conviction, in view of the decision as here formulated, is that AT&T will *not* so choose. Why should it, in effect, subsidize its own competition—and competition operating under a protective umbrella at that—by [863] filling idle satellite capacity with the only substantial traffic now available?

There is, in all candor, no ideal solution to this problem. Our job is to come up with the best alternative *available*—and I make no apologies for thus relying on marketplace realities in an effort to bring to the consuming public some immediate benefits of a new technology. In my view, the answer is to be found in an approach that affirms in essence the AT&T/Comsat contractual arrangement but then attaches to it one critical condition: namely, that Comsat, with its unique technical and managerial expertise, also provide satellite service to those entities who, lacking the initial nucleus of assured traffic, might be unwilling or unable to risk the huge investment necessary to launch satellite facilities of their own. As an alternative, Comsat should be free to elect the route of an end-to-end retailer.

The majority attempts to "structure" behavior largely by recourse to penalties and blue-sky "models" of pure competition. But the proposal before use, in my judgment, suffers from two fatal flaws: it may retard the evolution of satellite technology, not get it going, and it may thus withhold realistic benefits to the public. The Commission can and must do better than that.

(Commissioners Reid and Wiley join with Chairman Burch in this Dissenting Statement.)

CONCURRING OPINION OF COMMISSIONER NICHOLAS JOHNSON

The Commission now arrives at the denouncement of this seven year old proceeding. An examination of the plot of this story, and its several acts, gives a revealing insight to the policymaking process at the FCC.

Domestic satellites became a policy question at the FCC, not because of Commission action, but with the filing of a proposal for domestic satellite television network interconnection by ABC in September 1965. To examine the important policy questions before taking definitive action, the Commission returned the ABC application and instituted an inquiry. 31 F.R. 3507 (March 2, 1966).

In response to the inquiry, the Ford Foundation filed a proposal in August 1966 linking the financing of public broadcasting to the institution of domestic satellite service. Under the Ford plan, the savings in interconnection costs would be used to finance public broadcasting as a "people dividend" from the $40 billion of public expenditures to develop the space technology that made the satellite system possible. This was a proposed alternative use of the savings—rather than flowing them through to networks' profits, or lower costs to users and their customers. J. Dirlan and A. Kahn, "The Merits of Reserving the Cost-Savings from Domestic Communications Satellites for Support of Educational Television," 77 Yale L.J. 494 (1968).

The FCC responded with a further notice of inquiry. 31 F.R. 13763 (October 20, 1966). In February 1967 President Lyndon Johnson proposed the legislation that later became the Public Broadcasting Act of 1967. And in April 1967 Comsat proposed a pilot domestic satellite system to demonstrate the potential and benefits of satellites, including their use for public broadcasting.

[864] On August 14, 1967, President Johnson announced the formation of a Task Force to review a variety of telecommunications policy questions, including domestic satellites. This began what was to become a three year review by the Executive Branch of important policy questions before the FCC in this area. By late 1968 the Johnson Task Force had completed its work with a recommendation that a Comsat-directed pilot program be authorized. In early 1969 the FCC was prepared to authorize such a pilot program. A

report and order had been drafted, and tentative expressions of the position of each Commissioner had been made.

Before issuing it, however, then-Chairman Hyde took the document to the White House to inform the White House staff of the action the Commission was to take. In the interim there had been a change in Administration, and the information-providing trip resulted in a request that the Commission hold any action while the White House once again examined the policy questions.

The White House recommendations, for an "open-entry" policy, came in a January 1970 memorandum from Peter Flanigan to Chairman Dean Burch. In March 1970 an FCC Report and Order, 22 F.C.C. 2d 86, concluded that no decision could be made on the appropriate policy for domestic satellite entry and specific proposals from potential entrants were requested. The next Commission order, and the staff's recommended decision came in March 1972.

Today's action seems to signal the end. Open entry is adopted with certain modifications. The benefits to be realized by public broadcasting are, at this point, speculative.

There are several interesting conclusions to be drawn about the Commission's role in policymaking at least for domestic satellites.

(1) The Commission has relied heavily on the parties appearing before it for the analyses and proposals it has considered. Although there is no readily available way to make an exact calculation, I suspect that most of the important parties appearing before the Commission have invested significantly more resources, each, on these policy questions than has the Commission in total. This seems particularly true for the Executive Branch. The Commission has been a "captive," responding to and arbitrating between the variety of forces which have attempted to move it.

(2) The relative congruence between Commission action and White House recommendation, occurring over periods of significant shifts in policy, is striking. The ability of the Commission to move in variance with White House positions on important policy questions (regardless of who is President) is very questionable.

(3) The effects, benefits and costs, of both regulation and delay would be worthy of a detailed analysis. Suppose any entrant, including ABC, had been able to launch a satellite system in 1965 by merely "purchasing" the needed resources, including spectrum. Suppose the Commission had gone ahead with a pilot program authorization in early 1969. What would have been the results of these—or other alternatives—on services, technology development, and so forth? Are we better off, or worse today? Should the domestic satellite question have been handled differently, and if so, what can we learn about [865] handling other policy questions before this and other governmental agencies that engage in an economic planning function?

(4) Over and over again the Commission meets the question of melding competitive and monopoly portions of the telecommunications common carrier industry. The issues were joined in the Telpak and other bulk offering and private line proceedings, and are still unresolved. They are met again in the relationships between monopoly landline telephone companies and miscellaneous carriers who offer a variety of land mobile services in competition and monopoly in communications equipment and interconnection. They are met in the pricing questions surrounding the entry of specialized competitive carriers. And they are met here in the treatment, particularly of AT&T and Comsat, of certain entrants for domestic satellite services. The issues remain unresolved.

Given these limitations, I believe the staff work and ultimate Commission position put forward today is much better than anyone had a right to expect. Accordingly, as a realist, I concur.

Because of the significance of the policy, however, perhaps a few more words regarding my own preferred approach to decision would be appropriate.

We are entering into a new area of communications. The next few years will be years of experimentation and gathering of experience. It's not that we don't know how to launch and operate a satellite. Comsat, NASA, the military, and numerous American companies have a great deal of expertise in this field.

But we have no experience with the non-technical aspects of this operation. Will the public tolerate the short delay, or echo effect, in voice communications by satellite? What new institutional (and possibly personal) uses of communications will evolve to use the peculiar qualities of satellite distribution systems (cheaper long-haul costs, possibility of multiple distribution points, and so forth)? What problems will arise in joint operations of satellites, or of earth stations? What new ratemaking or regulatory concepts and procedures will be needed?

(1) Accordingly, I still believe there is some merit to the idea of a pilot project at this stage. Rather than have it operated by a chosen company (Comsat, AT&T, some other present company, or a new entity), however, I would have it operated by NASA or some other entity of government. This is not such a radical idea. It is the way every other nation in the world has dealt with the problem. And most have resolved the issue long before us. It is the way, in fact, that we run our space program. It is the way we evolve new technology in many areas of the economy. And, even as to space communications satellites, the military and NASA have already operated such systems.

All I would propose is that for the first generation of experience (3 to 7 years) a public entity undertake the operation of America's first domestic communications satellite system for the benefit of all potential users and operators. Every effort would be made to test, at cost, any reasonable proposal from any American company, institution, or individual. The results of all tests would be made fully open to any interested party. Training opportunities would be made avail- [866] able to as many interested persons as possible. This would save a tremendous amount of money for American business, as well as the public, and open up the possibility of a great deal more use (and competition—if that's what we're *really* interested in) when the system or systems are finally established on a commercial basis.

I have made this proposal throughout my six year term at the Commission. It has never received the support of the White House or a majority of the Commissioners. There is little doubt in my mind that we would be much further down the road today if it had been adopted in 1966.

(2) If there is not to be an experimental system, there is much to be said for a chosen instrument. A single system operator can insure economies of scale, fair and open access to all comers, the lowest possible rates, and the most geographically disbursed system (including, for example, the best service to Alaska, Hawaii and so forth).

My preference would be to create a new entity—a Domsat—for domestic satellite services only, that would have every incentive to compete fully with AT&T. No carrier would be permitted to hold stock in the company or sit on the board (although, of course, individual shareholders could hold stock in AT&T and Domsat).

Another alternative would be to give AT&T a monopoly over domestic satellite service. AT&T is now having some growing pains even keeping up with expanding service on earth. But AT&T exclusive operation in space would have the advantage that all users— including the homeowner—would get *some* benefit from the new technology, which will now flow almost exclusively to large corporate users of satellites. If this were done, AT&T should probably want to be required to provide such service through a separate corporate entity for purposes of bookkeeping (as its current corporate practices would indicate it would probably want to do anyway).

Comsat could also be the chosen instrument. It does have the expertise. But it would not have the advantage just described that AT&T would have—virtually monopoly control of all U.S. communications on the ground for purposes of rate averaging. Moreover, Comsat has additional problems as an international operative. At one time I urged that Intelsat be encouraged to become a truly international communications carrier, supplying domestic communications services for the world as well as internationally. It seemed to me an appropriate, and symbolic, peaceful venture for nations in need of one. But that idea never caught on either. So now, it seems, we are doomed to a world in which every nation must have not only its own airline, merchant marine, and steel mill, but its own domestic satellite system as well. Given such a world, however, it seems to me inappropriate for Comsat—already carrying the burdens of Big Brotherism into its international meetings—to have to confront its world partners with the potential conflicts of interest (and division of managerial energies) involved in operating the world's most lucrative domestic satellite system.

(3) If we are not to have an experimental system or a chosen instrument, because of a deistic reverence for competition, then we ought to really have competition. I'm reminded of the children's riddle: [867] "Where does an 800-pound gorilla sleep?" And the answer: "Any place he chooses." True competition is one of the most highly regulated states of economic operation possible. That's what the antitrust laws are all about—when they're enforced. You either keep the 800-pound gorilla (in this case the $18 billion Bell) out of the canary cage entirely, or you tell him where to sleep.

If we're really serious about experimenting with the radical notion of free private enterprise, I'm all for it. But then there have to be some very meaningful restraints on AT&T and Comsat—at the very least in the initial stages. Otherwise, we're just kidding ourselves—though very likely nobody else.

If we want a competitive arena I would keep out AT&T and Comsat entirely. (AT&T has never been consistently enthusiastic about using space anyway.) Let anyone else in who wants in. Let them experiment with equipment and the search for services and markets. Try to maintain some conditions of fair competition. If after a few years the Commission wants to reassess this decision, and let AT&T into the business in ways consistent with maintaining this newly burgeoning industry, fine. But not until then.

(4) Finally, I cannot but bemoan our failure to provide expressly for—at least—free interconnection for the Public Broadcasting Corporation and other educational users. I always felt that the Ford Foundation had made a fairly persuasive case that more was called for. The American people, having invested more than $40 billion in the soaring growth stock called civilian space, are entitled, someday, to a little bit of a dividend. One has yet to be declared. Ford proposed that a proportion of the savings to the commercial networks from the use of space be passed on to the public in terms of a funding source for public broadcasting. It seemed to me a fair idea.

But all this is history. We're now in countdown. It's no time to dissent. I'm on board.

Document I-27

Document title: George M. Low, Deputy Administrator, NASA, "Personal Notes," December 23, 1972.

Source: George M. Low Papers, Rensselaer Polytechnic Institute, Troy, New York (used with permission).

George Low was NASA's deputy administrator from the fall of 1970 until 1976. During that period, he dictated "personal notes" on a regular basis to record his actions and thoughts. These notes provide a fascinating record of the events of the time. As the Apollo 17 astronauts explored the Moon during the last Apollo mission in December 1972, Low, NASA Administrator James Fletcher, and other top NASA officials had to divide their time between monitoring the lunar surface activity from

Houston and meeting on NASA's fiscal year 1973 and fiscal year 1974 budgets with officials of the Office of Management and Budget (OMB) in Washington. To meet the stringent budget cuts proposed by the Nixon administration's OMB, it was Low's idea to take NASA out of communications satellite research and development (R&D). NASA had had an Applications Technology Satellite (ATS) program since 1963 to follow up its support of the initial Relay and Syncom projects, and that program had helped develop many new technologies and capabilities in the communications satellite area. The NASA decision to withdraw from communications satellite R&D meant that the final ATS mission, the ATS-G, was canceled.

[1]   December 23, 1972

PERSONAL NOTES NO. 83

[2] Fiscal Year 1973 and 1974 Budget

Meanwhile back on earth, things weren't going quite as well. On Monday, December 11, approximately at the time of the lunar landing, we received a call from Bill Morrill asking Fletcher, Lilly, and me to meet with him to discuss the FY 1973 and 1974 budgets. We tried to get him to give us our mark by telephone, but he was unable to do so. As a result, Fletcher, McCurdy, Shapley, Lilly, and I traveled back to Washington on Wednesday, December 13, roughly between EVAs 2 and 3. (We were in Houston for the full period of both of these EVAs.)

We met in Bill Morrill's office with Morrill, Young, and Taft. (We had left Houston at 5:15 in the morning on a Jet Star for Andrews [Air Force Base] and arrived in Morrill's office at precisely 9:30, the time of the appointment.) Morrill informed us that the President was determined to bring the FY 1973 budget down to a $250 billion ceiling in outlays, and to have a not too much higher number for FY 1974. As a result, all departments and agencies had to take major cuts, both in FY 1973 and in FY 1974. The 1973 cuts were particularly difficult to sustain, since only one-half year was left for money savings. In effect then, any cut made in 1973 would have double the normal effect. In NASA's case, OMB had accepted the "submarginal submission" and made drastic cuts below that level. Within the submarginal budget, we had already cut out the aircraft engine retrofit work, most of the new starts, almost all of the nuclear work, and had cut back in many other areas. The OMB mark, in addition, canceled Viking, canceled QUESTOL, delayed the Shuttle, delayed ERTS-B [Earth Resources Technology Satellite or Landsat] (did not allow ERTS-C in the Interior budget), and made further across-the-board cuts. (I should [3] have mentioned that OSO-I was also cancelled in our submarginal submission.) The net result was a budget at approximately the $3 billion level in outlays for both FY 1973 and FY 1974. We were also told that the number of cuts were policy decisions approved by the President and not ours to change. These were particularly the major ones such as Viking, OSO [Orbiting Solar Observatory], nuclear work, QUESTOL, etc. In the area of minor cuts, we would be allowed to make adjustments. The President then also asked, we were told personally, that a fairly substantial number of dollars be included in the NASA budget on the supersonic transport, with the words that he felt that this was a mandatory development for the country and that NASA should take on the fight with the Congress.

Our meeting lasted for about an hour, and following that meeting, Lilly continued to meet with Young and Taft for approximately one more hour. We then got back on our airplane and returned to Houston. We held additional meetings on the plane on the way back to Houston, in Houston the next morning, and then returned to Washington immediately after the lunar rendezvous and docking for meetings on the following Friday, Saturday, Sunday, and Monday. By Sunday noon we had firmed up our position and Monday was spent in writing the position for a reclama submission to OMB.

Within NASA we were fairly unanimous in deciding that a Viking cut would be unacceptable. First of all, Viking is the only highly visible sign of space exploration in the middle 70's. Secondly, more than half of the $800 to $900 million on Viking has already been spent. Third, it would be almost impossible to sustain the support of the scientific community for the rest of the NASA program if Viking were cancelled. For all of these reasons and many more, we decided to do our best to try to get Viking back into the budget. My first inclination was to try to cancel the ASTP [Apollo-Soyuz Test Project] mission, since from NASA's point of view it contributed least to our overall program. However, after some discussion and after some G-2ing by Fletcher, it became very clear that the President considers ASTP the highest priority NASA mission, and that any suggestion on our part to cancel this flight would be totally unacceptable. The President also considers the Shuttle the second priority NASA mission, and, we were told, would not consider cancelling [4] that project. From NASA's point of view, of course, it was clear that at a $3 billion level we would not have started either the Shuttle or ASTP. Thus, we were in a major bind.

I then suggested that it might be time to phase out of the communications business. The reasoning here goes something like this: NASA has been a catalyst for space communications development in the early phase of the space program and until now. However, there now has developed a significant communications satellite capability in private industry. For example, COMSAT/INTELSAT is spending $14 million a year on advanced R&D. It is clear, therefore, that communications work will go on whether or not NASA participates. Of course, there are some areas, such as direct broadcasting, which will take much longer without federal government participation. In other areas of applications, such as earth resources, environmental work, etc., there exists no commercial/industrial capability that will carry on if the federal government gets out of it. I, therefore, reasoned that it would be best to do one applications area well instead of doing two major areas not nearly so well. Fletcher at first was quite reluctant to accept this reasoning, but after a day or so of thinking about it, enthusiastically supported it. As a result, we decided to propose cancellation of ATS-G, to carry out ATS-F because most of the money on it was already spent, but at the same time to phase down all in-house communications R&D so that by the time ATS-F flies we will completely phase out of this business. Incidentally, this may be a major first for a government agency to get out of an R&D business of its own volition.

In the Office of Space Science we decided to keep Viking, but suspend HEAO [High Energy Astronomy Observatory]. Suspending a program is something else that has never happened in NASA before. Basically, we would keep a skeleton team together, both in NASA and in industry, for a year or more while we reviewed HEAO to determine whether we can meet its objectives at, for example, half the costs. Naugle was in favor of outright cancellation, if this were the case, but my view was that through suspension we might be able to pick the project up again without again seeking a "new start." In space science also, OSO is no longer in [5] the program as we submitted it (I will come back to that later), and there were many across-the-board cuts.

In OAST [Office of Aeronautics and Space Technology], in our basic and first submission, QUESTOL and the engine refanning were out, almost all nuclear work was canceled, and there were additional cuts in SRT/ART. In the overall SRT/ART program, I established guidelines that 90% of this work should have a promise of being relevant within a period of seven years; and that only 10% of our SRT/ART work should be in the future beyond the seven-year period.

In Manned Space Flight, Skylab and ASTP were left as they were, and the Shuttle was cut back somewhat in costs and thereby delayed by a total of one year, considering the schedule changes already made by previous 1973 expenditure cuts on top of the present cuts.

OMB also suggested major cuts in personnel totaling 1880 with the bulk of these coming at Marshall and at Lewis/Plumbrook. We have, in effect, accepted the Lewis/

Plumbrook cut because this is where all the nuclear work was going on. However, we have indicated that until we can get things sorted out, we would not accept a cut at Marshall or elsewhere at this time. We stated, instead, that a number roughly approximating the 1880 would be coming out of NASA's budget, but exactly where these cuts would be made we will determine later. In the meantime, I want to make a major effort to see whether we can "sell" the excess NASA capabilities to agencies such as EPA [Environmental Protection Agency] (for Lewis) and the DOT [Department of Transportation] (for Marshall). This is different from what we attempted last year when indicated that we would make NASA capability available as a service to these other agencies. After trying for a year to make that work, it just is clear that it won't. Instead, our intention now is to "spin off" some of the capabilities directly to other agencies so that they can develop an in-house capability.

As I mentioned before, I spent Monday, December 18, writing our reclama letter to Weinberger, and, in addition, writing a letter to Kissinger soliciting his support on Viking. Copies of the drafts of these letters are attached. In the [6] meantime, Fletcher had been working with Whitehead, Anders, and Jon Rose to get their G-2 on what was really going on in the White House, and, at the same time, he also received their free advice. Jon, who is used to dealing within the White House, felt that the letters that I had written might make their mark with OMB but he really felt that they were needed with Erlichman and Flanigan and were not suitable for that purpose. Accordingly, he rewrote both letters just before Fletcher had a meeting with Weinberger and Morrill on December 19 (I was back in Houston at that time). A copy of their rewrite is also attached. There were no changes in substance with one exception: the engine refan program was back in the words but not back in the budget. This is a program where a great deal of pressure has been applied to the Vice President's office and Bill Anders would, therefore, like to see it back in the budget. We indicated to OMB that we would certainly undertake the project if additional money were added over and above the mark for this purpose. At the time of this writing, it is quite probable that this money will be added. I forgot to mention that Bill Anders met with us on the 18th, and that we engaged in a very significant philosophical argument with him. It is Bill's opinion (shared apparently by all White House staffers) that NASA's main objectives should be to explore and to provide launch services. Subjects such as applications and science we should only do as a service for others, and, therefore, should seek their funding, e.g., user agencies or NSF [National Science Foundation], for this purpose. Both Fletcher and I engaged in a fairly vehement argument with Anders on this point. Although I don't think we persuaded Anders, at least he knows where we stand.

Our budget submission as revised, was only approximately $50 million over the OMB mark for both FY 1973 and FY 1974. Weinberger was apparently quite pleased with our proposals, and it is quite probable that they will be accepted. However, at the time of Christmas weekend we have not yet heard positively that our proposals have been accepted or that the NASA budget is locked up. As a final afterthought, Fletcher went back to Weinberger and asked him whether it wouldn't be possible to reinstate OSO. The reasoning is that this might be a minor concession to make to the scientific community. [7] This reinstatement, of course, we could only make with additional funding. This, too, is an open item at the time of this writing.

Document I-28

Document title: Committee on Satellite Communications, Space Applications Board, Assembly of Engineering, National Research Council, "Federal Research and Development for Satellite Communications," 1977.

Source: National Research Council, National Academy of Sciences, Washington, D.C.

By 1975, critics were beginning to call into question NASA's decision to withdraw its support from the research and development (R&D) of communications satellites. Such critics argued that the nation needed a government program to investigate related technologies in their early stages of development, as well as to explore technologies and applications that were not clearly linked to private-sector objectives. In response, NASA contracted with the independent National Research Council to conduct a study on whether there was a justifiable federal role in communications satellite R&D. The study concluded that there was definitely a need for such a program (excerpts of the study report are included here). This conclusion formed one of the bases for NASA going ahead with the controversial Advanced Communications Technology Satellite (ACTS) program; ACTS was finally launched in September 1993.

[i]

Report of the
COMMITTEE ON SATELLITE COMMUNICATIONS
of the
SPACE APPLICATIONS BOARD
ASSEMBLY OF ENGINEERING
NATIONAL RESEARCH COUNCIL

Federal Research and Development
for Satellite Communications

Published by
NATIONAL ACADEMY OF SCIENCES
WASHINGTON, D.C.
1977 . . .

[no page number]   PREFACE

In January 1973, the National Aeronautics and Space Administration (NASA), faced with the necessity of reducing expenditures, examined its programs to determine what could be eliminated. While NASA made a number of reductions, one of interest to this study was the decision to essentially eliminate its satellite communications activities because this was felt to be a relatively mature field and NASA believed that R&D in support of future activities could be provided by the communications industry. Since January 1973, several organizations have assessed the consequences of that decision and have urged that the decision be re-examined.[1]

In late 1975, NASA asked and the National Research Council agreed to study further the question "Should federal research and development on satellite communications be resumed and, if so, what is the proper federal role in this field?" To undertake the study, a Committee on Satellite Communications (COSC) was formed under the auspices of the Space Applications Board (SAB). This report presents the Committee's findings; signifi-

---

1. "The Federal Role in Communications Satellite R&D," American Institute of Aeronautics and Astronautics, New York City, 1975; "The NASA R&D Program on Satellite Communications," A Position Paper of the Satellite Telecommunications Section, Communications and Industrial Electronics Division, Electronic Industries Association and the Government Products Division, Electronics Industries Association, Washington, D.C., 1974; untitled paper, Aerospace and Electronic Systems Group, The Institute of Electrical and Electronic Engineers, Inc., Washington, D.C., 1976.

cant background information and working papers assembled by the Committee during its deliberations will be published separately.[2]

[1]   INTRODUCTION

In the one hundred years since the invention of the telephone, telecommunications has become a pervasive part of the developed world. The telephone is in nearly every home and in every office in the United States, and there is about one telephone for every ten persons on earth. Radio broadcasting and other radio links have become common-place tools for providing both entertainment and services. Television provides entertainment, news, and educational services to most homes in the technologically developed countries of the world. There remain, however, some troubling limitations to further improvements in communications services. For example, the cost of providing telephone or TV service by conventional means is high in remote and sparsely populated regions. Thus, the Rural Electrification Administration has made and guaranteed about $650 million in federal loans annually to stimulate an extensive rural telephone service now serving 3.1 million subscribers in 47 states.

High frequency radio is widely used to span great distances but suffers from outages caused by solar disturbances of the ionosphere. As a result, ships and aircraft are frequently out of communication with their bases for long periods or during critical phases of their journeys. High frequency radio is also severely spectrum-limited and its use is largely confined to the provision of voice and low-speed data services. First steps in improving ship communications began in 1976 with the launch of COMSAT General's MARISAT satellites which now provide urgently needed, reliable services to U.S. Navy and commercial ships in the Atlantic and Pacific Ocean basins.

Nineteen years ago when the first satellites were launched, it was clear that they could serve as high-altitude relay stations and thus overcome some of the limitations of terrestrial communications systems. First efforts involved bouncing radio signals from orbiting balloons and even from earth's natural satellite, the moon. Another approach involved the use of a receiver-transmitter, called a transponder, in a satellite to relay signals from one distant point on earth to another. Early efforts using low-altitude satellites showed the feasibility of the transponder technique, but such satellites had short orbital periods, did not remain within sight of the earth stations at all times, and required that earth stations continuously track those satellites in view.

The promise of communications via satellite was realized with the use of satellites in geostationary orbits at an altitude of 36,000 km [kilometers]. At that height, the orbit period, synchronized with the earth's rotation, places the satellite in an essentially stationary position above a selected point on the equator and within line-of-sight of about one-third of the earth's surface. This possibility [2] for providing continuity of service and solving the tracking problem was pointed out by Arthur Clarke[3] in 1945 and first achieved by NASA's SYNCOM in 1963.

In 1963, the U.S. Congress established the Communications Satellite Corporation (COMSAT) to bring about a commercial international satellite communications system as quickly as possible and to represent the U.S. in the International Telecommunications Satellite Organization. International satellite communications service began in 1965 with

    2.   *Federal Research and Development for Satellite Communications: Working Papers.* Committee on Satellite Communications of the Space Applications Board, National Research Council. National Academy of Sciences, Washington, D.C., 1977.
    3.   Clark, A.C. "Extraterrestrial Relays," *Wireless World.* October, 1945, pp. 305–308.

INTELSAT I which could carry 240 telephone channels or one TV channel. INTELSAT II, III, IV, and IVA satellites were added in subsequent years. As of January 1977, the system provides telephone and TV links between the 94 countries that share in ownership of the system. There are also 13 non-owner countries that use the INTELSAT system.

Use of satellites for domestic communications within the U.S. was delayed by political and regulatory processes until 1974 when policy decisions were made about who would provide such services. Meanwhile, Canada's ANIK satellite system had become operational in early 1973, supplying some U.S. domestic services. Since then, a number of companies have entered the field and today satellites are being used to provide domestic telephone or TV services. Additional domestic satellites are planned for the U.S. and for other countries.

Since 1963, the United States has led the world in satellite communications. Initial experiments were conducted by the National Aeronautics and Space Administration and the Department of Defense. Transition from experimental to practical use of satellites was rapid for transoceanic telephone and TV services because there existed an infrastructure ready to exploit this new medium and because the number of new undersea cables was unable to keep pace with the demand. U.S. aerospace and electronic industries were able to capitalize on their own work as well as on the research and development funded in these industries by the federal government to develop a competitive advantage in the world market.

The private sector has continued to make advances in the technology for providing conventional telephone and TV services. The industry has taken some risks; for example, one company paid for launch vehicle improvements and incorporated much advanced technology, not previously proven in flight, in its satellite to improve performance. However, it became clear that the risk the private sector was willing (or could permit itself) to take was limited and that most private initiatives were being channelled to existing markets and to where technical risks were not perceived as unacceptably high. It is clear that even in the largest companies, prudent management requires that large investments in R&D not be made unless there is reasonable assurance that relatively short term payoffs will result. Furthermore, the risk of violating federal anti-trust and trade regulation statutes has led companies to refrain from entering into joint efforts that might permit them to share risk. As a result, following the withdrawal of the federal government from satellite communication R&D, there have been no commercial experimental satellites to test new techniques and concepts or to permit users to experiment with new services.

There are a number of potential communications services, such as for health care delivery, educational services, search and rescue, electronic mail, teleconferencing, and environmental data collection, which apparently cannot readily [3] or economically be provided using the technology available to the common carriers for producing conventional telephone and television services. If the option to initiate some of these services is to remain open in the future, then advances must be made in needed technology by undertaking research and development programs now.

There are examples of work which must be undertaken if new services are to be contemplated. These include technology for utilizing new portions of the radio frequency spectrum, employing larger and more sophisticated spacecraft antennas, utilizing a satellite as a switchboard in space, and advancing technology to drive down the cost of communications.

As time passed, many concerned with the development and the future of satellite communications came to realize that NASA's 1973 decision to reduce R&D in the field might indeed close options if advancements in technology such as those just cited did not become available. Mindful of this, NASA, in the fall of 1975, asked the NRC to conduct a study of the federal role in satellite communications research and development. The NRC agreed on October 7, 1975, to undertake the study and decided that the work should be done by a new Committee on Satellite Communications (COSC) under the NRC's Space

Applications Board. It was also agreed that the Committee should be constituted of technologists, communications system operators, satellite communications users, a communications policy specialist, and a regulatory economist. The members were selected with due regard for a balance in viewpoints. . . .

In its work, the Committee considered whether it is likely that satellites in geostationary orbits could make voice, video, and data communications attractive for a variety of public uses not presently provided. Such satellite systems should be able to provide new services to remote and distant places and to sparsely distributed users. For example, using the ATS-6 satellite, Brazil has experimented with delivering television broadcasts to some of its isolated populace. The U.S. has experimented with providing health care information and educational services to inhabitants of remote villages in Alaska, Appalachia, and the Rocky Mountain West. When the ATS-6 was withdrawn from such experiments to keep an international commitment to conduct similar demonstrations in India, a number of user groups testified to their need for the replacement satellite which NASA had planned to launch.[4] However, funds to complete and launch the replacement satellite were not appropriated and no individual user or combination of users was able to afford the estimated $45 million to $50 million to launch and operate it. While the cost-effectiveness of any single application of this type by a satellite may be questionable,[5] the use of multi-purpose satellites may open an increasing number of opportunities for public service, government, and commercial uses.

[4] Among the non-technical questions confronting the Committee, therefore, were these: Are there a large number of disaggregated, mainly public service users in remote places likely to need and want the capabilities of satellite communications? Is an experimental program, building on the experience of the curtailed ATS-6 experiments, warranted to permit users to evaluate the worth of such services and to demonstrate the market and the costs? If so, what should such a program comprise and what should be the respective roles of the government, the communications industry and the potential public service sector users?

Collectively in Committee meetings and individually outside of those meetings, the members of COSC: (1) reviewed the history and present status of satellite communications, (2) considered a number of important communications service needs expressed by potential users, (3) identified advances in technology required for meeting those needs, (4) judged which of those advances probably would, and which probably would not, be met by the private sector, (5) structured and evaluated several possible NASA roles in the advancement of technology, and (6) decided upon recommendations.

[5] PERCEIVED NEEDS AND REQUIRED TECHNOLOGY

PERCEIVED NEEDS

The government investment in research and development on multi-channel point-to-point satellite communications, which began with the space age and culminated in the formation of the Communications Satellite Corporation, clearly has borne rich dividends for the country. The revenues from this new industry currently exceed $200 million per year and are expanding rapidly. It was only after the Department of Defense (DOD) and

---

4.   U.S. Senate Committee on Aeronautical and Space Sciences. Hearings on S.3542, A Bill to Authorize Appropriations to the National Aeronautics and Space Administration for Research and Development Relating to the Seventh Applications Technology Satellite, July 23, 1974.

5.   See Educational Policy Center, *Instructional Television: A Comparative Study of Satellite and Other Delivery Systems.* Syracuse Research Corporation, Syracuse, New York, 1976.

NASA had developed the technology and demonstrated its practical use, however, that commercial firms were able to risk operational systems. Today the price of multi-channel point-to-point voice service has dropped to several thousand dollars per channel-year. Both transoceanic and domestic systems are in operation or planned in a large number of countries.

The situation for other classes of long-range satellite communications—for example, service to mobile platforms (ships and aircraft) or to widely distributed or remote ground locations—is much less favorable. Most users of such communication terminal installations feel they can afford only modest sized and low-cost antennas. The services so provided might include public activities such as education, mail, environmental monitoring, geophysical exploration, hazard warning, health care delivery, navigation aids, time and frequency dissemination, public safety, search and rescue, or wildlife monitoring.

The U.S. Department of Health, Education and Welfare and NASA have recently conducted experiments in Appalachia, the Rocky Mountain States, Alaska, and Washington State.[6] These experiments were designed to assess the value of service to remote locations and to assess the communications satellite as a means for providing it. For example, using television, voice, and a variety of data signals relayed by ATS-6 (Applications Technology Satellite 6), the experiments delivered health care and education services to thousands of Alaskans living in [6] areas too remote to reach readily in person or through ground-based communications.[7][8]

These experiments successfully demonstrated the capability to provide diagnostic consultative services between medical professionals and paraprofessionals, transmit and provide consultations on x-rays, and transmit and up-date medical records, all in real-time via satellites. As a result, the Alaska Native Health Board now assigns highest priority to development of the community health aide program and to improving the communications that provide the aides with professional back-up.[9]

The Public Service Satellite Consortium[10] has compiled the needs of numerous current and potential users similar to those portrayed in the Alaska example, but the fact is that most potential users cannot afford current communication service prices, much as the transoceanic point-to-point users could not afford early satellite communications systems before technology advances brought lower prices. If prices could be reduced, an increased market for such services might well develop.

REQUIRED TECHNOLOGY

The technical challenge in reducing costs for satellite service to small terminals is difficult, but it is no greater than that faced in originating satellite communications in 1958. The basic approach already can be envisioned.[11] To enable small antennas to be used at

---

6.   Marion H. Johnson, "ATS-6 Impact: A View from the Control Room," *National Library of Medicine News.* Vol. XXX, No. 10-11, October-November, 1975, pp. 3–7.

7.   Charles Brady, "Telemedicine Moves North to Alaska," *National Library of Medicine News.* Vol. XXX, No. 10-11, October-November, 1975, pp. 7–10.

8.   Martha R. Wilson and Charles Brady, "Health Care in Alaska Via Satellite," AIAA Conference on Communication Satellites for Health/Education Applications, AIAA Paper 75-898, New York, 1975.

9.   Subcommittee on Appropriations for the Department of the Interior and Related Agencies, U.S. House of Representatives. Testimony on behalf of the Alaska Native Health Board by Lillie H. McGarvey, May 13, 1975.

10.   The Public Service Satellite Consortium is a private organization dedicated to aggregating the public services satellite market. Its subscribers number more than 65 state, local and regional organizations currently conducting over 20 public service satellite communications experiments with the NASA ATS-series satellites and the NASA/Canadian Communications Technology Satellite.

11.   Walter E. Morrow, "Current and Future Communications Satellite Technology," Presentation to the International Astronautical Federation 26th Congress, Lisbon, September 1975.

earth terminals, high-gain satellite antennas must be employed. To be economical, these must be shared by large numbers of users at many locations. Many antenna beams from a single satellite will [7] be required, along with methods for accurately aiming the antenna and a means for switching signals from one beam to another by means of a switching system aboard the satellite.

High Gain Spacecraft Antennas

The possibility of high gain (large) spacecraft antennas seems antithetical to the notion of spacecraft weighing, at most, a few thousand kilograms. (The standard 25-meter ground antennas weigh hundreds of thousands of kilograms.) There is one large difference, however, between the surface of the earth and space; namely, in the absence of gravity and wind forces, large space antennas can be built using very light structures.

The NASA ATS-6 spacecraft incorporates a 10-meter parabolic antenna that weighs less than 100 kg [kilograms] and is operable to 10 GHz [gigahertz]. This antenna consists of a series of sheet aluminum ribs on which is stretched a metallized net. During launch, the antenna is packed into a small container by wrapping the ribs and mesh around a central hub. Upon reaching orbit, the ribs are released whereupon they unwind into their deployed position.[12] Other designs need investigation with the objectives of further reducing weight, increasing performance, and increasing size.

Multiple Beams

One difficulty with high gain spacecraft antennas is that they produce very narrow beams and therefore have limited coverage on the earth's surface. For instance, the ATS-6 10-meter antenna has a beamwidth of about 1° at one of the operating frequencies, 2.6 GHz. If such an antenna is to be usefully employed over the earth's surface visible to the satellite, it will be necessary to generate a total of about seventy-five beams and to share the spacecraft antenna aperture among these many beams.

As an example, the Massachusetts Institute of Technology's Lincoln Laboratory developed a 10 GHz lens antenna about 0.75 meter in diameter, illuminated by 19 feed horns and producing 19 beams—which in the case of this antenna will just cover that part of the earth visible from geostationary orbit. The satellite transmitter can be connected by command to any combination of the feed horns. The entire antenna system weighs less than 20 kg. Similar arrangements might be made for large parabolic reflector antennas. In that case, a cluster of antenna feeds would be located at the focus of the parabola. Further development of these concepts is needed both to achieve the proper performance over the required bandwidth and to minimize effects of the space environment such as extremes of temperature.

[8] Precision Antenna Aiming

With today's technology, aiming an antenna in space to a precision of 0.1° is relatively easy. However, the high gain antennas of anticipated future spacecraft will have beamwidths of 0.1° to 0.5° and will require a pointing precision of 0.01° or better. It is advantageous to attach the antenna rigidly to the spacecraft and aim the structure as a

---

12.   Computer Sciences Corporation. *NASA Compendium of Satellite Communications Programs.* Report of Work on Contracts NAS 5-24011 and NAS 5-24012. Computer Sciences Corporation, Silver Spring, Maryland, 1975, pp. 13-59 to 13-81.

whole. To point the beam accurately, the satellite's location in space must be known, the directional vector to the earth determined, and then pitch, roll and yaw maneuvers performed. The spacecraft location can be determined by means of a series of ground-based observations of satellite range and range rate or by means of an on-board sensor system. One on-board system, in a Lincoln Experimental Satellite, used a precision chronometer and visual and/or infrared sightings of the sun and the earth's edge. The satellite location was determined by noting the time at which the observed angle between the sun and earth reached a given value.

A spacecraft with a large antenna can be turned in space by means of an onboard momentum wheel or wheels. By speeding up or slowing down the wheel, pitch maneuvers can be made. Pivoting of the wheel axis can produce roll and yaw motions. The spacecraft must also be kept in proper orbital position. This is often accomplished by hydrazine-fueled thrusters. Ammonia thruster systems can also be used and electronically powered thrusters have been considered. Current aiming techniques need to be improved and additional research and development initiated to provide simple and accurate systems.

On-Board Message Switching

The use of multiple beam high-gain satellite antennas will permit the use of small terminals. On the other hand, the problem remains of how to interconnect users on different beams. One solution would be to collect the signals from the various beams and transmit them on a very wide-band downlink to a large ground terminal. The interconnection could then be made by conventional switching equipment and the signals returned to the spacecraft on a wide-band link with each signal addressed to the proper downlink beam. This solution, while permitting the complex switching equipment to be located on the ground, would require additional very wide-band channels in the already crowded radio frequency spectrum. Much more power would be required in the satellite and the existing 0.25 second time delay would be doubled.

Another solution would be to perform the switching in the satellite. On[-]board switching can be done in several ways. While switching at radio frequency would avoid the complexity of demodulation, time sharing in the use of the downlink transmitter would be very difficult.

An alternative is demodulation of the up-coming signals to identify on which beam the down-going signals must be placed to reach the intended recipients. Recent advances in high-speed digital signal processors offer encouragement that on-board switching is possible. Much research and development is needed to arrive at practical solutions and experimental verification in flight will be necessary before the communications industry can risk operational use.

[9] Higher Satellite Power

A way to increase satellite capacity or achieve a given capacity with low cost ground stations is to increase the satellite transmitter power. The transmitter power output is the product of the available prime power and the efficiency of the transmitters.

There is relatively little possibility of increasing the 60% efficiency of current satellite solid-state transmitters operating at frequencies up to 2.0 GHz. At frequencies above 2.0 GHz, travelling wave tubes with efficiencies of up to 40% are commonly used and improvements in efficiency should be possible.

Significant advances in the performance of prime power systems should be possible. Most current satellites employ silicon solar cell power systems having efficiencies as low as 10%. The lightest weight arrangement involves solar-oriented planar arrays having about

20 watts of power per kilogram. New designs having more efficient cells on lightweight flexible substrates should be able to produce 50 watts per kilogram.

It may also be possible to develop even higher power per unit weight by means of larger solar array structures or deployed parabolic solar concentrators which could be used with either solar cells or perhaps Brayton closed-cycle turboalternators. These means for achieving larger satellite capacities and thus lower earth station costs require new technology in prime power devices, in structural efficiency, and in the high power transmitter devices themselves.

Modulation Systems

Most contemporary systems employ analog frequency modulation voice and TV transmission. For FM voice systems, a 50 dB [decibel] power signal-to-noise ratio in a one-cycle band is required. Digital speech transmission systems operating at 2400 bits per second with very efficient modulation systems have been demonstrated to operate at power signal-to-noise ratios of about 40 dB. While currently these digital systems are far too costly to be used in inexpensive mobile terminals, recent advances in the reduction of the cost of digital equipment indicate the possibility of low-cost voice systems operating at significantly lower signal-to-noise ratios.

OTHER TECHNOLOGY AND PHENOMENOLOGY

Other improvements are needed in satellite support systems. Typical of these needs are those for lighter, longer life (nickel-hydrogen) batteries and station-keeping engines (ion engines). Better understanding is also needed of certain space phenomena such as static discharges at geostationary orbit and the effects of rain on the polarization of radio signals. It should be noted that AT&T's COMSTAR satellite carries radio propagation experiments at 18 GHz and 30 GHz. These experiments, although singular, are typical of the many experiments needed to better understand potentially limiting natural phenomena. . . .

[29]   CONCLUSIONS AND RECOMMENDATIONS

The Committee, in its deliberations, reviewed a number of future communications needs which potentially could be satisfied by satellite systems. These included needs in fields such as education, health care delivery, hazard warning, navigation aids, search and rescue, electronic mail delivery, time and frequency dissemination, and geophysical exploration. Many of these are public service needs which might be satisfied by satellite communications systems using high power and a high-gain antenna in the space segment, permitting low-gain, low-cost earth stations. To make such systems possible, technological advances in multibeam spacecraft antennas, low-cost earth stations, large satellite power systems, high-speed spacecraft communications switches, and spacecraft supporting technology may be required. If costs can be reduced by the application of new technology, many potential public service users may benefit from new satellite communications services.

The Committee concludes that the technology to meet such needs is often not provided by the private sector because of the technical and cost risks involved. The Committee therefore concludes that there is an appropriate federal role and that NASA should resume the research and development activities needed to provide the new technology. . . .

As discussed earlier in this report, it became clear as the Committee progressed through its deliberations that it would be neither possible, nor appropriate, for a part-time, short-duration committee to undertake an exhaustive study of the future needs of the country in satellite communications and then to make detailed recommendations on the basis of such a comprehensive study. Instead, the Committee focused upon classes of possible NASA programs (called "options" in this report) and, accordingly, the Committee's conclusions and recommendations are focused on the options considered.

The Committee concludes that the current NASA satellite communications program (Option 1) is inadequate, both in terms of meeting NASA's statutory advisory obligations and in terms of meeting the country's needs in satellite communications research and development. Some members, but not all, felt that if this option were the only one that the nation was willing to support, NASA should drop out entirely of the satellite communications research and development business, and that legislation should be sought which would terminate NASA's statutorily mandated advisory responsibilities in satellite communications.

The Committee believes that the extra funding required to support an expanded NASA satellite communications technology program (Option 2) is not likely [30] to produce enough returns of value to the country to make it worthwhile pursuing, and therefore recommends against it.

Option 3, a satellite communications technology flight-test support program, has considerable appeal in that it is directed at removing a major roadblock in the way of increased private sector investment in satellite communications research and development. Such a program would face many difficulties in deciding fairly who should be provided such opportunities and in resolving questions of access to results, patent protection, government rights, and proprietary rights, to name a few. The Committee therefore is skeptical of the likely efficacy of such a program and recommends against pursuing it— even if undertaken in conjunction with Option 1.

> *The Committee recommends that NASA implement an experimental satellite communications technology flight program (Option 4) using the safeguards provided by the first two phases of the decision process discussed in the preceding section.*

That procedure is intended to ensure that the communications technology program is responsive to the perceived needs of the entire satellite communications community, including, in particular, potential users of the services. In addition, it is believed that following this procedure will help foster better transition of the experimental results into subsequent operational systems.

It seems clear to the Committee that there are a number of potential public service satellite communications systems which should be investigated in detail for possible implementation. However, as discussed in the preceding chapter, the Committee also believes firmly that NASA should pursue such a program only if one or more potential user groups are involved from the start of the program through its finish, and only if the estimated costs and benefits are thoroughly investigated and the balance indicates the pursuit of the program is worthwhile.

> *The Committee recommends that NASA implement an experimental public service satellite communications system program (Option 5), provided that the program is carried out using the entire four-phase decision process discussed in the preceding section.*

The Committee concludes that the arguments against an operational public service satellite communications system program (Option 6) are compelling, that such an option is inappropriate for NASA, and recommends against it.

In summary, the Committee on Satellite Communications concludes that there might well be a number of public service communications needs which satellite communications systems of the future could help satisfy. Some of these services and systems may require the development of technology such as multi-beam spacecraft antennas, low-cost earth stations and on-board signal switching—technologies which do not readily derive from current or anticipated future activities of the private communications common carriers. In addition, because of the disaggregated nature of those who need these services, the private sector often cannot find a ready market which justifies the risk of expansion into the provision of these new services. There is, then, an appropriate federal role in [31] assisting the development of needed technology and in demonstrating new public services for a sufficient period that their users may be perceived as a viable market by the private sector. The most appropriate supplier of the needed technology is NASA.

> *The Committee recommends that as soon as possible, NASA, with the participation of appropriate user groups, begin conceptual definition of both the needed technology (Option 4) and the public service experiments themselves (Option 5).*

These initiatives are the first steps in the implementation of the Committee's Options 4 and 5 which have been described earlier in this report. The report also describes a process of checks and balances which the Committee believes are essential to channel the expanded NASA role in the needed direction.

Document I-29

Document title: John J. Madison, Legislative Affairs Specialist, NASA, Memorandum for the Record, "Advanced Communications Technology Satellite (ACTS) program meeting, October 13, 1983."

Source: NASA Historical Reference Collection, NASA History Office, NASA Headquarters, Washington, D.C.

Beginning in 1980, NASA reentered the communications satellite research and development area, first with a technology development effort and then with a proposal for a satellite mission that would demonstrate various new technologies and their ability to work together as a system. This mission, known as the Advanced Communications Technology Satellite (ACTS), was controversial within the government. The Reagan administration believed that it was the private sector's responsibility to invest in technology and demonstrate its capabilities in areas where the primary payoffs would be commercial. For several years during the mid-1980s, President Reagan refused to approve funding for NASA to develop ACTS. Hughes, the world's leading builder of communications satellites, also opposed the program on the grounds that it represented inappropriate government competition with the private sector. Congress, with a Democratic majority, believed in a partnership between the public and private sectors would assure the nation's continued leadership in the communications satellite sector. Each year, for several years, Congress restored funding for the mission to NASA's budget. It was not until early 1987 that the Reagan administration reversed its opposition to the program, allowing it to move forward. ACTS was finally launched in 1993. This memorandum captures the early NASA justification for the program, as presented in a meeting of two staff members of the space subcommittee of the House of Representatives (Rad Byerly and Tim Clark), head of the ACTS program at NASA Headquarters (Robert Lovell), and NASA's legislative affairs specialist (John Madison). The program's rationale and design underwent continual change until a program concept acceptable to both the executive branch and Congress was developed.

[no pagination]

Memorandum for the Record

SUBJECT: Advanced Communication Technology Satellite (ACTS) program meeting, October 13, 1983

PRESENT: Committee staff: R. Byerly, T. Clark;
          NASA: R. Lovell and J. Madison

The purpose of the meeting was to review the status of the NASA ACTS program.

Dr. Byerly started the meeting by asking a number of questions about the rational for NASA's recommitment to develop a second generation satellite communications technology base for industry. The following points were established by the ensuing discussions:

- The current global shift form [sic] an industrial to an information-based economy is creating a rapidly increasing demand for capacity that cannot be met by the satellite communication technology base developed by NASA and industry during the period of 1962 to 1973.
- The satellite communications industry is not monolithic; individual sectors like the hardware manufacturers, the common carriers, the antenna manufacturers, the entrepreneurs who buy transponders one-at-a-time and resell them and other[s] have had little interest in maintaining an advanced technology base.
- The U.S. competitive edge in the world market has been substantially eroded by the transfer of technology to foreign manufacturers; in the U.S., only two out of five former leaders in the world market remain competitive.
- To provide the capacity to meet the forecasted demand, NASA and industry over the past five years focused research on the precursor technologies to an experimental system like ACTS; the technologies include frequency reuse through spot beams, on-board switching and regeneration, data compression, modulation and demodulation and beam hopping; there is now a good understanding of the technical risks related to most of the technologies.
- The ACTS program provides for the testing of many of these technologies in an experimental network that could be applied to the next generation of geostationary communications satellites. It's [sic] objective is to restore the preeminence of the U.S. industry in satellite communications.

Dr. Byerly and Mr. Lovell discussed the NASA effort that supports the ACTS program. It consists of two activities. One involves fundamental research aimed at developing the devices and processes that support an advanced components development activity and some highly sophisticated components which are beyond the technical level of ACTS. The second is directed toward developing components that will reduce some of the technical risk related to the ACTS experimental flight systems.

Dr. Byerly inquired into the status of the ACTS program. Mr. Lovell indicated that NASA was still involved in the source selection process. One proposal was submitted in response to the RFP [request for proposals] which was issued in March 1983. The proposed industry team is composed of RCA, TRW, COMSAT, Hughes and Motorola. RCA would act as the prime contractor responsible for the satellite bus and the integration of the ACTS payload. The total estimated cost of the ACTS program is $354.0 million. Industry will contribute to the cost of the program.

Some discussion about the incentives for industry to participate in the ACTS program followed. The principal motivators are: a $10-$15 billion per year commercial communications market in the 1990's, an opportunity to be the beneficiary of a good technology transfer mechanism and good protection of proprietary data amid an activity including a number of competitors.

No additional items were reviewed. Dr. Byerly requested a two page programmatic description of the ACTS program.

John J. Madison
Legislative Affairs Specialist

Document I-30

Document title: William Schneider, Under Secretary of State for Security Assistance, Science, and Technology, and David J. Markey, Assistant Secretary of Commerce for Communications and Information, "A White Paper on New International Satellite Systems," Senior Interagency Group on International Communication and Information Policy, February 1985.

Source: NASA Historical Reference Collection, NASA History Office, NASA Headquarters, Washington, D.C.

This report sets forth the reasoning behind the November 28, 1984, determination by President Ronald Reagan that "separate international communications satellite systems are required in the national interest." (A copy of the determination is included in the excerpts from this report.) This decision marked the end of an era during which it was U.S. policy to protect the position of INTELSAT as the only provider of global point-to-point communications via satellite.

A White Paper on New International Satellite Systems

Senior Interagency Group
on International Communication
and Information Policy

William Schneider, Jr.
Under Secretary for Security
   Assistance, Science, and Technology
U.S. Department of State

David J. Markey
Assistant Secretary for
   Communications and Information
U.S. Department of Commerce

February 1985

[1]   Introduction

Since 1983, several U.S. firms have filed applications with the Federal Communications Commission (FCC) to establish international communications satellite systems in addition to the global system owned by the [International] Telecommunications Satellite Organization (INTELSAT). Orion Satellite Corporation, International Satellite, Inc. (ISI), and Cygnus Corporation propose new transatlantic communications systems, and RCA American Communications, Inc. (RCA) has applied to use capacity on a U.S. domestic satellite to provide international service. Pan American Satellite Corporation (PanAmSat) proposes to establish a system which would serve Latin America. In addition to existing and planned regional satellite systems independent of INTELSAT, other transoceanic satellite systems are under consideration abroad. Approved and proposed transatlantic submarine cable communications facilities, many of which are actually or potentially competitive with INTELSAT, are pending as well.

Focus of Report

The filing of U.S.-based satellite system applications with the FCC prompted action by the Executive branch, which has special responsibilities in this field under the Communications Satellite Act of 1962, as amended (47 U.S.C. 701 et seq.) including the responsibility to determine whether additional U.S. international satellite systems are "required in the national interest." The Senior Interagency Group on International Communication and Information Policy (SIG) reviewed U.S. international satellite policy to determine whether, and under what conditions, authorizing satellite systems and services in addition to INTELSAT would be: (a) consistent with prevailing U.S. law, practice, and international treaty obligations; (b) compatible with sound foreign policy and telecommunications policy goals; and, (c) in the U.S. national interest.[1]

[2] The Executive agencies represented on the SIG undertook a study and reached a unanimous position in favor of new entry, subject to certain limitations. A recommendation subsequently was made to the President by the Secretaries of State and Commerce. The President determined on November 29, 1984, that international satellite systems separate from INTELSAT were required in the U.S. national interest, subject to certain conditions. Specific criteria relating to the President's determination were then forwarded to the FCC by the Secretaries of Commerce and State jointly. See Appendixes A and B.

This report provides background information regarding the President's determination, and it also provides information on important regulatory and other parallel measures which are desirable to ensure that the Executive branch's fundamental policy goal—an efficient and responsive international communications environment—is achieved. The discussion here focuses on the major communications and information policy issues raised by the applications before the FCC. It addresses commercial, trade, and legal matters, and also examines major U.S. foreign policy interests and concerns.

This report does not seek to resolve all of the questions that have been raised regarding new international satellite systems nor to direct action by the FCC on specific pending applications. It does, however, consolidate much of the extensive analysis that has been undertaken by the Executive branch and sets forth the requirements applicable to any system the FCC may eventually authorize.

The Executive branch has concluded, in brief, that it is technically feasible, economically desirable, and in the national interest to allow new entry by U.S. firms into the international satellite field. Customers should be afforded both the new service options and the benefits of competition among customized service providers that new entry promises. This can be accomplished, moreover, while maintaining the technical integrity of the INTELSAT global system and avoiding significant economic harm to that system. U.S. foreign policy, and international communications and information policy, require a continued strong national commitment to INTELSAT as "a single global commercial telecommunications satellite [3] system as part of an improved global telecommunications network."[2] But our national commitment to INTELSAT and other important goals can be accommodated, provided that new international satellite systems and services are authorized and regulated along the lines discussed in this report.

---

1.   The SIG is composed of representatives of the Departments of State, Justice, Defense, and Commerce; the Offices of Management and Budget, Science and Technology Policy, Policy Development, and the U.S. Trade Representative; the National Security Council; the Central Intelligence Agency; the U.S. Information Agency (USIA); the Board for International Broadcasting; the Agency for International Development; and the National Aeronautics and Space Administration. Commerce and State co-chair the SIG and USIA serves as vice chair.

2.   Preamble, Agreement Relating to the International Telecommunications Satellite Organization "INTELSAT," TIAS 7532, 23 UST 3813, 3814 (1973).

Specifically, this report concludes that—

(a) Additional international satellite facilities should be permitted by the FCC, provided they satisfy conventional regulatory requirements, but the new entrants must be restricted to providing customized services, as defined in this report. When one or more authorities abroad authorizes use of such new systems, the United States with those authorities will enter into consultation procedures with INTELSAT under Article XIV(d) of the INTELSAT Agreement. Construction permits may be issued at the conclusion of regulatory proceedings to those applicants meeting the public interest requirements of the Communications Act. Final licenses and authorizations should not be issued, however, until after INTELSAT consultation is completed.

(b) The FCC should examine allowing U.S. carriers and users in addition to the Communications Satellite Corporation (Comsat) to have cost-based access to the INTELSAT space segment for customized services. This matter can be pursued on a parallel track, as the pending applications are being processed, however, and does not constitute a condition to FCC action on these applications.

(c) The United States should, and will, maintain its full commitment to INTELSAT, while permitting technology-driven competition in this important sector to evolve. . . .

[50]   Conclusion

The applications to establish additional international satellite systems now pending before the FCC presented four options. The Executive [branch] could have recommended (1) approval, (2) denial of the applications outright, (3) approval of the applications subject to specific qualifications, or (4) further study, with postponement of any decision for an indefinite period. The unanimous view among the member agencies represented on the SIG is that it would be in the U.S. national interest to allow new providers of international satellite facilities, provided INTELSAT were not exposed to significant economic harm. The President's determination reflects this view.

There is sufficient risk of significant adverse economic impact on INTELSAT to make blanket approval of unrestricted competition unwise. It would also be premature to take such a step until the results of cost-based access, new fiber optic cables, and new INTELSAT services are fully evaluated. Unrestricted entry could ultimately undermine the economic integrity of this important international enterprise, which would be inconsistent with the U.S. national interest.

[51] The case has not been made for flatly disapproving the existing applications. The new entrants have made a threshold showing that services they propose are not now available on comparable terms. Limited entry along the lines recommended would further U.S. international trade interests, promote technological progress, and be consistent with national defense and security interests as well. Given these limitations, and the restrictions likely to be placed on any new satellite system by telecommunications authorities abroad, the risk of any significant adverse impact on INTELSAT is exceedingly small.

Further study and resulting delay [are] unlikely to further the national interest. Over a year of extensive study and review by the Executive branch has already taken place. This review has not resulted in the submission of credible information supplied by anyone, including INTELSAT and Comsat, which demonstrates plausible adverse effects. There is no basis to assume such information will be forthcoming.

Satellite systems entail significant lead time. Time is required to secure the requisite spacecraft, to reach launch agreements, and to secure operating arrangements. U.S. regulatory procedures are generally more time consuming than those abroad, where decisions can sometimes be reached and implemented without the regulatory proceedings and protracted court appeals characteristic of U.S. regulation. Consultation with INTELSAT is also required. Even were the pending applications approved by the FCC immediately, service would not be available for some time.

Government should not stifle private entrepreneurial initiatives absent sound and compelling public policy reasons. Such initiatives should not be discouraged when the services proposed could prove of value to customers, improve their productivity and efficiency, and thus enable American firms to compete more effectively both at home and abroad. The public policy case for continuing the status quo and flatly prohibiting additional international satellite systems is weak. Simply the pendency of U.S. applications has caused INTELSAT to accelerate plans for special business-oriented services and has precipitated a beneficial review of competitive conditions in the international satellite field generally. Further study and inevitable delay are unlikely to yield public dividends commensurate with the economic costs imposed.

[52] It is the view of the Executive branch that the national interest will be furthered by approving additional international communications satellite systems subject to limitations designed to minimize adverse effects on INTELSAT. Specifically, additional systems should be restricted to providing services through the sale or long-term lease of transponders or space segment capacity for communications not interconnected with public-switched message networks (except for emergency restoration service). Consultation must be undertaken with INTELSAT pursuant to Article XIV(d) of the Definitive Agreement.

[53]   Appendix A

[54]   THE WHITE HOUSE
                                   WASHINGTON

                              November 28, 1984

                                                    Presidential Determination
                                                              No. 85-2

MEMORANDUM FOR   THE SECRETARY OF STATE
                 THE SECRETARY OF COMMERCE

By virtue of the authority vested in me by the Constitution and statutes of the United States, including Sections 102(d) and 201(a) of the Communications Satellite Act of 1962, as amended (47 U.S.C. 701(d), 721 (a)), I hereby determine that separate international communications satellite systems are required in the national interest. The United States, in order to meet its obligations under the Agreement Establishing the International Telecommunications Satellite Organization (INTELSAT) (TIAS 7532), shall consult with INTELSAT regarding such separate systems as are authorized by the Federal Communications Commission. You are directed jointly to inform the Federal Communications Commission of criteria necessary to ensure the United States meets its international obligations and to further its telecommunications and foreign policy interests.

This determination shall be published in the Federal Register.

                                                    [hand-signed: "Ronald Reagan"]

[55]   **THE SECRETARY OF COMMERCE**
                        Washington, D.C. 20230

                                                            November 30, 1984

Honorable George P. Shultz
Secretary of State
Washington, D.C. 20520

Dear George,

There are two matters regarding the President's determination on new international satellite systems that need to be clarified. First, the White House has directed our departments to examine the scope of INTELSAT's pricing flexibility. Second, our position on the related issue of direct access to INTELSAT should be made clear.

The executive agreement establishing INTELSAT generally requires uniform pricing for each service. Prices on heavily trafficked routes may now exceed costs while those on thin routes may be below costs. It is not clear whether INTELSAT could vary its prices under the agreement. If INTELSAT's prices on busy routes are artificially inflated, inefficient entry by new systems may be induced. INTELSAT should have pricing flexibility when confronted with actual or potential competition as long as the prices it charges cover its costs.

A related issue is direct, cost-based access to the INTELSAT space segment. Allowing users and carriers in addition to Comsat the option to deal with INTELSAT directly for competitive services would foster competition based on superior efficiency and foresight and tend to deter entry by inefficient systems.

We should express clear positions on these two important points in the filing we will soon be submitting jointly to the Federal Communications Commission. I have asked Dave Markey to work with Bill Schneider to ensure this is done.

                                Sincerely,

                                [hand-signed: "Mac"]
                                Secretary of Commerce

cc: Chairman Mark Fowler

[56]   **THE SECRETARY OF STATE**
                            WASHINGTON

                                                            December 20, 1984

Dear Mac:

Thank you for your letter of November 30 relating to the President's determination on international satellite systems separate from INTELSAT. Your understanding conforms with ours that the White House is interested in having us examine the issues of pricing flexibility in INTELSAT and direct access to INTELSAT by users other than COMSAT.

We have received, and are reviewing, the draft paper prepared by NTIA [National Telecommunications and Information Administration] which might be sent jointly to the FCC.

The Office of the Coordinator for International Communication and Information Policy, together with others concerned with the issue, are working with your staff on these and additional issues emanating from the Presidential determination.

Sincerely yours,

[hand-signed: "George"]
George P. Shultz

The Honorable
  Malcolm Baldridge,
    Secretary of Commerce.

cc: Chairman Mark Fowler

[57]   Appendix B

[58]   THE DEPARTMENT OF COMMERCE
                                Washington, D.C. 20230

November 28, 1984

Honorable Mark S. Fowler
Chairman
Federal Communications Commission
Washington, D.C. 20554

Dear Mr. Chairman:

The President has determined that separate international communications satellite systems are required in the national interest. He has also directed that we inform the Federal Communications Commission of criteria necessary to ensure the United States meets its international obligations and to further its telecommunications and foreign policy interests. Prior to final authorization by the Commission of any systems, to assure that the United States meets its obligations as a Party to the Agreement Establishing the International Telecommunications Satellite Organization (INTELSAT) (TIAS 7532):

(1) each system is to be restricted to providing services through the sale or long-term lease of transponders or space segment capacity for communications not interconnected with public-switched message networks (except for emergency restoration service); and,

(2) one or more foreign authorities are to authorize use of each system and enter into consultation procedures with the United States Party under Article XIV(d) of the INTELSAT Agreement to ensure technical compatibility and to avoid significant economic harm.

The President's determination, its conditions, and these criteria are premised on our review of the issues prompted by the applications now before the Commission. If proposals substantially different are forthcoming, further Executive Branch review may be required.

The Commission should afford interested parties an opportunity to submit timely comments on the pending applications in view of these Executive Branch recommendations.

A memorandum of law concerning Article XIV of the INTELSAT Agreement is enclosed.

Sincerely,

[hand-signed: "George P. Shultz"]   [hand-signed: "Malcolm Baldridge"]
Secretary of State   Secretary of Commerce

Enclosure . . .

Chapter Two

Observing the Earth From Space

by Pamela E. Mack and Ray A. Williamson[1]

Programs that apply the capabilities of space technology to needs such as telecommunications and Earth observation have brought society many concrete benefits. However, developing projects to realize those benefits has not been easy, particularly for Earth observations. Applications programs have neither the glamor and high-profile political impact of human spaceflight nor the well-organized advocacy community of the space scientists. NASA, which is primarily a research and development agency, has had an ambivalent relationship with the application of space technology to Earth-bound needs. While the space agency welcomes opportunities to prove its value in concrete ways, it recognizes that an applications program that has completed development and entered the operational phase must usually be transferred from NASA to another agency or to a private sector user; not surprisingly, NASA staff have often preferred to work on those programs that do not have to be "given away."

Although scientific and technological feasibility and accomplishment are essential to space applications, they are only part of the story. Tensions between NASA, as developer of space capabilities, and the organizations or experts who actually distribute or use the services or data provided by applications satellites also play an important part in the success or failure of applications programs. To give a few examples, some scientists were excited about the data that meteorological satellites could provide, but their enthusiasm played a smaller role in the origin of the Television Infrared Operational Satellite (TIROS), the first meteorological satellite project, than did military needs. The Kennedy administration created the Communications Satellite Corporation (Comsat) as an innovative way of bringing a new form of international telecommunications into being, but traditional communications corporations steadily increased their role in satellite communications.[2] NASA predicted large benefits from crop surveys using data from Earth resource satellites, but agricultural scientists took a different approach to using the data than the one NASA had developed.

This broad theme—that different players have different goals and expectations—also has been played out in specific controversies in different applications projects. Usually at about the time of the launch of an initial satellite, programs have often experienced disputes over whether to conduct further research or to develop an operational program immediately. Knowing that the research satellite would set much of the pattern for their operational program, users have often sought more control over the initial development of an applications satellite than NASA wanted them to have. Finally, programs have suffered from major controversies over the proper role of the government in their development and operations. Communications satellite systems became the province of private industry, but only after a bitter debate concerning whether or not to turn the fruits of government research over to private profit. Congress in the early 1980s rejected in no

---

1.   The authors thank Don Blersch, Russ Koffler, Rob Masters, and Brent Smith for providing information and Frank Eden for his review of a draft of this essay.
2.   Communications satellites are discussed in Chapter One; they are mentioned in this chapter only in terms of their relationship to Earth observation satellite programs.

uncertain terms proposals that meteorological satellites be commercialized. The debate about turning Earth resource satellites over to private industry has been long and controversial, and by the time privatization occurred, the U.S. system had fallen behind the state of the art in important aspects. This debate also contributed to the slow development of commercial remote-sensing satellite systems. (The "value-added" business for data from meteorological and land remote-sensing satellites has been more commercially successful.)

Meteorological Satellites

Today's widespread familiarity with satellite images used by television weather forecasters encourages the assumption that meteorological satellites were an eagerly awaited breakthrough in the technology underpinning weather forecasts. In fact, at the start of the space age, meteorologists were not certain satellite data would prove useful. One of the pioneers of meteorological satellites, Harry Wexler, wrote in 1954:

> To predict the future of the atmosphere, the meteorologist must know its present state—as defined by the three-dimensional distribution of pressure, temperature, wind, humidity. . . . Knowing the present state of the atmosphere and past motions of the storms enables a prediction to be made by extrapolation and other techniques.[3]

Wexler pointed out that a satellite could provide only a "bird's eye" view, not the three-dimensional data meteorologists needed. Therefore, a satellite would "serve principally as a 'storm patrol.'"[4] [II-1] A warning of a severe storm obviously would be of great practical value, but most of the practice of meteorology addressed more routine situations. A meteorologist's desire for three-dimensional measurements of many variables was one of the arguments for developing more sophisticated weather satellites in the 1960s and 1970s.

Because the value of gathering weather data from satellites was not immediately obvious to civilian meteorologists, early meteorological satellite proposals emphasized military uses. Ground stations and hurricane patrol airplanes provided acceptable storm warnings for the continental United States and nearby waters, but the Navy needed storm warnings in whatever remote areas ships might be operating, and the Air Force had similar needs for worldwide forecasts.

Planning for a U.S. space program began with a 1946 Project RAND report titled "Preliminary Design of an Experimental World-Circling Spaceship."[5] This report emphasized various military applications of a satellite; it noted that "perhaps the two most important classes of observation which can be made from such a satellite are the spotting of the points of impacts of bombs launched by us, and the observation of weather conditions over enemy territory."[6] In the section of the report discussing the scientific uses of a satellite, the authors commented that observations of cloud patterns "should be of extreme value in connection with short-range weather forecasting, and tabulation of such data over a period of time might prove extremely valuable to long-range weather forecasting."[7]

3.   Dr. Harry Wexler, "Observing the Weather from a Satellite Vehicle," *Journal of the British Interplanetary Society* 7 (September 1954) 269–76; see Document II-1.

4.   *Ibid.*

5.   For the history of Rand's role in early space planning, see Merton E. Davies and William R. Harris, *RAND's Role in the Evolution of Balloon and Satellite Observation Systems and Related U.S. Space Technology* (Santa Monica, CA: The RAND Corporation, 1988).

6.   Douglas Aircraft Company, Inc., "Preliminary Design of an Experimental World-Circling Spaceship," Report No. SM-11827, May 2, 1946, p. 11, Space Policy Institute Documentary History Collection, Washington, DC.

7.   *Ibid.*, p. 13.

Later RAND studies sought to tackle the problem of whether cloud images alone would be of much benefit to meteorologists. An April 1951 RAND report titled "Inquiry into the Feasibility of Weather Reconnaissance from a Satellite Vehicle" considered whether the data meteorologists wanted could be derived from cloud images. [II-2] The report stated the following:

> *It is obvious that in observing the weather through the "eye" of a high-altitude robot almost all of the regular quantitative measurements usually associated with meteorology must fall by the wayside. It is impossible to make more than an intelligent guess at the values of temperature, pressure, humidity, and the remaining quantitative meteorological parameters. . . . Clouds, being the objects most easily discernable [sic] from extremely high altitudes, become the important item and must be utilized to the utmost in forming a synoptic picture. It is apparent that from clouds alone it will be impossible to tell everything about the current synoptic situation. Combined, however, with both theoretical knowledge and that gained through experience, accurate cloud analysis can produce surprisingly good results.*[8]

Starting in 1947 with imagery taken from V-2 rockets fired at White Sands, New Mexico, scientists sought to classify clouds and to deduce weather parameters from historical data and cloud patterns. They judged their results to be quite successful, but they argued that new approaches would be needed to make the best use of the data. For example, they wanted a new method of classifying clouds, rather than the traditional classification method based solely on appearance.[9]

An analysis of similar images by the Naval Research Laboratory a few years later provided more evidence of the value of weather-related observations from space. For example, Otto Berg discovered that images taken by a Navy Aerobee rocket in October 1954 had shown a major hurricane in the Gulf of Mexico—a storm that then hit the United States with no advance warning from weather stations. He argued that satellites would be immediately useful for providing storm warnings. Berg suggested that "in the more distant future, these techniques of rocket reconnaissance will be applied to investigation of other meteorological phenomena."[10]

The first weather satellite project, TIROS, resulted not just from the perceived usefulness of storm warnings but also from the existence of many different groups in the Department of Defense (DOD) that wanted a hand in space. The Air Force sponsored a number of studies to explore technology for reconnaissance satellites, leading eventually to a development contract with Lockheed for what eventually became the Satellite Military Observation System (SAMOS) reconnaissance satellite. The RCA Corporation, one of the unsuccessful bidders, then approached the Army Ballistic Missile Agency (ABMA) with a proposal to develop a satellite with a television camera for either meteorology or surveillance.[11] The ABMA initiated Project Janus to test the television concept for the purpose of reconnaissance, but in mid-April 1958, DOD assigned the satellite reconnaissance mission exclusively to the Air Force. The ABMA then changed the mission of what had become the Janus II project from reconnaissance to meteorology.[12]

---

8.   S.M. Greenfield and W.W. Kellog, "Inquiry into the Feasibility of Weather Reconnaissance from a Satellite Vehicle," The RAND Corporation, R-365, August, 1960, p. 1; see Document II-2. This is the unclassified version of RAND Report R-218, April 1951.

9.   *Ibid.*, p. 22.

10.   Otto E. Berg, "High-Altitude Portrait of Storm Clouds," *Office of Naval Research Reviews,* September 1955, Space Policy Institute Documentary History Collection, Washington, DC.

11.   Richard LeRoy Chapman, "A Case Study of the U.S. Weather Satellite Program: The Interaction of Science and Politics," Ph.D. Diss., Syracuse University, 1967, pp. 20–24.

12.   *Ibid.*, pp. 30–33.

In May 1958, Janus II was transferred within DOD from the ABMA to the Advanced Research Projects Agency (ARPA), a new organization established a few months earlier and intended to centralize military (and, temporarily, civilian) space research under tighter control by the secretary of defense. After another reconfiguration to take advantage of a larger booster, ARPA changed the name of the project from Janus II to TIROS (for Television Infrared Observation Satellite) and committed funds to final design and construction for a planned launch in the summer of 1959.[13] The U.S. Signal Research and Development Laboratory at Fort Monmouth, New Jersey, managed a contract with RCA for construction of the satellite.

The Civilian Program

NASA took over the TIROS project, upon its creation later in 1958, with the understanding that the space agency would cooperate with the Weather Bureau. However, DOD interest in weather satellite data continued, complicating the process of planning an operational program. Later in the 1960s, the Air Force began developing a separate Defense Meteorological Satellite Program (DMSP) to meet specific military needs for data to support its operations.[14] Early satellites in the DMSP differed little from TIROS, but later satellites in the program provided quantitative radiometric data designed specifically to support DOD requirements. Civilian meteorological satellites continued to be used by both civilian and military meteorologists; the eventual convergence of the two programs is discussed in the next section.

In July 1958, after President Dwight D. Eisenhower had decided that all space programs that were not clearly military should be transferred to the new civilian space agency, the White House assigned TIROS to NASA. Arranging the actual transfer posed difficulties because the program was so far along in its development, but a number of scientists and engineers agreed to move from DOD to NASA along with the project, and NASA arranged for the Weather Bureau to provide research support in meteorology.[15] A transfer agreement was signed in April 1959. [II-3]

Despite the difficulties of the transfer, NASA launched the experimental TIROS I on April 1, 1960—a spin-stabilized satellite carrying two television cameras. The results generated so much excitement among meteorologists that NASA soon set up a system to transfer the resulting cloud cover information onto standard weather maps and to send them to weather stations and to the military services.[16] Although meteorologists found satellite data difficult to integrate into the forecasting process, because their models required data on temperature, pressure, and wind speed and direction, they found that satellite images showed large-scale weather patterns so clearly that they were immediately useful.[17] Satellites also demonstrated their value for storm warning. In September 1961, a TIROS satellite helped track an extremely dangerous hurricane, Carla, bearing down on the Gulf Coast. Warnings led to the evacuation of more than 350,000 people.[18] Also in

---

13.   *Ibid.,* pp. 36–54, 61–62.

14.   In addition, the highly classified CORONA photoreconnaissance program was jointly managed by the Air Force and the Central Intelligence Agency.

15.   Chapman, "A Case Study," pp. 60–64.

16.   Janice Hill, *Weather from Above: America's Meteorological Satellites* (Washington, DC: Smithsonian Institution Press, 1991), pp. 9–16.

17.   For a discussion of the resistance of meteorologists to the use of satellite data, see Margaret Eileen Courain, "Technology Reconciliation in the Remote-Sensing Era of the United States Civilian Weather Forecasting, 1957–1958," Ph.D. Diss., Rutgers University, 1991.

18.   Patrick Hughes, "Weather Satellites Come of Age," *Weatherwise* 37 (April 1984): 68–75.

September 1961, a fully developed hurricane, Esther, was located through satellite images. This was the first hurricane to be identified by a satellite before being observed by then-conventional means.

Even before these successes, the usefulness of the TIROS satellites led to pressure to transform the experimental project into an operational system. To address this issue, NASA called together an interagency Panel on Operational Meteorological Satellites in October 1960. Disagreements over the future of the program quickly appeared. The Weather Bureau sought more control than NASA wanted to give up; it asked for complete authority over the operational system, including launching, data retrieval, and final decisions on the design of new operational satellites. [II-4]

The panel issued a compromise plan in April 1961 calling for a national operational meteorological satellite system (based on a second-generation satellite already under development by NASA) to be managed by the Weather Bureau.[19] This plan did not go as far as the Weather Bureau had originally proposed; NASA would maintain control of launch services and ground support and would develop and procure spacecraft under contract to the Department of Commerce (of which the Weather Bureau was a part). President Kennedy's May 25, 1961, speech to Congress titled "Urgent National Needs," in which he urged funding a program to land an American on the Moon before the end of the decade, also requested funds to put the weather satellite plan into operation. Congress approved the funding Kennedy had requested, despite continuing controversy over what would be the best division of responsibility between NASA and the Weather Bureau for operating the system.[20]

Meanwhile, NASA was working on the second-generation meteorological satellite, Nimbus, as a prototype for the operational system. Nimbus was a more sophisticated spacecraft than TIROS—stabilized so that it always pointed toward the Earth rather than continuously rotating. It was to be launched into a Sun-synchronous polar orbit so that it could collect data from the whole Earth at the same local time each day. The satellite would carry not only more sophisticated television cameras, but also a high-resolution infrared radiometer that used thermal infrared sensors to map temperature. (A simpler experimental sensor of the same type had been carried on later TIROS flights.) Plans for Nimbus also included a variety of more sophisticated sensors—most significantly, sounding instruments providing data that could be used to determine temperatures at various levels in the atmosphere.[21] These new instruments reflected the efforts of NASA scientists to meet the continuing demand from research meteorologists for basic numerical data in addition to cloud images.[22]

While researchers wanted the more sophisticated data Nimbus would provide, the Weather Bureau became concerned about its increasing cost and delays in launch. NASA and the Weather Bureau also had differing perspectives on what decisions should be made

---

19. U.S. National Coordinating Committee for Aviation Meteorology, Panel of Operational Meteorological Satellites, *Plan for a National Operational Meteorological Satellite System* (Washington, DC: U.S. Government Printing Office, 1961).

20. These issues are discussed in "The National Meteorological Program," preliminary staff report of the Committee on Science and Astronautics, 87th Cong., 1st sess., July 13, 1961. For a discussion of the process of approving the plan, see Science Policy Research Division, Congressional Research Service, *United States Civilian Space Programs: Volume II, Applications Satellites,* report prepared for the Subcommittee on Space Science and Applications, U.S. House of Representatives, 98th Cong., 1st sess., May 1983, pp. 198–99; Chapman, "A Case Study," pp. 107–27.

21. Chapman, "A Case Study," p. 161.

22. Indeed, the Nimbus series continued through Nimbus 7 and provided the basic information needed to develop later research satellite systems, including the Upper Atmosphere Research Satellite and the Earth Observing System. H.F. Eden, B.P. Elero, and J.N. Perkins, "Nimbus Satellites: Setting the Stage for Mission to Planet Earth," EOS, *Transactions, American Geophysical Union* 74 (June 29, 1993): 281, 285.

about the operational system and on which agency should make them.[23] This became a serious problem as NASA's plans for Nimbus increasingly diverged from Weather Bureau priorities. NASA had agreed to extend the TIROS program from the original two satellites to ten so as to provide continuous data for those users who already depended on the data, but delays in the Nimbus program still created the likelihood of gaps in coverage. The possibility of experiencing a period of months with no meteorological satellite data available particularly worried DOD.[24] Critics also raised questions about the reliability of Nimbus because it constituted such a large leap in sophistication over TIROS.[25] In addition, the predicted annual cost of an operational system based on Nimbus had nearly doubled since the original plan had been submitted to Congress.[26] The Weather Bureau would be responsible for funding the operational system, but it was a small agency with a limited budget and with little chance of getting that budget expanded substantially.[27]

By the late summer of 1963, differences between NASA and the Weather Bureau had hardened into an impasse.[28] On September 27, 1963, the Weather Bureau notified NASA that it was pulling out of existing interagency agreements and pursuing an interim operational satellite system based on TIROS technology. The Weather Bureau was able to make such a stand only because it had found a partner; DOD had agreed to provide launch services for the operational TIROS system the Weather Bureau wanted.[29]

As a result, NASA found itself in a weak position. The space agency could not justify developing advanced satellites for a user that did not want them, and it could not afford to have DOD as a competitor in providing launch services for civilian satellites. NASA compromised and agreed to give the Weather Bureau a larger voice in shaping the next-generation meteorological satellite system. The compromise resulted in decisions to build a TIROS Operational System with funding provided by the Weather Bureau, to continue the Nimbus program on a purely experimental basis, and to sign a new formal agreement for cooperation between the two agencies. [II-5]

The agreement called for NASA to develop and launch the initial version of any new instrument or spacecraft and for the Weather Bureau to provide funding for operational versions. This agreement remained in force until 1982, when NASA decided to withdraw from providing operational improvements as a continuing obligation. At that time, the National Oceanic and Atmospheric Administration (NOAA), the successor agency to the Weather Bureau, assumed responsibility for development, as well as operation, of all civilian meteorological satellites. The agency lacked the capability and funding necessary for such development, and eventually it returned to informal and then formal cooperation with NASA.

The satellites of the TIROS Operational System (renamed ESSA 1 through 9[30] after launch) were less capable than Nimbus, but they did involve significant improvements over the original TIROS satellites. Some of these spacecraft used a higher resolution camera first tested on Nimbus. Others provided real-time data to users around the world through the Automatic Picture Transmission system developed for Nimbus, but first tested on TIROS-VIII in 1964.

---

23.   James E. Webb to J. Herbert Hollomon, June 28, 1962, with attached memo: Abraham Hyatt to the Administrator, "Weather Bureau Plan," June 25, 1962, Space Policy Institute Documentary History Collection, Washington, DC.

24.   Chapman, "A Case Study," pp. 155–71.

25.   *Ibid.*, pp. 210–11.

26.   *Ibid.*, p. 192.

27.   *Ibid.*, pp. 217–29.

28.   *Ibid.*, pp. 229–42.

29.   *Ibid.*, pp. 244–59.

30.   The Weather Bureau had become part of a new organization, the Environmental Science Services Administration (ESSA), established on July 13, 1965.

Once they had made decisions about the scope of the initial operating system, NASA and the Weather Bureau were more free to think about experimental satellites to test out new instruments that might be incorporated into future generations of operational satellites. Nimbus 1, launched April 28, 1964, experienced a number of problems and operated for only a month; Nimbus 2, launched May 15, 1966, was much more successful, testing out improved cameras. NASA also tested meteorological satellite technology as part of the Applications Technology Satellite (ATS) project to orbit experimental geosynchronous satellites (used for communications experiments as well as meteorological ones). In December 1966 and November 1967, ATS-1 and -3 explored the possibility of observing weather with line scan imagers, a possibility conceived by Vernon Soumi, a professor at the University of Wisconsin; the resulting continuous coverage (images of the full Earth disc every thirty minutes) proved extremely valuable for tracking storms and even showed short-lived cloud patterns correlated to tornadoes.[31] Continuous coverage from geosynchronous orbit made it possible to observe the motion of clouds and deduce wind speed at the level of the clouds—a significant step toward the three-dimensional quantitative data meteorologists wanted. Three other satellites in the series, ATS-2, -4, and -5, also carried meteorological experiments, but all suffered launch problems.[32] ATS-6, launched in May 1974, carried a new cloud-imaging radiometer along with a more powerful transmitter that made it possible for anyone with an easy-to-build ground station to receive the images.[33]

While the ATS program tested ideas for weather satellites in geosynchronous orbit, the Nimbus program continued to test advanced instruments in low-Earth orbit, with five launches between 1969 and 1978. Nimbus 3, launched April 1969, carried five new sensors. These included the first sounding instruments using remote sensing to furnish measurements of temperature and other variables at different levels of the atmosphere for providing numerical data for climate models. The sounding instruments on Nimbus 3 measured temperature, water vapor, and ozone content of various atmospheric levels; later Nimbus satellites carried sounding instruments to measure other variables. The sensors worked well, but the data proved much less useful for weather prediction than scientists had expected. Meteorologists had hoped that data on temperature, wind speed, and other factors could be plugged into a model of how the atmosphere worked to provide weather predictions. Satellite sounding instruments provided much of the data needed with reasonable accuracy, but existing climate models were not designed to assimilate these data easily. Meteorologists discovered that they needed to perform much more research before they could use data acquired by satellite to improve the accuracy of weather predictions.[34]

In August 1966, the Weather Bureau stated the following as its objectives for an operational satellite system: "(1) [t]he establishment and maintenance of a satellite system to obtain global observations on a regular basis, (2) meteorological observations from synchronous altitude, and (3) global observations of atmospheric structure needed for numerical weather forecasting." [II-6] The first objective was met by continuing improvements in the TIROS series of low-altitude satellites, which were flown from 1966 to 1969.

---

31. The ATS cameras provided pictures every thirty minutes, compared to once or twice a day from the TIROS Operational System. For a discussion of the usefulness of continuous coverage, see W.L. Smith *et al.*, "The Meteorological Satellite: Overview of 25 Years of Operation," *Science* 231 (January 31, 1986): 455–62.

32. Hill, *Weather from Above*, pp. 23–26, 29–32.

33. *Ibid.*, pp. 33–35.

34. James C. Fletcher to Stuart Eizenstat, "Possible Initiatives," February 16, 1977, suggests that NASA hoped that sounding instruments would lead to a major new research initiative. Pamela E. Mack, "Cloudy Seeing: Developing New Sensors for Weather Satellites," paper presented at the Society for the History of Technology's annual meeting, London, England, August 1996.

NASA launched an Improved TIROS Operational Satellite (ITOS) in January 1970.[35] Starting in 1972 with the third satellite in the ITOS series, NASA replaced the television cameras that had been carried on all earlier flights with a two-channel scanning radiometer providing visible and infrared imagery.[36] This infrared imagery was used to monitor nighttime cloud cover and to produce sea-surface temperature maps. This same satellite (designated NOAA-2) carried the first operational sounding instruments, which provided vertical temperature profiles through the atmosphere. These instruments provided the data needed to meet the third of the objectives established in 1966. Additional improvements to the low-altitude satellites made in the late 1970s resulted in the TIROS-N design, carrying a finer resolution radiometer and sounder as well as a data collection platform and a solar energetic particle monitor.[37]

ATS-1 and -3 provided data to meet the second of the 1966 objectives, but budgetary constraints delayed the operation of a geosynchronous meteorological satellite system until the mid-1970s. In the interim, NASA funded two prototype Synchronous Meteorological Satellites, launched in May 1974 and February 1975. The space agency then launched the first Geostationary Operational Environmental Satellite (GOES) on October 16, 1975. Sounding instruments were also included in improved GOES satellites, starting with GOES-4 in September 1980.

While NASA developed new capabilities for meteorological satellites and the National Weather Service integrated the resulting data into the operational weather forecasting system, budgetary pressures continued to grow. The Reagan administration wanted to transfer operational space systems to private industry to cut the federal budget. In early 1981, Comsat proposed taking over both the Landsat (see below) and the meteorological satellite systems; officials in the Reagan administration responded with enthusiasm. Congress, however, disagreed strongly with the idea of privatizing meteorological satellites; members argued that the government properly provided weather forecasts as a public good and therefore should retain control of the production of meteorological satellite data.[38] Late in 1983, Congress passed and President Reagan signed an appropriations bill that included a specific prohibition against the sale of the meteorological satellite system to private industry.[39] However, the issue of charging users for weather satellite data arose again in the 1990s as a result of cooperative programs with other countries that took such an approach.

Clearly, the balance among technological possibilities, user needs, and financial limitations shaped not only the origins but also the continuing development of the meteorological satellite system. Weather forecasts improved, although not as much as meteorologists had predicted when they looked forward to the new capabilities various satellite technologies would provide. Part of the problem was that the path from a good idea to its incorporation into the operational system was inevitably slow and rocky. Probably, however, the more important factor was that predicting weather was, and continues to be, a problem of much greater complexity than scientists had anticipated.

---

35.   The first TIROS satellite was an operational prototype; subsequent satellites in the series were to be renamed ESSA 10, 11, and so on, after launch. However, the new National Oceanic and Atmospheric Administration replaced ESSA, and the satellites were named NOAA. It is at this point that the Weather Bureau was renamed the National Weather Service.

36.   Hill, *Weather from Above*, pp. 37–38.

37.   *Ibid.*, pp. 49–51.

38.   Press Release, Senate Commerce, Science and Transportation Committee, September 20, 1983; "Weather Satellites," *Congressional Record*, S. 14367, October 20, 1983; "Transfer of Civil Meteorological Satellites," *Congressional Record*, H.R. 9812-9822, November 14, 1983. See also Hill, *Weather from Above*, p. 60.

39.   Office of Technology Assessment, U.S. Congress, "Remote Sensing and the Private Sector: Issues for Discussion," *Technological Memorandum* (March 1984): 22.

Converged Polar-Orbiting Meteorological Satellite Systems

Although the civilian and military polar-orbiting meteorological satellite programs have followed separate paths, there have been several attempts to bring them together over the years. Officials within several administrations kept hoping that a merged system could meet the requirements of both NOAA and DOD (because each had a need to acquire imagery of clouds) while providing an overall savings to the government. However, NOAA and DOD weather systems acquire varying kinds of data at different times of the day to support distinct types of uses. For example, DOD is interested in cloud image data acquired in the early morning to support tactical and strategic operations; NOAA is more interested in atmospheric soundings in the early afternoon, which the National Weather Service feeds into its predictive weather models. Furthermore, until the 1980s, DMSP data were not shared with civilian users.

In 1973, a national space policy study led by the Office of Management and Budget and the National Security Council[40] examined the fiscal and policy implications of conducting separate DOD and NOAA operational weather satellite systems. This study based its assessment of the technical feasibility and costs of a converged system on NOAA, NASA, and DOD analyses, concluding that no option could maintain performance levels and also reduce costs significantly. In addition, policy concerns regarding the open distribution of weather data useful to potential adversaries argued for separate programs.[41] The 1973 review did, however, result in the Nixon administration directing NOAA to use the DMSP Block 5D spacecraft bus, then under development by the Air Force, as the basis for the next-generation series of polar-orbiting satellites. In addition, NOAA and DOD were instructed to coordinate more closely the management of the separate programs.

On seven other occasions since 1972, the Department of Commerce and DOD studied the potential for integrating their programs. These studies did not lead to merged programs, but they did result in a number of modest economies, including the use of similar spacecraft with numerous common subsystems and components. In addition, both programs have used a common launch vehicle and have shared responsibility for creating products derived from the data. The two programs have also worked together closely on research and development efforts and provided complementary environmental information. Most of the sensors, however, remained under the design and control of each agency (see Table II–1).

Despite these efforts, until the early 1990s, foreign policy and national security concerns precluded full program integration. By that time, the drive to reduce the federal budget and increase government efficiency led a number of observers to suggest again consolidating the two systems. In addition, in October 1992, NASA and NOAA had begun to explore the potential for consolidating aspects of NOAA's Polar-orbiting Operational Environmental Satellite (POES) system and NASA's Earth Observing System satellite, EOS-PM. The latter is an afternoon equator-crossing satellite that will gather data similar to the POES afternoon satellite, but of much higher quality and complexity.

---

40.   "The Meteorological Satellite Analysis Study (MSAS)," Office of Management and Budget, 1973. This study was begun in 1972.

41.   The United States had pledged to maintain an open civilian weather satellite system. Also, NOAA's environmental satellites demonstrated the U.S. "open skies" policy and satisfied long-standing U.S. obligations to exchange Earth data with the meteorological agencies and scientific organizations of other nations.

Table II–1

Key Sensors and Priorities for NOAA's and DOD's Polar Meteorological Programs

Agency and Data Acquired	Sensor	Attributes
NOAA		
Multispectral imagery (cloud, vegetation)	Advanced Very High Resolution Radiometer (AVHRR)	Calibrated, multispectral imagery
Temperature and humidity (initialize numerical weather prediction model)	TIROS Operational Vertical Sounder (TOVS)	High spatial resolution, cross-track scanning (PM equator crossing)
DOD		
Visible and infrared cloud imagery (cloud-detection forecast, tactical imagery dissemination)	Operational Linescan System (OLS)	Constant field of view; low-light (early AM crossing)
Microwave imagery (ocean winds, precipitation)	Special Sensor Microwave/Imager (SSM/I)	Conical scan
Temperature and humidity (electro-optical propagation, initialize numerical weather prediction models)	Special Sensor Microwave/ Temperature Sounder (SSM/T-1); Special Sensor Microwave/ Water Vapor Sounder (SSM/T-2)	Low spatial resolution, cross-track scanning

Source: Office of Technology Assessment, U.S. Congress, *Civilian Satellite Remote Sensing: A Strategic Approach,* OTA-ISS-607 (Washington, DC: U.S. Government Printing Office, September 1994), p. 79.

In February 1993, Representative George Brown, then chair of the House Committee on Science, Space, and Technology, sent a letter to NOAA Administrator D. James Baker, requesting a review of the NOAA and DOD polar-orbiting programs to explore possible cost savings. [II-7] As a result of this initiative and similar interest within the Clinton administration, the two agencies began to examine the two programs once again. A few months later, Senator James Exon, chair of the Senate Subcommittee on Nuclear Deterrence, Arms Control and Defense Intelligence, sent a similar request to Commerce Secretary Ron Brown. [II-8] A report of the U.S. Congress Office of Technology Assessment also offered consolidation of the two programs as an option for reducing federal spending.[42]

By July 1993, the two major convergence studies were consolidated into a single tri-agency study involving DOD, NASA, and NOAA. With input from this study, by September 1993, Vice President Al Gore's National Performance Review made a firm proposal to

---

42.   Ray A. Williamson, "NASA's Mission to Planet Earth," Statement before the Space Subcommittee of the House Committee on Science, Space, and Technology, May 6, 1993; U.S. Congress, Office of Technology Assessment, *The Future of Remote Sensing: Civilian Satellite Systems and Applications,* OTA-ISC-548 (Washington, DC: U.S. Government Printing Office, July 1993), p. 16.

integrate the two systems. [II-9] This proposal estimated that the government would save $300 million through the year 2000 and $1 billion over a decade by creating a converged environmental satellite system. The National Performance Review also recommended that NASA

> *assist in ongoing efforts to converge U.S. operational weather satellites, given the benefits of streamlining the collection of weather data across the government... By considering [Mission to Planet Earth] research activities in context with operational weather satellite programs, cost savings are possible through convergence of the current operational satellite fleets. Convergence of the National Oceanic and Atmospheric Administration (NOAA) Polar Metsat and NASA's EOS-PM (Earth Observing System Afternoon Crossing [Descending] Mission) will eliminate redundancy of measurements, enhance the capability of NOAA's data set and potentially result in cost savings.[43]*

After further study, Vice President Gore's initial proposition resulted in a plan detailed in a May 1994 Presidential Decision Directive (also known as NSTC-2) on the "Convergence of U.S.-Polar-orbiting Operational Environmental Satellite Systems." [II-10] This decision directive called for an Integrated Program Office (IPO) that will "be responsible for the management, planning, development, fabrication, and operations of the converged system." The IPO was set up in October 1994. NOAA gained the lead responsibility for operations, with DOD leading systems acquisition and NASA leading new development and the insertion of new technologies.

One of the important considerations in exploring the exact terms of convergence was prior interest at NOAA in cooperating more closely with Europe in NOAA's polar-orbiting program. Europe had been contributing instruments to NOAA's POES spacecraft since 1978. During the early 1980s, the Reagan administration attempted to reduce the two-satellite constellation (one morning-crossing, one afternoon-crossing) to a single afternoon-crossing spacecraft. NOAA officials became extremely concerned that maintaining only a single orbiter would greatly reduce the reliability of data delivery. Hence, the agency began discussions with other countries, forming the International Polar Orbiting Meteorological Satellite Group (IPOMS) to promote a more equitable sharing of the burden of maintaining polar-orbiting meteorological satellites. Membership in IPOMS included most of the major remote-sensing satellite operators. Within IPOMS, the European Space Agency (ESA) together with the newly created European Organisation for the Exploitation of Meteorological Satellites (EUMETSAT) expressed interest in providing a European polar orbiter that could replace one of NOAA's spacecraft.[44]

Under the strategy developed during the 1980s, NOAA planned to provide several instruments for a European orbiter, which would replace the morning-crossing NOAA satellite. Originally, this satellite was to be a large ESA spacecraft carrying both global change research instruments and operational meteorological instruments.[45] By 1992, this plan had evolved into one in which Europe would orbit two spacecraft: an ESA global

---

43.   Office of Vice President, *National Aeronautics and Space Administration, NASA05: Clarify the Objectives of the Mission to Planet Earth Program, in Accompanying Report of the National Performance Review* (Washington, DC: Office of the Vice President, September 1993).

44.   Minutes of the Fourth IPOMS Plenary and First Administrative Working Group Tokyo, Japan, November 12–13, 1987, IPOMS Reports, National Environmental Data and Information System, National Oceanic and Atmospheric Administration, 1987, Space Policy Institute Documentary History Collection, Washington, DC.

45.   This spacecraft was originally conceived to be one of an international fleet of large research and operational polar orbiters, launched and serviced by the Space Shuttle. With the loss of *Challenger* in January 1986 and the subsequent change of U.S. policy toward the use of the Space Shuttle, these plans were abandoned.

change research satellite, Envisat, and a EUMETSAT meteorological operational satellite, METOP. NOAA and EUMETSAT began to develop explicit plans to operate the METOP satellite series as morning-crossing spacecraft carrying three U.S. instruments: the Advanced Very High Resolution Radiometer (AVHRR), the High-Resolution Infrared Sounder (HIRS), and the Advanced Microwave Sounding Unit. On May 6, 1994, NOAA formally invited EUMETSAT to participate in the converged U.S. system. [II-11, II-12]

Under the convergence plan, by about 2005 or 2007, the United States will keep two satellites in orbit at all times: a polar orbiter that will cross the equator early in the morning to obtain early cloud data of particular interest to DOD and an afternoon-crossing orbiter that will provide the atmospheric soundings that the National Weather Service needs to support data inputs to its predictive models. METOP-1 will cross the equator in the late morning to collect data of particular interest to EUMETSAT's European data users.

One important issue that had to be decided was the data policy for METOP. In keeping with its long-standing U.S. data policies articulated in Office of Management and Budget Circular A-130, the United States has insisted that data from its sensors not be controlled even if they fly on the spacecraft of other nations.[46] In keeping with its data policy, EUMETSAT wishes to control the data from the satellite to assure that the countries benefitting from the system contribute to its funding. On the other hand, the United States wants the power to deny data delivery to an adversary in times of crisis or war.

The three U.S. agencies will need the long period between the convergence decision and operational status of the system to resolve several technical, programmatic, and institutional issues:

1. *Developing new sensors and spacecraft.* Because DOD and NOAA have different data requirements, it will be challenging to meet all the primary data requirements and still reduce program costs. The IPO, for example, may find it difficult to satisfy DOD's need for images of constant resolution across the field and maintain the radiometric quality that NOAA desires. NOAA also has a requirement for sounding data of higher quality than DOD's.

2. *Incorporating new capabilities in operational sensors.* NASA's involvement in the integrated program could lead to interesting opportunities to incorporate improved sensors or new data types in DOD's and NOAA's operational programs, derived from experience with NASA's Mission to Planet Earth research efforts. However, the sensors designed to tackle research problems are generally much more expensive and capable than those designed for routine data collection. Hence, making the transition to operational status also presents special challenges to designers in making cheaper instruments capable of meeting operational requirements. Data users will also have to adjust their operations to make efficient use of more complex, but more useful data.

3. *Maintaining institutional collaboration among U.S. agencies.* The three agencies in the IPO have worked diligently to establish a collaborative working relationship. However, each agency's budget is subject to radically different priorities and is overseen within the Office of Management and Budget by different examiners. In addition, each receives oversight and its appropriation from different congressional committees and

---

46. International Space University, *Toward an Integrated International Data Policy Framework for Earth Observations: A Workshop Report,* ISU/REP/97/1 (Illkirch, France: International Space University, January 1997).

subcommittees. Hence, continued progress in maintaining the IPO will depend on the determination of several elements of the administration and Congress to follow through on their program commitments.

4. *Maintaining close international cooperation.* Cooperating with EUMETSAT and in time possibly other nations in operating a fleet of operational environmental satellites poses additional challenges for the three agencies. EUMETSAT policies and funding mechanisms will continue to be driven by the needs, philosophies, and funding mechanisms of its member weather organizations, which are likely to be different than those of the IPO. Maintaining the system and high-quality data delivery over time will require continued flexibility on the part of the IPO in negotiating system upgrades. Adding additional organizations to the system will further complicate such negotiations, although this could increase system capability and further reduce U.S. costs.

Despite these challenges, the attempt to consolidate NOAA's and DOD's meteorological programs is more likely to succeed than past efforts because of the confluence of several factors. First, continuing pressures to maintain reduced agency budgets will encourage agency officials to continue to seek program efficiencies. Cost savings are also an important factor in resolving possible frictions among congressional oversight and appropriations committees over programmatic aspects of the converged system. Second, earlier plans by NOAA and DOD to upgrade both the DMSP and POES instruments and spacecraft shortly after the turn of the century will support technical convergence. Third, the changed international security environment will cause DOD analysts and managers to continue to moderate their historical objection to shared military-civilian systems. In addition, the opportunity to involve EUMETSAT and perhaps other nations or organizations in providing environmental data could further reduce overall program costs. Finally, including NASA explicitly in the partnership provides the opportunity to plan ongoing innovation and the transition of research instruments built in support of NASA's Earth Science activities to operational status for the converged system.

Earth Resource Satellites

Earth resource satellites suffered many of the same kinds of controversies in their transition from research to operations as meteorological satellites, with worse results. Part of the tension over this transition resulted from worries about whether a program to collect images of the Earth for civilian purposes would threaten the secrecy surrounding DOD's reconnaissance satellite programs. Another more important source of difficulty was a fragmented data user community; the images taken from Earth resource satellites were useful for geologists, hydrologists, agricultural scientists, city and regional planners, geographers, and people from other disciplines. Yet, a system serving all these disciplines was unable to meet the needs of any one extremely well. Even within the federal government, two agencies with different interests, the Department of the Interior and the Department of Agriculture, sought to shape the program. Another problem was that Earth resource satellite projects started later than meteorological satellites. With a first launch in 1972, the Earth resource satellite program began only after enthusiasm for Apollo had waned; at this point, NASA's budget was subject to much more intense scrutiny from the Office of Management and Budget and Congress.

Research Program

The idea for a civilian Earth resource satellite had two sources. DOD had an active reconnaissance satellite program growing out of experience with reconnaissance aircraft dating back to World War I. Scientists who investigated new technology for the classified reconnaissance program often had training in geology or geography, and they saw much potential for civilian use of the classified data they studied. In addition, NASA hired significant numbers of geologists to prepare the scientific program for Apollo, and some of them became interested in looking at the Earth as well as the Moon from space.[47] In 1965, NASA started to investigate the potential of studying Earth resources from space using instruments flown in its own aircraft. [II-13] The space agency wanted to involve the U.S. Geological Survey (a branch of the Department of the Interior) and the Army Corps of Engineers in remote-sensing research.

NASA proceeded slowly, testing a variety of sensors from aircraft before planning an experimental satellite. In a pattern similar to the debate over Nimbus, the space agency's initial plans for Earth resource satellites called for a large, sophisticated experimental satellite. [II-14] Meanwhile, scientists at the Department of the Interior had become convinced of the value of satellite data for applications and wanted an early operational satellite instead of elaborate experiments. [II-15] Impatient with NASA's lack of action, a group of scientists at the U.S. Geological Survey persuaded Secretary of the Interior Stewart L. Udall to announce in September 1966 that the Department of the Interior would start its own operational satellite program.[48] [II-16] When the Weather Bureau pulled out of its meteorological satellite agreement with NASA, it made an alternative alliance with DOD. The Department of the Interior, on the other hand, was unable to find a partner with space expertise; thus its announcement was more a bureaucratic maneuver than a realistic plan. Nevertheless, the resultant publicity forced NASA to commit to faster action on an experimental project to build the kind of small satellite the Department of the Interior wanted.[49] [II-17] NASA initially called the project the Earth Resources Technology Satellite (ERTS), but it changed the name to Landsat in 1975. General Electric won the prime contract for both the Nimbus and Landsat programs, and the Nimbus platform that had been developed and flown by NASA was also used for Landsat.

Many problems remained after the agreement on what kind of satellite NASA would build. The interested agencies continued to disagree over a variety of management and technical issues and over the proper balance between an experimental and an operational program.[50] The Department of the Interior and the Department of Agriculture wanted different kinds of sensors; Interior preferred a return-beam vidicon (a type of television cam-

---

47.   Pamela E. Mack, *Viewing the Earth: The Social Construction of the Landsat Satellite System* (Cambridge, MA: MIT Press, 1990), pp. 31–42. To see how an Earth observation program grew out of research on the other planets, see Peter C. Badgley, "The Applications of Remote Sensors in Planetary Exploration," paper presented at the Third Annual Remote Sensing Conference, Ann Arbor, Michigan, October 14, 1964.

48.   The Department of the Interior called its program Earth Resources Observation Satellites (later Systems), or EROS.

49.   W.T. Pecora, Director, Geological Survey, to Under Secretary, Department of the Interior, "Status of EROS Program," draft, June 15, 1967, Space Policy Institute Documentary History Collection, Washington, DC. For more details, see Mack, *Viewing the Earth*, pp. 56–65.

50.   Peter C. Badgley, Program Chief, Earth Resources Survey, to Distribution, "Meeting of Earth Resources User Agency Representatives with Space Applications Staff Members and Advanced Manned Missions Staff Members, April 20, 1967," May 4, 1967; Jacob E. Smart, Assistant Administrator for Policy, NASA, to Dr. Seamans, "Meeting with Representatives of Department of Agriculture and Interior Earth Resources," September 11, 1967, with a confidential second memo of same subject and date; Jacob E. Smart, Assistant Administrator for Policy, NASA, to Dr. Mueller, *et al.*, "Earth Resources Survey Program," October 3, 1967; Edgar M. Cortright for George E. Mueller, Associate Administrator for Manned Space Flight, Memorandum to Assistant Administrator for Policy, "Earth Resources Study Program," November 17, 1967; Harry J. Goett to Daniel G. Mazur, December 3, 1967. All of these documents can be found in the Space Policy Institute Documentary History Collection, Washington, DC.

era), while Agriculture desired a multispectral scanner. Both sensors involved relatively untested technology.[51]

Data processing and distribution provided even more serious challenges: a satellite taking pictures of the entire Earth, even at a spatial resolution of about 100 meters, would quickly produce an overwhelming amount of data. Effective use, particularly coverage of large areas and repeated coverage of the same scene to observe changes, would require analysis by computer rather than by a human photointerpreter. However, technology for such large-scale image processing had not yet been developed in the civilian world.[52]

Budgetary constraints proved even more serious than technical problems. Facing declining support for the space program once NASA reached its Apollo goal, the space agency's leaders attempted to capitalize on the usefulness of space to promote applications programs.[53] [II-18, II-19, II-20] The strategy did not work; the Bureau of the Budget repeatedly deleted the ERTS project from the budgets of NASA and the Department of the Interior. [II-21] In fact, the strategy of promoting usefulness may have backfired: the Bureau of the Budget (Office of Management and Budget after 1970) repeatedly asked NASA to prove that the benefits of Landsat would exceed the costs. NASA sponsored the required studies and also appealed cuts in the project's budget directly to the president.[54] The Department of the Interior obtained funding to build a data processing and distribution center only with the help of Republican Senator Karl Mundt of South Dakota. Senator Mundt had become a major supporter of the project when the Department of the Interior decided to locate its data processing center in Sioux Falls, South Dakota.[55] [II-22]

Despite all the discord, the first satellite proved a technical success. NASA launched Landsat 1 on July 23, 1972, and scientists quickly found many uses for the data. Prior to the satellite's launch, NASA had received more than 600 proposals from scientists and others requesting funding to investigate uses of data from the satellites. More than 200 proposals from U.S. investigators and 100 from overseas were funded; these scientists stood ready to use Landsat data as soon as they became available.

The use of Landsat data raised a number of issues. Developing countries had initially worried about the misuse of data gathered without their consent, but when satellite data began to arrive, they found it of considerable value in providing information on areas that were inadequately mapped. [II-23] Despite these benefits, debate continued over international political and legal issues associated with remote sensing. [II-24] Landsat data proved useful to scientists of many sorts, for everything from searching for oil to mapping ice.[56] Yet unlike meteorological satellites, whose data proved more useful to weather forecasters than to research scientists building models, Landsat data were quickly used by researchers, and much more slowly such data found widespread operational use. NASA

---

51. Mack, *Viewing the Earth*, pp. 66–79. NASA was not allowed to use better tested technology that had been developed for reconnaissance satellites.

52. Badgley to Distribution, "Meeting of Earth Resources User Agency Representatives, April 20, 1967," May 4, 1967; Mack, *Viewing the Earth*, pp. 107–18.

53. Leonard Jaffe to the Record, "Commentary Delivered by Mr. Leonard Jaffe at the Airlie House Planning Seminar, June 1966," July 8, 1966, Space Policy Institute Documentary History Collection, Washington, DC.

54. Mack, *Viewing the Earth*, pp. 80–93. This issue continued even after launch. See Paul A. Vander Myde to George M. Low, February 25, 1975, Space Policy Institute Documentary History Collection, Washington, DC.

55. Press release from the office of Senator Karl E. Mundt, March, 30, 1970, Space Policy Institute Documentary History Collection, Washington, DC; David L. Stenseth, "EROS—The Local Story," *IDEL Earth Trak* 1 (June 1974): 4–5; Mack, *Viewing the Earth*, pp. 132–45.

56. Anthony J. Calio to Director, Johnson Space Center, "Earth Resources Briefing for Petroleum Industry Representatives," November 12, 1973; Charles D. Centers to Mathews, February 15, 1973. Both documents are located in the Space Policy Institute Documentary History Collection, Washington, DC.

leaders discovered that if the space agency did not find ways to convince potential users to apply the data, then the project would not bring the benefits promised.[57] [II-25]

NASA therefore became involved not only in promoting the use of Landsat data, but also in supporting research to develop approaches to their use. One major project involved agricultural surveys, because better prediction of harvests was one use of Landsat data that the space agency had predicted would bring significant benefits.[58] The Large Area Crop Inventory Experiment (LACIE) involved NASA, the National Weather Service, and the Department of Agriculture in an attempt to develop and test a crop forecasting system using Landsat data. The data, it was claimed, could provide two things: information on how much land had been planted in a given crop (assuming one could differentiate among crops in the data) and information on crop health (because badly stressed vegetation reflected less infrared light). The project achieved reasonable success, but the problem of identifying and differentiating crops from satellite imagery turned out to be much more difficult than expected.[59] [II-26] Furthermore, the Department of Agriculture chose not to adopt the system developed by the LACIE project for operational use.[60] Several other experiments were more successful, but use of Landsat data for operational applications continued to develop more slowly than its promoters had hoped.[61] A NASA-sponsored technology transfer and utilization program involving universities, state and local governments, and industry could not reverse this reality.

Operational Landsat Program

The development of an operational program became the subject of a series of political debates. Different groups proposed at least three different alternatives. First, the federal government could develop an operational Earth resources program on the model of the operational weather satellite program, providing satellite data for the public good. Second, a private company might take over the existing Landsat system and run it as a business, an option usually called privatization. Third, a private company might develop and launch a new and separate Earth observations satellite, an option usually called commercialization.

Despite the somewhat disappointing growth of Landsat data use, the Department of the Interior wanted an early transition to a government-sponsored operational satellite program, but the agency faced a number of obstacles. [II-27] First, the Office of Management and Budget questioned whether Landsat had yielded enough benefits to jus-

57.   James C. Fletcher to Frank E. Moss, February 20, 1973; Hans Mark to Clifford E. Charlesworth, July 31, 1973; William E. Stoney to Distribution, "Summary Thoughts on Earth Resources Transfer Meeting," June 27, 1975. All of these documents are in the Space Policy Institute Documentary History Collection, Washington, DC.

58.   Carroll G. Grunthaver, U.S. Department of Agriculture, to Leonard Jaffe, Deputy Associate Administrator, NASA, August 22, 1973, discusses a new research and development study to develop a computer-based system for spring wheat yield estimation.

59.   James L. Mitchell, Office of Management and Budget, to Richard E. Bell and Don Paarlberg, Department of Agriculture, "Large Area Crop Inventory Experiment," October 8, 1976, Space Policy Institute Documentary History Collection, Washington, DC.

60.   Mack, *Viewing the Earth*, pp. 146–58.

61.   For a story with a successful outcome see John P. Erlandson, Army Corps of Engineers, to R. B. MacDonald, Johnson Space Center, March 16, 1976, Space Policy Institute Documentary History Collection, Washington, DC; Mack, *Viewing the Earth*, p. 130. For the reasons of slow adoption, see letter from Sally Bay Cornwell, National Conference of State Legislatures, to Allen H. Watkins, EROS Data Center, December 3, 1977, and letter from Allen H. Watkins to Sally Bay Cornwell, December 22, 1977, Space Policy Institute Documentary History Collection, Washington, DC.

tify a continued government-funded program.[62] [II-28] This issue posed a dilemma: potential users did not want to invest in the information systems necessary to process and analyze Earth resources satellite data until they knew that the data would continue to be available in the future, while the Office of Management and Budget did not want to fund an operational program until it was clear that enough users would participate to justify it.[63] As delays mounted, Landsat technology became increasingly out of date; in 1986 France launched an Earth resources satellite named SPOT (*Satellite Pour l'Observation de la Terre*) that in some ways was more sophisticated.

Instead of choosing which federal agency would house an operational program, an alternative approach gained increasing attention: that private industry should take over Earth resource satellites because it had communications satellites (instead of retaining the program as a government function as had been done with meteorological satellites). The impetus for the idea came primarily from those interested in reducing the federal budget; unlike communication satellites, the potential profitability of Earth resource satellites was so uncertain that private industry had only limited interest in taking over the whole system.[64] [II-29, II-30] In October 1978, President Jimmy Carter issued Presidential Decision 42, which asked NASA and the Department of Commerce to find ways to encourage private industry participation in civilian remote sensing (including Landsat, weather satellites, and ocean observation satellites).[65]

Some type of decision about an operational Landsat program had to be made, but privatization raised many difficult questions. Presidential Decision 42 led to the creation of an interagency task force to study the problems of and potential for private-sector participation in remote sensing, with a particular focus on Landsat. The task force report addressed issues ranging from cost to potential international sensitivity about private-sector control of data. It concluded that privatization of the whole system or of the space segment was premature, but that private industry should be encouraged to make proposals for investment in any part of the system.[66] This resulted in Presidential Decision 54 in November 1979, giving NOAA temporary responsibility for managing an operational Landsat system and asking NOAA to study ways to encourage private participation. Presidential Decision 54's long-term goal was eventual operation by the private sector.[67] [II-31, II-32]

62.   James C. Fletcher to James T. Lynn, Office of Management and Budget, September 15, 1976, Space Policy Institute Documentary History Collection, Washington, DC.

63.   Bruno Augenstein, Willis H. Shapley, Eugene B. Skolnikoff, "Earth Information from Space by Remote Sensing," report prepared for Dr. Frank Press, Director, Office of Science and Technology Policy, June 2, 1978; see Document II-30. This document also addresses another obstacle: controversy over which agency would control the operational system. See Mack, *Viewing the Earth*, pp. 204–05.

64.   Privatization meant the development of something similar to Comsat—a private corporation that would purchase satellites, pay for launch services, manage the working spacecraft, and process and sell data, all at its own risk. This is different from the small industry that had developed to sell analysis and special processing of Landsat data. That industry did not want the government providing too many services, but it did want the government to provide the basic data. See J. Robert Porter, Jr., President, Earth Satellite Corporation, "Statement before the House Subcommittee on Space Science and Applications," June 23, 1977, Space Policy Institute Documentary History Collection, Washington, DC.

65.   Also in October, Senator Harrison Schmitt introduced a bill calling for the creation of an Earth Resources Information Satellite Corporation modeled on Comsat. No action was taken on the bill. Science Policy Research Division, Congressional Research Service, *United States Civilian Space Programs. Volume II, Applications Satellites,* prepared for the Subcommittee on Space Science and Applications of the Committee on Science and Technology, U.S. House of Representatives, May 1983, pp. 249–50.

66.   "Private Sector Involvement in Civil Space Remote Sensing," draft, June 4, 1979, Space Policy Institute Documentary History Collection, Washington, DC.

67.   *United States Civilian Space Programs, Volume I,* pp. 238–42. For issues relating to the transition to an operational system, see the letter from Richard D. Lamm to George S. Benton, NOAA, April 30, 1980, with attached: "Recommendations of the National Governor's Association, National Conference of State Legislatures, Intergovernmental Science, Engineering & Technology Advisory Panel, National Resources & Environment Task Force, for the Final Transition Plan for the National Civil Operating Remote Sensing Program (first draft April 10, 1980)," April 30, 1980.

A lengthy debate followed concerning whether and how privatization might take place.[68] During that time, two additional Landsat satellites were launched to provide continuity of data delivery, but the development of a more advanced operational system was put on hold. These satellites did carry one new instrument, the Thematic Mapper (TM), which was a significant improvement over the Multispectral Scanner that had been carried on the initial satellites. The most provocative proposal to come from private industry was from Comsat in July 1981 to take on full responsibility for both Earth resource and meteorological satellites.[69] [II-33] Many observers had doubted that commercialization could protect the public interest in meteorological satellite data; similar concerns were voiced with respect to Earth resource satellites.[70] [II-34]

President Reagan proved an even stronger supporter of the transfer of government projects to private industry than President Carter had been. In March 1983, Reagan announced a decision to transfer Landsat, the meteorological satellites, and future ocean observation satellites to private industry.[71] As already mentioned, Congress rejected the idea of transferring the meteorological satellite program to private industry. [II-35] However, the Department of Commerce proceeded with a request for proposals from private industry for operational control of the Landsat system.[72] Congress passed a bill setting the terms for transfer in 1984, and the Earth Observation Satellite Company (EOSAT), a joint venture of Hughes and RCA, won the competition. [II-36]

This 1984 legislation supported the concept of providing sufficient subsidy to continue Landsat operations while EOSAT built a market for data. Department of Commerce officials envisioned that with government help, EOSAT would build Landsats 6 and 7. Eventually, administration and congressional supporters believed, the data market would grow large enough to support entirely private ownership and operation of future Landsat systems. NOAA's 1985 Commercialization Plan called for continued government funding of $250 million to build Landsats 6 and 7.[73] To assist in this process, EOSAT began building its own operations control and receiving station in Norman, Oklahoma.

In the fall of 1985, EOSAT complicated negotiations over the amount of subsidy by proposing to fly the TM on a spacecraft designed to be launched by the Space Shuttle.[74] Despite the loss of the orbiter *Challenger* in January 1986, NOAA agreed to the proposal

68.   William H. Gregory, "Free Enterprise and Landsat," *Aviation Week and Space Technology* 113 (July 14, 1980): 13; Ed Harper to Craig Fuller and Martin Anderson, "Resolution of Issues Related to Private Sector Transfer of Civil Land Observing Satellite Activities," July 13, 1981; National Oceanic and Atmospheric Administration, "Transfer of the Civil Operational Earth Observation Satellites to the Private Sector," draft, January 19, 1983. The last two documents are located in the Space Policy Institute Documentary History Collection, Washington, DC.

69.   Communications Satellite Corporation News Release, "Comsat President Proposes Bold Restructuring of Earth Sensing Satellite Systems," July 23, 1981, Space Policy Institute Documentary History Collection, Washington, DC.

70.   David A. Stockman, Office of Management and Budget, to Malcolm Baldrige, Secretary of Commerce, May 9, 1983. See Pamela E. Mack, "Commercialization, International Cooperation, and the Public Good," in Daniel S. Papp and John R. McIntyre, eds., *International Space Policy: Legal, Economic, and Strategic Options for the Twentieth Century and Beyond* (Westport, CT: Quorum Books, 1987), pp. 195–202.

71.   "Statement by Dr. John V. Byrne, Administrator, National Oceanic and Atmospheric Administration, U.S. Department of Commerce," March 8, 1983, Space Policy Institute Documentary History Collection, Washington, DC.

72.   U.S. Department of Commerce, "Request for Proposals for Transfer of the United States Land Remote Sensing Program to the Private Sector," January 3, 1984.

73.   Landsat proponents argued that nearly double this amount was necessary to ensure adequate support for the commercialization process, but David Stockman, director of the Office of Management and Budget under President Reagan, would agree only to the $250 million.

74.   At the time, NASA envisioned being able to launch the Shuttle into polar orbit from Vandenberg Air Force Base in California. Landsat satellites then would have been serviced in orbit by Shuttle crews.

in March 1986; in August of the same year, the Reagan administration issued a decision limiting Shuttle payloads to those requiring the unique characteristics of the Shuttle.[75] This caused NOAA to direct EOSAT to prepare for launch on an expendable launch vehicle. Other disagreements between the administration and Congress delayed a decision to fund the Landsat system until the spring of 1988.[76] By that time, it had become fully apparent that the subsidy ($219 million) would cover only the development and construction of one spacecraft. The Reagan administration and Congress nearly terminated EOSAT's operation of Landsats 4 and 5 several times for lack of a few million dollars of operating funds.[77]

Part of the difficulty arose because, in the 1980s, proponents of land remote sensing faced the same problem they had experienced in the 1960s. No single agency was willing to commit funding ($15–30 million per year beyond EOSAT's revenue from data sales) to continue system operations. Unlike the weather satellites, which NOAA operated to provide data for its own National Weather Service, the Department of Commerce had no internal constituency for collecting remotely sensed land data. The Carter administration had selected NOAA because of the agency's experience in operating the weather satellite systems. Congress expressed only lukewarm interest in supporting NOAA's long-term operation of Landsat. This lack of commitment to a continuously operated remote-sensing system undermined what little confidence data customers had in the Landsat system. Relatively few customers were willing to develop the necessary processing infrastructure and training programs or make other investments that depended on the routine delivery of Landsat data.

NOAA and EOSAT expected to launch Landsat 6 in 1992, with the federal government providing most of the funding for building and launching the satellite. However, even if Landsat 6 successfully reached orbit and operated as designed for five years, this plan still left the United States with the prospect of entering the late 1990s with no capability to collect Landsat data. It soon became clear that even if the data market doubled or tripled, EOSAT would not earn sufficient revenue to build Landsat 7. To resolve growing concerns over the future of the Landsat program, President Bush "directed the National Space Council and the Office of Management and Budget to review options with the intention of continuing Landsat-type data collections after Landsat 6." [II-37]

The Land Remote Sensing Policy Act of 1992

By the early 1990s, several circumstances led to the decision to return Landsat system operation to the government. First, the U.S. military made extensive use of Landsat and SPOT data to create maps used in planning and executing U.S. maneuvers during the 1991 Gulf War.[78] Second, Landsat proponents worried that failing to develop Landsat 7 would give SPOT full control of the international market for multispectral satellite data. Third, global change researchers began to appreciate that the twenty-year Landsat data archive would allow them to follow environmental change on parts of Earth's surface. Fourth, the attempt to commercialize the Landsat system had faltered badly, and policy

75. The White House, "Statement on the Space Shuttle," August 15, 1986, Space Policy Institute Documentary History Collection, Washington, DC.

76. U.S. Congress, Congressional Research Service, "Future of Land Remote Sensing System (Landsat)," 91-685-SPR, pp. 6–7, Space Policy Institute Documentary History Collection, Washington, DC.

77. M. Mitchell Waldrop, "Landsat Commercialization Stumbles Again," *Science* 246 (January 9, 1987):155–56; Eliot Marshall, "Landsat: Cliff-hanging, Again," *Science* (October 20, 1989): 321–22.

78. B. Gordon, Statement before the U.S. House of Representatives, Joint Hearing of the Committee on Science, Space, and Technology and the House Permanent Select Committee on Intelligence, Scientific, Military, and Commercial Applications of the Landsat Program, Hearing Report 102-61, June 26, 1991.

makers began to feel that no private company was soon likely to be able to provide equivalent data on the scale needed by federal agencies.[79] Finally, the advent of the geographic information system (GIS) and the development of other information technologies, such as high-powered computers, inexpensive storage devices, and the Internet, promised to reduce the costs and complexities of processing Landsat data.[80]

As a result of these and other pressures to continue collecting Landsat data, in 1992 the administration, with the strong support of Congress, moved to place operational control of Landsat 7 and beyond to DOD and NASA. [II-38] Under the Landsat management plan negotiated between DOD and NASA, DOD agreed to fund the development of the spacecraft and its instruments, while NASA agreed to fund the construction of the ground data processing and operations systems, to operate the satellite, and to provide for Landsat data distribution. [II-39] The Land Remote Sensing Policy Act of 1992, signed into law in October, codified the management plan and authorized approximately equal funding from each agency for the operational life of Landsat 7. [II-40]

Landsat 6 was to carry an Enhanced Thematic Mapper (ETM) having better radiometric calibration than previous TM sensors, along with an additional "sharpening" panchromatic band of fifteen-meter resolution, allowing it to deliver data with resolution nearly equivalent to SPOT data. NASA had studied this capability in the mid-1970s but dropped any plans to build higher resolution instruments as a result of national security restrictions on the sharpness of data from civilian satellites. By the 1990s, other countries had started selling fine-resolution data, so those national security concerns had become moot (see below).

Initial NASA and DOD plans called for Landsat 7 to include an ETM Plus, an improved version of the ETM under development for Landsat 6. Later, DOD began to consider adding a new multispectral sensor to the satellite, the High Resolution Multispectral Stereo Imager (HRMSI), capable of collecting five-meter resolution data particularly useful for mapping. NASA and DOD analysts estimated that developing, launching, and operating Landsat 7 for five years would equal $880 million (1992 dollars). NASA considered the additional instrument optional; in the course of discussions, DOD decided that it should be an operational requirement. However, the HRMSI sensor and additional ground operations equipment would have cost an additional $400 million. The high data rates expected for the HRMSI nearly doubled the overall required system data rate and would have added significant costs to NASA's yearly operations budget for Landsat 7.

In September 1993, Landsat 6 was launched but failed to reach orbit, raising additional concerns about the loss of data continuity. That same month, NASA officials concluded that the costs of operating Landsat 7 with HRMSI were too large, given other strains on the space agency's budget. In December 1993, DOD decided not to fund the resulting Landsat 7 budget shortfall. As a result of disagreement over the Landsat 7 requirements and budget, DOD decided to drop out of the agreement altogether. [II-41, II-42, II-43, II-44] That left NASA to fund the development of Landsat 7, carrying only the planned thirty-meter-resolution ETM Plus. After some discussion, DOD transferred

---

79. In 1990, John Knauss, Commerce Under Secretary for Oceans and Atmosphere, stated: "Our experience with the Landsat program . . . [has] led us to the conclusion that commercialization of Landsat, as had originally been envisioned, is not possible." J. Knauss, Under Secretary for Oceans and Atmosphere, Department of Commerce, Statement before the Senate Committee on Commerce, Science, and Transportation, June 12, 1990, Space Policy Institute Documentary History Collection, Washington, DC.

80. Ray A. Williamson, "The Landsat Legacy: Remote Sensing Policy and the Development of Commercial Remote Sensing," *Photogrammetric Engineering and Remote Sensing* 63 (July 1997): 877–85.

$90 million to NASA to assist in developing the satellite and sensor because DOD would be a major customer of data from Landsat 7.

In early 1994, the question of which agency would actually operate Landsat 7 had not yet been resolved. NASA planned to use Landsat data to support its research into land use and land change as part of the U.S. Global Change Research Program. Landsat 7 is formally now part of NASA's Mission to Planet Earth. These data will also support many federal government operational programs and the data needs of state and local governments, the U.S. private sector, and foreign entities.

In May 1994, the Clinton administration resolved the outstanding issue of procurement and operational control of the Landsat system by assigning it jointly to NASA, NOAA, and the Department of the Interior. Under this plan, NASA will procure the satellite, NOAA will manage and operate the spacecraft and ground system, and the Department of the Interior will archive and distribute the data at the marginal cost of reproduction. [II-45] NASA has scheduled the launch of Landsat 7 for 1999. However, the future of government-funded land remote-sensing satellites beyond Landsat 7 is still uncertain.

The Beginning of Commercial Remote Sensing

Having failed in successfully transferring the Landsat system to private ownership and operation, government programs and policy were nonetheless in part responsible for making commercial remote sensing possible. The Land Remote Sensing Policy Act of 1992 included Title II, which sets out the terms for government licensing of private operators of remote-sensing satellite systems. Title IV of the 1984 act had included identical wording, requiring that potential private operators of remote-sensing satellites acquire an operating license from the federal government in accordance with international obligations, but until 1992, no company had taken advantage of that provision. The legislation assigned to the secretary of commerce the responsibility for considering and granting such licenses, requiring that the secretary act on such applications within 120 days, "in consultation with other appropriate United States Government agencies. . . ."[81]

In October 1992, shortly after President Bush signed the 1992 act, WorldView, Inc., applied for a license to operate a commercial remote-sensing system. WorldView's plans called for building a system capable of collecting stereo panchromatic data of three-meter resolution and multispectral data of fifteen-meter resolution in green, red, and near-infrared spectral bands, although with a narrow field of view.[82] WorldView's sensor was designed to collect stereo pairs along the satellite track as well as sideways off track, enabling a rapid revisit of areas of particular interest. Technology developed as part of Lawrence Livermore Laboratory's ballistic missile defense program supplied the instrumental basis for a commercial system. Perhaps the greatest innovation, however, was a data marketing plan based on commercial objectives, rather than on meeting government requirements. WorldView's officials judged that the ultimate market for these data was the information industry, which was planning to use the Internet, CD–ROM, and other information technologies to reach customers quickly and efficiently.[83] Such plans depended on the ability of WorldView and other companies that followed to build and operate a satellite at much lower cost than Landsat. A commercial data marketing plan involves

---

81.   The Land Remote-Sensing Policy Act of 1992 (15 U.S.C. 5621, Sec. 201 (c)).

82.   The Early Bird satellite will be capable of gathering panchromatic data along a three-kilometer swath and multispectral data along a fifteen-kilometer swath.

'83.   Williamson, "The Landsat Legacy," pp. 883–84.

collecting data only of sufficient quality and quantity to meet the needs of most customers.[84] This also reduces costs compared to a system such as Landsat 7, which is designed to collect as much data as possible to provide a global archive for the future needs of scientists.

Department of Commerce officials coordinated the license application with DOD, the Central Intelligence Agency, and the Department of State. By late 1992, national security planners were more inclined than ever before to ease earlier restrictions on the resolution limits of civilian data. Their decisions were moved in part by the knowledge that the French were planning to improve the resolution of their SPOT system, and the Indian Space Agency was also moving to higher resolution instruments. Furthermore, the Russian firm Soyuzkarta had begun to market high-resolution multispectral photographic data (two-meter resolution) from the formerly secret Russian KVR-1000 sensor. On January 4, 1993, the Department of Commerce sent a license to WorldView, allowing it to operate a three-meter satellite system. [II-46, II-47, II-48].

Other companies soon filed their own applications for systems that would achieve even greater sharpness. In June 1993, Lockheed, Inc., filed with the Department of Commerce for a license to operate a system capable of achieving one-meter resolution. Shortly after, Orbital Sciences Corporation, in partnership (later dissolved) with GDE Systems and Itek, also filed a similar license request. The proposal to collect higher resolution data caused the Clinton administration to reconsider desirable policy for commercial remote sensing. Although the sale of such data abroad posed no threat of the transfer of critical technology, in the view of some, one-meter data were too close to the reconnaissance capabilities of high-flying aircraft and classified satellites.[85] Others, while recognizing the risk of marketing these data worldwide, have argued that data of high resolution can moderate potential conflict if they are available to all sides.[86]

Nevertheless, one-meter data, delivered in a timely manner, are of significant security utility for surveillance, military planning, and the creation of the up-to-date maps needed to fight battles effectively. When combined with the geolocational capabilities of the global positioning system, these data also make it possible for belligerent nations to target specific locations for cruise missile and other precision attacks. Hence, intelligence officials argued, if the data were sold globally, there would have to be some sort of control over distribution. Ultimately, after several months of discussion, officials decided that the benefits of keeping such data under the control of U.S. suppliers were greater than the risks posed by possible data misuse.

In March 1994, eight months after receiving the license application, the White House released a policy statement concerning licenses for commercial remote-sensing systems. The policy required the satellite operator to maintain satellite tasking records and to make them available so that the federal government could determine who purchased what data, if necessary. It also authorized the government to cut off or restrict the flow of data during times of crisis to protect national security interests. [II-49] The Department of Commerce has granted several licenses based on this policy, including one to Lockheed, Inc. [II-50]

---

84.   For EOSAT, the operator of Landsats 4 and 5, this meant collecting fewer scenes than government operators would have collected. NASA and NOAA were interested in gathering as many scenes as possible to file the archive of Landsat scenes.

85.   V. Gupta, "New Satellite Images for Sale: The Opportunities and Risks Ahead," Center for Security and Technology Studies, Lawrence Livermore National Laboratory, UCRL-ID-118140, 1994, Space Policy Institute Documentary History Collection, Washington, DC.

86.   B. Gordon, "The Moderating Effects of Higher Resolution Civil Satellite Imaging on International Relations," paper presented at the 1996 AFCEA Conference, Washington, DC, June 1996.

Conclusion

One might anticipate that space applications programs would have been the least controversial aspects of the space program because they would seem to be the most obviously beneficial. An examination of their history, however, suggests that applications satellites raised difficult institutional policy issues, resulting particularly from the large number of interested organizations involved. A project such as Apollo served primarily public and political interests in a space race. For basic space science, NASA had a clear constituency of scientists. In the 1970s, NASA sought to control research and development for satellite applications in the same way it controlled space science, but the space agency found that user agencies expected to direct research to meet their own perceived needs. In addition, the technological potentials of the various applications fields that scientists found most interesting were not necessarily the ones with the most short-term practical value.

During the 1980s, government-funded applications satellite systems faced an increasingly difficult budgetary climate. Continuing development of satellite technology made it possible to offer more and more sophisticated services, but in a time of tremendous pressure on the federal budget, the government has been reluctant to fund more expensive systems, even if they resulted in better services.

Commercial interests in land remote sensing and international cooperation in meteorological observations have helped invigorate these two applications. As history demonstrates, land remote-sensing applications have proven more difficult to integrate into existing systems than meteorological or communications satellites had been. The obstacle was not primarily a lack of usefulness of the data produced by the satellites; rather, proponents of the Landsat program faced an intense debate over the proper role of government in developing and operating a system that benefits both public and private data users.[87] If the operation of commercial remote-sensing satellites proves successful, it may resolve not only the long-standing tensions between research and operational uses of remotely sensed Earth observation data, but also the question of the proper role of government and the private sector in supplying them.

In contrast to land remote sensing, most observers continue to support the public provision of meteorological data. However, pressure to reduce satellite system costs has endangered the robustness of NOAA's system. A changed political environment resulting in a merged civil-military system and increased international cooperation should improve the ability of the government to continue to provide high-quality meteorological data while reducing system costs.

Document II-1

Document title: Dr. Harry Wexler, "Observing the Weather from a Satellite Vehicle," *Journal of the British Interplanetary Society* **7 (September 1954): 269–276.**

This article was originally presented by Harry Wexler, Chief of the Scientific Services Division of the U.S. Department of Commerce Weather Bureau, as a speech at the Third Symposium on Space Travel, held at the Hayden Planetarium in New York City on May 4, 1954. It is one of the earliest inquiries into the possible uses of satellites in forecasting weather. While correct in anticipating a satellite's utility in observing large-scale weather patterns, it is interesting that Wexler nevertheless dismisses a satellite's potential for what are now routine observations of pressure, temperature, and humidity. The two

---

87. Philip J. Hilts, "Landsat Satellites Termed Incapable of Profitable Operation This Century: Substantial Demand Not Enough to Sustain Business, Reports Say," *Washington Post*, March 12, 1989, p. A4.

figures accompanying the article are omitted here, but their captions are included, along with the description of Figure 1. Note the British-style spellings, such as the word "centre."

[269]

Observing the Weather from a Satellite Vehicle*

By Dr. Harry Wexler,

Chief Scientific Services Division, U.S. Department of Commerce,
Weather Bureau

Introduction

To predict the future of the atmosphere, the meteorologist must know its present state—as defined by the three-dimensional distribution of pressure, temperature, wind, humidity, clouds, precipitation, etc. To do this, at hundreds of stations throughout the Northern Hemisphere, the atmosphere is probed by balloon-borne instruments which radio back to Earth values of pressure, temperature, humidity, and whose paths can be translated into wind direction and speed of the various layers through which the balloon ascends. These observations expressed as numbers or symbols, plus auxiliary information of clouds, precipitation, etc., are plotted on weather charts and synthesized into an instantaneous picture of the atmosphere which, however, is presently incomplete because of lack of observations in large portions of the atmosphere, specially over oceans and unpopulated areas. Knowing the present state of the atmosphere and past motions of the storms enables a prediction to be made by extrapolation and other techniques.

A satellite vehicle traveling about the Earth outside the atmosphere would not assist in portraying the pressure, temperature, humidity, and wind fields by direct measurement. However, by a "bird's-eye" view of a good portion of the Earth's surface and the cloud structure, it should be possible by inference to identify, locate, and track storm areas and other meteorological features. The vehicle would then serve principally as a "storm patrol." There exists under normal conditions a characteristic cloud condition for a "typical" extra-tropical storm.

A plan view of a typical mid-latitude storm shows cold and warm fronts, whose low-pressure centre is at their vertex, and their accompanying cloud systems.

A major cyclonic storm-cloud system visible from above will be the warm front cloud from which the major portion of the storm's precipitation usually falls. In west-east vertical sections, the cloud at the extreme right is composed of high-level (5–10 miles) tenuous cirrus or cirrostratus clouds which change to denser altostratus and altocumulus and finally to thick precipitating nimbostratus as one approaches the storm. If the ascending warm air above the warm front is unstable enough, cumulonimbus or thunderstorm clouds will penetrate above the top of the nimbostratus cloud.

In the warm sector, or the area between the warm front and cold front, there will be stratus and fog, if the surface is colder than the air, or cumulus clouds, if it is warmer. Approaching the cold front the higher altocumulus clouds will [text continued on page 271 after Figure 2]

---

* Presented at the Third Symposium on Space Travel, American Museum, Hayden Planetarium, New York, May 4, 1954.

[270] [figure omitted] FIG. 1. The Earth from an altitude of 100 miles.

This picture show the Earth's curvature and more than 200,000 square miles of the U.S.A. and Mexico, and was taken from a V2 on March 7, 1947. The view stretching to the horizon is a distance of about 900 miles, and the dark body of water near the top of the picture is the Gulf of California, about 65 miles wide. The picture also shows rivers, islands in the Gulf of California, the peninsula of lower California and part of the Pacific Ocean. The two cameras were installed amidship in the rocket as part of 2,000 lb. of scientific instruments. They operated automatically taking pictures through an infra-red filter, used to cut the haze. The time of flight from launching to the break-up of the rocket was 6 1/2 minutes.

[271] [figure omitted] FIG. 2. Diagram of area included in fig. 1.

appear, closely followed by a narrow band of cumulonimbus and then scattered fair-weather cumulus in the cold air mass well behind the cold front.

The characteristic features of the cold and warm front cloud systems plus the adjacent air mass clouds should enable unique identification of a cyclonic storm, either in its maturing or fully developed stages. The incipient, or embryonic storm will be more difficult to detect because of lack of fully developed cloud systems. However, because of the tendency of cyclonic storms to form in "families," arrayed in a southwest-northeast line with the older storms located farther northeast and with a known average spacing between storms, it may be possible to detect an incipient storm by its position relative to the more noticeable mature storms and possible clues from the cloud system.

In Fig. 1 is shown an actual cloud photograph taken from a V2 rocket at a height of 100 miles above White Sands, New Mexico, on March 7, 1947. Unfortunately there was no mature extra-tropical storm within the field of [272] view of the camera, and the clouds shown are mostly "fair-weather" clouds caused mainly by heating of the ground and lifting of the air by the mountains. The most prominent clouds are thousands of bright cumuli—arrayed in roughly parallel bands, called "cloud streets," which usually indicate direction of the wind. These clouds usually occur two to eight miles above the surface, the higher cloud tops being associated with thunderstorms. The fuzzy clouds, so transparent that the cumulus clouds are visible through them, are the high-level cirrus clouds found at heights eight to ten miles. Far to the west, off the California cost, are patches of the characteristic low California stratus clouds (height one to two miles) with parts of the ocean surface visible.

The most that a meteorologist could obtain from such a cloud view would be the negative knowledge that no major storm is present plus some indication of the wind direction at cloud height and possibly the distribution of thunderstorms.

In order to reconnoitre the weather most effectively, the Satellite Weather Station should have the following properties:—

(a) It should be located far enough away to have an instantaneous field of view comparable to North America and adjacent ocean areas—similar to the area covered by the forecaster's "working" chart.

(b) It should not be so high that cloud areas and geographical features are not readily identifiable.

(c) It should move in such a manner as to have the same cloud system in the field of view at least twice in a 12-hour period to obtain a track of the storm associated with the cloud system.

(d) It should not move so fast that individual cloud systems cannot be located accurately with respect to known ground features.

(e) It should cover the entire Earth in daylight at least once daily.

(f) It should have a westward component of motion relative to the Earth's surface so as to detect quickly new storms which usually move from west to east.

Such a vehicle is one which is located at 2•01 Earth's radii from the Earth's centre or about 4,000 miles from the Earth's surface and which has a period of rotation about the Earth of exactly 4 hours. If the Earth were not rotating the vehicle would move in the same meridional plane through the North and South Poles. But since the Earth does rotate as the vehicle moves, its path relative to the Earth's surface is a series of curves.

Let us assume that at noon on March 21 the vehicle is directed poleward from the Equator at the 95th meridian west, at "0" hour. Assuming no external perturbations, the orbit of the vehicle is always maintained in a plane parallel to its initial orbitary plane, but attached to the centre of the Earth in its motion through space. The Earth rotates under the vehicle in such a way that as the vehicle proceeds northwards, it crosses all latitudes at exactly noon and after one hour it passes over the North Pole; afterwards it then moves southward at all latitudes at exactly midnight. At 2 hours it is at the Equator, at 3 at the South Pole, after which it enters into the daylight hemisphere again crossing all latitudes at exactly noon in its northward passage. At 4 hours, it crosses the [273] Equator at the 155th meridian, west, and repeats a similar path on the Earth's surface, but displaced westward from its initial path. In 24 hours it returns to its initial point of departure after having made both a daylight (noon-time) and night (midnight) surveillance of the entire Earth's surface.

Twenty minutes after its departure on its first leg, when the vehicle has moved over Amarillo, Texas, its horizon will enclose an area almost identical to the weather chart used in preparation of weather forecasts for North America and adjacent oceans.

What would be seen from the vehicle at some 4,000 miles above Amarillo, Texas, at exactly noon on June 21? An attempt has been made to portray the scene below under the assumption that the Sun is directly overhead. In drawing a chart before sketching in the clouds, an attempt was made to indicate the surface features of the Earth, taking into account its normal colour and reflectivity (albedo) of sunlight, and the scattering and depleting effects on the passage of light through the Earth's atmosphere in the following way:—

(a) Normal illumination values at the surface were first entered in the chart according to zenith distance of the Sun.

(b) Next, values of the apparent illumination or "brightness" were obtained by taking the product of the surface albedoes and the illuminations. For simplicity only two albedo figures were used: 4 per cent. for water and 15 per cent. for land. This then gives the brightness field of the Earth before passage of the light up through the atmosphere.

(c) Next the Earth's surface brightness was computed after depletion by the atmosphere, values for which are known from the incoming sunlight.

(d) Next was computed the atmospheric contribution to the brightness field at the vehicle. This was done by estimating from available observations, the portion of radiation coming from the sky to the ground (i.e. the downward radiation or "skylight") and by assuming that the same fraction of illumination is scattered upward. This procedure assumes that the atmosphere is a "uniform diffuse reflector" of the brightness shown.

(e) The two brightness values—from the Earth's surface and from the atmosphere—are added together to give a total brightness.

To distinguish the over-all brightness contrast between ocean and land, for example,

the fractional contrast $F = 2 \dfrac{B_L - B_O}{B_L + B_O}$ must be larger than $1/10$. The computed values of F (not shown) are considerably larger than this value, except near the periphery, indicating that for most of the observed area land can be readily distinguished from ocean.

Colour contrasts of objects on the ground tend to be suppressed in at least two ways: selective scattering of the bluer components by atmospheric molecules, and "dilution" of colours by the white diffuse component contributed by the non-selective foreign particle scattering in the atmosphere.

The effect of the first type of scattering is to deplete the blue colours relatively more than the longer wave-lengths. The over-all effect is to emphasize red [274] colour components of the objects on the ground, as compared to the blue components; and to screen both with diffuse light consisting of a relatively large blue component mixed with white light. The over-all result would be to give a bluish tinge to what is seen, since the blue scattered from the incident solar beam would more than make up for the blue-depletion of light coming from the ground.

As to the colour of the sky on the horizon, we might expect that there would be a grey layer, corresponding to the atmosphere say in the lower 10,000 ft., with an upper thin blue region in the region of substantial Rayleigh scattering, and black above that.

Thus, as a result of all these calculations, a reasonable picture was obtained of the surface features of the Earth under normal conditions of June 21 ground cover illumination, albedo and atmospheric effects, but without clouds. Over this chart was sketched a hypothetical cloud pattern normally associated with certain atmospheric disturbances. Albedo values were assigned to various cloud types and their brightness as computed. These "disturbances" included the following:—

(a) A cyclone family of three storms in various stages of development extending from Hudson Bay south-westward to Texas.

(b) The north-eastern part of another such cyclone family whose oldest member is in the Gulf of Alaska, the remaining members to the southwest being invisible.

(c) A fully developed hurricane embedded in "streets" of trade cumuli in the West Indies.

(d) The Intertropic Convergence Zone (or Equatorial Front)—a zone of interaction between the north-east trades of the northern hemisphere and the south-east trades of the southern hemisphere—extending west of Isthmus of Panama to the mid-Pacific.

(e) A "line-squall"—favourite breeding-ground of severe wind storms and tornadoes—in the eastern U.S. moving ahead of the cold front and surrounded on both sides by the cauliflowerlike cumulus congestus.

(f) Scattered cumulus clouds of varying thicknesses over the heated land areas—especially in the mountains and other areas where dynamic effects encourage the lifting of air in vertical columns.

(g) Altocumulus lenticularis or lens-shaped clouds formed by lifting of layers of moist air over mountains and usually found where the "jetstream" crosses mountains, as over the northern Canadian Rockies.

(h) Low stratus and fog found off the southern and lower California coasts, over the Great Lakes, the Newfoundland area, formed by passage of warm moist air over cold surfaces.

The cumulus cloud systems over the oceans will tend to fall in fairly regular patterns or "streets"—even more so than was observed over the rough terrain in the V2 picture. The regularity of the ocean cloud systems in the present sketch is probably exaggerated, but its breakdown into a more irregular pattern over land is believed to be real. The centres of the anticyclonic or "high pressure" areas are marked by little or no cloud.

[275] This then is the hypothetical picture visible from the 4,000-mile high vehicle over Amarillo, Texas. Some of these clouds, such as the Trade Cumuli, could undoubtedly be observed on almost any day and others, such as the hurricane, seen only rarely. The cyclone families would be observed daily, but their location, the number of individual

storms, size and intensity would vary geographically. A meteorologist given a clear picture of the cloud distribution, as here portrayed, could without difficulty sketch in a very useful weather chart showing location of the various stormy and fair weather areas; in fact, he would have a much better idea of the large-scale weather distribution than his Earth-bound colleague, who is forced to rely on scattered observations taken at or near the Earth's surface.

As for obtaining two or more "fixes" on storms within 12 hours, this would be possible as the vehicle makes successive passages toward the Poles—and the closer to the Pole the storm is located the more such fixes could be made. For example, the large fully developed storm depicted over Hudson Bay would be visible first on one leg of the path, again 4 hours later on the next leg, and again 4 hours later on a third leg. This would, by reference to known surface features, enable tracking of the storm in the 12-hour interval. A word of caution is necessary since the clouds which on one hand make possible the visual identifications of the storm will hinder its location with respect to known surface features. Nor would trying to track the storm by observing the edge of its cloud shield necessarily give an accurate track, since the changing cloud pattern associated with such large, usually dissipating storms may give spurious motions, as they form on one side and dissipate on the other. Thus, there will not be too good an accuracy for tracking these large storms, but this is not too important since their speed of motion is usually slow anyway. On the other hand, the incipient or developing storm, so important for future weather developments, is faster moving and has a less extensive cloud system associated with it so that more accurate fixes should be possible. The hurricane, with its cloud bands, similar to the arms of a spiral nebula, and its open "eye" at the centre, will be a much easier storm to detect and follow accurately. Cloud systems associated with cold fronts and squall-lines will also lend themselves to accurate tracking.

As the days pass, however, and the Earth moves in its orbital motion about the Sun, the vehicle will cross each latitude about 4 minutes earlier than the preceding day. Thus, if motion northward is started at noon on March 21, this will change on June 21 to 6 a.m. moving north, and 6 p.m. moving south; in this case, the field of view in daylight will be mostly to the east (going north) and to the west (going south) and the efficiency of the vehicle as a cloud patrol will have diminished considerably. On September 21 its efficiency will increase again as it moves south at noon and north at midnight. However, on December 21 its efficiency will drop again—and to its lowest point as far as the Northern Hemisphere is concerned. It will move north at 6 p.m. and south at 6 a.m.—but because of the low solar declination at this time and consequent lack of daylight hours, its usefulness as a cloud and storm detector will be greatly impaired. This is a serious defect because the winter season is the busiest period for storms in the Northern Hemisphere. This suggests as a better [276] solution that the preceding plan, the initial movement northward or southward at noon on December 21. This will then give optimum conditions for winter weather patrol—excluding the Arctic and some distance south where little or no daylight will prevail.

This visual cloud reconnaissance might be taken automatically by a television camera in an unmanned vehicle and relayed to Earth to various collection centres for study, analysis and exchange with other forecast offices to obtain a truly global weather picture. If the vehicle could be properly manned and equipped, then other valuable geophysical and solar data could be obtained as follows:—

(a) *Temperature* of the Earth's surface and a rough average temperature of the intervening atmosphere by observing the infrared spectrum.

(b) *Precipitation Areas* (rain, snow, etc.) could be detected by radar as well as the heights of their formation above the surface; also the height of the freezing level which shows as a bright band in the radar scope.

(c) *Thunderstorm Areas*—by location of lightning either visually (at night) or electronically (at day).

(d) *Solar Radiation* measurements, particularly in the ultra-violet, to correlate with weather changes in an attempt to see if unusual spells of weather are solar-controlled.

(e) *Albedo Measurements*—to keep a global account of the day-to-day changes in reflectivity of the Earth's surface to solar radiation from the ground and water surfaces (including snow and ice cover), clouds, atmospheric turbidity. Long-time variations in the Earth's albedo could be correlated with similar variations in climate. For example, it has been estimated that a one-point drop in Earth's albedo from its average value of 35 per cent. would lead to an average world-wide warming of 1° C. The fraction of sky covered by clouds is of such critical importance in albedo changes that it has been estimated a variation in the average world cloudiness from 0•4 to 0•6 would explain the whole range of climatic changes—from ice ages to the intervening warm periods.

(f) *Meteoric Dust*—samples could be obtained to test a recently proposed theory that these particles may serve as cloud-seeding agents, thus causing increases in rainfall, especially after meteoric showers. Samples of the dust to test in cold boxes, together with measurements of their natural concentration would shed direct evidence on a problem which heretofore has only been possible to treat statistically.

In summary, it can be stated without question that a satellite vehicle, moving about the Earth at the proper height and manner would be of inestimable value as a weather patrol for short-range forecasting and as a collector of basic research information for solar and geophysical studies, including long-term weather changes and climatic variations.

Document II-2

Document title: S.M. Greenfield and W.W. Kellog, "Inquiry into the Feasibility of Weather Reconnaissance from a Satellite Vehicle," The RAND Corporation, R-365, August 1960, pp. v–vi, 1–23, 31.

Source: NASA Historical Reference Collection, NASA History Office, NASA Headquarters, Washington, D.C.

One of the possible uses of an orbital satellite that Project RAND had addressed in its 1945 report "Preliminary Design of an Experimental World-Circling Spaceship" was predicting the weather by observing cloud patterns on a large scale. Recognizing the potential of such a capability, but unsure of how useful cloud images alone would be in predicting the weather, the Air Force commissioned Project RAND to conduct a further study to determine more precisely what useful information could be gained from high-altitude observations. The fourteen figures accompanying the report are omitted here, but their captions are included.

Inquiry into the Feasibility of Weather Reconnaissance from a Satellite Vehicle

S. M. Greenfield and W. W. Kellogg

August, 1960

NOTE

Originally published as classified Report R-218
of the same title, dated April, 1951. Amended
and released for open publication, August, 1960.

[v]

SUMMARY

The value of observing the weather over inaccessible areas by aerial weather reconnaissance has been recognized for many years. An alternative method of obtaining broad
coverage of the weather, however, is thought to lie in the use of a special satellite vehicle
which could observe cloud patterns. It is obvious that any meteorological reconnaissance
utilizing only observations from such a high-altitude "eye" cannot provide quantitative values for the parameters normally associated with standard weather observation and forecasting techniques. In determining the feasibility of such a system, therefore, the
questions that must be answered are (1) What extent of coverage can be expected from a
satellite viewing system? (2) In terms of resolution and contrast, what can be seen from
the satellite? (3) Given proper coverage and resolution, what can actually be determined
regarding the synoptic weather situation from this information?

General considerations of ease of satellite launching and photographic coverage suggest an orbiting altitude of 350 to 500 mi.[1] For the purpose of the present study, however,
only the 350-mi altitude was considered to any extent. At this altitude, a vehicle would
have an orbital velocity of about 24,870 ft/sec and would make one complete circuit of its
orbit in 1.6 hr. Assuming that any regressive motion of the satellite's orbit owing to the
spatial motion and oblate shape of the earth is corrected for, and that the area it is desired
to observe is in daylight during the vehicle passage for an extended period, this area will
be covered and televised in a grid fashion once every 24 hr. It is visualized that, by means
of mechanical scanning transverse to the path of the satellite, a continuous strip whose
width is equal in order of magnitude to the altitude of the vehicle will be viewed. As an
example of the sort of coverage which could be provided at middle latitudes, with a satellite at a 354.6-mi altitude the fraction of the area between 45° and 50° latitude which can
be covered grid-fashion with a 100-mi wide scanning path in 24 hr is one-third, and if the
width of the path is increased to 450 mi, the 24-hr coverage is complete.

Utilizing photographs from recent vertically fired rockets (V-2), an estimate of the
dimensions of the smallest increment necessary for proper cloud identification was made.
This was found to be approximately 500 ft and is termed the "usable resolution" in this
report: Entering Tables 1, 2, and 3, which give resolution versus contrast for various values of frame speed, aperture size, and various types of illumination, showed that it was possible to obtain this value of resolution in sunlight illumination with contrast between

---

1.   Additional information concerning the problems of satellite operation is given in the RAND Report
R-217, April 1951 (out of print).

cloud and background of less than 10 per cent. An examination of the albedos from typical background objects, as presented in Fig. 2 . . ., compared with cloud albedos seems to indicate that 10 per cent contrast is available over a wide range of [vi] possible cloud-ground and cloud-cloud combinations. This, therefore, appears to establish the feasibility of cloud identification from high altitudes, at least from the standpoint of contrast and resolution.

Owing to the lack of quantitative measurements, the clouds must be utilized to their utmost in determining the synoptic weather picture. Experience and statistical climatological values play their part in forming this picture, and the process involves a "hunting technique" that oscillates between the three main tools at the analyst's command. Some detailed estimates of various parameters are possible from the visual cloud characteristics. Items, such as moisture content, temperature gradient, stability, magnitude or direction of vertical pressure gradient, wind shear, and wind direction[,] all show promise of yielding good estimates of the actual values to this type of analysis and of helping to clarify the final estimated synoptic picture. For any future operational use, this study has shown that such things as a cloud atlas of clouds viewed from above, complete climatological material on the area in question (including a possible statistical survey of fluctuations from the normal of the various parameters as attributed to synoptic systems and broken down into small regions of similar climate and topography), and perimeter weather will immeasurably help the job of the observer and analyst. An aid to getting a "feel" for the problem involved, photographs from three rocket flights were analyzed and the synoptic situation was estimated. These results and the actual weather for the corresponding times are presented in the section entitled "Results of Three Attempts at Analysis. . . ." In an attempt to correlate further the rocket photographs with the actual synoptic picture, Dr. J. Bjerknes, of U.C.L.A., independently made an analysis of photographs taken on a flight on July 26, 1948. In this analysis, all other synoptic meteorological data available for that date were utilized. . . .

[1]   **INTRODUCTION**

The foundation of all meteorological forecasting systems is the weather-observing network. Whether the forecast is "local" or for the entire Northern Hemisphere, the starting point must be an appraisal of the synoptic weather picture. Since storm systems at middle latitudes generally move from west to east, a meteorologist who does not have good observations from a rather wide area (particularly to the west) is at a disadvantage; and such is often the case for coastal regions, since weather reporting over the oceans is often inadequate.

Although ship reports and weather reconnaissance by aircraft help to some extent to fill the gap, there has long been a need for extending weather observations over the oceans and inaccessible areas. A solution to this problem may lie in weather observations made by means of a television camera placed in an unmanned satellite vehicle. Such a method has the advantage of providing a means of observation of the over-all picture of the wide-scale weather situation that is lacking in normal daily weather observations, and should give new insight into the behavior of the atmosphere.

It is obvious that in observing the weather through the "eye" of a high-altitude robot almost all of the regular quantitative measurements usually associated with meteorology must fall by the wayside. It is impossible to make more than an intelligent guess at the values of temperature, pressure, humidity, and the remaining quantitative meteorological parameters. Because of this, the analyst must rely on the visible components of meteorology to ascertain to some usable degree the synoptic weather situation.

Clouds, being the objects most easily discernible from extreme altitudes, become the important item and must be utilized to the utmost in forming a synoptic picture. It is apparent that from clouds alone it will be impossible to tell everything about the current synoptic situation. Combined, however, with both theoretical knowledge and that gained through experience, an accurate cloud analysis can produce surprisingly good results.

The purpose of this report is to present methods of attack on the above problem, to show what may be actually seen from high-altitude photographs (primarily a discussion on necessary resolution and area coverages), to discuss what may be determined from these photographs (both directly and indirectly), and to give some results obtained. Although all the present analysis is based on data obtained from vertically fired rockets, the experience gained therefrom permits recommendations on possible methods of forming a synoptic picture from satellite-missile photographs.

[2]   **THE SATELLITE VEHICLE**

Owing to the ever-changing pattern of the atmosphere, the need for almost constant surveillance must be foremost in any plan to trace synoptic weather situations. Any vehicle designed for such a purpose must therefore have the ability to make many trips over the area in question. These traverses, moreover, must be made in such a fashion that they not only cover a representative portion of the area, but also complete their cycle often enough to enable an observer to notice any significant change or shift in the cloud systems.

Such a vehicle is the satellite. Flying high above the sensible part of the atmosphere, so that atmospheric drag becomes negligible, the satellite becomes an unparalleled instrument for weather reconnaissance when scope of view is considered. For the purpose of simplicity, all calculations and performance considerations in this report will be based on a satellite assumed to be circling the earth at an altitude of about 350 mi.[2] At this altitude such a vehicle[3] would have an orbital velocity equal to 24,870 ft/sec[4] and would make one complete circuit of its orbit in 1.6 hr. Also, because of the fact that this missile is theoretically moving in a stable orbit around the earth, the globe turning under the vehicle causes the trajectory of the satellite to appear to "creep" over the face of the earth, thereby increasing the area observed.[5] Depending on the efficiency of the power plant, the order of magnitude of the time period for which the vehicle could be kept operating is thought to be 1 yr. However, in attempting to decide the satellite's full worth for weather reconnaissance, the questions that must be considered are as follows: Can enough be seen from such altitude to enable an intelligent, usable, weather (cloud) observation to be made, and what can be determined from these observations?

---

2.   *Ibid.*

3.   The actual altitude to which these figures apply is 354.6 mi.

4.   The actual velocity of a projection of the satellite's image over the face of the globe is really a variable resulting from the change in angular velocity from latitude circle to latitude circle.

5.   It should be noted that the concept of "repetitive traverses" is in itself complicated in that, regardless of the stability of the satellite orbit, the spatial movement and the oblate shape of the earth impart a regressive motion to the vehicle relative to fixed points on the earth. From a satellite at approximately 350-mi altitude in an orbit set tangent to a latitude of 56°, 78 days will be required for it to appear twice over the same point on earth at exactly the same time. This regressive motion can be partially corrected by an adjustment of the speed (through altitude change) of the satellite. It further imparts a limitation on successful viewing in that for approximately half of the 78-day period (assuming 12 hr of photographable time out of every 24) the desired area will have night at the time of the satellite's passage. For a complete discussion of regression of the orbit, the interested reader is referred to RAND Report R-217 (see footnote 1 . . .).

[3]   **WHAT CAN BE SEEN**

Naturally, any estimate of the amount that can be seen from an extreme altitude must be a function of both the resolving power of the camera system and the area that can be scanned and recorded (or televised) and still retain usable data. Much of the discussion and most of the figures in this section are the result of previous RAND studies conducted by Dr. R. S. Wehner.

AREA COVERAGE

Fig. 1—Viewing system

Using the relation (see Fig. 1)

$$\frac{W}{F} = \frac{w}{d} = 2 \tan \frac{a}{2},$$

where W = sensitive element width, in.
       w = width of surface pictured per frame, mi.
       F = focal length of camera, in.
       d = optical range, mi.
       a = angle of view, deg.,

and using Tables 1, 2, and 3, it is possible to compute the width of square surface viewed and the angle of view for any given camera and aperture. This has been done and is summarized in Table 4. As can be seen, if a limiting resolution[6] of 500 ft is [4] set, it is still possible to obtain this resolving power under sunlight illumination with a contrast as low as 2.5 per cent (with a 5.0-in. aperture). Under moonlight, however, this resolution is possible only with 100 per cent contrast, a very fast f/1.4 lens, and a minimum exposure time of 0.25 sec; under light of the night sky illumination it is not possible at all. Assuming, then, that the chosen limiting resolution is correct, the probability of obtaining identifiable cloud photographs under any but sunlight illumination appears to be small.

---

6.   The term "limiting resolution," as used in the television field, refers to the greatest possible resolution attainable by a given TV pick-up tube and is wholly dependent on the structural make-up of the tube itself. As used in this report, limiting, minimum, or usable resolution is a quantity depending on scene contrast signal-to-noise ratio, aperture, f number of camera, etc., and is chosen to pick up the smallest object that it is desired to view.

Table 1*

RESOLUTION OF CLOUDS BY SUNLIGHT

(Image orthicon f/10 camera operated at 20:1;[+] signal-to-noise ratio
at rates of 40, 10, and 4 exposure frames/sec; and a satellite height of 350 mi)

Minimum Resolvable Surface Dimension

Contrast (%)	Aperture (in.)	40 Frames/Sec (ft)	10 Frames/Sec (ft)	4 Frames/Sec (ft)
100	0.5	200±	100±	64±
	1.0	100±	50±	32±
	2.0	50±	25±	16±
	5.0	20±	10±	6±
25	0.5	800	400±	250±
	1.0	400	200±	125±
	2.0	200	100±	64±
	5.0	80	40±	25±
10	0.5	2,000	1,000	640
	1.0	1,000	500	320
	2.0	500	250	160
	5.0	200	100	64
2.5	0.5	8,000	4,000	2,500
	1.0	4,000	2,000	1,250
	2.0	2,000	1,000	640
	5.0	800	400	250
1	0.5	20,000	10,000	6,400
	1.0	10,000	5,000	3,200
	2.0	5,000	2,500	1,000
	5.0	2,000	1,000	640

*   The material contained in this table was prepared by Dr. R. S. Wehner and is includ-
    ed in RAND Report R-217 (see footnote 1 . . .).
+   It should be noted that this table (and also Tables 2 and 3) is unrealistic in that the
    20:1 signal-to-noise ratio is applicable only to 25 per cent contrast. For 10 per cent
    contrast, a signal-to-noise ratio of 50:1 is required. This would mean a required trans-
    mitter power increase by a factor of 2.5 (assuming a 2-in. aperture 1000 TV lines, and
    a frame frequency of 10 sec). This is still not prohibitive but does become so with a
    substantial increase in either the number of TV lines or the frame frequency.
±   Values of computed resolution smaller than realizable with present commercial image
    orthicons.

[5]   **Table 2***

RESOLUTION OF CLOUDS BY SECOND- AND THIRD-QUARTER MOONLIGHT

(Image orthicon f/1.4 camera operated at 20:1; signal-to-noise ratio
at rates of 40, 10, and 4 exposure frames/sec; and a satellite height of 350 mi)

Minimum Resolvable Surface Dimension

Contrast (%)	Aperture (in.)	40 Frames/Sec (ft)	10 Frames/Sec (ft)	4 Frames/Sec (ft)
100	5	1.08	0.54	0.34
	10	0.54	0.27	0.17
	20	0.27	0.14	0.09
25	5	4.32	2.16	1.36
	10	2.16	1.08	0.68
	20	1.08	0.54	0.34
10	5	10.8	5.4	3.4
	10	5.4	2.7	1.7
	20	2.7	1.35	0.85

*   The material contained in this table is included in RAND Report R-217 (see footnote 1. . .).

Table 3*

RESOLUTION OF CLOUDS BY LIGHT OF THE NIGHT SKY

(Image orthicon f/0.7 camera operated at 20:1; signal-to-noise ratio
at rates of 40, 10, and 4 exposure frames/sec; and a satellite height of 350 mi)

Minimum Resolvable Surface Dimension

Contrast (%)	Aperture (in.)	40 Frames/Sec (ft)	10 Frames/Sec (ft)	4 Frames/Sec (ft)
100	10	4.3	2.15	1.36
	20	2.15	1.08	0.68
	40	1.08	0.54	0.34
25	10	17.2	8.6	5.4
	20	8.6	4.3	2.7
	40	4.3	2.15	1.36
10	10	43.0	21.5	13.6
	20	21.5	10.8	6.8
	40	10.8	5.4	3.4

*   The material contained in this table is included in RAND Report R-217 (see footnote 1. . .).

[6]   **Table 4**

POSSIBLE PERFORMANCE CAPABILITIES FOR VARIOUS IMAGE ORTHICON CAMERAS, TAKING INTO ACCOUNT ILLUMINATION SOURCES

Ratio of Focal Length to Aperture Diameter, and Illumination Source*	Aperture (in.)	Focal Length (in.)	Approx. Minimum Contrast Necessary to Give at Least 500-Ft Resolution (%)	Maximum Number of Frames/Sec	Computed Angle of View of Each Frame (deg)	Computed Width of Square Surface Viewed in Each Frame+ (mi)
f/10 camera,	0.5	5	25	10	11.44	70
clouds	1.0	10	10	10	5.74	35
illuminated by	2.0	20	10	40	2.86	17.5
sunlight	5.0	50	2.5	10	1.14	7
f/1.4 camera,	5	(†)	(†)	(†)	(†)	(‡)
clouds	10	(†)	(†)	(†)	(†)	(‡)
illuminated by 2nd- and 3rd-quarter moons	20	28	100	4	2.05	1.25
f/0.7 camera, light of night sky illumination	...	(**)	(**)	(**)	(**)	(**)

NOTE: For the purpose of computation, in the relation written on p. 3:
     W = width of the target in inches, which is taken to be equal to 1 in., the size of the commercial RCA image orthicon target
     d = optical range, which is taken to be equal to 350 mi (the height of the satellite).
All computations made assuming a minimum signal-to-noise ratio of 20:1.
     *   All cameras mentioned here refer to those using an image orthicon tube.
     +   Since the curvature of the earth was not taken into account, the figures in this column are lower than the actual figures.
     †   No resolution of the order of 500 ft or less.
     **   No resolution of the order of 500 ft or less, regardless of aperture.

[7] Calculations must also be performed to arrive at the possible area coverage. Since it is apparent that cloud observations, to be at all useful, have to be made over a wide-enough strip (at least as wide as the height of the satellite), it should be considered that the camera will be mechanically scanned. This may be accomplished by means of a 45° plane mirror rotatable about the axis of the camera. The mirror actually does the "looking" and scanning for the camera, which is mounted horizontally, its axis being parallel to the axis of the missile. Taking a sequence of 20 nonoverlapping frames will produce a strip 350 mi long, transverse to the trajectory of the satellite, and 17.5 mi wide. If the camera is set to take 5 frames/sec and the rotatable mirror is fixed with a fast snap-back device, the system will then be in position to take a second strip by the time the satellite has moved ahead approximately 17.5 mi relative to the earth. (The speed of the missile relative to the earth's surface is about 4.4 mi/sec.) This will produce a continuous 350-mi-wide strip around the earth with each complete traverse of the missile.

The daylight camera with an f/10 lens and an image orthicon television tube[,] and whose performance is summarized in Table 4, should have a 2-in. objective to give the proper ground coverage per frame. This combination allows a 500-ft object to be resolved with only 10 per cent contrast,[7] which is reasonably small. It should be emphasized that these figures are presented here merely to give some *examples* of performance of viewing systems and not as a description of the *optimum* system performance.

Some calculations of the efficiency of coverage of an inaccessible area such as an ocean were also made by direct measurements (assuming different strip widths) on a grid map. On this map were projected complete cycles[8] of traverses for two proposed satellite trajectories. Once again, curvature of the earth was neglected. The results obtained are as follows:

1. For a satellite with 24-hr complete cycle (354.6-mi altitude, angular velocity 15 times that of earth, and trajectory tangent to lat. 56°N.)—

    Assuming a 100-mi-wide scanning band (50 mi on either side of path): In the vicinity of lat. 45°–50°N., we find that in 24 hr the surface has been covered in a grid fashion such that about one-third of its area has been scanned and presumably televised.

    Assuming a 200-mi-wide scanning band (100 mi on either side of path): As may be expected, doubling the scanning band does not quite double the area covered. This is owing to some overlapping of the bands. (It can be shown that, [8] to cover the area completely, a scanning band approximately 450 mi in width is needed.)

    As a result of the grid-like coverage, the 100-mi-wide band, at its worst, should pick up at least portions of the largest, most active weather disturbances and enough of the remaining cloud coverage to orient the system in relation to the ground.

2. For a 48-hr complete cycle (altitude 453.3 mi, angular velocity 14.5 times that of the earth, and trajectory tangent to lat. 56°N.)—

    Assuming a 100-mi-wide scanning band (50 on either side of path): In the vicinity of lat. 45°–50°N., we find that in 48 hr it has been covered, grid fashion, so that two-thirds of its area has been scanned and presumably televised. (It can be shown that to cover the area completely in a 48-hr cycle, a scanning band approximately 250–300 mi in width should be required.)

The results so obtained give an idea of the areas which can be covered (or scanned) from a vehicle in an orbit 350 mi above the surface of the earth. The 350-mi-wide strip discussed in the first part of this section will therefore cover in 24 hr a large percentage of the area between 45°N. and 56°N.[9] with considerable overlapping of scanning, particularly

---

7.   This resolution and contrast represents the maximum needs of satellite weather observation. This is obtainable with 25 percent contrast when using an f/10 lens with a 2-in. objective in sunlight illumination (see footnote to Table 1 marked (+).

8.   Initially, the trajectory of the satellite is set tangent to a given latitude. Owing to the relative difference in the angular velocities between the satellite and the earth and to the relative stability of the orbit of the missile, the vehicle's trajectory appears to "creep" over the surface of the globe. A complete cycle is the time it takes for the trajectory of the satellite once again to become tangent to the original point. (This "creep" causes the traverses to become more widely dispersed as the trajectory approaches the equator.)

9.   A large percentage of the area should be covered in the 24-hr trajectory, and almost all should be scanned in the 48-hr trajectory.

around the 56th parallel. In any event, the coverage, as mentioned here, if achieved with any measure of success, should produce good weather reconnaissance results.

RESOLUTION AND LIMITING CONTRASTS

Since it is now obvious that clouds will be the chief meteorological element directly observable from high altitude photographs, it must be ascertained how closely these clouds may be identified and what may be determined from them, either directly or indirectly.

As can be seen from Tables 1, 2, and 3, when a set of conditions such as aperture, illumination, exposure time, and focal length-aperture diameter ratio of a given camera have been established, the remaining factor for determination of the minimum resolution attainable is the contrast value. In cloud photography of the type to be attempted from the satellite, one is unable to choose the surrounding photographic conditions. Features such as background, lighting at time of observation, etc., are examples of the uncontrolled variables, and, as a consequence, any system of data gathering by photographic means must be flexible enough to give adequate results over a wide range of limiting factors. The question is: If the camera and optical system are chosen,[10] and if the various conditions of lighting, background, etc., are assumed to remain within the limits providing [9] usable resolution, will the resulting limiting contrast values still enable one to observe the weather under a wide-enough range of actual conditions?

Before endeavoring to answer this question it is desirable to define the term "usable resolution." It was thought that details of cloud structure as small as several hundred feet in diameter might possess significance when an attempt was made to form a synoptic picture by means of cloud analysis. This was borne out when high-altitude rocket photographs were examined. Further reasons for asserting this to be the appropriate minimum size to be resolved were found when a simple test was conducted on these photographs. (The heights at which these pictures were taken varied between 50 and 70 mi.) Using an adjustable viewer, the photograph was taken slowly out of focus until it was impossible to identify definitely the forms of clouds other than by saying that they were widespread or were in small clusters. For example, beyond this point it was impossible to distinguish between closely packed cumulus and a deck of altocumulus, and also between a dense layer of stratus or altostratus and the fibrous texture of cirrostratus. A study of other parts of the photograph, where recognizable or measurable objects were located at ranges about equal to those of the clouds, showed that the limiting resolution at which the clouds lost their distinguishability was from 500 to 1000 ft. This is what is meant by "usable resolution." As may be imagined, this is at best only a rough approximation, but because of its apparent agreement with previously estimated values it should serve very well as a working basis.

It was mentioned above that in order to obtain a known, usable resolution, once the camera and lighting conditions are chosen, the limiting contrast value must also be spec-

---

10.   Previous studies at RAND have shown that one of the best available television cameras for use in the satellite would be one employing an *image orthicon pick-up tube*. The characteristics of this tube approach those of the human eye over part of its operating range, it has a greater sensitivity than earlier types, and it is capable of stable operation without adjustment over a wide range of illumination intensity. Since it is not the purpose of this report to delve too deeply into the technical aspects of the the problems of a television viewing system, only results of resolution computation of the image orthicon tube are presented here. These are summed up in Tables 1, 2, and 3. For technical information and equations involved, see R. B. James, R. E. Johnson, and R. S. Moore, *RCA Review,* Vol. 10 July 1949, pp. 191–223; and A. Rose, "Television Pickup Tubes and the Problem of Vision," *Advances in Electronics,* Vol. 1, Academic Press, New York, 1948.

ified. It is obvious that if, for various combinations of cloud-ground and cloud-cloud, albedo differences are such that their contrast values fall below the limiting contrast, these combinations cannot be observed by high-altitude weather reconnaissance.

Hewson, in an article in a meteorological journal[11] and in his book (written in collaboration with Longley) on theoretical and applied meteorology,[12] calculated and tabulated diffuse-reflection coefficients for clouds of various thicknesses. In doing so, as a result of the extensive variation of cloud liquid-water densities and cloud droplet radii, he was forced to choose one set of values for these two parameters. Those on which his figures are based are a density of 1.0 gm of liquid water per cubic meter of cloud and a droplet radius of 5×10-4 cm. Owing to the fact that these values probably apply to a large percentage of the usable clouds observable from extreme altitudes, they may be reasonably [10] employed in making estimates for this study. These values are plotted in Fig. 2, the ordinate and abscissa being contrast and background albedo, respectively. Each curve represents a particular albedo applicable to a particular cloud thickness. According to the definition of contrast,

$$C = \frac{P_b - P_d}{P_b}$$

where P_b = brightness (albedo) of the brightest thing viewed (either object or background)
         P_d = brightness of darkest object viewed (albedo)
         C = contrast between the two.
From the above definition, each curve may be represented by the following relation:

$$C \begin{cases} 1 - \dfrac{A_b}{A_t}, & \text{for } A_b < A_t \\[2ex] = 0, & \text{for } A_b = A_t \\[2ex] 1 - \dfrac{A_t}{A_b}, & \text{for } A_b > A_t \end{cases}$$

where C = contrast between object and background
        A_b = albedo of background
        A_t = albedo of clouds of various thicknesses.

It is therefore seen that, except for the small range of albedo combinations around the point of discontinuity on the curves, a large majority of possible cloud-background albedo combinations fall within the range of at least 10 per cent contrast. As can be seen from Table 1, assuming at least a 2.0-in. aperture and sunlight illumination, an f/10 camera will permit at least 10 per cent contrast for approximately 500-ft resolutions.[13] Table 5

---

11.   E. W. Hewson, *Quart. J. Roy. Met Soc.*, Vol. 69 (1943), p. 47.
12.   E. W. Hewson and R. W. Longley, *Meteorology, Theoretical and Applied*, John Wiley & Sons, Inc., New York, 1944, pp. 73–75.
13.   Using the equation (for altitude of 350 mi)

$$C = \frac{1600}{ga\sqrt{t}}$$

where C = contrast
           a = aperture
           t = exposure time (or time of one frame)
           g = minimum resolvable surface dimension,
it is possible to calculate the contrast (minimum) needed to obtain at least 500 ft resolution under the conditions given in the example of ground coverage which assumed full daylight illumination. This value turns out to be 3.56 per cent. Owing to the unrealistic power requirements necessary to transmit 3.56 per cent contrast, this value has been raised to 10 per cent.

. . . gives the albedos for various ground covers. Applying these values to Fig. 2, it can be seen that, except for the case of newly fallen snow combined with clouds thicker than 600 meters and the case in which the background albedo approaches very close to cloud albedo, 500-ft resolutions are obtainable over a wide range of conditions.

There is one other factor that might limit contrast and, therefore, resolution. This is aerial haze between the camera and the ground. As has recently been shown in several [11] [original placement of "Fig 2—Available contrast with varying cloud and background albedos"] V-2 photographs, this problem is almost completely solved by use of an infrared filter in the optical system.

Fig. 2—Available contrast with varying cloud and background albedos

From the foregoing section we may conclude that, from the standpoint of area coverage and resolution, weather observations from a satellite are a definite possibility.

[12]   **Table 5***

SURFACE ALBEDO AND SCENE CONTRAST OF CLOUDS
AGAINST VARIOUS BACKGROUND SURFACES

Ground Surface	Albedo[+]	References[±]
Fresh snow	.80-.93	1, 3, 4
Old snow, sea ice	.40-.60	3, 4
Brown soil	.32	1
Grass	.10-.33	4
Green leaves	.25	1
Sandy loam	.24	2
Sand	.13-/18	3
Asphalt paving	.15	2
Dry earth	.14	4
Rock	.12-.15	4
Moist earth	.08-.09	2, 4
Cultivated soil, vegetable	.07-.09	3
Smooth sea surface		
Solar elev 5 deg	.40	3
Solar elev 10 deg	.25	
Solar elev 20 deg	.12	
Solar elev 30 deg	.06	
Solar elev 40 deg	.04	
Solar elev 50-90 deg	.03	

*   This table was prepared by Dr. R. S. Wehner and is included in the RAND general report on the satellite (see footnote 1 . . .).

   +   Values of albedo apply to illumination by "white" light or sunlight.

  ±   References:

    1.   *International Critical Tables,* 1929 ed., Vol. 5, p. 262.

    2.   *Handbook of Chemistry and Physics,* 1942 ed., pp. 2147–2148.

    3.   H. Landsberg, *Handbook of Meteorology,* McGraw-Hill Book Company, Inc., New York, 1945, p. 932.

    4.   J. Charney, *Handbook of Meteorology,* McGraw-Hill Book Company, Inc., New York, 1945, p.296.

[13]   **LIMITATIONS OF THE ANALYSIS**

It is a known fact that the reliability of any form of synoptic meteorological analysis depends on the experience of the analyst. An analysis of the type dealt with in this report is no exception. If anything, it is even more dependent on analytical experience because of the sparseness of data and the difficulties in interpretation. To date, the meteorological cloud atlas has been built up almost entirely from ground observations. The changeover to "looking down" upon the clouds means that the dominant features which served to identify types of clouds when observed from the ground are no longer to be seen. The halo and corona that served so well to classify cirrostratus and altostratus, respectively, are absent. Also, the upper surface of large scale cloud decks is, for the most part, completely different in appearance from the lower surface. Therefore a completely new concept of cloud-identification features must be formed, and only those experienced on these new concepts will be able to make an intelligent analysis.

There is also danger of an incorrect interpretation of the cause for the clouds, which might lead to a completely erroneous analysis. Take, for example, the case in which the entire picture under consideration exhibits one complete deck of clouds. In this case the deck of clouds might be stratus caused by radiational cooling and so might constitute an entirely local phenomenon. An analyst looking at this situation might jump to the conclusion that the clouds in question were of frontal origin, possibly altostratus, and might forecast accordingly. It is evident that a forecast made from such an erroneous assumption of the cause would be completely incorrect. (Methods of attack on this problem of analysis are treated more fully in a later section of this report.)

There are also many definite advantages to be gained in the analysis of weather by this method; chief among these is the fact that extremely large areas may be visually observed in a relatively short period of time. The disadvantages of large gaps (between stations) on the usual weather map and the comparatively limited field of view of each ground observer are eliminated. What is obtained is, in effect, the cloud pattern integrated over a wide area. From many points of view this is highly desirable, owing to the fact that, for the first time in the history of synoptic meteorology, the classical models of various weather situations may be examined *in toto*.[14]

[14] **WHAT CAN BE DETERMINED FROM HIGH-ALTITUDE OBSERVATIONS**

                    CLOUD IDENTIFICATION

Assuming, from the previous section, that cloud shapes of the order of 500 ft or more in diameter are distinguishable from an altitude of 350 mi, the problem of identifying these clouds can be treated. As stated previously, attributes and/or phenomena that served to establish the classification of clouds when viewed from the ground are almost completely different when these same clouds are viewed from above. The question is, What can actually be done to tell the various cloud forms apart?

The solution to this problem may lie in a new classification system formed by means of close correlation of observations of clouds viewed from above with observations of these same clouds viewed from below. In this manner, an atlas of identifying cloud features

---

14.   This idea of "the over-all look" was first described by Major D. L. Crowson, USAF, in a recent article in *B. Amer. Meteorol. Soc.*, Vol. 30, No. 1, January, 1949. His primary object was the use of vertically fired rockets in conjunction with the regular meteorological observations as a supplement rather than as a possible replacement. In this regard, his analysis of rocket photographs is very similar to that presented by Dr. J. Bjerknes in Appendix I.

as scanned from extreme altitudes might be built up. Using this information, a trained observer should have little trouble in establishing the identity of almost any visible cloud. The importance of such an atlas cannot be over-emphasized, because the degree of confidence in a synoptic picture formed from this type of observation or in the subsequent forecast becomes extremely small if the identity of the clouds cannot be established. An attempt along these lines has been made, utilizing several series of photographs taken from V-2's fired at White Sands, New Mexico. It should be kept in mind, however, that this attempt was made using data which were not originally gathered for this purpose, and the necessary ground observations are therefore not available for positive identification purposes. Because of this, the results presented here are a classification and identification based on the writer's observational experience.

From a study of the above-mentioned photographs it was observed that two general cloud forms stand out from each other under the usable resolution conditions. Since these two forms are also two types of cloud formations, most other clouds can be considered as being a special form or combination of these and may be so categorized. This is partially attempted in the table [below].

It is noticed that certain formations of clouds very often assemble in over-all patterns peculiar to these formations. Clouds, therefore, may also be partially categorized according to pattern. In the case of clouds formed by globular masses joined together to produce a single layer, the pattern is still apparent to an observer on the ground as a result of the differences in light intensities caused by the variations in cloud thickness. It is likely, therefore, from the section on cloud contrast, that these patterns will also be visible to an observer stationed above the layer, owing to the difference in albedo values caused by cloud sections of different thicknesses. These patterns are very useful in cutting down the overlap present in the following table.

[15] A Vertically Developed	Remarks	B Stratiform	Remarks
1. Cumulus } 2. Cumulonimbus }	Varying degrees of vertical development	1. Stratus } 2. Altostratus }	In some forms may be very similar.
		3. Cirrostratus	May be distinguished because fibrous texture is visible even when viewed from above.
		4. Nimbostratus	When there is no vertical development on top this form may appear to be very similar to Nos. 1 and 2.

Combinations of A and B (forms similar in appearance are bracketed)

Cloud Formations	Remarks
[Altocumulus] [Cirrocumulus]	Altocumulus cloud elements may exhibit vertical development, or there may be just closely packed globular masses. In the first case, the altocumulus may seem to be very similar to altostratus or nimbostratus that have vertical development in their tops, although the layer may retain some semblance of orderly pattern.

[Altostratus]   Very often these formations contain considerable vertical development.
[Nimbostratus]   This seems to be especially true when these forms are associated with
the passage of a front.

It is clear that any attempt to formulate an atlas of cloud appearance as seen from high altitude is a major undertaking. The work done on the subject in this report represents, at best, the beginning of the work that must be accomplished to make high-altitude cloud photographs a usable weather tool.

THE ANALYSIS

Having once established the identity of almost all the clouds viewed, the formation of the synoptic weather picture becomes the next problem. The following question arises: Given an over-all cloud picture, what, in fact, can be determined, either directly or indirectly?

According to conventional meteorological practice, the various parameters, such as pressure, temperatures, etc., are plotted on a map, and the subsequent analysis of these quantities produce the synoptic picture. Almost the reverse is true in the case at hand. Here the synoptic situation must first be established, and the various parameters must be estimated from it. Actually, it is not quite so straightforward a procedure. Rather, it becomes a "hunting" technique, in which one makes a first approximation to the over-all weather situation, using the clouds, and, from this, a first estimate of the value of temperature, pressure, humidity, etc. This picture of the weather is then modified to fit [16] obvious deviations of the estimated values from those indirectly observed. This process continues until a satisfactory situation is evolved that appears to fit all existing conditions (an attempt being made to satisfy both physical and theoretical considerations); from this, final estimations of the various parameters are made. (Several possible approaches to the problem of approximating the synoptic picture are discussed in the section entitled "Suggested Methods of Attack on the Problem of Determining the Synoptic Situation.". . .)

The normal observable meteorological parameters may be divided into two main categories, viz., those that may be estimated in some measure directly from observations of the clouds and/or ground, and those that require a knowledge of the over-all weather patterns before an estimate can be made. In the first category may be listed wind, humidity, precipitation, and a variable not normally considered by itself as such—degree of stability. In the second listing may be found pressure (and pressure tendency) and temperature. Before an attempt at its analysis can be made, a considerable amount of experience and general knowledge of the workings of the atmosphere is required concerning each item, regardless of which category it comes under. It is found that this estimation method is neither a quick nor a simple process, regardless of the qualifications of the analyst. Rather, each of the items requires a very careful study and the weighing of all the possible influencing conditions before approximate values can be assigned.

As a result of this pilot study, several suggested methods of estimating the various meteorological parameters were evolved and are discussed as follows:

Wind

1. From the established meteorological models it is assumed that certain definite weather situations will produce certain sequences of clouds preceding or following them. This will therefore tend to orient the situation with respect to the ground. Once this orientation has been established, the wind direction may be approximated through a knowledge of the theoretical circulation associated with a given synoptic weather situation.

2. It has been noticed in several photographs that, in the presence of strong upper winds, cumulus clouds that have formed in mountainous country appear to form to the

lee of the mountains rather than to their windward side. In the presence of very light winds, it was noticed that the cumulus tended to form on the peak of the mountain. This phenomenon requires further study before its degree of usefulness as an observational tool can be determined.

3.   Owing to the fact that cumulonimbus clouds extend from as low as 1600 ft up to 40,000 ft, their slope becomes a good indication of the vertical shear within the layer. It was first thought that this direction of slope would be an indication of the direction of the upper winds. However, although the wind velocity normally increases with altitude, it is obvious that for any given case one should not disregard the possibility that the wind velocity might decrease with height or that the direction and velocity distribution in the vertical might be of such a nature as to cause the cumulonimbus to slope into the upper wind. When this slope is combined with other factors that indicate wind direction [17] at one particular level, it may be possible to construct a picture of the change of wind direction with height in the layer under consideration.

4.   A further indication of wind direction (in the lower levels) was observed when small, detached clouds were seen to form in line, stretching from a mountain top. These could be due to moist air being forced upward by the mountain and then moving down-slope on the lee side, causing the formation of small "rotors" or individual cellular eddies each capped by small cumulus clouds and extending for a considerable distance down-wind from the mountain. This phenomenon is known as a "standing wave" and is often accompanied by other standing clouds at higher altitudes.

5.   It has also been observed in a layer of stratus overlying mountainous terrain that air funneling down a valley and spreading out in a relatively flat section produced lines and swirls in the top of this cloud layer that closely matched the path the air must have taken. This action may be very useful in determining wind direction in sections completely covered by sheet-type clouds and may be found to be of further use over areas that are not particularly mountainous. Although photographs of large flat areas were not available for analysis, it is thought that wind-direction determination in these sections may still be accomplished in the lower levels. This may be done by utilizing many of the above methods and several others that could be an outgrowth of such an analysis. One such method might use the inherent uniform structure of a stratus sheet. In this case it is thought that if a sheet passes over flat ground on which there are isolated protuberances projecting into the sheet, a wake will be produced in the cloud that may also show up when viewed from above and that will stretch downwind from the object.

Temperature

The starting point for any determination of temperature must be the statistical normal for that time of the year. The first estimation may then be modified by the various affecting conditions. The prevailing weather situation provides the first modifying influence. This estimation is, of course, dependent on the analyst's ability to estimate the synoptic conditions with a degree of accuracy that will answer the question, Is the sector under observation being affected by relatively cold or warm air? Cloud systems, wind directions, and even forms of ground cover (snow, etc.) will help in deciding this. This is the first indication of the over-all complexity of this type of analysis and serves as an actual illustration of the "hunting" technique mentioned above.

Upper air temperatures may be estimated in the same manner, clouds indicating the boundaries between air masses (fronts). A further help in estimating this quantity is the fact that, once having decided on a ground temperature, the degree of stability (indicated by vertical development in clouds) and the presence or absence of intervening fronts will enable one to construct an applicable temperature lapse rate. (The degree of stability will determine the departure from an adiabatic lapse rate, while the degree of cloudi-

ness (moisture) will help an analyst to decide whether to use the moist or the dry adiabatic lapse rate as the limiting one.)

Vertically developed cloud will also aid in determining the temperature gradient of
the surrounding area. This is true because of the fact that the vertical shear, as indicated
[18] by the slope of towering cumulus and cumulonimbus clouds, orients the direction of
the higher and lower temperatures in the areas. This method is employed by taking the
direction of vertical shear as being from the low levels toward the high levels. If one then
faces in the direction of shear in the northern hemisphere, the lower temperature will be
on the observer's left and the higher on the observer's right (see Fig. 3).

This relationship holds for the northern hemisphere. In the southern hemisphere the
directions of decreasing and increasing temperature in relations to vertical shear are
reversed.

Fig. 3—Vertical wind shear—temperature gradient relationship

Pressure

It is apparent that no quantitative values of pressure are forthcoming from this analysis. Furthermore, it is virtually impossible even to make a quantitative estimate other than
to state whether the area is thought to be under the influence of a high- or a low-pressure
system. Charts of average pressures for various times of the year in different areas of the
world are available. Using these and the weather situation at the time, trends of pressure
may be established. This information when applied in conjunction with known weather
may be a very useful tool for forecasting purposes. Little work has been attempted on this
subject in this pilot study, and the above should serve only as a possible starting point for
any detailed research along these lines.

C. F. Brooks[15] points out some further pressure information that may be obtained
from clouds. He says, in effect, that, since in the presence of any constant vertical shear
the cumulus clouds will tend to lean or slope (the amount of departure from the vertical
being a resultant of the vertical velocity and the rate of change of wind velocity with
height), any cloud that has a uniform rate of vertical growth and a 90° slope throughout
is an indication of the "uniformity of wind velocity in all layers pierced." This indicates a
decrease of horizontal pressure gradient with height. (This can be shown very simply by
an examination of the geostrophic wind equation

$$V_g = \frac{1}{p} \frac{(\partial p)}{(\partial n)} \frac{1}{\lambda},$$

where V_g = the geostrophic wind velocity
       p = density o air
  $\partial p/\partial n$ = horizontal pressure gradient
       λ = Coriolis parameter.

[19] It can be seen that since λ, which depends on the sine of the latitude, will remain
constant and p decreases with height, $\partial p/\partial n$ must also decrease *for* V_g, to remain constant.) This decrease turns out to be very small when actual values are used. In the case of
a uniformly growing cumulus that slopes in its lower layers and then straightens or even
bends back on itself with increasing height, the decrease of the horizontal pressure

---

15.   C. F. Brooks, "Clouds in Aerology and Forecasting," *B. Amer. Meteorol. Soc.*, Vol. 22, November, 1941,
pp. 335–345.

gradient with height is (as Brooks also points out) much stronger than in the previous case. If one assumes that the slope of vertically developed clouds may be observed from 350 mi altitude (at least at the edges of scanning strip), further pressure data may be gathered.

Degree of Stability

As has been mentioned above, the degree of stability in a given layer may be estimated by the amount of vertical development present in clouds. In any mechanism of vertical development, the stability of the air plays a major part. Convective, orographic, or upslope lifting may produce clouds in the absence of instability, but, for any large-scale vertical build-up of clouds, a great tendency for the atmosphere to "overturn" must be present. [original placement of Fig. 4] (Absolute instability is taken to mean that the decrease of temperature with height is greater than the dry adiabatic lapse rate. In the presence of unsaturated water vapor, the dry adiabatic lapse rate is about 9.8°C/km, whereas, in the presence of saturated water vapor, the smaller saturated adiabatic lapse rate with a non-linear variation of temperature is used.) In the presence of water vapor, the latent heat (energy) of condensation that is released when the air is forced to rise and its moisture forced to condense may be sufficient to continue independently the upward motion. This motion indicates a condition of instability where none may have existed at the beginning of the process. Continuation of this motion, therefore, indicates the instability of the air in the presence of saturated water vapor and is evidenced in towering cumulus or cumulonimbus. If, on the other hand, condensation occurs but the ascending air is not provided with a sufficiently large amount of heat so as to warm it to a higher temperature than that of the surrounding air, the layer is considered absolutely stable and may be characterized by smooth, flat-topped cloud forms, usually arranged in layers or sheets. This is also true when a small layer of instability is "capped" by an inversion (increase of temperature with height). This concept of absolute stability, absolute instability, and conditional instability (unstable or stable depending on whether the water vapor present condenses or not) is presented graphically in Fig. 4.

Fig. 4—Graphical representation of degrees of stability as given by lapse rate of temperature

It may be said that, in the presence of vertically developed clouds, a dry adiabatic lapse rate (or very close to it) exists below the base of the cloud, a relatively steep lapse rate exists within the cloud, and a relatively stable lapse rate exists above the cloud. In [20] the case of flat-topped or sheet-type clouds, it may be that, although instability may exist in a small layer comprising the cloud, an inversion layer of very stable air exists immediately above, causing the cloud to stop its vertical growth.

In his paper on clouds, Brooks[16] suggests the following further refinements on this:

1. Detached, lumpy cloud with a flat base and rounded top has (a) adiabatic lapse rate below it, (b) greater than saturated-adiabatic lapse rate (unstable) within the cloud, and (c) almost the same lapse rate as (b) (unstable) from its top to the height that the cloud will grow.
2. Towering, sharply-bounded cumuliform cloud: The diameter of cloud at different levels is an indication of the relative steepness of the lapse rate (except in the presence of large wind shear). "The narrower such a cloud or cloudlet is, relative to its height, the greater the lapse rate of the surrounding air."

16.   *Ibid.*

This provides one with very rough criteria for estimating the degree of stability of the air.

To sum up, water vapor in the air is a latent source of heat energy. When moist air is carried rapidly upward, the water vapor condenses in the form of liquid droplets and the latent heat of condensation is released to the surrounding atmosphere. It is this source of latent heat that feeds thunderstorms and other types of vertically developed clouds. Cumulus clouds are an indication of moisture and relative instability, and, conversely, when there is moisture in the air there will be a greater tendency toward convection and turbulence.

Moisture

Clouds, being composed of water droplets, naturally indicate the presence of moisture in the atmosphere (see the above section). Resulting from the difference in formation conditions, cloud types can give a further breakdown of moisture distribution. For example, cumulus and cumuliform clouds of vertical development require the entrainment of continuous supplies of moist air to prevent their complete evaporation shortly after forming. It can therefore be said that with this type of cloud we may associate fairly moist air near the surface. In like manner, positioning of the moisture in the atmosphere may be associated with other cloud forms, and an over-all estimate may be made from visual observations. Once the synoptic picture has been established, closer estimates may be made utilizing the other meteorological parameters, and the value of moisture content may be worked into the "hunting" technique previously mentioned.

Precipitation

Although it will not be possible to observe any form of precipitation directly, it is known that the largest amounts usually fall from two main types of clouds: cumulonimbus (showers—rain, snow, etc.) and nimbostratus (steady precipitation, sleet, etc.). Furthermore, the probability of precipitation in one form or another, which arises whenever these types are present, is higher than for any other types of clouds. Further infor-
[21] mation may be obtained from the fact that it may be possible to distinguish between newly fallen snow and old snow, owing to a difference in albedos (see Table 5 . . .), and the new snow may then be connected with the proper form of cloud observed downwind from it.

[22]   **SUGGESTED METHODS OF ATTACK ON THE PROBLEM OF
         DETERMINING THE SYNOPTIC SITUATION**

From the above discussion it can be seen that the analysis is based primarily on cloud observations. During the course of this study several systematic methods of accomplishing these presented themselves. Although neither time nor proper data were available for a complete study of these possibilities, the most promising were considered and are presented herewith as a guide to any more intensive study.

1.   It is suggested that a typing of clouds as to *cause* rather than *appearance* will greatly facilitate the identification of the synoptic situation. Classification into two main categories would constitute a possible breakdown, as follows: (*a*) Regional clouds (those caused by purely local conditions), and (*b*) migrating cloud systems (clouds that appear to move as a unit). The breakdown might then be coupled with a knowledge of the clouds associated with various weather phenomena to complete the synoptic picture.

2.   It is a recognized fact that similar synoptic situations occurring under different climatic and/or topographic conditions may produce radically different weather. A statistical analysis is therefore suggested, in which (*a*) the desired area is divided into small

regions of similar climate, geography, etc., and (*b*) a statistical survey of cloud types and associated weather found with various weather situations (fronts, etc.) in each region is made.

3. Owing to the fact that identification of fronts as fronts may be very difficult, it is suggested that it may be possible to identify air masses from high-altitude pictures and to utilize them in the formulation of the synoptic picture. Since general classifications of air masses include as integral identifying features the stability of the air, the moisture, and the type of clouds produced in a given air mass, this should not be too difficult, in many cases. An air-mass identification has the further advantage of establishing more closely the possible limits of the various meteorological parameters.

4. A major advantage of satellite weather observations is the repeated broad spatial coverage. Such broad coverage provides the meteorologist with an essential element for his analysis, which is generally referred to as *continuity*. It permits him to follow a given system as it moves and develops over a period of days. It is a relatively simple matter to identify a system once it is known that such a system is present. Once a weather situation is so identified, it can be earmarked from high-altitude pictures, and not only may it then be tracked across an inaccessible area like an ocean, but any over-all changes or modifications that affect the visible parameters may be almost immediately noticed. It is also likely that, having a complete analysis of the surrounding territory on land, where observations are plentiful, and many satellite observations of the unknown area (through which it is possible to get fixes on systems and to examine visually the over-all weather [23] picture), a complete analysis of the desired region will become a much simpler thing to construct.

Each of the above suggestions affords excellent possibilities of providing the required information. It should he kept in mind, however, that these suggestions appear to offer the best solution when systematically used together. . . .

[31]   **CONCLUSION**

In the section entitled "What Can Be Seen," . . . it was shown that, given at least 500-ft resolution, it was possible to differentiate between the various types of clouds. Under "Limitations of the Analysis," . . . the possible limitations to the type of analysis to be studied were indicated. Given the identity of virtually all the cloud forms viewed, it was further shown, in the section entitled "What Can Be Determined from High-Altitude Observations," . . . that it may be possible to estimate the various meteorological parameters under certain conditions and assumptions. The main assumption was that some estimate of the over-all synoptic situation could be made initially and a "hunting" technique could be applied. Several suggested methods of estimating the synoptic picture were presented and discussed.

This report has attempted to show what is thought to be necessary in the making of such an analysis. It is obvious, however, that, with the limited data available, many important points may inadvertently have been overlooked. An inquiry of this type can therefore serve only as a guide to a full-scale study of the subject, in which every suggestion and method is put to a full test and is either accepted, modified, or discarded.

The development of all the suggested methods mentioned in this report appears to be feasible. As any analysis depends on its integral parts for its accomplishment, from this standpoint, if from no other, the analysis of synoptic weather from satellite observations is also feasible.

Document II-3

Document title: Hugh L. Dryden, for T. Keith Glennan, NASA, and Roy W. Johnson, Department of Defense, "Agreement Between the Department of Defense and the National Aeronautics and Space Administration Regarding the TIROS Meteorological Satellite Project," April 13, 1959.

Source: NASA Historical Reference Collection, NASA History Office, NASA Headquarters, Washington, D.C.

Among the military space projects President Eisenhower transferred in 1958 to the newly created National Aeronautics and Space Administration was the Television and Infrared Operational Satellite (TIROS) meteorological satellite project, previously controlled by the Department of Defense's Advanced Research Projects Agency. Although the Defense Department continued to participate in the TIROS project and made excellent use of the information returned from operational satellites, this agreement marked the beginning of a permanent split in military/civilian meteorology that led for more than three decades to both the military and the civil sector designing and operating their own meteorological satellite systems.

[1]

Agreement Between the Department of Defense and the National Aeronautics and Space Administration Regarding the TIROS Meteorological Satellite Project

1. Effective April 13, 1959, the National Aeronautics and Space Administration (NASA) shall assume technical and management direction of the meteorological satellite project designated Project TIROS, as set forth in Order No. 10-59, dated July 25,1958, and Task No. 1 of Order No. 17-59, dated September 4, 1958, of the Advanced Research Projects Agency (ARPA) of the Department of Defense (DOD).

2. In order to insure the complete availability to DOD and NASA of all information developed under Project TIROS and to insure that the respective interests of both are fully recognized in carrying out the Project, the following arrangements are agreed to:

    a. A committee will be established under NASA chairmanship, with representation from both DOD and NASA, to advise NASA on technical matters related to Project TIROS, including DOD requirements, and to make any necessary arrangements for the close cooperation and full exchange of information between NASA and DOD.

    b. Copies of all NASA directives issued to agencies of DOD in connection with Project TIROS will be furnished to ARPA for information.

    c. Copies of all documents pertinent to the conduct of Project TIROS in the possession of ARPA will be furnished to NASA.

3. Contracts under Project TIROS to be funded by DOD will continue to be placed and administered by procuring activities of DOD, subject to the technical and management direction of NASA, and any facilities, equipment and personnel of DOD currently assigned to Project TIROS will remain available to NASA to carry the Project to completion.

[2] 4. ARPA will fund Project TIROS up to a total of $11,649,000. An amount of $6,711,000 has already been committed under ARPA Order No. 10-59, and $2,000,000 under Task No. 1, ARPA Order No. 17-59, leaving a balance of $2,938,000 which will be set aside for obligation by DOD on Project TIROS at the request of NASA. These funds are not, however, available for the construction of facilities. NASA will provide any funds required for Project TIROS in excess of the $11,649,000 provided by ARPA.

5. Equipment acquired for Project TIROS will remain available to the Project until its conclusion. The disposition of any such equipment at the conclusion of the Project will be as mutually agreed upon by NASA and DOD.

[hand-signed: "Hugh L. Dryden for"]   [hand-signed: "Ray W. Johnson"]
T. Keith Glennan   Roy W. Johnson
for NASA   for Department of Defense

Document II-4

Document title: U.S. Department of Commerce, Weather Bureau, "National Plan for a Common System of Meteorological Observation Satellites," Technical Planning Study No. 3, Preliminary Draft, October 1960, pp. 1–3.

Source: NASA Historical Reference Collection, NASA History Office, NASA Headquarters, Washington, D.C.

When NASA's first TIROS satellite proved highly useful in studying large-scale weather systems, the U.S. Weather Bureau began planning for a fully operational national weather satellite system. The Weather Bureau proposed to remove NASA from its overall lead position to the role of performing research and development in support of the operational system. In October 1960, NASA organized the first meeting of an interagency panel to discuss the issue of an operational system. At the meeting, the Weather Bureau brought forth its plan; the foreword and first section of the study appears here. Predictably, NASA objected to giving up as much control over the program as the U.S. Weather Bureau desired. The result was a compromise plan issued in April 1961.

National Plan for a Common System
of Meteorological Observation Satellites

Washington, D.C.
October 1960 . . .

[no page number]

FOREWORD

The present report is a summary of planning that commenced shortly after the successful launching and operation of TIROS I, April–June 1960. The results of this remarkably successful meteorological satellite clearly show that satellites must be included as an integral part of a comprehensive, world-wide weather observing system. Their ability to give complete global coverage, to look at familiar meteorological phenomena from a new vantage point and to reveal organized motions and processes over a great range of dimensions will influence virtually all phases of meteorological development and operations.

Representatives of the government departments directly interested met at the National Aeronautics and Space Administration Headquarters on October 10, 1960 for discussions of how to proceed with an operational meteorological satellite program. Need for a national plan indicated at this meeting prompted issuance of this report at the present time. It represents an effort to utilize results of studies made since 1954, including a 1959 report to the World Meteorological Organization, and experience gained from Explorer VII and TIROS I. This report delineates a first approach to the design of the sys-

tem, the required organization and a method of implementation for obtaining and utilizing meteorological satellite data as soon as possible for daily weather forecasting and
storm warnings.

This plan, although preliminary, represents a starting point in formalizing a program,
portions of which are already in operation. The plan is circulated among meteorological
groups for the purpose of inviting comments and cooperation on how best to take advantage of this epochal new means for observing certain meteorological phenomena and to
assist in planning more effectively for future programs.

[hand-signed: "F. W. Reichelderfer"]
F. W. Reichelderfer
Chief of Bureau

November 3, 1960 . . .

[1] I.   Goals of a National Plan
The TIROS I Weather Satellite has brought the objective of meteorologists for a world
wide observational network a long step toward fulfillment. Conceived initially as a
research project, TIROS I demonstrated immediate limited operational value. This mon-
umental scientific achievement is a manifestation of the policy declared by the Congress
in the National Aeronautics and Space Act of 1958—"that activities in space should be
devoted to peaceful purposes for the benefit of all mankind." The President of the United
States in his address to the United Nations General Assembly, September 22, 1960 proposed that "We press forward with a program of international cooperation for constructive uses of outer space under the United Nations. Better weather forecasting, improved
world wide communications . . . are but a few of the benefits of such cooperation."

Several additional weather satellite experiments are planned for the next two years.
In view of the impact of the increasing operational aspects of the experiments, a national
plan leading to a fully operational system is necessary for making maximum use of meteorological satellite data at the earliest possible time.

It is the purpose of this study to formulate a national plan for a common system of
meteorological observation satellites. This plan would have the following goals:
1.   Complete global coverage.
2.   Uninterrupted continuity in time.
3.   Design for maximum national and international utilization for the benefit of all
     mankind.
4.   Adequate readout stations to insure timely receipt of all the data.
5.   Complete communication facilities to transmit data from the readout stations to
     the National Meteorological Center and other user points.
6.   Analysis of the data received at the National Meteorological Center.
[2] 7.   Depiction of these analyses in forms suitable for transmission via adequate
     communication facilities.
8.   Communication of processed data to all domestic (civil and governmental) and
     international users for application to their particular requirements.
9.   Intensive research to improve weather forecasts through the application of satel-
     lite data.
For planning purposes it has been assumed that:
1.   With respect to the Operational System:
     a.   The U. S. Weather Bureau as the National Meteorological Service will have
          program responsibility for the operational meteorological satellite observing
          and data processing system. This would include equipment procurement,
          launching, data retrieval and processing, and dissemination to users.

   b. An organization to perform all activities related to the operational meteoro-
       logical satellite observing system would be established as a self-contained enti-
       ty reporting to the Chief of the Weather Bureau.
   c. Coordination will be accomplished by a "Civil-Military Liaison Committee"
       and resident liaison personnel.
   d. The operational system will be started by the end of Fiscal Year 1962.
       (Adequate funding is required to accomplish this.)
2. With respect to <u>Research and Development:</u>
   a. NASA has the responsibility for the equipment design and development,
       launching and data retrieval associated with experimental satellites. The
       design of the operational space craft will be based on the results of this work.
   b. The Weather Bureau has the responsibility for the data analysis and meteo-
       rological research.
3. In <u>operations</u> and <u>research</u> and <u>development</u>, it is assumed that the Department
   of Defense would be responsible for:
   a. Military application of satellite data and National Meteorological Center
       products.
[3]   b. Specialized communication systems and other facilities to meet unique needs
       not covered by the National Satellite Meteorological Program.
4. International participation will be developed by existing international bodies
   such as the World Meteorological Organization. . . .

Document II-5

**Document title: Hugh L. Dryden, Deputy Administrator, for James E. Webb, Administrator,
NASA, and Luther H. Hodges, Secretary of Commerce, "Basic Agreement Between U.S.
Department of Commerce and the National Aeronautics and Space Administration
Concerning Operational Meteorological Satellite Systems," January 30, 1964.**

**Source: NASA Historical Reference Collection, NASA History Office, NASA
Headquarters, Washington D.C.**

*When the Interagency Panel on Operational Meteorological Satellites released a plan in April 1961
calling for NASA to develop an operational weather satellite system to be managed by the U.S. Weather
Bureau, the bureau was concerned by the lack of influence it had over the developmental stages of this
next generation of weather satellites, named Nimbus. Some of these fears were realized as the NASA-
managed Nimbus program suffered numerous delays as well as dramatic costs increases. On
September 27, 1963, frustrated with NASA's performance, the Weather Bureau announced that it
would develop its own satellite system based on the TIROS design, with the Department of Defense providing launch services. Outmaneuvered, NASA agreed to helping the Weather Bureau develop its
TIROS-based system, as well as granting it an increased role in developing the Nimbus satellite, which
became an experimental rather than an operational system. This agreement codified the NASA-
Weather Bureau relationship.*

[1]

Basic Agreement Between U.S. Department of Commerce and the National Aeronautics and Space Administration Concerning Operational Meteorological Satellite Systems

SECTION 1. AUTHORITY AND PURPOSE

Recognizing the success of the National Aeronautics and Space Administration (NASA) research and development (R&D) meteorological satellite program and the utilization by the Department of Commerce - Weather Bureau (DOC - WB) of satellite data in weather analysis and forecasting; and

Taking note that Congress, also recognizing this success, provided in the Supplemental Appropriation Act of 1962, and thereafter, funds authorizing DOC - WB to establish and operate a system for the continuous observation of world-wide meteorological conditions from space satellites and for the reporting and processing of the data obtained for use in weather forecasting . . .": and

Recognizing the broad responsibilities of NASA under the Space Act for continuing a research and development program for the development of spacecraft technology and meteorological satellite systems for (1) application to operational systems and (2) research activity in the atmospheric sciences; and

Taking note that Congress appropriated separate funds to NASA for the purpose of supporting such an R&D program of spacecraft technology and meteorological satellite systems;

[2] It is, therefore, the purpose of this agreement to define the relationship between, and the functions to be performed by the DOC - WB and NASA (1) in the conduct of operational meteorological satellite programs, and (2) in the development of supporting technology for operational meteorological satellite programs.

SECTION II. ESTABLISHMENT AND OPERATION OF THE NATIONAL OPERATIONAL METEOROLOGICAL SATELLITE SYSTEM

A. Objective

The primary objectives of the National Operational Meteorological Satellite System (NOMSS) is to provide meteorological information for prompt and effective use by the various national meteorological services in weather analysis and prediction. The operation of NOMSS will be based on, but not be limited to, the technology produced in the NASA R&D Meteorological Satellite Program and will satisfy the meteorological requirements of user agencies, subject to limitations of budget, resources of law. The DOC - WB may modify NOMSS as appropriate and in accordance with the terms of this agreement to accommodate changes in meteorological requirements and developments in technology.

[3] B. Basic Responsibility and Functions

1. The DOC - WB by law has the basic responsibility for the establishment and operation of the NOMSS, which includes obtaining necessary funds.

2. Each agency agrees to perform the functions and follow the management duties and procedures set forth in paragraphs D and E of this section.

C. Funding
    The DOC - WB will submit requests for annual appropriations to carry out NOMSS. The DOC - WB will develop the plans and budget estimates and the justification thereof, with the assistance and support of NASA and user agencies, as appropriate. At the beginning of each fiscal year, or as soon thereafter as funds are appropriated, or as program changes require, the DOC - WB will issue a reimbursement order to NASA for the full amount of funds required for the execution of the NASA portion of the program in accordance with the approved project proposal prepared under this agreement. NASA will account to DOC - WB for the funds so transferred. NASA also will provide necessary reports to the DOC - WB regarding the proposed and actual commitments, obligations and expenditure of these funds, so that DOC - WB [4] may meet its fiscal responsibilities with respect to the funds appropriated to it or otherwise received.

D. Functional Responsibilities
    The functions of the DOC - WB and NASA in the conduct of NOMSS are as follows:
1. The Department of Commerce - Weather Bureau shall:
    a. Determine overall meteorological program requirements (including cost and schedule).
    b. Specify quantities to be measured by satellite meteorological instruments.
    c. Approve Project Development Plan and changes involving schedules, resources, interfaces, and performance.
    d. Monitor the performance of the system for meeting meteorological requirements.
    e. Determine the need for replacing a spacecraft that has experienced marginal failure in providing meteorological data.
    f. Operate the Weather Bureau Command and Data Acquisition [CDA] Stations, including control of the operational satellite after NASA has determined that the satellite is ready for operational use.
[5]   g. Manage meteorological data analysis activities at the CDA stations.
    h. Communicate operational data from CDA stations to [the National Weather Service Center and the Goddard Space Flight Center (GSFC)] and others, as appropriate.
    i. Process data for integration into weather analyses.
    j. Disseminate data, analyses and forecasts.
    k. Archive the information (processing, storage, retrieval).
    l. Use the data for research and climatological purposes.
    m. Conduct system studies as required to meet its responsibilities.
2. The National Aeronautics and Space Administration shall:
    a. Prepare the Project Development Plan.
    b. Design, engineer, procure and qualify flight spacecraft.
    c. Select and procure launch vehicles.
    d. Maintain and operate launch sites.
    e. Design, construct, and insure initial operational status of Command and Data Acquisition stations.
    f. Prepare the pre-launch of spacecraft and launch vehicle.
    g. Conduct launch operations.
[6]   h. Track and determine basic orbit during the useful life of the satellite.
    i. Monitor the engineering status of the satellite and command the satellite during initial time in orbit, and, as requested, during periods of malfunction using the Weather Bureau Command and Data Acquisition stations.
    j. Consult, as appropriated, on technical matters.

E. Management Responsibilities and Procedures
    1.   The management of the functions of the Department of Commerce portion of this agreement is a responsibility of the Weather Bureau under the authority delegated by the Secretary of Commerce to the Chief of the Weather Bureau under Department Order No. 91 of May 23, 1963. The Weather Bureau shall provide or obtain the necessary DOC resources for NOMSS and shall serve as the official DOC contact for this program.
    2.   The management of the NASA portion of this agreement is the responsibility of Headquarters, Office of Space Science and Applications. It shall provide the necessary NASA resources to NOMSS and shall serve as the official NASA contact for this program.
    3.   The following specific management functions and procedures are agreed upon:

[7]
    a.   The DOC - WB will forward mission requirements to NASA for review and acceptance.
    b.   NASA will forward the basic plan of approach to the DOC - WB for review and approval.
    c.   NASA will forward the Project Proposal to the DOC - WB for review and approval.
    d.   NASA will forward requests for proposals to the DOC - WB for review and comment.
    e.   A representative of the DOC - WB will be assigned to the NASA GSFC Project Office and will participate in review for source evaluation, definitization of statements of work and project status reviews.
    f.   NASA will make the final source selection, and will negotiate with and be the single interface with the contractor.
    g.   NASA will submit to DOC - WB for review and approval the definitized contract work statement, schedules and cost.
    h.   NASA will forward the final Project Development Plan to the DOC - WB for review and approval.
    i.   Major changes involving schedules, costs and system performance will be forwarded by NASA to the DOC - WB for review and approval.
[8]
    j.   All changes affecting the interface between NASA provided equipment and DOC - WB equipment will be forwarded to the DOC - WB for review and approval.

F. Interagency Relationships
    The DOC - WB will furnish a statement of mission requirements to NASA, and will ensure that such requirements and the resulting project plans meet the needs of DOD and other user agencies.

SECTION III. THE DEVELOPMENT OF SUPPORTING TECHNOLOGY
FOR OPERATIONAL METEOROLOGICAL SATELLITES

A. Scope
    This section deals only with the development of space technology which is specifically identified as applying to NOMSS.

B. Basic Responsibility and Functions
    1.   NASA has the basic responsibility for supporting civilian satellite technology.
    2.   The DOC - WB will submit to NASA estimates of future meteorological satellite requirements and the DOC - WB estimates of present technological limitations to meeting them. NASA will draw up its R&D plans with due consideration of the stated Weather Bureau requirements and will keep DOC fully informed on R&D program plans and developments.

[9]   3.   The DOC - WB may conduct sensor development but will maintain close liaison with NASA to ensure compatibility with future spacecraft configurations.

C.   Funding
NASA will fund for the supporting technology for operational meteorological satellite development programs.

D.   Management Procedures
1.   Supporting technology for Operational Meteorological Satellite programs is the responsibility of the Office of Space Science and Applications, in NASA Headquarters.
2.   NASA will maintain a coordinating mechanism whereby the contributions of the Weather Bureau and other competent agencies can be considered in the development program. It will consist of five members, two nominated by NASA (Chairman), two by the Weather Bureau, and one representing the non-Government meteorological community.
3.   The NASA, after considering the advice of the coordinating mechanism, will choose, and allocate space in meteorological satellites for flight tests of experimental meteorological sensors. NASA will fund for these tests, including the costs of flight hardware beyond the preprototype stage, but the execution will be the responsibility of the experimenting agency.

[10] E.   Data
1.   Data from proven meteorological sensors flown in NASA research and development meteorological satellites, such as TIROS camera data and Nimbus AVCS data, will be made available at the request of the Weather Bureau for operational use on a cost reimbursable basis, for such added costs as may result from the operational requirement.
2.   Data from experimental meteorological sensors flown in NASA R&D meteorological satellites, such as the Nimbus HRIR and other new sensor developments of potential operational use, will be made available to the Weather Bureau as soon as practicable on a non-interference basis to NASA missions for the conduct of operational experiments. In the case of these data, the experimenting agency retains exclusive publication rights for a period of eighteen months, but the Weather Bureau may conduct operational experiments during this period with the proviso that dissemination of these data is restricted to such purposes and that scientific publication will not result without the concurrence of the experimenter. NASA will be reimbursed for all additional costs incurred in making such data available to the Weather Bureau.

SECTION IV. METEOROLOGICAL SATELLITE PROGRAM REVIEW BOARD

A Meteorological Satellite Program Review Board is hereby established. [11] The Board is composed of two members each from NASA and DOC - WB with the Associate Administrator for Space Science and Applications of NASA and the Chief of the Weather Bureau serving as co-chairmen. The Board will meet quarterly or at the request of either co-chairman to review the program and consider any substantive issues which may arise. It may make recommendations to the DOC - WB on the resolution of issues concerning the operational programs, and to the NASA concerning the responsiveness of the NASA R&D program to the needs of NOMSS. Either chairman may refer any issue to the Associate Administrator of NASA and to the Assistant Secretary of Commerce for Science and Technology for resolution.

SECTION V. AMENDMENTS

This agreement may be amended at any time by the mutual consent of the Agencies concerned. The agreement will be reviewed formally for necessary changes at least once every two years from the date of the agreement or as required at the request of either agency. For particular programs, a Memorandum of Understanding may be used at the working level to clarify any of the functional responsibilities and procedures.

SECTION VI. RELEASE OF PUBLIC INFORMATION

Release of public information on the operational and R&D programs may be initiated by either the Weather Bureau or by NASA. Before any [12] release is issued to the public, however, clearance and final approval must be given by the agency having the assigned function listed in Section 2 D or Section 3 B. Coordinated or joint releases should be issued where appropriate.

SECTION VII. INTERNATIONAL RELATIONS

1.   Regarding the international aspects of meteorology and space satellites, international negotiations may be carried out by either agency according to its basic responsibilities and functions as defined in this agreement, with due regard to the provisions 2 and 3 below and subject to normal State Department policy guidance.

2.   Where such negotiations imply obligations or place commitments upon the other agency, that agency will be consulted in advance of international agreement or commitment.

3.   The design of operational meteorological systems will give due consideration to commitments already expressed or implied by the United States.

Hugh L. Dryden
Deputy Administrator
for JAMES E. WEBB   Luther H. Hodges
ADMINISTRATOR, NASA   LUTHER H. HODGES
January 30, 1964   SECRETARY OF COMMERCE
                                                                     January 30, 1964

Document II-6

Document title: Robert M. White, Administrator, Environmental Science Services Administration, National Environmental Satellite Center, U.S. Department of Commerce, to Dr. Homer E. Newell, Associate Administrator for Space Science and Applications, NASA, August 15, 1966.

Source: NASA Historical Reference Collection, NASA History Office, NASA Headquarters, Washington, D.C.

Transcending the turf battles that marked its relationship with NASA in the early 1960s, the U.S. Weather Bureau made progress during the later half of the decade in refining its requirements for an operational meteorological satellite program. Using information gathered from three separate satellite programs—TIROS, Nimbus, and the Advanced Technology Satellite (ATS)—in August 1966, the Weather Bureau issued to NASA this statement outlining its objectives for an operational meteorological satellite program.

[1]   [rubber stamped: "AUG 15 1966']
Dr. Homer E. Newell
Associate Administrator for Space
    Science and Applications
National Aeronautics and Space
    Administration
Washington, D.C. 20546

Dear Homer:

I would like to thank you for the briefing given by NASA on its proposed FY 1968 meteorological research and development program at the 1 July 1966 meeting of the Meteorological Satellite Program Review Board (MSPRB). As in the past two years our comments are directed toward an evaluation of the effectiveness of your program plans in the development and improvement of operational meteorological satellite systems.

In arriving at our comments on the NASA R&D program, we have reviewed our letters of the past two years, based on similar briefings. We find that significant progress was made last year with respect to interagency program development and effective use of available resources in attempting to attain program objectives. Our primary objectives have not changed and remain as follows: (1) The establishment and maintenance of a satellite system to obtain global observations on a regular basis, (2) meteorological observations from synchronous altitude, and (3) global observations of atmospheric structure needed for numerical weather forecasting. Following are specific comments with respect to each of these.

1.   Global observations on a regular basis.
The initial deployment of the Tiros Operational Satellite (TOS) system, with the launch of the ESSA 1 and ESSA 2 satellites, has been quite successful and is providing very useful data to the meteorological community throughout the world. Launch of ESSA 3 is now imminent due to a recent camera failure on ESSA 1. We are very pleased with the effort, direction and progress being made in the "TIROS M" program. We look forward to this development solving the major problem raised under this objective in our letter last year. The program review showed a line item for "improved HRIR day/night imaging and higher resolution," to be accomplished by a two-channel radiometer (visible and 11 microns) of high resolution. Also, funds were shown for development of a [2] multichannel radiometer under the Nimbus B flight program and in the TOS improvement program. [handwritten underlining on original] We would like to determine whether or not additional radiometer development activity is needed to meet the TOS system requirements. Also, there is a strong requirement to increase the resolution of satellite cloud pictures to one mile (photo resolution). We would like to examine with you the possible technical approaches to meeting this requirement within the framework of the TOS system and the probable costs of doing so. If a reasonable approach can be found, from the point of view of both cost and technology, we would want to proceed with those steps needed to provide this improved operational capability. We consider these programs essential to the full attainment of the first objective and hope that it will be possible for NASA to devote additional resources and increased priority in the NASA R&D program to them.

2.   Meteorological observations from synchronous altitude.
The continued progress on meteorological experiments with the ATS series is most gratifying. The data relay experiments being planned jointly by NASA and ESSA, in conjunction with the spinscan camera, will make a major contribution to the development of the World Weather Watch.

We are pleased with the promptness and thoroughness with which the Goddard Space Flight Center conducted the feasibility study for a Synchronous Operational Meteorological Satellite which could be built and launched in a short time period should strong National interest dictate such a move. It now appears likely that the earliest we will obtain funds for such a system will be in our regular FY 1968 budget. Therefore, with the extra time available for planning studies, it would seem wise to examine the trade-offs of spacecraft and ground station cost and performance as a function of system design, especially with regard to frequencies and data format.

We consider the continued strong emphasis and support of NASA in this area to be very important.

3.   Global observations of atmospheric structure needed for numerical weather forecasting.
As cited in last year's letter this goal continues to carry the highest priority to ESSA from a meteorological point of view, because of the importance of describing the atmosphere adequately in terms required for numerical weather prediction. The schedule of development of the World Weather System is critically dependent upon progress in this area. Therefore, we support ongoing Nimbus flights and those of other advanced satellites in the NASA program [3] which are devoted to attaining this objective. Because of the critical importance of this portion of the NASA program, we are hopeful that adequate priority will continue to be supplied in support of this program element with respect to others in the National budget. We hope NASA will continue, and if possible expand its support of the development and flight test of the new sensors and supporting subsystems in its R&D program in order to provide the technology needed to meet this objective.

The Environmental Science Services Administration is now examining how its present and future operational satellites can satisfy environmental data requirements in other areas than meteorology. Undoubtedly there will be a need for R&D support from NASA in these new areas. I suggest we review this matter after our needs are established and discussions have been held under the leadership of Messrs. Jaffe and D. S. Johnson. We have been most pleased with the joint effort this past year in resolving problems and in allocating available resources to meet operational and R&D meteorological satellite needs. We are looking forward to the continuation of this excellent cooperation.

Sincerely yours,

Robert M. White
Administrator

Document II-7

Document title: George E. Brown, Jr., Chairman, Committee on Science, Space, and Technology, U.S. House of Representatives, to D. James Baker, Acting Under Secretary for Oceans and Atmosphere, U.S. Department of Commerce, February 22, 1993.

Document II-8

Document title: Jim Exon, Chairman, Subcommittee on Nuclear Deterrence, Arms Control and Defense Intelligence, U.S. Senate, to Ron Brown, Secretary of Commerce, June 2, 1993.

Source: Both in NASA Historical Reference Collection, NASA History Office, NASA Headquarters, Washington, D.C.

These two letters, one from Congressman George E. Brown, chair of the House Committee on Science, Space, and Technology, and the other from Nebraskan Senator Jim Exon, chair of the Senate Subcommittee on Nuclear Deterrence, Arms Control and Defense Intelligence, urged National Oceanic and Atmospheric Administration (NOAA) Administrator D. James Baker and Commerce Secretary Ron Brown to evaluate the concept of converging NOAA's Polar-orbiting Operational Environmental Satellite System with the Department of Defense's Defense Meteorological Satellite Program. These letters helped establish the political basis within Congress for the convergence to take place.

Document II-7

February 22, 1993

Dr. D. James Baker
Acting Under Secretary for Oceans and Atmosphere
U.S. Department of Commerce
Washington, D.C. 20230

Dear Dr. Baker:

I want to congratulate you on your selection to become the next National Oceanic Atmospheric Administration (NOAA) Administrator. I look forward to working with you on the very important programs NOAA has planned for the future. The Committee on Science, Space, and Technology intends to consider a number of dual-use technology and defense convergence issues during the 103rd Congress. In that regard, we believe that the issues related to the convergence of the NOAA Polar Satellite Program with the Defense Meteorological Satellite Program (DMSP) should be re-examined to look for potential opportunities for reducing overall costs. As you know, recent significant changes in agency requirements and policies may be more conducive to a merged system than was the case in the past.

In addition, I feel that the relationship between NOAA's continuing operational global observing system and NASA's planned 15-year Earth Observing System should also be examined to determine the potential benefits and liabilities of closer cooperation on these programs.

I believe that it would be prudent to examine these opportunities for convergence and possible cost savings at the same time rather than separately. Therefore, I ask that NOAA, and as soon as appropriate, NASA and DOD, jointly study and assess the possible benefits and mechanisms for merging all or parts of the three programs and provide a jointly developed plan to the Committee on Science, Space and Technology.

Thank you for your cooperation.

Sincerely,

GEORGE E. BROWN, JR.
Chairman

Document II-8

[1]   June 2, 1993

The Honorable Ron Brown
Secretary of Commerce
US Department of Commerce
Herbert C. Hoover Building
14th Street and Constitution Avenue, N.W.
Washington, D.C. 20230

Dear Secretary Brown:

I plan to make a statement on the Senate floor soon about the weather satellite systems operated by the Department of Defense (DoD) and the National Oceanographic and Atmospheric Administration (NOAA). I do not believe that two separate U.S. Government weather satellite systems can be justified any longer given the budget problems we face.

The DoD operates a constellation of two weather satellites called the Defense Meteorological Satellite Program (DMSP). These satellites are flown in sun-synchronous polar orbits (meaning that they cross the same point above the Earth twice a day at the same times every day). They are built by the General Electric Corporation. They are equipped with sensors for imaging clouds, determining moisture and temperature in the atmosphere, and for measuring ocean currents. The data from DMSP is broadcast to tactical users over an encrypted link and at the same time the data is remotely relayed to Omaha's Weather Central for comprehensive analysis. Historically, the DoD system's primary customer was a classified intelligence gathering program.

The National Oceanographic and Atmospheric Administration (NOAA) of the Department of Commerce also operates a constellation of two weather satellites called TIROS. These satellites are also flown in sun-synchronous polar orbits, are built by General Electric, and have sensors for imaging clouds and taking readings of moisture and temperature in the atmosphere. TIROS data also is broadcast directly to users around the world as well as to a central processing location in the United States. TIROS data, however, is completely unclassified.

In terms of capacity, the United States does not need four weather satellites in orbit. In last year's defense authorization act, the conferees directed the Secretary of Defense to develop a comprehensive space investment strategy. As [2] part of this effort, the conferees directed DoD to examine anew the potential for greater cooperation between civil and military weather satellite programs in light of changes in the world and budget pressures.

Merging the two satellite programs will take time—time to design a common system, to determine management arrangements between DoD and NOAA, to build new satellites, and to launch them. Both DoD and NOAA will obviously have to continue to launch and operate their own systems until the new system can be deployed.

Ideally, DoD and NOAA would run out of their current satellites at precisely the same time and precisely when the new system became operational. It appears that this is possible, but not without some planning.

Last year Congress directed NOAA to procure two more TIROS satellites. If this happens, NOAA will have 7 satellites, which could last until 2005, and DoD's inventory of 9 DMSP satellites will last until 2007 or longer. That would mean waiting 12 to 15 years to deploy a common, merged satellite system and waiting several years before starting development work on a new, common satellite. The Government ought to examine whether it would make more sense to speed things up.

I propose that DoD consider transferring to NOAA two DMSP buses, which NOAA could then modify for the TIROS configuration and add to it the TIROS sensors. This would give DoD and NOAA 7 satellites each. It would save money in the short term, some of which could be used to fund development of a common satellite system. This is important because budgets are so tight for both DoD and NOAA that neither may be able to afford modernization on their own. It would also mean that DoD and NOAA would likely use up their inventories at about the same time, for a smooth transition to a new common system.

I am writing to urge you and Deputy Secretary of Defense Perry to create a formal working group under appropriate senior officials to attempt to resolve any outstanding issues standing in the way of merging the two government polar-orbiting weather satellite systems.

What is required is leadership from both agencies to resolve issues of data encryption; management of a merged system; potential transfer of DoD assets to NOAA; integration with European meteorological satellite efforts; and cooperation with NASA on the Earth Observing System polar platform.

[3] To take one example, without the personal involvement of you and Secretary Perry, subordinate DoD officials will continue to insist that data must be encrypted and NOAA will insist that encryption is not acceptable. My suspicion is that the case for encryption rests on weak arguments, given the availability of geosynchronous satellites, and European, Chinese, Russian, and NOAA systems all broadcasting in the clear. On the other hand, NOAA may be shortsighted in disregarding other national interests that might justify some form of encryption capability. Resolving this type of problem requires creativity from top policymakers.

The nation cannot afford to maintain and modernize two satellite weather constellations. Working together, however, DoD, NOAA and NASA could pool resources, achieve efficiency and improve capabilities at reduced cost to the taxpayer.

I look forward to hearing your views.

Sincerely,

Jim Exon
United States Senator

cc: Deputy Secretary of Defense William Perry
     Vice President Al Gore
     NASA Administrator Goldin

Document II-9

Document title: National Performance Review, Department of Commerce, "Establish a Single Civilian Operational Environmental Polar Satellite Program," September 30, 1993.

Source: NASA Historical Reference Collection, NASA History Office, NASA Headquarters, Washington, D.C.

This document spells out the financial advantages of achieving the consolidation of the National Oceanic and Atmospheric Administration's Polar-orbiting Operational Environmental Satellite System and the Defense Department's Defense Meteorological Satellite Program. It also points out that a consolidated or converged system with NASA involvement could make efficient use of NASA's development of new Earth observation instruments. This document was part of the Clinton administration's National Performance Review.

Department of Commerce

Accompanying Report of the National Performance Review
Office of the Vice President
Washington, DC

September 1993

DOC12: **Establish a Single Civilian Operational Environmental Polar Satellite Program**

Background

The United States is committed to an operational environmental polar satellite program because of the critical value of the data the satellites collect.(1) Polar satellites collect temperature and moisture measurements (key inputs to computer weather prediction models generating all national three- to five-day weather forecasts); measurements of the Antarctic ozone levels; long-term environmental measurements used to support global climate change studies; sea surface temperature measurements; and global cloud-cover images. Polar satellites also provide other valuable support missions, such as monitoring emergency distress beacons to aid search and rescue missions and worldwide data collection to support a variety of activities, such as endangered species monitoring.

However, at present, the nation maintains two polar-orbiting meteorological satellite systems: (1) the National Oceanic and Atmospheric Administration (NOAA) Polar-orbiting Operational Environmental Satellite (POES) for civil forecasting and research purposes; and (2) the Department of Defense (DOD) Defense Meteorological Satellite Program (DMSP) for national security purposes.

In addition to these programs, the National Aeronautic and Space Administration (NASA) has initiated a climate research program called Mission to Planet Earth (MTPE). A key portion of this effort is the Earth Observing System (EOS), a series of six different satellites measuring various parameters critical to understanding global climate change. One of these satellites is called the EOS-PM (PM indicating that the satellite passes over the equator in the afternoon). The climate monitoring instruments on EOS-PM are basically more modern versions of the meteorological instruments currently flying on the NOAA weather satellites. In essence, the nation will have three different satellite systems with very similar capabilities.

Over the past 20 years, the POES and DMSP programs have made numerous attempts to converge to the greatest extent possible.(2) The programs have similar spacecraft, use a common launch vehicle, share products derived from the data, provide complementary environmental data to the nation, and work closely together on research and development efforts. In all, the programs achieved substantial commonality, but national security concerns have precluded full convergence.(3)

DOD has stated it would manage a converged system, but a single program run by DOD was and still is unacceptable given international concern over the militarization of space.(4) Today, however, with the end of the Cold War, the issues which have precluded complete convergence seem to have diminished in importance.(5) With both programs planning a new satellite design, the time is appropriate to consolidate their efforts.

The EOS-PM climate research satellite is being designed with the idea that many of the instruments can be used by NOAA within the POES program. This continues a historical NOAA-NASA relationship wherein NASA develops new technology and demonstrates prototype hardware, and NOAA buys identical units for continued operational support.(6) However, current plans involve flying EOS-PM for 15 years, during which time

POES also will have operational satellites.(7) Over most of this period, both programs would be flying duplicate instruments. The nation would be more efficiently served if NASA would develop and fly the prototypes once and then transfer the systems to NOAA's operational program for future flights.

Convergence studies began in 1972 and have continued ever since.(8) NOAA recently performed an internal study of the opportunities available through convergence of the programs.(9) Recently, initial talks have begun among the three agencies with the goal of performing another study of convergence opportunities among the three programs.(10) What is needed, however, is a clear decision to create a single, civilian polar satellite program.

Currently, the NOAA POES program, the DOD DMSP program, and the NASA EOS-PM program all are in various stages of developing new spacecraft and instruments. In the next 10 years, the estimated total cost for these three efforts exceeds $6 billion in development, production, and operations costs. However, many policy makers feel that the nation cannot afford to develop three separate satellite systems with such similar missions.

For example, Congressman George Brown of California has stated that a converged system seems more achievable than in the past. He therefore has directed NOAA to work with DOD and NASA to "jointly study and assess the possible benefits and mechanisms for merging all or parts of the three programs."(11) Senator James Exon of Nebraska was more direct in his letters to DOD and Commerce: "The nation cannot afford to maintain and modernize two satellite weather constellations."(12) Recently, at the National Space Outlook Conference, Air Force General Charles Horner, Commander United States Space Command, stated: "How you do convergence is really the question, not if you do convergence."(13)

A single operational polar satellite program could meet the needs of all users by incorporating key DOD requirements into the NOAA POES program. Furthermore, the synergy achieved through DOD and NOAA cooperation could allow both agencies to meet critical operational requirements (such as collecting oceanographic and global tropospheric wind data) which neither agency has been able to afford alone. The converged operational program could save additional costs by using the NASA EOS program's state-of-the-art spacecraft and instruments instead of forcing NOAA to design and build its own. The result would be a single development program (compared to the three planned today) and minimal overlap between NASA's climate research and the NOAA-DOD converged operational meteorological missions.

The difficulty will be to successfully incorporate DOD requirements into the program. Based upon historical studies, key areas requiring consideration are data deniability, orbit selection, international cooperation, and adequate oversight to ensure DOD concerns are adequately met.(14) The following summarizes how each of these can be addressed:

Data deniability. The satellite must broadcast data free to everyone but also have the capability to deny data to specific adversaries. New technology, such as that used to deny cable-TV pay channels to non-subscribers, makes this task easier.

Orbit selection. Currently, the DOD desires the capability to change its satellite orbits depending on mission requirements. Past studies have identified a three-satellite constellation as sufficient for meeting all orbit needs.(15) Allowing DOD to influence orbits selection should alleviate their concerns.

International cooperation. A NOAA-led system could easily maintain and even improve international cooperation in environmental data exchange. However, since NOAA plans to use foreign satellites as part of the converged program, DOD may be reluctant to rely upon foreign satellites for important data. This concern could be alleviated by maintaining one or more ground spare U.S. satellites at all times that could be launched if a foreign-controlled satellite ever became unreliable.

Oversight. DOD will require some mechanisms to ensure their requirements continue to be met. Possible implementation details could involve including DOD user and acquisition experts in the NOAA program offices and operations facilities, allowing DOD to fund and manage DOD-unique parts of the program, and establishing an interagency oversight group to which the program would have to report periodically to ensure that all agency requirements were adequately met. Such oversight mechanisms should not be difficult to achieve. The driving force behind this effort is clearly the desire to reduce costs.

Further cost reduction could be achieved through greater international participation. According to Dr. Ray A. Williamson of the Office of Technology Assessment: "Greater international coordination and collaboration on sensors and systems . . . will eventually be needed in order to reap the greatest benefit from the world-wide investment in remote sensing."(16)

NOAA is already working on such arrangements in its POES program by asking the Europeans to assume a greater role. An agreement in principle has been reached between NOAA and the European Organization for the Exploitation of Meteorological Satellites (EUMETSAT) whereby EUMETSAT will purchase, launch, and operate one of the two current POES missions beginning in the year 2000. This will save the U.S. more than $100 million for each launch of one of these satellites. Such cooperation with the Europeans is an important component of cost-efficient operations and is the first step to a truly international environmental satellite observing system.

Action

Legislation should be enacted to establish a single environmental polar satellite program under the direction of NOAA.

Congress should enact legislation to establish a single environmental polar satellite under the direction of NOAA. The legislation should direct NOAA, NASA, and DOD to undertake activities to establish this effort within their existing programs.

Implications

The proposed changes would allow for a more efficient, less-costly global satellite observation program. A strong, efficient U.S. polar environmental monitoring program would be the foundation for a cooperative international system. The Europeans already plan to increase funding for an element of this system. With a solid, unified U.S. national program in place, other countries may align their programs to complement the basic system. The result will be additional environmental data collected at minimal cost to the nation. The convergence concept provides a feasible and cost-effective opportunity to accurately monitor and predict the impact of the environment on the world's societies.

The greatest difficulty in the proposal will be to ensure that a single, national program under civilian leadership will be responsive to national security needs. However, these concerns can be met much more easily now than they could have in the past.

Fiscal Impact

Cost savings over ten years would total about $1.3 billion. This is based on a three-satellite system (with European participation) relying on NASA to develop new hardware.

Budget Authority (BA) and Outlays (Dollars in Millions)

	1994	1995	1996	1997	1998	1999	Total
			Fiscal Year				
BA	0.0	0.0	-75.0	-75.0	-75.0	-75.0	-300.0
Outlays	0.0	0.0	-50.0	-70.0	-75.0	-75.0	-270.0
Change in FTEs	0	0	0	0	0	0	0

Endnotes

1.   See Hussey, John W., "Economic Benefits of Operational Environmental Satellites," reprinted from A. Schnap (ed.), *Monitoring Earth's Ocean, Land, and Atmosphere from Space-Sensors, Systems, and Applications,* Vol. 97 of Progress in Astronautics series (Washington, D.C.: American Institute of Astronautics and Aeronautics, 1985).

2.   See Blersch, Donald, *DMSP/POES Convergence Materials Handbook,* STDN-91-18. 2nd ed. (Arlington, VA: Analytic Services, Inc., October 1991).

3.   See U.S. Department of Commerce, *Comparison of the Defense Meteorological Satellite Program (DMSP) and the NOAA Polar-orbiting Operational Environmental Satellite (POES) Program,* Envirosat-2000 Report (Washington, D.C., October 1985).

4.   See U.S. Department of Commerce, National Oceanic and Atmospheric Administration (NOAA), *International Implications of Converging the DOD and DOD Polar Orbiting Meteorological Satellite Systems* (Washington, D.C., 1987).

5.   See Blersch, Donald, *DMSP/POES Convergence: A Post Cold War Assessment (A Re-Examination of Traditional Concerns in a Changing Environment)* (Arlington, VA: Analytic Services, Inc., June 1993).

6.   See U.S. Department of Commerce and National Aeronautic and Space Administration, "Basic Agreement between U.S. Department of Commerce and the National Aeronautics and Astronautics Association Concerning Operational Environmental Satellite Systems of the Department of Commerce," 1973.

7.   See National Aeronautics and Space Administration, *1993 Earth Observing System Reference Handbook* (Washington, D.C., March 1993).

8.   See Blersch, *DMSP/POES Convergence Materials Handbook.*

9.   See U.S. Department of Commerce, "Report of the Working Group on NOAA Polar Satellite Convergence Opportunities" (Washington, D.C., June 1993). (Draft.)

10. See U.S. Department of Commerce, National Oceanic and Atmospheric Administration, National Aeronautical and Space Administration, "Terms of Reference for Joint Defense Meteorological Satellite Program (DMSP), Polar-orbiting Operational Environmental Satellite (POES) and Earth Observing System (EOS) Program Assessment," June 29, 1993.

11. See Letter from George E. Brown, Jr., Chairman of the House Committee on Science, Space and Technology, to Dr. D. James Baker, NOAA Administrator, February 22, 1993.

12. See Letters from Senator James Exon to Secretary of Commerce Ron Brown and Deputy Secretary of Defense William Perry, June 2, 1993.

13. "Horner Supports Converged System," *Space News* (June 27, 1993), p. 4.

14. Blersch, *DMSP/POES Convergence Handbook,* p. II-2.

15. See U.S. General Accounting Office, *Economies Available to Converging Government Meteorological Satellites* (Washington, D.C.: U.S. General Accounting Office, 1986).

16. U.S. Congress, House Committee on Science, Space and Technology, Subcommittee on Space, testimony by Dr. Ray A. Williamson, Office of Technology Assessment, May 6, 1993.

Document II-10

Document title: Presidential Decision Directive/NSTC-2, The White House, "Convergence of U.S. Polar-orbiting Operational Environmental Satellite Systems," May 5, 1994.

Source: NASA Historical Reference Collection, NASA History Office, NASA Headquarters, Washington, D.C.

This document, the result of a recommendation from the Clinton administration's National Performance Review, lays out a broad plan for the convergence of the U.S. Polar-orbiting Environmental Satellite System, operated by the National Oceanic and Atmospheric Administration (NOAA), with the Defense Meteorological Satellite Program's polar-orbiting system, operated by the U.S. Air Force. It calls for the establishment of an Integrated Program Office by October 1, 1994, to be operated jointly by the Departments of Commerce and Defense and by NASA. It gives NOAA the lead responsibility for operating the converged system. The Department of Defense would be responsible for major systems acquisition, and NASA would lead in "facilitating the development and insertion of new cost effective technologies" into the system.

[no pagination]

THE WHITE HOUSE WASHINGTON   May 5, 1994
PRESIDENTIAL DECISION DIRECTIVE/NSTC-2

TO: The Vice President
      The Secretary of State
      The Secretary of Defense
      The Secretary of Commerce
      The Director, Office of Management and Budget
      The Administrator, National Aeronautics and Space Administration
      The Assistant to the President for National Security Affairs
      The Assistant to the President for Science and Technology
      The Assistant to the President for Economic Policy

SUBJECT: Convergence of U.S. Polar-orbiting Operational Environmental Satellite Systems

I. Introduction
    The United States operates civil and military polar-orbiting environmental satellite systems which collect, process, and distribute remotely-sensed meteorological, oceanographic, and space environmental data. The Department of Commerce is responsible for the Polar-orbiting Operational Environmental Satellite (POES) program and the Department of Defense is responsible for the Defense Meteorological Satellite Program (DMSP). The National Aeronautics and Space Administration (NASA), through its Earth Observing System (EOS-PM) development efforts, provides new remote sensing and spacecraft technologies that could potentially improve the capabilities of the operational system. While the civil and military missions of POES and DMSP remain unchanged, establishing a single, converged, operational system can reduce duplication of efforts in meeting common requirements while satisfying the unique requirements of the civil and national security communities. A converged system can accommodate international cooperation, including the open distribution of environmental data.

II. Objectives and Principles
    The United States will seek to reduce the cost of acquiring and operating polar-orbiting environmental satellite systems, while continuing to satisfy U.S. operational requirements for data from these systems. The Department of Commerce and the Department of Defense will integrate their programs into a single, converged, national polar-orbiting operational environmental satellite system. Additional savings may be achieved by incorporating appropriate aspects of NASA's Earth Observing System.
    The converged program shall be conducted in accordance with the following principles:
— Operational environmental data from polar-orbiting satellites are important to the achievement of U.S. economic, national security, scientific, and foreign policy goals.
— Assured access to operational environmental data will be provided to meet civil and nation security requirements and international obligations.
— The United States will ensure its ability to selectively deny critical environmental data to an adversary during crisis or war yet ensure the use of such data by U.S. and Allied military forces. Such data will be made available to other users when it no longer has military utility. The implementing actions will be accommodated within the overall resource and policy guidance of the President.

III. Implementing Actions
    a. Interagency Coordination
        1. Integrated Program Office (IPO)
    The Departments of Commerce and Defense and NASA will create an Integrated Program Office (IPO) for the national polar-orbiting operational environmental satellite system no later than October 1, 1994. The IPO will be responsible for the management, planning, development, fabrication, and operations of the converged system. The IPO will be under the direction of a System Program Director (SPD) who will report to a triagency Executive Committee via the Department of Commerce's Under Secretary for Oceans and Atmosphere.
        2. Executive Committee (EXCOM)
    The Departments of Commerce and Defense and NASA will form a convergence EXCOM at the Under Secretary level. The members of the EXCOM will ensure that both civil and national security requirements are satisfied in the converged program, will coordinate program plans, budgets, and policies, and will ensure that agency funding commitments are equitable and sustained. The three member agencies of the EXCOM will develop a process for identifying, validating, and documenting observational and system requirements for the national polar-orbiting operational environmental satellite system. Approved operational requirements will define the converged system baseline which the

IPO will use to develop agency budgets for research and development, system acquisitions, and operations.

b. Agency Responsibilities

1. Department of Commerce

The Department of Commerce, through NOAA, will have lead agency responsibility to the EXCOM for the converged system. NOAA will have lead agency responsibility to support the IPO for satellite operations. NOAA will nominate the System Program Director who will be approved by the EXCOM. NOAA will also have the lead responsibility for interfacing with national and international civil user communities, consistent with national security and foreign policy requirements.

2. Department of Defense

The Department of Defense will have lead agency responsibility to support the IPO in major system acquisitions necessary to the national polar-orbiting operational environmental satellite system. DOD will nominate the Principal Deputy System Program Director who will be approved by the System Program Director.

3. National Aeronautics and Space Administration

NASA will have lead agency responsibility to support the IPO in facilitating the development and insertion of new cost effective technologies that enhance the ability of the converged system to meet its operational requirements.

c. International Cooperation

Plans for and implementation of a national polar-orbiting operational environmental satellite system will be based on U.S. civil and national security requirements. Consistent with this, the United States will seek to implement the converged system in a manner that encourages cooperation with foreign governments and international organizations. This cooperation will be conducted in support of these requirements in coordination with the Department of State and other interested agencies.

d. Budget Coordination

Budgetary planning estimates, developed by the IPO and approved by the EXCOM, will serve as the basis for agency annual budget requests to the President. The IPO planning process will be consistent with agencies' internal budget formulation.

IV. Implementing Documents

a. The "Implementation Plan for a Converged Polar-orbiting Environmental Satellite System" provides greater definition to the guidelines contained within this policy directive for creating and conducting the converged program.

b. By October 1, 1994, the Departments of Commerce and Defense and NASA will conclude a triagency memorandum of agreement which will formalize the details of the agencies' integrated working relationship, as defined by this directive, specifying each agency's responsibilities and commitments to the converged system.

V. Reporting Requirements

a. By November 1, 1994, the Department of Commerce, the Department of Defense, and NASA will submit an integrated report to the National Science and Technology Council on the implementation status of the national polar-orbiting operational environmental satellite system.

b. For the fiscal year 1996 budget process, the Departments of Commerce and Defense and NASA will submit agency budget requests based on the converged system, in accordance with the milestones established in the Implementation Plan.

c. For fiscal year 1997 and beyond, the IPO will provide, prior to the submission of each fiscal year's budget, an annual report to the National Science and Technology Council on the status of the national polar-orbiting operational environmental satellite system.

Document II-11

Document title: D. James Baker, Under Secretary for Oceans and Atmosphere, U.S. Department of Commerce, to John Morgan, Director, EUMETSAT, May 6, 1994.

Document II-12

Document title: D. James Baker, Under Secretary for Oceans and Atmosphere, U.S. Department of Commerce, to Jean-Marie Luton, Director, European Space Agency, May 6, 1994.

Source: Both in NASA Historical Reference Collection, NASA History Office, NASA Headquarters, Washington, D.C.

These letters, which follow from the Presidential Decision Directive of the previous day (May 5, 1994), respectively invite EUMETSAT to join the converged satellite system and formally inform the European Space Agency of the invitation.

Document II-11

[rubber stamped: "May 6, 1994"]

Mr. John Morgan
Director, EUMETSAT
Am Elfengrund 45
D-64242 Darmstadt-Eberstadt
Germany

Dear John:

I am pleased to invite the European Organisation for the Exploitation of Meteorological Satellites (EUMETSAT) to consider expanded cooperation as an important partner in the United States converged, polar-orbiting operational environmental satellites program.

This week, the President has directed the National Oceanic and Atmospheric Administration (NOAA), the Department of Defense (DOD) and the National Aeronautics and Space Administration (NASA) to work together to implement a converged system which integrates NOAA and DOD systems while capitalizing on NASA technology. Building on longstanding plans to cooperate with European partners in this area, the U.S. Government's preferred option for such cooperation includes the METOP satellite series assuming U.S. missions requirements for such cooperation can be achieved.

Cooperation with the METOP satellite series and our EUMETSAT and ESA partners is critical to our efforts to enhance further development of a global operational observing system. Inclusion of METOP as one of three elements in the preferred converged satellite constellation underscores the importance we place on environmental satellite cooperation with our European partners.

Recognizing these important benefits in cooperation, we propose that EUMETSAT join us in exploring the accommodation of converged system mission requirements in the joint polar system planning that is already underway.

Sincerely,

[hand-signed: "Jim"]
D. James Baker

cc: Jean-Marie Luton, ESA

Document II-12

[rubber stamped: "May 6, 1994"]

Mr. Jean-Marie Luton
Director, European Space Agency
8-10, rue Mario-Nikis
75738 Paris Cedex 15
France

Dear Jean-Marie:

I am writing to you in recognition of the important role of the European Space Agency (ESA), together with the European Organisation for the Exploitation of Meteorological Satellites (EUMETSAT), in the METOP satellite series.

This week, the President directed the National Oceanic and Atmospheric Administration (NOAA), the Department of Defense (DOD) and the National Aeronautics and Space Administration (NASA) to work together to implement a converged system which integrates NOAA and DOD systems while capitalizing on NASA technology. Building on our longstanding plans to cooperate with European partners in this area, the U.S. Government's preferred future satellite constellation includes the METOP satellite series.

Cooperation with the METOP satellite series and our EUMETSAT and ESA partners is critical to our efforts to enhance further development of a global operational observing system. Our long-term understanding is that METOP-related cooperation will be addressed in a NOAA/EUMETSAT Agreement closely associated with an Agreement between EUMETSAT and ESA. Our desire to include METOP as one of three elements in the converged satellite constellation underscores the importance we place on environmental satellite cooperation with our European partners.

Recognizing the important benefits to cooperation, we are proposing that EUMETSAT join NOAA in exploring the accommodation of converged system mission requirements into the joint United States/European polar system planning that is already underway.

Sincerely,

[hand-signed: "Jim"]
D. James Baker

cc: John Morgan, EUMETSAT

Document II-13

Document title: Peter C. Badgley, Program Chief, Natural Resources, NASA, "Current Status of NASA's Natural Resources Program," *Proceedings of the Fourth Symposium on Remote Sensing of Environment held 12, 13, 14, April 1966* **(Ann Arbor, MI: University of Michigan, 1966), pp. 547–558.**

Source: NASA Historical Reference Collection, NASA History Office, NASA Headquarters, Washington, D.C.

NASA began its natural resources program in 1965 with the goal of studying the Earth from space. Unsure of what observational technique offered the greatest utility, the agency conducted a number of experiments from aircraft in an attempt to determine optimal instrument design for satellites. Peter Badgley, head of the Natural Resources Program at NASA, presented the results of the experiments at a symposium on remote sensing sponsored by the Department of the Navy's Office of Naval Research and the Air Force Cambridge Research Laboratories. The results helped clarify issues about coordination between NASA and the Department of the Interior, which was interested in beginning an Earth resource survey program of its own.

[547]

Current Status of NASA's Natural Resources Program

Peter C. Badgley
Program Chief, Natural Resources
National Aeronautics and Space Administration Headquarters
Washington, D.C.

ABSTRACT

The National Aeronautics and Space Administration (NASA) is supporting research activities in those areas of remote sensing from Earth-orbiting spacecraft which are related to the study of natural cultural resources. These sensors are believed to possess a number of unique advantages for the discovery, inventory, evaluation, development and conservation of such resources. Many Government agencies, universities, and research institutions are cooperating with NASA in this effort. The current status of this program is described in this paper.

1. INTRODUCTION AND OBJECTIVES OF NASA'S NATURAL AND CULTURAL RESOURCES PROGRAM

Natural resources are defined as those naturally occurring materials, such as mineral deposits, timberstands, and fresh water, which are of value to mankind. Cultural resources are defined as those items of value to man which result from his own activities and are in general derived from the natural resources.

Since World War I airborne mapping by photographic means has been used extensively for the study of natural and cultural resources. Radar and infrared sensors have been used to a lesser extent. Historically, the development and use of such techniques has been fostered by the military, but in recent years there have been widespread applications beyond the military field. During the past three decades civil and commercial interests have also used airborne imaging devices very successfully. In addition, gravity, magnetic, and radioactive measuring instruments have been applied to the search for mineral and petroleum deposits. In the past few years imaging sensors in unmanned and manned spacecraft have been employed to provide the first true synoptic coverage of the lithosphere, hydrosphere, and atmosphere.

The objectives of the NASA Natural Resources Program are as follows:

1. To determine those natural and cultural resource data which can be acquired best from spacecraft for the benefit of mankind.
2. To test and develop the best combination of observational procedures, instruments, subsystems, and interpretive techniques for the acquisition and study of terrestrial, lunar, and planetary natural and cultural resource data from spacecraft.
3. To determine how the increased frequency and synoptic coverage uniquely afforded by spacecraft observations can aid the study of time variant and relatively unchanging phenomena on the surface of the Earth.
4. To develop improved methods of displaying and disseminating space-acquired natural and cultural resource data on a global basis suitable for utilization [548] by scientific, technical, and commercial interests.
5. To determine which natural and cultural resource data can be most effectively and economically obtained by manned spacecraft, unmanned satellite, interrogation of surface sensors, or the means currently being used.
6. To discover, by virtue of trained scientists in spacecraft, what unforeseen natural and cultural resources or geoscience phenomena may be observable from the overview available at orbital altitudes.

A large number of potential users having interests in a variety of geoscientific problems and applications have been identified:

1. Agriculture/Forestry Resources,
2. Geography (Cultural Resources),
3. Geology/Hydrology (Mineral and Water Resources),
4. Oceanography (Marine Resources).

2. POSSIBLE PHENOMENA WHICH MAY BE OBSERVED AND RECORDED ADVANTAGEOUSLY FROM SPACECRAFT

Figure 1 gives a partial listing of phenomena which can be advantageously "mapped" from space and those types of sensors that may be used, based on the state-of-the-art. As new or better sensors are developed, the listings of observable phenomena will undoubtedly grow. The word "mapped" is used here to mean that certain natural and cultural resources phenomena are observed from space and recorded on photographs, images, tapes, or other data storage media. After these raw data are recovered and analyzed, the pertinent information is plotted on appropriate bases which become thematic maps. These thematic maps, together with written reports, constitute one of the principal end-products expected of the Natural Resources Program.

3. UNIQUE ADVANTAGES OF SPACE FOR NATURAL RESOURCE STUDIES

There are many advantages to obtaining imagery of the entire Earth or major parts of it by means of spaceborne geoscience sensing systems. These systems encompass a number of instruments and techniques applicable to many disciplines, both cultural and natural, and of use to scientific and applications users. These systems have the unique utility of complementing one another in their results, hence their broad applications.

For sizable areas within the field of view of the sensor, spacecraft coverage is truly synoptic because the high altitude and speed of the spacecraft permit the scientist to obtain information of large areas at a single instant of time. This is of great advantage to research work in the Earth sciences and in natural resources, which have been hampered by the time and space scales that arise in the measurement of the variable under investigation. Further, there is information in the whole pattern of an integrated structure which can neither be derived from elements of the whole nor considered simply as the sum of the elements.

An important advantage of satellite photography is the aspect of real-time data acquisition. With this characteristic remote areas of special significance could be canvassed on short notice, thus providing information on impending disasters such as tsunamis, and forest fires, and studying disaster areas which result from storms, earthquakes, etc. Many of these problems such as earthquakes, volcanoes, air-sea interactions, fish migrations, crop growth, and disease, etc., are global in nature and are consequently best studied by globe encircling data gathering systems.

[549] For complete aerial photographic coverage over large areas many technical problems pertaining to data reduction exist, for example, the assembling of broad-scale mosaics. Here the photos must be matched, and corrections in density, scale, or color reproduction be made, and finally the joining lines must be reduced as much as possible. Using space photography would reduce these tasks to a minimum since one space photograph, depending on scale, will cover many times the area as most aerial photographs.

The long duration of spaceflights and all-weather operations are highly advantageous aspects of remote sensing from space. There are many regions of the Earth that are covered with clouds for long periods of time. The cloud cover not only absorbs and reflects a large part of the radiation from the Sun to the Earth, but also absorbs radiation from the Earth out into space. In a manned orbiting satellite the problem can be partially overcome since the scientist on board the craft will be able to see when an area is clear enough to make observations and to also employ those sensors which can penetrate certain types of overcast. A less obvious but highly significant advantage results from the clearer images possible when the observation is made from far above the turbulent refracting and diffusing layer which often seriously degrades aerial observations.

With orbital sensing, it appears that the costs will be considerably less than even a single synoptic global coverage with aircraft. This is true because of the great amount of data that can be rapidly acquired, of the more complete coverage, and of the superior quality of some of the data which greatly reduces the effort needed for processing and analyzing.

Low-altitude photography applied to natural resource surveys and exploration has proven to be of great value. However, up-to-date and comprehensive data require frequent overflights and near blanket coverage; thus extensive aircraft acquisition is prohibitively costly. Since the resolving power of remote-imaging instruments from satellite altitudes will be sufficient to permit identification of many different parameters of Earth resources, a potential means of economically acquiring such data on a world basis is offered.

A further advantage of orbital sensing is that global coverage can be obtained by uniform types of equipment and methods of calibration and measurements. This will insure that data will be collected under controlled conditions and will not be subjected to these uncertainties. Obvious technical and operational advantages result from the precise regularity of spacecraft motion, from the lack of vibration, and from the high rate of speed.

The Earth-orbital missions are also of great value for the experience gained and the testing of sensors and techniques prior to the conducting of lunar and planetary orbital missions.

World-wide resource management through the use of operational spacecraft will provide a combination of scientific, sociological, political, and economic benefits. Through resource management, man is able to monitor the total resource availability, make efficient use of existing resources, protect existing resources against damage or loss, and uncover new resources. To be effective, action must be taken well in advance of the depletion of available resources. Thus accurate data on current inventory and rate of depletion furnishes the basic information required to anticipate forthcoming pressures on resources and to indicate appropriate steps to be taken.

4. ACCOMPLISHMENTS TO DATE

With the assistance of disciplinary groups in several Federal agencies and institutions, program definition activities have been initiated in the four disciplinary areas. Many of the phenomena which each natural resource discipline wishes to observe and record from space have been identified (Figure 1). The instruments and their frequencies needed to gather these data have also been [550] identified as closely as possible (Figure 1).

The coordinated requirements of the several natural resource disciplines for photography and radar on initial flights have been compiled in document form. A document of infrared instrument requirements is being prepared. Albums of imagery acquired by this aircraft program together with spacecraft-acquired imagery (from Gemini, Nimbus, etc.) of value to natural resource scientists are being compiled. An atlas which analyzes the potential of this data for natural resource scientists is in preparation.

One of the principal tasks of this Program is the determination of the best combination of instruments and the best resolutions for observing natural resource phenomena. These are currently being identified. However, until several generations of instruments have been flown in space and the data analyzed, it will be impossible to be completely precise on such instrument specifications.

The various remote sensing instruments recommended by the disciplinary groups are being flown over carefully selected test sites with aircraft. The data obtained from such test site overflights are then studied to determine the best combination of instruments for spaceflight and the best analytical processes for acquiring the maximum amount of information from the data. Accomplishments in these areas are described in detail below.

4.1 AIRCRAFT DATA GATHERING SYSTEM

The Natural Resources Program together with the MSC [Manned Space Center] Engineering and Development Directorate is presently engaged in gathering data over test sites with a number of airborne electronic and electro-optical remote sensors for a number of user agencies and cooperating scientists.

This program has been set up to obtain precursor data for the calibration of instruments over known features and for the development of the best observational and interpretive techniques in the period 1965–1968 preceding the earliest (1968) natural resource spaceflight missions. Further, the costs of developing such a data gathering system initially with airborne instruments is substantially less than proceeding directly to a spaceborne system. The experience gained in this aircraft phase (1965–1968) is already providing a solid basis for planning of the spaceflight testing phase.

It is expected that aircraft-acquired data will also be obtained over a number of key test sites simultaneously with the initial spaceflight data in order that the spaceflight instruments may be calibrated and in order that the aircraft and spacecraft data may complement each other to the maximum extent.

4.2 STATUS OF THE AIRCRAFT PROGRAM

It can be seen from Figure 1 that each sensor has multiple uses, and in most cases a combination of sensors is desirable to provide complete data on any particular observed feature.

To carry out the remote sensing program in a satisfactory manner, it is necessary to conduct aircraft testing from several altitudes (low, intermediate, and high) over a period of several years. The low-altitude phase, altitudes up to 20,000 feet, is presently being conducted using the NASA-MSC Convair 240A (NASA 926) over sites in the Continental United States. It is planned to conduct the second phase at altitudes up to 40,000 feet over the same test sites and over several overseas sites. The third phase should be conducted at high altitudes, possibly up to 80,000 feet over the same test sites. Ultimately, it may be highly desirable during the early Earth-orbital missions to have the ground truth teams on the site, and the aircraft overhead at the time the orbiting spacecraft overflys the area.

[551] <u>Convair 240A</u>
Implementation of this airplane was initiated in October 1964, and initial survey operations over geologic test sites got underway in early December 1964. First flights were made using only the camera systems; other instruments were installed as they became available. The airplane is now scheduled to its full capability of sensors and no additional ones are being planned at this time for this aircraft. Following is a list of those sensors installed on board:
1. Microwave radiometer,
2. Metric camera system,
3. Multiband camera system,
4. Ultraviolet imager,
5. Recon IV imaging IR,
6. Redop scatterometer,
7. Doppler chirping radar.

<u>Lockheed P-3A (Electra)</u>
This airplane was acquired in December 1965 and is ideally suited as a remote sensor test aircraft for altitudes up to 40,000 feet and contains much of the basic instrumentation necessary to meet the objectives of the Natural Resources Program. The navigation system contains the following items:
1. APN 153 doppler navigator,
2. ASA 47 air-mass computer,
3. LN 12 attitude reference system,
4. Inertial platform,
5. ASQ 80 weather and ground point radar,
6. APN 70 LORAN overwater navigation system.
This equipment will provide the flight parameters such as roll, pitch, yaw, ground speed, heading, altitude, position, etc., for sensor operation, data correlation, and navigation of the aircraft. The aircraft has both a large cabin area and a bomb-bay area with a number of radomes and instrument mounting provisions in which to install sensor systems and other experiments. With the installation of an auxiliary power unit, operation without the use of ground-based starting equipment and other ground support equipment will be possible. The P-3A is now being implemented to receive a compliment [sic] of sensors basically as described in Figure 2 and is planned to be in operation by July or August of 1966.

<u>Douglas A-3B</u>
This airplane contains a Westinghouse AN/APQ 97 (XE-1) side-looking radar system. It has completed seventy-two hours of flight time acquiring radar imagery over a number of test sites and other areas of interest. An additional sixty hours of flight time has been requested (twenty hours in FY 1966; forty hours in FY 1967).

[552] 4.3 OBJECTIVES OF THE TEST SITE PROGRAM

The prime objectives of the Natural Resources Test Site Program are calibration of the remote sensors and the development of a capability for supporting the use of remote sensors in performing natural resources investigations. To meet this objective, scientists are currently gathering data with electronic and electro optical remote sensors in and over areas of specific interest. Examples of aircraft and Gemini-acquired data are shown in Figures 3–7. [Figures 3–7 omitted] The data thus acquired will then be used to:

1. Advance our knowledge of the effects of terrain parameters on sensor data.
2. Provide a means of calibrating data return from sensors in aircraft and spacecraft.
3. Define sensor operational parameters and spacecraft integration requirements.
4. Develop data handling and interpretation techniques.
5. Define sensor systems to meet the scientific objectives of Earth-orbital missions.
6. Provide a group of scientists skilled in application of remotely sensed data to natural resources investigations.

4.4 TYPES OF TEST AREAS

Experience in studies to date have indicated that two types of test areas or sites are necessary. The first type is designated as an "Instrument Calibration Test Site," and for abbreviation purposes is referred to as a "Test Site." The second type is designated as a "Natural Resources Applications Area," and for abbreviation purposes is referred to as an "Applications Area." Figures 8–12 give a complete list of the currently proposed Applications Areas and Test Sites and the names of the scientists responsible for data analysis at these sites. [Figures 8–12 omitted]

A Test Site is an area where studies are conducted in the calibration of the instruments. These studies will test instrument response to well-defined preselected conditions. Tasks will include the development of interpretation and correlation techniques, and investigations of the response of the remote sensors in terms of biological, chemical, and physical conditions in the area. Applications Areas are areas where extensive investigations are conducted using fully-developed instruments to gather and interpret data in terms of the area's known conditions and features, e.g., agricultural, geographic, geologic, hydrologic, and oceanographic. These Applications Areas tentatively include a number of international sites which have been chosen principally to provide data on problems in the various Earth sciences, that are global or continental in scope and to promote international cooperation in line with NASA's policy.

Test Sites should have the following characteristics:

1. They must satisfy the requirements of the specific instrument to be tested at whatever development stage it exists.
2. There should be available an extensive amount of ground data so they do not require extensive basic study.
3. They should be as uniform as possible, commensurate with the purpose for which selected, so as to permit identification of the remote sensor response with a single (or minimum number of) features.

Applications Areas should be:

1. Areas in which studies by the participating agency (or one of its cooperating agencies or institutions) are taking place or have taken place in the recent past.
[553] 2. Areas of broad natural resources or scientific interest, with scientific resources problems whose solution will contribute to the progress in the Natural Resources Program.
3. Areas with well-documented features and for which there is an active scientist prepared to analyze sensor data and report on results in a competent manner.

International Applications Areas should:

1. Contain features and conditions not well-developed or available in the U.S.
2. Be readily accessible to all accredited scientists of other countries involved in the Program and should not be located in countries where political instability could adversely affect the Program.
3. Be areas in which studies can be adequately supported and which present no logistic problems.

4.5 SELECTION OF SITES

Prospective Test Sites are proposed by the instrument and/or disciplinary scientists with approval being vested in an ad hoc "Test Site and Aircraft Committee." This Committee is chaired by the Chief, Natural Resources Program, or his deputy, and is composed of the program managers of the several natural and cultural resources disciplines and the chairmen of the "teams." Actual investigations are carried out by scientists affiliated with the instrument teams, in the participating Federal agencies, and at universities under NASA contract.

4.6 CURRENT STATUS OF TEST SITE ACTIVITIES

Studies of Test Sites started in 1964 with infrared studies at the Pisgah Crater Area, California, for geology and at the Purdue Farm for agriculture. The number of studies increased in 1965 with additional studies in Western Kansas; Mono Craters, California; Davis, California; Weslaco, Texas; and Willcox Playa, Arizona. Concomitantly, the work at the original was broadened to include other sensors.

Work at the Test Sites is being carried out principally by scientists of the instrument teams working under NASA contracts or grants. Current status of "Test Site" studies is given in Figure 13. [Figure 13 omitted]

Many of these measurements and problems require the development of new study concepts. For example, statistical sampling programs are being developed by geologists at Northwestern University. Geoscientists from the University of Nevada are working with the instruments and user scientists to determine the influence of the not usually measured parameters on remote sensor data, as well as providing very detailed geological, mineralogical, and micrometeorological data. Similar studies are being carried out in the other disciplinary areas.

Airborne remote sensor data are now being acquired over a number of Applications Areas, and are being studied and evaluated by participating user agency scientists. A number of preliminary reports have been written describing the uses of the data. Although this program is in its infancy, it appears that the objectives of the Applications Areas are being met.

4.7 DATA PROCESSING AND DISTRIBUTING UNIT

The Data Processing and Distributing Unit (Data Unit) has been established to handle the data recorded by several kinds of remote sensors, both electromagnetic and force-field, onboard NASA-conducted aircraft and other cooperating flights over selected geoscience test sites. Additional calibration and ground-reference data is also collected at some specific ground-site installations using contact or short-range sensors for correlation and corroboration of the airborne remote-sensor data. The purpose of such collections is to aid in evaluating the usefulness of apparatus and data analysis techniques for remote sensing of [554] natural and cultural resources by means of spaceborne instrumentation. This involves a number of different data formats (film, paper, tapes, charts, etc.) providing records which present the data in a variety of forms, i.e., digital, analog, alphanumeric, etc.

The Data Unit will perform the following functions in support of the NASA Natural Resources Program:

1. Process, reproduce, catalog, classify, index, disseminate, store, and retrieve geoscience data (original, reduced and/or analyzed, including preliminary, intermediate, and final reports) received from NASA-supported, or cooperating remote sensing of natural resources investigators.

2. Provide supplemental support by maintaining an adequate supply of charts and maps relating to test sites and by conducting file searches and related services. Such charts and maps will be supplied from standard sources.

3. Compile and furnish periodic accession lists of data to cooperating investigators. Accession lists will indicate type and size of format, originator, sensor type, geographic area, altitude, time and date of acquisition by originator. This information may be in the form of computer print-outs.

4. Design and supply check lists to investigators for submission of data to the Data Unit. These check lists will contain minimum terms, descriptions and information required to provide a basis for data cataloging and entry into a computer or other retrieval systems.

5. Design a format for queries to the Data Unit and make copies available to investigators for their use.

6. Process, reproduce, associate with related sensor data, catalog, classify, index, disseminate, store and retrieve all ground control data from test site supplied by investigators who support the NASA Natural Resources Program.

5. ECONOMIC BENEFITS TO BE DERIVED

It is difficult to establish the economical importance or yield of a system which is virtually untried and untested. Economic benefits, however, cannot be derived from a natural resource until that resource is located and until sufficient knowledge and understanding of that resource and its environment are obtained to permit efficient exploitation.

As examples of areas of potential economic benefit, the National Academy of Sciences, National Research Council (NASCO Publication 1228) cited the following:

Shipping—The shipping industry currently transports about $40 billion per year in cargo. It is expected to rise about 48 per cent by 1970. By 1975 the shipping bill for transporting this cargo will be about $5 billion with the added costs of $500 million per year for new construction. Even a small improvement in ship routing techniques resulting from increased knowledge derived from spacecraft concerning high waves, shoaling of channels, existence of uncharted shoals, icebergs, and pack ice distribution, etc., will contribute a significant dollar savings when compared to the expected overall cost.

Fisheries—U.S. fisheries production has increased less than 10 per cent in the past decade, yet the importing of fisheries products into the United States has resulted in an adverse balance of payments of approximately $500,000,000 in 1965. In fact, improved domestic development of fisheries in our near shore area could result in doubling domestic production in the next ten to fifteen years and bring about a marked reduction in the adverse gold flow problem. Although dependent on many factors, any aid provided by spacecraft oceanography to locating, delineating, predicting the productivity of fishing grounds globally could produce large dollar payoffs to the entire industry and hence the nation.

[555] Another example of interest relates to the field of water resources:

A single, medium-sized, Canadian hydroelectric plant saves $1 million for each 1 per cent increase in accuracy in predicting April to August flow. This amount of power would otherwise be lost because of the need to waste water to provide room for unanticipated flood conditions.

Base and thematic mapping by conventional aerial surveys is an extremely costly operation. It has been estimated that over $1 billion is spent annually to obtain the aircraft-acquired data from which such maps are made, and yet only a very small percentage of the Earth's resources are so covered in any one year. The Earth's surface consists of 504 million square kilometers, of which 146 million is land area. The following tabulation gives an idea of the costs involved to cover the land and adjacent shallow sea areas (150 million square kilometers) once by aircraft.

ESTIMATED COST OF AERIAL SURVEYS

Types of Surveys	Cost/km^2	Cost to Cover Land and Adjacent Areas
A. Small Scale Mapping and Supplementary Photography and Infrared	$6.00	$900,000,000
B. Magnetic-Gravity	5.00	750,000,000
C. Side-Locking Radar	1.00	150,000,000
		$1,800,000,000

If the remaining ocean areas were covered only in a synoptic manner using magnetic-gravity, infrared, microwave, and selective photography at an estimated cost of $10 per square kilometer, this would add $3.5 billion to the land coverage figures. Thus, a one-time aerial survey of the Earth and its force fields would cost about 7.5 billion. However, many of the phenomena affecting resources are time variant and repeated coverage is needed. Even if repetitive annual coverage were limited to about 20 per cent of the world (100 million square kilometers) and if only synoptic sensing totaling $10 per kilometer were performed, the program maintenance cost would be $1 billion annually. These figures cover only the cost of data acquisition. Data reduction and dissemination would involve costs far exceeding those of the acquisition phase.

The cost of mounting an operational resource-sensing space program cannot be accurately determined at this time. Parameters, such as the payload, power, mode (manned or unmanned) have not been fully defined. However, if one assumes that these parameters will be compatible with one or more of the space vehicles being developed, then the costs attributed to the Natural Resources Program for space-acquired data would be reasonable.

Comparing the costs of a space program to one of conventional aerial surveys does not provide the total answer. Many aircraft surveys will still be required and some types of anticipated surveys might not prove practical by either aircraft or spacecraft. However, indications are that, where coverage of a global repetitive nature is required and obtainable by both modes, a space system has unquestionable economic advantages. It appears that the cost will be on the order of a magnitude less for the space mode. The potential economic advantages of utilizing space for resource analysis is not limited to the acquisition phase, but it is also important in the data reduction phases. Since space-acquired data will be of a uniform and systematic nature, its conversion into maps and statistics will be enormously simplified when compared to conventional methods.

The question of whether the Program is worth the cost of a spaceborne data gathering system must be considered on the basis of future demands for natural resources. Current indications strongly support the need for new revolutionary means, such as orbiting spacecraft, to meet rapidly increasing demands for natural resources. The data of scientific value from such a program will also be enormous, but unfortunately cannot have a price tag put on it at this time. The full extent of the economic benefits to accrue from this program can be properly evaluated only by the scientists and economists associated with the various natural and cultural resources. This evaluation has been initiated as an [556] integral part of this Program, and there are indications that the potential benefits will greatly exceed the cost of the Program. However realistic experiments and additional research must be conducted before any specific dollar values can be placed on these benefits.

6. FUTURE PROGRAM

Ideally, the Program should proceed through the following phases:

Feasibility—This phase (in progress) is basically one of experimentation from aircraft to determine signatures of natural resource phenomena in terms of assumed spacecraft sensor resolution. During this phase, carefully selected and controlled test sites are being utilized. Using available aircraft instrumentation, the correlation and relative value of each sensor to the phenomena in question are being studied. Suborbital and orbital flights also will be utilized to obtain some limited sensor responses. These are being analyzed and used as a basis for relating aircraft to spacecraft obtained signatures. As instrumental value is established, design and procurement of space hardware are being initiated. Facilities for data handling and reduction must also be established during this stage. This includes provisions for space-acquired data as well as aircraft-acquired data. Aircraft testing of sensors is expected to be a continuing activity, and will continue beyond this phase to provide data from both aircraft and spacecraft simultaneously during the spacecraft testing phase.

Spacecraft Testing—During the 1968–1972 period the first flights with the primary purpose of sensing the Earth's resources are expected. These are expected to include flights of the Apollo Applications Program, where manned spacecraft will carry a sizable number of sensors which can be directed at various parts of the Earth simultaneously. On these initial spaceflights, coverage will be concentrated over areas such as the United States where ground controls may be used to verify the conclusions derived during the feasibility stage. Arctic, tropical, and other representative test sites will also be included. As a result of these flights, it is expected that sufficient information will be available to determine the optimum:

1. Mode—unmanned, manned or man-serviced;
2. Orbital configuration and flight duration;
3. Extent and variety of sensors;
4. Mode of data recording and return to Earth;
5. Methods of data reduction and dissemination.

During this stage, the basic economics of resource sensing from space must be determined. This will involve weighing the benefits as opposed to the costs of the program entering into an operational stage. Although not an operational phase, it is expected that considerable data of economic importance will be obtained in addition to a large amount of scientific information.

Operational—The existence and extent of this stage will depend on the economic analysis made during the previous stage. Indications are that it will be multidiscipline in nature, global in extent, and more or less continuous since many of the important phenomena associated with resources are time variant. Operational flights may well begin while the orbital testing stage is still in progress—perhaps during 1971 or 1972. By 1972 it is expected that testing and operational spaceflights will be combined.

[557]

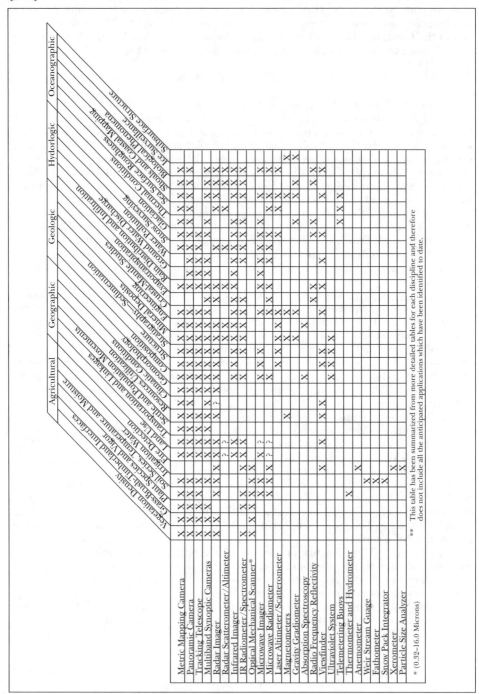

*Figure 1. Anticipated Applications** of Earth Orbital Natural Resources Data Gathering System*

[558]

Instrument	Part of Spectrum	Information Obtained	Format	Resolution or Instantaneous Field of View	Field of View
RC8 Camera, Color IR*	Visible 0.4–0.7μ	Color Distribution Qualitative	Film 9x9	70 lines/mm	74° Edge-Edge
T11 Camera, Black & White*	Visible 0.5–0.7μ	Conventional Photo Information	Film 9x9	35 lines/mm	74° Edge-Edge
Multispectral Camera*	Visible Near IR 0.9 Near UV 0.3–0.4μ	Color Distribution Quantitative	Film 70mm	70 lines/mm	18° (not stereo)
IR Imager #1	IR 4.5–5.5	Radiant Temperature Distribution	Single Channel 35mm Split Channel 16mm	***	78°
IR Imager #2	IR 8–13	Radiant Temperature Distribution	Film 70mm	***	90°
Passive UV	2900–4000 Å	UV Reflectance	Single Channel 35mm Split Channel 16mm	***	78°
IR Spectrometer	5–15	Spectral Distribution of Emitted Energy (Gross Chemistry of Surface)	Line Trace	Spectral Resolution 0.1 of Spectral Range	1°²
Microwave Radiometer	33mm; 8.5mm; 19mm; 13.5mm	Radiant Temperature Distribution to Some Depth	Line Trace	1° Kelvin	Several Degrees
Ryan Scatterometer	13.3Gh	Radar Backscatter (∇σvsθ)	Magnetic Tape	0.1° Beam Width	60° Fore 60° Aft 30° Wide
Side Looking Radar **	35Gh	Radar Image	Strip Photo	***	***
Broad Band IR Radiometer**	8–14	Radiant Temperature of Surface	Line Trace	1° Kelvin	1°²
Passive Microwave Imager**	Broad Band Centered at 9Gh	Distribution of Radiant Temperatures to Some Depth	Film 70mm	1°²	90°
Imaging Radar***	***	Radar Image of Reflectivity of Surface and Near Surface	9-1/2" Film Split Channel	~15m	10-mile Swath

*	Two of these operable simultaneously
**	Not available at present
***	Separate aircraft
****	Classified
	Gh = (Gigahertz)=10⁹ cycles per second

Figure 2. Characteristics of Remote Sensing Instruments for the CV-240A, P3A, and Other NASA Aircraft

Document II-14

Document title: "Prepared by Jaffe and Badgley at Seamans' Request: NASA Natural Resources Program," May 13, 1966.

Source: NASA Historical Reference Collection, NASA History Office, NASA Headquarters, Washington, D.C.

A continual problem in the NASA Earth Resources Program involved conflicts between the developers and the users of technology regarding spacecraft design. NASA wanted to develop large, complicated satellites, which required considerable time between inception and flight. The various user communities, on the other hand, were willing to settle for less advanced hardware so as to fly earlier. This report, requested by NASA Deputy Administrator Robert C. Seamans, Jr., and prepared by Director for Space Applications Leonard Jaffe and Natural Resources Program Chief Peter Badgley, demonstrates NASA's bias toward large experimental satellites and includes a rather lengthy instrument "wish list."

[1]

Prepared by Jaffe and Badgley at Seamans' Request
NASA Natural Resources Program

[rubber stamped: "May 13 1965"]

Objective

To conduct an experimental program to determine and develop the feasibility of the use of satellite borne instrumentation to make measurements of and assist in surveying the earth's natural and cultural resources (i.e. atmospherics, agriculture, forestry, hydrology, geology, geography, oceanography, and so forth).

Scope

1.   Studies by user agencies of the applicability of spaceborne instrumentation to their needs.

2.   Accumulation of data from an instrumented aircraft flight test program over calibrated test sites.

3.   Orbital flight testing of instrumentation and development of required data analysis techniques to determine feasibility and practicability of this space application.

Urgency

User agencies have expressed an urgent need for improvements in their observational and surveying capabilities to help maintain an adequate supply of natural resources. If vigorously pursued, the technology of spaceborne systems could be provided to fill many of the expressed needs by 1975.

Requirements of the Experimental Program

The specific and ultimate requirements of the various users must necessarily be a result of the experimental program. This application will require data collection and analysis systems of the most advanced types with an optimum mix of high resolution (small area detailed data samples of resolutions below 20 meters) and moderate resolution (broad areas, resolutions above 20 meters) coverage. Instruments recording data from many portions of the electromagnetic spectrum are being considered including optical (visible), infrared, ultraviolet, and microwave radar. (Reference: "Objectives, Instrumentation, and Flight Time Recommendation of User Agencies and Cooperating Scientists Involved in the NASA Natural Resources Program").

The current NASA experimental program began in 1964 with the initiation of discussions with and studies by the user agencies to assess their requirements and the applicability of space derived data by analysing [sic] available results from existing programs (Tiros, Nimbus, Mercury, and Gemini) and simulated results from instrumented aircraft tests.

[2] The instrumented Aircraft Flight Test Program began in mid 1964 with flights of an infrared imaging system over calibrated agricultural ground test sites near Purdue University. Since then we have added multispectral cameras, active imaging and scatterometer radars, passive microwave radiometers, and UV line scanners in 1965 and plan

to add more advanced multispectral cameras, metric and panoramic cameras, more advanced infrared scanners, infrared spectrometers and passive microwave imagers in 1966. The studies and the aircraft program are planned to be carried out on a continuing basis to provide for economical testing of instrumentation over sites of known characteristics prior to orbital testing.

As a result of these activities the user community has identified and specified candidate instruments for orbital test. The initial and most urgent of these are a wide range spectral scanner (0.3–13 microns) covering the near UV, visible and part of the IR spectrum (190 meter resolution) and a multispectral synoptic photographic system (of approximately 30 meters resolution). NASA is currently planning incorporation of these instruments into an already planned flight in 1969.

A rather completely instrumented natural resources satellite is being considered for flight in 1970 or 1971 involving the following instruments:

Metric Cameras
Panoramic Cameras
Multispectral Tracking Telescope
Multispectral Synoptic Cameras
Imaging Radars (8.0 gc)
Radar Altimeter/Scatterometers (0.4 and 0.8 gc)
Wide Range Spectral Scanner (0.3–13 microns)
Infrared Long Wavelength Spectrometer (18–16 microns)
Infrared Short Wavelength Spectrometer (0.4–2.5 microns)
Infrared Radiometer (10–12 microns)
Passive Microwave Imager (9.0 gc)
Passive Multichannel Microwave Radiometers (0.4–21 cm)
UV Imaging Spectrometer (3900–4900 Angstroms and 5800–6800 Angstroms)
Laser Altimeter/Scatterometer
Absorption Spectrometer (UV, visible, IR)
Chirp Radar System (75–450 mc's)
Gravity Gradiometer
Magnetometer System (triaxial fluxgate and rubidium vapor)

Flight test of this entire group of instruments simultaneously is highly desirable because the cross correlation of data from various portions of the electromagnetic spectrum acquired under similar lighting and weather conditions will yield far more information than data acquired at separate times by individual instruments.

If successful and needed, a repeat of this satellite is being considered for flight in 1971–72 to continue the development of the technology and the ability to handle and analyse [sic] accumulated data.

[3] Orbital Requirements

During the experimental phase of this program (pre-1975) the natural resource users require data to be collected periodically over a number of natural and cultural resources test sites. These sites must be readily accessible and available for study by well trained scientists. Many of the sites essential to this program fall within the United States both because of the subject matter involved (such as land use in metropolitan areas, water pollution in Great Lakes, water resources in northeast and arid west, geothermal power sites in Pacific Northwest etc.) and because of the ready availability of trained personnel for ground control studies. Orbital inclinations of at least 480 are therefore extremely important during the experimental phases of this program. It should be emphasized however, that a number of important users (Canadian government, Arctic Institute, several oceanographic groups, the National Science Foundation and the U.S. Geological Survey) have expressed strong recommendations for polar orbits. The U.S. Geological Survey and the National Science Foundation requirements relate to their work in Antarctica. Although requirements for polar orbits are not mandatory during the experimental phases, they are nevertheless highly desirable. However, data acquired during such flights can be controlled so that it is collected only over well studied end well specified test sites.

Conclusions
    1.   The discourse with the user scientists is well underway.
    2.   The development of an operational capability requires that experience with orbital data be built up over a long period.
    3.   In order to make progress in this field it is necessary that we have the widest participation of competent scientists and this can only be accomplished by unclassified access to data and the pertinent characteristics of the instrumentation and through availability to the program of already developed technology and hardware for incorporation in the NASA experimental flight program.
    4.   Many of the data collection instruments required for the Natural Resources Program are also vital to the broad exploration of the moon and planets and therefore should be made available to NASA in any case.

Document II-15

Document title: Leonard Jaffe, Director, Space Applications Programs, OSSA, to Deputy Administrator, thru Homer S. Newell, Associate Administrator for Space Science and Applications, "Meeting at the U.S. Geological Survey (USGS), August 25, 1966, regarding Remote Sensing and South America," August 31, 1966, with attached: Robert G. Reeves, For the Record, "Meeting at the U.S. Geological Survey (USGS), 10 a.m., August 25, 1966," August 31, 1966.

Source: NASA Historical Reference Collection, NASA History Office, NASA Headquarters, Washington, D.C.

As different government and nongovernmental institutions became interested in the potential for a satellite Earth resources program, NASA came under pressure to deliver user-oriented working satellites rather than large experimental spacecraft. The State Department, as one of these interested institutions, along with its Agency for International Development, viewed resource satellites as a way to assist third-world nations. Interested in providing a resource satellite for South America, the Departments of State, Agriculture, and the Interior used the threat of a non-NASA satellite program based on a proposal by the Radio Corporation of America (RCA) to prod NASA into a more responsive attitude.

[no pagination]

                                                   [rubber stamped: "SEP 6 1966"]

        AD/Deputy Administrator
THRU: S/Associate Administrator for
           Space Science and Applications
         SA/Director, Space Applications Programs, OSSA

Meeting at the U.S. Geological Survey (USGS), August 25, 1966, regarding Remote Sensing and South America

The enclosed is a fairly complete memorandum on the subject meeting, but following comments are in order:

(1) USGS is apparently the prime mover in soliciting the Agency for International Development (AID) to promote South American use of orbital remote sensing techniques in general support of the Rostow Report recommendations approved as the National Security Action Memorandum 349 (NSAM 349) enclosed.

(2) USGS would like to budget for an "operational" satellite as soon as possible to establish jurisdiction.

(3) Plan based on the Radio Corporation of America (RCA) proposal to develop Delta class satellite with TV camera of 5,000 line resolution.

(4) USGS and Department of Agriculture agree that the South American need is urgent and that the system proposed by RCA would be useful.

(5) The RCA proposed camera is far beyond the Nimbus/TIROS state of the art.

(6) We made note at the meeting that there is some agreement between Department of State and the National Aeronautics and Space Administration (NASA) that NASA should establish and chair an interagency committee to develop governmental thoughts on the subject.

(7) I suggested that the agencies pull together requirements for such a lower resolution system and convey them to NASA. NASA will then determine how such a development might proceed from a technical standpoint and in the interests of overall economy in the space budget.

(8) I suggest that the Deputy Administrator (AD), NASA, organize the committee recommended in the Administrator's memorandum to AD, dated August 12, 1966, (See also NSAM 349) and that the agencies' requirements and the NASA suggested approach to development be submitted to that committee. This will insure that NASA is the focal point for advice during the developmental period. This is most appropriate—because of the multi-agency interest, NASA is in the best position to serve all of the interests of the government.

[rubber stamped: "Leonard Jaffe"]
Leonard Jaffe

Enclosures
NSAM 349
The Frontiers of South America
Memo For the Record, from SAR/Reeves, dated August 31, 1966
SA/LJaffe/mc 8/31/66

bcc: S/Newell
        AXC
        Concurrence: _____
                        Homer S. Newell
                        Associate Administrator for
                        Space Science and Applications

I/Morrison

[handwritten note: "9/6/66"]

****

[1] For the Record

[rubber stamped "AUG 31 1966"]

SAR/Robert G. Reeves

Meeting at the U.S. Geological Survey (USGS), 10 a.m., August 25, 1966

A meeting was called by Mr. Edgar L. Owens, Chief, Planning Division, Bureau for Latin America, Department of State, Agency from International Development (AID), and was held in the Director's Conference Room, U.S. Geological Survey. Attendees were:

Edgar L. Owens, State/AID
Kenneth Milow, State/AID
Leonard Jaffe, NASA/Code SA, Director
James R. Morrison, NASA/Code I
Robert G. Reeves, NASA/Code SAS (USGS)
Arch B. Park, USDA/Agricultural Research Service
James Bailey, U.S. Naval Oceanographic Office (USNOO)
Leo V. Strees, USNOO
William A. Fischer, USGA/Office of the Director
Montie R. Klepper, USGS/Associate Chief Geologist
Charles J. Robinove, USGS/Hydrology Program Manager
John Place, USGS/Geography Program Manager (Acting)
Robert Peplies, USGS/Geography Program (East Tennessee State University)

The subject of the meeting was to discuss the possibility of using Remote Sensors for resources surveys in the AID program, Latin America. Specifically, preparation of a paper explaining the techniques of remote sensing, suitable for use by higher officials State was requested (See enclosed letter to W.A. Fischer from E.L. Owens).

[2] The meeting opened with a brief review of the recommendations of the Rostow Report, which includes formation of an interagency committee to, among other things, investigate use of space for resources studies in Latin America. The Department of State (Assistant Secretary for Latin American Affairs) has action, according to NSAM 349, to convene such a committee. The National Aeronautics and Space Administration (Dr. Seamans) will probably chair the committee. A "working group" drawn probably from among those in attendance at this meeting (and including others, no doubt) will probably be constituted; results of this meeting will probably influence the decision whether or not to recommend constitution of a formal working group.

Mr. Jaffe cautioned about overenthusiasm on part of underdeveloped nations for new and exotic techniques, with concomitant exclusion of proven methods. He gave as an example the reliance on satellites for communication in Latin America. Work on a microwave net connecting South American capitals was stopped when communication satellites made their appearance; however, the South American countries have not been able to use satellites, and neither do they have a microwave link.

Political sensitivity of remote sensing was briefing discussed. Mr. Morrison mentioned the Brazilian proposal, and I reviewed a recent memorandum from Dr. Fernando de Mendonca (Technical Director, Brazilian National Space Activities Commission) stating that the undertaking of the proposed Brazilian Remote Sensing Project has been approved by the Brazilian National Security Council.

Mr. Fischer reviewed the USGS briefing to Secretary Udall outlining the USGS views on and needs for an "evolutionary" Earth Resources Observational Satellite (EROS).

The RCA proposal was briefly discussed. It was agreed that it is 1) a commercial proposal, therefore some performance claims and cost data are suspect 2) possibly beyond present state of the art.

Dr. Park stated need on part of agricultural/forestry scientists in Latin America and other "developing" countries for more and improved data.

He also stated that USDA has a need for a "tropical" test site and other activities to get remote sensor signatures and correlations and develop interpretative techniques. This is necessary to fulfill USDA's mission of keeping track of worldwide agricultural activities.

Mr. Fischer stated that USGS has same need as USDA for data, to fulfill its missions of supporting AID in assistance programs in underdeveloped countries and also studying geologic features and phenomena and mineral resources that are global in contact.

[3] Mr. Robinove stated that more data, of the type possibly obtainable from space, are necessary for hydrologic studies. These too are global in space.

It was generally agreed that more and improved data are necessary to assist in Latin American studies; commonly the platform is of no interest to the user scientists, only good data. However, the platform does effect the characteristics of data and it seems likely that data from space may provide at least some of the necessary resources, information in a form superior to aircraft data; and that these data are needed *now.*

To provide the necessary "background" materials various documents already in being or in preparation were discussed—chiefly the OSSA-SA-SAR prospectus and summer study prospectus. Certain key items are needed—in my opinion, it may be better to prepare a separate summary. These items are:

(1) Statement of usefulness of remote sensors for resources investigations (discuss both airborne and spaceborne).

(2) What is underway in development and use studies of remote sensors for resources investigation.

(3) What needs to be done, in addition to work already under way.

(4) Competence and abilities of Latin America scientists and organizations to undertake this project. (Also, desires of various countries should be commented on—for example, Argentina, in person of the Director of their Geological Survey, Dr. Felix Bonorino Gonzales, is very interested in this program.)

(5) Cost of "natural resources" satellite program and how funded.

Mr. Jaffe made the following points:

(1) The user agencies should come up with their data requirement: resolution, areal [sic] coverage, spectral range, etc. Especially needed are requirements that might be met by a system such as that proposed by RCA.

(2) NASA will examine requirements and determine how, from a technical stand point, they can best be met; especially, how they can be melded with on-going programs, with minimum cost.

(3) That, although from a technical stand point, No. 2 could be done in a few weeks, certain decisions are being made which would greatly affect any examination of 1)[;] therefore, it would be better to await the results of these decisions. These decisions are expected in about a month.

[4] Mr. Fischer agreed to pull together USDA, USNOO, and USGS requirements into one document, for review by the user agencies and submission to NASA for engineering examination.

Mr. Morrison stated that it is NASA policy for foreign countries to be "self supporting," although NASA assists in providing some equipment, training, and technical service. If the foreign countries have to pay their share of the costs of space programs, they then make every effort to ensure that the programs are properly carried out. Mr. Owens

generally agreed with this NASA concept, although AID does have limited technical cooperation funds to assist where absolutely necessary. Mr. Fischer made the point that it's better to tie any space resources study projects to on-going program, to take advantage of
United States and counterpart scientists already working in an area or on a project. I suggested that even if scientists are already at work in an area, it will probably be necessary to
furnish scientific/technical advice on the application of remote sensors to the problems at
hand.

The meeting pointed up the following:

(1) USDA and USGS agree on the immediate need for data, which might be obtained
from spaceborne remote sensors, for domestic and foreign resources investigations.

(2) The less costly, the more the chances of success of any program involving the use
of remote sensors for resources investigations will be.

(3) High resolution is still a very touchy problem. An "intermediate" resolution evolutionary system has a good chance of being accepted, whereas a program using
high resolution instruments might run into great political difficulties.

Robert G. Reeves

Enclosure

SARR/RGReeves:mc 8/31/66
bcc: S/Newell
       AXC
       SA/Jaffe
       SAD/Tepper
       USGS/Fischer
       USDA Park
       NAVOCEANO/Alexiou
       SAR/Colvo
       SAR/Badgley
       I/Morrison

Document II-16

**Document title: Office of the Secretary, U.S. Department of the Interior, "Earth's
Resources to be Studied from Space," News Release, September 21, 1966.**

Source: Department of the Interior Library, Department of the Interior, Washington, D.C.

*When the Department of the Interior issued this press release announcing Project EROS (Earth
Resources Observation Satellites) on September 21, 1966, it not only lacked a satellite supplier, but
also funds had yet to be allocated for the project. Proposing the EROS project was actually a bold move
by the Interior Department to force NASA into providing an initial Earth resource survey satellite in
short order. Although NASA moved to comply, budget battles between both organizations and the
Bureau of the Budget delayed launch of the first satellite until July 1972.*

[1]   For release: SEPTEMBER 21, 1966

Earth's Resources to be Studied from Space

Project EROS was announced today by Secretary of the Interior Stewart L. Udall. EROS (Earth Resources Observation Satellites) is a program aimed at gathering facts about the natural resources of the earth from earth-orbiting satellites carrying sophisticated remote sensing observation instruments.

"Project EROS," said Udall "is based upon a series of feasibility experiments carried out by the U. S. Geological Survey with NASA, universities, and other institutions over the past two years. It is because of the vision and support of NASA that we are able to plan project EROS."

Udall said that "this project will provide data useful to civilian agencies of the Government such as the Department of Agriculture who are concerned with many facets of our natural resources. The support of these agencies is vital to the success of the program."

The Interior Secretary said that "the time is now right and urgent to apply space technology towards the solution of many pressing natural resources problems being compounded by population and industrial growth."

Udall said that the Interior Department program will provide us with an opportunity to collect valuable resource data and use it to improve the quality of our environment." [2] "Facts on the distribution of needed minerals, our water supplies and the extent of water pollution, agricultural crops and forests, and human habitations, can be obtained on a global basis, and used for regional and continental long-range planning," he said.

Secretary Udall named Dr. William T. Pecora, Director of the U. S. Geological Survey, to head the program.

"A team of knowledgeable scientists and resource data users will guide government and private agencies in making their data needs known, and to help plan a major effort in the exploration of the earth for human benefit," Udall said.

Pecora and his earth science colleagues described space-sensing of the earth as "the ability to 'see' more easily beneath the water and forest or soil cover, and the ability to view areas of the earth repetitively at various times and seasons. Another basic advantage is the fact that comparable observations can be made all over the earth."

"Although we are now gaining valuable information from existing satellites," Pecora said, "none are capable of providing global coverage of the type required for successful resource application."

"We visualize EROS as an evolutionary program," said Pecora, "beginning with television cameras flown in an orbit that will cover the entire surface of the earth repeatedly, under nearly-identical conditions of illumination."

Pecora said that "we plan to fly the first satellite in 1969," and that "the cost of launching the first EROS vehicles is not expected to exceed $20 million—far less than the cost of photographing the earth by conventional aerial means."

"What we have learned from photographs taken recently from orbiting spacecraft," the Survey Director said, "indicates that the lands can be examined, evaluated, and mapped, and the type and vigor of plants can be determined. In addition to the cameras that will provide the photographic record, the first vehicle will also have a small telecommunications unit so that we may relay data to and from ground stations that will aid in interpreting the television images. These relayed ground data will include seismic and other information that, hopefully, will enable us to predict some natural disasters."

Pecora explained that "future sensing systems will employ heat-measuring devices to monitor the earth's volcanoes and search for sources of geothermal power, radar that will 'see' beneath the clouds, and eventually cameras with sufficient resolving power to permit timely up-dating of our national topographic map series."

[3] "In addition to savings in the cost of updating these maps," said Pecora, "the availability of updated maps will result in a savings of over $100 million annually to the American public. Applied on a global basis, the savings would exceed a billion dollars a year."

The earth scientist emphasized the importance of feasibility experiments that have been carried out by his agency with NASA and other research and technical agencies. "These experiments enable us to start the EROS program with confidence in its useful application for the benefit of man," he said.

In announcing the EROS program, Secretary Udall pointed to the huge national requirements for natural resources needed to feed out technologic society as well as the need to conserve the Nation's lands. "We must insure that we use our resources wisely," he cautioned, adding that "the information gained from EROS vehicles will be synthesized and made generally available; it will help us achieve maximum use of our resources with minimum waste."

"We firmly believe," said the Interior Secretary, "that the use of the Earth Resources Observation Satellite will provide technological support for the continuation of our society of 'plenty' for generations to come. EROS will be just the beginning of a great decade in land and resource analysis for a burgeoning population."

Document II-17

Document title: Charles F. Luce, Under Secretary, U.S. Department of the Interior, to Dr. Robert C. Seamans, Jr., Deputy Administrator, NASA, October 21, 1966, with attached: "Operational requirements for global resource surveys by earth-orbital satellites: EROS Program."

Source: Department of the Interior Library, Department of the Interior, Washington, D.C.

Shortly after the September 21, 1966, Department of the Interior press release announcing Project EROS, NASA served notice that it was willing to develop the sort of resource survey satellite that the U.S. Geological Survey required. The Department of the Interior wasted little time, quickly sending NASA its specific requirements for the initial satellite. Although the Department of the Interior hoped to be receiving data from a satellite within two years, budgetary and management disputes delayed the launch until 1972.

[no pagination]

October 21, 1966

Dear Bob:

In my letter to you of October 7, I indicated that the Geological Survey was prepared to submit a document of "performance specifications" of the EROS Program for evaluation by your staff. I am pleased to transmit the enclosed document which sets forth Interior's operational requirements as a basis for extensive discussions.

Since our meeting, Dr. Pecora has attended two meetings at NASA preparatory to specific recommendations on the Mexico-Brazil agreements and state-of-the-art analyses.

Secretary Udall is very happy to know that our coordination is progressing so favorably and hopes these discussions will lead to areas of early agreement and action in the context of the EROS concept.

Sincerely yours,

Charles F. Luce
Under Secretary

Dr. Robert C. Seamans, Jr.
Deputy Administrator, Code AD
National Aeronautics and
    Space Administration
Washington, D.C. 20546

Enclosure

****

[1]

Operational Requirements for Global Resource Surveys by Earth-Orbital Satellites EROS Program

The EROS Program of the Department of the Interior requires sensors to be placed in orbit to obtain systematic synoptic and repetitive imaging of natural and cultural features whose description and understanding is vital to Interior Department missions in many disciplines. The Interior Department is convinced that the earliest possible acquisition of satellite data is of great importance and therefore outlines the following requirements for the first EROS satellite on a "performance specification" basis. All presently available qualified satellite platforms, sensors, and facilities should be evaluated for their capabilities to fulfill Interior Department requirements. We hope to begin acquiring resource data from an operational system by the end of 1969.

The first satellite, planned as an optimum general purpose data collector, should be followed by a series of EROS satellites carrying more sophisticated sensors using other regions of the electromagnetic spectrum and also measuring force fields. A tentative priority for the later sensors is 1) high resolution infrared imager, 2) radar imager and scatterometer, 3) ultraviolet luminescence sensor, 4) microwave radiometer, and 5) gravity gradient and magnetic sensors. Specific requirements for these sensors are not stated at [2] this time, but will be developed as the engineering of the sensors and interpretive capability develops of users of data.

I. Basic Requirements
    A. Near global coverage
    B. Repetitive observation at same local time
    C. Photographic or imaging data
    D. Vertical viewing required in normal operation but oblique viewing desirable for supplementary coverage
    E. Slight overlap of images in direction of flight and sidelap at the equator
    F. Minimum operational life of one year
    G. Analog data return
    H. Capability for in-flight command programming

I.   Capability of carrying additional unspecified instruments with weight of 50–100
     pounds
J.   Data telemetering capability from ground-based instruments through satellite
     communications link to central data-reduction facilities
K.   Unclassified systems and data

II. System constraints imposed by basic requirements
     A. Sun-synchronous orbit with sun angle at 30° at 50° latitude, at vernal equinox at
         2.6 hours (local sun time) before or after true noon
[3] B. Power supply sufficient for sensors and data transmission and relay
     C. Stabilization sufficient to maintain a minimum of 1° pointing accuracy at vertical
     D. On-board data storage as required by relation of orbit altitude, and tracking stations
     E. Data rate commensurate with resolution and number of elements in final design

III. System requirements for optimum data acquisition
     A. Imagery
         1. Ground resolution of 100'–200' per resolution element (on a side)
         2. Field of view of about 100 statute miles on a side (square format)
         3. Spectral resolution
             a.   about 100 mμ spectral increment peaked at 510 mμ
             b.   150 mμ spectral increment peaked at about 700 mμ
         4. Ground recording in two modes (direct analog image and magnetic tape)
             with the least image degradation possible.
     B. Data communications relay
         1. Capable of relaying digital data from earth-based sensors to central data-
             reduction and computing facilities.
[4]   2. Required data transmission to be on a basis of at least daily readout of data
             stored at ground stations and collected at intervals of 1–5 hours. Data rate
             and volume would be low.
         3. Ground sensors may include, but not be limited to:
             a.   insolation meters
             b.   stream gages and water-quality recorders
             c.   tiltmeters
             d.   thermal probes
             e.   seismometers
             f.   strain gages
             g.   displacement meters

[5]   References

    Hackman, R. V., 1966, Time, shadows, terrain and photo-interpretation: U.S. Geol.
Survey Tech. Ltr. NASA 22. 8p., 10 figs.
    U.S. Geol. Survey, 1966, Detailed plan and status report of U.S. Geol. Survey research in
remote sensing under the Natural Resources Space Applications Program (internal report)

Document II-18

**Document title: Irwin P. Halpern, Director, Policy Staff, NASA, Memorandum for General
Smart, "Earth Resources Survey Program," September 5, 1967.**

**Source: NASA Historical Reference Collection, NASA History Office, NASA
Headquarters, Washington, D.C.**

In 1967, with no clear mission for the post-Apollo era, some NASA officials began to talk about space activities that could be justified in terms of public benefit. As the head of the Policy Staff, Irwin Halpern pointed out to Assistant Administrator for Policy Jacob E. Smart in this memorandum that the Earth Resources Survey represented the sort of justifiable program the space agency needed.

[1] NATIONAL AERONAUTICS AND SPACE ADMINISTRATION
WASHINGTON, D.C. 20546

September 5, 1967

MEMORANDUM for General Smart - E

Subject: Earth Resources Survey Program

I would strongly recommend, from a policy point of view, that the Earth Resources Survey [ERS] Program and its place in the overall space budget now be reassessed.

It is imperative, in my judgment, that NASA make plain that it is doing important things for people. This is what the President and the Congress are concerned about. This is the test to which the taxpayer is putting Federal programs. This is why, in essence, our budget has been cut so severely.

The ERS program would score well in this test: with proper emphasis and direction, it could provide major demonstrable economic returns in the near as well as distant future. The need now, it seems to me, is to capitalize upon this asset, to pursue the program with a greater level of effort than now planned, and thereby to seize an important opportunity (while there is still flexibility in the FY '68 budget) to strengthen the basis for public support for the entire space program in an election year.

We should, in my judgment, take the lead of the Vice President (who recently voiced concern that too little was being done in the ERS Program) and the recent Woods Hole Conference (which called for proceeding apace with an aircraft as well as satellite ERS development program) and ensure that this program is pursued with a higher level of effort. The Departments of Agriculture, Interior and Navy have asked NASA for support for R&D in the ERS area. We are responding by providing only 25 percent of the $8 million or so that is required. Such responsiveness to the interests of potential consumer agencies may well give rise to charges that NASA, instead of leading the assault on a whole new area of technology, is in the van dragging its feet. In my view, finding goals in common with other Federal agencies and programs is tantamount to finding new bases of support for the space program. This would argue, in turn, for increased responsiveness to the interests of other agencies.

[2] Despite the low level of effort mounted to date, the NASA ERS Program has already made a substantial contribution to the economy in stimulating a national awareness of the potential of remote sensing. It is probably fair to say that NASA has served as a catalyst in the field of exploiting the whole electromagnetic spectrum for economic uses. A new industry is aborning [sic]: Teledyne is planning to establish an ERS service company that may include from six to twenty twin-engine aircraft equipped with multiban camera systems; Westinghouse will soon hold a conference on the use of side-looking radar for natural resource exploitation. In my view, NASA has the opportunity and the obligation to follow through and remain at the cutting edge of the new technology (to the extent security considerations permit), in order to give new impetus to industry's advance in this field.

Furthermore, what we do or fail to do in this area may importantly affect our world position politically and economically. The ERS Program offers an opportunity to substantially expand international cooperation (Brazil and Mexico have already sought to work with us in this area). Encouraging the development of ERS capabilities at aircraft levels in the course of developing a satellite capability would afford the United States rich opportunities to assist less developed countries in ways that are very meaningful and beneficial to them in the near term. At the same time, by shirking leadership in the ERS field, we risk being upstaged by the Soviets, who have lately indicated an interest in the economic uses of satellite photography. On the security side, the NASA program holds out the promise of contributions to a more stable world by helping to gain wide acceptance for "open skies."

In conclusion, I believe that by strengthening the Earth Resources Survey Program, even at the expense of certain scientific or hardware projects, we would strengthen the total space program and protect its future. Such a step would draw public attention to meaningful post-Apollo activity and possibly even away from the moon mission itself. Above all, it would offer the President an opportunity to single out a tangible, indeed exciting, example of how the space program is supporting other important national goals, such as the war on hunger, and thereby helping people both at home and abroad.

[hand-signed: "Irwin P. Halpern"]
Director, Policy Staff

Document II-19

Document title: Jacob E. Smart, Assistant Administrator for Policy, NASA, Memorandum for Dr. Mueller, *et al.*, "Earth Resources Survey Program," October 3, 1967, with attached: Draft Memorandum for Mr. Webb, Dr. Seamans, Dr. Newell, "Issues Re: The Earth Resources Survey Program."

Source: NASA Historical Reference Collection, NASA History Office, NASA Headquarters, Washington, D.C.

Throughout 1967, NASA debated making the Earth Resources Survey (ERS) program a high priority for the post-Apollo era, perhaps involving the activities of a human crew. Although this program offered a number of potential payoffs for the agency, it contained several liabilities as well. Desiring agency-wide consensus on the proper approach to the ERS program, Assistant Administrator for Policy Jacob E. Smart sent this assessment of key issues involved with a request for feedback to Associate Administrator for Manned Space Flight George E. Mueller and other high-level NASA officials.

[no pagination]

NATIONAL AERONAUTICS AND SPACE ADMINISTRATION
WASHINGTON, D.C. 20546
October 3, 1967

OFFICE OF THE ADMINISTRATOR

MEMORANDUM for Dr. Mueller - M   Mr. Dembling - G
Dr. Adams - R   Mr. Scheer - F
Mr. Cortright - S   Mr. Allnutt - C
Mr. Finger - D   Mr. Lilly - B
Mr. Mathews - ML
Mr. Jaffe - SA
Mr. Wyatt - P
Mr. Frutkin - I
Admiral Boone - W
Dr. Eggers - R
General Cabell - OY

Subject: Earth Resources Survey Program

For some time, this office has been attempting to understand the full ramifications of the policy and procedural aspects of the ERS matter. Our purpose is to identify and articulate issues, to assist in establishing procedures to ensure that the important issues are recognized, evaluated, and viewed in perspective by NASA's decision makers and by the specialist and generalist staff members whose function it is to advise decision makers. To grasp the significance of earth resource sensing, of its complexities, and attendant problems, requires the best talent that NASA can muster. We solicit your personal assistance and the cooperation of your office in the development of policy on this matter.

Attached is a first draft appreciation of the issue as seen by this office. We request that you review and constructively criticize, either verbally or in writing, this draft and continue to share with us a sense of responsibility for the development of policy guidance with respect to NASA's internal and external activities.

> [hand-signed: "Jacob E. Smart"]
> Assistant Administrator
> for Policy

****

[1]   NATIONAL AERONAUTICS AND SPACE ADMINISTRATION
WASHINGTON, D.C. 20546

OFFICE OF THE ADMINISTRATOR

DRAFT

MEMORANDUM for Mr. Webb - A
    Dr. Seamans - AD
    Dr. Newell - AA

Subject: Issues Re: The Earth Resources Survey Program

The question of NASA's involvement in the Earth Resources Survey Program has reached a critical juncture at which fundamental decisions and overall guidance now seem imperative. On the one hand, ERS constitutes an important technical, political and economic challenge in which NASA could play a formidable role. On the other hand, diverse and complex inter-agency issues which bear directly on security and other key national policies, as well as the requirement to arrive at internal managerial decisions, need to be weighed very carefully before any broad NASA go-ahead is given.

The positive aspects of the program break down into three: compelling political realities, potential economic and social contributions, and fairly widespread scientific, technical and industrial interest. ERS offers NASA a unique and timely opportunity to furnish the President and the Congress with demonstrable evidence that the space program can be effectively applied to critical national and international problems such as poverty, overpopulation, urban stabilization and the enhancement of natural [2] resources. If it is to survive in today's budgetary environment, NASA must be responsive to these very real problems, as well as continue with its vital efforts to reach out and explore the solar system. With respect to the future, these problems will also haunt the next Administration— and the next. The U.S.S.R. also stands in the wings, readying itself to exploit aerospace science and technology for political and economic gain. At home, other governmental agencies, the Space and Marine Councils, the National Academy of Sciences and other bodies already are embarked on what adds up to a national ERS program. And segments of industry are becoming increasingly interested in the prospects of such activity and are beginning to invest their own resources in this field of endeavor. Clearly, there is no question of turning or holding back the technology, at least as far as aircraft-borne sensing is concerned. Moreover, the unclassified technology, including the release of Gemini pictures, has made plain that the state-of-the-art is ready for such a development.

Constraints: The controlling issues which must be resolved fall into the categories of (1) security and foreign policy implications, (2) multi-agency management and funding, (3) the need for a sound technical feasibility program to determine precisely what aerospace system[s] can and cannot accomplish, and (4) the several alternate organizational options which NASA could pursue in the event a go-ahead is decided.

[3] The conflict with DOD and other national agencies extends beyond the security issue—itself a knotty one—into the eventual roles and missions of the military and civilian space agencies. Moreover, the State Department has been chary to date in its endorsement of an international ERS, even on bilateral lines. At issue is whether the objections of those powerful agencies can be overcome and whether a mutually agreed-upon program, delineating respective responsibilities and soliciting mutual cooperation can be evolved. It must be made clear, in addition, that many of the constraints are policy constraints reflecting a dated policy environment that needs to be examined in the light of the environment of today and tomorrow.

With respect to other governmental agencies such as the Agriculture and Interior, the essential problem is definition of NASA's role: whether to behave as the lead agency, at least through the R&D phase, utilizing an inter-agency Program Review Board chaired by NASA as a coordination instrument, or to consider multi-agency funding and management with all its attendant problems as the program proceeds. An example of a longer-range consideration is what NASA's continuing role should be: Should we phase out as soon as an operational system is developed? Should we participate in operations beyond providing launch services?

In the technical arena, while much research in selected areas has been accomplished, the practicability of remote sensing of aerospace systems remains to be demonstrated. We must also examine the political, economic and social [4] consequences of ERS, as well as the technological aspects, in much greater depth, lest we prematurely build up public expectations. The fact is, however, that NASA and others have already proclaimed the program's promise.

Finally, ERS looms as a major undertaking for NASA if the above issues can be met. This warrants top-level review of Headquarters control and Center participation within NASA and a firm inter-agency arrangement to meet the myriad of problems and objections which occur.

We come then to the primary points at issue: At what pace should this nation pursue an Earth Resources Program utilizing aerospace systems and what should NASA's role be? What opportunities exist to extend our demonstrated capabilities in communications, weather and navigation satellites to the ERS area in order to be more responsive to problems of crucial concern to the Executive and Legislative Departments? Should NASA continue to be the major source of non-defense funds and resources? And in the same vein, should NASA fulfill a central role in a national civilian program? Should there not be a clear national policy which allocates responsibilities for an integrated approach?

Document II-20

Document title: Edgar M. Cortright for George E. Mueller, Associate Administrator for Manned Space Flight, Memorandum to Assistant Administrator for Policy, "Earth Resources Survey Program," November 17, 1967.

Source: NASA Historical Reference Collection, NASA History Office, NASA Headquarters, Washington, D.C.

More so than most other programs, NASA's Earth Resources Survey (ERS) program was an interagency endeavor. The Departments of Agriculture, the Interior, and State all had active stakes in Earth resource surveys. With the possibility for turf disputes in mind, NASA approached program development with a mixture of confidence and caution. Responding to an assessment of ERS issues by Assistant Administrator for Policy General Jacob E. Smart, Associate Administrator for Manned Space Flight George E. Mueller suggested that NASA should take a firm leadership position. Given Mueller's position as head of the portion of NASA in charge of humans in space, it is not surprising that Mueller seems to assume that NASA's remote-sensing activities would be conducted aboard occupied platforms.

[1]   UNITED STATES GOVERNMENT

Memorandum

To: E/Assistant Administrator   DATE: NOV 17 1967
           for Policy

FROM: M/Associate Administrator
           for Manned Space Flight

SUBJECT: Earth Resources Survey Program

REFERENCE: Your memorandum of October 3, 1967, same subject

Thank you for the opportunity to comment on your draft memorandum regarding an Earth Resources Survey Program. The draft states the issues and brings them into focus quite well. I therefore have a limited number of comments which are applicable to the draft as such.

Accepting your invitation for assistance I have extended my remarks in several areas to propose some solutions for consideration. Alternatively, I have suggested some factors which I feel must necessarily be accounted for in any approach to resolution of several questions raised in your memorandum. These suggestions are based on three years of study activity conducted by the Advanced Manned Missions Program Office, study in which participation of potential users has been encouraged. This work, which is continuing, has sharpened our intuition and provided some knowledgeability which I am pleased to share with you.

Scope of the Memorandum

Although the term "resources" is being used by some NASA spokesmen in a broad sense to include cultural resources as well as natural resources, I suggest that the scope of the memorandum be explicitly broadened by choice of other terms in title and text. It is not obvious, for instance, that a scheme for malaria control is within purview of the paper, and there are enough people who use the terms "resources" and "natural resources" interchangeably that confusion is likely over the extent of the subject matter involved. A title such as "Earth Applications Survey," for instance, would have wider implications although I am sure this can be improved.

The use of the terms "aerospace systems" and "aerospace technology" may also be misleading since I believe you are principally concerned here with those aspects of such programs which utilize space-borne technology.

[2] Presumably, aircraft work performed by NASA as a phase in the development of space technology does not pose policy problems of the same genre at all. NASA developmental work directed at air-borne systems as such does not appear to run the same gamut of policy considerations as the space-borne work, nor is it apparent too that, in areas of overlap, the Agency will necessarily wish to approach resolution of the issues in the same way. I suggest that issues relevant to air-borne systems be separated from those relevant to space-borne systems. The following commentary has been developed in the context of problems pertaining to space-borne systems.

Present Status

The assertion on page 2 that the U.S. virtually has a national ERS program requires comment. While it is true that a number of agencies, departments, councils, etc., are more or less actively engaged, the collection of efforts lacks the cohesiveness and leadership characteristic of a program. Moreover, of particular relevance in this context, the ongoing efforts are not heavily dependent on space technology. In addition, most of those projects oriented toward the application of space technology are being stimulated by NASA, usually with NASA funds. Thus, NASA is in a unique and strong position in the application of space-borne technology to earth oriented requirements.

NASA Position vis a vis Other Agencies

In terms of assumed role, as well as in terms of expertise, NASA is in the de facto position of lead agency for the space application of aerospace-borne sensors. It is logical and natural, as well as administratively superior, for NASA to continue in this role—at least through the R&D phases. With continued cooperation from the Departments of Agriculture, Interior, and others, this arrangement would greatly facilitate a strong NASA lead in demonstrating the benefits of exploiting the advantages of this new medium. It would also be the most efficient way to rapidly bring the new technology along on behalf of potential users who are without strong R&D orientations.

Since it is very unlikely on the face of it that a separate satellite system will become a reality for every user, the multi-purpose space platform is certain to play a major part in a future operational time period. Furthermore, the fact that the cost per application ought to go down markedly as the total number of activities on board the platform increases weighs heavily in favor of such an approach. The establishment of a lead agency, at least through the R&D phase, is essential to carrying this kind of enterprise forward and it would be a mistake for NASA to turn aside from the lead position already established.

[3] Although NASA might accept R&D funding from potential users, it is desirable to continue the practice of NASA funding. This is particularly true in cases where potential users lack R&D funding or management experience. They cannot be expected to crank up programs where the beneficial outcome to them is not reasonably assured, whereas it is well within NASA's charter to expend money for endeavors where a good measure of faith in the expectation of favorable results is required.

In the later operational period, after practical, earth-oriented applications of space-borne systems have been demonstrated cost-effective, users should be expected to pay their own way with NASA bearing responsibility for launching, tracking, data acquisition, and data dissemination.

Security and the Department of Defense

It may be that the Department of Defense must be considered a special case in the sort of situation envisaged above but it would seen neither necessary nor desirable for NASA to abandon the field. NASA's charter is clear in authorizing military support and, properly handled, such undertakings will enhance public confidence in the worth of NASA programs. NASA's predecessor, the NACA [National Advisory Committee for Aeronautics], supported the military without diminution of its public image or international reputation. In addition, workable arrangements exist with respect to classified undertakings and these can be extended into the new arena.

Proposals have been advanced that NASA be restricted from using some readily available technology as, for example, cameras larger than 6 inch aperture for gathering earth data from orbit. Such blanket proscriptions must never be allowed to become established policy since this would render some very promising remote sensing techniques inaccessible to anyone but the DOD. The existence of the technology which would be useful, and its capabilities, are openly known, hence security policy should concentrate on the collection and utilization of certain data and not on the equipment. Moreover, NASA has no need for access to special technologies which are of uniquely military character and which are therefore not at issue.

International Considerations

The reluctance in various sectors to consider international space applications programs must be broken down if we are not to forego an excellent opportunity to foster genuine international cooperation as well as to greatly enhance our prestige around the world. There is ample successful precedent for international cooperation in space and it would be a mistake not to expand on it. Whatever peculiar problems are posed by an effort like an Earth Resources Survey, our experience with international efforts in research, in meteorology, in communications, and in manned flight offers an excellent basis upon which to proceed. It may be possible to initiate viable efforts on a national basis, but the full potential of "benefits for all mankind" will not be achieved without substantial international participation.

[4] Feasibility, Development, and Operations

It is quite important to differentiate between these phases of program effort since the policy to be pursued, as well as the promise of results, may be quite different in each case. The current aircraft program and early Apollo Applications flights are aimed at establishing the feasibility of remote sensing from orbit in the many spectral bands of interest.

Based on Gemini [handwritten insertion in the original: ", meteorological satellites,"] and aircraft work to date, however, I believe that sufficient technical feasibility has been established to warrant laying out a development program. Such a program planning effort would provide focus and perspective for on-going and future feasibility efforts. It would furnish a structure into which the results of these efforts could be fitted, thereby hastening the day on which a development program could be initiated. Potential user agencies should officially participate in the developmental program planning which would delineate program objectives, the benefits which can be expected, and the cost effectiveness of different ways of achieving the benefits, all as related to an eventual operational program.

Much groundwork directly applicable to such program planning has been laid in the normal course of our advanced studies program. These studies, which embraced both technical and programmatic aspects, include the Manned Orbital Research Laboratory (NAS1-3612), an ORL Experiment Program (NASw-1215), and Spent Saturn S-IVB Utilization (NAS8-21064). In conjunction with these studies, we have examined how such programs could be implemented organizationally, emphasising [sic] the use of NASA in-house capability and the application of special, user-agency capabilities. We are currently pursuing a more definitive description of potential economic benefits through contract NASw-1604 with the Planning Research Corporation. This study is expected to provide greater understanding of this difficult area and yield a methodology which could support a [Planning Programming Budgeting System] type of analysis. These types of studies must be continued and expanded or we cannot hope to compete successfully in the Planning Programming Budgeting Systems with quantitative assessments of promising applications.

In closing, I should like to request that you keep me informed of your progress in this matter. You may contact the Director, Advanced Manned Missions Program, for information and assistance in support of your efforts.

signed by Edgar M. Cortright
for George E. Mueller

cc: See attached sheet

Document II-21

**Document title: Interior Department, "Appeal of 1971 Budget Allowance: EROS,"
November 25, 1969.**

**Source: Record Group 255, Records of the National Aeronautics and Space
Administration, Federal Records Center, Suitland, Maryland.**

Despite overcoming technical and boundary problems, and successfully selling the utility of an Earth resources survey system to a variety of users, NASA and the Department of the Interior encountered almost intractable opposition to the initiative within the Nixon administration's Bureau of the Budget. The Budget Bureau nearly eliminated the Earth Resources Technology Satellite (ERTS) program, but it relented when Senator Karl Mundt, pleased that the Interior Department had decided to locate the data-processing facility in his home state of South Dakota, led a drive to save the program.

[no pagination]

Appeal of 1971 Budget Allowance
EROS

Requested increase   - $7,500,000

Allowed increase   -   <u>minus</u> $3,900,000

The most temperate comment that can be made about this allowance is that it must have been based on incredibly bad advice.

Follow intensive review in the early days of this Administration President Nixon proposed in his 1970 budget start of a major earth resources satellite experiment. In his speech before the General Assembly of the United Nations on September 18, 1969, the President reaffirmed his commitment to an earth resources satellite program.

Ambassador Yost has undertaken further initiative with Secretary General U Thant to implement the President's commitment.

During November 1969, Dr. DuBridge discussed with Canadian representatives the basis of their collaboration with the United States in the earth resources satellite program.

Dr. Paine's support for the program as a major high payoff component of our space effort was publicly stated on a National TV appearance within the past ten days.

The EROS-ERTS program has exceptionally strong support within the Congress. Interior 1970 budget proposals were not only approved but increased. The House Committee on Science and Astronautics terms the program ". . . perhaps the best possible opportunity to achieve tangible economic returns from the substantial investment already made by the American taxpayer in the U.S. space program." The Committee has been vigorous in its advocacy of the earth resources concept and has urged expedited action on the EROS-ERTS programs.

The program has attracted widespread professional, public, industrial, scientific, and academic interest throughout the Nation and the World.

The abrupt change in Administration policy proposed by [the Bureau of the Budget] cannot help but become a major embarrassment to the Administration. The Budget Bureau has advanced no adequate rationale in support of its action. Accordingly there is no basis for its defense by myself and my staff before the Congress or the public.

I understand funds for the NASA ERTS program were cut because [the Bureau of the Budget] claimed the resolution capability of the proposed system did not meet needs of the use[r] agencies. This is not the case. Dr. Pecora and other user agency authorities wrote NASA clarifying this point and I believe these letters have been furnished to your examining staff. We support the technical discussion provided by NASA in defense of the ERTS program. The questions posed in the Interior "passback" have been previously answered by the numerous independent appraisals of the earth resources concept and program and the Issue Papers provided by this Department in prior years.

The course of action proposed by the Bureau of the Budget affects our '70 program as well as our 1971 budget. For this reason a decision must be made as soon as possible. If, after consideration of this appeal, you are still unable to recommend that the program proceed as planned, I propose that the issue should be discussed by myself and Administrator Paine with the President at his earliest convenience.

Document II-22

Document title: Robert P. Mayo, Director, Bureau of the Budget, to Honorable Walter J. Hickel, Secretary of the Interior, April 14, 1970, with attached: "Statement for Senator Mundt."

Source: Record Group 51, Records of the Office of Management and Budget, Federal Records Center, Suitland, Maryland.

Although the Bureau of the Budget wanted to block spending on an Earth resources survey system, NASA and the Department of the Interior maintained moderate program support through congressional pressure as well as through direct appeals to the president. NASA was given the go-ahead to work on two experimental satellites, while the U.S. Geological Survey in the Department of the Interior would build a data-processing facility in Sioux Falls, South Dakota, but only if the satellites' performance demonstrated the need for it. The latter development was made possible with the intervention of South Dakota Senator Karl Mundt, who lobbied strenuously for ERTS funding in the Department of the Interior after Sioux Falls was designated as the preferred location for the data-processing facility.

[1]   APR 14, 1970

Honorable Walter J. Hickel
Secretary of the Interior
Washington, D.C.

Dear Wally:

I am concerned that you and your staff feel that you have experienced embarrassment in justifying the Administration's position on the EROS program. I want to reassure you that the President and the Bureau carefully reviewed this program and we believe that the approach developed is a realistic one for the development of this new technology.

In our informal communications to your staff on the FY 1971 allowances, and also in material prepared for your review with the President on the FY 1971 budget, the rationale for this approach was explained. We note from your congressional justifications for the FY 1971 budget that you have supported the Administration's decision on this program by emphasizing the fact that the program is still experimental and that no substantial investments will be made now for facilities and equipment which would be needed for a possible follow-up operational system. We hope that this letter will further clarify the Administration's position and be helpful to you in any necessary strengthening of your presentation to Congress. It is important for all of us that the interested members in Congress clearly understand the Administration's position.

This Administration is placing high priority on the development of practical applications of space technology. Surveying earth resources is one of the applications which appears to be potentially productive. In an effort to explore this potential, the National Aeronautics and Space Administration is seeking authorization from Congress to proceed with the development of two experimental Earth Resources Technology Satellites. However, the space program of the Seventies will not be characterized by the pursuit of a single goal. We are attempting to develop a space program with a balanced emphasis on exploration, science and applications.

We recognize that the FY 1971 budget allowance for the EROS program is somewhat smaller than your request. In our judgment, a constructive program of preparation for useful application of ERTS data can be developed within the resources provided. I am sure that you are aware [2] that NASA's request was also reduced. While an increase has been allowed for actual satellite development, the NASA budget does not provide for increases requested for other parts of their earth resources program.

We must emphasize that this program is an experimental effort. We do not believe that it is prudent to invest substantial resources at this time in preparing for an operational system capable of analysis of all ERTS data. Present funding for NASA will focus on satellite development while Interior and Agriculture will concentrate on continued research relating to potential applications of ERTS data and preparation for use of the data once it becomes available.

The Administration does not want to move beyond the experimental phase of this program until we are confident that the benefits of the program are more than the expected costs. Moreover, the evidence is unclear as to the technological capability of ERTS to provide pictures with sufficient resolution to give the payoff claimed in some studies that have been undertaken thus far. Until we have greater assurances on these points, the financial commitment to the programs should remain at a minimum level.

We have previously furnished to your staff a copy of the enclosed statement which was supplied to Senator Mundt's office. The statement outlines the Administration's position with respect to location and construction of facilities.

I know that you and your staff recognize that the need for fiscal restraint requires selectivity and great care in our commitment to claims on our limited future budget resources. I appreciate the continuing support you are giving to the President's program. If you desire to discuss this matter further I and my staff will make arrangements.

Sincerely,

Robert P. Mayo
Director

Enclosure

****

[no pagination]

Statement for Senator Mundt

In 1972, the National Aeronautics and Space Administration plans to launch two experimental satellites designed to conduct experiments in the survey of agricultural and geological resources through the use of remote sensors on the spacecraft. If these experiments are successful, this program may have considerable potential for future applications in the survey of earth resources.

A review of the facility requirements for the Government's earth resources survey programs has indicated that Sioux Falls, South Dakota, will [on the original, there is an edited portion where the word "will" has been crossed out and replaced with the word "would"] be a desirable geographic location for a data processing and distribution facility for this program.

The Administration will propose initial funding to be used for site selection and design of the proposed facility. The Sioux Falls earth resources data processing and distribution facility would be designed to process data from later satellites if the initial experiments are successful. For the initial Earth Resources Technology Satellites (ERTS), NASA will supply data to interested governmental and private organizations. Actual facility construction at Sioux Falls will depend on the results of the experimental satellites.

The city of Sioux Falls has agreed to supply the land and construct the building under a leasing agreement with the Government agencies involved if the ERTS satellite experiments prove promising. The facility will be designed to meet the specifications of the Government agencies involved so as to serve the data reduction and analysis requirements of private and Government organizations which contemplate the use of the earth resources survey data. Future management arrangements for the Sioux Falls facility will be developed by the Federal agencies currently involved in this experimental program.

Document II-23

Document title: Arnold W. Frutkin, Memorandum to Dr. Fletcher, Administrator, NASA, *et al.*, "Some Recent International Reactions to ERTS-1," December 22, 1972.

Source: NASA Historical Reference Collection, NASA History Office, NASA Headquarters, Washington, D.C.

One persuasive argument for creating an orbital Earth resources survey system was that this system would provide significant benefits to third-world nations in areas such as accurate mapping and resource location. NASA was suitably pleased, therefore, with favorable international responses to these types of data from the first ERTS satellite.

---

[1]   NATIONAL AERONAUTICS AND SPACE ADMINISTRATION
                           WASHINGTON, D.C. 20546

                                                              DEC 22, 1972

MEMORANDUM TO   A/Dr. Fletcher
                 AD/Dr. Low
                 AA/Dr. Newell
                 ADA/Mr. Shapley
                 E/Mr. Mathews

SUBJECT: Some Recent International Reactions to ERTS-1

We are listing below a few of the early foreign reactions to ERTS-1 on the basis that they may be of some interest.

a.   Thailand: "The best of the (ERTS) scenes received are excellent; they appear to demonstrate clearly that our decision to participate in the ERTS programme from its inception was wise, and that there is much to be learned and much work to be done in order to exploit the new technology fully. . ." (from November 23, 1972 letter to NASA from Dr. Pradisth Chesosakul, Secretary General, Thai National Research Council).

b.   Mali: "The Malians expressed warm appreciation for the opportunity to participate in the ERTS and Skylab programs. Minister of Industrial Development N'Daw was particularly pleased with the information already being derived from ERTS imagery. Schweitzer (US/AID) observed that the Malians take a very practical attitude toward ERTS results, which they already are beginning to utilize to make maps of remote areas, for guiding water exploration efforts and for deciding routing of new roads. They have quickly recognized the experimental nature of the ERTS project and are proud to be participating in this pioneering effort. . . . The ERTS project is a wonderful example of how the world's most advanced country can cooperate with the poorest to our mutual benefit. For two air tickets (AID purchased airline tickets for the Malian Principal Investigator), the U.S. Government has gained a million dollars worth of Malian political mileage. This is an exciting project which should have increasingly important benefits for Mali and should also be applicable to other [2] developing countries. It deserves every bit of support we can give it, now and in the follow-up period." (November 17, 1972 State Department telegram from the U.S. Embassy, Bamako, Mali.)

c.   Canada: "I have just seen the ERTS imagery for the first time and I wish to congratulate NASA on this fine achievement. We consider it an important breakthrough in providing data for the understanding of our environment. I would also like to express our appreciation for having an opportunity to participate in this experiment." (August 2 telegram to Dr. Fletcher from Mr. Jack Austin, Deputy Minister of Energy, Mines and Resources, Government of Canada.)

d.   Brazil: Dr. Fernando de Mendonca, the Director-General of the Brazilian Space Agency, reported at the IAF on a wide range of valuable information provided by ERTS-1 in an experiment relating to mapping of the Amazon system. He has summarized the first preliminary findings in a letter to NASA. The following excerpts are of relevance for illustrating the value of ERTS-1:

"1. The course of the tributaries of the Amazon River are very different from the ones shown in the most recent available charts. The difference in position is sometimes off by 20 km or more and the difference in direction is sometimes off by 90 degrees.

"2. Islands with more than 200 km^2 exist which are not shown on maps.

"3. Some lagoons which are shown on maps as 20 km long are in reality more than 100 km long.

"4. Small villages and towns are located wrongly on the maps by several tens of kilometers.

"5. The drainage systems of some areas are entirely wrong and this has caused among other things, the construction of roads (Manaus - Porto Velho for instance) with extra expenditures for bridges. In fact, the mentioned road is placed wrongly (by more than 20 km) in recent maps (1971).

[3]   "6. Large unsuspected geological features have been detected, which might provide new insights into the formation of the basin.

"7. Large abandoned river meanders are shown which were not present in existing maps.

"8. Even with high percentage (75%) of cloud coverage in some images one can still make good use of the obtained information for correcting maps.

"9. Unmapped lineaments and fractures have been discovered.

"10.   The entire Amazonian region was covered last year with a Side Looking Airborne Radar (SLAR). The completed controlled photo-mosaics will not be ready for at least another year. Over 150 people are working on the SLAR project which has cost Brazil about 20 million dollars. Since the region is rather flat the ERTS-1 MSS channel 6 provides practically the same information as the SLAR imagery. If one considers the other MSS channels, then one has substantially more information from ERTS than the SLAR. This without mentioning the repeatability of ERTS imagery. The cost of ERTS imagery per square kilometer is about two orders of magnitude less than the SLAR if the satellite operates for the expected lifetime of one year."

e.   FAO (Food and Agriculture Organization of the United Nations): "NASA deserves FAO's compliments on a successful launching of the satellite which will remotely sense earth resources for the benefit of mankind for the first time." (letter of August 3 to NASA from Mr. Juan Yriart, Assistant Director-General, Development Department, FAO.)

f.   Iran: "We have located several lakes which do not appear on the Watershed Map of Iran. This phenomenon is presumably due to this year's relatively abundant rainfall. . . . In the extreme southeast part of Iran (near the Pakistan border) several igneous bodies have been observed which do not figure on the . . . geological map of Iran. . . . By comparing images taken from the extreme southeast part of the Caspian Sea with a map of the region prepared in 1945, it is quite noticeable that the shape of the [4] Bandar Shah peninsula has changed. This is possibly due to lowering of the Caspian Sea by evaporation which exceeds the inflow of stream waters. (November 12 letter to NASA from the Iranian Principal Investigator.)

g.   United Nations Secretariat: "It also gives me great pleasure at this time to extend my congratulations to you on the successful orbiting of this the first dedicated earth surveying satellite. It is a cause of particular satisfaction to us here at the UN that NASA and the other United States agencies involved in the programme have, by their imaginative approach, laid a sound basis for the international cooperation which will be such a fundamental requirement in order that this new application of space technology may serve, as we all hope, for the benefit of all mankind." (July 28 letter to NASA from A.H. Abdel-Ghani, Chief, Outer Space Affairs Division, UN.)

h.   Egypt: (Meguid) called ERTS a "significant technical achievement" in the UN General Assembly's First Committee during the week of October 30, 1972.

i.   Ghana: (Boaten) commended the Food and Agricultural Organization in cooperation with NASA for applying space technology to problems of desert locusts and food
in Africa, Asia and Latin America in the same meeting.
j.   The All-African Seminar held in Addis Ababa in August 1971 made a recommendation on the inventorying of natural resources reading in part as follows:

"Recommends:
–   that a complete inventory of natural resources, such as water, soils, vegetation,
    wild life, be undertaken everywhere in Africa, and that particular attention be
    given to this recommendation at both national and regional levels;
–   that the most modern techniques be used to achieve this aim, such as remote
    sensing through satellites; in particular noting that the two earth resources tech-
    nology satellites will be launched in 1972 and 1973."

[hand-signed: "AWF"]
Arnold W. Frutkin

Document II-24

**Document title: James V. Zimmerman for Arnold W. Frutkin, Assistant Administrator for
International Affairs, to Dr. John V.N. Granger, Acting Director, Bureau of International
Scientific and Technological Affairs, Department of State, September 12, 1974, with
attached: "Foreign Policy Issues Regarding Earth Resource Surveying by Satellite: A Report
of the Secretary's Advisory Committee on Science and Foreign Affairs," July 24, 1974.**

**Source: NASA Historical Reference Collection, NASA History Office, NASA
Headquarters, Washington, D.C.**

*Potential benefits to developing nations had been a key component in the NASA campaign to obtain
the Nixon administration's approval for an Earth resources survey system satellite. However, a number of nations objected to having little control over the dissemination of satellite information gathered
about their country. NASA proceeded with ERTS without a formal international regime, basing its
position on a broad interpretation of the previously established "open skies" doctrine. The Advisory
Committee on Science and Foreign Affairs of the Secretary of State suggested in its July 1974 report
that additional U.S. action was required to avoid conflicts, particularly at the United Nations, over
U.S. remote-sensing activities. NASA's response argued that the space agency had anticipated and
was effectively addressing any foreign policy repercussions of the ERTS program.*

[no pagination]
REPLY TO
ATTN OF   I/PP   SEP 12 1974

Dr. John V.N. Granger
Acting Director
Bureau of International Scientific
    and Technological Affairs
Department of State
Washington, DC 20520

Dear John:

This is in response to Herman Pollack's letter to Dr. Fletcher of August 2, 1974 forwarding the remote sensing policy report prepared by the State Department's Advisory Committee on Science and Foreign Affairs. The report, in our judgment, is a welcome contribution to the on-going discussions of policy alternatives in this important area of space activity. There are, however, a number of points arising out of the report which we suggest be brought to the attention of the Advisory Committee.

Page 2—The Advisory Committee's assumption that the cost and technical sophistication of the appropriate data analysis is generally beyond the reach of developing countries does not accord with our developing country experiences. In effect, through one device or another (U.S. or other aid), use of the data is in fact being made. We estimate, for example, that 19 African countries are directly or indirectly involved with the analysis of earth sources imagery of their territory.

Page 3 and 17—We suggest the report reflect a distinction between ERS data themselves, which the U.S. makes freely available, and analyses of these data, which could be held proprietary as desired. This distinction could apply to all potential ERS data users—governmental and private, domestic and foreign.

Page 5—It is important to note that Article I of the 1967 Outer Space Treaty provides that outer space shall be free for exploration and use by all States without discrimination of any kind. The "open skies" principle, therefore, can be considered a "cornerstone" of international as well as U.S. policy; no further international political commitment (discussed on page 1) is necessary.

Page 6—In view of the overall developments at the February-March 1974 U.N. Remote Sensing Working Group meetings, as well as the recent positions taken at the U.N. by the representatives of Canada and Sweden, we find it difficult to draw the conclusion that there is increased advocacy in the U.N. of restrictions on either the acquisition or dissemination of ERS data. Such advocacy continues to have a very narrow base (the USSR and France plus Argentina, Brazil and Mexico).

Pages 14 and 16—We believe the most effective means of enhancing international participation in U.S. earth resources programs and acceptance of the "open skies" concept is through the establishment of additional foreign ERTS data acquisition ground stations. To date such facilities have been established in Canada and Brazil. This spring an agreement providing for a similar facility was signed with Italy. Discussions are currently underway with a number of countries including Iran, Venezuela and the Federal Republic of Germany. Foreign governmental agencies establishing a ground station sign a memorandum of understanding with NASA which includes an open data dissemination provision; they are, in effect, ratifying the "open skies" principle.

Page 16—The Advisory Committee should know that a U.S. offer to provide (sell) to an agreed international distribution center a master copy of all ERTS imagery was made to the U.N. in January 1973. We understand that the FAO is currently studying the feasibility of establishing a world-wide ERS data storage, processing and dissemination center which could utilize the U.S.-offered imagery.

We hope the above comments prove useful and would welcome the opportunity to further discuss NASA's remote sensing activities with members of the Advisory Committee.

Sincerely,

[James V. Zimmerman]

[for] Arnold W. Frutkin
Assistant Administrator
for International Affairs

cc: Dr. James C. Fletcher

****

[1]   July 24, 1974

Foreign Policy Issues Regarding
Earth Resource Surveying by Satellite

A Report of the Secretary's Advisory
Committee on Science and Foreign Affairs

Summary

This report considers the options for U.S. foreign policy regarding the acquisition and dissemination of earth resource surveying data obtained from satellites (ERS). Foreign policy issues have arisen primarily in the course of debate in the Outer Space Committee of the UNGA, where the Brazilians, Soviets, French and others have introduced "principles" which would limit State's rights to acquire ERS data from space or to disseminate such data without the prior assent of the countries affected. While ERTS-1 experiments have been accepted under the principle of freedom for unrestricted space observations, often called "Open Skies," there is no international political commitment to "Open Skies" in the context of "operational" ERS systems.

For a variety of reasons, including national security, the U.S. (with the tacit support of the USSR) has insisted on the unencumbered right to *acquire* data from space. This posture must be maintained.

The present paper deals with appropriate future directions for the ERS program, and in particular with the policy of dissemination of ERS data. The authors point out that [2] experience with ERTS-1 and Skylab have indicated that space technology has great promise for generating useful data bearing on agriculture, forestry, fisheries, the location of natural resources, land use planning and in many other areas. However, experience to date is wholly inadequate to establish the economic value of these data or the cost-benefit character of its space acquisition as compared with other means. Extensive further experimentation is required to develop techniques for the interpretation and analysis of ERS data and for optimally combining that with data from other sources before its economic potential can be reliably assessed. The cost and technical sophistication of the appropriate data analysis, at this point, is generally beyond the reach of individual developing countries. Cooperative projects with LDC's could be important elements in U.S. strategy to develop for the needed applications of R&D that are the pacing elements in future progress. Eventually a viable commercial activity in data analysis may mature.

With respect to the policy issues, the authors cite three U.S. alternatives with regard to data acquisition and dissemination:
- assertion of unilateral rights;
- negotiation of internationally acceptable principles, offering technical cooperation and assistance as an incentive; and
[3] - abandonment of an open program and reliance on classified data.

The authors conclude that the further work necessary to establishing the economic utility of ERS requires the cooperation of other nations, and U.S. participation in and technical support of their efforts to develop data analysis techniques suited to their situation.

We should continue to assert the right to acquire and disseminate primary data. But to encourage applications R and D, the [U.S. government] should be willing to permit another country which so requests to restrict joint research to those applications the results of which they are willing to publish. Thus NASA and other agencies would continue the policy of full disclosures of both primary and secondary data in which the [U.S. government] is involved, but we would respect the right of another country to obtain the primary data tape or read the satellite directly if for our own purposes we energize it over their territory, and make what use of the data they will.

U.S. policy must therefore focus on the distinction between primary data, and information available after processing. Policy should be directed to obtaining international acceptance of freedom of acquisition and freedom of dissemination of primary data tapes if the potential benefits [4] of this technology are to be realized. The strategy for advancing these goals should be based on recognition that ERS should not be prematurely described as "operational" and in no event unless and until international acceptance of its potential benefits is obtained.

[5] Introduction

For more than a decade international acceptance of "Open Skies"—the right of any country to examine the earth from outer space without prior restraint—has been a cornerstone of U.S. space policy and should be continued. A number of factors reinforce the importance of retaining this freedom:

(a) SALT agreements force increasing reliance on national technical means of verification.

(b) Growing interest in the possibility that Earth Resources Satellites (ERS) might provide economic benefits in the future makes the continued viability of Open Skies of special interest. Since foreign customers for satellite data analysis must find the initial acquisition of data acceptable, either Open Skies policy must be maintained or else agreements or same form of internationalization of the space segment may be required.

(c) Many new space technologies will have to call for some observations from space. Restriction on open observations will create barriers to effective operation.

"Open Skies" is the preferred alternative, since it is necessary to sustain the legitimacy of remote sensing for national security purposes. In addition, it provides the minimum impediment to the development of ERS as a potential economic asset.

[6] While ERTS-1 has been accepted under the Open Skies principle, there is no international political commitment to Open Skies in the context of "operational" systems. However, as advocates of ERS become more vocal with the view that "operational" systems are economically viable, there is increased advocacy in the U.N. of restrictions on the acquisition and dissemination of ERS data and in the [U.S. government] on the possible value of restrictions on ERS data in order to capture more of the economic benefits.

The policy question requiring resolution is: What posture toward acquisition and dissemination of ERS data is optimal today in the light of the present state of evolution of ERS experience?

The answer depends substantially on an evaluation of that state of evolution and an analysis of the requirement to bring the ERS program to economic viability.

Experience with ERTS-1

A little over one year of experience with ERTS-1, plus some data from manned space flight programs and considerable commercial experience with airborne photography, have established both politically and commercially motivated interest in civil applications of remote sensing. From a technical point of view, the capability for image acquisition is well advanced and the potential for improvement established by very expensive national security programs.

[7] The spatial resolution achieved with multi-spectral scanning sensor in ERTS-1 is adequate to produce raw imagery that suggests a wide range of commercial/civil applications. It is not the case, however, that ERTS-1 and Skylab/Apollo experiments have given us the experience to permit the design of a remote sensing satellite appropriate to an economically viable, operational system. This is so primarily because of inadequate experience with application-specific image processing and interpretation. But even the space segment is in an early stage of development.

ERTS-1 is deficient in resolution for applications such as land use planning. Frequency of observation of a given point is not sufficient given the random interruption of cloud cover, for time-dependent problems in flood control, agricultural monitoring, iceberg tracking, etc. As a result, turnaround time between observation and analysis is too long for many purposes. Although technology exists to remedy these deficiencies, we do not have enough systems experience, with both satellite configurations and image data processing and analysis, to make the tradeoffs between these functional attributes and systems costs.

One must remember that ERS applications are in their infancy. Although the technical capability to acquire images, generate geometric and radiometric corrections, and extract information by image enhancement have been [8] demonstrated, the majority of applications research projects to date involve manual (visual) processing of primary images. These images are available several weeks after data acquisition.

A fully operational system would have to provide digital primary imagery to widely dispersed customers in approximately real time. This primary imagery is only the first step. The ground-based image filtering and contrast adjustment, information extraction and interpretation impose heavy data processing requirements. Correlation with ground truth data and final interpretation are tasks requiring very high levels [of] professional expertise, and extensive experience with end-use problems, institutions and customs. Since these interpretation services are the key to obtaining the benefits from remote sensing technology[,] decisions on the major international political issues and on the character of the follow-on space projects must be made with the desired institutional structure for the service delivery system clearly in mind.

The first step is the study of the economics with a view to evaluating the commercial potential for developing the services. At this time the institutional structure for exploiting an operational remote sensing system does not exist. A few commercial companies with substantial support from government R&D contracts are hoping to create a service business out of proprietary satellite photo interpretation. Several not-for-profits are developing experience, also on government contracts.

[9] It would appear that the conditions required to make information services based on remote sensing technology economically viable call for:

(a) Realistic evaluation of the total costs of processing, interpreting and marketing information as well as its initial acquisition. On a per-image basis, the space segment of the system may prove the least expensive.

(b) Identification of applications with very high economic leverage for which remote sensing from space is more economical than ground or airborne observation and for which a market can be aggregated.

(c) Determining the extent to which the economic leverage depends on exclusivity of access to either the raw data or to the processed and interpreted data in order to permit a proprietary advantage,

(d) Definition of [U.S. government] policies affecting the applications system economics—as well as international issues—and the establishment of the point of responsibility in the [U.S. government].

(e) Definition of the optimum satellite system with respect to economic and other considerations.

The [Office of Management and Budget] sponsored studies now under way may shed some light on those matters. However, the level of development of application services is so rudimentary at this time that much reliance will have to be placed on judgment—based on experience with similar new technologies.

[10] ERS Applications

It seems likely that ERS applications will have several characteristics:

(a) Satellite data will usually have to be combined with data from many more conventional sources before useful, commercially saleable [sic] results are obtained. Thus it is unlikely that many applications will be found that do not require complex interfaces with existing service organizations—both in the region under study and in the client's community.

(b) Exclusivity of the information product may be achieved through the quality of interpretation and amalgamation with other data, but exclusivity based on primary imagery requires either turning the satellite operation over to a private monopoly or finding legal justification and an administrative mechanism for exclusive licensing of imagery by the government. In any case, exclusivity is a will-o-the-wisp, since alternative platforms exist for surveying most areas of interest, and primary imagery alone will often prove inadequate.

(c) Some applications may prove cost effective in the developing countries before they do in the U.S., because of deficiencies in logistic and communications infrastructure, extensive poorly studied areas, lack of institutional structures for ground-based data collection, etc. (Flood management is an example.)

[11] (d) The most valuable applications are probably still undiscovered, and may depend heavily on very sophisticated automated image processing. In general, the computing requirements would appear to pose serious problems for developing countries and very small organizations. A number of years of subsidized operations will probably be required before commercial incentives will produce the required development investment.

Policy Issues

In light of this background, the policy issues requiring clarification are:

(a) What is the "critical path" impeding progress? Other than the obvious necessity for a follow-on program to ERTS-1, which is inadequate in resolution, re-observation rate, and is nearing end of useful life, it is experience in information extraction and use under realistic economic conditions that is most needed. A follow-on experimental satellite is well justified, and would permit several years of applications R&D, without which no viable commercial activity can occur. Thus, end-user programs would appear to deserve government priority. The second element in the critical path is the resolution of concerns of foreign nations with respect to acquisition and dissemination of data.

(b) Is there a relationship between the well established acceptability of meteorology satellites and the viability of Open Skies for ERS systems? U.S. policy in the meteorology [12] field is firmly and necessarily committed to free international exchange, as well as unilateral rights of observation. This is well accepted internationally because the benefits to all depend on information interchange. Many ERS applications have a similar dependency on global access, as well as exchange of "ground truth" data. If our ERS policy moves away from commitments to international cooperation and disclosure of space data, characteristics of most NASA programs to date, the possible impact on the U.S. weather satellite program must be clearly foreseen.

(c) How can we take best advantage of the U.S. lead in most of the professional end-user skills (oil exploration, plant disease studies, airborne mapping, etc.) plus the lead in digital image processing? These assets may offer sufficient "exclusivity" to adequately protect national economic self interest if we move ahead to develop and apply these skills. To facilitate this process, we should make imagery generally available to encourage further development of both institutions and technology.

(d) Should primary data tapes from ERS systems be made fully available to anyone at very low cost by government, or should the operating entity be empowered to restrict dissemination, and offer exclusive access to economically motivated customers? Which approach would develop most rapidly the technology, institutions, and capital and market structures to deliver useful services? The answer hinges on [13] whether the work to date would justify private investment in a satellite and ground facility system, in which case proprietary rights in primary data might encourage investment. In our judgment, the work to date would not justify private investment; particularly since the character of most applications is such that the benefits are not capturable [sic] by individuals. When the product of an enterprise is an indivisible public good, then it is certain that governments will be the primary customers. Examples are: land-use planning, flood control, crop estimation, iceberg tracking, pollution monitoring, etc.

(e) What policy regarding acquisition and dissemination of imagery over other countries has the best chance of leading to a stable basis for global access by the U.S., maximum cooperation in the exchange of ground data, and best opportunities for U.S. firms to market services abroad? Obvious alternatives include:

(1) Unilateral assertion of satellite data acquisition rights by the U.S., coupled with either an assertion of the right not to disseminate, or a commitment to do so.

(2) Proposal for an international network (like Intelsat), with the first U.S. operational bird as the first prototype. This would doubtless produce serious delays. This is also the most aggressive form of internationalization.

[14]   (3) Effort to negotiate international acceptance of "Open Skies" without asserting the unilateral right, perhaps with the commitment to share primary "(Master)" data fully and an offer of help to LDC's in image processing and interpretation as an incentive.

(4) Cancel the NASA program altogether and rely upon classified sources for U.S. national requirements.

(f) What is required to maintain the credibility of a unilateral U.S. program in which we offer to disseminate primary data internationally? If, occasionally, data are withheld, what will be the political consequences? Is there any merit to an extension of the "Master Tape" proposal to include the production and dissemination of photographs from the tape by a U.N. agency, in parallel with U.S. government distribution domestically? How can the U.S. enhance the sense of participation and equity of other countries in the new technology, as a foundation for viable international machinery in the future?

In an understandable desire to sustain or accelerate the public investment in ERS, its supporters are allowing policy authorities to underestimate the technical, institution and economic obstacles to be overcome before a widespread, cost-effective usage of remote sensing in civil applications can be sustained. The temptation to call the follow-on [15] program to ERTS-1 "operational" reflects this tendency. Political sensitivity of other nations to unilateral remote sensing both grows on and enhances this optimism. In fact, a great deal more research and, more important, more applications experience in a realistic economic environment is needed before an economically viable "operational" ERTS program can emerge. Requirements for this further experience include:

(a) Continued access to remote sensing and ground truth data on as close to a global basis as possible, clearly calling for cooperation by other nations.

(b) Continued broad availability of primary digital data at a subsidized cost from a NASA R&D system in which operational continuity is assured, and in which systems configuration tradeoffs can be evaluated.

(c) Expanded public support for user application and image processing technology development.

(d) A public commitment to the policy basis for future systems, should they prove economically viable, on the basis of which international cooperation and private investment will be forthcoming.

Thus we urge that policy reflect the fact that, while further development of ERS systems—especially the ground segment—is desirable, the U.S. is not prepared to move now to an "operational" system. In accordance with (e) 3 above, [16] the U.S. should seek to establish *de facto* acceptance of Open Skies through the inevitable gradual development of the technology and a private sector institutional base for delivering ERS services both here and abroad.

The pacing elements in commercial exploitation of the ERS technology are image processing, interpretation, and service marketing. Here the U.S. enjoys the most important elements of technological exclusivity. Rapid progress in improving and disseminating the basic image data distribution will maximize the opportunity to take advantage of this uniqueness. Thus a unilateral offer of the primary data tapes to an appropriate international body may well achieve the maximum political advantage of involvement of other countries and the UN at an early stage at a minimum cost to the development of the system. In addition, we should consider some form of expanded help to developing countries in the image processing and interpretation area. In the U.S. we should develop these activities in the private sector, initially encouraged by government contracts.

This policy framework could lead to acceptable negotiated agreements with other nations. We should continue to assert the right to acquire and disseminate primary data. But to encourage applications R and D, the [U.S. government] should be willing to permit other countries to restrict joint research [17] with the U.S. to those applications that they are willing to publish. Thus NASA and other agencies would continue the policy of full disclosures of both primary and secondary data in which [the U.S. government] is involved. We would respect the right of other countries to obtain the primary data tape or read the satellite directly, if for our own purposes we energize it over their territories, and make what use of the data they will.

Finally, in order to maximize private sector confidence in the continuity of this program and to minimize political problems, the most rigid separation possible should be maintained between the organizational and managerial environments for civil ERS systems and national security systems. At the technology level only commonality in both space and ground segments can be considered to the extent that security requirements and economics permit.

Document II-25

Document title: Clinton P. Anderson, Chairman, Committee on Aeronautical and Space Sciences, U.S. Senate, to Dr. James C. Fletcher, Administrator, NASA, October 14, 1972.

Source: NASA Historical Reference Collection, NASA History Office, NASA Headquarters, Washington, D.C.

Because threatened budget cuts kept the future of NASA's ERTS program in a state of doubt, the space agency was under steady pressure to demonstrate the program's utility. NASA had campaigned for the program on the grounds that it would provide solid benefits to various user communities. However, in the months following the launch of the first ERTS, NASA found that providing the specific data required by the users was in many ways more difficult than building the hardware. NASA's limited knowledge of many fields hindered its ability to specifically tailor data. Nontechnical communities, particularly state agencies, did not understand the technology well enough to realize its capabilities and limitations. Many potential users, unable to effectively communicate their precise needs to NASA and unaware of what was available, were not profiting from ERTS data. A note on Senator Anderson's letter to NASA Administrator James Fletcher refers to a "Chuck" and "Frank." "Chuck" was Charles Matthews, the NASA Associate Administrator in charge of the ERTS program, and "Frank" was Senator Anderson's aide Frank DiLuzio.

---

[1]   October 14, 1972

Dr. James C. Fletcher, Administrator
National Aeronautics and Space Administration
Washington, D.C. 20546

[handwritten note: "Chuck, I'm inclined to think Frank is right here. $ is a problem but I'm afraid NASA has to take the leadership here—JGF"]

Dear Jim:

For quite some time, this office has been following the progress of the ERTS-A Program, and, specifically, the transferring of information gathered by the cameras and sensors on board ERTS to end users. It is obvious from testimony, news stories and public news releases of NASA that the initial users of the data will be the Corps of Engineers, the Environmental Protection Agency and the Departments of the Interior, Commerce and Agriculture. These are Government agencies with statutory mission assignments that can utilize the data. It is also true that to a great degree, they have within their agencies the needed expertise to interpret and apply such data to specific programs. This is a talent which I do not believe exists in other potential user groups which need assistance and which are not being given the proper attention. These groups include state planning agencies, state regulatory agencies, regional compacts, state and regional industrial and educational groups and resource planning groups.

I firmly believe that full and profitable utilization of ERTS data will only be accomplished when it is made available to individual states, regional compacts and consortia. I further believe that unless there is a more concentrated effort to make the information available to the end users, nothing dramatic is going to happen regardless of how valuable and useful the collected data is for land use planning, flood plan control, forest land management, resources identification and development, etc.

This problem was perhaps best illustrated at a recent meeting of the Federation of Rocky Mountain States in Boise, Idaho. It was very evident that while NASA has done a fair job in setting up sources to which users could write for "imagry," [sic] the locations and capabilities of these centers were not well known. Also, the potential users did not seem to know what was available and, therefore, could not ask for specific "imagry" [sic] with a precise reference.

[2] I do not mean to criticize NASA, but merely to reiterate that it has always been very difficult to take new technology and make it easily available in understandable and useful forms to the end users. This has been true with many new technologies, and it also is true with ERTS. It is very difficult to initially inform and keep informed the several different ultimate users, and it also is difficult at times to even identify the ultimate user of such data.

I have been informed that there is an agreement between NASA and the several Governmental departments involved as to their specific roles. However, I firmly believe that there must be a lead agency in the application of any new technology, or it has a tendency to drift and dribble rather than follow a logical process of application which proves out its value to the country.

I would appreciate it very much if NASA would take a look at what is now being done in terms of making the "imagry" [sic] and data available; and perhaps with the assistance of Interior, Agriculture, Commerce, EPA and the Corps of Engineers, devise a more efficient method of accomplishing the following:

1.   A clear identification of information available, disseminated in a more effective way;

2.   An opportunity for states and regional users to participate in identifying the character of information needed and its form—not in lieu of, but in addition to the Principle Investigator process;

3.   A system whereby local users can obtain consultation with knowledgeable people on a regional, decentralized basis as to what is available and in what shape or form;

4.   Some direct technical assistance to the states and to potential and actual regional users as to how this data can be applied in their planning and problems; and

5.   Some method whereby state and regional people can be trained so that they, in turn, can assist others.

It seems to me that it is imperative that some Governmental department which already has either [a] regional or state offices network be designated [3] as the focal point through which all services can be provided in the initial application and evaluation of ERTS data. I am certain that as more and more data is accumulated and states and local users become familiar with the characteristics of the system, there will evolve better and more efficient channels of transmission and assistance to the local end users.

It may be that the local Geological Survey office of the Department of the Interior might well serve as the initial focal point if it were adequately staffed and its file of ERTS data were maintained on a current basis. I know that they have the expertise to provide the advice and interpretations necessary for the early applications of ERTS data.

Unless something is done along the lines that I have been outlining, we may fail to use this new capability which ERTS is giving us for constructive purposes. Further, we may lose more support for the space program because of the inability to transmit space-gathered information on a timely basis to users in order to help solve specific earth problems. The exchange of information may well lead to improved ERTS instrumentation and techniques.

I propose to talk to the Chairmen of the Interior and Commerce Committees about joining me in obtaining greater cooperation from the other respective Government agencies involved. I do not intend that the initial setup which I am recommending is in any way a precedent setter, but it may buy time for the Federal agencies to evolve a more efficient system of disseminating ERTS information in a form usable by the end user until such time as a final communication system has been developed by the states and agencies involved. One such approach might be the regional setup proposed by the group from Hobbs, New Mexico, with which you are familiar. That plan has great merit, and even NASA staff members agree with it.

In view of the fact that the Congress will adjourn prior to any action on this letter, I would appreciate it very much if you would keep Frank DiLuzio current on your thinking and your reactions to this letter, as he will be in Washington at least through December.

Sincerely yours,

[hand-signed: "Clinton Anderson"]

Document II-26

Document title: Walter C. Shupe, Chief, GAO Liaison Activities, NASA, Memorandum to Distribution, "GAO Report to Congress 'Crop Forecasting by Satellite: Progress and Problems,' B-183184, April 7, 1978," April 21, 1978.

Source: NASA Historical Reference Collection, NASA History Office, NASA Headquarters, Washington, D.C.

In its efforts to promote Landsat, NASA placed much emphasis on improved crop forecasting. Together with the Department of Agriculture and the Weather Service, NASA embarked on a project to demonstrate the system's ability to accurately predict crop yield and quality. The project, titled LACIE (Large Area Crop Inventory Experiment), experienced some success, but in general it failed to meet expectations.

---

[1] National Aeronautics and Space Administration
Washington, D.C. 20546

[rubber stamped: "APR 21 1978"]

LGG-19

MEMORANDUM

TO: Distribution

FROM: LGG-19/Chief, GAO Liaison Activities

SUBJECT: GAO Report to Congress "Crop Forecasting by Satellite: Progress and
           Problems," B-183184, April 7, 1978

This office made advance distribution of the enclosed report via route slip on April 11, 1978. Additional copies were obtained and are now being distributed to those who did not receive one at that time. A copy of the Comptroller General's transmittal letter, dated April 7, 1978, is enclosed, together with a separate copy of the report Digest.

GAO reported that USDA, NASA, and NOAA are currently planning a joint research program that will deemphasize wheat and expand LACIE techniques to other crops and applications, such as early warning of crop damage and crop condition assessment. GAO recommended that the Secretary of Agriculture provide cognizant congressional committees with periodic assessments of the LACIE project, the experimentation with other crops, and the experiments: with early warning of crop damage and crop condition assessment.

Inasmuch as the report recommendation is not addressed to the Administrator, it is not necessary for NASA comments to be provided for the OMB, for GAO, or for Congressional committees, unless requested at a later date.

[hand-signed: "Walter C. Shupe"]

Enclosures

Distribution attached

[2] <u>Distribution</u>
*A-1/Dr. Robert A. Frosch (Attn: Executive Secretary)
*AD-1/Dr. Alan M. Lovelace
ADB-1/Gen. Duward L. Crow
*B-1/Mr. William Lilly
*BR-2/Mr. Jeff Barber
*BR-2/Mr. Dick Midgett
*E-1/Dr. Anthony Calio
*E-1/Mr. Leonard Jaffe
*ER-2/Mr. Pitt Thome
*EPR-3/ Mr. Pinkler
G-1/Mr. Neil Hosenball
*L-1/Mr. Kenneth R. Chapman
*LC-5/Dr. Joseph Allen
*LG-2/Mr. Edward Z. Gray
LGW-2/Mr. Dick Stone
LI-15/Mr. Norman Terrell
N-1/Mr. Ray Kline
*P-1/Dr. John E. Naugle
R-1/Dr. James Kramer (Attn: Mr. R. Nysmith)
S-1/Dr. Noel Hinners (Attn: Mr. C. Wash)
T-1/Mr. Norman Pozinsky (Attn: Mr. R. Stock)

<u>Center</u>
*Dr. C. Kraft, Director, JSC (Attn: Messrs. P. Whitbeck & L. Sullivan)

****

[i]

COMPTROLLER GENERAL'S REPORT TO THE CONGRESS

Crop Forecasting by Satellite: Progress and Problems

<u>DIGEST</u>

A 3-year, three-agency Federal project is developing technology to improve estimates of foreign wheat crops. It is called the Large Area Crop Inventory Experiment (LACIE) and is carried out as follows:
— The National Aeronautics and Space Administration's (NASA's) Earth Resources Technology Satellite (Landsat) provides data for estimating wheat acreage.
— The National Oceanic and Atmospheric Administration provides information to assist in estimating crop yield under various weather conditions.
— The Department of Agriculture—the user of wheat crop estimates—provides historical data and defines requirements for the project.

<u>IMPORTANCE OF WORLD CROP INFORMATION</u>

To date, LACIE has had mixed success in achieving its performance goals. Farmers, importers, exporters, agribusiness companies, Federal and State policymakers, foreign governments, and international organizations use foreign agricultural information. But if more accurate and timely information were available, these parties could better achieve their goals by making improved decisions on planting, fertilizing, harvesting, storing, and exporting. (See p. 5.)

Agriculture initially planned to implement an operational wheat-forecasting system if LACIE technology could produce cost-beneficial, improved estimates. However, this emphasis on wheat has changed, and Agriculture is planning a research program which will define the potential of the LACIE technology for other crops and applications. (See p. 2.)

[ii] The Congress should be kept aware of the results of this program and of the experimentation with other crops and applications.

LACIE PERFORMANCE

The LACIE project is developing new technology and, as to be expected with new technology, has had some technical problems, such as
— difficulty in distinguishing spring wheat from other grains (see p. 12),
— slow progress in developing methods for machine classification of wheat areas to reduce the need for heavy manual involvement in identifying wheat-growing areas (see p. 13), and
— using current yield models which use highly aggregated weather inputs that are not fully, responsive to weather changes occurring for short periods over localized areas. (See p. 14.)

LACIE performance needs to be improved to meet its goals of 90-percent accurate production estimates, 9 out of 10 years. In the most important test country, the Soviet Union, the LACIE production estimate was close to the official estimate; however, this resulted from offsetting errors; i.e., the wheat area estimate was high by over 12 percent, and the wheat yield estimate was low by nearly 15 percent. (See p. 8.)

LACIE COSTS

LACIE and related efforts planned through fiscal year 1978 will cost about $54 million, not including NASA personnel costs. The total costs of the follow-on research program involving the three agencies have not been determined. However, Agriculture is investing a substantial amount of funds in computer equipment, and in programs and related items to establish a facility near the Johnson Space Center, where much of the research will be carried out. (See pp. 16 and 17.)

[iii] CONTINUING EFFORTS

Agriculture planned to implement the Application Test System—designed to test LACIE wheat-estimating techniques in an operational environment. It has, however, decided to extend experimentation to other crops and applications, such as early warning of crop damage and crop condition assessment. It will also defer a Landsat-based wheat information system until further experimentation and evaluation is completed. (See p. 17.) Project plans in 1974 called for the performance of a cost/benefit analysis to evaluate the usefulness and cost-effectiveness of a LACIE-type system in providing foreign crop information. The analysis will assess benefits based on expected improvements in timeliness and accuracy of information from a LACIE-type system for forecasting wheat. Accordingly, the reasonableness of the benefits set forth should be carefully examined if the analysis is used in deciding whether a crop-forecasting system based on LACIE technology should be carried out. The analysis will be carried out in 1978 but will not be completed by the end of the LACIE project in July 1978. (See p. 18.)

RECOMMENDATION TO THE SECRETARY OF AGRICULTURE

Since there have been technical problems in reaching LACIE objectives and the research direction has changed, GAO recommends that the Secretary of Agriculture provide cognizant congressional committees with periodic assessments of the LACIE project, the experimentation with other crops, and the experiments with early warning of crop damage and crop condition assessment. (See p. 21.)

AGENCY COMMENTS

The issues in this report have been discussed with LACIE officials in the three participating agencies, and their comments have been incorporated as appropriate. NASA believes that LACIE area and yield estimates for the Soviet Union should not be compared to the Soviet's figures for area and yield because [iv] the latter are suspect. However, the LACIE project makes this comparison, and Agriculture reports the figures in its regular periodic reports. . . .

Document II-27

Document title: Charles J. Robinove, Director, EROS Program Office, Geological Survey, U.S. Department of the Interior, Memorandum to Staff of the EROS Program, "Optimism vs. pessimism *or* where do we go from here? (some personal views)," December 10, 1975.

Source: NASA Historical Reference Collection, NASA History Office, NASA Headquarters, Washington, D.C.

In bargaining with the Bureau of the Budget to gain approval for an Earth resource survey satellite system, the Department of the Interior was forced to accept an initial provisional program rather than the fully operational program it sought. To become fully operational, the program would have to prove its utility. As time passed and progress was slower than anticipated, this bargain hung like Damacles' sword over the program. In an effort to boost severely sagging morale in the EROS program office, Charles J. Robinove prepared this memorandum.

[1]   December 10, 1975

Memorandum

To: Staff of the EROS Program

From: Charles J. Robinove

Subject: Optimism vs. pessimism *or* where do we go from here?
             (some personal views)

In the past months the staff of the EROS Program has shown increasing pessimism and discontent regarding the future of the EROS Program, the thrust of research and operations, programmatic and administrative problems, and the place of the scientific staff. All of us, from top to bottom, have at times been discouraged in our individual progress and in the progress of the program—this is only natural in a growing and evolving area of science within the Federal bureaucracy.

Certainly many things have progressed more slowly than any of us would like. In the ten-year span since the EROS Program was announced, only two satellites have been launched, no commitment has been made to an operational system, organizational progress has seemed to be stifled, and budgets have never been large enough to do the things that we have all felt need to be done.

Pessimism in the face of slow progress and setbacks is a natural human condition, but why is it so prevalent now? Probably it is because all of us believe in the EROS Program as a soundly based scientific and technical framework that is good for the United States and the world, remote sensing is a scientifically exciting field to work in, our contacts with others in a vast area of scientific and resource management disciplines is a broadening and continuing learning experience, and through our efforts we can build a sound and respected scientific reputation both individually and for the program.

Let us look at the progress that has been made. Landsats 1 and 2 are operating at performance levels higher than expected, Landsat-C is planned for launch in 1977, and Landsat-D is being planned. Other more specific missions such as Magnetometer Satellite (Magsat) and Heat Capacity Mapping Mission (HCMM) are being planned.

[2] Users of remote sensing data are increasing rapidly, and applications of data to resource and environmental problems are proving to be practical and beneficial. The EROS Data Center is an operating reality with increasing numbers of customers for data and training. The individual freedom of the staff to do research in both basic and applied remote sensing is a privilege envied by many outside the Survey. But in spite of these positive factors, pessimism seems greater than optimism.

Scientific progress is like individual freedom—both must be fought for day by day. Neither can be taken for granted. Regardless of the quality of our work or the necessity for such work to benefit the country and the world, there will always be those who say our endeavors are of little value, that our concepts won't work, and who battle against us in private and in public, either to eliminate the EROS Program entirely or to reduce it to an unworkable level.

Are these reasons for pessimism? No—they are reasons to increase our efforts and to convince our detractors, both by our results and our statements, that what we do is of value. This must be done continually. It is not enough to know we are making progress and helping others. We must show it and tell it and be proud of it.

In accomplishing the things that we have done, we have solved many problems—scientific, administrative, and political; we have built a constructive and useful phase of science and technology; and we have overcome many obstacles in doing so. There is every reason to continue to do so and every ability available to continue and increase our progress. In the context of pride in our purpose and our accomplishments, we can put up with enemies, misunderstandings, and continual nitpicking in small matters.

Optimism derives from our purpose and our accomplishments; pessimism derives only from frustration in reaching our goals. I believe we can overcome our pessimism by looking at our purpose and our goals, recognizing the hard and often frustrating scientific, administrative, and political roadblocks in our way, and trying, day by day, to overcome them.

If we remain firmly committed to our broad goals and confident of our ability, then optimism will aid us in achieving our goals and in overcoming the pessimism that can lead us to defeat ourselves.

[hand-signed: "Charles J. Robinove"]

Document II-28

Document title: James C. Fletcher, Administrator, NASA, to Mr. John C. Sawhill, Associate
Director, Office of Management and Budget, October 19, 1973.

Source: NASA Historical Reference Collection, NASA History Office, NASA
Headquarters, Washington, D.C.

*The Bureau of the Budget was given expanded responsibilities and renamed the Office of
Management and Budget (OMB) in 1971. This name change did not result in a more welcome attitude toward the ERTS program. The second satellite, ERTS-B, was originally scheduled for launch
in 1976. Fearing that a failure in ERTS-A might cause a potentially damaging gap in data, NASA
requested permission to advance the launch date by two years. OMB resisted, believing that an earlier launch might imply a premature decision to declare the program operational and thereby increase
funding. In this letter to OMB Associate Director John C. Sawhill, NASA Administrator James
Fletcher presents NASA's case.*

---

[1]   NATIONAL AERONAUTICS AND SPACE ADMINISTRATION
Washington, D.C. 20545

OCT 19 1973

Office of the Administrator

Mr. John C. Sawhill
Associate Director
Office of Management and Budget
Executive Office of the President
Washington, D.C. 20503

Dear John:

I am writing to clarify and explain in greater detail the reasons to accelerate the
launch of ERTS-B from 1976 to 1974, as communicated in my letter to Roy Ash of
September 5, 1973.

First, let me address three points on which discussions with your staff indicate there
may have been some misunderstanding.

1. The acceleration of ERTS-B is required for continued experimental work and
does not depend on approval of the interagency pilot project I suggested to
George Shultz in my letter of September 5, 1973. Even if it is decided not to
undertake such a global agricultural experiment, the earlier launch of ERTS-B is
required to continue current experimentation with the use of the ERTS data systems for applications in agriculture and other fields to provide a better base of
experience for the planning and design of possible future operational systems.

2. The launch of ERTS-B in 1974 does not require or necessarily imply an earlier
decision to commit to an operational ERS satellite system than envisaged with the
previously approved 1976 ERTS-B launch date, nor does it require or necessarily
imply a decision to launch another experimental satellite in 1976 to maintain continuity of data collection between the accelerated ERTS-B and an eventual operational satellite system. NASA and the user agencies generally [2] believe that an
earlier commitment to an operational satellite system could be made, and I have

recommended that continuity of data collection be maintained with another experimental satellite (ERTS-C) until an operational satellite system becomes available. However, even if neither an operational system nor ERTS-C is approved at this time, ERTS-B should nevertheless be launched in 1974 to extend the experimental operations conducted with ERTS-1. If there has to be a gap, it would be preferable to get a stronger experimental base sooner and accept a hiatus between a more meaningful experimental phase and the operational system, rather than introduce an unnecessary hiatus into the middle of the experimental phase.

3. The acceleration of ERTS-B, even if combined with a decision to proceed with ERTS-C for launch in 1976, as I have separately recommended, does not have budgetary implications that go beyond previously planned committed levels for space research and technology. The additional cost of ERTS-C can easily be accommodated within a total NASA budget well below the planning level of $3.2–3.4 billion (1971 dollars) agreed to in January 1972, as shown in my FY 1975 budget letter of September 28, 1973.

Now let me explain more fully the reasons for shifting the launch date of ERTS B to 1974, independent of the related considerations discussed above.

R&D Strategy

The research and development strategy for the ERTS program, for which NASA and the associated technical agencies are responsible, must be based on recognition of the fact that a multipurpose development program like ERTS cannot be treated like a classical hardware systems development with clearly identifiable decision points for separate phases of research, development, test, evaluation, and operations. We are dealing with a system and many potential applications [3] in which the experimental phase has to include (in different ways for each application) a whole variety of activities, including among others:

- Development and test of sensor performance capabilities.
- Development and test of satellite hardware capabilities.
- Development and test of data collection and relay capabilities.
- Development and test of ground data processing to provide data in useful form for each of many applications.
- Development of data analysis aids and methodologies to convert data into useful information.
- Test of information utility in each application for management and decision-making.

Two key points must be stressed: First, the satellite data requirements for experimentation necessary to achieve an adequate basis for evaluation are different for each class of potential application; and second, some of the most important potential applications require a data base of repetitive coverage extending over several years and/or quasi-operational testing using current, near-real-time data.

There are two major areas of ERTS applications that clearly appear to have great near-term potential value and which illustrate the importance of the continuity of data an ERTS-B 1974 launch would provide. These are the areas of vegetation and water boundary discrimination. It has been demonstrated, for example, that ERTS data can be used to identity and separate different crops, monitor crop vigor, end measure crop acreage. To be useful in the development of systematic crop yield [4] predictions, these techniques need to be applied to current growing seasons that are influenced by current climatology. The same holds true for measurement of forest stress under insect attack and for exact delineation of coastal wetlands by grass species discrimination: the dynamic nature of the problem requires current data for current use. For hydrology applications like flood mapping, seasonal lake and pond assessments, water sedimentation and pollution measurement, and sea ice monitoring, near-real-time data is even more critical, since the dynamic phenomena themselves are short-lived and the period for action in response to observation is also short. An early ERTS-B launch will provide assurance that the immediate values in these application areas can be realized.

These two examples illustrate how the first phase of ERTS investigations has been spent discovering and understanding the full extent of the informational content of space data; the next phase now needs to focus on making information extraction routine and on learning how the new information can be integrated into management and decision processes. This next phase requires experimental support and continuity in the flow of ERTS data. These are the considerations that led the Interagency Coordination Committee for the Earth Resources Survey Program in its annual report to your office submitted on October 4 to identify the acceleration of ERTS-B to 1974 as the best step in the overall transition from R&D to operations.

A first-order analysis of the tradeoffs between the two launch dates shows the following:

* The earlier launch maintains the momentum developed in the investigator and user communities, a momentum that would have to be reestablished at a cost in money and performance if a serious gap in data flow were permitted.
[5] * The earlier launch is likely to provide earlier definitive experience among many diverse users and applications upon which decisions as to operational systems timing, configuration, and benefits can be reached.
* The earlier launch of ERTS-3 means that the fifth (10–14 micrometer) MSS channel would not be carried until a later time; the development schedule for this thermal IR channel would not be affected and the instrument could be flown in 1976 if a suitable spacecraft is available. A two-year discontinuity in ERTS data availability is considered more serious than the potential delay in the flight test of the additional experimental channel.
* The earlier launch date provides the possibility of near-continuous coverage following ERTS-1 which has exceeded its design life and can be expected to fail at any time. A later gap in data flow following ERTS-B in the period 1975–76 could be filled with a follow-on experimental or operational system as I have already recommended; however, this is a matter for a separate decision which can be made on its merits in the FY 1975 budget process, independently of the decision on the ERTS-B launch date.
* The earlier launch results in lower FY 1975 and total program costs.

I would like to reiterate a basic point: the objective of the ERTS program, to identify whether an operational approach to ERS is warranted, is earliest and most economically served by accelerating ERTS-B.

[6] Congressional Direction

The four Committees charged with the NASA authorization and appropriation have clearly and unambiguously gone on record on the ERTS-B matter. The FY 1974 authorization act includes $8 million specifically targeted for this purpose. Quotes are: "The [House] Committee, however, places the highest priority on the ERTS project." ". . . the [House] Committee believes that ERTS-B should be prepared for launch as soon a practicable." ". . . the [Senate] Committee concurs with the House in adding $7 million to the Space Applications Program to bring ERTS B into a ready status for launch in its present configuration . . . ." ". . . the [House] Committee urges NASA to reprogram the necessary funds to launch ERTS-B as early as possible." [bracketed material in the original]

The basis for Congressional interest is not political in the usual sense. Data from ERTS-1 is beginning to be used extensively by many different State agencies, and a serious gap in data availability would cause these States to abandon their analysis teams—many of them before they have had the opportunity of working with real-time repetitive data. Perhaps the States were premature in building up their capabilities in response to an experimental satellite, but, on the other hand, it seems that ERTS-1 has more values of an operational character then [sic] had been anticipated.

In addition, I have been personally informed by a number of both House and Senate members from both sides of the aisle that failure to accelerate ERTS-B would be considered a serious disregard of Congressional intent and would result in political and programmatic repercussions it would be in the best interests of the Administration to avoid. While the appropriation bill was still under review, it has been possible to point to that fact as a limitation on NASA's programming flexibilities; now that it has passed, I am immediately accountable to the Committees on the matter of the FY 1974 program content.

[7] International Considerations

International interest, on the part of US agencies and firms, as well as numerous foreign governments, demonstrates a wide-spread positive evaluation of the ERTS program and a very strong concern for its continuity. The following are examples:

- President McNamara of the World Bank looks forward to an early launching of ERTS-B, has established a special Bank office for applications of space surveys, and has requested the Bank to be designated a special (user) agency.
- AID is utilizing ERTS data in specific projects of technical assistance and has requested continuity of coverage for some of these.
- The Department of State's Bureau of African Affairs has given special attention to ERTS data for the drought-stricken Sahelian area and there has been a positive response, specific to the ERTS program, from the Governments of Niger and Mali.
- Canada, Brazil, and Italy have committed to substantial investments in ERTS ground stations and data processing facilities, implying a clear expectation of the continuity of ERTS-type data from whatever source.
- Brazil convoked a meeting in May of this year in which some eleven countries expressed their interest in the continuity of ERTS data.
- More than a dozen countries have inquired regarding the establishment of ERTS stations.
- The Bendix company has informed us that US commercial advantage in overseas sales of equipment for ERTS data reception and analysis depends heavily upon an early start for ERTS-B.

[8] I can provide you further data on the international aspects if you desire. However, my own assessment is that the US stands to gain much by the earlier launch of ERTS-B and stands to lose, in both the near and the long term, if this launch is deferred for reasons which ignore US international policy interests.

In conclusion, I have decided to request apportionment of FY 1974 funds on the basis of a 1974 ERTS-B launch. Time is already becoming critical if we are to assure the smoothest and most economical contractual and schedule changes.

The separate question of whether to dedicate some parts of the ERTS-B capabilities to a special pilot project should be decided on its own merits in the context of national needs and national policy; it should be noted that the pilot project would not interfere with domestic data acquisition or with use of the satellite by those countries investing in ground stations. I am confident that this proposal is inherently sound and I urge your assistance in helping assure that it receives appropriate consideration within the Administration.

Sincerely,

James C. Fletcher
Administrator

bcc: A/Dr. Fletcher
          AD/Dr. Low
          AA/Dr. Newell   Prepared by: AAA:DWJr:djs:10/19/73
          ADA/Mr. Shapley
          B/Mr. Lilly
          E/Mr. Mathews
          AAA/Mr. Williamson
          AXM-3/Files

Document II-29

**Document title: Christopher C. Kraft, Jr., Director, Johnson Space Center, to Associate
Administrator for Applications, NASA Headquarters, "Private Sector Operation of
Landsat Satellites," March 12, 1976.**

**Source: NASA Historical Reference Collection, NASA History Office, NASA
Headquarters, Washington, D.C.**

*In January 1975, NASA renamed the ERTS program Landsat. By 1976, with two Landsat satellites
in orbit, another scheduled for launch, and a fourth under construction, many people associated with
the Earth resources/Survey program felt the experimental phase had continued long enough, and it
was time at least to begin to discuss where the program's ultimate home should be. Many thought it
would be best to transfer the entire program to private industry. Others, including Christopher Kraft,
Director of NASA's Johnson Space Center, argued that it was premature to decide on an operational
structure for Landsat, given the character of the market for its products to date.*

_____

[rubber stamped: "MAR 12 1976"]

TO: NASA Headquarters
           Attn: E/Associate Administrator for Applications

FROM: AA/Director

SUBJECT: Private Sector Operation of Landsat Satellites

The following comments are provided in response to your request at the recent APIB
Meeting regarding possible non-government operation of the Landsat system.

I have no doubt that private industry will eventually play a major role in the operational earth resources survey system. I believe this will make for a healthy situation and
should be retained as a future option. However, just what that role will be is not at all clear
at this time.

It appears likely that for the next 5–10 years the Federal Government will be the major
user of Landsat-type data. Although many others use the data for a variety of purposes,
they probably would not be willing at this time to contribute significantly to supporting
an operational Landsat system. Possible roles for private industry would be to build and
operate Landsat as a commercial venture or under a long-term lease back to the Federal
Government.

Considering the high risk attendant to the currently uncertain commercial market, it
is doubtful that industry would be willing to underwrite the Landsat development and
operation. The prospects for negotiating a mutually acceptable long-term lease arrangement—considering the budgetary process—are also highly uncertain at this time.

Therefore, I would suggest that no changes be made to present plans for operation of Landsat C and D. As operational data use develops in the next few years, this climate may change and a broader-based demand for Landsat-type data may emerge. The option for private sector operation of the Landsat system should be reassessed at that time.

Christopher C. Kraft, Jr.

HD/JFMitchell:nmm2/27/76:3751

cc:
BD/E. B. Stewart
HB/O. Smistad
HC/R. A. Hoke

Document II-30

Document title: Bruno Augenstein, Willis H. Shapley, and Eugene B. Skolnikoff, "Earth Information From Space by Remote Sensing," report prepared for Dr. Frank Press, Director, Office of Science and Technology Policy, June 2, 1978, pp. ii–iv, 1–14.

Source: NASA Historical Reference Collection, NASA History Office, NASA Headquarters, Washington, D.C.

When this report came out in June 1978, the United States had been engaged in an "experimental" Earth resources observation program since 1966. Although by 1978 the program's ultimate viability was still not clear, most involved felt it was time to make a policy decision regarding whether or not the program should be canceled, declared operational, or maintained on an experimental basis until further results made a choice more obvious. Among the issues facing decision makers was deciding which government agency should take the lead, if indeed an operational system was desired. This report from the Office of Science and Technology Policy of the Executive Office of the President examines the issues at hand and recommends NASA as the appropriate lead agency. The three authors of the report were consultants to Presidential Science Advisor Frank Press. Bruno Augenstein was a RAND Corporation scientist with early involvement in Earth observation systems. Willis Shapley was a long-time top staff official at the Bureau of the Budget and NASA. Eugene Skolnikoff was a political science professor who specialized in science, technology, and foreign policy issues at the Massachusetts Institute of Technology. Included here are the executive summary and first two chapters of their report.

Earth Information From Space by Remote Sensing

by

Bruno Augenstein
Willis H. Shapley
Eugene B. Skolnikoff

June 2, 1978 . . .

[ii]   EXECUTIVE SUMMARY

1.   The aim of the consultants in this report has been to develop policy recommendations and options for U.S. civil remote sensing activities.

2. There is an urgent need for a clarification of policies on the future evolution of U.S. activities in civil remote sensing. Executive branch policy has supported technological R&D and some experimental applications, but has up to now deferred as premature commitments to operational uses and decisions on policies for operational systems. Congressional committees have urged that at a minimum the Executive Branch prepare clear policies to guide future civil uses of remote sensing from space.

3. The report addresses first the basic question of the U.S. policy attitude toward civil remote sensing from space, and concludes that U.S. policies should be based on acceptance of the proposition that the U.S. should and will continue to be actively involved in civil remote sensing from space for the indefinite future. This conclusion is based on the manifold U.S. interests served by civil remote sensing, which include a wide variety of technical, public, economic, and international interests. The consultants recognize that there have been exaggerated claims for benefits and the times by which they can be achieved, but are nevertheless convinced that the many potential values fully justify a continuing U.S. effort to achieve them.

4. One important consequence of this conclusion is that the U.S. should, regardless of the program level approved, make a policy commitment to data continuity. This is essential to reduce present uncertainties among prospective users and to help forestall the growing threat of foreign competition.

5. Policies on U.S. civil remote sensing must recognize the dynamic nature of the technologies involved and support strong continuing research and development efforts. Operational systems, however, should take advantage of existing and low cost technologies when they are adequate to meet the needs of the users.

6. A top priority should be given to the preparation and periodic updating of a comprehensive plan, covering the expected technical, programmatic, and institutional evolution of U.S. civil remote sensing for 10 to 15 years in the future.

7. An equally high priority should be given to the designation of a lead agency. This is needed now to develop the initial version of the comprehensive plan to guide preparation of the FY 1980 budget and the accompanying legislative proposals. For the long term, it [iii] is essential to have a qualified single agency with authority and responsibility for leadership and management of U.S. activities in the collection and dissemination of civil remote sensing data and for serving as the interface with the private sector and international interests in civil remote sensing. Federal mission agencies would continue to be responsible for their user interests in remote sensing information.

8. After considering the criteria that should be applied, the consultants have concluded that the lead agency responsibility should be assigned to NASA, provided NASA is reconfigured to ensure that a user-oriented service outlook is given an equal footing with NASA's important present R&D missions. Lower ranked alternatives would be, first, the National Oceanic and Atmospheric Administration (NOAA) in the Department of Commerce, and second, the Department of the Interior.

9. The report notes that U.S. civil remote sensing activities unavoidably have significant international aspects, because of the global nature of remote sensing from space and because of the strong existing foreign and international interests in the technology and its uses. The consultants believe that there are many potential values to the U.S. (e.g., in dealing with lesser developed countries problems) in actively seeking constructive international involvement. The preferred institutional alternatives appear to be a U.S. owned and operated system serving international users or the establishment of an international consortium in which the U.S. and other nations would participate.

10. The consultants propose a two-tiered concept for the U.S. National System for civil remote sensing. This would include (a) a core segment consisting of the assets and services required for a multipurpose open access system which would serve international

as well as domestic users and which might become the nucleus of an international system if one is established, and (b) a segment composed of specific national means, which may be proprietary, for adding to and enhancing data and knowledge useful to interests of the U.S. This approach permits flexibility for international arrangements while assuring the availability of data services needed by U.S. interests.

11. There is a variety of modes in which the U.S. private sector can participate, although such participation, especially as it relates to data collection, will generally be under the supervision and authority of the U.S. Government. Private sector roles may include major systems management functions, systems engineering, or delegated operational functions, in addition to normal contractor roles in developing and producing system hardware and software.

[iv] 12. The final chapter of the report gives four alternative scenarios which indicate the range of options for the continuing involvement of the U.S. in civil remote sensing within the terms of the policy conclusions of the report. The scenarios differ in the level of U.S. commitment and the degree of international involvement sought. Scenario "A" is a "minimum" scenario, providing some degree of institutionalization of current activities and a policy commitment to data continuity. Scenario "B" provides for some significant commitments to a quasi-operational or operational system in the mode of a tiered U.S. national system permitting international participation in the core segment. Scenario "B" is the same as "A" except that it includes a U.S. invitation to other countries to participate in an International System. Scenario "C," a "maximum" scenario, goes beyond "B" to include active U.S. efforts to bring an international system into being.

13. Finally, the consultants recommend approval of the conclusions and recommendations of the report and the selection of Scenario "B" as the general guides for the future conduct of U.S. civil remote sensing activities. They further note the importance of announcing this approval no later than the summer of 1978 to permit the lead agency to prepare plans in time for incorporation in the FY 1980 budget decisions and the accompanying legislative program to be submitted to the Congress.

[1]   CHAPTER I - INTRODUCTION

A. Aim and Scope

1.   The aim of the consultants in preparing this report has been to develop policy recommendations and options to guide decisions on United States policy with respect to civil remote sensing from space. The topics addressed include the future involvement of the U.S. in civil remote sensing, institutional arrangements in the Federal Government, international and private sector participation, and a proposed concept for a U.S. national system compatible with several policy options.

2.   The report is intended to apply to civil remote sensing systems using current and future sensors of the Landsat type and other types of systems for obtaining information on or from the earth from space. National security interests are taken into account but not discussed; the policies, options, and recommendations in the report are consistent with current statements of applicable National Security Policy. The policies, options, and recommendations in the report are also compatible with current policies and arrangements regarding meteorological satellites; questions of possible changes in these arrangements are not addressed.

B. Background

1.   Official central policy in the Executive Branch has been somewhat ambivalent on civil remote sensing from space. There has been support for technological R&D and some experimental applications. But operational uses have been under a cloud—decisions on "operational systems" and even on policies for a transition to an operational mode have been regarded, up to now, as premature. On the Congressional side, however, and in other sectors there has been support for an early transition to an operational mode and strong demands that at a minimum the Executive Branch propose clear policies to guide the future civil uses of remote sensing from space.

2.   There now appears to be a consensus that the time has come to formulate and adopt policies to govern future U.S. involvement in civil remote sensing from space. Actions to reformulate overall national policy in space are underway. The Director of the Office of Science and Technology Policy has agreed to present an administration policy on civil remote sensing to the Congress later this year. The consultants' report is intended as one of the steps leading to decisions on what these policies should be.

3.   Some of the concerns which underline the urgency of the need for policy decisions on civil remote sensing are the following:

[2] a.   The absence of a U.S. policy on future directions and institutional arrangements has created an atmosphere of uncertainty and frustration. The U.S. agencies concerned and other prospective users, domestic and foreign, are finding it difficult to make sensible decisions on their future plans and commitments. No single agency has had authority and responsibility for leadership in developing national policy on an overall plan.

b.   Lead-times are running out. Decision-making on the systems and services to follow Landsat D should start this year; significant opportunities and options may otherwise be lost.

c.   Foreign competition is becoming a real threat. U.S. indecision in the face of strong, technologically advanced efforts by Japan, France, or the European Space Agency could result in loss of U.S. leadership in civil remote sensing and the opportunities that would go with it.

d.   Existing U.S. policies are now having some counterproductive effects. The limitation of civil remote sensing from space to R&D and experimentation, combined with the practice of incremental program decisions without a long-term plan for operational services, is calling into question the future of U.S. involvement in civil remote sensing. This is adding to the uncertainties, discouraging prospective users, and encouraging foreign competition. Decisions on uses beyond R&D are no longer premature; they are timely, even overdue.

e.   Finally, there is a widespread feeling that the time has come to make a stronger effort to realize benefits of value from civil remote sensing.

C. Definitions and Distinctions

1.   In this report, we will be dealing primarily with "Civil Remote Sensing from Space" activities and systems, by which we mean the totality of activities and systems required for the collection, production and initial dissemination of data on the earth obtained from space by civil remote sensing systems. These activities and systems are generally regarded as having a space segment and a ground segment. The space segment consists of the production activities and systems required for the collection, initial processing, and delivery of earth data from space. The ground segment consists of dissemination activities and systems required for making the processed data available in appropriate formats to users and for maintaining archives of remote sensing data.

2. To a lesser extent, the report will refer to the <u>analysis</u> activities or systems required by users for the analysis, enhancement, or display of remote sensing data and its consolidation with data from other sources. These are the activities and systems that convert processed remote sensing <u>data</u>—the output of the production and dissemination systems—into <u>information,</u> the product that is used by users and beneficiaries.

[3] 3. Each of the above types of systems (production, dissemination, analysis) may be operated in one of three modes:

a. <u>Experimental,</u> for R&D, test, or demonstration purposes (beneficial operational uses may also be made of data from such systems).

b. <u>Operational,</u> for the provision of services on a continuing basis in accordance with a stated plan.

c. <u>Quasi-operational,</u> for the combined or simultaneous conduct of experimentation and provision of operational services.

4. The availability of the data output of remote sensing production systems may be:

a. <u>Open access,</u> when all interested users are given access to the data on an equitable and non-discriminatory basis.

b. <u>Limited access,</u> when access to the data may be limited by the system operator on a national or proprietary basis.

[4]   CHAPTER II - U.S. INVOLVEMENT IN CIVIL
                       REMOTE SENSING FROM SPACE

A. <u>The Basic Policy Questions</u>

1. In this chapter, we address the basic question of the U.S. policy attitude toward a continuing future U.S. involvement in civil remote sensing from space. Policies and options on the conduct of civil remote sensing—to be discussed in succeeding chapters—are heavily dependent on the intentions of the U.S. with regard to a continuing future involvement, the reasons for such involvement, and expectations regarding benefits and other consequences.

2. The fundamental question is whether U.S. policy should be based on the premise that the U.S. will continue to be actively involved in civil remote sensing from space, in some mode and at a scale to be determined, for the indefinite future. If the basic policy attitude is that there are expected to be significant benefits, advantages, and opportunities of value to the U.S. from such involvement, then institutional and other policies should be directed at the best way of realizing these values. If, as a second alternative, the policy attitude is that this is a technology of uncertain value to U.S. interests that should remain indefinitely in an experimental mode, policies should nevertheless still be directed at achieving the maximum values for U.S. interests from the continuing experimental program. Finally, if the policy attitude is that the expected values do not justify recognition of a long-term U.S. involvement in civil remote sensing, then the implications for all U.S. interests should be squarely faced.

3. To arrive at the appropriate policy attitude toward civil remote sensing at this time and the implications of the alternatives, we will, in the following sections, (a) review the wide variety of U.S. interests that may be served by civil remote sensing from space, (b) give the consultants' views on the considerations that should determine the U.S. policy attitude, (c) state the conclusions on U.S. policy attitudes reached by the consultants, and (d) indicate the consequences of these conclusions which will be discussed in the remainder of the report.

B.   The Manifold U.S. Interests in Civil Remote Sensing

1.   The U.S. interests that can be served by civil remote sensing from space are of many different kinds. We will group them under the headings of technical, public, economic, and international interests.

2.   Technical Interests
   a.   Basic scientific knowledge and understanding. Remote sensing from space provides a means to learn about the earth on a global or regional basis, update such information, and provide data for basic research on various aspects and features of the earth. Current administration policy properly stresses the national importance of basic research. Remote sensing from space is a unique tool for a wide range of scientific research interests.
[5] b.   Global capacities. Remote sensing from space has unique capabilities to provide information needed to deal on a global basis with national and international problems, such as food production, energy and mineral resources, water availability, and others, especially in relation to developing countries.
   c.   Support for U.S. decisionmaking. Remote sensing from space can provide information and data useful for U.S. decisions, such as current information on crop production in major agricultural countries.
   d.   Technology development. Continuing R&D in remote sensing systems can provide a vehicle for achieving technological advances of significance in other fields as well as in remote sensing.

3.   Public Interests
   a.   Specific Federal needs and functions. Earth information derived by remote sensing from space can make significant contributions to the needs and functions of Federal agencies and programs in many different areas, such as weather, crops, climate, geological resources, topographic mapping, land use, and environmental monitoring. Significant contributions by remote sensing information systems have been demonstrated or are clearly foreseen in these and other fields.
   b.   Public interest needs and benefits. The broader public interests of the United States, as distinguished from specific Federal programmatic interests, are served by making available earth information from space to States, localities, universities, and the public at large, in usable form at a reasonable cost. State, regional, and local authorities and universities and other public interest groups can use such information in a variety of public interest functions, such as land use planning, environmental monitoring, demographic studies, etc.

4.   Economic Interests
   a.   Economic interests of U.S. private sector. The availability of earth information derived by remote sensing from space can be used to the overall economic advantage of the United States by private enterprise in a variety of fields, oil and mineral exploration being two outstanding examples.
   b.   U.S. competitive position in space technology. With the emergence in Europe and Japan of strong competitive capabilities and interests in space technology, including remote sensing systems, it is clearly in the U.S. economic (as well as political) interest to maintain a leadership position in civil remote sensing. Competitive areas include the manufacturing and servicing of satellites, sensors, and ground equipment; the dissemination of data; and the provision of technical services and assistance in the analysis, enhancement, [6] and interpretation of remote sensing data. In each of these fields, U.S. industry and private companies can undoubtedly compete successfully if the U.S. Government maintains a continuing active involvement in civil remote sensing.

    c.   <u>Reduce some U.S. information costs.</u> Civil remote sensing systems in a routine operational mode may result in reduced data collection costs for some U.S. government, State and local, and private information needs. Such possibilities can be meaningfully explored only if U.S. policy calls for continuing active U.S. involvement in civil remote sensing.

    d.   <u>Contribute to general economic growth.</u> The availability of civil remote sensing data and its appropriate use in resource planning and other fields can be expected to contribute to economic growth, especially in the less developed countries, which should have a positive effect on general economic growth.

    e.   <u>Return on space investment.</u> Civil remote sensing is an area in which there are opportunities for realizing returns, in the form of economic and other kinds of benefits, on the very large national investment in space of the past two decades, for relatively very small additional investments.

5.   <u>U.S. International Interests</u>

    a.   <u>Support of U.S. foreign policy.</u> U.S. involvement in civil remote sensing from space can provide opportunities and a vehicle for the support of many U.S. foreign policy objectives. The strong interest of many foreign countries in earth information derived by remote sensing, combined with the current U.S. leadership role in providing such information and related services, make this an area of positive potential in U.S. foreign relations.

    b.   <u>Maintain U.S. leadership.</u> Civil remote sensing is an area in which the U.S. has developed and demonstrated a benign technology of obvious potential benefit. Active continuing U.S. involvement—exploiting a dramatic space capability clearly developed by the U.S.—would help maintain the fact and image of U.S. leadership in space. Conversely, if other nations take over leadership in this field, the overall leadership position of the U.S. would be impaired.

    c.   <u>Support U.S. position with LDC's.</u> U.S. civil remote sensing activities can provide a constructive way to support U.S. commitments of technology to assist less developed countries. They can buttress the U.S. position in North/South dialogues, in the U.N. and other forums. For example, initiatives involving civil remote sensing might provide a constructive opportunity for U.S. leadership at the U.N. 1979 Conference on Technology for Developing Countries.

[7] d.   <u>Support international cooperation in space.</u> U.S. commitments to service foreign Landsat ground stations, while contractually limited to support from U.S. experimental satellite systems, have resulted in foreign investments and expectations based on the assumption that U.S. Landsat or other generally compatible satellite data will continue to be available. It is generally in the U.S. interest to fulfill rather than disappoint these expectations.

    e.   <u>Promote openness.</u> U.S. civil remote sensing policies and activities can be used to support the general objective of treating open information as an international good, and contribute to the development of international law to that end. They can also continue to support the U.S. policy to preserve without limitation the legitimacy of remote sensing from space, especially in the context of an active U.S. effort to permit and help other countries share in the benefits.

    f.   <u>General international cooperation.</u> The development of international working arrangements in the field of remote sensing from space can provide the U.S. with useful opportunities to encourage international cooperative arrangements to deal with broader international concerns. It also offers opportunities to innovate in building international institutions to develop models of effective working bodies that also meet requirements of participation and equity.

C.   The Guiding Considerations

1.   The consultants believe that the considerations presented below should guide and do
in fact largely determine the basic policy attitude that should be taken toward a continuing U.S. involvement in civil remote sensing.
2.   The Basis for Assessment
    a.   The assessment of the value of the contributions of a continuing U.S. involvement
       in civil remote sensing to each of the 19 U.S. interests listed above is necessarily a
       matter of judgment. Few cases at most are susceptible to quantitative or other pre-
       cise analysis; even in such cases the assessments also depend on judgments regard-
       ing objectives, criteria, and future eventualities. An overall assessment of the
       combined value of remote sensing to all 19 U.S. interests is all the more a matter
       of judgment, a judgment that must take account of the impact of continuing or
       not continuing U.S. involvement in civil remote sensing on each of the different
       U.S. interests identified.
    b.   The final assessment would be simplified if there had already emerged a dramat-
       ic single beneficial use of civil remote sensing which provided an overriding jus-
       tification for an operational system that was clear to all concerned (or if there
       were a clear overriding reason for discontinuing U.S. civil remote sensing
       [8] activities). Given the present situation, however, in which there is room for
       disagreement on the values and benefits that can be achieved, it is important to
       recognize that remote sensing systems from space are generally multipurpose in
       nature; in fact, their utility arises in good measure from the wide variety of uses to
       which the data and information they provide may be put. The final assessment
       has to take the whole range of prospective uses into account.
3.   A Positive Assessment
    a.   Taking into account the manifold U.S. interests and the variety of beneficial uses
       that can be served, the consultants believe that there is ample justification for a
       continuing active U.S. involvement in civil remote sensing from space. The poten-
       tial technical public interest, and economic and international benefits and oppor-
       tunities, while not provable or even fully definable in advance, now seem clearly
       to justify the continuing federal investment and operating costs likely to be
       required (see paragraph C-5-b. below).
    b.   In making this assessment, the consultants recognize and have taken account of
       the fact that exaggerated claims have been made for benefits of civil remote sens-
       ing systems. It will take time—many years—to begin to realize the full potential.
       "Overselling," especially with respect to how soon experimental demonstrations
       can produce definitive results, has frequently been the response to critical bud-
       getary policies which threatened the extension of R&D programs while demand-
       ing an early determination of the value of operational uses. Policy acceptance of
       a continuing U.S. program would have the effect of removing such a threat to
       "survival" and encourage more realistic long-term planning.
    c.   The difficulties of unambiguous justification must be noted. Estimates made to date
       of the economic benefits properly attributed to use of remote sensing have pro-
       duced wide assessment variances. In part this is due to use of different baselines,
       accounting differences, and the need to predict future user markets from a starting
       position of a currently highly fragmented user community. Other problems arise
       from the supposition that some users may tend to disguise or conceal their own
       authoritative benefit estimates for commercial reasons. Finally, there is no agreed
       method for estimating the value of public services where remote sensing can pro-
       vide new or improved responses to hitherto inadequate service availability.

We consider that there are unique attributes of remote sensing systems, and that their utility resides in a complex mix of: (1) direct benefits presumably quantifiable by conventional cost benefit analysis, and perhaps less direct benefits accruing from an increased tax base resulting from new ventures prompted by the availability of remote sensing services and products; (2) general benefits for [9] better decision making by informed societies; (3) the payoffs to the public good by basic investments in information services; and (4) in some cases, by fundamental structural changes in economic and social welfare not capturable [sic] by conventional cost benefit analysis (e.g., here a general equilibrium approach is relevant, vice the simpler and much less encompassing cost benefit assessments).

We believe that there is both merit in and opportunities for additional analyses by which the utility, in the broadest sense, of remote sensing systems can better be measured, particularly in the case of the latter three factors of the preceding paragraph. For the immediate future, decisions to pursue remote sensing must in part be founded on intuitively based social and political rationales. The current lack of fully quantifiable utility assessments does not outweigh the preponderance of evidence that remote sensing systems should be pursued.

4.   Consequences of not continuing. The consultants have also considered the implications of alternative U.S. policy attitudes, e.g., a decision not to continue indefinitely a U.S. involvement in civil remote sensing or a decision to defer still longer the making of a decision on a continuing future U.S. involvement.

a.   A decision not to continue would mean loss of the benefits and opportunities that are the basis of the favorable assessment given above. The consultants believe that the potential of many of these benefits is widely recognized both in the U.S. and abroad. Any U.S. policy implying or presaging U.S. withdrawal, now or in the future, from an active role in civil remote sensing from space would undoubtedly be met by the early development of foreign systems designed to meet U.S. as well as foreign needs. The consultants believe that such a loss of benefits, opportunities, and U.S. leadership is unacceptable.

b.   A decision to defer a decision would permit further erosion of U.S. opportunities and leadership and could lead to results like those cited immediately above. It would also perpetuate the current unsatisfactory situation of general uncertainty and generate further dissatisfaction in Congress and among State, local, private and international users of remote sensing data from space. The consultants are convinced that the time has come to make a positive decision.

5.   Some concerns addressed. Two recurring concerns that have played a part in policy consideration of the U.S. involvement in civil remote sensing need to be noted and discussed briefly:

[10] a.   Reliance on a market test of value. In the determination of the Executive Branch policy toward U.S. involvement in civil remote sensing there has been a tendency to judge the value of possible operational uses in terms of the willingness of the user to pay for the establishment and operation of the system, and, therefore, to defer a policy commitment to continuity of data services until the users are willing and able to commit themselves to providing the funds required.

The logic and appropriateness of this approach have been criticized on many grounds. Thus, it has been pointed out that in the private sector individual customers are not expected to finance in advance the investment costs necessary to produce a product or establish the capability for providing a service. It is also pointed out that making data continuity dependent on demonstrated benefits and user commitments to operational use tends to place the program in a "Catch 22" situation because of the lead times involved: the decision on data continuity has to be faced before the demonstrations needed for a commitment have been completed.

Another criticism relates to the tendency to regard the principal federal user agencies as the "market" for civil remote sensing and to judge the need for an operational system based on their willingness to budget for the cost. As shown by our listing of the many U.S. interests in civil remote sensing, the direct mission interests of federal agencies represent only a fraction of the overall U.S. interest. No federal agency has mission interests broad enough to represent the total "market." In any case, the federal budget process is at best a very imperfect "market place"; to give one example, federal agencies do not have the options for separate financing of capital investment that are normally followed in the private sector.

The consultants agree that users of civil remote sensing should pay a reasonable charge for the data products they receive (transparencies, prints tapes, etc.), and would expect such charges to cover the out-of-pocket costs of producing them. The question of further recoupment of costs should receive further study; such study must, however, give full recognition to the public interest benefits of civil remote sensing for which there may be not identifiable customers to charge and to the broader national U.S. interests served by civil remote sensing that are not the budgetary responsibility of any federal department or agency. The basic policy decision on continuing U.S. involvement in remote sensing should be made on an overall national policy basis and not depend on user funding commitments or resolution in advance of the complex questions of user charges and cost recoupment.

[11] b. <u>Scale of funding and other commitments.</u> The budgetary costs of a continuing U.S. involvement in civil remote sensing are a matter of legitimate concern and deserve careful consideration; an early priority should be given to the development of cost projections for the principal programmatic options.

Current rough NASA projections (which require validation and refinement) suggest that a constructive evolution of operational civil remote sensing data services, including production and dissemination of data to primary users and analysis centers, could be accommodated within a budget averaging somewhere in the range of $150 to $300 million per year (FY 1979 dollars) over the next decade. In addition, there would be a need for strong continuing R&D and experimental applications efforts (where some savings might result from the availability of operational services) and some added user analysis costs required as federal agencies learn to take full operational advantage of remote sensing and to combine these data with data from their conventional sources.

Costs of this magnitude, while considerable, do not constitute a major or "uncontrollable" budget threat. Insofar as they go beyond R&D, they can be regarded conceptually as a necessary new element in the Nation's continuing overall investment in space technology—an essential step for realizing benefits from this investment that would otherwise be lost.

The implications of an "operational commitment," in the sense of an assurance of continuity of data services, may sometimes be exaggerated. On the space segment side, the provision of continuous data services would not, in principle, have to represent a major expansion over a continuing R&D effort. Reasonable data continuity has been maintained—in fact if not by policy—by R&D satellites since the launching of LANDSAT-1; the additional costs required in the future can be minimized by careful integration of R&D and operational planning. On the user side, the necessary expenditures by each using agency can be decided on a case-by-case basis, since each use has its own timetable of operational need and readiness. Policy decisions assuring data continuity and a continuing future U.S. involvement in civil remote sensing should be backed up by the budgetary support required. However, the commitment level for a continuing U.S. involvement and the commitments of each federal user can be controlled through the regular budget process.

D. Conclusions and Consequences

1. On the basis of the considerations discussed above, the consultants' conclusions on the basic question of U.S. policy attitude toward civil remote sensing are as follows:

[12] a. U.S. policy should accept that the U.S. should and will be actively involved in civil remote sensing from space for the indefinite future.

   b. The Federal Government should establish and support affirmative policies in its continuing involvement in civil remote sensing from space, directed at realizing the potential benefits and taking advantage of the opportunities, both domestic and international.

2. Six significant policy consequences the consultants see as flowing from the above conclusions on the basic policy attitude are outlined below. The first three are discussed briefly below; the last three require more extended discussion and are addressed in the succeeding chapters of the report.

   a. The need for data continuity. Using interests require, in varying degrees, reasonable assurances on the nature, frequency, and other characteristics of remote sensing information that will be available and on the period of time in the future for which it will be available. As noted above, it has been unreasonable to expect significant investment or other operational type commitments by using interests in the absence of a clear expectation on their part that the information or data needed will be available for a period that will justify the commitments they have to make. User lead times for operational preparations, and remote sensing systems lead times for maintaining flow of data, dictate the necessity of a long-term plan for data and continuity.

   The consultants have concluded that (1) without a long-term (periodically updated) plan for data continuity, U.S. remote sensing from space activities are not likely to generate the user commitments needed to realize the potential benefits, and (2) a policy commitment to such a plan is an essential cornerstone of any continuing U.S. involvement in civil remote sensing. The planning of the technical characteristics of the data to be provided should take account of the needs of Federal, State, local, private, and international users. The consultants note that in the nature of things there cannot be an absolute or permanent commitment, and that a policy commitment can always be rescinded in the event that a future zero-base review shows that it is no longer warranted.

   b. The need for continuing R&D. It must be recognized that remote sensing systems are based on a very dynamic technology and that new possibilities for data collection and beneficial uses will continue to be discovered in the years ahead. This makes it essential that and the consultants conclude that:

[13]   (1) Strong and imaginative programs of technological R&D and experimental applications must continue to receive a high priority.

   (2) Remote sensing information systems should not become frozen to a particular technology.

   (3) At the same time, the strong focus on advancing technology must not be permitted to obstruct the use of existing or lower technologies in operational systems when it is economically or technically advantageous to do so.

   (4) A close coupling must be maintained between R&D and operational activities in remote sensing information systems. Provision should be made whenever feasible for operational uses of remote sensing information produced by R&D systems or in experimental applications. Conversely, the use for experimental purposes of operational data and systems should be encouraged when operational uses will not be unacceptably downgraded.

c. <u>Need for a comprehensive plan.</u> Given acceptance of the fact of a continuing U.S. involvement in civil remote sensing, the consultants <u>conclude</u> that a high priority should be given to the development, approval, and periodic updating of a <u>comprehensive plan</u> to guide U.S. Government activities in civil remote sensing. Preparation of an initial version of the comprehensive plan should be the first order of business once there has been (a) policy acceptance of continuing U.S. involvement in civil remote sensing and (b) a lead agency has been designated (see next section). In the consultants' view these two actions can and must be taken prior to the development of an initial comprehensive plan, because a meaningful plan cannot be developed without a decision on the U.S.'s basic policy attitude toward civil remote sensing, as previously discussed, or without a lead agency with the necessary authority and competence, as will be discussed below and in Chapter III. Essential characteristics of the comprehensive plan should include the following:

(1) The plan should cover programmatic, technical, and procedural plans for the collection and dissemination of civil remote sensing data and for providing assistance and other services to users. It should cover R&D and experimental activities as well as the provision of services on an operational basis.

(2) The plan should also cover institutional plans for the conduct and appropriate evolution of Federal activities related to civil remote sensing, including provision for effective participation by all federal agencies concerned and other user [14] interests (State, local, private, and international) in the decision process on the technical and other characteristics of the data services to be provided.

(3) The plan must be as realistic as possible, both in technical expectations and projected schedules.

(4) The plan should reflect a phased approach to future decisions and commitments, i.e., it should avoid predetermining matters that can be left to future decision and should indicate as clearly as possible the timing, extent, and implications of the commitments required.

d. <u>Need for designation of a lead agency.</u> Acceptance of the policy that there will be a continuing U.S. involvement in civil remote sensing underscores the urgent need—one that has been evident for some time—for the designation at the earliest possible date of a lead agency for U.S. civil remote sensing activities. Chapter III below is devoted to a discussion of the needs for a lead agency, the functions it should perform, the criteria and options for its selection, and the consultants' conclusions on the agency that should be designated. Chapter V includes some further discussion of proposed lead agency activities.

e. <u>The need for attention to international involvement.</u> A continuing future U.S. involvement in civil remote sensing will clearly require attention to international interactions that are (1) necessary because of the global nature of remote sensing, existing U.S. international commitments, and actions that have been or may be taken by other countries, the U.N., ESA, etc., or (2) desirable for U.S. interests or for general international interests as seen by the U.S. Alternatives for international institutional arrangements are discussed in Chapter IV. The proposed concept for a U.S. national system presented in Chapter V provides for international involvement under several scenarios.

f. <u>The need for a U.S. system concept.</u> Finally, there is a need to develop a viable conceptual framework for the continuing future involvement of the U.S. in civil remote sensing. Chapter V presents a proposed concept of a U.S. national system that is consistent with the policy conclusions of the preceding chapters, discusses some policy issues involved, and outlines alternative scenarios for implementing such a system.

Document II-31

Document title: Zbigniew Brzezinski, The White House, Presidential Directive/NSC-54, "Civil Operational Remote Sensing," November 16, 1979.

Source: NASA Historical Reference Collection, NASA History Office, NASA Headquarters, Washington, D.C.

When the Nixon administration's Bureau of the Budget agreed to finance an experimental multi-agency Earth resources observation program in 1970, it was on the condition that such a system would have to prove its effectiveness before any commitment to an operational program. In 1978 and 1979, the Carter administration commissioned studies to determine what should be done with the Landsat program. The first study (Document II-30) concluded that the federal government should move to an operational system run by NASA. The second study, however, focused on the feasibility of turning control of the Landsat system over to private industry. It was this second study that carried more weight in the deliberations leading to Presidential Directive/NSC-54. This document outlines the November 1979 decision by President Jimmy Carter to name the Department of Commerce's National Oceanic and Atmospheric Administration (NOAA) the lead agency for the Landsat program. Also, NOAA would be responsible for exploring ways to increase private-sector involvement.

November 16, 1979

Presidential Directive/NSC-54

TO: The Secretary of State
        The Secretary of Defense
        The Secretary of Interior
        The Secretary of Agriculture
        The Secretary of Commerce
        The Secretary of Transportation
        The Secretary of Energy
        The Director, Office of Management and Budget
        The Assistant to the President for Domestic Affairs and Policy
        The Administrator, Agency for International Development
        The Director, Arms Control and Disarmament Agency
        The Chairman, Joint Chiefs of Staff
        The Director of Central Intelligence
        The Administrator, National Aeronautics and Space Administration
        The Administrator, Environmental Protection Agency
        The Director, Office of Science and Technology Policy
        The Director, National Science Foundation

SUBJECT: Civil Operational Remote Sensing

    The President has approved the civil space policy discussed below. The policy amplifies that established in PD/NSC-37—National Space Policy and PD/NSC-42—Civil and Further National Space Policy. . . .
[in the original, there was still a "blacked-out" classified area of the document in this position]

2. <u>LAND PROGRAMS.</u> The National Oceanic and Atmospheric Administration (NOAA) of the Department of Commerce is assigned the management responsibility for

civil operational land remote sensing activities in addition to its ongoing atmospheric and oceanic responsibilities. Initially, the operational land remote sensing system from space will be based on LANDSAT technology. Commerce's initial responsibility—in coordination with other appropriate agencies—will be to develop a time-phased transition plan covering: (1) a Program Board (discussed below); (2) organization for management and regulation; (3) system financing including pricing policies for the users['] sharing of costs; (4) technical programs; (5) establishment of private and international participation; (6) identification of facilities (including the EROS data center), hardware, and personnel that should be transferred; and (7) identification of actions such as executive orders and legislation required. Commerce will submit to OMB a preliminary implementation plan by December 15, 1979, covering any required FY 1981 budget adjustments and a final transition plan by June 1, 1980.

a. Federal Management Mechanism. Commerce will establish and chair a Program Board for continuing federal coordination and regulation with representatives from the involved federal organization (e.g., Defense, Interior, Agriculture, Transportation, Energy, State, NASA, CIA, AID, EPA, and Executive Office of the President). Organizations such as the National Governors' Association and National Conference of State Legislatures will be asked to participate as necessary. The Board will forward recommendations on unresolved policy issues to the Policy Review Committee (Space) for consideration and action.

b. Private Sector Involvement. Our goal is the eventual operation by the private sector of our civil land remote sensing activities. Commerce will budget for further work in FY 1981 to seek ways to enhance private sector opportunities (e.g., joint venture with industry, a quasi-government corporation, leasing etc.). Commerce will be the contact for private industry on this matter and with the Program Board will analyze any proposals received prior to submitting policy issues to the Policy Review Committee (Space) for consideration and action.

c. International Participation. The United States will generally support non-discriminatory direct readout to foreign ground stations to continue our present policy and to provide data to foreign users under specified conditions. Pricing policies must be developed that are consistent for foreign and domestic users. We will promote development of complementary nationally operated satellite systems so as to limit US program costs, but protect against unwarranted technology transfer.

3. WEATHER PROGRAMS. Defense and Commerce will maintain and coordinate dual polar orbiting meteorological programs. We will continue procurement of current satellite systems with Defense and Commerce each operating separate satellites to meet the differing needs of the military and civil sectors. When any new polar orbiting satellites are justified they will be jointly developed and procured by Defense, Commerce and NASA to maximize technology sharing and to minimize cost. An appropriate coordination mechanism will be established to assure effective cooperation and to prevent duplication.

4. OCEAN PROGRAMS. If a decision is made to develop oceanographic satellites, joint Defense/Commerce/NASA development, acquisition and management will be pursued. A Committee will be established, with the above representation expanded to include State, CIA, and NSF. The Committee will forward recommendations on policy issues to the Policy Review Committee (Space) for consideration and action.

Zbigniew Brzezinski

Document II-32

Document title: David S. Johnson, Chairman, Satellite Task Force, *Planning for a Civil Operational Land Remote Sensing Satellite System: A Discussion of Issues and Options* (Rockville, MD: U.S. Department of Commerce, National Oceanic and Atmospheric Administration, June 20, 1980), pp. 1–16.

Source: National Oceanic and Atmospheric Administration, Rockville, Maryland.

Established by Presidential Decision Directive (PDD) 42, an interagency task force charged with study-ing options for privatizing all or part of the nation's remote-sensing systems concluded that such action was premature, but that more private-sector involvement was appropriate. Consequently, in November 1979 Presidential Directive 54 decreed that NOAA temporarily manage the Landsat sys-tem while also studying ways to increase private-sector involvement. The resulting study was released in June 1980, but its recommendations were swept aside as the Reagan administration entered office with its own agenda for Landsat. What follows is the executive summary of that study.

Planning for a Civil Operational Land Remote Sensing Satellite System: A Discussion of Issues and Options

June 20,1980

Satellite Task Force
David S. Johnson, Chairman

[1]   <u>EXECUTIVE SUMMARY</u>

This document discusses the issues and options relating to a national civil operational land remote sensing satellite system pursuant to the President's decision to assign to the National Oceanic and Atmospheric Administration of the Department of Commerce the management responsibility for civil operational land remote sensing satellite activities.[1] This document, prepared by the Commerce Department (Commerce), in coordination with other interested agencies,[2] discusses the issues involved in implementing an opera-tional land remote sensing system from space, initially based on Landsat technology, with the goal of eventual private sector ownership and operation of the system. Some policy and technical options related to implementing an operational system are contained in this document, but decisions on these options will, for the most part, await the Administration's FY 1982 budget review and subsequent actions.

A land remote sensing satellite system provides information about the condition of the Earth's surface by a process of sensing radiation from objects on the Earth. The sys-tem uses sensors located on satellites which transmit the data to ground receiving stations

---

1.   The White House Press Release of November 20, 1979, announcing this decision is appended to this Summary as Attachment A.
2.   The National Aeronautics and Space Administration, the Departments of Agriculture, the Interior, Energy, State, and Defense, the Agency for International Development, the Environmental Protection Agency, and the Director of Central Intelligence.

for processing into usable data products. The current system is largely an experimental program called Landsat managed by the National Aeronautics and Space Administration (NASA). Information from the system has proven of value to a variety of public and private sector users in the United States and abroad for helping to make decisions related to such areas as agricultural crop forecasting, rangeland and forest management, mineral and petroleum exploration, mapping, urban and regional land use planning, water quality assessment and disaster assessment.

[2] Background

The issuance of the President's decision regarding civil operational remote sensing from space culminated a two-year Administration review of the nation's space policy. During this period, the Policy Review Committee (Space) was established and national policy on space programs was clarified. In May 1978, the President announced that the United States will encourage domestic commercial exploitation of space capabilities under appropriate U.S. authorization and supervision. Further, in October 1978, the President made a commitment to continue the availability of data from the Landsat program for all classes of users. In his March 27, 1979, Science and Technology Message, the President reiterated his Administration's commitment to the continuity of land remote sensing satellite data over the coming decade. Subsequently, Dr. Frank Press, the President's Science Advisor, in Administration testimony before the Senate Subcommittee on Science, Technology, and Space on April 9, 1979, stated that "the Administration is committed to an operational remote sensing system, although yet undefined."

From October 1978, through the summer of 1979, Executive Branch agencies examined the potential for integrating U.S. civil remote sensing satellite programs and for private sector involvement in U.S. civil space activities. They recommended that all U.S. civil operational remote sensing programs be managed by a single agency. The agencies also reported that the private sector would be interested in assuming more responsibility for land remote sensing from space if Federal policy and market uncertainties were clarified.

In November 1979, the President provided the framework within which a civil operational land remote sensing satellite system should be implemented, and assigned to the National Oceanic and Atmospheric Administration (NOAA) in Commerce the management responsibility for civil operational land remote sensing activities in addition to its ongoing atmospheric and oceanic responsibilities. NOAA's related ongoing responsibilities include managing the national civil operational meteorological satellite program and the Commerce Department's responsibilities for a joint operational demonstration by the Department of Defense (DoD), NASA and Commerce of a National Oceanic Satellite System (NOSS).

[3] The Executive Branch's review of remote sensing satellite programs and policies was paralleled by a series of Congressional hearings during the 96th Congress on operational land remote sensing from space, including hearings before the House Subcommittee on Space Science and Applications of the Committee on Science and Technology and the Senate Subcommittee on Science, Technology and Space of the Committee on Commerce, Science and Transportation. Two bills before the 96th Congress focused on operational land remote sensing: S. 663, introduced by Senator Adlai E. Stevenson, which proposed the establishment of an Earth Data and Information Service in NASA, and S. 875, introduced by Senator Harrison Schmitt, which proposed the creation of a for-profit Earth Resources Information Corporation.

Assumptions

This document was developed in accordance with the following assumptions, which reflect the policies, established in the President's decision on civil operational remote sensing and previous space policy pronouncements, and the prerequisites to their achievement:

- The Federal government will ensure continuity of data during the 1980s;
- A national civil operational land remote sensing satellite system should ensure continuity of data and the appropriate reliability and timeliness of standard data products;
- User requirements, projected levels of demand and the cost of meeting these requirements should determine the design of the operational system;
- The Administration's goal is eventual private sector ownership and operation of the operational system, which includes the assumption of financial risk, as well as operational control by the private operator;
- Prices for land remote sensing satellite products should be set at levels that ensure maximum recovery of system costs consistent with the public good;
- The practice of the widest practical dissemination of Landsat data on a public nondiscriminatory basis will be continued for the data and standard data products from the Interim and Fully Operational Systems in accordance with prevailing U.S. national policies;

[4] • Eventual private sector ownership and operation of the U.S. program will be conducted under Federal government regulation, consistent with U.S. policies and international obligations;

- The civil operational land remote sensing satellite program is a national program responsive to Federal interests and U.S. user requirements. Due regard will be given to foreign user interests and to foreign participation in the U.S. program;
- NOAA will manage the operational system until a new institutional framework is established.

The Present Landsat System

The existing Landsat system consists of one satellite, Landsat 3, launched in 1978, which covers the Earth once every 18 days and transmits sensed data from an on-board multi-spectral scanner (MSS) and two return beam vidicon (RBV) cameras back to Earth, either directly to U.S. or foreign ground stations or indirectly from an on-board tape recorder which stores data until the satellite is within range of a U.S. ground station. NASA's Goddard Space Flight Center controls the satellite and performs the initial preprocessing of the data transmitted to Goddard from U.S. ground stations via domestic communications satellite (DOMSAT).

At the Department of the Interior's EROS Data Center in Sioux Falls, South Dakota, the Goddard preprocessed high density digital tapes are archived and further processed into standard data products (either computer compatible tapes or photographic images) for dissemination to domestic and foreign users at the cost of processing the order and reproduction. Similar preprocessing, processing, archiving and dissemination functions are performed by the nine foreign ground stations that now receive data direct from Landsat 3.

Two additional satellites, Landsat D and D', currently are under construction, with Landsat D tentatively planned for launch in 1982. The Landsat D series of satellites is designed to carry a new sensor, the Thematic Mapper (TM), which will provide 30m resolution[3] for the first time, as well as the MSS, and to use the Tracking and Data Relay [5] Satellite System (TDRSS) for relay of data direct from Landsat to a single U.S. ground station at White Sands, New Mexico. To provide continuity with data from previous Landsats, the multispectral scanner (MSS), which provides 80m resolution, will continue to be deployed on Landsat D and D'. Direct readout of sensor data to foreign ground stations will be continued.

Because of difficulties in developing the TM and the associated ground data processing system, NASA is considering launching Landsat D without TM in 1982, to be followed by Landsat D' with TM later.[4] Current estimates for the operational preprocessing of Landsat D and D' data at Goddard are 200 MSS scenes per day beginning no earlier than 1983 and up to 50 TM scenes per day when the TM system becomes operational possibly no earlier than 1985.

The Interim and Fully Operational Systems

A fully operational land remote sensing system that meets optimal performance standards can be implemented at the earliest in 1989, given best estimates of the state of the art advances in sensors and the time required for Federal contracting procedures if they are used. Until that time, extension of the Landsat D system can ensure that, after 1983, the commitment to continuity of data during the decade of the 1980s is met.

From a technical standpoint, the following performance standards have been identified as applicable to a high quality operational system:

• Sensors designed to generate data meeting a broad range of user requirements at a reasonable price;

• Assured continuity of satellite coverage without break, with one backup satellite in orbit at all times and another on the ground;

• 95% confidence that, averaged over a two-day period, all data will be processed and made available from the ground station within 48 hours of receipt; and

• Ability to identify and process certain data out of order to meet urgent user needs.

[6] However, the extent to which these compabilities [sic] are pursued will depend upon their full capital and operating costs and the demonstrated existence of an adequate private and Federal market to justify such costs.

While sensors specifically designed to generate data meeting a broad scope of user requirements cannot be provided until the late 1980s, the Landsat D sensors can be used as the basis for an interim system which will help to ensure continuity of data during the 1980s and meet many user needs.

The Administration is currently reviewing the Landsat D system to see where improvements may be required to ensure data continuity during the 1980s. For instance, the current Landsat system includes no satellites after Landsat D'. Anticipated gaps in spacecraft coverage of several years between about 1986 and the initiation of a fully operational system may have to be filled by the construction of one or more satellites or by the refurbishment of Landsat D. In addition, changes in the Landsat D ground segment may be required to minimize the risk of losing some data or having an excessively long delay in processing some data. The Landsat D system, with any follow-on satellites and ground system improvements, has been designated the "Interim Operational System."

The earliest possible date by which all four performance standards for a high quality operational system could be met is 1989, when the R&D necessary for the new solid state,

---

3.   The term "resolution," as used in this document, refers to the instantaneous field of view (IFOV).

4.   The Administration is also considering other alternatives such as delaying the launch of Landsat D until 1983 when the TM sensor will be ready.

multilinear array sensors should have been completed, and the sensors will have been fab-
ricated, tested, and incorporated into either an existing multi-mission modular spacecraft
(MMS) or a new spacecraft. The Landsat D system so modified is designated the "Fully
Operational System."

A decision on when to implement the Fully Operational System requires careful
examination of the Federal government's priorities, needed financial assistance, private
sector willingness to invest in and take over the system, user demands during the interim
system and the potential risk of foreign satellite systems obtaining a portion of the domestic and foreign land remote sensing market.

Management Arrangements for the Interim Operational System

Certain changes in management responsibility will take place as the Interim
Operational System is implemented. Although the exact dates for transferring managerial responsibility to NOAA are subject to changes in NASA's schedule for Landsat D, NOAA
plans to assume the following responsibilities from NASA and Interior on the following
schedule:
[7] • NOAA will assume responsibility from NASA in FY 1983 for the command and control of the system and will begin providing MSS data on an operational basis after the successful launch and check-out of Landsat D and the MSS ground system and after NASA
has demonstrated that the system is operational. NOAA will assume responsibility for TM
data when that portion of the system reaches an initial operational level of performance;
• NOAA will assume responsibility from NASA and the EROS Data Center in FY
1983–84 for the generation and dissemination of data and standard data products.
Assuming it is cost-effective, a new facility would be co-located with the Landsat D preprocessing facility at Goddard and would be the sole sales outlet in the United States
of data and standard data products from the Interim Operational System; and
• NOAA will take title to the Landsat archival material at Goddard and the EROS Data
Center in FY 1984 and will be responsible for archival and dissemination functions for
the Interim Operational System.

During the interim operational phase based on the Landsat D series of satellites,
NOAA will manage the system in coordination with an interagency Assistant Secretary
level Program Board. In addition, the Secretary of Commerce will establish a Land
Remote Sensing Satellite Advisory Committee with representatives of state and local governments, other domestic non-Federal users, and interested domestic private sector
groups. Within NOAA, a new major line component, the National Earth Satellite Service,
has been proposed to have managerial responsibility for the civil operational land remote
sensing satellite program.

User Requirements for the Fully Operational System

User requirements should determine the design of the fully operational land remote
sensing satellite system. A survey of governmental and private users indicates a wide range
of possible requirements, depending on the type of application being considered, which
could justify differing types of satellite systems.

To assist NOAA or an eventual private owner to develop a responsive operational system, a preliminary survey of possible user requirements was made. This survey indicated,
[8] for example, that agencies that are interested primarily in renewable resource applications such as agricultural crop assessment want frequent observations, delivery of data
within 48 hours in certain circumstances, spectral bands that discriminate between various
types of vegetation and resolution higher than that provided by the current Landsat sys-

tem. State and local governments, requiring data for land use management and protecting environmental quality, request higher resolution over urban and suburban areas and time-series analyses to detect detailed changes. The U.S. mineral extraction and related industries call for stereoscopic[5] capabilities, global coverage, thirty to forty meter resolution and processing of data within a few weeks. Foreign users['] interests appear to be similar to those of their U.S. counterparts, although area coverage requests obviously differ.

Further analysis and sorting of these requirements with respect to resolution, spectral bands, stereo coverage, frequency of observation and timeliness of product delivery will be necessary as plans are developed for the operational system.

Performance Options for the Fully Operational System

Hypothetical system performance options have been identified to meet some or most of the preliminary user requirements identified above. These options range from designing a system with capabilities similar to the Landsat 3 with MSS only, at an estimated 10-year cost of $1 billion, to building a new system which meets most of the currently stated user requirements, including two meter resolution, at an estimated 10-year maximum cost of $10 billion.[6] Stereo coverage can be provided at an additional cost of up to $700 million.

A final decision on the system design to be pursued for the Fully Operational System can be reached only after further analysis of user requirements, technical options, cost comparisons, system financing, and the effect of potential foreign competition.

[9] Revenues, Pricing Policies and Financial Assistance

Reliable projections of revenues from sales of standard data products, and from the direct reception fees to be paid by foreign ground station operators cannot be made at this time since the characteristics of the Interim and the Fully Operational Systems, the users' level of demand at various prices, the impact of a market expansion program and the impact of foreign competition are not now known. Tentative projections indicate that this system may not and probably will not be self-financing before the end of the century. Therefore, continued Federal financial contributions to support of the system likely will be necessary for the foreseeable future.

System revenues, generated by the sale of standard data products and foreign ground station access fees, now amount to only $6 million[7] a year.[8] Current fees consist of a nominal $200,000 access fee for foreign ground stations and cost of reproduction charges for standard data products—$200 for a computer compatible tape and between $8 and $50 for various types of Landsat images. The projected costs of the Fully Operational System range from $100 to $400 million a year. To achieve the objectives for the sharing of costs by users, and for the eventual ownership and operation by the private sector, prices must be increased to cover, over time, the capital and operating costs of the system and the data and data products treated in a proprietary manner.

The system's manager could charge three types of fees for data and standard data products:

---

5.   As used in this context, stereoscopic means two or more images, taken from different angles, to permit inference of the relative height of various topographic features.

6.   All costs are in FY 1980 dollars.

7.   All revenues are in FY 1980 dollars.

8.   This figure includes $2.7 million from sales, $1.8 million from foreign ground station access fees and $1.3 million attributed to the value of the data distributed without charge to Federal agency users.

- Basic Fee. A fee paid by each user on each standard data product it purchases from the U.S. system operator. These fees would vary in proportion to the costs incurred in producing that product. They would be paid by users of both real-time and retrospective data. Other factors such as timeliness, the placing of special orders and special handling could be reflected in a surcharge schedule.

[10] • Royalty Fee. A fee paid by each U.S. and foreign user and foreign ground station operator on the reproduction or resale of Landsat standard data products.

- Direct Reception Fee. One or more fees paid by foreign ground station operators receiving data directly from U.S. land remote sensing satellites. Examples of such fees are: (1) an annual access fee like the $200,000 fee per station per year currently being paid by Landsat station operators, and (2) a transmission fee paid by foreign ground station operators for data transmitted to and received by the foreign ground stations. This latter fee would be based on the amount of data requested.

Upon the completion of pricing studies, a proposed pricing schedule will be developed based on these types of fees, and possibly others, for consideration by the Program Board and the Land Remote Sensing Satellite Advisory Committee.

Since a substantial shortfall is projected between annual revenues and the estimated annual costs of running an operational system of between $100 and $400 million per year, Federal financial assistance likely will be required. In this event, the Federal government could provide various types of capital and operating assistance to a private or government corporation, whichever institutional option is eventually chosen. Such Federal capital assistance could include grants, equity guarantees, and Federal loan and loan guarantees. Federal operating assistance could include Federal support of research and development, purchase guarantees, appropriations, free services and tax incentives.

Whether for the Interim or Fully Operational System, three possible options for Federal agencies to share in the costs of financing the operational land remote sensing system are under consideration:

- NOAA could budget for all "core"[9] and special system costs;
- NOAA could budget for "core" system costs and user agencies would budget for special system capabilities;[10]

[11] • User agencies could fund individually a predetermined portion of all "core" and special system costs.

A decision on the preferred financing option will weigh, on the one hand, the benefits of having a mechanism that forces agencies to make trade-offs between land remote sensing data and other sources and, on the other hand, the advantages of focusing responsibility for the program and budgeting in one agency.

Institutional Approaches to Eventual Private Sector Ownership and Operation

1. Institutional Alternatives

Several institutional options exist for achieving the goal of eventual ownership and operation by the private sector of our civil land remote sensing satellite activities. The four principal institutional options discussed in the document are:

---

9.   The "core" system includes the space and ground segment elements necessary to meet the common needs of the majority of users.

10.   Special system capabilities include stereoscopic coverage.

(1) A private corporation (or consortium) selected competitively to own and operate all or part of the civil operational land remote sensing satellite system and to sell data to Federal agency users under a guaranteed purchase contract;

(2) A for-profit private corporation, authorized by Federal legislation, with private equity and privately and publicly appointed Board members;

(3) A wholly-owned government corporation authorized by Federal legislation, with Government equity, reporting to the Secretary of Commerce, with provision for subsequent transformation to a private stock corporation as system revenues warrant; and

(4) Federal agency ownership with private contractor operation, and provision for subsequent transfer to a private sector owner as system revenues warrant.

Options 1 and 2 offer the earliest possibilities of private sector ownership and assumption of risk. Options 3 and 4 delay implementation of private sector ownership until the next decade.

[12] These options will be examined by the Administration over the next several months to evaluate which alternative best serves the Federal, state and local government and private sector interests in having an operational land remote sensing satellite program.

2. Establishment of Federal Policy to Encourage Private Sector Investment

Several policies impact the likelihood or willingness of the private sector to own the operational system. For example, under present policy, a system owner has no ownership rights in the Landsat data and standard data products. Without a change in this policy, a private owner would be denied the opportunity for profitability; therefore the Federal government would have to authorize the private sector to own and sell civil operational land remote sensing satellite data and standard data products on terms that eventually permit a reasonable return on investment. Other factors that affect private sector investment are competition from ongoing Federally funded R&D land remote sensing satellite systems and the duration of the Federal government's financial commitment to the land remote sensing satellite program. Conversely, a private system owner should be required to abide by the government policy of widest practical dissemination of data and standard data products on a public nondiscriminatory basis at prices that are consistent for domestic and foreign users.

3. Regulation of Private Sector Operation

A private owner of the land remote sensing satellite system could enjoy a monopoly. To protect the national interest, the private owner's activities should be regulated to the extent necessary to conform to national space and other domestic and foreign policy objectives. A private or government entity owning the operational system should be required, for example, to comply with international treaties such as the Outer Space Treaty for the conduct of peaceful activities in outer space; continue the widest practical dissemination of data and standard data products on a public nondiscriminatory basis; meet the needs of U.S. government users; and refrain from misuse of insider knowledge obtained from the land remote sensing satellite data.

Market Expansion

The system manager should undertake a market expansion program to increase revenues, reduce required Federal [13] financial assistance, and enhance decision-making through the use of land remote sensing satellite data. An important element of this program is assuring continuity of land remote sensing data.

A market expansion program for the operational system can build on the types of training and technology transfer activities now being conducted by NASA and the

Department of the Interior. NOAA could arrange for reimbursable training programs, enter into joint applications demonstration projects with users in all sectors, encourage university land remote sensing instructional programs and work with domestic and international assistance agencies to promote new opportunities for American business in the land remote sensing satellite field. As part of its ongoing R&D responsibility, NASA could continue to develop and demonstrate to users new techniques and technologies for using land remote sensing satellite data.

International Aspects

The United States should continue to encourage international participation in the U.S. civil operational land remote sensing satellite program by further developing an international community of data users and by continuing discussions with prospective foreign land satellite system operators to explore the prospects for encouraging complementary and compatibility among future operational land satellite systems.

The United States should ensure that data from the Interim and Fully Operational Systems are made available to foreign users through sales of standard data products on a nondiscriminatory basis. NOAA, working closely with the Department of State and other interested agencies, should take the following actions:
• Consider foreign user requirements in planning the Fully Operational System;
• Conclude agreements with those foreign agencies wishing to receive data directly from the Interim and Fully Operational Systems;
• Establish pricing policies for data sales and direct reception fees that are consistent for domestic and foreign users; and
[14] • Continue the Landsat Ground Station Operations Working Group as a forum for the exchange of technical information.

The land remote sensing satellite systems being developed by other countries offer the prospect of both competition and cooperation with the U.S. The competitive challenge to U.S. technologies leadership is likely to occur in such areas as the development of multilinear array sensor technology, and sales of ground equipment, services and data products. NOAA, working closely with the Department of State and other interested agencies, should encourage the expansion of world-wide markets for U.S. equipment, services and data products, and pursue prospects for complementary with foreign satellite operators in order to develop complementary system characteristics (e.g., orbits, coverage patterns and repeat cycles) and compatible system outputs (e.g., standard data product formats).

Legislation for the Operational System

Legal authority in four principal areas may be required in order to implement a civil operational land remote sensing satellite system:
1. Authorization for NOAA to develop, own and manage the civil operational land remote sensing satellite system until the responsibility is transferred to a private or other entity;
2. Establishment of the institutional structure, financial assistance and transition to private sector ownership and operation of the U.S. civil land remote sensing satellite system;
3. Establishment of a regulatory system to ensure that a private sector owner's activities are in compliance with U.S. laws, policies and international obligations; and
4. Establishment of proprietary interests in operational land remote sensing data and standard data products.

Summary of Issues

The following is a summary of the issues that have to be addressed as the Federal government moves toward an operational land remote sensing satellite system:

[15] 1. Continuity of Data in the 1980s

    a. Operations
- Whether to fund, construct and launch additional Landsat D series satellites with tape recorders to provide continuity in the acquisition of data from space until a Fully Operational System can be deployed?
- Whether to improve the existing Landsat D ground segment at the Goddard Space Flight Center to provide continuous processing of the acquired data into timely and reliable standard data products?
- Whether to transfer responsibility for command and control of the Landsat D space and ground segments from NASA to NOAA?
- Whether to transfer responsibility for archiving and disseminating land remote sensing satellite standard data products from the Department of the Interior to NOAA, and whether to co-locate these functions with the satellite command and control and preprocessing facilities at the Goddard Space Flight Center?

    b. Management
- When to submit to Congress an Administration bill that authorizes NOAA to own and manage an operational land remote sensing satellite system until that system is transferred to another entity?

2. Initiation of a Fully Operational System
- How to validate user requirements and their priorities?

- When to establish a Fully Operational System utilizing new sensors that meet a broad range of user needs?

[16] 3. Pricing Policies and Financial Assistance
- How to establish initial price increases for direct reception and for data and standard data products that are consistent for foreign and domestic users, provide adequate advance notice of price increases, and encourage potential users to invest in support equipment and reduce use of competing methods of data collection?
- When to implement price increases?
- How to fund the capital and operating costs of the Interim and Fully Operational Systems that exceed revenues?

4. Institution for Private Sector Involvement

- What, if any, institutional framework for private sector ownership should be submitted to Congress?
- What mechanisms for regulating and providing Federal financial assistance to the private sector should be provided in any bill authorizing an institutional framework for private sector involvement?
- What policies should control the activities of any private sector owner for ownership of data and standard data products, for conditioning their dissemination on the payment of appropriate fees, for making possible the users' sharing of system

costs beyond the costs of reproduction, and for requiring consistent pricing and ensuring nondiscriminatory availability of standard data products.

5. Market Expansion
   • What market expansion should be authorized for the Federal system manager?

6. International Aspects
   • How to encourage the growth of worldwide markets for U.S.-produced equipment, services and land remote sensing satellite data and standard data products?

Document II-33

Document title: Ed Harper, Office of Management and Budget, Memorandum to Craig Fuller/Martin Anderson, "Resolution of Issues Related to Private Sector Transfer of Civil Land Observing Satellite Activities," July 13, 1981.

Source: NASA Historical Reference Collection, NASA History Office, NASA Headquarters, Washington, D.C.

The new Reagan administration, eager to reduce the federal budget and to transfer as many government functions as possible to the private sector, quickly reversed the key elements of President Carter's approach to creating an operational framework for remote sensing and sought to commercialize the program as soon as possible. In response, Comsat proposed that the government transfer the operation of both weather satellites and remote-sensing satellites to the private sector, arguing that the profits from selling weather imagery back to the government could be used to finance the long-term commercial development of remote sensing. The White House formed a Cabinet Council working group to consider this proposition.

July 13, 1981

MEMORANDUM TO: Craig Fuller/Martin Anderson

FROM.:   Ed Harper

SUBJECT: Resolution of Issues Related to Private Sector Transfer of Civil Land Observing Satellite Activities

The purpose of this memo is to <u>request that a working group</u> within the Cabinet Council system be established to consider <u>the following two</u> issues related to <u>private sector transfer</u> of civil land observing satellite activities:
   –   What is the best <u>mechanism to implement the current policy</u> of <u>transfer of civil land</u> remote sensing systems (LANDSAT) to the private sector as soon as possible?
   –   Should the Administration consider <u>simultaneously private sector transfer of both civil weather</u> and <u>land remote sensing systems</u>?

Background

With the <u>revisions to the 1982 Budget</u> the Administration explicitly stated its <u>intention to hand-off operational responsibilities</u> for land remote sensing to the <u>private sector</u> in the mid-1980's or sooner, if possible. This policy reflected the judgment that the Federal investments in the LANDSAT program contained in the revised budget were sufficient to evaluate the usefulness of this data and that, <u>if the operational uses were significant,</u> the <u>private sector would provide follow-on</u> satellites—there would be <u>no need for the Federal</u> Government to purchase additional satellites beyond the two new NASA budgeted satellites (i.e., LANDSAT D and D1). Thus, the <u>Administration withdrew the Carter commitment to data continuity through the end of the decade</u> and decided that additional satellites beyond the two new NASA <u>satellites</u> would depend on the private sector's willingness to invest in and operate follow-on satellites. <u>We are not asking the Cabinet Council to revisit this policy.</u>

The Department of Commerce (NOAA) is currently developing draft legislation designed to facilitate private sector transfer of land observing satellite activities. This legislation needs to be consistent with the policy decisions on the issues being referred to the Cabinet Council.

A potential <u>private sector owner/operator</u> has <u>requested</u> that the Administration consider <u>transferring simultaneously both the civil weather and land remote sensing satellite systems</u> to the private sector, and that selection of a private sector proposal or combination of proposals be based on the merits of the total package.

- What <u>is the best</u> mechanism to implement the current policy of private sector transfer, as soon as possible? The options available to the Administration seem to be the following:

Laissez-faire approach—continue NOAA operation of satellites consistent with current policy and do nothing to encourage or discourage independent private sector initiatives.

A decision to consider transferring the current Government inventory of civil remote sensing satellites and ground equipment to a private corporation or consortium of private corporations in return for cash and/or future considerations.

A decision to provide some form of subsidy or long-term data contract (details to be specified consistent with budget of user agencies) in order to facilitate private sector transfer.

A combination of the two previous options.

A decision to establish a federally chartered for-profit private corporation to own and operate a civil, land remote sensing satellite system—along the lines envisioned in the Schmitt Bill introduced in the previous Congress.

- Should the Administration <u>simultaneously consider private sector transfer of both civil weather and land remote sensing systems</u>?

Transfer of the civil weather satellite program to the private sector would place more emphasis on the private sector and market forces in determining the level and scope of these satellite activities. However, the assertion that such a transfer could reduce the Federal budget and increase the Federal tax base without incurring significant additional Federal risks has not yet been validated.

The Administration probably will not be able to determine if such a private sector transfer can be achieved on terms acceptable to the Government until proposals are received and evaluated.

- The <u>sub-issues</u> that will need serious review and consideration include:

  What type of Federal commitment, if any, would be appropriate for purchase of either weather and/or land satellite data? To what extent should the Federal Government continue related technology development (e.g., R&D on advanced sensors)?

  What type of relationship should exist between the Government and any potential private sector owner/operator?

  What Federal assets and data rights should the Government consider transferring to the private sector?

<u>Assumptions</u>

- In light of the need for fiscal restraint, an increase in the Federal commitment to land remote sensing from space should be considered only to the extent that user agencies are <u>willing to make tradeoffs against previously approved activities for 1983</u> and beyond in order to facilitate an expanded Federal commitment.
- Since there are <u>other options for reducing the Federal expenditures for</u> needed weather satellite data (e.g., combining civil/military polar-orbiting satellites, reducing the number of civil weather satellites in orbit, and placing weather sensors on commercial communications satellites), it should be <u>assumed that the 1983–86 budget projections for civil weather satellites may be revised downward.</u>

  The agencies affected include:

Agency	Area Affected
Department of Commerce	NOAA operation of weather and land satellite systems.
Department of Agriculture	Agriculture forecasting based on weather and land satellite data.
Department of Defense	Data from civil weather satellites (in addition to data from military weather satellites).
Department of Interior	Geological, mineral and land management activities use land satellite data.
Department of State	International agreements on satellite remote sensing.
Central Intelligence Agency	National security.
National Aeronautics and Space Administration	R&D using satellite data and new sensor development for weather and land satellites.

Document II-34

Document title: Government Technical Review Panel, "Report of the Government Technical Review Panel on Industry Responses on Commercialization of the Civil Remote Sensing Systems," November 10, 1982, pp. 1–25.

Source: NASA Historical Reference Collection, NASA History Office, NASA Headquarters, Washington, D.C.

While the Reagan administration was quite intent on transferring control of the nation's civilian Earth observation satellites to private industry, implementing this policy was a formidable task. Members of Congress raised a number of objections, particularly to the suggestion of selling off weather satellites, as well as the Landsat system. Private industry, the supposed beneficiary of this proposal, was less than enthusiastic, as evidenced by this report of the Government Technical Review Panel established to review various options for establishing a new remote-sensing policy. This report is based on corporate responses to a Department of Commerce request for information concerning the transfer of remote-sensing satellites to private industry. Respondents were placed into four categories based on their degree of support for the concept of the privatization of remote sensing.

Report of the Government Technical Review Panel on Industry Responses on Commercialization of the Civil Remote Sensing Systems . . .

November 10, 1982 . . .

[1] I. <u>OVERVIEW</u>

The panel convened on 26 October and reviewed fourteen responses to the Request for Information [RFI] that appeared in the Commerce Business Daily (CBD) on September 10, 1982. No attempt was made to solicit additional information or clarification from respondents.

The responses varied in scope and sophistication from a handwritten postcard to a fairly comprehensive, all inclusive submission. Criteria for evaluation could not, therefore, be applied uniformly to all proposals. However, the following general criteria were used, as applicable:

    (1) Responsiveness to federal needs;

    (2) Continuity of data services;

    (3) Feasibility; and

    (4) National security and foreign policy concerns.

Responses were grouped into four (4) natural categories reflective of their basic thrust:

    (1) Those favoring near-term commercialization of existing civil remote sensing capabilities, entirely or in part.

    (2) Those espousing independent entrepreneurial interests and advocating a climate conducive to free market competition.

    (3) Those favoring government retention of the existing system, at least for the immediate future.

    (4) Other.

[2] II. SUMMARY AND OBSERVATIONS

It could fairly be stated that a simple evaluation of responses to the RFI would fulfill the charter of this panel and that further comment is gratuitous. Nonetheless, our study of this issue and the responses produced a consensus which we would be remiss not to surface.

Insofar as the responses are positive toward the issue of commercialization, they tend to assert rather than demonstrate an ability to satisfy whatever criteria we might establish. Nonetheless, the panel harbors significant doubt as to whether all U.S. government interests could be satisfactorily protected if the approach is simply to substitute one monopolistic organization for another. Perhaps, it will take an RFP [Request for Proposals] to answer the toughest questions.

The RFI elicited more interest than might have been expected and surfaced a strong body of opinion that urges restraint and caution in proceeding with commercialization. There is an underlying theme common to the submissions from several large, responsible, and knowledgeable entities that commercialization now could inhibit the free market process. They suggest continued government operation of the system while fostering an environment conducive to an expansion of free enterprise activities.

One of the concerns which permeated our discussions of a non-government monopoly environment, was the potential lack of vigor in the R&D effort and lack of incentive to adopt improvements which may materialize. This has been the case in the satellite communications field. It is our belief that the best answer to the emerging foreign competition lies in the continuation of a dynamic U.S. government R&D system.

It is also the belief of the panel that there is considerable financial, policy and program risk to the government in commercializing weather satellites and that there is no clear policy or financial benefit to be realized. Too, there is no clear consensus among the respondents as to the desirability or feasibility of commercializing any of our civil remote sensing systems at this time.

[3] Additionally, creation of a single, government-chartered, subsidized firm for this purpose would seem antithetical to the underlying economic philosophy of the United States and, in particular, this Administration, as we understand it. If regulated, it would result in the creation of a "utility" without the competitive incentives for reducing operating costs or increasing efficiency. If unregulated, the chartered entity would tend to assume the characteristics of a legislated monopoly.

Finally, the following general national security concerns exist in commercialization of remote sensing from space even though not specifically addressed in the individual evaluations:

(1) There is some potential for military and intelligence application of current data products, and

(2) with possible system improvement under private sector control these concerns would increase.

(3) There are technology transfer issues which might be exacerbated if a private sector operator became the world wide supplier of remote sensing equipment and spares.

(4) Controls over data dissemination, and provision for DOD emergency use would require very careful stipulation in any transfer of civil remote sensing activities.

[4] III. EVALUATION OF THE RESPONSES

1. CATEGORY ONE - Those favoring near-term commercialization of existing civil remote sensing capabilities, entirely or in part.

COMSAT, Environmental Satellite Data, Inc., and Control Data Corporation are willing to proceed now in assuming at least part of the existing land and/or weather satellite systems. COMSAT proposes to take over both systems in their entirety. Control Data Corporation proposes a phased take-over of the Landsat ground processing and distribution system, while Environmental Satellite Data, Inc. suggests operating specific segments of the GOES ground system. Although the magnitude of the private sector take-over varies substantially among these firms, sufficient details were provided to permit an evaluation under all four general criteria. A fourth respondent, the <u>American Science and Technology Corporation,</u> proposes to take over the command and control of the Landsat satellites. However, the thrust of this response focuses on new entrepreneurial interests and is therefore reported under Category 2.

A. <u>Communications Satellite Corporation, Comsat General Corporation (COMSAT)</u>
COMSAT has submitted the only proposal advocating total commercialization of civil remote sensing. Additionally, COMSAT emphasizes a "concern for urgency" in such a transfer. While COMSAT presents the most detailed proposal (due to the magnitude of the transfer), it reiterates its earlier position of requiring both the civil weather and land remote sensing satellites to insure future commercial viability.
(1) <u>Responsiveness to Federal Needs:</u>
The COMSAT concept involves private sector purchase of the current assets in the government's land and civil weather systems. These would be enhanced in the future by incorporating additional sensors upon identification [5] of the specific user needs. The federal government would pay most of the incremental costs. For example, COMSAT suggests adding sensors that would collect oceanic data on water color, winds, ice, and wave conditions.
It is not clear that Landsat's coarse (80 meter) multispectral scanner data, which are used extensively for making agricultural assessments, would be included as part of the basic data collection package that COMSAT suggests for the post-Landsat D' era of the late 1980's. However, if this need is identified as a continuing federal user requirement COMSAT would provide this capability, <u>at additional cost,</u> if it were not part of the projected array of imagery collection capabilities. Thus, from the viewpoint of system technical capabilities, this concept would be more than fully responsive to the current and future level of federal user requirements with reference to timeliness, extent and frequency of coverage, imagery characteristics and data formats, and timeliness.
However, the proposal would not appear to meet another critical user requirement—assurance of data availability at reasonable cost. The COMSAT concept would require federal data purchases at an annual level of about $315–330 million per year. For the government to meet this amount, it appears that there will have to be either a substantial increase in the cost of land and/or weather data or there will be substantial direct subsidy payments to COMSAT.
Another factor that is less significant than either of the two preceding factors is the matter of proprietary rights. The COMSAT concept also calls for the system owner/operator to have copyright and proprietary rights over the data that are collected. Such rights, while desirable from the viewpoint of helping make the system self-financing, have the disadvantage of inhibiting the use of the collected data.

[6]   (2) <u>Continuity of Service:</u>
Implementation of the COMSAT concept would appear to satisfy the major consideration of maintaining continuity of data flow. However, the concept does afford the system operator a loophole for (a) restricted liability only if the "best efforts" are not made, and (b) also for performance being contingent upon the federal government meeting extensive financial commitments for a 15 year period.

Should changes occur in the international price structure which adversely effect profitability, i.e., undercutting of U.S. commercial data prices by a foreign competitor, a commercial operator might elect to abandon the enterprise, or use the "best effort" principle to demand increased federal price subsidies.

(3) Feasibility

This proposal would replace existing weather ground systems, which need technical improvement, and would use current Landsat facilities and equipment. In addition, a centralized facility is proposed for both land and environmental data based on Landsat-type hardware. Centralization is technically valid, but does not exploit what is currently known about the advantage of distributed processing architecture, insofar as service to users is concerned[,] i.e., throughput, availability, and accessibility of data. The proposed technology to be employed on COMSAT's LANDSTAR, the successor to LANDSAT, includes linear array focal planes and on-board data compression capability. These are needed improvements if high resolution solid-state sensors are adopted. Not mentioned, but very much needed on any follow-on sensor, is the addition of cooled focal planes for short-wave and thermal infrared data, and a capability of off-nadir pointing for more frequent coverage of the same scene. The addition of synthetic aperture radar sensors for data set merging is very appealing as is the integration of weather and land remote sensing systems for [7] more efficient programming of data acquisition and low-orbit collection of environmental data. COMSAT recognizes that guaranteed progress in technological advances and maintenance of U.S. technological leadership can only be realized if government retains an active role in advanced technology development, either alone or in some joint role with the operator of the systems.

(4) National Security and Foreign Policy Concerns:

The COMSAT proposal is based on a "best effort" principle which does not commit the Corporation to provide continuous services over the lifetime of the contract. A disruption of land remote sensing and weather services could have substantial foreign policy implications in terms of traditional U.S. international data exchange policies which emphasize continuity and nondiscrimination.

COMSAT states its intentions to broaden the primary data market by limiting secondary reproduction and distribution of data. This could ultimately jeopardize the continuous and reciprocal international exchange of meteorological data upon which this country is vitally dependent.

If a national emergency required disruption of commercial service in Metsats and/or Landsats, the government would have to reimburse the commercial operator for lost revenues. This would entail an additional government expense not incurred under government ownership and would constitute an additional complicating factor in making national security and foreign policy decisions.

COMSAT proposes to sell "surplus" DCS (data collection system) capacity to commercial users which could be used for non-environmental purposes. However, by agreement in the International Telecommunications Union (ITU), DCS frequencies are to be used exclusively for environmental monitoring.

No contractual or legislative stipulation should preclude the government from developing its own satellite systems for national security purposes.

[8] B.   Environmental Satellite Data, Inc. (ESD)

ESD does not address Landsat or the overall weather satellite system. Instead the company offers to assume responsibility for a small segment of the GOES data processing while focusing on the distribution of GOES imagery to commercial users. The proposal would terminate the current GOES-TAP "no-cost" service and replace it with a larger, more efficient system requiring a commercial user contractual fee. This concept could be implemented under existing policies and regulations.

(1) Responsiveness to Federal Needs:
Within the limited scope of this proposal ESD appears to satisfy federal needs and to be responsive to increasing numbers of users.

(2) Continuity of Service:
Phased implementation should assure continuity of service.

(3) Feasibility:
This proposal contains no technical detail, but it does assert that the Visible and Infrared Spin-Scan Radiometer earth locating program can be improved to the point that one pixel location accuracy of the grid on the satellite imagery can be obtained within six hours after completion of maneuver of the satellite, vice 10 pixels and 24 hours. Since larger and faster computers than those currently used are available, this should be readily achievable. However, no conclusions on capability are possible without specifics on computers and data management.

(4) National Security and Foreign Policy Concerns:
ESD proposes to distribute only a small part of the total GOES weather satellite imagery. So long as any government agreement with ESD includes provision for equal, non-discriminatory access to data, no foreign policy issues are raised. No national security problems exist.

[9] C. Control Data Corporation, CDC
CDC proposes a time-phased, joint government/industry venture beginning with ground processing of Landsat data by 1984. This would lead to economic validation of a transition from a subsidized, government service operation to a product-based, profit making venture. The feasibility of transfer of the space segment for remote land sensing would be evaluated during the transition. Weather data could be included in such a ground data processing system.

(1) Responsiveness to Federal Needs:
CDC's joint venture concept could be designed to meet specific federal needs throughout the transition to private ownership and operation.

(2) Continuity of Service:
Not specifically addressed, but preceding comments are applicable.

(3) Feasibility:
CDC's stated experience with data processing and analysis and with DOD space systems would attribute to CDC a technical credibility sufficient to warrant serious consideration of their suggestions on commercialization. However, no details of a technical nature were provided. CDC's concept represents a conservative, low technical risk approach to eventual commercialization. CDC, along with COMSAT and Terra-Mar, explicitly recognizes the greatly improved information extraction potential in merged or "fused" data sets from different parts of the electromagnetic spectrum. CDC recognizes that technological progress can only be assured if the government maintains an active role in advanced technology development, either alone or in some joint role with the operator of the systems.

The data provided on concepts for user services facilities illustrates at least a basic understanding of how such facilities should function in data processing, archiving, and distribution. The highly buffered system concept presented represents state-of-the-art thinking on high-rate, high-throughput, functional requirements.

[10]   (4) National Security and Foreign Policy Concerns:
         CDC's concept does not pose a foreign policy concern if the suggested user service scheme satisfies present policies on data access and distribution, and if a discriminatory pricing system is not imposed. However, the response fails to address these issues. No national security concerns are raised.

[11] 2. CATEGORY TWO - Those espousing independent entrepreneurial interests and advocating a climate conducive to free market competition.
         Three firms—Terra-Mar, American Science and Technology Corporation and Space Services, Inc.—advocate a free enterprise environment which permits a natural evolution and competitive development of the private sector remote sensing industry. Their premise is that successful commercialization will occur only in a competitive market where government regulations and guarantees are held to a minimum. System development would be driven by market forces and user requirements rather than technological capabilities. While Terra-Mar looks at the philosophy underlying the development of free-market remote sensing, AS&T and SSI outline their respective entrepreneurial concepts and future plans concerning satellite remote sensing, satellite launch and associated services.

   A. Terra-Mar
         Terra-Mar is developing an earth resources data service aimed primarily at commercial clients. The company's data service plan is based on extensive in-house market research. Terra-Mar states that the exploitable market in remote sensing is based on computer and information technology. They advocate an open market for data within the value-added industry. This company is concerned that the immediate transfer of the existing Landsat assets could be as much a hindrance to commercialization as a benefit, unless the government takes prudent steps to smooth a gradual transition to private operation. Terra-Mar is opposed to near-term transfer of weather satellites because of the vital nature of weather information in serving the national interest.

         (1) Responsiveness to Federal Needs:
             Not specifically addressed

[12]   (2) Continuity of Service:
             Not addressed.

         (3) Feasibility:
             No specific systems were recommended or discussed in detail. Comments regarding the desirability and feasibility of a distributed processing system, and the probable advance of computer technology are well within the current consensus of industry and government on this technology. The efficiencies of a distributed system regarding throughput and availability of data are recognized. In addition, a distributed system can provide many levels of complexity in analysis capability, which will permit superior tailoring of product and information extraction capabilities.

         (4) National Security and Foreign Policy Concerns:
             No concerns were noted.

B. American Science and Technology Corporation (AS&T)

AS&T proposes to take over command and control of Landsat-4 and D' while pursuing the development of their own remote sensing satellites. They do not plan to process or to distribute Landsat data, but will build their own ground segment to serve future AS&T space platforms. AS&T does not express any interest in civil weather satellite systems. Additionally, they do not believe further government regulation or legislation is required or necessary for the implementation of this proposal.

NOTE: AS&T has been working in conjunction with SSI in planning the launch of AS&T remote sensing satellites as early as 1984.

[13]   (1) Responsiveness to Federal Needs:
AS&T would assume operational control (not ownership) of Landsat-4 and Landsat D'. These satellites would be integrated with AS&T's own low-cost Advanced Earth Resources Observation Satellite (AEROS) earth remote sensing satellites, which would provide complementary data. Thus, many user data needs, e.g., data format, compatibility, resolution, frequency of coverage, would be initially satisfied by the combined Landsat-AS&T system. AS&T's proposed flat fee for access to sensors and data flows is attractive, and on the surface, very cost competitive. However, data from non-U.S. areas would have to be relayed via TDRSS, or obtained from foreign ground stations—with potentially high additional costs. Foreign coverage, lacking TDRSS capability and foreign stations, may prove inadequate.

The stable of sensors which AS&T proposes to build and launch by 1985 cannot, by themselves, satisfy current and projected federal requirements for multispectral data since the specifications do not include the spectral coverage or spectral resolutions required. Spatial resolutions which are comparable to Landsat do not solve the problem, since the large majority of analysis is with spectral, not spatial, information extraction.

(2) Continuity of Service:
Continuity of service appears to be assured, provided AS&T can maintain development, launch and operational schedules.

(3) Feasibility:
AS&T proposes the take-over of existing Landsat-4 ground subsystems relating to command, control, and maintenance of the health of the spacecraft. However, AS&T apparently does not possess the broad range of experience and expertise necessary to maintain highly complex spacecraft such as Landsat, nor is there a personnel and facility resource extant in AS&T upon which the company could rely in any spacecraft emergency.

[14]   Statements made previously by AS&T in public fora, but not included in this submittal, have alluded to probable costs of replacement or complementary sensors which are unrealistically low, in our estimation. Technical risks associated with space activities to be undertaken by companies new to this activity are very high, and translate directly into the necessity for large cash contingencies. In addition, the probability that complex sensors can be acquired, integrated, launched and checked out in three years is very low, as discussed in the comparable section on feasibility for SSI.

(4) National Security and Foreign Policy Concerns:
AS&T does not wish to pay for TDRSS. Unless the U.S. government continues to support TDRSS for Landsat, it is unlikely that a current Landsat global data base could be maintained for U.S. or foreign users. This has both national security and foreign policy implications.

The AS&T proposal raises the following additional foreign policy and national security concerns: (a) Possible assumption of U.S. foreign policy obligations by a private firm; and (b) Data distribution in a nondiscriminatory manner to all customers.

C. Space Services Incorporated of America (SSI)
SSI is developing launch services for space activities and is not interested in the ownership or operation of Landsat, related data acquisition or data distribution. SSI believes land and weather satellites should not be considered simultaneously for transfer to private industry, and further that weather systems should not be commercialized at this time. SSI states that any transfer to private industry should not create a monopoly, but allow opportunity to compete in the free marketplace.
[15]   NOTE: SSI has been working in conjunction with AS&T in planning the launch of AS&T remote sensing satellites as early as 1984.

(1) Responsiveness to Federal Needs:
Not specifically addressed.

(2) Continuity of Service:
See AS&T proposal.

(3) Feasibility:
This proposal states an intent to launch private sensors as early as 1984 which would "complement" or "overlap" existing Landsat capabilities. The development and acquisition of free-flying space remote sensors typically take six to eight years. However, disregarding government delays and procedures, and assuming state-of-the-art technology, the fabrication, test, integration and launch might take three to five years given the most optimistic estimate. In addition, the inclusion of cooled focal planes aboard the sensor, which are important in geology and other space applications, are unlikely to be achieved in this time frame, thereby eliminating the majority of federal and other sophisticated users of thermal data. Lastly, SSI has not demonstrated a capability to launch a 1000–2000 Kg payload into orbit, nor does SSI appear to have the capability to command and control the spacecraft, check out on-board systems, or process sensor data. SSI makes the statement that it "supports the concept of allowing the private sector to use government facilities and equipment already in place." However, existing facilities and equipment can only be used by expert, experienced personnel with access to a very broad range of resources to solve problems, maintain system capabilities and provide continuity. SSI does not appear to possess these attributes

[16]   (4) National Security and Foreign Policy Concerns:
SSI recognizes government responsibilities to authorize and supervise private remote sensing activities in accordance with the 1967 Outer Space Treaty and national security interests. However, there is concern whether further commercial space launch services, from within the U.S., be permitted until a well-defined national policy and regulatory framework dealing with such activities have been established and approved.

[17] 3. CATEGORY THREE - Those favoring government retention of the existing system, at least for the immediate future.
The University of Massachusetts, RCA, Hughes Aircraft, General Electric and Ocean Routes Inc., support continued government operation and ownership of the remote sensing system through the next ten years or more. They present alternatives that could lead

to eventual commercialization of Landsat, but these involve extensive government participation through various management, joint-operator or financing options. These respondents are opposed to commercialization of the weather satellite programs, as well as the
creation of any monopolistic commercial entity for land sensing. It should be noted that
three of the nation's major aerospace industries (RCA, GE, and Hughes) share this basic
position.

A.   Remote Sensing Center, University of Massachusetts (UMass)
       The UMass proposal is somewhat ambiguous but looks to an evolutionary, step-
by-step approach leading toward commercialization. However, they believe the current
systems would require extensive federal support if a transfer to private industry were
attempted in the near-term. In the UMass plan, individual segments of the current systems
would be modified and gradually replaced by the private sector until full commercialization of the land sensing system was achieved. They doubt that total private-sector owner-
ship/operation of the weather sensing system is feasible now or in the foreseeable future.

       (1) Responsiveness to Federal Needs:
             Insufficient information is provided to permit an evaluation; however, a
       phased take-over should fulfill user needs initially.

       (2) Continuity of Service:
             Information provided is indeterminate for evaluation; comments in the pre-
       ceding paragraphs apply.

[18]   (3) Feasibility:
             The plan for stepwise commercialization can be implemented within current
       technology. The subsystems of data receive/record, command and control, image
       processing, assessment and analysis, and communications and distribution are
       readily separable. The distributed processing proposed is subject to the same com-
       ments as for Terra-Mar. The (apparent) desire for a central archive, as well as local,
       limited archives, is technically feasible; it requires only a data management deci-
       sion. Since the "existing systems would be unaffected," federal global modeling
       capacities would not be affected in the near-term (8–10 years). However, the pro-
       posal postulates eventual replacement of government sensors by private sensors
       which could acquire "Landsat-like" data. It seems unlikely that the requirements
       for more advanced sets of multispectral data can be satisfied by Landsat-like pri-
       vate sensors. For example, geological researchers are finding that spectral resolu-
       tions on the order of 10–20 nanometers within the short wavelength region
       (1.1–2.6 micrometers) and 30–50 nanometers within the thermal region (8–14
       micrometers) are showing enormous promise for extracting unique signatures of
       surficial minerals, thereby allowing inferences on subsurface content and struc-
       ture. These capabilities are beyond the spectral resolutions on Landsat.

       (4) National Security and Foreign Policy Concerns:
             This response mentions some aspects of national security and foreign policy
       issues—e.g., providing priority weather service to DOD. However, the entire con-
       cept is drawn up in the context of U.S. coverage only, with no provision for meet-
       ing federal needs for foreign area data, or for serving foreign users.

B. RCA (RCA Astro-Electronics, RCA American Communications and RCA Service Company)

RCA believes that Landsat is a candidate for future transfer to private industry only if there are significant government commitments and financial [19] guarantees. While stating that both weather systems (polar-orbiters and geostationary) must remain in the government, RCA offers an alternate financing proposal, i.e., leasing of GOES and TIROS spacecraft, tracking, data reception and data distribution. The leasing option does not provide cost savings over the long term and does not include provisions for R&D, data processing, or launch services.

(1) Responsiveness to Federal Needs:
Not adequately addressed for evaluation, although a leasing arrangement could stipulate those requirements to be met.

(2) Continuity of Service:
Landsat is not addressed. Continuity of metsat data is assured for the duration of a lease.

(3) Feasibility:
RCA has well-established technical credentials to perform as they have outlined. The proposed changes in operations and system upgrade do not represent technical risk or significant technical development. The postulated 3-axis stabilized version of GOES appears reasonable, but not enough information is given to assess the weight margin or implied lifetime. The proposed Satellite Operations Control Center (SOCC) functional flow is credible and achievable, but whether or not higher automation will result in "minimum on-site support" cannot be ascertained. A particularly attractive aspect was the approach to distributed processing architecture, single-point failure recovery, and redundancy. Somewhat questionable was the claim made for optimized real-time versus off-line partitioning for telemetry analysis, since it is not clear what may be optimum for SOCC functional allocations.

[20]   For data distribution, the proposal indicates that satellite communications are the best form of consolidation of National Weather Service (NWS) requirements, but the L-band frequency allocation would require a variance from the FCC. The relay capability proposed for the GOES satellite requires new design work plus development effort on the Marisat traveling-wave-tube, not an attractive prospect from a technical risk standpoint. The electronic specifications given for power, frequency, gain, bandwidth and sidelobes were not analyzed in detail, but appear reasonable.

(4) National Security and Foreign Policy Issues:
There are no national security or foreign policy implications in the RCA response.

C. Hughes Aircraft Company (Hughes)
Hughes does not propose to participate in commercialization activities and states that both land and weather programs should remain in the federal government for at least ten years. Hughes recognizes the national importance of land remote sensing data in international trade and global strategic resource inventories. Therefore, they propose that if transfer of land sensing should proceed, then: (a) The land and weather programs be addressed separately; (b) Vendor selection be done competitively; and (c) No dominant entity or monopoly be created.

(1) Responsiveness to Federal Needs:
Not applicable.

(2) Continuity of Services:
Not applicable.

[21]   (3) Feasibility:
Not applicable.

(4) National Security and Foreign Policy Concerns:
Not applicable.

D. General Electric Company, Space Systems Division (GE)
GE states, "The Landsat program continues to perform as a significant national asset." After extensive analysis, GE finds that private sector takeover of Landsat has unacceptably high business risks at this time. However, ". . . continuation of Landsat under government sponsorship is imperative to protect the $1 billion investment . . . and to permit eventual private sector involvement if it proves economically feasible in the future." Weather satellite systems are not addressed, and there is no basis for substantive evaluation.

(1) Responsiveness to Federal Needs:
Not applicable.

(2) Continuity of Service:
Not applicable.

(3) Feasibility:
Not applicable.

(4) National Security and Foreign Policy Concerns:
Not applicable.

[22] E. Ocean Routes Inc. (OCEANROUTES)
This company strongly opposes the concept of transferring weather satellites to the private sector, since such an enterprise would have to be regulated and run as a monopoly with government subsidies or guarantees. They do not address land remote sensing.

(1) Responsiveness to Federal Needs:
Not applicable.

(2) Continuity of Service:
Not applicable.

(3) Feasibility:
Not applicable.

(4) National Security and Foreign Policy Concerns:
Not applicable.

[23] 4. CATEGORY FOUR - Other Responses.
Three additional responses were received which do not fit the criteria used in this report or explicitly address the questions outlined in the CBD announcement of

September 10, 1982. Comments on these responses are included to express the additional interest generated by the Commerce announcement.

A.   Autometric, Inc.

Autometric favors the move to commercialize civil remote sensing systems but believes a near-term move is premature, i.e., industry is being placed in a position of bidding on an unknown entity. If industry miscalculates the market, the government may have to ". . . bail them out." Autometric does propose a quantitative evaluation of the relative merits of the Landsat 4 Thematic Mapper, the French SPOT [Haute (High) Resolution Visible on SPOT] HRV sensor, and the Large Format Camera that will be flown on the Space Shuttle. The evaluation would be used to ascertain the commercial value of the Thematic Mapper data. No commercialization proposition or comments are provided.

B.   Computer Sciences Corporation (CSC)

CSC has been involved with the Landsat program since its inception. They do not comment directly on private sector transfer, but state a keen interest in the future of land sensing and request inclusion in further discussions which may be held with industry.

C.   Robert Georgevic

This individual described himself as a University Professor from southern California. He did not respond in detail, but advocates retaining civil remote sensing systems in the U.S. government.

[24] IV. LIST OF RESPONSES

Section   Response

1.A   Report, "Commercialization of Civil Remote Sensing," Communications Satellite Corporation (COMSAT), October 22, 1982.

1.B   Unsolicited Proposal, Environmental Satellite Data, Inc., September 17, 1982.

1.C   White Paper, "Civil Operational Remote Sensing From Space," Control Data Corporation, October 1982.

2.A   Document TMA 10-011-82, "Civil Operational Remote Sensing From Space," Terra-Mar, October 1982.

2.B   Report, "Response to the Request for Information With Respect to Civil Operational Remote Sensing From Space," American Science and Technology Corporation, October 22, 1982.

2.C   Letter to Dr. John H. McElroy, Re: "Civil Remote Sensing Satellites; Request for Information," Space Services Incorporated of America, October 22, 1982.

3.A   Document, "A Plan for Commercialization of the U.S. Civil Remote Sensing System," Remote Sensing Center - Hasbrouck, University of Massachusetts, October 21, 1982.

3.B   Report R-4412, "Private Sector Involvement in Civil Operational Remote Sensing from Space," RCA, October 22, 1982.

3.C			Report, "Commercialization of United States Civil Remote Sensing Satellite Systems," Hughes Aircraft Company, October 21, 1982.

[25] 3.D		Letter to Dr. John H. McElroy, "Civil Operational Remote Sensing from Space," Space Systems Division, General Electric Company, October 21, 1982.

3.E			Letter to Dr. John H. McElroy, Response to Secretary's Request for Information, Ocean Routes Incorporated, October 15, 1982.

4.A			Precis, "A Quantitative Evaluation of the Landsat/TM, the SPOT/HRV and the Large Format Camera," Autometric, Incorporated, October 20, 1982.

4.B			Letter to Dr. John H. McElroy, Computer Sciences Corporation, October 21, 1982.

4.C			Postcard to Dr. John H. McElroy from Dr. Robert Georgevic, San Diego, California, October 12, 1982. . . .

Document II-35

Document title: "Transfer of Civil Meteorological Satellites," House Concurrent Resolution 168, November 14, 1983.

Source: NASA Historical Reference Collection, NASA History Office, NASA Headquarters, Washington, D.C.

While Congress was willing to consider transferring control of the Landsat system to the private sector, it did not want to do the same with the weather satellites, citing the argument that weather services were clearly a public good. Therefore, House member Don Fuqua of Florida introduced House Concurrent Resolution 168 in November 1983; the resolution would effectively exclude weather satellites from the privatization process. Because the issue of weather satellites was impeding progress on the transfer of Earth observation satellites to the private sector, President Reagan signed the resolution, thus opening the door for the Land Remote Sensing Commercialization Act of 1984, which was passed just seven months later.

H 9812			**CONGRESSIONAL RECORD—HOUSE**			**November 14, 1983**

[no pagination] TRANSFER OF CIVIL METEOROLOGICAL SATELLITES

Mr. FUQUA. Speaker, I move to suspend the rules and agree to the concurrent resolution, (H. Con. Res. 168) expressing the sense of the Congress that it is not appropriate at this time to transfer ownership or management, of any civil meteorological satellite system and associated ground system equipment to the private sector.
	The clerk read as follows:

H. Con. Res. 168

Whereas the Federal Government has traditionally provided weather forecasts which rely significantly upon data gathered by civil meteorological satellites;
	Whereas within the United States the Federal Government is the principal user of data gathered by civil meteorological satellites:

Whereas the Federal Government has the responsibility for providing forecasts and warnings regarding severe weather in order to protect property and public safety;

Whereas the United States has engaged for over one hundred years in the free international exchange of meteorological data;

Whereas civil meteorological satellite systems and associated ground system equipment are essential components in ensuring the national security of the Nation through their use in conjunction with satellites operated by the Department of Defense;

Whereas transfer to the private sector of ownership or management of any civil meteorological satellite system and associated ground system equipment would likely create a Government-subsidized monopoly and jeopardize the cost efficiency and reliability of data gathered by civil meteorological satellites;

Whereas it is highly unlikely that, under the current plan for transfer of civil meteorological satellites, any significant new commercial venture involving marketing of weather data would develop;

Whereas skepticism in the Congress about the transfer of the civil meteorological satellite system could complicate and delay the pressing decision about the future of the civil land remote sensing satellite system; and

Whereas no satisfactory explanations or proposals have been advanced for the transfer of ownership or management of any civil meteorological satellite and associated ground system equipment to the private sector: Now, therefore, be it

Resolved by the House of Representatives (the Senate concurring), that it is the sense of the Congress that it is not appropriate at this time to transfer ownership or management of any civil meteorological satellite system and associated ground system equipment to the private sector.

The SPEAKER pro tempore. Pursuant to the rule, a second is not required on this motion.

The gentleman from Florida (Mr. FUQUA) will be recognized for 20 minutes and the gentleman from New York (Mr. CARNEY) will be recognized for 20 minutes.

The Chair recognizes the gentleman from Florida (Mr. FUQUA).

Mr. FUQUA. Mr. Speaker, I yield myself 2 minutes.

(Mr. FUQUA asked and was given permission to revise and extend his remarks.)

Mr. FUQUA. Mr. Speaker, House Concurrent Resolution 168 expresses the sense of the Congress that it is not appropriate at this time to sell this Nation's civilian weather satellites to the private sector. The purpose of the concurrent resolution is to halt the administration's effort to sell these satellites by sending a clear and unambiguous signal to the administration and to U.S. industry that the Congress does not consider such a transfer to be in the national interest.

Mr. Speaker, our committee is of the view, after reviewing both the public testimony on this issue and the administration's plans for transfer, that the proposal to transfer weather satellites is not sound and that considerable time, effort, and resources will be saved by halting this proposal immediately. Our committee reached this conclusion for several reasons:

The public service nature of weather services;

The danger of establishing a federally protected monopoly;

National security considerations; and

The need to concentrate the debate on commercializing land remote-sensing satellites.

In sum the committee concurred with the view presented in a November 10, 1983 Joint NASA/Department of Defense study:

There is considerable financial, policy, and program risk to the Government in commercializing weather satellites and there is no clear policy or financial benefit to be realized.

Mr. Speaker, House Concurrent Resolution 168 currently has over 150 co-sponsors. The resolution received strong bipartisan support in our committee, and an identical resolution passed unanimously in the other body last month. This has not been, and it should not be, a partisan issue in the Congress, and I urge my colleagues to support passage of the resolution.

Mr. Speaker, I yield to the gentleman from New York (Mr. SCHEUER), the chairman of the subcommittee that handled this concurrent resolution.

Mr. SCHEUER. I thank the gentleman for yielding.

Mr. Speaker, the chief Sponsor of this measure, the gentleman from Texas (Mr. ANDREWS), has done an outstandingly fine leadership job in promoting this resolution and in developing the support that the chairman just mentioned. I yield such time as he may consume to the gentleman to explain the resolution, what motivated him, what the resolution means, and for any further explanation the gentleman might like to make.

(Mr. ANDREWS of Texas asked and was given permission to revise and extend his remarks.)

Mr. ANDREWS of Texas. I thank the gentleman for yielding time to me.

Mr. Speaker, I rise today to speak on an issue that I suspect many thought fell by the wayside long ago. Back in March of this year when the President announced his intention to commercialize our Nation's four weather satellites, there rose from the public and from Congress an outcry that was so loud, so overwhelmingly clear and devoid of partisanship that I would have thought the administration would have understood that this ill-conceived proposal would never gain the approval of Congress. In hearing after hearing held this year by the Science and Technology Committee, we have received testimony and reports which have been consistently negative with respect to this proposed sale—from sources which include the President's own private sector survey on cost control, NASA, the Department of Defense, three congressionally chartered panels, a Commerce Department Advisory Committee, and the World Meteorological Organization. Representative of the testimony was the conclusion of a joint DOD-NASA study:

There is considerable financial policy, and program risk to the Government in commercializing weather satellites and there is no clear policy or financial benefit to be realized.

Despite the testimony, despite the fact that the Senate has passed not only a resolution identical to this one but language actually prohibiting the use of Commerce funds to write a bid document including the weather satellites, despite the introduction of this resolution in the House; despite its 150 co-sponsors and its overwhelming endorsement by the Science and Technology Committee, the administration is moving ahead full speed. The Department of Commerce is drafting a request for proposal that includes weather satellites and that document is due to be released to the private sector in final form in December.

Selling the weather satellites does not make economic sense. The National Weather Service accounts for 95 percent of the first use of our weather data. By what logic should the Government sell its $1.6 billion weather satellite system to a private company at a greatly reduced rate, only to sign a long-term monopolistic contract for data services which could end up costing taxpayers more than $100 million per year?

Furthermore, such a sale would have serious national security implications. Selling these satellites would necessitate significant, and perhaps unwieldy, oversight and regulation by the DOD which relies on civil weather satellites both in its routine operations and in military emergencies.

The sale would threaten the quality of our weather data since a private operator would have little incentive—other than price-gouging perhaps—to improve services. Stagnant technology would hurt everyone who relies on weather information: the farmer, the pilot and the citizen dependent on Federal tornado and hurricane warnings.

In addition, while some weather information is commercially marketed now, the sale of our satellite system would not enhance that market one iota. In reviewing the initial draft of Commerce's RFP, the DOD insisted that a provision be included allowing the free and open distribution of weather data to our allies around the world, as we have traditionally done. The DOD insisted on this provision because of their fear that should the United States begin selling weather data to foreign governments instead of offering it free, those foreign governments would limit the data they provide to the United States, data which is crucial to U.S. military operations. This leads inescapably to the conclusion that the transfer of these satellites is neither militarily prudent nor commercially viable. For if weather data from U.S. satellites is made available internationally, it can be transmitted on public airwaves to the United States, thus destroying not only the international market, but the domestic market as well. The demands of our military and the requirements of commercialization are irreconcilable.

But perhaps the most compelling reason to disassociate the weather satellites from the RFP process now is that the extended consideration we are giving to their sale only delays the decision to commercialize land remote sensing satellites. The continued coupling of the land and weather satellites only guarantees that the land satellite issue will not be resolved by Congress until as late as mid-1985 at which point we may have forfeited to our competitors overseas what promises to be a multibillion domestic market.

Thanks to the Bateman amendment to the NASA bill, the administration will have to come to Congress sooner or later to seek approval for that RFP. And when they do, I feel sure the weather satellite portion of it will not survive the scrutiny of this Congress. My concern is that the private sector be adequately forewarned. It is disingenuous and unfair to ask American companies to spend their valuable time and resources responding to a proposal that has no hope of getting by Congress. It is time that we send a clear message to the administration and to the boardrooms of those companies that would bid our weather satellites. They are not for sale. That is the sole purpose of this resolution. It deserves to pass and I urge its prompt approval.

Mr. SCHEUER. Mr. Speaker, I yield myself such time as I may consume.

(Mr. SCHEUER asked and was given permission to revise and extend his remarks.)

Mr. SCHEUER. Mr. Speaker, as the gentleman from Texas (Mr. ANDREWS) indicated, this proposal to sell the weather satellites is truly a nonstarter. It is truly a proposal whose time will never come. And it is a dangerous proposal, not just a silly proposal. It is a dangerous proposal because it is diverting our attention from the main game, from the main competitive global arena in which U.S. enterprise really can play a dynamic successful role and that is in the area of Landsat, in the area of Landsat that is appropriate for commercialization. And if we dilly and dally with this absolutely preposterous idea of commercializing weathersat which 13 out of 14 Government-industry respondents have told us is intrinsically a Government function and should never be commercialized, if we dilly and dally, fiddling about with this, we are going to continue to let our position in Landsat erode to the point where the French and the Japanese are going to beat us to the punch and preempt our leadership in space in the very arena where private capital could pay a dynamic and successful role.

Mr. Speaker, 1 would like to expand upon a point raised by the gentleman from Florida (Mr. FUQUA)—that pursuing the sale of weather satellites needlessly complicates the urgent need to maintain U.S. leadership in land remote sensing.

The U.S. Government has operated civilian land-remote-sensing satellites, or Landsat, since 1972.

Since that time, Landsat has provided a wealth of information on: Natural resources, mineral deposits, and agricultural productivity.

At low cost to: Federal agencies, to private industry, and over 40 nations around the world.

The program has been an enormous scientific success, and it has been a great source of international good will for our Nation.

Landsat, unlike weather satellites, might profitably be operated by the private sector, since the data that it produces are of use not only to scientific researchers and to Government agencies, but also to a number of private companies, including oil and mineral exploration firms and various agricultural interests.

I agree with the administration that the time is right to investigate our options for commercializing this emerging technology, especially since the United States will face stiff competition in the field of land remote-sensing from the French and the Japanese by the mid- to late-1980's.

I disagree strongly, however, with the administration's perverse approach to this international challenge—namely, to kill immediately all funding for the Landsat program and to put our civilian weather satellites, an inherently governmental system, up for sale.

The administration pursued these policies under the assumption that U.S. industry would rush into competition with subsidized French and Japanese systems by launching fully private land remote-sensing systems.

Predictably the gamble failed.

No U.S. industry could possibly enter this field at the present time without some form of temporary Government subsidy.

This subsidy is a necessary evil if we hope to avoid giving up another potentially lucrative commercial opportunity to our competitors.

The irony of our deteriorating situation, of course, is that this commercial opportunity exists only because the U.S. taxpayer, over the past 12 years, has supported a Landsat program which has convincingly demonstrated the feasibility of commercial land remote sensing.

Mr. Speaker, many of us share a deep concern that the United States is about to fritter away its technological leadership in an area where, over the next 20 years, we expect to see tremendous global commercial expansion.

In this case, our industrial expansion and ability to compete in global markets is threatened not by a lack of resources, nor by a lack of creativity, nor by a lack of productivity.

We are threatened solely by Government inaction.

I believe we still have time to preserve our remote-sensing industry, but only if we act expeditiously on two fronts.

First, we need to halt this absurd proposal to commercialize weather satellites. It is a silly diversion from the main global game—Landsat. We must get our eye on the ball.

Next, we need to agree on a policy which will effect a rational transition between Government operation of land remote-sensing systems and fully private operation of these systems.

Last week, the gentleman from Missouri (Mr. VOLKMER) and I held 2 days of joint hearings on draft legislation aimed at effecting such a transition.

Concurrent with our consideration of this draft legislation, the administration has been preparing a request-for-proposal (RFP) to solicit industry bids on present and future land remote-sensing systems.

I was greatly heartened by the testimony that we received on the legislation, both from witnesses representing U.S. industry and from the administration's witness, Mr. Ray Kammer, who is chairman of the Source Evaluation Board for Civil Remote Sensing.

Mr. Kammer found many parallels and few major inconsistencies between his efforts and the provisions of the legislation, and I look forward to working with Mr. Kammer and others in the administration to insure that, as Congress articulates a coherent policy for commercializing land remote-sensing activities over the next 6 months, the administration aggressively implements that policy.

It would be truly unfortunate to see a repetition of the circumstances surrounding the weather-satellite issue, wherein despite repeated signals from the Congress, the administration continued to pursue a policy which had little, if any, support in either House of Congress.

Mr. Chairman, I strongly support the concurrent resolution sponsored by the gentleman from Texas (Mr. ANDREWS), both because selling weather satellites is a bad idea and because we need to proceed quickly toward the resolution of the real issue of this debate—namely, how to transfer, in a rational way, our land remote-sensing capabilities to the private sector.

I urge my colleagues to support House Concurrent Resolution 168.

1640

Mr. McGRATH. Mr. Speaker, I yield myself such time as I may consume.

(Mr. McGRATH asked and was given permission to revise and extend his remarks.)

Mr. McGRATH. Mr. Speaker, I rise in support of this resolution, but I do so with some reservations. Right now I do not feel that the transfer to the private sector of the weather satellites is in the national interest. However, I would like to withhold final judgment on this issue until the RFP process currently being conducted by the Department of Commerce is complete.

After the responses to the RFP are in, we will have a great deal more information about whether or not to go forward with this transfer. We will have the factual basis for determining whether the weather-satellite system is suitable for transfer to the private sector or if, as I believe will be the case, they are not suitable.

Mr. Speaker, I have been following this issue very carefully for the past 2 1/2 years. The civilian weather satellites are essential to the protection of public health and safety, and they serve as a backup to the defense meteorological system. The weather satellites are inherently governmental, as evidenced by the fact that over 95 percent of the market for the data from these satellites is the U.S. Government. For these reasons, I do not think it would be wise to transfer this system to the private sector.

In the Committee on Science and Technology, we held five hearings on the commercialization of the land and weather satellites, and while the consensus was that the land system should be commercialized, strong opposition to the transfer of the Metsat system was expressed by a wide range of witnesses from both the public and private sectors.

Mr. Speaker, it is my hope that once the controversy of the Metsat transfer has been removed from consideration, the Congress and the administration will be able to concentrate on the commercialization of the land remote-sensing system which is more rationally in the public interest.

Mr. Speaker. I yield 2 minutes for myself.

(Mrs. SMITH of Nebraska asked and was given permission to revise and extend her remarks.)

Mrs. SMITH of Nebraska. Mr. Speaker, I rise in strong support of this legislation expressing the sense of Congress with regard to weather satellites.

I am a cosponsor of this bill which puts Congress on record in opposition to the transfer of ownership or management of the Nation's weather satellites to the private sector at this time.

In March of this year, the Department of Commerce proposed to commercialize civil weather and land remote-sensing satellites. This proposal has had remarkably little support especially with respect to weather satellites. It is no wonder. Over 99 percent of the data generated by the weather satellites is in fact used by civil and military agencies of the U.S. Government.

The proposed sale of weather satellites would not save the Government any money. The Defense Department estimates that the sale would cost tax-payers about $800 million more over 10 years than it would cost the Government to continue to operate them.

Congressional hearings have brought out considerable opposition to selling off the Government's satellites. The administration's own private sector survey on cost control, the Department of Defense, NASA, and other Government groups all find that there is no advantage to commercializing the weather satellites.

My main concern about the satellite sale proposal is the effect it would have on our farmers and ranchers, pilots, and other citizens who depend on accurate and timely reports on the weather. The future of the present weather reporting systems would be in doubt with the sale of our "eyes in the sky." There are no answers to the questions about future availability of weather information, the national security implications of selling our satellites, or the safety implications because this idea has not had enough study. Let us adopt this simple resolution and give the whole issue of weather satellites and Landsat thorough study.

Mr. McGRATH. Mr. Speaker, I yield 2 minutes, for debate only, to the gentleman from Virginia (Mr. BATEMAN).

Mr. BATEMAN Mr. Speaker, I appreciate the gentleman's yielding this time to me.

I would like to associate myself with the remarks that have been made by the distinguished Members who preceded me who spoke in favor of the resolution and against the proposition of commercializing our weather satellite systems.

As the gentleman from Texas (Mr. ANDREWS) observed, earlier this year I offered an amendment which was adopted and has passed both Houses of the Congress which would have made the prior consent of the Congress as to the commercialization of the weather satellites a condition of such a step being taken. In view of the fact that the other body has conclusively and unanimously indicated that it is not disposed and would not consent to any commercialization of the weather satellites, it is my feeling, which I share with the gentleman from Texas, that private industry and the Government itself should not be put through a frivolous exercise, with the expenditure of a great deal of money, research, investigation, and study, all to come to absolutely naught.

For those reasons Mr. Speaker, I think this resolution is well conceived and should be supported.

Mr. McGRATH. Mr. Speaker, I reserve the balance of my time.

Mr. FUQUA. Mr. Speaker, I yield such time as he may consume to the distinguished gentleman from Florida (Mr. NELSON), a member of the committee.

Mr. NELSON of Florida. Mr. Speaker, I thank the chairman of the committee for yielding this time to me.

I just want to say that this resolution sponsored by the gentleman from Texas (Mr. ANDREWS) is a resolution that we should not have to be debating here. But here is constitutional government working at its best, one branch balancing off against the other branch. The executive branch has simply made a mistake. And they have to be brought back onto the correct course by the congressional branch.

With security at risk, we should not be sending our weather satellites to the commercial sector where in a time of emergency the Government may not have access to that weather information.

So, Mr. Speaker, I certainly urge a yes vote on this resolution.

Mr. FUQUA. Mr. Speaker, I yield such time as he may consume to the distinguished delegate from Puerto Rico (Mr. CORRADA).

Mr. CORRADA. Mr. Speaker, I thank the gentleman for yielding this time to me.

Mr. Speaker, I join in voicing my strong support for House Concurrent Resolution 168 which expresses the sense of the Congress that the civil meteorological satellite system should not be sold or transferred to the private sector.

The important data and work collected by these satellites underscore the need for maintaining them in Government hands. The collection and dissemination of weather information is crucial to every person across the Nation, from the farmers in Nebraska to vacationers in Puerto Rico. The need for precise and reliable access must be protected and preserved to ensure not only the health of our population but also our continued economic vitality.

The plan to transfer the weather satellites to the private sector is misguided and inimical to the public good. I urge my colleagues to vote in favor of this resolution and thus send a clear signal to the administration about our position on this important issue.

Mr. McGRATH. Mr. Speaker, I yield 2 minutes, for debate purposes, to the gentleman from Pennsylvania (Mr. WALKER).

Mr. WALKER. Mr. Speaker, in his book "The High Road," Ben Bova talked about the proposition that there are two types of people that approach policy in this country or throughout history. They are the Promethians and the Luddites. The Promethians are essentially those people who look toward the future, who try to find ways in which to use that which is in the present to promote the future. The Luddites are those people who are wedded to the past, who try to hold on to exactly what is without having any concept of the future at all.

I submit, Mr. Speaker, that this is a Luddite bill. I say that by stating that we are unwilling to move forward toward commercialization of weather satellites; we are exempting one large area from commercialization that has the potential of being commercialized immediately. If there is one thing that we should be all about in our space program, it is moving out of research and development toward commercialization. The more commercialization we promote as a nation in outer space, the more chance we have of reaping the economic rewards that come from that.

1650

These are not minor economic rewards. There are some people who are farsighted enough to believe that commercialization of outer space over the next 20 to 25 years, if given the proper investment attitude, could reap a trillion dollar economy from outer space, a spaced-based economy of a trillion dollars. That in terms of 1983, that is the equivalent of 35 million jobs.

We often sit on this House floor and we hear debating about the fact, where are the jobs going to come from? How are we going to provide jobs for the future? Where are the jobs for people who do not have them? With hi-tech emerging, where are the jobs going to come from?

One of the places they are going to come from is by properly industrializing and commercializing outer space. When we pass bills of this type, when we say that we are going to take as a matter of public policy and X-out of our consideration weather satellites, we are taking the first step toward limiting the amount of investment that will ever be made in outer space and the amount of jobs that can be created.

I think that is wrong. I think this is a sad bill and I hope people will vote against it.

Mr. BROOKS. Mr. Speaker, this past March the President announced his intention to commercialize our Government's weather and land remote sensing satellites. It was clear from the beginning that the offer to sell the weather satellites was simply an attempt to make the sale of the land remote sensing satellite more appealing.

I find this entire commercialization effort most bothersome and have been vocally opposed to it from the very beginning. While many of us may disagree with the desirability of retaining a land remote sensing satellite program within the Government, I know of no one who disagrees with the wisdom of retaining Government control over our Nation's weather satellites. Many of us have made this fact known to the administration during the past few months, and yet the process to commercialize them continues. At risk, if this process continues to completion, is the very security and well-being of our Nation, for we depend daily upon our ability to gather and analyze weather data from around the country as well as the globe.

I urge all of you to support this resolution, of which I am a cosponsor, and hope that the administration will financially understand and accept that our Nation's weather satellites will not be sold.

Mr. McGRATH. Mr. Speaker, I yield back the balance of my time.

GENERAL LEAVE

Mr. FUQUA. Mr. Speaker, I ask unanimous consent that all Members may have 5 legislative days in which to revise and extend their remarks on House Concurrent Resolution 168.

The SPEAKER pro tempore. Is there objection to the request of the gentleman from Florida?

There was no objection.

Mr. FUQUA. Mr. Speaker, I have no further requests for time and I yield back the balance of my time.

The SPEAKER pro tempore. The question is on the motion offered by the gentleman from Florida (Mr. FUQUA) that the House suspend the rules and agree to the concurrent resolution, House Concurrent Resolution 168.

The question was taken.

Mr. CARNEY. Mr. Speaker, on that I demand the yeas and nays.

The yeas and nays were ordered.

The SPEAKER, pro tempore. Pursuant to the provisions of clause 5 of rule I and the Chair's prior announcement, further proceedings on this motion will be postponed. . . .

1810

So (two-thirds having voted in favor thereof) the rules were suspended and the concurrent resolution was agreed to.

The result of the vote was announced as above recorded [yeas 377, nays 28, not voting 29].

A motion to reconsider was laid on the table.

Document II-36

Document title: "Land Remote-Sensing Commercialization Act of 1984," Public Law 98–365, 98 Stat. 451, July 17, 1984.

Source: NASA Historical Reference Collection, NASA History Office, NASA Headquarters, Washington, D.C.

The Land Remote-Sensing Commercialization Act of 1984 was the culmination of many years of debate over who should control the Earth remote-sensing system operated through the Landsat satellites. Although the Reagan administration had advocated the outright sale of the entire Landsat system as well as the nation's weather satellites, the Remote Sensing Act of 1984 was much more limited in scope. Specifically, it gave the Secretary of Commerce authority to contract to private industry the marketing of unenhanced Landsat data. Subsequently, the contract was awarded to the Earth Observation Satellite Company (EOSAT), a joint venture between RCA and Hughes Aircraft Corporation.

[no pagination]   PUBLIC LAW 98–365—JULY 17, 1984

                                                              98 STAT. 451

Public Law 98–365
98th Congress

An Act

To establish a system to promote the use of land remote-sensing satellite data, and for other purposes. [citation in margin: "July 17, 1984 (H.R. 5155)."]

Be it enacted by the Senate and House of Representatives of the United States of America in Congress assembled, That this Act may be cited as the "Land Remote-Sensing Commercialization Act of 1984." [citation in margin: "Land Remote Sensing Commercialization Act of 1984. Communications and telecommunications. 15 USC 4201 note."]

TITLE I—DECLARATION OF FINDINGS, PURPOSES, AND POLICIES

FINDINGS

SEC. 101. The Congress finds and declares that— [citation in margin: "Congress. 15 USC 4201."]

(1) the continuous civilian collection and utilization of land remote-sensing data from space are of major benefit in managing the Earth's natural resources and in planning and conducting many other activities of economic importance;

(2) the Federal Government's experimental Landsat system has established the United States as the world leader in land remote-sensing technology; [marginal note: "Landsat system."]

(3) the national interest of the United States lies in maintaining international leadership in civil remote sensing and in broadly promoting the beneficial use of remote-sensing data;

(4) land remote sensing by the Government or private parties of the United States affects international commitments and policies and national security concerns of the United States; [marginal note: "Defense and national security."]

(5) the broadest and most beneficial use of land remote-sensing data will result from maintaining a policy of nondiscriminatory access to data;

(6) competitive, market-driven private sector involvement in land remote sensing is in the national interest of the United States;

(7) use of land remote-sensing data has been inhibited by slow market development and by the lack of assurance of data continuity;

(8) the private sector, and in particular the "value-added" industry, is best suited to develop land remote-sensing data markets;

(9) there is doubt that the private sector alone can currently develop a total land remote-sensing system because of the high risk and large capital expenditure involved;

(10) cooperation between the Federal Government and private industry can help assure both data continuity and United States leadership;

(11) the time is now appropriate to initiate such cooperation with phased transition to a fully commercial system;

(12) such cooperation should be structured to involve the minimum practicable amount of support and regulation by the Federal Government and the maximum practicable amount of competition by the private sector while assuring continuous availability to the Federal Government of land remote-sensing data;

(13) certain Government oversight must be maintained to assure that private sector activities are in the national interest and that the international commitments and policies of the United States are honored; and

(14) there is no compelling reason to commercialize meteorological satellites at this time.

PURPOSES

SEC. 102. The purposes of this Act are to— [citation in margin: "15 USC 4202."]

(1) guide the Federal Government in achieving proper involvement of the private sector by providing a framework for phased commercialization of land remote sensing and by assuring continuous data availability to the Federal Government;

(2) maintain the United States worldwide leadership in civil remote sensing, preserve its national security, and fulfill its international obligations; [marginal note: "Defense and national security."]

(3) minimize the duration and amount of further Federal investment necessary to assure data continuity while achieving commercialization of civil and land remote sensing;

(4) provide for a comprehensive civilian program of research, development, and demonstration to enhance both the United States capabilities for remote sensing from space and the application and utilization of such capabilities; and

(5) prohibit commercialization of meteorological satellites at this time.

POLICIES

SEC. 103. (a) It shall be the policy of the United States to preserve its right to acquire and disseminate unenhanced remote-sensing data. [citation in margin: "15 USC 4203."]

(b) It shall be the policy of the United States that civilian unenhanced remote-sensing data be made available to all potential users on a nondiscriminatory basis and in a manner consistent with applicable antitrust laws.

(c) It shall be the policy of the United States both to commercialize those remote-sensing space systems that properly lend themselves to private sector operation and to avoid competition by the Government with such commercial operations, while continuing to preserve our national security, to honor our international obligations, and to retain in the Government those remote-sensing functions that are essentially of a public service nature. [marginal note: "Defense and national security."]

DEFINITIONS

SEC. 104. For purposes of this Act: [citation in margin: "15 USC 4204."]

(1) The term "Landsat system" means Landsats 1, 2, 3, 4, and 5, and any related ground equipment, systems, and facilities, and any successor civil land remote-sensing space systems operated by the United States government prior to the commencement of the six-year period described in title III.

(2) The term "Secretary" means the Secretary of Commerce.

(3) (A) The term "nondiscriminatory basis" means without preference bias, or any other special arrangement (except on the basis of national security concerns pursuant to section 607) regarding delivery, format, financing, or technical considerations which would favor one buyer or class of buyers over another.

(B) The sale of data is made on a nondiscriminatory basis only if (i) any offer to sell or deliver data is published in advance in such manner as will ensure that the offer is equally available to all prospective buyers; (ii) the system operator has not established or changed any price, policy, procedure, or other term or condition in a manner which gives one buyer or class of buyer de facto favored access to data; (iii) the system operator does not make unenhanced data available to any purchaser on an exclusive basis; and (iv) in a case where a system operator offers volume discounts, such discounts are no greater than the demonstrable reductions in the cost of volume sales. The sale of data on a nondiscriminatory basis does not preclude the system operator from offering discounts other than volume discounts to the extent that such discounts are consistent with the provisions of this paragraph.

(C) The sale of data on a nondiscriminatory basis does not require (i) that a system operator disclose names of buyers or their purchases; (ii) that a system operator maintain all, or any particular subset of, data in a working inventory; or (iii) that a system operator expend equal effort in developing all segments of a market.

(4) The term "unenhanced data" means unprocessed or minimally processed signals or film products collected from civil remote-sensing space systems. Such minimal processing may include rectification or distortions, registration with respect to features of the Earth, and calibration of spectral response. Such minimal processing does not include conclusions, manipulations, or calculations derived from such signals or film products or combination of the signals or film products with other data or information.

(5) The term "system operator" means a contractor under title II or title III or a license holder under title IV.

TITLE II—OPERATION AND DATA MARKETING OF LANDSAT SYSTEM

OPERATION

SEC. 201. (a) The Secretary shall be responsible for— [citation in margin: "15 USC 4211."]

(1) the Landsat system, including the orbit, operation, and disposition of Landsat system, including the orbit, operation, and disposition of Landsats 1, 2, 3, 4, and 5; and

(2) provision of data to foreign ground stations under the terms of agreements between the United States Government and nations that operate such ground stations which are in force on the date of commencement of the contract awarded pursuant to the title.

(b) The provisions of this section shall not affect the Secretary's authority to contract for the operation or part or all of the Landsat system, so long as the United States Government retains—

(1) ownership of such system;

(2) ownership of the unenhanced data; and

(3) authority to make decisions concerning operation of the system.

CONTRACT FOR MARKETING OF UNENHANCED DATA

SEC. 202. (a) In accordance with the requirements of this title, the Secretary, by means of a competitive process and to the extent provided in advance by appropriation Acts, shall contract with a United States private sector party (as defined by the Secretary) for the marketing of unenhanced data collected by the Landsat system. Any such contract— [citation in margin: "15 USC 4212."]

(1) shall provide that the contractor set the prices of unenhanced data;

(2) may provide for financial arrangements between the Secretary and the contractor including fees for operating the system, payments by the contractor as an initial fee or as a percentage of sales receipts, or other such considerations;

(3) shall provide that the contractor will offer to sell and deliver unenhanced data to all potential buyers on a nondiscriminatory basis;

(4) shall provide that the contractor pay to the United States Government the full purchase price of any unenhanced data that contractor elects to utilize for purposes other than sale;

(5) shall be entered into by the Secretary only if the Secretary has determined that such contract is likely to result in net cost savings for the United States Government; and

(6) may be reawarded competitively after the practical demise of the space segment of the Landsat system, as determined by the Secretary.

(b) Any contract authorized by subsection (a) may specify that the contractor use, and, at his own expense, maintain, repair, or modify, such elements of the Landsat system as the contractor finds necessary for commercial operations.

(c) Any decision or proposed decision by the Secretary to enter into any such contract shall be transmitted to the Committee on Commerce, Science, and Transportation of the Senate and the Committee on Science and Technology of the House of Representatives for their review. No such decision or proposed decision shall be implemented unless (A) a period of thirty calendar days has passed after the receipt by each such committee of such transmittal, or (B) each such committee before the expiration of such period has agreed to transmit and has transmitted to the Secretary written notice to the effect that such committee has no objection to the decision or proposed decision. As part of the transmittal, the Secretary shall include information on the terms of the contract described in subsection (a). [marginal note: "Congress."]

(d) In defining "United Stated private sector party" for purposes of this Act, the Secretary may take into account the citizenship of key personnel, location of assets, foreign ownership, control, influence, and other such factor.

CONDITIONS OF COMPETITION FOR CONTRACT

SEC. 203. (a) The Secretary shall, as part of the advertisement for the competition for the contract authorized by section 202, identify and publish the international obligations, national security concerns (with appropriate protection of sensitive information, domestic legal considerations, and any other standards or conditions which a private contractor shall be required to meet). [citation in margin: "15 USC 4213."]

(b) In selecting a contractor under this title, the Secretary shall consider—

(1) ability to market aggressively unenhanced data;

(2) the best overall financial return to the Government, including the potential cost savings to the Government that are likely to result from the contract;

(3) ability to meet the obligations, concerns, considerations, standards, and conditions identified under subsection (a);

(4) technical competence, including the ability to assure continuous and timely delivery of data from the Landsat system;

(5) ability to effect a smooth transition with the contractor selected under title III; and

(6) such other factors as the Secretary deems appropriate and relevant.

(c) If, as a result of the competitive process required by section 202(a), the Secretary receives no proposal which is acceptable under the provisions of this title, the Secretary shall so certify and fully report such finding to the Congress. As soon as practicable but not later than thirty days after so certifying and reporting, the Secretary shall reopen the competitive process. The period for the subsequent competitive process shall not exceed one hundred and twenty days. If, after such subsequent competitive process, the Secretary receives no proposal which is acceptable under the provisions of this title, the Secretary shall so certify and fully report such finding to the Congress. In the event that no acceptable proposal is received, the Secretary shall continue to market data from the Landsat system. [marginal note: "Report."]

(d) A contract awarded under section 202 may, in the discretion of the Secretary, be combined with the contract required by title III, pursuant to section 304(b).

SALE OF DATA

SEC. 204. (a) After the date of the commencement of the contract described in section 202(a), the contractor shall be entitled to revenues from sales of copies of data from the Landsat system, subject to the conditions specified in sections 601 and 602. [citation in margin: 15 USC 4214."]

(b) The contractor may continue to market data previously generated by the Landsat system after the demise of the space segment of the system.

FOREIGN GROUND STATIONS

SEC. 205. (a) The contract under this title shall provide that Contractor shall act as the agent of the Secretary by continuing to supply unenhanced data to foreign ground stations for the life, and according to the terms of those agreements between the United States Government and such foreign ground stations that are in force on the date of the commencement of the contract. [citation in margin: "15 USC 4215."]

(b) Upon the expiration of such agreements, or in the case of foreign ground stations that have no agreement with the United States on the date of commencement of the contract, the contract shall provide—

(1) that unenhanced data from the Landsat system shall be made available to foreign ground stations only by the contractor; and

(2) that such data shall be made available on a nondiscriminatory basis.

TITLE III—PROVISION OF DATA CONTINUITY AFTER THE LANDSAT SYSTEM

PURPOSES AND DEFINITION

SEC. 301 (a) It is the purpose of this title— [citation in margin: "15 USC 4221."]

(1) to provide, in an orderly manner and with minimal risk, for a transition from Government operation to private, commercial operation of civil land remote-sensing systems; and

(2) to provide data continuity for six years after the practical demise of the space segment of the Landsat system.

(b) For purposes of this title, the term "data continuity" means the continued availability of unenhanced data—

(1) including data which are from the point of view of a data user—

(A) functionally equivalent to the multispectral data generated by the Landsat 1 and 2 satellites; and

(B) compatible with such data and with equipment used to receive and process such data; and

(2) at an annual volume at least equal to the Federal usage during fiscal year 1983.

(c) Data continuity may be provided using whatever technologies are available.

DATA CONTINUITY AND AVAILABILITY

SEC. 302. The Secretary shall solicit proposals from Unites States private sector parties (as defined by the Secretary pursuant to section 202) for a contract for the development and operation of a remote-sensing space system capable of providing data continuity for a period of six years and for marketing unenhanced data in accordance with the provisions of sections 601 and 602. Such proposals, at a minimum, shall specify— [citation in margin: "Contracts with U.S. 15 USC 4222."]

(1) the quantities and qualities of unenhanced data expected from the system;

(2) the projected date upon which operations could begin;

(3) the number of satellites to be constructed and their expected lifetimes;

(4) any need for Federal funding to develop the system;

(5) any percentage of sales receipts or other returns offered to the Federal Government;

(6) plans for expanding the market for land remote-sensing data; and

(7) the proposed procedures for meeting the national security concerns and international obligations of the United States in accordance with section 607.

AWARDING OF THE CONTRACT

SEC. 303. (a)(1) In accordance with the requirements of this title, the Secretary shall evaluate the proposals described in section 302 and, by means of a competitive process and to the extent provided in advance by appropriation Acts, shall contract with the United States private sector party for the capability of providing data continuity for a period of six years and for marketing unenhanced data. [citation in margin: "15 USC 4223."]

(2) Before commencing space operations the contractor shall obtain a license under title IV.

(b) As part of the evaluation described in subsection (a), the Secretary shall analyze the expected outcome of each proposal in terms of—

(1) the net cost to the Federal Government of developing the recommended system;

(2) the technical competence and financial condition of the contractor;

(3) the availability of such data after the expected termination of the Landsat system;

(4) the quantities and qualities of data to be generated by the recommended system;

(5) the contractor's ability to supplement the requirement for data continuity by adding, at the contractor's expense, remote-sensing capabilities which maintain United States leadership in remote sensing;

(6) the potential to expand the market for data;

(7) expected returns to the Federal Government based on any percentage of data sales or other such financial consideration offered to the Federal Government in accordance with section 305;

(8) the commercial viability of the proposal;

(9) the proposed procedures for satisfying the national security concerns and international obligations of the United States;

(10) the contractor's ability to effect a smooth transition with any contractor selected under title II; and

(11) such other factors as the Secretary deems appropriate and relevant.

(c) Any decision or proposed decision by the Secretary to enter into any such contract shall be transmitted to the Committee on Commerce, Science, and Transportation of the Senate and the Committee on Science and Technology of the House of Representatives for their review. No such decision or proposed decision shall be implemented unless (1) a period of thirty calendar days has passed after the receipt by each such committee of such transmittal, or (2) each such committee before the expiration of such period has agreed to transmit and has transmitted to the Secretary written notice to the effect that such committee has no objection to the decision or proposed decision. As part of the transmittal, the Secretary shall include the information specified in subsection (a). [marginal note: "Congress."]

(d) If, as a result of the competitive process required by this section, the Secretary receives no proposal which is acceptable under the provisions of this title, the Secretary shall so certify and fully report such finding to the Congress. As soon as practicable but not later than thirty days after so certifying and reporting, the Secretary shall reopen the competitive process. The period for the subsequent competitive process shall not exceed one hundred and eighty days. If, after such subsequent competitive process, the Secretary receives no proposal which is acceptable under the provisions of this title, the Secretary shall so certify and fully report such finding to the Congress. Not earlier than ninety days after such certification and report, the Secretary may assure data continuity by procurement and operation by the Federal Government of the necessary systems, to the extent provided in advance by appropriation Acts. [marginal note: "Report."]

TERMS OF CONTRACT

SEC. 304. (a) Any contract entered into pursuant to this title— [citation in margin: "15 USC 4224."]

(1) shall be entered into as soon as practicable, allowing for the competitive procurement process required by this title;

(2) shall, in accordance with criteria determined and published by the Secretary, reasonably assure data continuity for a period of six years, beginning as soon as practicable in order to minimize any interruption of data availability;

(3) shall provide that the contractor will offer to sell and deliver unenhanced data to all potential buyers on a nondiscriminatory basis;

(4) shall not provide a guarantee of data purchases from the contractor by the Federal Government;

(5) may provide that the contractor utilize, on a space-available basis, a civilian United States Government satellite or vehicle as a platform for a civil land remote-sensing space system, if—

(A) the contractor agrees to reimburse the Government immediately for all related costs incurred with respect to such utilization, including a reasonable and proportionate share of fixed, platform, data transmission, and launch costs; and

(B) such utilization would not interfere with or otherwise compromise intended civilian Government missions, as determined by the agency responsible for the civilian platform; and

(6) may provide financial support by the United States Government, for a portion of the capital costs required to provide data continuity for a period of six years, in the form of loans, loan guarantees, or payments pursuant to section 305 of the Federal Property and Administrative Services Act of 1949 (41 U.S.C. 255).

(b)(1) Without regard to whether any contract entered into under this title is combined with a contract under title II, the Secretary shall promptly determine whether the contract entered into under this title reasonably effectuates the purposes and policies of title II. Such determination shall be submitted to the President and the Congress, together with a full statement of the basis for such determination.

(2) If the Secretary determines that such contract does not reasonably effectuate the requirements of title II, the Secretary shall promptly carry out the provisions of such title to the extent provided in advance in appropriation Acts.

MARKETING

SEC. 305. (a) In order to promote aggressive marketing of land remote-sensing data, any contract entered into pursuant to this title may provide that the percentage of sales paid by the contractor to the Federal Government shall decrease according to stipulated increase in sales levels. [citation in margin: "15 USC 4225."]

(b) After the six-year period described in section 304(a)(2), the contractor may continue to sell data. If licensed under title IV, the contractor may continue to operate a civil remote-sensing space system.

REPORT

SEC. 306. Two years after the date of the commencement of the six-year period described in section 304(a)(2), the Secretary shall report to the President and to the Congress on the progress of the transition to fully private financing, ownership, and operation of remote-sensing space systems, together with any recommendations for actions, including actions necessary to ensure United States leadership in civilian land remote sensing from space. [citation in margin: "15 USC 4226."]

TERMINATION OF AUTHORITY

SEC. 307. The authority granted to the Secretary by this title shall terminate ten years after the date of enactment of this Act. [citation in margin: "15 USC 4227."]

TITLE IV—LICENSING OF PRIVATE REMOTE-SENSING SPACE SYSTEMS

GENERAL AUTHORITY

SEC. 401. (a)(1) In consultation with other appropriate Federal agencies, the Secretary is authorized to license private sector parties to operate private remote-sensing space systems for such period as the Secretary may specify and in accordance with the provisions of this title. [citation in margin: "15 USC 4241."]

(2) In the case of a private space system that is used for remote sensing and other purposes, the authority of the Secretary under this title shall be limited only to the remote-sensing operations of such space system.

(b) No license shall be granted by the Secretary unless the Secretary determines in writing that the applicant will comply with the requirements of this Act, any regulations issued pursuant to this Act, and any applicable international obligations and national security concerns of the United States.

(c) The Secretary shall review any application and make a determination thereon within one hundred and twenty days of the receipt of such application. If final action has not occurred within such time, the Secretary shall inform the applicant of any pending issues and of actions required to resolve them. [marginal note: "Review date."]

(d) The Secretary shall not deny such license in order to protect any existing licenses from competition.

CONDITIONS FOR OPERATION

SEC. 402. (a) No person who is subject to the jurisdiction or control of the United States may, directly or through any subsidiary or affiliate, operate any private remote-sensing space system without a license pursuant to section 401.

(b) Any license issued pursuant to this title shall specify, at a minimum, that the license shall comply with all of the requirements of this Act and shall—

(1) operate the system in such manner as to preserve and promote the national security of the United States and to observe and implement the international obligations of the United States in accordance with section 607;

(2) make unenhanced data available to all potential users on a nondiscriminatory basis;

(3) upon termination of operations under the license, make disposition of any satellites in space in a manner satisfactory to the President;

(4) promptly make available all unenhanced data which the Secretary may request pursuant to section 602;

(5) furnish the Secretary with complete orbit and data collection characteristics of the system, obtain advance approval of any intended deviation from such characteristics, and inform the Secretary of any unintended deviation;

(6) notify the Secretary of any agreement the licensee intends to enter with a foreign nation, entity, or consortium involving foreign nations or entities;

(7) permit the inspection by the Secretary of the licensee's equipment, facilities, and financial records;

(8) surrender the license and terminate operations upon notification by the Secretary pursuant o the section 403(a)(1); and

(9) (A) notify the Secretary of any "value added" activities (as defined by the Secretary by regulation) that will be conducted by the licensee or by a subsidiary or affiliate; and

(B) if such activities are to be conducted, provide the Secretary with a plan for compliance with the provisions of this Act concerning nondiscriminatory access.

ADMINISTRATIVE AUTHORITY OF THE SECRETARY

SEC. 403. (a) In order to carry out the responsibilities specified in this title, the Secretary may— [citation in margin: "15 USC 4243."]

(1) grant, terminate, modify, condition, transfer, or suspend licenses under this title, and upon notification of the licensee may terminate licensed operations on an immediate basis, if the Secretary determines that the licensee has substantially failed to comply with any provision of this Act, with any regulation issued under this Act, with any terms, conditions, or restrictions of such license, or with any international obligations or national security concerns of the United States;

(2) inspect the equipment, facilities, or financial records of any licensee under this title;

(3) provide penalties for noncompliance with the requirements for licenses or regulations issued under this title, including civil penalties not to exceed $10,000 (each day of operation in violation of such licenses or regulations constituting a separate violation);

(4) compromise, modify, or remit any such civil penalty;

(5) issue subpenas [sic] for any materials, documents, or records, or for the attendance and testimony of witnesses for the purpose of conducting a hearing under this section;

(6) seize any object, record, or report where there is probable cause to believe that such object, record, or report was used, is being used, or is likely to be used in violation of this Act, or the requirements of a license or regulation issued thereunder; and

(7) make investigations and inquiries and administer to or take from any person an oath, affirmation, or affidavit concerning any matter relating to the enforcement of this Act.

(b) Any applicant or licensee who makes a timely request for review of an adverse action pursuant to subsection (a)(1), (a)(3), or (a)(6) shall be entitled to adjudication by the Secretary on the record after an opportunity for an agency hearing with respect to such adverse action. Any final action by the Secretary under this subsection shall be subject to judicial review under chapter 7 of title 5, United States Code. [citation in margin: "5 USC 701 *et seq.*"]

REGULATORY AUTHORITY OF THE SECRETARY

SEC. 404. The Secretary may issue regulations to carry out the provisions of this title. Such regulations shall be promulgated only after public notice and comment in accordance with the provisions of section 553 of title 5, United States Code. [citation in margin: "15 USC 4244."]

AGENCY ACTIVITIES

SEC. 405. (a) A private sector party may apply for a license to operate a private remote-sensing space system which utilizes, on a space-available basis, a civilian United States Government satellite or vehicle as a platform for such system. The Secretary, pursuant to the authorities of this title, may license such system if it meets all the conditions of this title and— [citation in margin: "15 USC 4245."]

> (1) the system operator agrees to reimburse the Government immediately for all related costs incurred with respect to such utilization, including a reasonable and pro-portionate share of fixed, platform, data transmission, and launch costs; and
>
> (2) such utilization would not interfere with or otherwise compromise intended civilian Government missions, as determined by the agency responsible for such civilian platform.

(b) The Secretary may offer assistance to private sector parties in finding appropriate opportunities for such utilization.

(c) To the extent provided in advance by appropriation Acts, any Federal agency may enter into agreements for such utilization if such agreements are consistent with such agency's mission and statutory authority, and if such remote-sensing space system is licensed by the Secretary before commencing operation.

(d) The provisions of this section do not apply to activities carried out under title V.

(e) Nothing in this title shall affect the authority of the Federal Communications Commission pursuant to the Communications Act of 1934, as amended (47 U.S.C. 151 *et seq.*). [citation in margin: "47 USC 609."]

TERMINATION

SEC. 406. If, five years after the expiration of the six-year period described in section 304(a)(2), no private sector party has been licensed and continued in operation under the provisions of this title, the authority of this title shall terminate. [citation in margin: "15 USC 4246."]

TITLE V—RESEARCH AND DEVELOPMENT

CONTINUED FEDERAL RESEARCH AND DEVELOPMENT

SEC. 501. (a)(1) The Administrator of the National Aeronautics and Space Administration is directed to continue and to enhance such Administration's programs of remote-sensing research and development. [citation in margin: "15 USC 4261."]

> (2) The Administrator is authorized and encouraged to—
>
> > (A) conduct experimental space remote-sensing programs (including applications demonstration programs and basis research at universities);
> >
> > (B) develop remote-sensing technologies and techniques, including those needed for monitoring the Earth and its environment; and
> >
> > (C) conduct such research and development in cooperation with other Federal agencies and with public and private research entities (including

private industry, universities, State and local governments, foreign governments, and international organizations) and to enter into arrangements
(including joint ventures) which will foster such cooperation.

(b) (1) The Secretary is directed to conduct a continuing program of—

(A) research in applications of remote-sensing;

(B) monitoring of the Earth and its environment; and

(C) development of technology for such monitoring.

(2) Such program may include support of basic research at universities and demonstrations of applications.

(3) The Secretary is authorized and encouraged to conduct such research, monitoring, and development in cooperation with other Federal agencies and with public and private research entities (including private industry, universities, State and local
governments, foreign governments, and international organizations) and to enter into
arrangements (including joint ventures) which will foster such cooperation.

(c) (1) In order to enhance the United States ability to manage and utilize its renewable and nonrenewable resources, the Secretary of Agriculture and the Secretary
of the Interior are authorized and encouraged to conduct programs of research
and development in the applications of remote sensing using funds appropriated
for such purposes.

(2) Such programs may include basic research at universities, demonstrations of
applications, and cooperative activities involving other Government agencies, private sector parties, and foreign and international organizations.

(d) Other Federal agencies are authorized and encouraged to conduct research and
development on the use of remote sensing in fulfillment of their authorized missions,
using funds appropriated for such purposes.

(e) The Secretary and the Administrator of the National Aeronautics and Space
Administration shall, within one year after the date of enactment of this Act and biennially thereafter, jointly develop and transmit to the Congress a report which includes (1) a
unified national plan for remote-sensing research and development applied to the Earth
and its atmosphere; (2) a compilation of progress in the relevant ongoing research and
development activities of the Federal agencies; and (3) an assessment of the state of our
knowledge of the Earth and its atmosphere, the needs for additional research (including
research related to operational Federal remote-sensing space programs), and opportunities available for further progress. [marginal note: "Report."]

USE OF EXPERIMENTAL DATA

SEC. 502. Data gathered in Federal experimental remote-sensing space programs may be
used in related research and development programs funded by the Federal Government
(including applications programs) and cooperative research programs, but not for commercial uses or in competition with private sector activities, except pursuant to section
503. [citation in margin: "15 USC 4262."]

SALE OF EXPERIMENTAL DATA

SEC. 503. Data gathered in Federal experimental remote-sensing space programs may be
sold en bloc through a competitive process (consistent with national security interests and
international obligations of the United States and in accordance with section 607) to any
United States entity which will market the data on a nondiscriminatory basis. [citation in
margin: "15 USC 4263."]

TITLE VI—GENERAL PROVISIONS

NONDISCRIMINATORY DATA AVAILABILITY

SEC. 601. (a) Any unenhanced data generated by any system operator under the provisions of this Act shall be made available to all users on a nondiscriminatory basis in accordance with the requirements of this Act. [citation in margin: "Public availability. 15 USC 4263."]

(b) Any system operator shall make publicly available the prices, policies, procedures, and other terms and conditions (but, in accordance with section 104(3)(C), not necessarily the names of buyers or their purchases) upon which the operator will sell such data.

ARCHIVING OF DATA

SEC. 602. (a) It is in the public interest for the United States Government— [citation in margin: "15 USC 4272."]

(1) to maintain an archive of land remote-sensing data for historical, scientific, and technical purposes, including long-term global environmental monitoring;

(2) to control the content and scope of the archive; and

(3) to assure the quality, integrity, and continuity of the archive.

(b) The Secretary shall provide for long-term storage, maintenance and upgrading of basic, global, land remote-sensing data set (hereinafter referred to as the "basic data set") and shall follow reasonable archival practices to assure proper storage and preservation of the basic data set and timely access for parties requesting data. The basic data set which the Secretary assembles in the Government archive shall remain distinct from any inventory of data which a system operator may maintain for sales and for other purposes.

(c) In determining the initial content of, or in upgrading, the basic data set, the Secretary shall–

(1) use as a baseline the data archived on the date of enactment of this Act;

(2) take into account future technical and scientific developments and needs;

(3) consult with and seek the advice of users and products;

(4) consider the need for data which may be duplicative in terms of geographical coverage but which differ in terms of season, spectral bands, resolution, or other relevant factors;

(5) include, as the Secretary considers appropriate, unenhanced data generated either by the Landsat system, pursuant to title III, or by licensees under title IV;

(6) include, as the Secretary considers appropriate, data collected by foreign ground stations or by foreign remote-sensing space systems; and

(7) ensure that the content of the archive is developed in accordance with section 607.

(d) Subject to the availability of appropriations, the Secretary shall request data needed for the basic data set and pay to the providing system operator reasonable costs for reproduction and transmission. A system operator shall promptly make requested data available in a form suitable for processing for archiving.

(e) Any system operator shall have the exclusive right to sell all data that the operator provides to the United States remote-sensing data archive for a period to be determined by the Secretary but not to exceed ten years from the date the data are sensed. In the case of data generated from the Landsat system prior to the implementation of the contract described in section 202(a), any contractor selected pursuant to section 202 shall have the exclusive right to market such data on behalf of the United States Government for the duration of such contract. A system operator may relinquish the exclusive right and consent to distribution from the archive before the period of exclusive right has expired by terminating the offer to sell particular data. [marginal note: "Marketing."]

(f) after the expiration of such exclusive right to sell, or after relinquishment of such right, the data provided to the United States remote-sensing data archive shall be in the public domain and shall be made available to requesting parties by the Secretary of prices reflecting reasonable costs of reproduction and transmittal. [marginal note: "Public availability."]

(g) In carrying out the functions of this section, the Secretary shall, to the extent practicable and as provided in advance by appropriations Act, use existing Government facilities.

NONREPRODUCTION

SEC. 603. Unenhanced data distributed by any system operator under the provisions of this Act may be sold on the condition that such data will not be reproduced or disseminated by the purchaser. [citation in margin: "15 USC 4273."]

REIMBURSEMENT FOR ASSISTANCE

SEC. 604. The Administrator of the National Aeronautics and Space Administration, the Secretary of Defense and the heads of other Federal agencies may provide assistance to system operators under the provisions of this Act. Substantial assistance shall be reimbursed by the operator, except as otherwise provided by law. [citation in margin: "15 USC 4274."]

ACQUISITION OF EQUIPMENT

SEC. 605. The Secretary may, by means of a competitive process, allow a licensee under title IV or any other private party to buy, lease, or otherwise acquire the use or equipment from the Landsat system, when such equipment is no longer needed for the operation of such system or for the sale of data from such system. Officials of other Federal civilian agencies are authorized and encouraged to cooperative with the Secretary in carrying out the provisions of this section. [citation in margin: "15 USC 4275."]

RADIO FREQUENCY ALLOCATION

SEC. 606. (a) Within thirty days after the date of enactment of this Act, the President (or the President's delegee [sic], if any, with authority over the assignment of frequencies of radio stations of classes of radio stations operated by the United States) shall make available for nongovernmental use spectrum presently allocated to Government use, for use by United States Landsat and commercial remote-sensing space systems. The spectrum to be so made available shall conform to any applicable international radio or wire treaty or convention, or regulations annexed thereto. Within ninety days thereafter, the Federal Communications Commission shall utilize appropriate procedures to authorize the use of such spectrum for nongovernmental use. Nothing in this section shall preclude the ability of the Commission to allocate additional spectrum to commercial land remote-sensing space satellite system use. [citation in margin: "President of the U.S. 15 USC 4276."]

(b) To the extent required by the Communications Act of 1934, as amended (47 U.S.C. 151 *et seq.*), an application shall be filed with the Federal Communications Commission for any radio facilities involved with the commercial remote-sensing space system. [citation in margin: "47 USC 609."]

(c) It is the intent of Congress that the Federal Communications Commission complete the radio licensing process under the Communications Act of 1934, as amended (47 U.S.C. 151 *et seq.*), upon the application of any private sector party or consortium operator of any commercial land remote-sensing space system subject to this Act, within one hundred and twenty days of the receipt of an application for such licensing. If final action has not occurred within one hundred and twenty days of the receipt of such an application, the Federal Communications Commission shall inform the applicant of any pending issues and of actions required to resolve them.

(d) Authority shall not be required from the Federal Communications Commission for the development and construction of any United States land remote-sensing space system (or (component thereof), other than radio transmitting facilities or components, while any licensing determination is being made.

(e) Frequency allocations made pursuant to this section by the Federal Communications Commission shall be consistent with international obligations and with the public interest.

CONSULTATION

SEC. 607. (a) The Secretary shall consult with the Secretary of Defense on all matters under this Act affecting national security. The Secretary of Defense shall be responsible for determining those conditions, consistent with this Act, necessary to meet national security concerns of the United States and for notifying the Secretary promptly of such conditions. [citation in margin: "Defense and national security. 15 USC 4277."]

(b) (1) The Secretary shall consult with the Secretary of State on all matters under this Act affecting international obligations. The Secretary of State shall be responsible for determining those conditions, consistent with this Act, necessary to meet international obligations and policies of the United States and for notifying the Secretary promptly of such conditions.

(2) Appropriate Federal agencies are authorized and encouraged to provide remote-sensing data, technology, and training developing nations as a component of programs of international aid.

(3) The Secretary of State shall promptly report to the Secretary any instances outside the United States of discriminatory distribution of data.

(c) If, as a result of technical modifications imposed on a system operator on the basis of national security concerns, the Secretary, in consultation with the Secretary of Defense or with other Federal agencies, determines that additional costs will be incurred by the system operator, or that past development costs (including the cost of capital) will not be recovered by the system operator, the Secretary may require the agency or agencies requesting such technical modifications to reimburse the system operator for such additional or development costs, but not for anticipated profits. Reimbursements may cover costs associated with required changes in system performance, but not costs ordinarily associated with doing business abroad.

AMENDMENT TO NATIONAL AERONAUTICS AND
SPACE ADMINISTRATION AUTHORIZATION, 1983

SEC. 608. Subsection (a) of section 201 of the National Aeronautics and Space Administration Authorization Act, 1983 (Public Law 97–324; 96 Stat. 1601) is amended to read as follows: [citation in margin: "15 USC 1517 note."]

"(a) The Secretary of Commerce is authorized to plan and provide for the management and operation of civil remote-sensing space systems, which may include the Landsat 4 and 5 satellites and associated ground system equipment transferred from the National Aeronautics and Space Administration; to provide for user fees; and to plan for the transfer of the operation of civil remote-sensing space systems to the private sector when in the national interest."

AUTHORIZATION OF APPROPRIATIONS

SEC. 609. (a) There are authorized to be appropriated to the Secretary $75,000,000 for fiscal year 1985 for the purpose of carrying out the provisions of this Act. Such sums shall remain available until expended, but shall not become available until the time periods specified in sections 202(c) and 303(c) have expired. [citation in margin: "15 USC 4278."]

(b) The authorization provided for under subsection (a) shall be in addition to moneys [sic] authorized pursuant to title II of the National Aeronautics and Space Administration Act, 1983. [citation in margin: "15 USC 1517."]

TITLE VII—PROHIBITION OF COMMERCIALIZATION OF WEATHER SATELLITES

PROHIBITION

SEC. 701. Neither the President nor any other official of the Government shall make any effort to lease, sell, or transfer to the private sector, commercialize, or in any way dismantle any portion of the weather satellite systems operated by the Department of Commerce or any successor agency. [citation in margin: "President of U.S. 15 USC 4291."]

FUTURE CONSIDERATIONS

SEC. 702. Regardless of any change in circumstances subsequent to the enactment of this Act, even if such change makes it appear to be in the national interest to commercialize weather satellites, neither the President nor any official shall take any action prohibited by section 701 unless this title has first been repealed. [citation in margin: "15 USC 4292."]

Approved July 17, 1984.

LEGISLATIVE HISTORY—H.R. 5155:
HOUSE REPORT No. 96-647 (Comm. on Science and Technology).
SENATE REPORT No. 98-458 (Comm. on Commerce, Science, and Transportation.
CONGRESSIONAL RECORD Vol. 130 (1984);
    Apr. 9, considered and passed House.
    June 8, considered and passed Senate, amended.
    June 28, House concurred in Senate amendment with an amendment.
    June 29, Senate concurred in House amendment.
WEEKLY COMPILATION OF PRESIDENTIAL DOCUMENTS, Vol. 20, No. 29 (1984):
    July 17, Presidential statement.

Document II-37

Document title: Office of the Press Secretary, The White House, "Statement by the Press Secretary," June 1, 1989.

Source: NASA Historical Reference Collection, NASA History Office, NASA Headquarters, Washington, D.C.

By mid-1989, EOSAT had been operating Landsats 4 and 5 for more than three years, but EOSAT's income from data sales did not quite equal, let along exceed, its operating costs. Hence, EOSAT still relied on the support of several million of dollars from the federal government to continue to collect Landsat data. No single government agency was willing to provide this relatively small amount of money. Because of this dispute, Landsat operations were in danger of being closed down. This presidential decision settled the immediate future of Landsat and set up a mechanism for later reexamination of its status.

[no pagination]
For Immediate Release   June 1, 1989

Statement by the Press Secretary

The President today announced he had approved funding for continued operations of Landsat satellites 4 and 5 and for the completion and launch of Landsat 6. The President's action endorsed a recommendation from the National Space Council chaired by Vice President Dan Quayle. The President also directed the National Space Council and the Office of Management and Budget to review options with the intention of continuing Landsat-type data collections after Landsat 6.

Landsat, which takes detailed photographs of the earth, is the U.S. Government's civil, space-based, land remote sensing program. Landsat-type imagery data is important for such applications as global change research, environmental monitoring, law enforcement, natural resource estimates, national security and a variety of private sector uses. In addition, Landsat provides a visible symbol of the U.S. commitment to, and leadership in, the use of space for the common good.

Over recent years, it has become increasingly evident that commercializing the entire Landsat program would not be feasible until at least the end of the century. Since earlier government planning was based on commercializing the entire program, the absence of near-term commercial viability threatened continuity of Landsat and jeopardized continuity of Landsat data. The National Space Council, at its first meeting on May 12, recommended the action endorsed by President Bush today.

Continued operation of Landsats 4 and 5 will require and additional $5 million in FY 89 and $19 million in FY 90. Cost of completion and launch of Landsat 6 by 1991 has already been included in the Commerce Department budget.

Document II-38

Document title: Office of the Press Secretary, The Vice President's Office, "Vice President Announces Landsat Policy," February 13, 1992, with attached: "Landsat Remote Sensing Policy."

Source: NASA Historical Reference Collection, NASA History Office, NASA Headquarters, Washington, D.C.

The National Space Council's detailed reexamination of the Landsat program in late 1991 prompted the Bush administration to release this policy statement about the Landsat system's future. The plan called for transferring the development and operations of Landsat 7 back to the government. Landsat 6, on the other hand, would still be launched and operated by EOSAT, which also would be responsible for overseeing Landsats 4 and 5 until Landsat 6 became fully operational.

[no pagination]
For Immediate Release   February 13, 1992

Vice President Announces Landsat Policy

The Vice President announced today that President Bush has approved a National Space Policy Directive which reaffirms the importance of Landsat-type multispectral imaging and provides a plan for maintaining continuity of Landsat coverage into the 21st century.

Landsat is an important satellite program which provides multispectral pictures of the Earth. It supports U.S. government needs, including those related to national security and global change research, and benefits the U.S. private sector. In May 1989, President Bush directed that continuity of Landsat-type remote sensing data be maintained, and approved a series of near term actions to implement this policy. The new National Space Policy Directive, which was developed by the National Space Council chaired by Vice President Quayle, establishes a comprehensive, long range strategy and assigns agency responsibilities for the future.

A key element of this strategy is the assignment of management and funding responsibility for the next satellite, Landsat 7, to the agencies which have the primary requirements for the data, NASA and the Department of Defense. The strategy seeks to minimize the cost of Landsat-type images for U.S. government uses, calls on agencies to eliminate unnecessary regulations governing private sector remote sensing activities, and fosters development of advanced remote sensing technologies to reduce the cost and improve the performance of future satellites.

#####

Attachment

Landsat Remote Sensing Strategy

I.   Policy Goals

A remote sensing capability such as is currently being provided by Landsat satellites 4 and 5 benefits the civil and national security interests of the United States and makes contributions to the private sector which are in the public interest. For these reasons, the United States government will seek to maintain continuity of Landsat-type data. The U.S. government will:

    a)   Provide data which are sufficiently consistent in terms of acquisition geometry, coverage characteristics, and spectral characteristics with previous Landsat data to allow comparisons for chance detection and characterization;

    b)   Make Landsat data available to meet the needs of national security, global change research, and other federal users; and,

    c)   Promote and not preclude private sector commercial opportunities in Landsat-type remote sensing.

II.   Landsat Strategy

    a.   The Landsat strategy is composed of the following elements:

        (1) Ensuring that Landsat satellites 4 and 5 continue to provide data as long as they are capable of doing so, or until Landsat 6 becomes operational.

        (2) Acquiring a Landsat 7 satellite with the goal of maintaining continuity of Landsat-type data beyond the projected Landsat 6 end-of-life.

(3) Fostering the development of advanced remote sensing technologies, with the goal of reducing the cost and increasing the performance of future Landsat-type satellites to meet U.S. government needs, and potentially, enabling substantially greater opportunities for commercialization.

(4) Seeking to minimize the cost of Landsat-type data for U.S. government agencies and to provide data for use in global change research in a manner consistent with the Administration's Data Management for Global Change Research Policy Statements.

(5) Limiting U.S. government regulations affecting private sector remote sensing activities to only those required in the interest of national security, foreign policy, and public safety.

(6) Maintaining an archive, within the United States, of existing and future Landsat-type data.

(7) Considering alternatives for maintaining continuity of data beyond Landsat 7.

b.   These strategy elements will be implemented within the overall resource and policy guidance provided by the President.

III. Implementing Guidelines

a.   The Department of Commerce will:

(1) Complete and launch Landsat 6.

(2) In coordination with OMB, arrange for the continued operation of Landsat satellites 4 and 5 until Landsat 6 becomes operational.

b.   The Department of Defense and the National Aeronautics and Space Administration will:

(1) Develop and launch a Landsat 7 satellite of at least equivalent performance to replace Landsat 6 and define alternatives for maintaining data continuity beyond Landsat 7.

(2) Prepare a plan by March 1, 1992, which addresses management and funding responsibilities, operations, data archiving and dissemination, and commercial considerations associated with the Landsat program. This plan will be coordinated with other U.S. government agencies, as appropriate, and reviewed by the National Space Council.

(3) With the support of the Department of Energy and other appropriate agencies, prepare a coordinated technology plan that has as its goals improving the performance and reducing the cost for future Landsat-type remote sensing systems.

c.   The Department of the Interior will continue to maintain a national archive of Landsat-type remote sensing data.

d.   Affected agencies will identify funds, within their approved fiscal year 1993 budget, necessary to implement this strategy.

IV. Reporting Requirements

U.S. government agencies affected by these strategy guidelines are directed to report by March 15, 1992, to the National Space Council on the implementation of this strategy.

Document II-39

Document title: Department of Defense and National Aeronautics and Space Administration, "Management Plan for the Landsat Program," March 10, 1992.

Source: NASA Historical Reference Collection, NASA History Office, NASA Headquarters, Washington, D.C.

This management plan spells out how the Bush administration planned to manage the continuation of the Landsat program. It roughly split the financial responsibility for Landsat development and operations evenly between the Department of Defense and NASA over the projected lifetime of the satellite. According to the agreement, the Department of Defense was to procure the satellite and NASA was to build and operate the data reception and distribution facility.

---

[1]

Management Plan for the Landsat Program

Introduction

The Landsat Program benefits a wide community of users, including the private sector, the global change research community, national security and other Government users. The National Aeronautics and Space Administration (NASA) and the Department of Defense (DoD) agree that the program provides a unique capability that should be continued. The two agencies will therefore cooperate in the continuation of the Landsat program, including the development and operation of a Landsat follow-on (Landsat 7) satellite, as well as in planning for future operations and advanced technology development with other appropriate agencies.

This plan responds to the President's National Space Policy Directive 5 on Landsat Remote Sensing Strategy, dated February 1992. It outlines an integrated approach to the management, development and operation of a newly structured Landsat program tailored to be more responsive to national security and global change research needs through the year 2002 and potentially beyond.

To implement this plan, the involved agencies will work with the Congress to obtain any necessary enabling legislation.

Concept

DoD, representing the national security community, and NASA, representing the U.S. Global Change Research Program and the civil/private Landsat use community in general, will divide the management responsibilities and costs for the program with approximate equality.

General Description

The program will:
- Be consistent with the following goals:
    - Maintain Landsat program data continuity beyond Landsat 6 by:
[2]
    - Seeking to launch Landsat, approximately 5 years after the launch of Landsat 6
    - Continuing to provide data which are sufficiently consistent in terms of acquisition geometry, calibration, coverage characteristics and spectral characteristics with previous Landsat data to allow comparisons for global and regional change detection and characterization
    - Continue to make such data available for U.S. civil, national security, and private sector uses
    - Seek to expand the use of such data for global change research and national security purposes
- Acquire a Landsat 7 satellite which is, as a minimum, functionally equivalent to the Landsat 6 satellite, with the addition of a Tracking and Data Relay Satellite [System] (TDRSS) communications capability. Additional improvements will be

sought if they do not increase risk to data continuity, and are attainable within agreed-to funding. Potential improvements could include features such as improved spatial or spectral resolution, stereoscopic viewing and other capabilities that could improve the operational utility of the data.

- Evaluate the need, and alternative means, for implementing follow-on satellite systems and improvements beyond Landsat 7. This would include evaluation of potential changes in program management, funding responsibilities, data management/utilization, system configuration and operational concepts, as well as use of advanced technologies to improve performance and reduce cost.

Program Management Responsibilities

- DoD will have the lead responsibility for the acquisition and launch of the Landsat 7 satellite, and with NASA and Department of Energy (DoE) participation, will prepare a technology demonstration plan for post-Landsat 7 satellites. In addition, the DoD Project Office will provide general systems level engineering and integration services in support of both the NASA and DoD Project Managers. The DoD portion of the program will be administered under the Director, Defense Support Project Office, as part of the Defense Reconnaissance Support Program. NASA will provide appropriate participation in the responsible DoD project office as required.

[3] • NASA will have the lead responsibility for the development and operation of the Landsat ground system, including data processing, archiving, distribution, user support and mission operations management. The NASA portion of the program will be administered under the Director, Earth Science and Applications Division, Office of Space Science and Applications in coordination with the Mission to Planet Earth Program. DoD will provide appropriate participation in the responsible NASA project office as required.

- A jointly chaired Landsat Coordinating Group (LCG) will be formed, with appropriate representation from both NASA and DoD. The group will be responsible for coordinating top-level program plans, budgets and policies; handling interagency matters related to the program; staffing any issues requiring adjudication at senior departmental levels; and coordinating reports to Congress and other tasks related to the program. Participation of other government agencies in LCG activities will be sought as appropriate.

- The Assistant Secretary of Defense for Command, Control, Communications and Intelligence and the NASA Associate Administrator for Space Science and Applications will be the senior agency officials responsible for program oversight and issue resolution. In addition, the Director, Defense Research and Engineering (DDR&E), the NASA Associate Administrator of Aeronautics and Space Technology, and the Director of the DoE Office of Space will be consulted on matters related to advanced technology.

Funding Responsibilities

- NASA and DoD will each fund that portion of the program for which it is responsible. Thus DoD will fund the procurement and launch of the Landsat 7 satellite, and NASA will fund satellite operations, data processing, archiving, and data distribution (including any ground hardware and facilities that are required). A mutually acceptable cost baseline will be developed, with each agency's total funding responsibility approximately equal as spread across the development and operational life of Landsat 7. Any significant funding disparities incurred in program planning or execution will be resolved through mutually acceptable funding adjustments as agreed to by the Deputy Secretary of Defense and the NASA Administrator (the cost baseline, reflecting this approach, is provided in Attachment 1).

- Any improvements over a Landsat 6 functional equivalent capability for Landsat 7 will be funded by the sponsoring agency, if the required funding exceeds the baseline defined above. If it is agreed that improvements benefit the interests of both agencies, they would be funded based upon a mutually [4] acceptable sharing arrangement approved by the Deputy Secretary of Defense and the NASA Administrator.
- NASA and DoD will coordinate their interactions with Congress with regard to the Landsat program.
- Agency funding and management responsibilities for subsequent Landsat satellite(s) would be the subject of a separate agreement.

Data Management

- **Data Access and Acquisition for Landsat 7:**
  - U.S. Government (USG) civil, national security, commercial and noncommercial users, including global change research users, will have near-real time and/or archival access to all data acquired. Collection scheduling for such users will be accomplished jointly by NASA and DoD, through the NASA project office.
  - Commercial users will be given input into collection scheduling and access to data through NASA.
  - USG users will have unrestricted rights of redistribution within the USG.
- **Data from Landsats 1–6.** NASA will seek to negotiate an agreement such that data from Landsats 1–6 are made available to USG civil, national security, commercial and non-commercial users, including global change research users in a manner similar to the arrangements for data access and acquisition for Landsat 7, described above.
- **Data Pricing.** The program will seek to limit the cost of data for USG civil, national security, and global change research use to the marginal cost of fulfilling the specific user request. In doing so, it will make such data available to the global change research community in a manner consistent with the Administration's Data Management for Global Change Research policy statements. Data will be provided for commercial use, with the goal of encouraging Landsat remote sensing commercialization and economic growth. Prices, policies, procedures and other terms and conditions for the distribution and sale of unenhanced Landsat data will be made publicly available.
- **Data Archiving.** NASA will work with the Department of the Interior to develop and maintain a permanent national archive for all Landsat data.
[5] • **National Security and Foreign Policy Considerations.** As a general principle, all Landsat data will remain unclassified. Special data prioritization, distribution procedures or restrictions might be necessary under certain national security and foreign policy conditions, or if future system improvements substantially increase the national security or foreign policy sensitivity of some Landsat data. DoD and NASA will develop procedures to minimize the impact of potential restrictions on Landsat system users.

Other Considerations

- **International Cooperation.** NASA will have the lead responsibility, with DoD support, for evaluating opportunities for international cooperation and utilization of

Landsat. NASA will have the lead responsibility for arranging for foreign ground station operations.

- **Commercialization.** NASA will have the lead responsibility, with support from DoD and other agencies, for promoting and periodically assessing U.S. commercial opportunities, to the extent feasible in the Landsat program.
- **Advanced Technologies for Landsat 8 and Beyond.** NASA, DoD and other USG agencies are pursuing advanced technologies that hold significant promise for future land remote sensing systems. Conducting an advanced technology demonstration and evaluation effort, with the goal of allowing for technology insertion at an appropriate point in the program, is desirable. Accordingly, with the support of NASA, the DoE and other Federal agencies, DoD will have the lead responsibility for preparing a coordinated technology plan that has as a goal improved performance and reduced cost for future Landsat-type systems. The plan will identify relevant agency activities and funding that can contribute to this goal.

[6] Approved:

[hand-signed: "Aaron Cohen for"]
Richard H. Truly
Administrator
National Aeronautics and
Space Administration

[hand-signed: "Donald J. Atwood"]
Donald J. Atwood
Deputy Secretary of Defense

__3/10/92__
Date Approved

__3/20/92__
Date Approved

[7] Attachment 1

Cost Baseline (Then Year $M)[1]

DoD Costs:

FY	92	93	94	95	96	97	98	99	00	01	02	Total
	30	80	158	134	52	6	2	2	2	2	2	470

Includes:

- Development of One Landsat 6-Equivalent Performance Satellite
  - Enhanced Thematic Mapper-class Sensor Performance as a Minimum
  - Baseline includes TDRSS Communications
- Launch
  - Planned for FY 1997
  - Titan II-class launch vehicle from West Cost (Vandenberg AFB)
- Program Support/General Systems-Level Engineering and Integration (SE &I)

NASA Costs:

FY	92	93	94	95	96	97	98	99	00	01	02	Total
	7	25	59	61	48	30	32	34	36	38	40	410

---

1. Subject to enabling legislation and contract negotiations.

Includes:

- Ground System Development
    - Enhanced Command/Control/Telemetry System
    - Enhanced Data Processing/Product Generation Capability
    - Archival Restoration
- Mission Operations
    - Landsat 4–6 Operations, with Landsat 7 Operations Beginning in FY 1997
    - Landsat 4–7 Data Processing, Archival, and Distribution Beginning mid-1993
    - Program Support/Ground Segment System Engineering
    - Mission Operations Management
- TDRSS Link Added to Landsat 7 Satellite

Document II-40

Document title: "Land Remote-Sensing Policy Act of 1992," Public Law 102–555, 106 Stat. 4163, October 28, 1992.

Source: NASA Historical Reference Collection, NASA History Office, NASA Headquarters, Washington, D.C.

This law codified the substantial changes in policy toward the Landsat program that had developed in the early 1990s. It returned the development and operation of the Landsat program to the government at the end of the operational life of Landsats 4, 5, and 6. It also reiterated the federal government's willingness, first noted in the Land Remote-Sensing Commercialization Act of 1984, to grant an operating license to operators of private remote-sensing satellites.

[no pagination]

PUBLIC LAW 102–555—OCT. 28, 1992   106 STAT. 4163

Public Law 102–555
102d Congress

An Act

To enable the United States to maintain its leadership in land remote sensing by providing data continuity for the Landsat program, to establish a new national land remote sensing policy, and for other purposes. [citation in margin: "October 28, 1992, H.R. 6133"]

Be it enacted by the Senate and House of Representatives of the United States of America in Congress assembled, [citation in margin: "Land Remote Sensing Policy Act of 1992. National defense. 15 USC 5601 note"]

SECTION. 1. SHORT TITLE.

This Act may be cited as the "Land Remote Sensing Policy Act of 1992."

SEC. 2. FINDINGS.

The Congress finds and declares the following:
(1) The continuous collection and utilization of land remote sensing data from space are of major benefit in studying and understanding human impacts on the global envi-

ronment, in managing the Earth's natural resources, in carrying out national security functions, and in planning and conducting many other activities of scientific, economic, and social importance.

(2) The Federal Government's Landsat system established the United States as the world leader in land remote sensing technology.

(3) The national interest of the United States lies in maintaining international leadership in satellite land remote sensing and in broadly promoting the beneficial use of remote sensing data.

(4) The cost of Landsat data has impeded the use of such data for scientific purposes, such as for global environmental change research, as well as for other public sector applications.

(5) Given the importance of the Landsat program to the United States, urgent actions, including expedited procurement procedures, are required to ensure data continuity.

(6) Full commercialization of the Landsat program cannot be achieved within the foreseeable future, and thus should not serve as the near-term goal of national policy on land remote sensing; however, commercialization of land remote sensing should remain a long-term goal of United States policy.

(7) Despite the success and importance of the Landsat system, funding and organizational uncertainties over the past several years have placed its future in doubt and have jeopardized United States leadership in land remote sensing.

(8) Recognizing the importance of the Landsat program in helping to meet national and commercial objectives, the President approved, on February 11, 1992, a National Space Policy Directive which was developed by the National Space Council and commits the United States to ensuring the continuity of Landsat coverage into the 21st century.

(9) Because Landsat data are particularly important for national security purposes and global environmental change research, management responsibilities for the program should be transferred from the Department of Commerce to an integrated program management involving the Department of Defense and the National Aeronautics and Space Administration.

(10) Regardless of management responsibilities for the Landsat program, the Nation's broad civilian, national security, commercial, and foreign policy interests in remote sensing will best be served by ensuring that Landsat remains an unclassified program that operates according to the principles of open skies and nondiscriminatory access.

(11) Technological advances aimed at reducing the size and weight of satellite systems hold the potential for dramatic reductions in the cost, and substantial improvements in the capabilities, of future land remote sensing systems, but such technological advances have not been demonstrated for land remote sensing and therefore cannot be relied upon as the sole means of achieving data continuity for the Landsat program.

(12) A technology demonstration program involving advanced remote sensing technologies could serve a vital role in determining the design of a follow-on spacecraft to Landsat 7, while also helping to determine whether such a spacecraft should be funded by the United States Government, by the private sector, or by a international consortium.

(13) To maximize the value of the Landsat program to the American public, unenhanced Landsat 4 through 6 data should be made available, at a minimum, to United States Government agencies, to global environmental change researchers, and to other researchers who are financially supported by the United States Government, at the cost of fulfilling user requests, and unenhanced Landsat 7 data should be made available to all users at the cost of fulfilling user requests.

(14) To stimulate development of the commercial market for enhanced data and value-added services, the United States Government should adopt a data policy for Landsat 7 which allows competition within the private sector for distribution of unenhanced data and value-added services.

(15) Development of the remote sensing market and the provision of commercial value-added services based on remote sensing data should remain exclusively the function of the private sector.

(16) It is in the best interest of the United States to maintain a permanent, comprehensive Government archive of global Landsat and other land remote sensing data for long-term monitoring and study of the changing global environment.

SEC. 3. DEFINITIONS. [citation in margin: "15 USC 5602"]

In this Act, the following definitions apply:

(1) The term "Administrator" means the Administrator of the National Aeronautics and Space Administration.

(2) The term "cost of fulfilling user requests" means the incremental costs associated with providing product generation, reproduction, and distribution of unenhanced data in response to user requests and shall not include any acquisition, amortization, or depreciation of capital assets originally paid for by the United States Government or other costs not specifically attributable to fulfilling user requests.

(3) The term "data continuity means the continued acquisition and availability of unenhanced data which are, from the point of view of the user—

(A) sufficiently consistent (in terms of acquisition geometry, coverage characteristics, and spectral characteristics) with previous Landsat data to allow comparisons for global and regional change detection and characterization; and

(B) compatible with such data and with methods used to receive and process such data.

(4) The term "data preprocessing" may include—

(A) rectification of system and sensor distortions in land remote sensing data as it is received directly from the satellite in preparation for delivery to a user;

(B) registration of such data with respect to features of the Earth; and

(C) calibration of spectral response with respect to such data, but does not include conclusions, manipulations, or calculations derived from such data, or a combination of such data with other data.

(5) The term "land remote sensing" means the collection of data which can be processed into imagery of surface features of the Earth from an unclassified satellite or satellites, other than an operational United States Government weather satellite.

(6) The term "Landsat Program Management" means the integrated program management structure—

(A) established by, and responsible to, the Administrator and the Secretary of Defense pursuant to section 101(a); and

(B) consisting of appropriate officers and employees of the National Aeronautics and Space Administration, the Department of Defense, and any other United States Government agencies the President designates as responsible for the Landsat program.

(7) The term "Landsat system" means Landsats 1, 2, 3, 4, 5, and 6, and any follow-on land remote sensing system operated and owned by the United States Government, along with any related ground equipment, systems, and facilities owned by the United States Government.

(8) The term "Landsat 6 contractor" means the private sector entity which was awarded the contract for spacecraft construction, operations, and data marketing rights for the Landsat 6 spacecraft.

(9) The term "Landsat 7" means the follow-on satellite to Landsat 6.

(10) The term "National Satellite Land Remote Sensing Data Archive" means the

archive established by the Secretary of the Interior pursuant to the archival responsibilities defined in section 502.

(11) The term "noncommercial purposes" refers to those activities undertaken by individuals or entities on the condition, upon receipt of unenhanced data, that—

(A) such data shall not be used in connection with any bid for a commercial contract, development of a commercial product, or any other non-United States Government activity that is expected, or has the potential, to be profitmaking;

(B) the results of such activities are disclosed in a timely and complete fashion in the open technical literature or other method of public release, except when such disclosure by the United States Government or its contractors would adversely affect the national security or foreign policy of the United States or violate a provision of law or regulation; and

(C) such data shall not be distributed in competition with unenhanced data provided by the Landsat 6 contractor.

(12) The term "Secretary" means the Secretary of Commerce.

(13) The term "unenhanced data" means land remote sensing signals or imagery products that are unprocessed or subject only to data preprocessing.

(14) The term "United States Government and its affiliated users" means—

(A) United States Government agencies;

(B) researchers involved with the United States Global Change Research Program and its international counterpart programs; and

(C) other researchers and international entities that have signed with the United States Government a cooperative agreement involving the use of Landsat data for non- commercial purposes.

SEC. 4. REPEAL OF LAND REMOTE-SENSING COMMERCIALIZATION ACT OF 1984.

The Land Remote-Sensing Commercialization Act of 1984 (15 U.S.C. 4201 *et seq.*) is repealed.

TITLE I—LANDSAT

SEC. 101. LANDSAT PROGRAM MANAGEMENT. [citation in margin: "15 USC 5611"]

(a) ESTABLISHMENT.—The Administrator and the Secretary of Defense shall be responsible for management of the Landsat program. Such responsibility shall be carried out by establishing an integrated program management structure for the Landsat system.

(b) MANAGEMENT PLAN.—The Administrator, the Secretary of Defense, and any other United States Government official the President designates as responsible for part of the Landsat program, shall establish, through a management plan, the roles, responsibilities, and funding expectations for the Landsat Program of the appropriate United States Government agencies. The management plan shall—

(1) specify that the fundamental goal of the Landsat Program Management is the continuity of unenhanced Landsat data through the acquisition and operation of a Landsat 7 satellite as quickly as practicable which is, at a minimum, functionally equivalent to the Landsat 6 satellite, with the addition of a tracking and data relay satellite communications capability;

(2) include a baseline funding profile that—

(A) is mutually acceptable to the National Aeronautics and Space Administration and the Department of Defense for the period covering the development and operation of Landsat 7; and

(B) provides for total funding responsibility of the National Aeronautics and Space Administration and the Department of Defense, respectively, to be approximately equal to the funding responsibility of the other as spread across the development and operational life of Landsat 7;

(3) specify that any improvements over the Landsat 6 functional equivalent capability for Landsat 7 will be funded by a specific sponsoring agency or agencies, in a manner agreed to by the Landsat Program Management, if required funding exceeds the baseline funding profile required by paragraph (2), and that additional improvements will be sought only if the improvements will not jeopardize data continuity; and

(4) provide for a technology demonstration program whose objective shall be the demonstration of advanced land remote sensing technologies that may potentially yield a system which is less expensive to build and operate, and more responsive to data users, than is the current Landsat system.

(c) RESPONSIBILITIES.—The Landsat Program Management shall be responsible for—

(1) Landsat 7 procurement, launch, and operations;

(2) ensuring that the operation of the Landsat system is responsive to the broad interests of the civilian, national security, commercial, and foreign users of the Landsat system;

(3) ensuring that all unenhanced Landsat data remain unclassified and that, except as provided in section 506 (a) and (b), no restrictions are placed on the availability of unenhanced data;

(4) ensuring that land remote sensing data of high priority locations will be acquired by the Landsat 7 system as required to meet the needs of the United States Global Change Research Program, as established in the Global Change Research Act of 1990, and to meet the needs of national security users;

(5) Landsat data responsibilities pursuant to this Act;

(6) oversight of Landsat contracts entered into under sections 102 and 103;

(7) coordination of a technology demonstration program, pursuant to section 303; and

(8) ensuring that copies of data acquired by the Landsat system are provided to the National Satellite Land Remote Sensing Data Archive.

(d) AUTHORITY TO CONTRACT.— The Landsat Program Management may, subject to appropriations and only under the existing contract authority of the United States Government agencies that compose the Landsat Program Management, enter into contracts with the private sector for services such as, but not limited to, satellite operations and data preprocessing.

(e) LANDSAT ADVISORY PROCESS—

(1) ESTABLISHMENT.— Landsat Program Management shall seek impartial advice and comments regarding the status, effectiveness, and operation of the Landsat system, using existing advisory committees and other appropriate mechanisms. Such advice shall be sought from individuals who represent—

(A) a broad range of perspectives on basic and applied science and operational needs with respect to land remote sensing data;

(B) the full spectrum of users of Landsat data, including representatives from United States Government agencies, State and local government agencies, academic institutions, nonprofit organizations, value-added companies, the agricultural mineral extraction, and other user industries, and the public; and

(C) a broad diversity of age groups, sexes, and races.

(2) REPORTS.—Within 1 year after the date of the enactment of this Act and biennially thereafter, the Landsat Program Management shall prepare and submit a report to the Congress which—
    (A) reports the public comments received pursuant to paragraph (1); and
    (B) includes—
        (i) a response to the public comments received pursuant to paragraph (1);
        (ii) information on the volume of use, by category of data from the Landsat system; and
        (iii) any recommendations for policy or programmatic changes to improve the utility and operation of the Landsat system.

SEC. 102 PROCUREMENT OF LANDSAT 7. [citation in margin: "15 USC 5612"]

(a) CONTRACT NEGOTIATIONS.—The Landsat Program Management shall, subject to appropriations and only under the existing contract authority of the United States Government agencies that compose the Landsat Program Management, expeditiously contract with a United States private sector entity for the development and delivery of Landsat 7.

(b) DEVELOPMENT AND DELIVERY CONSIDERATION.—In negotiating a contract under this section for the development and delivery of Landsat 7, the Landsat Program Management shall—
    (1) seek, as a fundamental objective, to have Landsat 7 operational by the expected end of the design life of Landsat 6;
    (2) seek to ensure data continuity by the development delivery of a satellite which is, at a minimum, functionally equivalent to the Landsat 6 satellite; and
    (3) seek to incorporate in Landsat 7 any performance improvements required to meet United States Government needs that would not jeopardize data continuity.

(c) NOTIFICATION OF COST AND SCHEDULE CHANGES.—The Landsat Program Management shall promptly notify the Congress of any significant deviations from the expected cost, delivery date, and launch date of Landsat 7, that are specified by the Landsat Program Management upon award of the contract under this section.

(d) UNITED STATES PRIVATE SECTOR ENTITIES.—The Landsat Program Management shall, for purposes of this Act, define the term "United States private sector entities," taking into account the location of operations, assets, personnel, and other such factors.

SEC. 103. DATA POLICY FOR LANDSAT 4 THROUGH 6. [citation in margin: "15 USC 5613"]

(a) CONTRACT NEGOTIATIONS.—Within 30 days after the date of enactment of this Act, the Landsat Program Management shall enter into negotiations with the Landsat 6 contractor to formalize an arrangement with respect to pricing, distribution, acquisition, archiving, and availability of unenhanced data for which the Landsat 6 contractor has responsibility under its contract. Such arrangement shall provide for a phased transition to a data policy consistent with the Landsat 7 data policy (developed pursuant to section 105) by the date of initial operation of Landsat 7. Conditions of the phased arrangement should require that the Landsat 6 contractor adopt provisions so that by the final phase of the transition period—
    (1) such unenhanced data shall be provided, at a minimum, to the United States Government and its affiliated users at the cost of fulfilling user requests, on the condition that such unenhanced data are used solely for noncommercial purposes;

(2) instructional data sets, selected from the Landsat data archives, will be made available to educational institutions exclusively for noncommercial, educational purposes at the cost of fulfilling user requests;

(3) Landsat data users are able to acquire unenhanced data contained in the collective archives of foreign ground stations as easily and affordably as practicable;

(4) adequate data necessary to meet the needs of global environmental change researchers and national security users are acquired;

(5) the United States Government and its affiliated users shall not be prohibited from reproduction or dissemination of unenhanced data to other agencies of the United States Government and other affiliated users, on the condition that such unenhanced data are used solely for noncommercial purposes;

(6) nonprofit, public interest entities receive vouchers, data grants, or other such means of providing them with unenhanced data at the cost of fulfilling user requests, on the condition that such unenhanced data are used solely for noncommercial purposes;

(7) a viable role for the private sector in the promotion and development of the commercial market for value-added and other services using unenhanced data from the Landsat system is preserved; and

(8) unenhanced data from the Landsat system are provided to the National Satellite Land Remote Sensing Data Archive at no more than the cost of fulfilling user requests.

(b) FAILURE TO REACH AGREEMENT.—If negotiations under subsection (a) have not, by September 30,1993, resulted in an agreement that the Landsat Program Management determines generally achieves the goals stated in subsection (b) (1) through (8), the Administrator and the Secretary of Defense shall, within 30 days after the date of such determination, jointly certify and report such determination to the Congress. The report shall include a review of options and projected costs for achieving such goals, and shall include recommendations for achieving such goals. The options reviewed shall include— [marginal note: "Reports"]

(1) retaining the existing or modified contract with the Landsat 6 contractor;

(2) the termination of existing contracts for the exclusive right to market unenhanced Landsat data; and

(3) the establishment of an alternative private sector mechanism for the marketing and commercial distribution of such data.

SEC. 104. TRANSFER OF LANDSAT 6 PROGRAM RESPONSIBILITIES. [citation in margin: "15 USC 5614"]

The responsibilities of the Secretary with respect to Landsat 6 shall be transferred to the Landsat Program Management, as agreed to between the Secretary and the Landsat Program Management, pursuant to section 101.

SEC. 105. DATA POLICY FOR LANDSAT 7. [citation in margin: "15 USC 5615"]

(a) LANDSAT 7 DATA POLICY.—The Landsat Program Management, in consultation with other appropriate United States Government agencies, shall develop a data policy for Landsat 7 which should—

(1) ensure that unenhanced data are available to all users at the cost of fulfilling user requests;

(2) ensure timely and dependable delivery of unenhanced data to the full spectrum of civilian, national security, commercial, and foreign users and the National Satellite Land Remote Sensing Data Archive;

(3) ensure that the United States retains ownership of all unenhanced data generated by Landsat 7;

(4) support the development of the commercial market for remote sensing data;

(5) ensure that the provision of commercial value-added services based on remote sensing data remains exclusively the function of the private sector; and

(6) to the extent possible, ensure that the data distribution system for Landsat 7 is compatible with the Earth Observing System Data and Information System.

(b) In addition, the data policy for Landsat 7 may provide for—

(1) United States private sector entities to operate ground receiving stations in the United States for Landsat 7 data;

(2) other means for data: access by private sector entities to unenhanced data from Landsat 7; and

(3) the United States Government to charge a per image fee, license fee, or other such fee to entities operating ground receiving stations or distributing Landsat 7 data.

(c) LANDSAT 7 DATA POLICY PLAN.—Not later than July 15, 1994, the Landsat Program Management shall develop and submit to Congress a report that contains a Landsat 7 Data Policy Plan. This plan shall define the roles and responsibilities of the various public and private sector entities that would be involved in the acquisition, processing, distribution, and archiving of Landsat 7 data and in operations of the Landsat 7 spacecraft. [marginal note: "Reports"]

(d) REPORTS.—Not later than 12 months after submission of the Landsat 7 Data Policy Plan, required by subsection (c), and annually thereafter until the launch of Landsat 7, the Landsat Program Management, in consultation with representatives of appropriate United States Government agencies, shall prepare and submit a report to the Congress which—

(1) provides justification for the Landsat 7 data policy in terms of the civilian, national security, commercial, and foreign policy needs of the United States; and

(2) provides justification for any elements of the Landsat 7 data policy which are not consistent with the provisions of subsection (a).

TITLE II—LICENSING OF PRIVATE REMOTE SENSING SPACE SYSTEMS

SEC. 201. GENERAL LICENSING AUTHORITY. [citation in margin: "15 USC 5621"]

(a) LICENSING AUTHORITY OF SECRETARY.—

(1) In consultation with other appropriate United States Government agencies, the Secretary is authorized to license private sector parties to operate private remote sensing space systems for such period as the Secretary may specify and in accordance with the provisions of this title.

(2) In the case of a private space system that is used for remote sensing and other purposes, the authority of the Secretary under this title shall be limited only to the remote sensing operations of such space system.

(b) COMPLIANCE WITH THE LAW, REGULATIONS, INTERNATIONAL OBLIGATIONS, AND NATIONAL SECURITY.—No license shall be granted by the Secretary unless the Secretary determines in writing that the applicant will comply with the requirements of this Act, any regulations issued pursuant to this Act, and any applicable international obligations and national security concerns of the United States.

(c) DEADLINE FOR ACTION ON APPLICATION.—The Secretary shall review any application and make a determination thereon within 120 days of the receipt of such application. If final action has not occurred within such time, the Secretary shall inform the applicant of any pending issues and of actions required to resolve them.

(d) IMPROPER BASIS FOR DENIAL.—The Secretary shall not deny such license in order to protect any existing licensee from competition.

(e) REQUIREMENT TO PROVIDE UNENHANCED DATA.—

(1) The Secretary, in consultation with other appropriate United States Government agencies and pursuant to paragraph (2), shall designate in a license issued pursuant to this title any unenhanced data required to be provided by the licensee under section 202(b)(3).

(2) The Secretary shall make a designation under paragraph (1) after determining that—

(A) such data are generated by a system for which all or a substantial part of the development, fabrication, launch, or operations costs have been or will be directly funded by the United States Government; or

(B) it is in the interest of the United States to require such data to be provided by the licensee consistent with section 202(b)(3), after considering the impact on the licensee and the importance of promoting widespread access to remote sensing data from United States and foreign systems.

(3) A designation made by the Secretary under paragraph (1) shall not be inconsistent with any contract or other arrangement entered into between a United States Government agency and the licensee.

SEC. 202. CONDITIONS FOR OPERATION. [citation in margin: "15 USC 5622"]

(a) LICENSE REQUIRED FOR OPERATION.—No person who is subject to the jurisdiction or control of the United States may, directly or through any subsidiary or affiliate, operate any private remote sensing space system without a license pursuant to section 201.

(b) LICENSING REQUIREMENTS.—Any license issued pursuant to this title shall specify that the licensee shall comply with all of the requirements of this Act and shall—

(1) operate the system in such manner as to preserve the national security of the United States and to observe the international obligations of the United States in accordance with section 506;

(2) make available to the government of a country (including the United States) unenhanced data collected by the system concerning the territory under the jurisdiction of such government as soon as such data are available and on reasonable terms and conditions;

(3) make unenhanced data designated by the Secretary in the license pursuant to section 201(e) available in accordance with section 501;

(4) upon termination of operations under the license, make disposition of any satellites in space in a manner satisfactory to the President;

(5) furnish the Secretary with complete orbit and data collection characteristics of the system, and inform the Secretary immediately of any deviation; and

(6) notify the Secretary of any agreement the licensee intends to enter with a foreign nation, entity, or consortium involving foreign nations or entities.

(c) ADDITIONAL LICENSING REQUIREMENTS FOR LANDSAT 6 CONTRACTOR.—In addition to the requirements of paragraph (b), any license issued pursuant to this title to the Landsat 6 contractor shall specify that the Landsat 6 contractor shall—

(1) notify the Secretary of any value-added activities (as defined by the Secretary by regulation) that will be conducted by the Landsat 6 contractor or by a subsidiary or affiliate; and

(2) if such activities are to be conducted, provide the Secretary with a plan for compliance with section 501 of this Act.

SEC. 203. ADMINISTRATIVE AUTHORITY OF THE SECRETARY. [citation in margin: "15 USC 5623"]

(a) FUNCTIONS.—In order to carry out the responsibilities specified in this title, the Secretary may—

(1) grant, condition, or transfer licenses under this Act;

(2) seek an order of injunction or similar judicial determination from a United States District Court with personal jurisdiction over the licensee to terminate, modify, or suspend licenses under this title and to terminate licensed operations on an immediate basis, if the Secretary determines that the licensee has substantially failed to comply with any provisions of this Act, with any terms, conditions, or restrictions of such license, or with any international obligations or national security concerns of the United States.

(3) provide penalties for noncompliance with the requirements of licenses or regulations issued under this title, including civil penalties not to exceed $10,000 (each day of operation in violation of such licenses or regulations constituting a separate violation);

(4) compromise, modify, or remit any such civil penalty;

(5) issue subpoenas for any materials, documents, or records, or for the attendance and testimony of witnesses for the purpose of conducting a hearing under this section;

(6) seize any object, record, or report pursuant to a warrant from a magistrate based on a showing of probable cause to believe that such object, record, or report was used, is being used, or is likely to be used in violation of this Act or the requirements of a license or regulation issued thereunder; and

(7) make investigations and inquiries and administer to or take from any person an oath, affirmation, or affidavit concerning any matter relating to the enforcement of this Act.

(b) REVIEW OF AGENCY ACTION.—Any applicant or licensee who makes a timely request for review of an adverse action pursuant to subsection (a)(1), (a)(3), (a)(5), or (a)(6) shall be entitled to adjudication by the Secretary on the record after an opportunity for any agency hearing with respect to such adverse action. Any final action by the Secretary under this subsection shall be subject to judicial review under chapter 7 of title 5, United States Code.

SEC. 204. REGULATORY AUTHORITY OF THE SECRETARY. [citation in margin: "15 USC 5624"]

The Secretary may issue regulations to carry out this title. Such regulations shall be promulgated only after public notice and comment in accordance with the provisions of section 553 of title 5, United States Code.

SEC. 205. AGENCY ACTIVITIES. [citation in margin: "15 USC 5625"]

(a) LICENSE APPLICATION AND ISSUANCE.—A private sector party may apply for a license to operate a private remote sensing space system which utilizes, on a space-available basis, a civilian United States Government satellite or vehicle as a platform for such system. The Secretary, pursuant to this title, may license such system if it meets all conditions of this title and—

(1) the system operator agrees to reimburse the Government in a timely manner for all related costs incurred with respect to such utilization, including a reasonable

and proportionate share of fixed, platform, data transmission, and launch costs; and

(2) such utilization would not interfere with or otherwise compromise intended civilian Government missions, as determined by the agency responsible for such civilian platform.

(b) ASSISTANCE.—The Secretary may offer assistance to private sector parties in finding appropriate opportunities for such utilization.

(c) AGREEMENTS.—To the extent provided in advance by appropriation Acts, any United States Government agency may enter into agreements for such utilization if such agreements are consistent with such agency's mission and statutory authority, and if such remote sensing space system is licensed by the Secretary before commencing operation.

(d) APPLICABILITY.—This section does not apply to activities carried out under title III.

(e) EFFECT ON FCC AUTHORITY.—Nothing in this title shall affect the authority of the Federal Communications Commission pursuant to the Communications Act of 1934 (47 U.S.C. 151 *et seq.*).

TITLE III—RESEARCH, DEVELOPMENT, AND DEMONSTRATION

SEC. 301. CONTINUED FEDERAL RESEARCH AND DEVELOPMENT. [citation in margin: 15 USC 5631]

(a) ROLES OF NASA AND DEPARTMENT OF DEFENSE.—

(1) The Administrator and the Secretary of Defense are directed to continue and to enhance programs of remote sensing research and development.

(2) The Administrator is authorized and encouraged to—

(A) conduct experimental space remote sensing programs (including applications demonstration programs and basic research at universities);

(B) develop remote sensing technologies and techniques, including those needed for monitoring the Earth and its environment; and

(C) conduct such research and development in cooperation with other United States Government agencies and with public and private research entities (including private industry, universities, non-profit organizations, State and local governments, foreign governments, and international organizations) and to enter into arrangements (including joint ventures) which will foster such cooperation.

(b) ROLES OF DEPARTMENT OF AGRICULTURE AND DEPARTMENT OF INTERIOR.—

(1) In order to enhance the ability of the United States to manage and utilize its renewable and nonrenewable resources, the Secretary of Agriculture and the Secretary of the Interior are authorized and encouraged to conduct programs of research and development in the applications of remote sensing using funds appropriated for such purposes.

(2) Such programs may include basic research at universities, demonstrations of applications, and cooperative activities involving other Government agencies, private sector parties, and foreign and international organizations.

(c) ROLE OF OTHER FEDERAL AGENCIES.— United States Government agencies are authorized and encouraged to conduct research and development on the use of remote sensing in the fulfillment of their authorized missions, using funds appropriated for such purposes.

SEC. 302. AVAILABILITY OF FEDERALLY GATHERED UNENHANCED DATA. [citation in margin: "15 USC 5632"]

(a) GENERAL RULE.—All unenhanced land remote sensing data gathered and owned by the United States Government, including unenhanced data gathered under the technology demonstration program carried out pursuant to section 303, shall be made available to users in a timely fashion.

(b) PROTECTION FOR COMMERCIAL DATA DISTRIBUTOR.—The President shall seek to ensure that unenhanced data gathered under the technology demonstration program carried out pursuant to section 303 shall, to the extent practicable, be made available on terms that would not adversely effect [sic] the commercial market for unenhanced data gathered by the Landsat 6 spacecraft. [marginal note: "President"]

SEC. 303. TECHNOLOGY DEMONSTRATION PROGRAM. [citation in margin: "15 USC 5633"]

(a) ESTABLISHMENT.—As a fundamental component of a national land remote sensing strategy, the President shall establish, through appropriate United States Government agencies, a technology demonstration program. The goals of such programs shall be to— [marginal note: "President"]

(1) seek to launch advanced land remote sensing system components within 5 years after the date of the enactment of this Act;

(2) demonstrate within such 5-year period advanced sensor capabilities suitable for use in the anticipated land remote sensing program; and

(3) demonstrate within such 5-year period an advanced land remote sensing system design that could be less expensive to procure and operate than the Landsat system projected to be in operation through year 2000, and that therefore holds greater potential for private sector investment and control.

(b) EXECUTION OF PROGRAM.—In executing the technology demonstration program, the President shall seek to apply technologies associated with United States National Technical Means of intelligence gathering, to the extent that such technologies are appropriate for the technology demonstration and can be declassified for such purposes without causing adverse harm to United States national security interests. [marginal note: "President"]

(c) BROAD APPLICATION.—To the greatest extent practicable, the technology demonstration program established under subsection (a) shall be designed to be responsive to the broad civilian, national security, commercial, and foreign policy needs of the United States.

(d) PRIVATE SECTOR FUNDING.—The technology demonstration program under this section may be carried out in part with private sector funding.

(e) LANDSAT PROGRAM MANAGEMENT COORDINATION.—The Landsat Program Management shall have a coordinating role in the technology demonstration program carried out under this section.

(f) REPORT TO CONGRESS.—The President shall assess the progress of the technology demonstration program under this section and, within 2 years after the date of enactment of this Act, submit a report to the Congress on such progress. [marginal note: "President"]

TITLE IV—ASSESSING OPTIONS FOR SUCCESSOR LAND
REMOTE SENSING SYSTEM

SEC. 401. ASSESSING OPTIONS FOR SUCCESSOR LAND REMOTE SENSING SYSTEM. [citation in margin: "15 USC 5641"]

(a) ASSESSMENT.—Within 5 years after the date of the enactment of this Act, the Landsat Program Management, in consultation with representatives of appropriate United States Government agencies, shall assess and report to the Congress on the options for a successor land remote sensing system to Landsat 7. The report shall include a full assessment of the advantages and disadvantages of— [marginal note: "Reports"]

(1) private sector funding and management of a successor land remote sensing system;

(2) establishing an international consortium for the funding and management of a successor land remote sensing system;

(3) funding and management of a successor land remote sensing system by the United States Government; and

(4) a cooperative effort between the United States Government and the private sector for the funding and management of a successor land remote sensing system.

(b) GOALS.—In carrying out subsection (a), the Landsat Program Management shall consider the ability of each of the options to—

(1) encourage the development, launch, and operation of a land remote sensing system that adequately serves the civilian, national security, commercial, and foreign policy interests of the United States;

(2) encourage the development, launch, and operation of a land remote sensing system that maintains data continuity with the Landsat system; and

(3) incorporate system enhancements, including any such enhancements developed under the technology demonstration program under section 303, which may potentially yield a system that is less expensive to build and operate, and more responsive to data users than is the Landsat system projected to be in operation through the year 2000.

(c) PREFERENCE FOR PRIVATE SECTOR SYSTEM.—If a successor land remote sensing system to Landsat 7 can be funded and managed by the private sector while still achieving the goals stated in subsection (b) without jeopardizing the domestic, national security, and foreign policy interests of the United States, preference should be given to the development of such a system by the private sector without competition from the United States Government.

TITLE V—GENERAL PROVISIONS

SEC. 501. NONDISCRIMINATORY DATA AVAILABILITY. [citation in margin: "15 USC 5651"]

(a) GENERAL RULE.—Except as provided in subsection (b) of this section, any unenhanced data generated by the Landsat system or any other land remote sensing system funded and owned by the United States Government shall be made available to all users without preference, bias, or any other special arrangement (except on the basis of national security concerns pursuant to section 506) regarding delivery, format, pricing, or technical considerations which would favor one customer or class of customer over another.

(b) EXCEPTIONS.—Unenhanced data generated by the Landsat system or any other land remote sensing system funded and owned by the United States Government may be made available to the United States Government and its affiliated users at reduced prices, in accordance with this Act, on the condition that such unenhanced data are used solely for noncommercial purposes.

SEC. 502. ARCHIVING OF DATA. [citation in margin: "15 USC 5652"]

(a) PUBLIC INTEREST.—It is in the public interest for the United State Government to—

(1) maintain an archive of land remote sensing data for historical, scientific, and technical purposes, including long-term global environmental monitoring;

(2) control the content and scope of the archive; and

(3) assure the quality, integrity, and continuity of the archive.

(b) ARCHIVING PRACTICES.—The Secretary of the Interior, in consultation with the Landsat Program Management, shall provide for long-term storage, maintenance, and upgrading of a basic, global, land remote sensing data set (hereinafter referred to as the "basic data set") and shall follow reasonable archival practices to assure proper storage and preservation of the basic data set and timely access for parties requesting data.

(c) DETERMINATION OF CONTENT OF BASIC DATA SET.—In determining the initial content of, or in upgrading, the basic data set, the Secretary of Interior shall—

(1) use as a baseline the data archived on the date of enactment of this Act;

(2) take into account future technical and scientific developments and needs, paying particular attention to the anticipated data requirements of global environmental change research;

(3) consult with and seek the advice of users and producers of remote sensing data and data products;

(4) consider the need for data which may be duplicative geographical coverage but which differ in term of season, spectral bands, resolution, or other relevant factors;

(5) include, as the Secretary of the Interior considers appropriate, unenhanced data generated either by the Landsat system, pursuant to title I, or by licensees under title II;

(6) include, as the Secretary of the Interior considers appropriate, data collected by foreign ground stations or by foreign remote sensing space systems; and

(7) ensure that the content of the archive is developed in accordance with section 506.

(d) PUBLIC DOMAIN.—After the expiration of any exclusive right to sell, or after relinquishment of such right, the data provided to the National Satellite Land Remote Sensing Data Archive shall be in the public domain and shall be made available to requesting parties by the Secretary of the Interior at the cost of fulfilling user requests.

SEC. 503. NONREPRODUCTION. [citation in margin: "15 USC 5653"]

Unenhanced data distributed by any licensee under title II of this Act may be sold on the condition that such data will not be reproduced or disseminated by the purchaser for commercial purposes.

SEC. 504. REIMBURSEMENT FOR ASSISTANCE. [citation in margin: "15 USC 5654"]

The Administrator, the Secretary of Defense, and the heads of other United States Government agencies may provide assistance to land remote sensing system operators under the provisions of this Act. Substantial assistance shall be reimbursed by the operator, except as otherwise provided by law.

SEC. 505. ACQUISITION OF EQUIPMENT. [citation in margin: "15 USC 5655"]

The Landsat Program Management may, by means of a competitive process, allow a licensee under title II or any other party to buy, lease, or otherwise acquire the use of equipment from the Landsat system, when such equipment is no longer needed for the operation of such system or for the sale of data from such system. Officials of other United States Government civilian agencies are authorized and encouraged to cooperate with the Secretary in carrying out this section.

SEC. 506. RADIO FREQUENCY ALLOCATION. [citation in margin: "15 USC 5656"]

(a) APPLICATION TO FEDERAL COMMUNICATIONS COMMISSION.—To the extent required by the Communications Act of 1934 (47 U.S.C. 151 *et seq.*), an application shall be filed with the Federal Communications Commission for any radio facilities involved with commercial remote sensing space systems licensed under title II.

(b) DEADLINE FOR FCC ACTION.—It is the intent of Congress that the Federal Communications Commission complete the radio licensing process under the Communications Act of 1934 (47 U.S.C. 151 *et seq.*), upon the application of any private sector party or consortium operator of any commercial land remote sensing space system subject to this Act, within 120 days of the receipt of an application for such licensing. If final action has not occurred within 120 days of the receipt of such an application, the Federal Communications Commission shall inform the applicant of any pending issues and of actions required to resolve them. [marginal note: "Licensing"]

(c) DEVELOPMENT AND CONSTRUCTION OF UNITED STATES SYSTEMS.— Authority shall not be required from the Federal Communications Commission for the development and construction of any United States land remote sensing space system (or component thereof), other than radio transmitting facilities or components, while any licensing determination is being made.

(d) CONSISTENCY WITH INTERNATIONAL OBLIGATIONS AND PUBLIC INTEREST.—Frequency allocations made pursuant to this section by the Federal Communications Commission shall be consistent with international obligations and with the public interest.

SEC. 507. CONSULTATION. [citation in margin: "15 USC 5657"]

(a) CONSULTATION WITH SECRETARY OF DEFENSE.—The Secretary and the Landsat Program Management shall consult with the Secretary of Defense on all matters under this Act affecting national security. The Secretary of Defense shall be responsible for determining those conditions, consistent with this Act, necessary to meet national security concerns of the United States and for notifying promptly the Secretary and the Landsat Program Management promptly of such conditions.

(b) CONSULTATION WITH SECRETARY OF STATE.—

(1) The Secretary and the Landsat Program Management shall consult with the Secretary of State on all matters under this Act affecting international obligations. The Secretary of State shall be responsible for determining those conditions, consistent with this Act, necessary to meet international obligations and policies of the United States and for notifying promptly the Secretary and the Landsat Program Management of such conditions.

(2) Appropriate United States Government agencies are authorized and encouraged to provide remote sensing data, technology, and training to developing nations as a component of programs of international aid.

(3) The Secretary of State shall promptly report to the Secretary and Landsat Program Management any instances outside the United States of discriminatory distribution of Landsat data. [marginal note: "Reports"]

(c) STATUS REPORT.—The Landsat Program Management shall, as often as necessary, provide to the Congress complete and update[d] information about the status of ongoing operations of the Landsat system, including timely notification of decisions made with respect to the Landsat system in order to meet national security concerns and international obligations and policies of the United States Government.

(d) REIMBURSEMENT.—If as a result of technical modifications imposed on a licensee under title II on the basis of national security concerns, the Secretary, in consultation with the Secretary of Defense or with other Federal agencies, determines that additional costs will be incurred by the licensee, or that past development costs (including the cost of capital) will not be recovered by the licensee, the Secretary may require the agency or agencies requesting such technical modifications to reimburse the licensee for such additional or development costs, but not for anticipated profits. Reimbursements may cover costs associated with required changes in system performance, but not costs ordinarily associated with doing business abroad.

SEC. 508. ENFORCEMENT. [citation in margin: "15 USC 5658"]

(a) IN GENERAL.—In order to ensure that unenhanced data from the Landsat system received solely for noncommercial purposes are not used for any commercial purpose, the Secretary (in collaboration with private sector entities responsible for the marketing and distribution of unenhanced data generated by the Landsat system) shall develop and implement a system for enforcing this prohibition, in the event that unenhanced data from the Landsat system are made available for noncommercial purposes at a different price than such data are made available for other purposes.

(b) AUTHORITY OF SECRETARY. —Subject to subsection (d), the Secretary may impose any of the enforcement mechanisms described in subsection (c) against a person who—

(1) receives unenhanced data from the Landsat system under this Act solely for noncommercial purposes (and at a different price than the price at which such data are made available for other purposes); and

(2) uses such data for other than noncommercial purposes.

(c) ENFORCEMENT MECHANISMS.—Enforcement mechanisms referred to in subsection (b) may include civil penalties of not more than $10,000 (per day per violation), denial of further unenhanced data purchasing privileges, and any other penalties or restrictions the Secretary considers necessary to ensure, to the greatest extent practicable, that unenhanced data provided for noncommercial purposes are not used to unfairly compete in the commercial market against private sector entities not eligible for data at the cost of fulfilling user requests.

(d) PROCEDURES AND REGULATIONS.—The Secretary shall issue any regulations necessary to carry out this section and shall establish standards and procedures governing the imposition of enforcement mechanisms under subsection (b). The standards and procedures shall include a procedure for potentially aggrieved parties to file formal protests with the Secretary alleging instances where such unenhanced data has been, or is being, used for commercial purposes in violation of the terms of receipt of such data. The Secretary shall promptly act to investigate any such protest, and shall report annually to the Congress on instances of such violations. [marginal note: "Reports"]

TITLE VI—PROHIBITION OF COMMERCIALIZATION OF
WEATHER SATELLITES

SEC. 601. PROHIBITION. [citation in margin: "15 USC 5671"]

Neither the President nor any other official of the Government shall make any effort to lease, sell, or transfer to the private sector, or commercialize, any portion of the weather satellite systems operated by the Department of Commerce or any successor agency.

SEC. 602. FUTURE CONSIDERATIONS. [citation in margin: "15 USC 5672"]

Regardless of any change in circumstances subsequent to the enactment of this Act, even if such change makes it appear to be in the national interest to commercialize weather satellites, neither the President nor any official shall take any action prohibited by section 601 unless this title has first been repealed.

Approved October 28,1992.

LEGISLATIVE HISTORY—H.R. 6133:
CONGRESSIONAL RECORD, Vol. 138 (1992):
    Oct. 5, considered and passed House.
    Oct. 7, considered and passed Senate.
WEEKLY COMPILATION OF PRESIDENTIAL DOCUMENTS, Vol. 28 (1992):
    Oct. 28, Presidential statement.

Document II-41

Document title: George E. Brown, Jr., Chairman, Committee on Science, Space, and Technology, U.S. House of Representatives, to John H. Gibbons, Assistant to the President, Office of Science and Technology Policy, August 9, 1993.

Document II-42

Document title: John Deutch, Under Secretary of Defense, to George E. Brown, Jr., Chairman, Committee on Science, Space, and Technology, U.S. House of Representatives, December 9, 1993.

Document II-43

Document title: John H. Gibbons, Director, Office of Science and Technology Policy, to George E. Brown, Jr., Chairman, Committee on Science, Space, and Technology, U.S. House of Representatives, December 10, 1993.

Document II-44

Document title: George E. Brown, Jr., Chairman, Committee on Science, Space, and Technology, U.S. House of Representatives, to John H. Gibbons, Assistant to the President, Office of Science and Technology Policy, December 14, 1993.

Source: All in NASA Historical Reference Collection, NASA History Office, NASA Headquarters, Washington, D.C.

Theses four letters highlight the policy and funding dispute over the Department of Defense (DOD) provision of the High Resolution Multispectral Stereo Imager (HRMSI) for Landsat 7, which DOD had originally proposed for the satellite but then backed away from when NASA refused to fund a substantial upgrade for the ground system to collect the data. U.S. Representative George Brown had played a major role in drafting and sponsoring the Land Remote-Sensing Policy Act of 1992 and therefore had a stake in seeing the controversy successfully resolved. Dr. John Gibbons, President Clinton's science advisor and the director of the Office of Science and Technology Policy, helped craft the Clinton administration's policy toward Landsat 7.

Document II-41

August 9, 1993

The Honorable John Gibbons
Assistant to the President
Office of Science and Technology Policy
Old Executive Office Building
Washington, DC 20500

Dear Jack:

As you know, NASA and the Department of Defense are roughly splitting the cost of the Landsat program in accordance with the Land Remote Sensing Policy Act (P.L. 102–555), which the Vice President played an important role in shaping. However, only DOD has requested funding for the High Resolution Multispectral Stereo Imager (HRMSI), a new sensor that would produce 5-meter resolution imagery; NASA's budget does not include its proposed funding share for HRMSI, which would be used to build the data handling system.

While I recognize the difficult choices that the Administration faced in formulating NASA's budget, I was greatly disappointed that this funding for HRMSI was not included. HRMSI is exactly what many users of remote sensing data have wanted for years. In addition to global change research, environmental applications that need HRMSI-type data include biodiversity ecosystem mapping, forest and coastal wetlands inventories, oil spill tracking, toxic waste sitting and monitoring, and land use planning.

The lack of NASA co-funding for HRMSI could lead to the cancellation of HRMSI if DOD is not able to obtain additional funding. Without HRMSI, DOD's interest in Landsat could diminish, possibly jeopardizing this valuable program that we have worked so hard to preserve. Even if DOD is able to fund the entire cost of HRMSI, most civil users may be unable to obtain imagery they need because DOD funding would not cover data acquisition and processing for civil users.

I hope that you will turn your attention to the need to establish a funding framework for HRMSI that ensures full civil participation and use of the data. I stand ready to assist you in any way possible.

Sincerely,

[hand-signed: "George"]
GEORGE E. BROWN, JR.
Chairman

Document II-42

December 9, 1993

Honorable George E. Brown, Jr.
Chairman
Committee on Science, Space and Technology
U.S. House of Representatives
Washington, D.C. 20515

Dear Mr. Chairman,

I promised to keep you informed about LANDSAT. Yesterday, Dan Goldin and I met to talk abut some program decisions in preparation for the FY95 budget. Basically, the decision is for NASA to go its own way with a 30 meter resolution thematic mapper, and DoD will consider if it will go forward with the 5 meter High Resolution Multispectral Instrument (HRMSI). DoD and NASA could not afford to go forward with the LANDSAT 7 and LANDSAT 8 program.

I am available to discuss this with you at anytime. Hope you had a productive trip to the Far East.

Best regards,

[hand-signed: "John"]
John Deutch

Document II-43

[1]   December 10, 1993

[handwritten note: "George"]

Dear Mr. Chairman:

Thank you for sharing your concerns on the loss of Landsat 6. I also share your disappointment with respect to the loss of Landsat 6 and your concern with the potential data gap that could result from this loss. The Office of Science and Technology Policy is taking an active role in reviewing the options for responding to this new situation. To this end OSTP is working with the Department of Defense, the National Oceanic and Atmospheric Administration, and the National Aeronautics and Space Administration to develop options for reducing the likelihood of a gap in critical remote sensing data.

The Administration recognizes that Landsat data is valuable to a broad user community as represented by Federal, state, and local governments, as well as the business, intelligence, and academic communities. Landsat is primarily valuable because its 20+ year long data base allows trends in land-surface characteristics to be determined, its radiometric measurement accuracy supports detailed characterization of the land-surface, and its combination of wide area coverage and 30 meter resolution allows broad areas to be assessed prior to initiating detailed studies requiring higher spatial resolution. Consequently, as the Administration develops a response to the loss of Landsat 6 we will take into consideration the requirements of the broad user community and the capabilities required to meet their needs.

Although we are in the early stages of assessing options, initial indications from EOSAT, the system's operator, are that Landsat 5 could continue to operate for several more years even though it is already 9 years old and well past its design life. Consequently, the likelihood of a significant data gap may not be as great as originally anticipated, assuming that Landsat 7 is launched on schedule. However, we do have serious concerns with the coverage of the earth and the site revisit time of Landsat 5. Landsat 6 would have addressed these concerns and added some additional capabilities that would have further increased the competitiveness and utility of the Landsat system.

OSTP in conjunction with the DOD, NOAA, and NASA is carefully considering options to address both the loss of Landsat 6 and the cost and risk of the Landsat 7 program. These two items are linked because the availability of Landsat 7 directly affects when or if a replacement spacecraft for Landsat 6 should be launched. As you are aware the current budgetary environment has made it necessary to examine options in terms of their cost and risk; therefore, I want to assure you that any alternatives to the current program will undergo careful interagency review in order to evaluate potential impacts relative to the Landsat program's diverse requirements and users. [2] Further, I would like to assure you that any proposal for responding to the loss of Landsat 6 or for significantly modifying the current Landsat program will address data continuity; impacts on agency budgets; schedule impacts; the acquisition, launch and operation of the necessary satellites and instruments; and the needs of the user communities.

The necessary interagency review will be expedited to meet the needs of FY95 budgetary decision timetables. However, it is important to make clear that no decision on program alternatives has been or will be taken prior to the completion of this review.

I look forward to working with you on this important issue and will provide a more detailed discussion of the alternatives being considered in the near future.

Sincerely yours,

[hand-signed: "Jack"]
John H. Gibbons
Director

The Honorable George E. Brown, Jr.
Chairman
Committee on Science, Space and Technology
U.S. House of Representatives
Washington, DC 20510

Document II-44

December 14, 1993

The Honorable John H. Gibbons
Assistant to the President
Office of Science and Technology Policy
Old Executive Office Building
Washington, D.C. 20500

[handwritten note: "Jack"]

Dear Dr. Gibbons:

Thank you for your letter concerning the Landsat program. I appreciate your keeping me informed about the review of Landsat issues, but the letter does not directly respond to the concerns about Landsat 7 outlined in my letter to you of [handwritten underlining] August 9. [handwritten note in margin: "(attached)"] Specifically, the letter made no mention of what I believe is the most critical issue—resolving the differences between NASA and DOD concerning funding for the High Resolution Multispectral Stereo Imager (HRMSI) on Landsat 7.

I would like to reiterate my strong support for the Landsat 7 program and my belief that HRMSI represents a valuable new capability for both national security and civilian users.

I would like to request again that you personally intervene to provide a solution that ensures continuation of the Landsat 7 program. Any solution should fulfill the requirements of the Land Remote Sensing Policy Act (Public Law 102–555), including an integrated management by NASA and the Department of Defense, procurement of Landsat 7 as quickly as practicable that is at least as capable as Landsat 6 would have been, and, where possible, incorporation of any improvement needed to meet U.S. Government needs.

As always, I stand ready to assist you in any way possible.

Sincerely,

[hand-signed: "George"]
GEORGE E. BROWN, JR.
Chairman

Document II-45

Document title: The White House, Presidential Decision Directive/NSTC-3, "Landsat Remote Sensing Strategy," May 5, 1994.

Source: NASA Historical Reference Collection, NASA History Office, NASA Headquarters, Washington, D.C.

After the Department of Defense decided to stop directly participating in the Landsat program, policy makers within the Clinton administration faced the task of determining how to proceed. After considerable discussion within the Clinton administration's National Science and Technology Council, with input from the National Security Council, President Clinton signed this directive, which gave NASA responsibility for procuring Landsat 7, NOAA the responsibility for managing operations, and the Department of the Interior the responsibility for distributing the data through the U.S. Geological Survey's EROS Data Center in Sioux Falls, South Dakota.

[no pagination]

<u>Presidential Decision Directive/NSTC-3</u>

TO: The Vice President
        The Secretary of Defense
        The Secretary of Interior
        The Secretary of Commerce
        The Director, Office of Management and Budget
        The Administrator, National Aeronautics and Space Administration
        The Assistant to the President for National Security Affairs
        The Assistant to the President for Science and Technology
        The Assistant to the President for Economic Policy

SUBJECT: Landsat Remote Sensing Strategy

I.   Introduction

This directive provides for continuance of the Landsat 7 program, assures continuity of Landsat-type and quality of data, and reduces the risk of a data gap.

The Landsat program has provided over 20 years of calibrated data to a broad user community including the agricultural community, global change researchers, state and local governments, commercial users, and the military. The Landsat 6 satellite which failed to reach orbit in 1993 was intended to replace the existing Landsat satellites 4 and 5, which were launched in 1982 and 1984. These satellites which are operating well beyond their three year design lives, represent the only source of global calibrated high spatial resolution measurements of the Earth's surface that can be compared to previous data records.

In the Fall of 1993 the joint Department of Defense and National Aeronautics and Space Administration Landsat 7 program was being reevaluated due to severe budgetary constraints. This fact, coupled with the advanced age of Landsat satellites 4 and 5, resulted in a re-assessment of the Landsat program by representatives of the National Science and Technology Council. The objectives of the National Science and Technology Council were to minimize the potential for a gap in the Landsat data record if Landsat satellites 4 and 5 should cease to operate, to reduce cost, and to reduce development risk. The results of this re-assessment are identified below.

This document supersedes National Space Policy Directive #5, dated February 2, 1992, and directs implementation of the Landsat Program consistent with the intent of P. L. 102–555, the Land Remote Sensing Policy Act of 1992, and P. L. 103–221, the Emergency Supplemental Appropriations Act. The Administration will seek all legislative changes necessary to implement this PDD.

II.   Policy Goals

A remote sensing capability, such as is currently being provided by Landsat satellites 4 and 5, benefits the civil, commercial, and national security interests of the United States and makes contributions to the private sector which are in the public interest. For these reasons, the United States Government will seek to maintain the continuity of Landsat-type data. The U.S. Government will:

(a) Provide unenhanced data which are sufficiently consistent in terms of acquisition geometry, coverage characteristics, and spectral characteristics with previous Landsat data to allow quantitative comparisons for change detection and characterization;

(b) Make government-owned Landsat data available to meet the needs of all users at no more than the cost of fulfilling user requests consistent with data policy goals of P. L. 102–555; and

(c) Promote and not preclude private sector commercial opportunities in Landsat-type remote sensing.

III. Landsat Strategy

a. The Landsat strategy is composed of the following elements:

(1) Ensuring that Landsat satellites 4 and 5 continue to provide data as long as they are technically capable of doing so.

(2) Acquiring a Landsat 7 satellite that maintains the continuity of Landsat-type data, minimizes development risk, minimizes cost, and achieves the most favorable launch schedule to mitigate the loss of Landsat 6.

(3) Maintaining an archive within the United States for existing and future Landsat-type data.

(4) Ensuring that unenhanced data from Landsat 7 are available to all users at no more than the cost of fulfilling user requests.

(5) Providing data for use in global change research in a manner consistent with the Global Change Research Policy Statements for Data Management.

(6) Considering alternatives for maintaining the continuity of data beyond Landsat 7.

(7) Fostering the development of advanced remote sensing technologies, with the goal of reducing the cost and increasing the performance of future Landsat-type satellites to meet U.S. Government needs, and potentially, enabling substantially greater opportunities for commercialization.

b. These strategy elements will be implemented within the overall resource and policy guidance provided by the President.

IV. Implementing Guidelines

Affected agencies will identify funds necessary to implement the National Strategy for Landsat Remote Sensing within the overall resource and policy guidance provided by the President. {In order to effectuate the strategy enumerated herein, the Secretary of Commerce and the Secretary of the Interior are hereby designated as members of the Landsat Program Management in accordance with section 101(b) of the Landsat Remote Sensing Policy Act of 1992, 15 U.S.C. 5602 (6) and 5611(b).} Specific agency responsibilities are provided below.

a. The Department of Commerce/NOAA will:

(1) In participation with other appropriate government agencies arrange for the continued operation of Landsat satellites 4 and 5 and the routine operation of future Landsat satellites after their placement in orbit.

(2) Seek better access to data collected at foreign ground stations for U.S. Government and private sector users of Landsat data.

(3) In cooperation with NASA, manage the development of and provide a share of the funding for the Landsat 7 ground system.

(4) Operate the Landsat 7 spacecraft and ground system in cooperation with the Department of the Interior.

(5) Seek to offset operations costs through use of access fees from foreign ground stations and/or the cost of fulfilling user requests.

(6) Aggregate future Federal requirements for civil operational land remote sensing data.

b. The National Aeronautics and Space Administration will:

(1) Ensure data continuity by the development and launch of a Landsat 7 satellite system which is at a minimum functionally equivalent to the Landsat 6 satellite in accordance with section 102, P. L. 102–555.

(2) In coordination with DOC and DOI, develop a Landsat 7 ground system compatible with the Landsat 7 spacecraft.

(3) In coordination with DOC, DOI, and DOD, revise the current Management plan to reflect the changes implemented through this directive, including programmatic, technical, schedule, and budget information.

(4) Implement the joint NASA/DOD transition plan to transfer the DOD Landsat 7 responsibilities to NASA.

(5) In coordination with other appropriate agencies of the U.S. Government develop a strategy for maintaining continuity of Landsat-type data beyond Landsat 7.

(6) Conduct a coordinated technology demonstration program with other appropriate agencies to improve the performance and reduce the cost for future unclassified earth remote sensing systems.

c.   The <u>Department of Defense</u> will implement the joint NASA/DOD transition plan to transfer the DOD Landsat 7 responsibilities to NASA.

d.   The <u>Department of the Interior</u> will continue to maintain a national archive of existing and future Landsat-type remote sensing data within the United States and make such data available to U.S. Government and other users.

e.   Affected agencies will identify the funding, and funding transfers for FY 1994, required to implement this strategy that are within their approved fiscal year 1994 budgets and subsequent budget requests.

V.   Reporting Requirements

U.S. Government agencies affected by the strategy guidelines are directed to report no later that 30 days following the issuance of this directive, to the National Science and Technology Council on their implementation. The agencies will address management and funding responsibilities, government and contractor operations, data management, archiving, and dissemination, necessary changes to P. L. 102–555 and commercial considerations associated with the Landsat program.

Document II-46

Document title: Gregory W. Withee, Acting Assistant Administrator for Satellite and Information Services, NOAA, to Walter S. Scott, President and Chief Executive Officer, World View Imaging Corporation, January 4, 1993.

Document II-47

Document title: Duane P. Andrews, Assistant Secretary of Defense, to Gregory W. Withee, Acting Assistant Administrator for Satellite and Information Services, NOAA, December 24, 1992.

Document II-48

Document title: Ralph Braibanti, Deputy Director, Office of Advanced Technology, Bureau of Oceans and International Environmental and Scientific Affairs, U.S. Department of State, to Michael Mignogno, Chief, Landsat Commercialization Division, NOAA, October 19, 1992.

Source: All in NASA Historical Reference Collection, NASA History Office, NASA Headquarters, Washington, D.C.

The letter from NOAA's Gregory Withee represents the first license to operate a commercial remote-sens-
ing satellite system granted under Title II of the Land Remote-Sensing Policy Act of 1992. The license
was granted to WorldView Imaging Corporation and subsequently transferred to EarthWatch, Inc.,
when WorldView and Ball Aerospace formed an alliance to build and operate remote-sensing satellites
and to market data from them.

Document II-46

[1]   JAN 4, 1993

Mr. Walter S. Scott
President and Chief Executive Officer
WorldView Imaging Corporation
7015 Elverton Drive
Oakland, California 94611-1111

Dear Mr. Scott:

This letter constitutes a license under Title II or the Land Remote-Sensing Policy Act of 1992, P.L. 102–555 (the Act), for WorldView Imaging Corporation (Licensee) to operate a private remote-sensing space system. The license is valid for a period of 10 years from the above date.

A. Approval is based on operational specifications set forth in paragraph (d) attached to the letter of application dated July 15, 1992, and additional information provided in the Licensee's letter dated August 20, 1992, in response to questions 2–4 and 8.

B. This license is subject to the following terms and conditions which apply to WorldView and any affiliate or contractor as appropriate.

1. Licensee will not change the operational characteristics described above in a way that would result in materially different capabilities than those described in the application information. Licensee will notify the National Environmental Satellite, Data, and Information Service (NESDIS) at least 60 days in advance of scheduled launch.

2. Licensee will operate the system in a manner that will preserve the national security of the United States through positive control of the spacecraft including safeguards to assure the integrity of spacecraft operations. In the event of a national security crisis, as defined by the Secretary of Defense, the Secretary of Commerce may, after consultation with the Secretary or Defense, require the Licensee to limit data collection and/or distribution by the system to the extent necessitated by the crisis. During such periods, the Licensee shall endeavor to provide system data requested by DOD and other national security agencies under reasonable cost terms.

[2] 3. Licensee will make available to the Government of any country (including the United States) unenhanced data concerning the territory under the jurisdiction of such Government as soon as such data are available and on reasonable terms and conditions.

4. Licensee will make available unenhanced data requested by the National Satellite Land Remote-Sensing Data Archive (the Archive) in the Department of the Interior on reasonable cost terms as agreed by the Licensee and the Archive. After a reasonable period of time as agreed with the Licensee, the Archive may make these data available to the public at the cost of fulfilling user requests.

Before purging any data in its possession, the Licensee shall offer them to the Archive at the cost of reproduction and transmission. The Archive may make these data available immediately to the public at the cost of fulfilling user requests.

5. Licensee will notify NESDIS of any agreement pertaining to operations under this license which the Licensee intends to enter into with a foreign nation, entity, or consortium involving foreign nations or entities, at least 30 days before concluding such agreement. Standard data purchase, distribution, and processing agreements and agreements incidental to the maintenance of sales offices in foreign countries are not included in this condition.

C.   Enforcement of the Act and this license will be carried out in accordance with section 203 of the Act. Any civil penalties authorized by section 203(a) (3) will be assessed in accordance with the procedures set forth in Subparts B and C of 15 CFR Part 904. Such civil penalties may be assessed in amounts up to $10,000 for any violation of the Act or any condition of this license with each day of violation constituting a separate violation.

D. The following require an amendment of the license;
   1. Assignment of the license;
   2. Any change in ownership of Licensee that would result in foreign individuals, entities, or consortia having an aggregate interest in Licensee in excess of 15 percent; and

[3]   3. Any operation outside the range of orbits and altitudes, the range of spatial resolution or the spectral bands approved above, provided that in the case of an emergency posing an imminent and substantial threat of harm to human life, property, environment, or the remote-sensing space system itself, Licensee shall not be required to obtain such amendment. If circumstances permit, however, Licensee shall attempt to obtain oral approval from NESDIS prior to making any substantial change.

I wish you well in your endeavors.

Sincerely,

Gregory W. Withee
Acting Assistant Administrator for
   Satellite and Information Service

Document II-47

December 24, 1992

Mr. Gregory W. Withee
Acting Assistant Administrator for Satellite,
   Data, and Information Services
National Oceanic and Atmospheric Administration
Department of Commerce
Washington, DC 20233

Dear Mr. Withee:

I have reviewed World View Imaging Corporation's answers to our questions provided in your September 16, 1992, letter. I am satisfied that the system and the terms and conditions in your proposed approval letter will provide the necessary protection to meet Department of Defense requirements. I recommend approval of their request.

There is currently considerable and growing international activity in remote sensing. Since this appears to be the first attempt to license a system under the LANDSAT Act, we are very interested and would like to track its progress. Please ask Mr. Scott to contact Colonel Pete Gill of my staff at (703) 697-9897 to arrange for a briefing to DoD representatives on how the system is developing.

Sincerely,

Duane P. Andrews

Document II-48

October 19, 1992

Mr. Michael Mignogno
Chief, Landsat Commercialization Division
NOAA/NESDIS, FB-4, room 3301E
U.S. Department of Commerce
Washington, D.C. 20233

Dear Mr. Mignogno:

We thank NOAA's Thomas Pyke for his September 16, 1992, letter to Richard Smith regarding the licensing of WorldView Imaging Corporation's proposed private remote-sensing satellite system.

Mr. Pyke's letter requested Department review of WorldView's August 20, 1992, response to earlier USG [U.S. Government] follow-up questions on WorldView's initial license application.

Based on the information provided, the WorldView license application appears to be consistent with U.S. international obligations, and this office recommends approval of the application.

We note in this connection that the "National Land Remote Sensing Policy Act of 1992" has codified the requirement for appropriate Government oversight (with appropriate interagency consultation) of private system activities affecting international obligations, foreign policy, and the national security of the United States. We look forward to participation in this process.

Sincerely,

Ralph Braibanti
Deputy Director
Office of Advanced Technology

cc: NOAA/NESDIS - Gregory Withee
    OES - Richard Smith

Document II-49

Document title: Office of the Press Secretary, The White House, "U.S. Policy on Licensing and Operation of Private Remote Sensing Systems," March 10, 1994.

Source: NASA Historical Reference Collection, NASA History Office, NASA Headquarters, Washington, D.C.

The June 1993 request of Lockheed Corporation's Space Systems, Inc., for a license to operate a remote-sensing system capable of receiving and marketing data of one-meter resolution caused the Clinton administration to review its policy toward the commercial operation of high-resolution satellites. After a thorough interagency review, the White House released this policy statement, which reaffirms support for the international sale of relatively high-resolution data and allows the Secretary of Commerce the right to limit data collection or distribution "during periods when national security or international obligations and/or foreign policies may be compromised." The policy is considerably restrictive toward the sales of systems to foreign entities, noting that each request will be reviewed on a case-by-case basis and will be made only "on the basis of a government-to-government agreement."

[no pagination]

For Immediate Release   March 10, 1994

U.S. Policy on Licensing and Operation of Private Remote Sensing Systems

License requests by US firms to operate private remote sensing space systems will be reviewed on a case-by-case basis in accordance with the Land Remote Sensing Policy Act of 1992 (the Act). There is a presumption that remote sensing space systems whose performance capabilities and imagery quality characteristics are available or are planned for availability in the world marketplace (e.g., SPOT, Landsat, etc.) will be favorably considered, and that the following conditions will apply to any US entity that receives an operating license under the Act.

1.   The licensee will be required to maintain a record of all satellite tasking for the previous year and to allow the USG access to this record.

2.   The licensee will not change the operational characteristics of the satellite system from the application as submitted without formal notification and approval of the Department of Commerce, which would coordinate with other interested agencies.

3.   The license being granted does not relieve the licensee of the obligation to obtain export license(s) pursuant to applicable statutes.

4.   The license is valid only for a finite period, and is neither transferable nor subject to foreign ownership, above a specified threshold, without the explicit permission of the Secretary of Commerce.

5.   All encryption devices must be approved by the US Government for the purpose of denying unauthorized access to others during periods when national security, international obligations and/or foreign policies may be compromised as provided for in the Act.

6.   A licensee must use a data downlink format that allows the US Government access and use of the data during periods when national security, international obligations and/or foreign policies may be compromised as provided for in the Act.

7.   During periods when national security or international obligations and/or foreign policies may be compromised, as defined by the Secretary of Defense or the
Secretary of State, respectively, the Secretary of Commerce may, after consultation
with the appropriate agency(ies), require the licensee to limit data collection and/or
distribution by the system to the extent necessitated by the given situation. Decisions
to impose such limits only will be made by the Secretary of Commerce in consultation
with the Secretary of Defense or the Secretary of State, as appropriate. Disagreements
between Cabinet Secretaries may be appealed to the President. The Secretaries of
State, Defense and Commerce shall develop their own internal mechanisms to enable
them to carry out their statutory responsibilities.
8.   Pursuant to the Act, the US Government requires US companies that have been
issued operating licenses under the Act to notify the US Government of its intent to
enter into significant or substantial agreements with new foreign customers.
Interested agencies shall be given advance notice of such agreements to allow them
the opportunity to review the proposed agreement in light of the national security,
international obligations and foreign policy concerns of the US Government. The
definition of a significant or substantial agreement, as well as the time frames and
other details of this process, will be defined in later Commerce regulations in consultation with appropriate agencies.

U.S. Policy on the Transfer of Advanced Remote Sensing Capabilities

Advanced Remote Sensing System Exports

The United States will consider requests to export advanced remote sensing systems
whose performance capabilities and imagery quality characteristics are available or are
planned for availability in the world marketplace on a case-by-case basis.
    The details of these potential sales should take into account the following:
  •   the proposed foreign recipient's willingness and ability to accept commitments to
      the US Government concerning sharing, protection, and denial of products and
      data; and
  •   constraints on resolution, geographic coverage, timeliness, spectral coverage,
      data processing and exploitation techniques, tasking capabilities, and ground
      architectures.
Approval of requests for exports of systems would also require certain diplomatic
steps be taken, such as informing other close friends in the region of the request, and the
conditions we would likely attach to any sale; and informing the recipient of our decision
and the conditions we would require as part of the sale.
    Any system made available to a foreign government or other foreign entity may be
subject to a formal government-to-government agreement.

Transfer of Sensitive Technology

The United States will consider applications to export remote sensing space capabilities on a restricted basis. Sensitive technology in this situation consists of items of technology on the US Munitions List necessary to develop or to support advanced remote
sensing space capabilities and which are uniquely available in the United States. Such sensitive technology shall be made available to foreign entities only on the basis of a government-to-government agreement. This agreement may be in the form of end-use and
retransfer assurances which can be tailored to ensure the protection of US technology.

Government-to-Government Intelligence and Defense Partnerships

Proposals for intelligence or defense partnerships with foreign countries regarding remote sensing that would raise questions about US Government competition with the private sector or would change the US Government's use of funds generated pursuant to a US-foreign government partnership arrangement shall be submitted for interagency review.

Document II-50

Document title: Robert S. Winokur, Assistant Administrator for Satellite and Information Services, NOAA, to Albert E. Smith, Vice President, Advanced Government and Commercial Systems, Lockheed Missile and Space Company, Inc., April 22, 1994.

Source: NASA Historical Reference Collection, NASA History Office, NASA Headquarters, Washington, D.C.

This letter, which "constitutes a license under Title II of the Land Remote Sensing Policy Act of 1992," spells out the requirements of government policy for the operation of Lockheed's one-meter system, consistent with the Clinton administration's policy on remote sensing, licensing, and exports of March 10, 1994 (see Document II-49).

---

[1]   [rubber stamped: "APR 22, 1994"]

Albert E. Smith
Vice President, Advanced Government
    and Commercial Systems
Lockheed Missiles and Space Company, Inc.
1111 Lockheed Way
Sunnyvale, California 94089-3504

Dear Mr. Smith:

This letter constitutes a license under Title II of the Land Remote Sensing Policy Act of 1992, P.L. 102–555 (the Act), for Lockheed Missiles and Space Company, Inc. (Licensee) to operate a private remote-sensing space system. The license is valid for a period of 10 years from the above date.

A.   Approval is based on the operational specifications set forth in your letter of application dated June 10, 1993, as amended by your letter dated August 13, 1993, and additional information provided in response to agency requests dated September 12, 1993.

B.   This license is limited to the operations of a land remote-sensing space system and subject to the following terms and conditions that apply to the Licensee and any subsidiary, affiliate, or contractor, as appropriate. The issuance of this license does not relieve the Licensee of the obligation to obtain export or other licenses from appropriate U.S. Government agencies pursuant to applicable statutes.

   1.   Licensee shall comply with the requirements of the Act, and any applicable regulations issued pursuant to the Act. The Licensee shall operate the system in a manner that preserves the national security and observes the international obligations and foreign polices of the United States. The Licensee shall at all times maintain positive control of the spacecraft including safeguards to ensure the integrity of spacecraft operations. The Licensee shall maintain and make available to the U.S. Government, as requested, a record of all satellite tasking operations, for the previous year.

During periods when national security or international obligations and/or foreign policies may be compromised, as defined by the Secretary of Defense or the Secretary of State, respectively, the Secretary of Commerce may, after [2] consultation with the appropriate agency(ies), require the Licensee to limit data collection and/or distribution by the system to the extent necessitated by the given situation. During those periods when, and for those geographic areas that, the secretary of Commerce has required the Licensee to limit distribution, the Licensee shall, on request, make the unenhanced data thus limited from the system available exclusively, by means of government furnished rekeyable encryption on the downlink, to the U.S. Government. The costs and terms associated with meeting this condition will be negotiated directly between the Licensee and DOD (for the U.S. Government) in accordance with Section 507(d) of the Act.

The Licensee shall ensure that all encryption devices used are approved by the U.S. Government for the purpose of denying unauthorized access to others during periods when national security or international obligations and/or foreign policies may be compromised.

The Licensee shall use a data downlink format that allows the U.S. Government access and use of the data during periods when national security or international obligations and/or foreign policies may be compromised.

The Licensee shall provide sufficient documentation to the U.S. Government on the Licensee's downlink data format to assure this access.

2. Licensee will make available to the Government of any country (including the United States) unenhanced data concerning the territory under the jurisdiction of such Government as soon as such data are available and on reasonable cost terms and conditions.

3. Licensee will make available unenhanced data requested by the National Satellite Land Remote Sensing Data Archive (the Archive) in the Department of the Interior on reasonable cost terms as agreed by the Licensee and the Archive. After a reasonable period of time, as agreed with the Licensee, the Archive may make these data available to the public at a price equivalent to the cost of fulfilling user requests.

Before purging any data in its possession, the Licensee shall offer such data to the Archive at the cost of reproduction and transmission. The Archive may make these data available immediately to the public at a price equivalent to the cost of fulfilling user requests.

[3] 4. Upon termination of operations under license, the Licensee will dispose of any satellite in space in a manner satisfactory to the President. To meet this condition and to deal with any circumstances involving [the] satellite's end of life/termination of mission, the Licensee shall obtain priori U.S. Government approval of all plans and procedures to deal with the safe disposition of the satellite (e.g., burn on reentry or controlled deorbit).

5. Licensee shall not change the operational specifications of the satellite system from the application as submitted, which would result in materially different capabilities than those described in the application, without filing an amendment as specified in paragraph D.3 of this license.

6. Licensee shall notify the National Environmental Satellite, Data, and Information Service (NESDIS) of any significant or substantial agreement the Licensee intends to enter with a foreign nation, entity, or consortium involving foreign nations or entities at least 60 days before concluding such agreement. Significant or substantial agreements include, but are not limited to, agreements which would provide for the tasking of the satellite and its sensors,

provide for real-time direct access to unenhanced data, or involve high-volume data purchase agreements. NESDIS, in consultation with the appropriate agencies, shall review the proposed agreement to ensure that it is consistent with the terms and conditions of this license. Specifically, the agreement shall require that the foreign entity will abide by the conditions in this license addressing national security, and international obligations and foreign policies. If NESDIS, in consultation with appropriate agencies, determines that the proposed agreement will compromise national security concerns or international obligations or foreign policy, NESDIS will so advise the Licensee.

C. Enforcement of this license will be carried out in accordance with section 203 of the Act. Any civil penalties authorized by section 203(a) (3) will be assessed in accordance with the procedures set forth in Subparts B and C of 15 C.F.R. Part 904. Such civil penalties may be assessed in amounts up to $10,000 for any violation of the Act or any condition of this license with each day of violation constituting a separate violation.

[4] D. Before the Licensee may take any of the following actions, NESDIS must grant an amendment to the license. NESDIS will consult with the appropriate Federal agencies as required by the Act before taking final action on the amendment. The Licensee must promptly file all relevant information with NESDIS if the Licensee anticipates the occurrence of any of the following conditions:

1. Assignment or transfer of the license;
2. Any change in ownership of the Licensee that would result in foreign individuals, entities, or consortia having an aggregate interest in the Licensee in excess of 25 percent; and
3. Any change in the orbital characteristics, performance specifications, or data collection and exploitation capabilities approved above. In the case of an emergency posing an imminent and substantial threat of harm to human life, property, environment, or the remote-sensing space system itself, Licensee shall not be required to obtain such amendment. If circumstances permit, Licensee shall attempt to obtain oral approval from NESDIS prior to making any such substantial change.

Sincerely,

Robert S. Winokur
Assistant Administrator for
  Satellite and Information Services

Chapter Three

Space as an Investment in Economic Growth

by Henry R. Hertzfeld

Introduction

The research and development investments that NASA has made have greatly affected the economy of the United States and the world. New industries have been created. New technologies have been advanced from the laboratory to the marketplace more quickly than if there had been no space program. Not only have jobs and income been created, but new ways of viewing the world now exist and other innovations that can be traced to NASA requirements and investments have improved the quality of life. Describing these advances is relatively easy. Measuring them is difficult. This chapter describes various economic methods that have been applied to the problem of the measurement of NASA investments, as well as the results of their use. It shows that economists are not in agreement in finding a clear and best approach to measurement. It is also clear that no one measure is a comprehensive indicator of NASA impacts and benefits.

This chapter also tracks two other issues. The first is the political and social need for NASA to measure its impact on the economy. From the beginning of the Apollo program until funding started to decrease in the late 1960s, NASA had no pressing need to justify its program from an economic perspective. Falling NASA budgets and very high national visibility greatly increased the need to explain to Congress and the public the usefulness of the space program. Also, with the growing budget deficit and social programs of the 1970s and 1980s, NASA had to compete for its share of the discretionary budget against many other national priorities. Finally, with the end of the Cold War in the late 1980s, the space race with the Soviet Union was over, and the pressure to view NASA and space investments from the perspective of a rate of return to the nation from its investments became paramount.

The second issue reflects an overall economic-related push within the United States to collect more and better data on research and development (R&D) and to expand the available methodological tools in economics to analyze those data. It is no coincidence that this trend also parallels the overall growth of R&D performed in large laboratories and institutions across the United States following World War II, including the establishment of the National Science Foundation, NASA, and the Department of Energy (which included the Atomic Energy Commission), as well as the very steady and rapid growth of the National Institutes of Health. Prior to World War II, successful government programs in technology development and transfer were limited to the Agricultural Extension Service and NASA's predecessor, the National Advisory Committee for Aeronautics. With the great expansion of R&D programs in government, there was an emphasis on documenting and measuring results to develop public support, convince Congress to continue and expand funding, and better understand the role of R&D and technological innovation in society. By the early 1990s, as described below, the mandate to develop performance measures for R&D had changed from a voluntary and ad hoc effort to one that is now mandated by congressional legislation.

NASA Activities

The cumulative investment that the United States has made in civilian space activities through NASA alone has been more than $200 billion over a period of nearly 40 years.[1] As with any federal expenditure, each year's budget outlays pump money into the economy, directly create jobs, and stimulate additional employment and income throughout the nation. Although the distribution among occupations and regions may differ for NASA-type expenditures, from the expenditures on new roads, housing, or welfare payments, the overall multiplier effects are similar. Also, if any of the programs are eliminated, these employment and income effects are eliminated.

What makes NASA and other R&D investments different is that they also stimulate the productive capacity of the economy through the development and introduction of new technology. New technology can make existing production methods more efficient, and it can also create new products and services that not only stimulate new markets, industries, and opportunities, but can also improve the way people do things and the overall quality of life. These benefits and impacts on the economy are not immediate—it may take years or decades for an idea to be transformed into a marketable product or service. Thus, the benefits from R&D investments in space are more difficult to measure and more unpredictable than the short-term benefits that accrue from the immediate jobs and income created by direct federal spending.

The measurement difficulties center around the unpredictable ways that technology is transferred and the problems inherent in tracing the progression of ideas and experiments into consumer products whose sales contribute to the gross domestic product. This process, even when the route to the marketplace is fast, may take years. For some fundamental research, the process may take decades. The process of technology transfer will include false leads, dead ends, and even products that are technical successes, but fail in the marketplace. However, it also may generate major commercial successes. In many cases, the R&D stage represents only a small fraction of the monetary investment necessary for bringing a product or service to the market. Also necessary for economic success are large-scale manufacturing technology, marketing efforts, advertising, product support, and so forth. Many government-developed technologies that are highly useful for their design objective in fulfilling a government mission need a large amount of additional R&D to make them optimal for commercial markets. Tracing over time the many paths of technology development and transfer and placing a value on the returns to the NASA R&D component of these innovations continue to be challenges to the economic community.

Certain "big" technologies for which NASA has been responsible and those that have been greatly stimulated through the space agency's programs have made their way into product and service markets with virtually no active government technology transfer program. Examples of these technologies are communications satellites, the miniaturization of electronic components, large management computer software applications, and advanced composite materials. Historically, industry has performed about two-thirds of NASA R&D under contract arrangements. Therefore, when a technology is developed that is basically a better or more efficient way of serving a proven and existing market, industry is already actively involved, knowledgeable, and often willing to commit the necessary funds to modify the government product or service for the commercial market. Of course, as time elapses, new consumer demands may emerge and grow from the original space-related technology. The rapid growth of new satellite telecommunications products and services offers an excellent example.

---

1.   If this were translated to constant 1997 dollars, the investment would be more than $450 billion.

However, discerning the economic impact of "little" technologies is another matter. Since NASA has been committed to increasing its impact on the economy through an aggressive technology transfer program, that program has primarily focused on funding demonstration projects that attempt to take ideas generated within NASA (or its contractors) and moving them toward usable end products. Although there have been some successes in this process, the technology transfer program has generated criticism over the years. NASA has been keenly aware of the vulnerability of the program as well as the importance of measuring its success through case studies and more aggregated metrics. In many ways, the agency has gained political support through the anecdotal accounts of useful products developed through the technology transfer process. These examples, particularly in the biomedical area, have provided a grounding for space technology that is understandable to Congress and the general population in human terms.

The focus of this chapter is twofold. First, it traces the history of macroeconomic measures of NASA's impact on the economy from both policy and economic methodology perspectives. Second, it discusses the efforts to develop measures of specific technology transfer activities. Together, these two trends are related to overall political and social forces that have influenced NASA's budget and the public's perception of space activities. This chapter does not review the short-term impact analyses or the impacts of the "big" technologies on the economy because these issues are well documented in other literature.[2]

Measuring NASA's Impact on the Economy

Three distinct approaches have been used to quantify the economic impacts of space R&D:

- An adaptation of a macroeconomic production function model estimates impacts of technological change attributed to R&D spending on the "gross domestic product" and derivative measures such as employment and earnings. The results of using this type of model can be expressed as a rate of return to a given investment or as a total value.
- A microeconomic model evaluates the returns to specific technologies through the use of benefit-cost ratios. Benefits derived from these studies are rarely additive to aggregate benefits across different technologies because of technical incompatibilities in data collection and economic assumptions underlying the models.
- An examination of data provides evidence of the direct transfer of technology from federal space R&D programs to the private sector. The results of these analyses tend to be reported in actual numbers measured (number of patents or inventions, value of royalties, value of sales, and so on). They are rarely compared to associated government expenditures because of the difficulty of linking specific funding to specific products or patents.

Until the very late 1960s, NASA did not have to worry much about defending its budget. During the early years of the Apollo era (which ran from 1961 to 1972) the United States was in the most heated part of the Cold War and had made a commitment to get a human on the Moon before the Soviet Union. The expense was almost secondary, and NASA was provided with a budget large enough to perform the job. At the same time, the United States had an overall budget surplus. This was before the impact of the expenses

---

2. The short-term direct impact of federal expenditures is a topic that is thoroughly discussed in any elementary economics textbook. Reference will be made in this essay to studies that have focused on the special distribution of NASA short-term employment and income effects.

of the Vietnam War were well understood and before the large social welfare programs passed by Congress during the administration of Lyndon Johnson attained rapid growth. This was also before the very high inflationary period of the 1970s and early 1980s, which propelled the federal government into a deficit and made the interest payments on the national debt one of the largest single components of the federal budget, dwarfing even the defense budget.

Early studies commissioned by NASA show that the space agency was not unaware of the implications of its spending on the economy. However, those impacts were secondary to NASA's primary mission of space exploration. A 1965 study by Jack Faucett Associates found that NASA's socioeconomic activities were "separate, uncoordinated, and almost incidental." The Faucett study recommended that NASA create a headquarters staff devoted to collecting economic data and coordinating various economic activities throughout the agency. These recommendations were ignored; to this day, the agency still addresses economic policy on an ad hoc and uncoordinated basis. [III-1]

The goal of stimulating economic growth through NASA technology was very much a side issue to the space agency during the 1960s and 1970s. For example, a 1968 NASA publication listing goals and objectives for the next decades in space does not mention economic growth at all, but only refers in passing to the facilitation of communications and navigation as an afterthought to developing space capabilities for managing the Earth's resources.[3]

During the 1960s, one of the important goals of NASA was to stimulate science and engineering at the universities. NASA sponsored social science and economic studies during this era, which were oriented primarily toward defining and identifying the impact of NASA expenditures on particular regions. One study, performed at the University of Pennsylvania in its Regional Science Department, was an input-output economic analysis of the Philadelphia economy that developed measures of NASA expenditures on the Philadelphia region.[4] There is no evidence that the results of this study were ever used by NASA to publicize the local benefits, nor is there any evidence that the results directly influenced NASA policy. The primary purpose and benefit of this study were to support university research activities and to advance knowledge in regional economic measurement techniques. NASA also sponsored another input-output study by William Miernyk of West Virginia University during this era.[5] Orr and Jones of Indiana University performed an industrial breakdown of NASA expenditures to measure national impacts.[6] Again, these were primarily academic theory-building studies that were not sponsored by policy-making offices at NASA, nor were the results extensively used by the space agency in support of its programs.

In 1974, Mary Holman of George Washington University published a comprehensive review of the ways in which NASA had an impact on the economy, based on her research at the agency during the late 1960s.[7] According to the preface in the book, this research was initiated by concerns that NASA's associate administrator for manned spaceflight, George Mueller, had about possible future "serious problems and distortions in several

3.   Space Task Group, *The Post-Apollo Space Program: Directions for the Future*, September 1969, published Document III-25 in John M. Logsdon, gen. ed., with Linda J. Lear, Jannelle Warren-Findley, Ray A. Williamson, and Dwayne A. Day, *Exploring the Unknown: Selected Documents in the History of the U.S. Civil Space Program, Volume I, Organizing for Exploration* (Washington, DC: NASA Special Publication (SP)-4407, 1995), 1:522–43.

4.   See Walter Isard, *Regional Input-Output Study: Recollections, Reflections, and Diverse Notes on the Philadelphia Experience* (Cambridge, MA: MIT Press, 1971).

5.   William H. Miernyk, *Impact of the Space Program on a Local Economy: An Input-Output Analysis* (Morgantown, WV: West Virginia University Press, 1967).

6.   L.D. Orr and D. Jones, "An Industrial Breakdown of NASA Expenditures," November 1969, Documentary History Collection, Space Policy Institute, George Washington University, Washington, DC.

7.   M.A. Holman, *The Political Economy of the Space Program* (Palo Alto, CA: Pacific Books, 1974).

sectors of the economy as a result of a probable peak in spending for the Manned Space Flight Program at that time." This book also includes a comprehensive bibliography of studies and books addressing the impacts of space and other federal programs during the 1960s and earlier.

The Stanford Research Institute looked at economic impacts of the space program in 1968. Although this report recognized the developing literature in the economics of R&D and new technology development, its emphasis was on the local and regional impacts on areas surrounding NASA field centers. These impacts fell generally into the category of short-term direct spending effects on per-capita income, construction, and so on, not on the lasting impacts of the new technologies created. [III-2]

By 1970, the NASA budget had fallen by nearly one-third from its 1965–1966 peak. A new era had arrived for NASA—one that meant a constant battle with Congress and the White House for money, programs, and new directions. Thus, it is no coincidence that NASA commissioned the first comprehensive economic analysis of its impact on the entire national economy at that time. The study was meant to be used as a lobbying and public relations tool for the agency. NASA officials hoped that the results of the study would show very robust impacts on the economy, proving the benefits from the investment in space.

A Midwest Research Center report in 1971 accomplished its purpose in two ways. [III-3][8] First, using an aggregate production function, it showed large long-run economic returns to R&D. Second, it documented a number of case studies of successful examples of NASA technology being used for commercial purposes.[9] It used a fairly new economic methodology developed by Robert Solow of MIT (who later received a Nobel Prize for his work on the economic impacts of technological change) and showed that there was a seven-to-one ratio of long-term economic benefits to expenditures.[10] This methodology and its application to NASA (a small subset of all R&D expenditures) were sharply criticized on both technical grounds and the interpretation of the data. While the calculation of a "bottom line" number provided NASA with some extra ammunition for its budget battles, in the long term, how successful this line of argument was in Congress is very debatable. NASA has always been funded because of the merit of its missions. However, the economic data added "window dressing" to the project mission requirements and made the funding decisions for the new programs of the 1970s and beyond easier to sell.

Following the 1971 Midwest Research Institute study, NASA commissioned several additional major macroeconomic studies of its R&D. The space agency hired Chase Econometrics in 1976 to conduct a macroeconomic simulation analysis. [III-4] Chase Econometrics performed a follow-on study in 1980. [III-5] Then in 1988, the Midwest Research Institute, under contract to the National Academy of Public Administration, performed an analysis that essentially replicated its 1971 study with updated data and econometric techniques. [III-6]

Although each of these four studies differed in time, technique, and reliability, they had much in common. Each was a rudimentary attempt to measure the overall returns to NASA in terms of national economic measures: gross national product, employment, and productivity. Each (with the exception of the Chase follow-on) calculated rates of return to NASA that were between seven-to-one and fourteen-to-one (which translated into discounted returns on investment between 30 and 43 percent). Also, each attempted to look at the lasting impact on the economy through technological changes that increased productivity.

---

8. Please note that the documents supporting this essay do not appear in chronological order but rather in the order in which they support the complex subject addressed in the essay.

9. Many of the case studies were taken from the aeronautics R&D program.

10. Robert M. Solow, "Technical Change and the Aggregate Production Function," *Review of Economics and Statistics* 38 (August 1957): 312.

These studies had their critics. The most pointed one was the report of the General Accounting Office (GAO), which examined the Chase Econometrics study of 1976. [III-7] The GAO analysis was sympathetic to the idea that NASA had generated technological returns, but it did not accept the Chase study metrics, nor did the GAO reviewers view the methodology of the Chase study as valid. They were probably correct because the Chase follow-on study of 1980 could not replicate the results of the 1976 study. There were a number of technical economic reasons for the lack of confidence in the Chase results. Perhaps the most telling reason was the belief by Chase that NASA needed to have "good" results from the study. Thus, Chase ran about sixty different simulations of the economy under different assumptions and chose the best results to present in the report.[11] The study did result in good public relations and press releases, but it was not well received by the professional economics community. Despite the critics' assessments, the results became benchmark figures that have been quoted time and again in speeches by NASA officials and in various testimony and publications.

During the mid-1970s, NASA also commissioned studies to attempt to develop measures of the aggregate benefits from specific successful technologies that the agency had supported. These studies were funded by the NASA headquarters office responsible for technology utilization and transfer, and they applied a case study approach to measuring benefits. By focusing on successful innovations, the selection of technologies was not random, nor did it adequately measure overall benefits to NASA.[12] These studies had the advantage of using more standard economic tools and of identifying concrete cases rather than abstract measures of gross domestic product. They were, in short, more easily understood than the macroeconomic approaches used by the Midwest Research Institute and Chase Econometrics.

In 1972, the Denver Research Institute studied NASA contributions to specific fields of technology. [III-8] It found that the major significance of NASA contributions was in causing technology advances in those fields to occur at an earlier time than they would have without NASA funding and support, that more than one-half of the technologies were employed in the aerospace and defense sectors, and that the technologies had only a moderate economic impact and relatively low scientific and social impacts. There was wide variation in the quantitative estimates from technology field to field. The study was based on interviews with the NASA engineers and scientists responsible for the innovations. Although the methodology was subjective and the sample biased because the data were collected from NASA officials, the general policy results are consistent with later studies that used more sophisticated methodologies.

In 1976, a study performed by Mathematica analyzed the contribution of NASA to only four technologies and found nearly $7 billion of economic impacts. [III-9] The $7 billion was more than the NASA budget of 1976, and the results of this study were used in connection with the Chase and Midwest Research Institute results to suggest the leverage that NASA funds have on the economy. This study focused on gas turbines, cryogenics, integrated circuits, and NASTRAN (a software program). The primary NASA benefits measured were the "speedup" of bringing these technologies into the marketplace, not

---

11.   This conclusion was related to the author in a conversation with Michael Evans, president of Chase Econometrics, several years after the study was completed. It is backed up by the research results of the 1980 follow-on study, which could not replicate the results for two reasons: (1) the calculation of the actual variables was not well documented in the original study, and the values assigned to some of the variables could not be verified; and (2) the statistical tests showed that the returns to NASA were not significantly different from zero. The second reason could also be explained by the very small percentage coming from the NASA budget of the R&D expenditures in the economy.

12.   In fact, by overlooking unsuccessful technologies, the cost side of these benefit-cost studies may have been significantly understated.

the development of the technologies themselves. Because this was a study of four cases and used the more traditional consumer surplus theory of microeconomics, the results were more readily accepted by the economics community than the results of the macro-economic studies of that era.

During the 1980s, NASA economic impact and benefit studies took on the additional role of creating the climate for large, new NASA programs, such as the space station, not of simply justifying the overall budget. The combination of the maturity of the space activities and the national budget deficit began to change the way the nation viewed the space program. Beyond the space race with the Soviet Union, it became apparent that practical uses of space and the role of the economic growth and benefits that space could provide were logical reasons for the government to invest in space and provide additional infrastructure and incentives for future business and the commercial development of space. Also, by the late 1980s and the 1990s, with the end of the Cold War, economic benefits became one of the more important justifications for a continued U.S. presence in space, at least from the civilian perspective. However, the evidence of economic benefits as measured by the existing studies was still relatively weak.

To build political support for the space station, several studies were performed to analyze its expenditures by industry and state.[13] The NASA Alumni League sponsored a study in 1983 that analyzed expenditures by standard economic industrial categories and by state. This study also attempted to measure the indirect (or multiplier) benefits by industry and state. It did not address the more interesting and more challenging task of measuring technological or productivity benefits. [III-10]

Major contractors also supported the space station with economic benefits analyses. Rockwell commissioned The WEFA Group (a merger between WEFA and Chase Econometrics had occurred) to perform a macroeconomic simulation of space station expenditures. [III-11] This was a more sophisticated attempt at measuring multiplier effects throughout the national economy, but it also did not take the further steps to analyze the productivity or technological changes that might be expected to occur with the space station R&D program.

The most recent study of economic benefits that looked at technology change and national growth stimulated by NASA was performed by the Midwest Research Institute (see Document III-6 above) through the National Academy of Public Administration under contract to NASA. This study repeated the methodology of the 1971 study and was conducted by the same researchers. It used updated econometric methods and more than fifteen additional years of data. The results were remarkably similar to the earlier study. They measured a nine-to-one rate of return to NASA R&D programs. This finding held up under Midwest Research Institute's sensitivity analysis. The institute also looked at case studies on a more qualitative basis. However, NASA never officially released the study, because the methodology used was still subject to many technical economic qualifications.[14]

---

13.   Internally, NASA procurement office reports even generated contracts and expenditures listed by voting districts as a method of influencing congressmen and senators of the importance of NASA on their constituents. However, no specific economic multipliers or analyses were performed internally to augment the raw procurement numbers.

14.   An October 5, 1988, internal NASA memorandum from Jim Bain to NASA Associate Administrator Willis Shapley, which commented on the Midwest Research Institute study, documented the reservations of in-house economists as well as some members of the National Academy of Public Administration's (NAPA) advisory panel concerning the results of the study. The memo stated: "The NAPA Advisory Panel letter to the NASA Administrator endorses the conclusion of positive R&D impacts on the economy, but does not reference endorsement of the study's conclusions on the magnitude of this impact. Discussions with individual members of the Advisory Panel have indicated that this was not an oversight, but the direct result of the unwillingness of some members to publicly endorse those numbers."

A detailed accounting of the difficulties of measuring macroeconomic benefits to NASA R&D is beyond the scope of this essay.[15]

Briefly, however, there were several major problems with these studies. First, the use of the production function approach implied that technological change is a function of R&D expenditures. In the usual statistical method of regression analysis, only an association between these parameters is measured; causation is not proven. Second, the macroeconomic studies (particularly the ones performed in the early to mid-1970s) had only about fifteen years of data with which to work. This skewed the results; the particular time span may have been measuring coincidental returns, because it was a time of expansionary economic growth in the United States that had little to do with the space program. The dramatic peak of NASA spending during the Apollo years was a one-time surge in the long-run trend of NASA's space expenditures. It has never been repeated. Third, although NASA R&D reached as much as one-third of all federal R&D in 1965 (one-fourth of all U.S. R&D), R&D in total comprised only 1.9 percent of the gross national product. The statistical errors ("noise") in macroeconomic measures such as the gross national product was larger than NASA R&D. In other words, the studies associate large impacts on gross national product from a relatively small component of the product. This relationship is extremely difficult to validate statistically, given the general accuracy problems with the national income data themselves. Finally, according to the accounting practices of the government during those years, there is no such thing as a government investment. All government expenditures are treated as outlays in the year spent. There is no imputed rate of return to the government expenditures and no government capital account.[16]

Therefore, the process of calculating and equating a rate of return to NASA (or any other government) expenditures with a rate of return to an equivalent private investment in a venture with high levels of risk is not accurate. Although there may well be robust returns to NASA R&D, using the type of methodology and statistics available when these studies were performed was much more of an exercise in doing research and experimentation on economic methodology than it was in measuring the benefits to NASA's programs.

National Policy, Commercial Space, and Economic Impacts

Economic growth and development and international competitiveness are goals of national policy. This is reflected in legislative and executive branch objectives as outlined in various laws and policy directives. The direct investment in space R&D and technology is one means to stimulate the economy of the United States.

The National Aeronautics and Space Act of 1958[17] calls for NASA to be a leader in technological development. In addition to NASA's charter for space and aeronautics activities, in the mid-1970s Congress added provisions chartering NASA to support R&D in civilian ground propulsion (including advanced automobile) systems.[18] The NASA budget

---

15.   A good description of the technical economic methodology and its use in measuring the benefits of NASA R&D can be found in "Measuring the Economic Returns to Space," in J. Greenberg, and H. Hertzfeld, editors, *Space Economics* (Washington, DC: AIAA Progress Series #44, 1992). Another perspective on government studies of R&D benefits can be found in P. Kochanowsky and H. Hertzfeld, "Often Overlooked Factors in Measuring the Rate of Return to Government R&D Expenditures," *Policy Analysis* 7 (Spring 1981): 16–27. Also, any standard textbook on public finance will include a detailed description of the problems with benefit-cost analyses.

16.   These practices have changed recently. The government now has begun to estimate a capital account, and special attempts are being made to more accurately measure R&D expenditures in the national income accounts.

17.   National Aeronautics and Space Act of 1958, Public Law 85–568, 72 Stat. 426.

18.   Subsection 102(e) was added by the Electric and Hybrid Vehicle Research, Development, and Demonstration Act of 1976, Public Law 94–413, September 17, 1976, section 15 (90 Stat. 1270). Also, Subsection 102(f) was added by the Department of Energy Act of 1978—Civilian Applications, Public Law 95–238, February 25, 1978.

during these years was comparatively low, and this was one method to employ NASA engineers in meeting the national need of developing more fuel-efficient energy systems as a result of the OPEC oil crises. It was also an example of one of the first direct links of NASA programs to civilian economic issues.

Other references to national economic growth policies began to become apparent in official space directives. In the Reagan administration's 1988 presidential directive on space policy, an entire section is devoted to commercial space sector guidelines, which outlined a number of government actions aimed directly at stimulating private-sector activity in space. [III-12] This was followed by the Bush administration's 1991 U.S. Commercial Space Policy Guidelines, which explicitly recognized the fast-growing commercial space sector and expanded government initiatives to encourage the space sector's growth. [III-13]

Finally, the most comprehensive statement of national commercial space policy is found in the Clinton administration's 1996 presidential space directive, which propels space activities directly into the mainstream of national economic policy and international competitiveness issues. [III-14] Two of the five goals listed in the introduction to this policy specifically mention both economic competitiveness and the stimulation of non-federal investment in space. The document also includes a list of guidelines aimed at enhancing commercial space. The significance of this document is the extent of coverage of commercial issues surrounding space activities. It is indicative of the maturity of the satellite communications industry coupled with the many industry proposals to develop commercial space systems. Prior presidential directives on space issues have evolved from barely addressing any commercial issues to those of the late 1980s and early 1990s, which initiated government studies and actions aimed at stimulating an emerging potential for the industrial use of space.

Economic forecasts have long played a part in NASA plans for commercial space. Some of the earliest were the projections of the demand for communications satellites, which was the major driver of the demand for expendable launch vehicles (and Space Shuttle flights before the *Challenger* accident). NASA was the owner and operator of these vehicles before the push to transfer expendable launch vehicles to the private sector, and the agency needed such forecasts for planning and budgeting purposes. In addition, the Space Shuttle was designed to launch commercial and industrial payloads as well as to perform NASA missions in space. Therefore, forecasts of potential users of the Space Shuttle vehicle provided information to NASA management that was important for developing a mission model, manifest, and pricing algorithm.

An example of the use of projections was the 1983 forecast of a $60 billion commercial space market fifteen years hence (2000) performed by the Center for Space Policy. This projection received a large amount of publicity and helped set the stage for the political and industrial support of the International Space Station.[19] However, any projection of economic markets many years in the future is so prone to error that economists and industry planners discount any such projections as premature and wishful thinking. In fact, the Center for Space Policy forecast was revised in 1984 with more industrial and economic detail, and it was presented as a range of possible markets in contrast to their controversial prior forecast. [III-15] Nevertheless, as with the earlier macroeconomic benefit studies, these and other projections of vast new markets had significant political impact and were influential in both stimulating new government initiatives in industrial space activities and in providing an underlying rationale for new big space programs such as the International Space Station.

---

19. *New York Times,* June 24, 1988, Section 3, p. 1.

A 1977 study by the Hudson Institute that did not receive a lot of publicity, but was extremely interesting and well balanced, made 100-year projections of space activities. The analysis provided a number of different scenarios ranging from space being dominated by military activities to space being an instrument of a "green" society. In each of the scenarios, the economic activities, costs, and benefits of space were major factors. The projections were based on a combination of economic growth and technology forecasting, superimposed on different assumptions concerning overall future societal and political perspectives. This is one of the first studies sponsored by NASA that gave equal weight to economic as well as technological trends and changes. The authors recognized that the purpose of the study was to provide NASA with both qualitative projections of the future for planning and public relations purposes and very broad guidelines to answering critics' questions about the value and long-term use of space. [III-16]

In the early 1980s, NASA embarked on a number of efforts to find new and better ways to commercialize space. NASA officials commissioned a task to review the agency's options. The Office of General Counsel prepared an in-depth legal and policy paper detailing the options available to NASA for stimulating commercial investments and opportunities. [III-17] The final series of reports of the task force was instrumental in establishing a headquarters office responsible for all commercial activities, ranging from technology transfer functions to direct support of R&D and joint projects with industry that had prospects of developing commercial space manufacturing activities. This office, as with most prior efforts to stimulate commercial uses of space, had some near-term success, but it failed to generate long-term changes in the operations or goals of the agency. The office (but not the commercialization functions) met with the fate of all prior NASA economics-oriented program offices; it was slowly dismantled in the early 1990s.

There were many reasons various internal NASA attempts at establishing and making economic analysis and economic stimulation programs were unsuccessful. Primarily, NASA is an agency managed and staffed by engineers and scientists. Historically, NASA officials and program managers have been recognized and rewarded by developing successful scientific or engineering programs. Cost management and economic stimulation were seen as important, but were not the yardsticks for promotion. Even though NASA's top management recognized the need and the potential for NASA to be an important element in economic policy for the United States, the message was not adequately transmitted to the program offices. The transition to a more business-like approach to space has been slow, but the trend, as evidenced in both legislative and presidential directives as well as NASA's own planning documents, is unmistakably toward emphasizing economic objectives as well as technological advances.[20]

In addition, during the 1983–1984 time period, the Reagan administration took a proactive stand on encouraging private-sector involvement in space. There had been a number of successful Space Shuttle flights, and experiments in materials processing in microgravity were beginning to show the promise of future business opportunities. In August 1983, business leaders and top-level executive branch officials attended a meeting on space commercialization and had lunch with President Reagan. [III-18] By April 1984, a memorandum from Craig Fuller on commercial space initiatives had been prepared by industry representatives that began a dialogue on the various incentives and changes in government activities and regulations that might be necessary to encourage more industrial participation in space. [III-19] An interagency working group was established under the Cabinet Council on Commerce and Trade to begin work on these commercialization issues.

---

20.   The motto, "faster, cheaper, better," which is the byline of Daniel S. Goldin's term as NASA Administrator, is indicative of this new culture being instilled within NASA.

International economic competition in space activities has developed as foreign nations have invested in space and used space and related technological R&D to stimulate their own industrial activities. Unlike the United States, one of the major reasons for European nations and many other nations of the world to invest in space has been an economic motivation.[21] Because of the different attitudes of the United States and the rest of the world concerning government subsidization of industrial development, a number of questions are frequently raised concerning the fair pricing of space products on the international market. The subject of international economic trade and competition in space goods and services is too large and complex a topic to address in detail in this essay.

NASA has made many attempts to encourage commercial uses of space.[22] Beginning in the late 1970s, and through the period of the development of the Space Shuttle and now the International Space Station, the government has commissioned space commercialization studies, given free access to microgravity facilities in space for R&D projects from universities, nonprofit institutions, and for-profit companies, and made personnel and terrestrial facilities available for testing. NASA officials have always hoped that commercial customers for the Shuttle, Space Station, and other space platforms will both provide the agency with revenue and the resources to encourage still more R&D and will also, in the longer run, provide the nation with large economic and social benefits.

One example of this type of stimulation of manufacturing in space is the production of new chemicals and drugs. Anticipating the operational phase of the Shuttle, NASA commissioned several studies of the feasibility of commercial space manufacturing. [III-20, III-21, III-22] These studies looked in detail at the many possible markets and opportunities that could be present for R&D in microgravity and eventual commercial products. The studies were comprehensive and included both a detailed technology assessment and a sample business plan with expected markets, development times, and potential rates of return. They assumed, of course, that space transportation and facilities would be available on a scheduled basis and would be competitively priced.[23] The study performed by McDonnell-Douglas in 1978, under contract to NASA's Marshall Space Flight Center, is particularly interesting. It outlines in great detail the potential technologies that could produce different types of new drugs in space, and it shows that a portfolio of such drugs could serve many markets and possibly be profitable. [III-23]

---

21. See, for example, the Convention for the Establishment of a European Space Agency (CSE.CD(73)19. rev. 7, Paris, May 30, 1975). Article VII (I) (b) states: "The industrial policy which the Agency is to elaborate and apply by virtue of Article II (d) shall be designed in particular to: . . . b) improve the worldwide competitiveness of European industry by maintaining and developing space technology and by encouraging the rationalisation and development of an industrial structure appropriate to market requirements, making use in the first place of the existing industrial potential of all Member States."

22. As mentioned in the introduction, this essay does not describe the "big" space commercial technologies. The communications satellite industry is the most obvious and most successful of all space ventures to date. Other examples include the global positioning system (a military space system) that has revolutionized land positioning systems, composite materials, and software management systems. These technologies are characterized by being new ways of providing services to existing markets. Only after the existing markets are satisfied and revenue streams are large, do the firms then innovate with new types of services and products that require significant market development. Most of the examples of both commercial space manufacturing and technology transfer discussed in the body of this essay are new goods and services that have no currently defined market. The risks of a new product coupled with the high risks of using space itself have not yet generated great commercial interest in space manufacturing. Therefore, the government has seen a role for itself in encouraging the "infant industry" of space manufacturing.

23. Most of these analyses also assumed that the Space Shuttle would be flying, according to early plans, seven orbiters, twenty to thirty flights per year, and so forth. This type of ready access did not materialize with the Shuttle program, and access to space today is limited to expendable launch vehicles for most commercial payloads.

McDonnell Douglas teamed with the pharmaceutical firm Johnson and Johnson (Ortho Division) in the early 1980s to build a space version of an electrophoresis instrument to conduct the R&D necessary to manufacture new drugs (the Electrophoresis Operations in Space (EOS) Program). NASA provided the Shuttle for testing the machine without charge, and it proved to be successful in separating chemicals and drugs in microgravity that could not easily be separated on Earth. An early, and perhaps overly optimistic, letter from John F. Yardley, president of McDonnell Douglas Astronautics Company, to NASA Administrator James M. Beggs testified to the success of these experiments and predicted commercial production in space by 1987.[24] [III-24] A 1983 McDonnell Douglas briefing details some of the commercial possibilities on which companies hoped to capitalize. [III-25]

Two major factors led to the end of the EOS Program. First, the *Challenger* accident in 1986 halted all flights on the Space Shuttle, and this eventually led to a national policy that eliminated most commercial payloads from the Shuttle. It also dramatically illustrated that space was a very risky place to do business. Access could be delayed or denied suddenly. These risks were greater than most normal terrestrial business risks. Not only were ongoing commercial programs halted, but the accident also put a long-term damper on companies considering new space business ventures (particularly those that depended on bringing back material from space).

The second factor is the unpredictable nature of innovation and new technological developments. The drug industry in the United States was undergoing change itself. Genetic engineering and a burgeoning biochemical industry were busy developing new drugs through new terrestrial methods. Also, some of those breakthroughs were in drugs that would directly compete with the ones with which McDonnell Douglas was working in space. Even though the space experiments in electrophoresis were extremely successful, other, less risky terrestrial methods to meet similar markets ended the private partnership to produce drugs in space. A little recognized spinoff benefit from this experiment was a significant improvement in the electrophoresis process on Earth.

The Centers for Commercial Development of Space programs of the 1990s was another attempt to stimulate private operations in space. [III-26] Through seed money to universities, NASA hoped to generate university-industry partnerships in space-related R&D leading to commercial operations.

Similar efforts by NASA to find commercial uses of space were undertaken for the International Space Station program as well as for remote sensing and other applications of space hardware and technology. To date, there are many proposals and ideas, but space manufacturing is still a tantalizing future business activity for industry. Perhaps after the International Space Station is operating, and transportation to and from space is cheaper and more reliable, a number of the possible business ideas for space manufacturing will materialize.

The Stimulation of Technology Transfer

One of the most historically successful technology transfer programs of the U.S. government has been the stimulation and development of aeronautics.[25] From 1915 to 1958, the agency directly responsible for R&D in aeronautics was the National Advisory

---

24.   This letter was also a direct positive and reinforcing response to the August 3, 1983, luncheon at the White House with industrial leaders and President Reagan. Yardley had attended the luncheon.

25.   The stimulation of aeronautics as an infant industry in the early part of this century involved not only the NACA, but also many other government activities, ranging from weather forecasting (aviation safety) to the postal service (creating a market for air cargo through mail delivery contracts).

Committee for Aeronautics (NACA). Founded in 1915, it was the predecessor agency to NASA. Although space activities involve more than 90 percent of the NASA budget, aeronautics R&D is still a very important activity for the agency. The NACA was organized so that industry and government personnel involved with aircraft worked together, in both directing the R&D program and making government facilities such as wind tunnels available to industry.

NASA inherited this tradition of creating technology transfer opportunities. However, it has discovered that transferring space technology to the civilian sector is different and more difficult than aeronautics technology. Aeronautics began as an infant industry serving the military and civilian government (post office) sectors. A civilian consumer market for passenger and cargo transportation grew rapidly. In contrast, space applications have yet to develop and capture a robust civilian market (outside of communications satellites) and have evolved over time as government programs carried out by aerospace contractors. Unlike aeronautics, finding ready markets in nonaerospace applications for the results of R&D has been difficult.

The 1958 Space Act recognizes, in a general way, the importance of the United States being a leader in technology.[26] This law also gives direction to NASA to disseminate information to the widest possible audience.[27] These sections of the law do not directly charter a technology transfer activity. The first section refers primarily to the Cold War technology race with the Soviet Union. The second was meant for the civilian effort to make widely available the data obtained from space for scientific and research uses. However, taken together, NASA has also used this legislative mandate to create, maintain, and expand its activities in moving technology from the NASA (and industry contractor) laboratories to benefit society.

An internal NASA study in 1969 recognizes that in the post-Apollo era, there was an important role for NASA to play in transferring its know-how and expertise to nonspace activities. The report recommended the establishment of a NASA office to coordinate efforts to transfer technology from NASA to other civil systems. There was little mention of industrial or commercial benefits from these proposed transfer activities. The emphasis was instead on supporting the social issues the nation faced, such as the design of new communities, the development of Earth resources survey systems, the application of technology to highway safety and traffic control, crime control, educational television systems, and so forth. [III-27]

The earliest NASA technology utilization and transfer programs did focus on the use of NASA management skills and technological know-how in other government (both national and state and local) applications. With the exception of the patent waiver program, there was a reluctance to directly involve industry in the programs because of the appearance of subsidies to particular firms.[28] Therefore, the majority of programs and projects of the technology transfer office focused on two areas: the dissemination of information and publications describing NASA advances in technology and the direct involvement in continued R&D on technologies that could be used to help solve specific social problems.

A 1971 evaluation of the NASA technology transfer information dissemination program concluded that the program had not been very successful.[29] Specifically, the study

---

26. National Aeronautics and Space Act of 1958.

27. *Ibid.*

28. Each year, NASA patented between 150 and 200 inventions, originating from both in-house R&D and contract research. Upon application and review by NASA, some patents' waivers were issued to individuals and to firms to permit them to develop inventions into commercial products.

29. S.I. Doctors, *The NASA Technology Transfer Program, an Evaluation of the Dissemination System* (New York: Praeger Publishers, 1971).

found that information dissemination concerning new inventions and innovations was not an effective way to encourage industry to use the information. More direct methods, such as interpersonal contacts, were found to be superior ways to transfer technology. The recommendations of the study centered around having more interactive exchanges between NASA and its regional development centers and giving the centers a longer time frame and more money to develop and focus the target technologies to their civilian uses.

There have been a number of attempts to measure the benefits of the transfer activities. As with the macroeconomic measures discussed earlier, the studies date from the mid-1970s, coinciding with the overall decline of NASA budget allocations. The most visible set of documents produced by the office of technology transfer has been the annual *Spinoff* publication.[30] The reports of many different applications of technology give the impression that the payoff from space investments has been very large. Indeed, it may have been. However, most of the reported technological successes in *Spinoff* are either demonstration projects (that is, not fully commercialized) or are public-sector uses of space technology. Public-sector applications may well have large social value, but that is not quite the same as a measure of productivity increases from privately produced and sold goods and services.[31] The benefits that accrue to commercial firms are also more easily measured, and in theory, some of the costs of the transfer activities of the government may be recouped through royalties and through eventual tax payments from the profits of the firms.

Studies of technology transfer activities in the mid-1970s were conducted for the same reasons as the macroeconomic analyses. NASA had a need to justify its budget by showing that there were more and longer term benefits than successful space missions. Agency officials were trying to answer the political questions raised by statements such as: "It's fine to walk on the Moon, but what have you done that improves the everyday life of the average American citizen?" The technology transfer studies were designed to provide concrete cases that illustrated advances such as heart pacemakers and other medical instrumentation, new materials such as anti-fog glasses, and new construction and building techniques.

In 1977, two studies were performed that attempted measures of these cases of technology transfer that could be traced to the specific efforts of the agency. One was the Denver Research Institute study of the NASA Tech Brief program. This study analyzed the users of the information publications that NASA produced that were made available to industry in the hope that technology developed by NASA would be adapted and extended by private firms. The results showed that there were few new products or commercial sales from these technologies, but that the information was particularly helpful to industry in improving the production process. [III-28]

Although the thinking in government was that this "free" information would be very valuable, the government overlooked three important factors. First, industry had to invest time and people to search for the information, which had a significant cost. Second, the published information was available to anybody, and there were no property rights that could be claimed by industry in the innovation, therefore making it risky to invest additional development funds in the technology. Third, very often government technologies

---

30.   Every year since 1976, this publication has been distributed by NASA. It is a glossy paper report that describes successful technology applications that have been fostered by NASA program offices and/or the office responsible for technology transfer. The discussion is descriptive in nature, and in some ways, the report looks like a corporate annual report, without the financial statements.

31.   One reason that the emphasis has been on public-sector benefits is that, until recently, NASA was reluctant to focus its transfer efforts on private companies because of the possibility of the allegation of an unfair competitive subsidy to one company or industry over another. In more recent years, NASA (and the government in general) has been more aggressive in helping firms and industries directly. One way this has been accomplished is through competitive awards and joint ventures with firms.

are not sufficiently optimized for the market, and a large financial commitment must be made to further develop and market the technology. Combined, it is not surprising that studies showed a relatively low economic return to the information publications.

The second study released in 1977 was done by MathTech (a successor company to Mathematica). This study looked at a number of successful technologies that the NASA technology transfer office had encouraged with additional R&D funds and personnel. [III-29] Technologies such as an improved firefighter's breathing apparatus and zinc paint coatings were analyzed from an economic benefits perspective. Benefit-cost ratios were calculated and in some cases were quite high. The major problem with those results is not that there were no significant benefits (there were), but that the cost figures were artificially low because they were primarily the costs of transferring the technology, not the costs involved with initially developing the technology. The selection of only a few successful cases ignores other transfers that were not successful and, again, tends to overstate the benefit-cost ratios. Nonetheless, an argument can be made that without the efforts of the added value contributed by the technology transfer process, either there would have been no benefits from these NASA innovations or the benefits would have come much later.

The Chapman study of 1989 is interesting because it selected 400 technologies mentioned in *Spinoff*, and researchers performed interviews with the companies to attempt to measure the cumulative benefits. They found that the benefits may have been as large as $21 billion (spread over twenty years). However, the report did not attempt to calculate the costs associated with the development and transfer of those technologies, nor did it determine when or if those technologies might have been forthcoming from the firms without NASA involvement. In addition, it did not attempt to separate the NASA investment or stimulation of the products from prior and future company investments. In essence, this report is a comprehensive update of the earlier studies that documented specific cases of technologies moving from NASA to industry. [III-30]

Another aspect of technology transfer was the realization in the early 1970s that the United States is unique among nations in the world in its openness concerning R&D results from civilian (unclassified) government-sponsored work. During the early years of the Nixon administration, an effort was made to provide U.S. government technology and information to U.S. firms first. Recognizing that under the then-current operating practices and laws of the nation, the United States could not easily restrict information from foreign nations and firms, a program named FEDD (For Early Domestic Distribution) was initiated. NASA participated in this program and made an attempt to get information to U.S. industry before it was openly published. However, no formal evaluation was ever made of the program, and a NASA white paper in 1978 essentially concluded that it was impossible to enforce this policy. [III-31]

The NASA technology transfer office also invested considerable funds in activities aimed at education and at encouraging new users for space data and products. To support these transfer programs, the office also sponsored economic studies that focused on specific program benefits. For example, there were a number of analyses performed in the mid-1970s on the benefits that might be attributed to having improved information from the Landsat series of remote-sensing satellites.[32] Because wheat is traded on the futures market and the prices are highly volatile depending on the size of the worldwide crops,

---

32. See, for example, ECON, Inc., *Economic Benefits of Improved Information on Worldwide Crop Production*, Report 76-243-1, November 15, 1976. A theoretical report on which the ECON studies were based was published by D. Bradford and H. Kelejian, *The Value of Information for Crop Forecasting with Bayesian Speculators: Theory and Empirical Results*, 1974, Documentary History Collection, Space Policy Institute, George Washington University, Washington, DC.

having better information about future yields is having information with a significant economic value. The extent that remote-sensing satellites can help provide that information and the subsequent influence that the information has on the futures market can be measured in terms of reducing the speculative swings on the market.

Recognizing that the NASA efforts to transfer technology have had a very uneven and often uncoordinated history, NASA established an internal task force in 1992 to study the problems and recommend changes. The final task force report had a series of recommendations to improve the system. These recommendations also included the development of measures to evaluate the performance of the components of the transfer process. [III-32] It is interesting to note that these findings and recommendations are not very different in thrust from those of the Doctors study of 1971.[33] The process of technology transfer and information dissemination still took too much time. Paperwork and other formal efforts remained cumbersome and slow. The money allocated to the program was insufficient. There was a lack of good data and feedback on many of the programs between NASA and industry, universities, and nonprofit organizations that use the technology transfer–generated information and facilities, and NASA employees generally did not get recognized for their efforts in fostering technology transfer.

The technology transfer programs at NASA are not unique in their frustrations and problems in implementing more effective programs. Other agencies face similar issues. There has been a general recognition of these problems, and Congress has passed a number of acts to help force the agencies to address the measurement and management of technology. One of the most important of these acts is the Government Performance and Results Act, which requires government agencies to develop new performance measures.[34] It is a clear signal from Congress that measuring economic activities and impacts is an important effort, not only to manage the programs better, but also to help justify continued government financial support of the programs. All R&D agencies are struggling with finding appropriate measures, and NASA is no exception.

Summary

Economics and commercial/industrial activities have never been the top priority of NASA space programs. NASA is an R&D agency, dedicated to advancing science and technology. That is its history and its culture. NASA's aeronautics activities have a very different history and a different relationship to industry than space activities, but they were never able to become models for the space side of NASA. The documents included in this chapter amply reflect the push and pull of economics. The push is to attempt to find uses of the innovative space technology in consumer applications. The pull is that if a robust set of market-driven uses for space can be developed, there will be a continued demand for resources, both public and private, to accelerate future space programs.

Various economic studies and forecasts have whetted the appetites of the public that there may actually be robust commercial uses of space that have been promised and that space spinoffs are very beneficial today. However, without the development of truly ongoing, profitable, and publicly visible commercial space ventures, the studies and projections will fall short of being convincing.

As space technology matures and as innovative products and services are developed using space technology, commercial and market developments will materialize. There are

---

33.   Doctors, *NASA Technology Transfer Program*. However, because the technology transfer program has changed, the specific recommendations are quite different.

34.   Government Performance and Results Act, Public Law 103–62, August 3, 1993.

still many hurdles for the space industry to overcome, but the tide is moving toward the stimulation of new space markets. The burden will have to shift from government-sponsored studies and analyses to business-sponsored proposals. In addition, these will have to reach beyond serving government markets and needs to serving the needs of the consumer.

In summary, the trends in government-led economic studies, technology transfer activities, and other stimuli toward business and market-driven space activities reflected the political needs of the times and the developing maturity of the space industry. Government will provide the infrastructure and will use space for its own civilian and security needs. The point is fast approaching when commercial space activities will either be able to develop and compete on a price basis or they will not materialize. The government can provide incentives, as it has in the past, but performing studies and pumping funds into marginal economic projects will have diminishing results.

Document III-1

Document title: Jack G. Faucett, President, Jack Faucett Associates, Inc., to Willis H. Shapley, Associate Deputy Administrator, NASA, November 22, 1965, with attachment omitted.

Source: NASA Historical Reference Collection, NASA History Office, NASA Headquarters, Washington, D.C.

The report that was attached to this letter to NASA is one of the first comprehensive analyses of the role economics could play in NASA management decisions. It points out that most of NASA's economic analyses, policies, and decisions are done on an ad hoc basis, without top management coordination. By 1965, with the Apollo program in full swing and NASA funding reaching its all-time high, the importance of economics as well as technology began to become evident to NASA management. The report concludes that an office for economics at NASA Headquarters be established to coordinate NASA economic policy. George Wright was an assistant to Willis Shapley.

November 22, 1965

Mr. Willis H. Shapley
Associate Deputy Administrator
National Aeronautics and Space
    Administration
Washington, D.C. 20546   Attention: Mr. George W. Wright

Dear Mr. Shapley:

The attached report, "A Preliminary Survey of Socio-Economic Capabilities for Meeting NASA's Future Policy and Program Analysis Requirements," represents the completion of NASA Contract NASW 1331, authorized by you on September 11, 1965.

As we see it, the study's principal usefulness to you is three-fold: (1) it identifies and describes the principal data collections, reports, and studies relevant to economic analysis in the NASA Headquarters; (2) it pinpoints the chief means by which your central staff capabilities can be improved for purposes of socio-economic analysis; and (3) it recommends a general course of action which, by pulling together the now largely uncoordinated activities, should serve to bring socio-economic analysis more effectively to bear on NASA's policy planning and decision processes.

A word of elaboration on point (3) is in order. It is not surprising to find that NASA's socio-economic activities are at this stage separate, uncoordinated, and almost incidental activities. Until now the highest national priority on NASA has been that of creating a large technical organization and getting on with the task of moving the nation toward the position of pre-eminence in space. Most of NASA's socio-economic studies are now carried out on an *ad hoc* basis by technical offices under the Associate Administrator. In each of these specialized staffs, the perspective and outlook are constrained by mission statement, and the analyses are almost never carried out by trained economists, statisticians, and other social scientists skilled in perceiving social implications. Moreover, the various reporting systems and data elements are not adequately integrated and coordinated from the point of view of economic analysis; they were developed almost without exception for other specialized management purposes and only fortuitously serve socio-economic analysis needs. To continue this approach would deny your office the strong central staff and perspective needed to analyze critical agency positions with respect to overall NASA programs and policies. A new professionalism is needed if NASA is to maintain leadership in the new national environment for goal selection and public policy formulation.

Our key recommendation is that your Chief Economist's responsibility of serving as a focal point for agencywide socio-economic analyses under the Associate Deputy Administrator be clarified, and his authority be spelled out in management instructions and other suitable media. Our detailed recommendations for achieving this are set forth in Chapter IV, especially section A, and the remainder of that chapter deals with other improvements. The NASA Economist should become involved in major NASA decision processes and selected interagency committees, and he should develop an information base to serve top agency needs just as the activities of the Management Information Systems Division now serve middle management and project management needs.

I do not want to close without noting that in our judgment successful implementation of our recommendations will require a number of additional positions with high enough grades to attract outstanding personnel, and also a period of about six months to assimilate the data collections and cited reports, and to study and begin to carry out the recommended improvements.

We wish to express our appreciation for the cooperation we received from NASA personnel and our gratitude for the opportunity to review for you the central socio-economic staff role in NASA.

I, of course, will be happy to discuss any points in this report at your convenience.

Yours sincerely,

[hand-signed: "Jack G. Faucett"]
President

Document III-2

Document title: Roger W. Hough, "Some Major Impacts of the National Space Program," Stanford Research Institute, Contract NASW-1722, June, 1968, pp. 1–2, 19–22, 36.

Source: NASA Historical Reference Collection, NASA History Office, NASA Headquarters, Washington, D.C.

During the 1960s, as economists began to develop methodologies designed to evaluate the impact of research and development (R&D) on the economy, NASA contracted for studies to document the space agency's specific impacts. These excerpts are from one of the early studies that focused on the impact of NASA field centers on their localities. The report did not address national economic impacts.

Some Major Impacts of the National Space Program

[1] <u>SUMMARY AND CONCLUSIONS</u>

Recent studies have begun to show conclusive evidence of a dramatic relationship between advancing technology and economic growth. Although such a relationship seems intuitively obvious, the proof has been, and still is, considerably elusive. No clear, definitive, quantitative theory yet exists in the economic literature, although much attention has been paid to the subject in the past 10 years.

Since NASA is exclusively a research and development agency of the government, it is important that the relationship between R&D and economic growth be understood. This report discusses some aspects of the relation-ship and indicates some of the required ingredients for economic growth. These are, for example, a more productive work force, gained largely through education; a continuous building up of the store of knowledge; greater utilization of knowledge by entrepreneurs; and a high rate of utilization of human capital, first by virtue of low unemployment in all occupational categories and second by a continual development and utilization of higher skills.

To find out how NASA contributes to these various elements, this study examined to some extent the role of the Agency in extending the quality of environment at its centers and production and test areas in the South. It was found that by contributing to improvements in local educational systems, NASA had effectively modified the direction of growth taking place in a number of locations. It is clear that this is more noticeable in small cities, towns, or counties than in large metropolitan areas. On the other hand, in a particular kind of environment, such as Houston, growth in both quantity and quality is also apparent if the immediate vicinity of a center is examined separately. Furthermore, the same elements that have been detected before with regard to scientific complexes in other cities, such as growth of graduate and higher education facilities, are noticeable in connection with NASA centers and other facilities in the South. In each of the areas—with the possible exception of New Orleans, where it may not be possible to detect such changes—NASA has contributed to those elements that constitute the ingredients for economic growth. It has upgraded the skills of the labor force, upgraded the level of education available to local inhabitants, decreased unemployment, and built up the store of knowledge by virtue of its scientific mission.

In summary, we find that:

1. NASA activities have had a positive and consequential influence on the localities in the South in which it has established research and development centers and production, testing, and launch facilities.

[2] 2. These influences have gone beyond those associated merely with the channeling of government funds into an area, primarily because of the research and development nature of the work.

3. R&D is different from other transfers of government funds because it requires more highly paid, highly educated workers who demand more in the way of quality of environment. This in turn affects the quality of education, for example, available to new residents of the community as well as to old ones, resulting in greater levels of achievement by all.

4. NASA and NASA-contractor personnel have contributed to this upgrading of the environment in each community in a variety of ways, from running for and being elected to local political offices to providing pressure through neighborhood and community organizations and volunteer, charitable, and religious groups.

Furthermore, it was found that, in many cases, a substantial portion of the teaching staff in local grade and high schools was made up of wives of engineers and scientists on NASA projects. These women are generally well educated, often from a more cosmopolitan environment than that found in many of the NASA locations in the South, and thus able to bring to school children a broader experience and a greater appreciation for education than they would have otherwise.

5. NASA's influence is also felt because of radical changes in per capita income that it brings about. Recent scholarly studies have indicated that the South must upgrade the productivity of its workers to achieve a position of economic (and social) health and well-being equivalent to that of the rest of the United States. Per capita income is the most reliable measure available to judge such progress, and this indicator has been affected greatly by NASA presence.

6. In certain cases, NASA has been a catalyst in stimulating other developments, particularly in New Orleans. In this case, the local economy was in a slump before the advent of space activities in the area. Uniform agreement was found among local business leaders, Chamber of Commerce officials, and others that the NASA presence was a critical influence in enlightening the community to new and progressive business opportunities.

7. The influence of NASA on education in the South is pertinent, above and beyond that mentioned above. In insisting on good educational facilities for their sons and daughters (and for themselves through college extension and graduate programs), NASA and NASA-contractor employees have laid the groundwork for a higher quality educational environment for all of the people in the communities where they reside. The South particularly needs such influences to enhance its own development. . . .

[19] IDENTIFIABLE NASA CONTRIBUTIONS

The impact on certain of the communities in which NASA operates in the South has been extensive. In three of these areas, the economic impact has been direct and substantial; in the other two areas, local economies have been affected only slightly, but the catalytic effect of the space program has stimulated businessmen and community leaders to think in even broader terms of expansion and utilization of local resources than they had before.

Much of this change might reasonably be attributed to an influx of, and an enthusiasm for, funds from the federal government—whatever their specific source and whatever their end use or purpose. On the other hand, it appeared during this study that there was something distinct about the infusion into a community of federal funds for research and development, as opposed to federal funds for other uses.

To investigate this hypothesis, the principal investigator in this study visited each of the NASA centers or bases of operations in the South. This chapter describes the results of those visits and discussions, together with background material drawn from previous studies along similar lines.

Huntsville, Alabama

In 1964, a Select Committee of the U.S. House of Representatives, 88th Congress, performed a study entitled, "Impact of Federal Research and Development Programs." In this investigation, impacts on communities, higher education, industry, and the economy and the nation generally were examined. The study revealed, in part, that:

1. Federal research and development programs make their impact on a given area in one or more of three ways: The federally owned or operated research and

development installation; the research and development contract . . .; and the expenditure of basic research funds . . . in institutions of higher learning.

2. Regardless of the channel, if significant numbers of Federal research and development activities, projects or dollars are involved, there is often a special impact on the locale, seemingly distinct from the spending of Federal dollars in other types of activities, which impact is especially noticeable in small cities.

3. Aspects of this impact include, on the one hand, the phenomenon of university and especially of graduate program expansion [20] following upon the location of a Federal research and development installation in a community; and, on the other hand, in a particular climate, the occasional phenomenon of certain types of research and development related industries growing up around those universities which are heavily endowed with Federal research and development grants and contracts.

4. One apparently consistent characteristic of Federal research and development activity, and especially any activity which involves significant numbers of scientists and other professional personnel, is that primary and secondary school systems are upgraded, often markedly. The select committee found striking examples of this in Huntsville, Alabama, and Tullahoma, Tennessee, where major Federal research and development installations are located; Stanford Research Institute found similar evidence of favorable changes in other communities.

These results of several years ago were substantiated in the present study. For example, discussions with members of the Executive Staff at Marshall Space Flight Center confirmed each of the findings of the select committee with regard to Huntsville, especially as to improvements in the school system, the active role played by the center in the establishment and continuity of the Research Institute, the establishment of an impressive research and industrial park, and the establishment and expansion of a branch of the University of Alabama at Huntsville.

Since these facts are important to the present discussion, it is appropriate to reiterate them here. First, the select committee found that educational facilities in Huntsville had been improved radically from 1950 to 1964:

> Total enrollment in primary and secondary schools increased from 3,138 in 1950 to 27,537 in 1964, with a consequent burden on local budgets. In 1956, Huntsville voters overwhelmingly approved an increase in ad valorem taxes, boosting the total tax rate (including city, county, and State levies) to $4.10 per $100. Revenues realized from this additional levy were earmarked for school construction programs.
>
> Since 1955 school construction in the Huntsville area has increased at the rate of approximately one new classroom a week. The number of public schools increased from 8 in 1956 to 28 in 1964, representing more than 800 classrooms. Because of the steady growth of facilities the community has not been forced to resort to double sessions at any time.
>
> Whereas in 1950 the figure for median school years completed by Huntsville residents 25 and older was well below the national average (7.5 years as against 9.3 years), by 1960 the Huntsville average was 0.2 years above the then national average of 10.6 years. Results of testing programs for students in grades [21] 1 through 12, showed that Huntsville students compared quite favorably with those in other States using similar tests. A majority of the scores are consistently in the upper 25 percentile. Comparable results show up on tests of the American College Testing program, the National Merit Scholarship program, and the college entrance examination program. Between 75 and 80 percent of all Huntsville secondary school graduates now enter college.

Some 350 spouses of Redstone personnel or of the local defense-related industry serve as teachers in the Huntsville system. Substantial salary increases in recent years, coupled with an actual decrease in student-teacher ratios contribute to the attractiveness of teaching opportunities. Of the more than 800 teachers in the city school system, approximately one-fourth hold master's degrees, a high percentage compared with other Alabama cities.

Significant improvements in curriculum content have been made since the initial influx of NASA and Missile Command personnel. Such courses as advanced biology and calculus have been added to the secondary school program.

The school system of Madison County (in which Huntsville is situated) consists of 30 schools with an enrollment of 12,860 students, about 1,500 of them "federally connected." Three private academies and two parochial schools serve an additional 1,800 students. An extension unit of the State vocational technical school was recently established at Huntsville, offering high school students the opportunity to complete the 11th and 12th grades with training in electronics, auto mechanics, and related technical fields. The effect of this new program has been to up-grade substantially the labor force for the entire area, and many of the graduates of the technical school find ready employment at the Redstone Arsenal.

The Huntsville Center of the University of Alabama and the university's Research Institute developed concurrently with the growth of the Redstone complex, largely as a result of concerted community effort. The university center has attracted an enrollment of more than 4,000 students. Of this number, 1,515 are estimated by Huntsville's superintendent of public schools to be dependents of Federal employees at Redstone and of those of related industries. The Center's drive toward expansion has been greatly accelerated since 1959 by the mushrooming demands placed upon it by scientific personnel in the area. The university allocated $250,000, matched by similar appropriations from both the city and Madison County, for a total of $750,000. In addition, the city and county donated 355 acres of land for the campus, and the county contributed the building of all necessary roads.

[22] The university center presently houses the largest graduate engineering school in the South. A further expansion of the center is currently underway, and another $750,000 is being raised to finance a new undergraduate program, first instituted in September 1964. Projected figures indicate that more than 6,000 students will be enrolled at the university center by 1966 in both graduate and undergraduate programs. This compares with a total enrollment of 1,500 in degree-granting institutions in the Huntsville area prior to 1958.

The Research Institute, founded in 1960, is adjacent to the university center, and with Research Park, constitutes the complex of research facilities bordering the arsenal. Many of the institute's staff participate as professors in the resident master's degree program offered by the university center, while the latter supplies the institute with graduate and undergraduate students who wish to participate in particular research projects. The institute is served by more than 200 full-time academic, research, and technical service personnel.

Second, the committee discussed the establishment of an R&D industry in Huntsville:

The Huntsville Industrial Expansion Committee, organized in the early forties, had long realized the vast opportunities which the Redstone Arsenal complex, in close proximity to the heart of the city on the one side and large, unused tracts of land, on the other, presented. Efforts to attract industry and educational institutions into the Huntsville area were intensified in the late 1950's and were centered around the two major projects. The expansion committee played an important role in raising funds for the establishment of the Research Institute, a branch of the University of Alabama created to provide research in the aerospace physical sciences. Dr. von Braun, shortly after being appointed Director of the Marshall Space Flight Center, helped by making an eloquent appeal before the Alabama State Legislature for funds to establish the institute. It was officially opened 3 months after Marshall Space Flight Center, on October 1, 1960, on an interim basis with personnel loaned by main campus departments. A State bond issue provided $3 million for buildings and equipment, and an additional $400,000 was pledged by the city of Huntsville and Madison County.

A concurrent development was the creation of Industrial Research Park by a nonprofit group known as Research Sites Foundation, Inc., a land holding arm of the industrial expansion committee. This organization leases and sells properties on a 2,000-acre tract adjacent to the arsenal to private firms and research groups at attractive rates; it is pledged to donate profits from these transactions to the Research Institute. . . .

[36] Because of the demand for additional courses of instruction, the University of Houston now teaches 15 graduate courses at MSC [Manned Space Center], including management and political science, as well as engineering, mathematics, and physics. The University is now establishing a new graduate school on a site adjacent to MSC on land donated by the Humble Oil Company. According to representatives of the center, this new facility represents an investment of more than $3 million at the outset, and is expected to serve a community of 80,000 people eventually in the Nassau-Clear Lake area.

Other examples exist such as the establishment of the Lunar Science Institute under joint sponsorship of Rice, NASA, and the National Science Foundation. All of these elements are representative of the pattern found in Huntsville. Although it may not be possible to credit NASA with the entire series of developments (in fact, it would be incorrect to do so), the influence of the establishment of the center at Houston goes much beyond mere numbers of people and a government payroll, just as it does elsewhere.

Finally, impacts on Houston and the Clear Lake Area are summarized in Table 7.

Table 7
Summary of Impacts on Houston-Clear Lake Area

Direct Economic Impacts*

	1960	1966
Population	6,500	33,000
School enrollment	1,900	6,700
School bonds (in millions)	$2.4	$ 7.4
Bank deposits (in millions)	$4.8	$30.9
Residential construction†	—	1,260

Other Impacts

Expansion of higher education facilities in general
Establishment of cross-disciplinary materials science laboratory at Rice University
Establishment of University of Houston Graduate School
Establishment of Lunar Science Institute
Dramatic change in character of Houston, expressed in pride in being a Space Center
Great expansion in number of firms locating in area, primarily to serve NASA, but expanding to other markets as well

---

*   As given in Reference 7.
†   New since 1961.

Document III-3

Document title: "Economic Impact of Stimulated Technological Activity," Final Report, Midwest Research Institute, Contract NASW-2030, October 15, 1971, pp. 1–11.

Source: NASA Historical Reference Collection, NASA History Office, NASA Headquarters, Washington, D.C.

This analysis was the first truly comprehensive national estimate of the returns from NASA R&D expenditures. It used a methodology developed by Robert Solow based on an aggregate national production function to estimate overall returns from R&D and then the subset of NASA R&D impacts. The results of a seven-to-one return on NASA expenditures and a projected 33-percent discounted rate of return from 1958 through 1987 were used for many years in press releases and public statements about the beneficial effects of NASA R&D. The following is Part I of what was a three-part report.

Economic Impact of Stimulated Technological Activity

[1] <u>ECONOMIC CONSEQUENCES OF STIMULATED TECHNOLOGICAL ACTIVITY</u>

"The nation's technological capacity, which is conceptually analogous to the capacity of its physical plant, is unquestionably a nation's most important economic resource. By the same token, the rate at which its technological capacity grows sets what is probably the most important ceiling on its long-term rate of economic growth.

The rate of growth of a nation's technological capacity depends jointly upon the rate at which it produces new technology and the rate at which it disseminates the old."

Jacob Schmookler
Invention and Economic Growth
1966

OVERVIEW

The degree to which a nation can satisfy its collective and individual wants is dependent upon the wealth of the nation and its citizens. The accumulation of economic wherewithal is obtained through combinations of labor, capital, and technology. All three inputs are essential but it is through technological progress that the productivity of labor and capital are increased to obtain more output per unit of input and, consequently, greater per capita wealth. The United States leads the world in the generation and application of technology. Our technological progress poses certain dilemmas, but is also the source of much of the economic power we are bringing to bear on societal deficiencies—deficiencies that many less wealthy nations cannot afford to consider, much less mount assaults upon.

This volume highlights the findings of a research inquiry into the relationships between technological progress and economic development, with emphasis on the several ways in which NASA research and development has aided in the accumulation and commercial application of new or improved scientific and technological knowledge.

[2] Scope of Research
The research undertaken had three separate, but related parts: Part I was an examination of the importance of technological progress in the generation of national eco-

nomic growth. The focus was on *aggregate* economic effects of technological progress—with technological progress being viewed *abstractly* as one of the principal growth-inducing forces operating in the economic milieu. Part I was concerned with *effects:* the economic effect of technological progress, the effect which R&D has on technological progress, and the effect of NASA on the nation's R&D spending. Specifically, this portion of the study was based on an econometric examination of the U.S. economy during the last 20 years to identify and measure the portion of growth which can be attributed to technological progress. Part I also examined the relationship between R&D and technological progress and, finally, made some tentative estimates of the relative effectiveness of NASA R&D expenditures in generating economic growth via technological progress.

Part II was a case study of the process whereby technology is developed and commercially applied. It was designed to undergird—*by example*—the findings of the econometric study. It was also intended to illustrate the extreme complexity of the application process—in particular, that any large technological undertaking produces both direct and indirect commercial applications, that these come in a wide variety of forms and types, that countless individual increments of technological progress are combined in any application, that there are many participants in the process—no one of whom can claim sole credit—and finally to examine the several roles that a mission-oriented research and development agency such as NASA plays in the application process.

The specific case study undertaken was of the R&D programs and application endeavors which have culminated in commercial communication via satellite.

Part III of the report was an *illustration* of ways in which a NASA undertaking has contributed to the nation's scientific and technical knowledge reservoir—the reservoir which is drawn upon and extended by any move toward application. The intent was to demonstrate that a large body of knowledge is accumulated in the process of satisfying mission-oriented program requirements and that this knowledge is retained for use by others for other purposes. The research procedure was again a case study. In this instance the focus was on what we had to learn to keep man alive and productive in space—with emphasis on those things which have relevance in one form or another to earthly problems.

[3] Thus, in three separate but interlocked studies, [Midwest Research Institute] attempted to touch upon major elements in the progression from science through technology to viable application in the economic realm: Part I measures the economic effect of technological progress. Part II illustrates the process whereby technology is developed and commercially applied (covering the invention/innovation portion of the spectrum). Part III shows that an inherent aspect of mission-oriented R&D is the generation of new or improved knowledge—in many fields: basic phenomena, applied science, engineering, design, materials, processing, etc. And, that this knowledge is added to the nation's knowledge bank for withdrawal when demand and the state of industrial practice evolve to the point where the technology will be applied.

[4]   PART I

OVERALL ECONOMIC IMPACT OF TECHNOLOGICAL PROGRESS—ITS MEASUREMENT

A. BACKGROUND

The central questions toward which this phase of the report was addressed are:
1.   What is the role of technological progress in national economic growth?
2.   What factors determine the rate of economic growth due to technological progress?

3.   Can the relationships between technological progress, its determinants, and subsequent economic growth be measured—quantitatively?

4.   And, how do the research and development activities of the space program tie into the preceding questions?

Before World War II, there was little need to ask such questions at the national level. Most development was performed by the individual inventor or by industrial laboratories supported by company funds. Choices as to whether or not to allocate resources to development and how to distribute resources among projects were made within individual companies. Most of the nation's research effort was performed at universities as an adjunct to graduate education. National priorities had little direct influence on the allocation of resources to R&D, and the scale of R&D was small enough that the formulation of precise relationships between R&D and the economy lacked urgency.

R&D grew dramatically following World War II under the stimulus of the Cold War and the race to combine atomic weapons with rocketry. Massive mission-oriented R&D programs were mounted, using as their model the Manhattan Project of World War II. All facets of research—basic and applied—as well as development and sophisticated production plus scientific and engineering education underwent huge federally funded expansions. A strong scientific and technological capability became an essential instrument for national survival—decisions to allocate resources to R&D were made on the basis of necessity.

[5] By the late 1950's, when the nation's first large-scale civilian mission-oriented R&D agency—NASA—was created, the economic effects of such undertakings were receiving explicit, if imprecise, recognition. At about the same time, the short-term and regional economic impacts of expanded R&D began to receive widespread recognition. Community after community strove to become another Route 128, or San Francisco Bay Area, or Huntsville. The immediate benefits of a local R&D complex were clear. Less clear were the processes whereby R&D led to new or improved processes, products, and services. But more important to the purposes of the present portion of this report, the theory, methodologies and empirical data needed to measure quantitatively the cumulative effect over time of the product and process advances were notably deficient.

During the 1960's a number of theorists and researchers undertook to improve our ability to measure the economic impact of technological advances, for it had become clear that technology was a large and powerful force in the accumulation of national wealth. Pioneering work by Solow, Kendrick, and Denison was amplified and extended by a number of others. Much progress has been made, but the fact remains that we got to the moon in a decade, but are, as yet, unable to fully measure the present and future economic impact of the science and technology accumulated on the way to the moon (or the aggregate effect of technological progress in general). Our present capability to measure the relationship between technological progress and R&D is even less precise.

Yet, national decisions with respect to the allocation of resources to and within R&D are being and will be made. These decisions cannot be postponed until precise measurements of their effects are possible. Thus, the intent of this part of the study was to provide—from within the existing state of the art—some measurements of technology's contribution to this nation's wealth during recent years and the role of R&D in generating growth through technological progress.

B.   RESEARCH APPROACH

The investigations were performed at the national economic level. We were exploring the aggregate effects of technological progress rather than those stemming from the individual inventions or innovations. Inadequacies in all existing macro-economic yardsticks forced the study to focus on the "cost savings" effects, i.e., increases in the productivity of labor and capital achieved through technological progress. The many improvements in the quality of goods and services due to research and development are not adequately reflected in existing aggregate economic series and cannot be directly measured.

[6] Given these restrictions on the scope of the study, six research tasks were performed:

First, we adopted a definition of technological progress that is consistent with how progress occurs and how it is generally perceived to occur. The definition presumes that all increases in output not attributable to added quantities of labor and capital are due to technological progress; i.e., all quality improvements in labor and capital are traceable to technological progress.

Second, within the framework of the definition of technological progress and neo-classical economic growth theory, a suitable macro-economic production function was structured.

The adopted production function states that technological progress acts in a multiplicative rather than an additive fashion in augmenting labor and capital in the output-generating process. The general form of the production function employed is:

$$Q_t = A_t f\,(K_t, L_t)$$

where:
Q_t = Output in time period t
K_t = Capital utilized in time period t
L_t = Labor expended in time period t
A_t = Level of technology applied in time period t.

Third, the technology index[1] implicit in the production function was used to assess quantitatively the impact of applied technology on economic growth and output.

Fourth, having determined the level of technology and resulting output, we related technological progress generating activities such as research and development, economies of scale, education, etc., in a mathematical model. Here, the determinants of technological progress were linked to the effect of their stimulus in terms of incremental economic output.

[7] With respect to growth in output in the private, non-farm sector of the economy traceable to R&D—which was denoted G(R&D)—we hypothesized the following relationship:

$$G(R\&D)_t = f(R_t)$$

where:
R_t = The weighted sum of past R&D expenditures for year.

Mathematically, the weights are expressed:

$$R_t = w_0 r_{t-0} + w_1 r_{t-1} + w_2 r_{t-2} + \ldots + w_i r_{t-i} + \ldots w_{18} r_{t-18}$$

where:
w_i = Weight for the ith year lag, and
r_{t-i} = R&D expenditures in the year t-i.

---

1.   The index, A_t, represents the technology being applied in the production process through time. It is arrived at through analysis of actual output and output possible with labor and capital quality—i.e., embodied technology—fixed at a base year.

Thus, R_t is a reflection of the current year's R&D activity plus the effective value of each of the past 18 years of R&D expenditures. Conceptually, R_t can be considered the effective investment in R&D "at work" in year t. The 18-year payout period and the payout pattern within the period were derived from several comprehensive and respected surveys of industry's pay-back expectations for R&D spending and new product lifetimes.

Fifth, through the use of statistical analysis, we empirically determined quantitative relationships existing between growth due to technological progress and determinants of technological progress.

Finally, within the preceding analytical framework, we examined the economic impact associated with the technological stimulus provided by the space program.

C. FINDINGS AND CONCLUSIONS

As have others before us, we found technological progress has been a powerful force in economic growth. Our study considered:
- That technology is one of the factors of production—along with labor and capital—with which the output requirements of the nation are satisfied;
- That what we term technological progress is responsible for improvements in the quality or productivity of labor and capital;
[8] - That technological progress results from the introduction of new or previously unused knowledge into the production process;
- That there are many mechanisms by which knowledge is productively applied, including: Improved worker skills, improved machine design, improved management techniques, and so on.

Measuring the effect of technological progress—so defined—during the 1949 through 1968 time period, we found that:
- *The technology added to the nation's production recipe after 1949 accounted for 40 percent of the real increase in private, non-farm output during the period.*
- *Cumulatively, total output for the period was about $8.2 trillion. If there had been no increase in the level of technology used after 1949, the stock of labor and capital applied would have only yielded a cumulative output of $6.9 trillion. Thus, the leverage on the other two factors of production by technological progress permitted almost 20 percent more output than might otherwise have been achieved with the same quantity of labor and capital.*
- *Throughout the period the technology factor in the production function increased at a compound rate of 1.7 percent per year. By the end of the period—in 1968—the compounding growth of technology had reached a point at which technological improvements beyond 1949 levels were accounting for 37 percent of output (Figure 1).*

Although it is possible to dissent on certain grounds about the exact amount of productivity gains due to technology, the major conclusion is clear. Without the increase of technology and its introduction into the production recipe, this nation would be substantially less wealthy than it is. Much of the economic wherewithal we are now attempting to apply toward the solution of pressing domestic problems is the product of applied technological progress. To expand this economic capacity for problem resolution, this nation must continue to allocate resources to enterprises which generate technological progress and encourage its productive utilization.

This brings us to the second set of findings—those related to the sources or determinants of technological progress. The theoretical and empirical foundation for these assessments is less definitive than for the preceding findings. However, there is general agreement on a list of forces important in the generation of technological progress. The forces are highly interactive but, for analytical reasons, were treated independently. Our findings indicated that most of these forces were of insignificant effect during the relatively short time period under study.

[9]

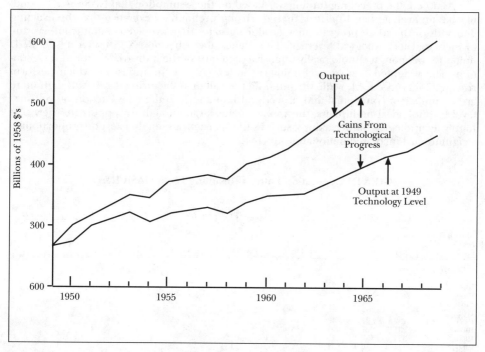

Figure 1. Output and Gains Resulting From Technological Progress (1949–1968)

[10] However, three factors—the sex mix of the workforce, education, and R&D—were found to be important determinants of economic gains through technological progress during the Post-World War II period. The first, sex mix, is the product of increasing participation by females in the workforce and increasing productivity by distaff employees. Improvements in this factor during the period accounted for 4 percent of the total gains due to technology. Improved worker productivity through higher educational levels contributed approximately 36 percent. The balance of the technology-induced gain—60 percent—was attributed to R&D after having ascertained that other possible determinants had no measurable or identifiable impact.

The relationship between R&D- and technology-induced economic gains was explored on a distributed-lag basis. Lag distributions between R&D expenditures and initial pay-back and final pay-out in the form of national economic gains were constructed from industry estimates and experience, but when subjected to statistical tests the relationships exhibited reasonably good explanatory power. The findings were that:

On the average—each dollar spent on R&D returns slightly over seven dollars in technologically induced economic gains over an 18-year period following the expenditure.

This finding leads to the strong conclusion that, on the average (including good, bad, and indifferent projects), R&D expenditures have been an excellent national investment.

The final set of findings relates to the economic impact—via technological progress—of NASA's R&D programs. Assuming that NASA's R&D expenditures had the same pay-off as the average, we found that:

The $25 billion, in 1958 dollars, spent on civilian space R&D during the 1959–1969 period has returned $52 billion through 1970 and will continue to produce pay-off through 1987, at which time the total pay-off will have been $181 billion (Table 1). The discounted rate of return for this invest-ment will have been 33 percent.

As noted, the preceding finding was based on the assumption that NASA R&D spending has an average pay-off effect; there is strong preliminary evidence that the exacting demands of the space program may produce greater than average economic effects due to increased technological leverage. This comes about because NASA allocates its R&D dollar to the more technologically intensive segments of the industrial sector of the economy. The weighted average technological index (At) of the industries which perform research for NASA is 2.1, while the multiplier for all manufacturing is 1.4. Although there are a number of conceptual and procedural limitations to the construction of industry-level technological multipliers, the spread seems large enough to support the view that highly technological undertakings, such as the space program, do exert disproportionate weight toward increased national productivity.

[11]

Table 1
G(R&D) Generation Pattern Resulting From NASA R&D

	1959	1960	1961	1962	1963	1964	1965	1966	1967	1968	1969	Annual G	Annual G
Annual NASA R&D	128	351	602	1,261	2,446	3,315	3,982	4,283	3,414	3,116	2,518		
G (R&D) Generated													
1959	1												
1960	6	2											
1961	21	16	4										
1962	48	57	28	8									
1963	84	132	97	58	16								
1964	118	231	227	204	113	22							
1965	138	324	397	475	395	153	26						
1966	138	378	556	832	922	536	184	28					
1967	120	378	649	1,165	1,614	1,250	644	198	22				
1968	93	331	649	1,359	2,259	2,187	1,501	692	158	21			
1969	64	256	567	1,359	2,636	3,572	2,627	1,615	552	144	17	12,898	35,167
1970	40	177	439	1,187	2,636	3,572	3,678	2,826	1,287	504	116	16,462	51,629
1971	22	109	303	919	2,303	3,572	4,291	3,956	2,253	1,175	407	19,311	70,939
1972	11	61	188	634	1,782	3,122	4,291	4,615	3,153	2,056	949	20,864	91,803
1973	5	31	105	393	1,231	2,415	3,750	4,615	3,679	2,878	1,661	20,764	112,566
1974	2	14	53	219	763	1,668	2,901	4,033	3,679	3,358	2,326	19,016	131,582
1975	1	5	24	110	426	1,034	2,003	3,121	3,215	3,358	2,713	16,010	147,592
1976		2	9	50	214	577	1,242	2,155	2,487	2,934	2,713	12,384	159,975
1977		1	3	20	96	290	693	1,336	1,718	2,270	2,371	8,797	168,773
1978			1	7	38	130	348	745	1,065	1,568	1,835	5,737	174,509
1979				2	13	52	156	374	594	972	1,267	3,430	177,940
1980					4	18	62	168	298	542	785	1,878	179,818
1981					1	5	21	67	134	272	438	939	180,757
1982						1	6	23	53	122	220	426	181,183
1983							1	7	18	49	99	174	181,358
1984								1	5	17	39	63	181,419
1985									1	5	14	19	181,438
1986										1	4	5	181,443
1987											1	1	181,444
TOTAL	914	2,506	4,298	9,002	17,462	23,665	28,427	30,576	24,372	22,245	17,976	181,444	

Document III-4

Document title: Michael K. Evans, "The Economic Impact of NASA R&D Spending," Executive Summary, Chase Econometric Associates, Inc., Bala Cynwyd, Pennsylvania, Contract NASW-2741, April 1976, pp. i–iii, 1–18.

Source: NASA Historical Reference Collection, NASA History Office, NASA Headquarters, Washington, D.C.

This study showed an overall seven-to-one return to NASA expenditures using a macroeconomic pro-
duction function approach. It differs from the 1971 Midwest Research Institute study (Document
III-3) in many technical ways, the most important of which was that it assumed that NASA R&D
was different from overall R&D. This study was the last in a series of major efforts in the mid-1970s
to justify NASA expenditures on the basis of their economic impacts. During the post-Apollo period of
the mid-1970s, NASA R&D budgets were less than one-half of those of the peak years of 1965–66;
NASA economic analyses were designed to buttress arguments for increased space budgets.

The Economic Impact of NASA R&D Spending

[i]   <u>Abstract</u>

In this study Chase Econometrics, Inc., has undertaken an evaluation of the economic impact of NASA R&D programs. The crux of the methodology and hence the results revolve around the interrelationships existing between the demand and supply effects of increased R&D spending, in particular, NASA R&D spending. The demand effects are primarily short-run in nature and have consequences similar to that of other types of government spending. The supply effects, which represent the results of a higher rate of technological growth manifested through a larger total productive capacity, are long-run in nature and have consequences very dissimilar to that of general types of government spending.

The study is divided into two principal parts. In the first part, the INFORUM Inter-Industry Forecasting Model is used to measure the short-run economic impact of alternative levels of NASA expenditures for 1975. The principal results of this part of the study are that a shift toward higher NASA spending within the framework of a constant level of total Federal expenditures would increase output and employment and would probably reduce the inflationary pressures existing in the economy. Hence, Chase concludes that NASA spending is more stabilizing in a recovery period than general government spending.

In the second part of the study, an aggregate production function approach is used to develop the data series necessary to measure the impact of NASA R&D spending, and other determinants of technological progress, on the rate of growth in productivity of the U.S. economy. The principal finding of this part of the study is that the historical rate of return from NASA R&D spending is 43 percent.

[ii] In the final part of the study, the measured relationship between NASA R&D spending and technological progress is simulated in the Chase Macroeconometric Model to measure the immediate, intermediate, and long-run economic impact of increased NASA R&D spending over a sustained period. The principal findings of this part of the study are that a sustained increase in NASA spending of $1 billion (1958 dollars) for the 1975–1984 period would have the following effects:

1)   Constant-dollar GNP would be $23 billion higher by 1984, a 2% increase over the "baseline," or no-additional-expenditure projections.

2)   The rate of increase in the Consumer Price Index would be reduced to the extent that by 1984 it would be a full 2% lower than indicated in the baseline projection.

3)   The unemployment rate would be reduced by 0.4% by 1984, and the size of the labor force would be increased through greater job opportunities so that the total number of jobs would increase by an additional 0.8 million.

4)   By 1984 productivity in the private non-farm sector would be 2.0% higher than indicated in the baseline projection.

Other simulations, of $100 to $500 million increases, show proportional results.

The large beneficial economic effects of NASA R&D programs, particularly the unique combination of increased real GNP and a lower inflation rate, stem from the growth in general productivity resulting from NASA programs. Growth in productivity means that less labor (and/or capital) is needed per unit of output. This results in lower unit labor costs and hence lower prices. A [iii] slower rate of inflation leads in turn to a more rapid rise in real disposable income, which provides consumers with the additional purchasing power to buy the additional goods and services made possible by the expansion of the economy's production possibility frontier. Finally, the increase in real consumer expenditure leads to an increase in demand for the services of labor.

[1]   Introduction

Chase Econometric Associates, Inc. has undertaken an evaluation of the economic impact of NASA R&D spending on the U.S. economy. This study reports on both the short-run and long-run effects of changing levels of spending. Both the Chase Econometrics macro model and input-output model are used to calculate the impact of different spending levels on the overall economy and on specific industries in the short-run part of the study. The long-run part of the study includes an estimate of the relationship between NASA R&D spending and the rate of technological growth. This relationship is used to determine how much higher spending levels would raise aggregate supply and increase the total productive capacity of the economy. The demand effects stemming from an increase in spending are not substantially different from traditional multiplier analysis and are primarily short-run in nature. The supply effects do not begin to have a significant effect on aggregate economic activity until five years later, but the ultimate effects are much larger and very different than the effects of most forms of government spending.

Short-Run Impacts of NASA R&D Spending

Description of Approach

The first part of the study deals with the short-term economic impact of NASA expenditures and attempts to answer the question of whether a higher level of NASA expenditures is more beneficial to the U.S. economy than a lower level during the year that the expenditures are made, holding the level of total [2] Federal spending constant. This analysis is useful in examining the effects of altering the level of NASA expenditures as part of an overall economic stabilization policy.

The economic impact was calculated by preparing two forecasts of the U.S. economy for 1975 using alternative levels of NASA expenditures, which we term NASAHI and NASALO. The NASALO forecast assumed an expenditure by NASA of $1.35 billion in 1971 dollars for goods and services (excluding NASA employee wages) during calendar 1975. The NASAHI forecast assumed an expenditure of $2.35 billion by NASA with other Federal government spending reduced by $1 billion, hence leaving the total level of government spending unchanged. Because of this, the aggregate economic impact shown for this shift is quite small.

In order to measure the differential industry effect of the NASAHI and NASALO expenditure levels, we utilized the INFORUM Inter-Industry Forecasting Model. This model, which was developed by the Interindustry Forecasting Project of the University of Maryland, has been expanded and modified by Chase Econometrics and has been linked to the Chase Econometrics Macroeconomic Forecasting Model to provide consistent economic forecasts for the industries included in the model. Through use of this model, it is possible to forecast the impacts on major economic indicators such as inflation, employment, GNP, and productivity of a shift in the Federal budget to a higher level of NASA spending.

Short-Run Results

The effects of the two alternative forecasts on the aggregate economy, as estimated through use of INFORUM, are shown in Tables 1 and 2. While the results are not dramatic, they do indicate that the direction of change in economic activity from an increase in the level of NASA expenditure is positive [3] [original placement of Tables 1 and 2] [4] and beneficial. The magnitudes are small because the total Federal expenditure has not been altered and these improvements result solely from a shift within total Federal expenditures. Nonetheless, these results do indicate that NASA expenditures are less inflationary than other Federal government expenditures, and that a shift toward higher NASA spending with a constant Federal expenditure is not inflationary in the present economy. Conversely, it would follow that a *shift* away from NASA to other Federal programs could be relatively inflationary in the present economy. Further, the employment effect of NASA expenditures is beneficial, although not large for this small change, and thus both goals of higher employment and lower rates of inflation would be hindered by a lower level of NASA expenditure.

Table 1
Macroeconomic Impact of NASAHI and NASALO Expenditures

	NASALO 1975	NASAHI 1975
Gross National Product	1529.9	1530.1
Gross National Product (1958$)	820.7	820.7
Consumer Price Index (% change)	10.5	10.5
Disposable Personal Income	1084.9	1085.0
Federal Government Deficit	17.0	16.9

All figures are in billions of dollars except where indicated otherwise.
NASAHI = NASA expenditures during 1975 of $2.35 billion in 1971 dollars.
NASALO = NASA expenditures during 1975 of $1.35 billion in 1971 dollars.

Table 2
Employment by Industries Affected by a NASA Spending Shift

Employment by Selected Industries			HI	LO	Diff
			(thousands)		
Industry Number	Industry	SIC Code			
5	Missiles and Ordnance	19	154	142	+12
59	Machine Shop Products	359	191	190	+ 1
67	Communication Equip.	366	404	402	+ 2
71	Aircraft		501	488	+13
	Total				+28

22	Logging and Lumber	241, 242	307	308	- 1
25	Furniture	25	543	544	- 1
27	Paper and Products	26	501	502	- 1
30	Printing & Publishing	27	688	689	- 1
31	Industrial Chemicals		295	296	- 1
72	Shipbuilding	373	169	171	- 2
	Total				-7

Net gain in Manufacturing Employment   +20
(thousands of jobs)

Thus in this section of the study we show that a shift to NASA expenditures from other Federal government spending will stimulate the economy without raising prices. In particular, we found the following effects of a shift of $1 billion in 1971 dollars.

1) A higher level of NASA expenditures would not have had an inflationary impact on the U.S. economy during 1975 and would probably have reduced the inflation pressures in the economy.

2) A shift of $1.0 billion in 1971 dollars, or $1.4 billion in 1975 estimated prices, from other Federal non-defense expenditures to NASA expenditures would have reduced the inflationary pressures in several key basic materials industries.

3) A shift to increase NASA expenditures would have increased employment by 25,000 in the missile and ordnance and aircraft industries. While it would have reduced employment in ten other industries, the net increase in the manufacturing sector would have been 20,000 jobs.

[5] 4) Output would have been stimulated in twenty-one industries. The principal industries which would have been affected had considerable excess capacity in 1975 and were producing at levels well below their peak years and in most cases below the average of the past five years.

The general conclusion reached in this section is that *a shift toward higher NASA spending within the framework of a constant level of total Federal expenditures creates jobs without raising the rate of inflation, and hence is more stabilizing in a recovery period than general government spending.*

The Impact of NASA R&D on the
Rate of Change of Technological Progress

Description of Approach

The second part of this study is an examination of the historical relationship between NASA R&D spending and the rate of technological progress. This examination requires two steps: (1) the construction of a time series to measure the rate of change of technological progress; and (2) an empirical investigation through regression analysis of the determinants of technological progress suggested by economic theory.

(1) Time Series for γ (gamma). The time series representing the rate of change in technological progress (γ) is a somewhat elusive measure, inasmuch as it requires developing a series for potential Gross National Product (GNP) as well as related series for labor and capital inputs. The series that was developed to measure γ is based on the methodology used by the Council of [6] Economic Advisers. In addition, an alternative series for γ was developed, following the methodology of E. F. Denison, to test the sensitivity of the results to a change in the formulation of the γ series.

Our formulation of γ is as follows:

$$\gamma = \frac{\Delta X}{X} - \alpha \frac{\Delta L}{L} - (1 - \alpha) \frac{\Delta K}{K}$$

where X = full capacity or maximum potential output (national income or GNP) in constant prices

L = maximum available labor force

K = capital stock, defined as $K = \sum_{i=0}^{N} \lambda^i I_{t-1}$ where λ is the rate of economic depreciation and I is fixed nonresidential investment.

α = share of potential output

γ = the rate of technological progress (that is, the rate of increase in full capacity real GNP that cannot be accounted for by a change in either the size and composition of the labor force or the size and composition of the capital stock).

(2) Determinants of γ. Economic theory and prior econometric studies suggest the following possible determinants for γ: (a) R&D spending; (b) an industry mix variable; (c) an index of capacity utilization; (d) an index of labor quality reflecting changes in age mix, sex mix, health levels, and educational levels of the labor force; and (e) an index of economies of scale. After considerable experimentation, we found the latter two determinants to be insignificant for the time period examined. The exclusion of economies of scale as an explanatory variable for γ can be justified on theoretical grounds since this variable is generally relevant to only firm [7] or industry or underdeveloped nation studies. The statistical insignificance of the labor quality variable may be partly explained by the fact that some of its characteristics are already reflected by the manner in which we constructed the labor force variable used to generate γ. Undoubtedly, the insignificance of the labor quality variable is also partly due to our inability to reflect significant improvements (variability) in labor education and training over an observation period as short as 15 years.

Hence, based upon both theoretical considerations and empirical investigation, we offer the following conclusions regarding the determinants of γ. First, R&D spending should be included as a determinant and should be subdivided into two explanatory variables, namely, NASA R&D spending and other R&D spending. Secondly, we found that both R&D variables could be closely approximated by a distributed lag structure that follows the general shape of an inverted U-distribution; that is, as a result of an increase in R&D spending in year 0, modest increase in the productivity growth rate begin in year 2, peak in year 5, and terminate in year 8. The actual distributed lag weights, determined by Almon method and used in the study, are given in Table 3. Thirdly, an industry mix variable should also be used in the equation that attempts to explain movements in γ. This specification is necessary to capture the impact on γ of shifts over time in resource allocation from high- to low-technology industries. Finally, the equation explaining γ should also include a capacity utilization variable to account for the fact that shortages and bottlenecks reduce productivity growth as the economy approaches full capacity.

[8]   **Table 3**
                         Distributed Lag Weights for R&D Spending

Time Lag (Yrs.)	Proportional Weight
0	0.0
1	0.0
2	0.061
3	0.164
4	0.220
5	0.232
6	0.200
7	0.123
8 and later	0.0

<u>The Measured Effect of R&D Spending on Productivity Growth</u>

(1) <u>The Regression Equation.</u> The final regression equation which was used to explain γ in this study is as follows:

$$\gamma = -1.81 + 0.426 \sum_{i=0}^{7} A_i (NRD)_{-i} + 0.074 \sum_{i=0}^{7} A_i (ORD)_i \frac{(1-Cp)}{(1-\overline{Cp})}$$
$$\quad\quad\quad (3.9) \quad\quad\quad\quad\quad\quad (2.0)$$

$$+0.031(IM - \overline{IM}) - 0.157(CP - \overline{Cp}) \quad\quad\quad\quad\quad \overline{R}^2 = 0.883$$
$$\quad (4.5) \quad\quad\quad\quad (3.1) \quad\quad\quad\quad\quad\quad\quad\quad\quad DW = 1.95$$
$$\quad\quad\quad\quad\quad\quad\quad\quad\quad\quad\quad\quad\quad\quad\quad\quad\quad\quad\quad\quad \text{Sample Period } 1960\text{–}1974$$

where: NRD = NASA R&D spending as a proportion of GNP
        ORD = other R&D spending as a proportion of GNP
        IM = industry mix variable, fraction
        Cp = index of capacity utilization, percent

[9] The numbers in parentheses below the regression coefficients represent t-statistics. As can be seen from the regression results, all coefficients are statistically significant and the overall fit of the equation to the data, as measured by the $\overline{R}^2$ value of 88.3 percent, is impressively high, especially for a first difference equation.

(2) <u>The NASA Contribution to γ.</u> Using the regression results above, we found that the increased levels of constant-dollar GNP stemming from a $1 billion increase in constant-dollar NASA R&D spending in 1975 are as given in Table 4. For purposes of this calculation we hold the baseline level of GNP constant and ignore all interactive and dynamic demand and supply multipliers. As will be explained later, the actual changes in GNP will be considerably larger once we do include the effect of these multipliers.

Table 4
Increase in GNP per Unit Increase in NASA R&D Spending
"Pure" Productivity Effects Only

Year	Cumulative Change in GNP
1975	0
1976	0
1977	0
1978	0
1979	0.26
1980	0.96
1981	1.90
1982	2.88
1983	3.74
1984 and succeeding years	4.26

[10] The rate of return on NASA spending may be found by substituting the results of Table 4 into the conventional rate of return formula. increase in spending, the appropriate expression would be

$$\frac{0.255}{(1=r)^5} + \frac{0.952}{(1=r)^6} + \frac{1.888}{(1=r)^7} + \frac{2.882}{(1=r)^8} + \frac{3.736}{(1=r)^9} + 4.261 \left[\frac{(1+r)^{10}}{1 - \frac{1}{1+r}} \right] = 1.00$$

where r is the rate of return. Solving this equation yields r = 43% to the nearest percent. If we resolve the equation by substituting $\frac{4.26}{(1=r)^{10}}$ for the last term, thus not assuming an infinite life, we find the rate of return diminishes to 38%.

Thus an increase of $1 billion in NASA R&D spending would increase productivity and total capacity of the U.S. economy by $4.26 billion in 1984 and each succeeding year. It should be stressed that this figure stems from a $1 billion increase in 1975 and then a return to previous spending levels. If spending were to remain $1 billion higher indefinitely, the first-order supply effects, i.e., disregarding interactive and dynamic effects, are shown in Table 5. As indicated above, the actual results are significantly larger because of the demand and multiplier effects calculated by simulating the Chase macroeconomic model.

[11]   **Table 5**
                          Cumulative Effect on GNP of a Sustained
                            Increase in NASA R&D Spending
                            "Pure" Productivity Effects Only

1975													
1976													
1977													
1978													
1979	0.26											=	0.26
1980	0.96	+	0.26									=	1.22
1981	1.90	+	0.96	+	0.26							=	3.12
1982	2.88	+	1.90	+	0.96	+	0.26					=	6.00
1983	3.74	+	2.88	+	1.90	+	0.96	+	0.26			=	9.74
1984	4.26	+	3.74	+	2.88	+	1.90	+	0.96	+	0.26	=	14.00

Macroeconomic Impacts of NASA
R&D-Induced Technological Progress

The third part of the study uses the relationship which has been developed between NASA R&D spending and the rate of technological progress to translate an increase in spending into a higher overall level of productivity for the U.S. economy. This section features a number of simulations with the Chase Econometrics macro model which determine the total effect of higher NASA R&D spending on the economy when interactive and dynamic effects are taken into account. These simulations consider the supply side of the economy as well as the demand side, and stress the fact that real GNP can be expanded by increasing productivity and lowering prices as well as by increasing government spending.

[12] Approach to Determining Macroeconomic Effects

Up to this point we have considered only the static supply or "pure" productivity effects of NASA R&D spending. We now employ the Chase Econometrics macro model to determine the effects of an increase of $1 billion in constant prices (1958 dollars) in NASA R&D spending. We assume that such spending is increased by this amount at the beginning of 1975 and remains in force throughout the next decade. There are two types of effects from this increased spending.

The first type of effect is the ordinary expenditure (demand) impact of increased government spending. The second type of effect—this effect being what really differentiates NASA R&D from other types of government spending—is the longer run impact of NASA R&D-induced changes in the rate of technological progress. These changes lead to an expansion in the productive capacity of the economy and ultimately lead to an increase in society's standard of living.

(1) The Expenditure (Demand) Impact of NASA R&D. In a period of economic slackness, an increase in government spending leads to increased real GNP and lower unemployment. These expenditure effects for NASA R&D are not markedly different than those experienced for most increases in other types of government spending or for the release of funds to the private sector for construction. It should be noted, however, that NASA R&D expenditure increases have a larger impact per dollar than similar spending on welfare or low productivity type job programs.

[13] (2) The Important Productivity Impacts of NASA R&D. The productivity impacts of NASA R&D generate social benefits in a somewhat more complex manner. We have

already shown above (Table 5) the magnitude of increase which will occur in the productive capacity of the economy for an increase in NASA R&D spending. However, there is no automatic increase in demand which will occur just because total supply is now higher, and until this newly created capacity is utilized through higher demand no social benefits are realized.

There is an economic mechanism through which increased supply does create its own demand. Greater R&D spending leads to an increase in productivity, primarily in the manufacturing sector. As a result of this increase, less labor is needed per unit of output. This in turn lowers unit labor costs, which leads to lower prices. Yet this decrease is not immediately transferred into higher output and employment. As prices are lowered (or grow at a less rapid rate), real disposable income of consumers increases at a faster rate. Consumers can then purchase a larger market basket of goods and services, which in turn are now available because the production possibility frontier has moved outward. Yet these decisions are not instantaneous and friction-less, as they would be in an oversimplified static model. We do not see significant effects of increased technology on aggregate demand until 1980.

Results of Macroeconomic Simulations

Once the increase in productive capacity has worked itself into aggregate demand through the mechanisms discussed above, real growth is then fairly steady as can be seen from Table 6. In particular, we find that real GNP rises near $5 billion per year faster than would be the case under [14] [original placement of Table 6] [15] the baseline simulation which does not include increased NASA R&D spending. Thus constant-dollar GNP is $6 billion higher in 1980, $10 billion in 1981, $14 billion in 1982, $18 billion in 1983, and $23 billion higher in 1984. If we were to continue this simulation farther into the future, we would find that the gap between GNP in the two simulations would continue to increase at approximately $5 billion per year—$28 billion in 1985, $33 billion in 1986, and so on.

[Table 6 originally placed here.]

As greater productivity is translated into higher demand, we find that the economy can produce more goods and services with the same amount of labor. This has two beneficial effects. First, unit labor costs decline, hence lowering prices. Second, lower prices enable consumers to purchase more goods and services with their income, hence leading to further increases in output and employment.

We find that the consumer price index grows at a slower rate with higher NASA R&D spending than without, and is a full 2% lower by 1984 than would otherwise be the case. Once again, this change does not occur in the early years of the simulation, but begins to become important in 1980.

One of the major effects of the higher level of real GNP and aggregate demand is the reduction in the unemployment rate of 0.4% by 1984. Since the labor force will be approximately 100 million strong by that date, this indicates, as a first approximation, an increase of 400,000 jobs. However, if we take into account the increase in the size of the labor force, the total will rise to 0.8 million new jobs. The increase in the labor force will occur for three principal reasons. First, the derived demand for labor will be greater because the marginal productivity of labor has increased. Second, the [16] supply of labor will rise because the real wage has increased. Third, and probably most important, the increase in aggregate demand will reduce the amount of hidden unemployment as more entrants join the labor force.

It is also important to note that labor productivity rises substantially as a result of the increased NASA R&D spending. The index of labor productivity for the private nonfarm

Table 6
Change in Selected Variables With an Increase in NASA R&D Spending of $1 Billion

	1975	1976	1977	1978	1979	1980	1981	1982	1983	1984
Gross National Product, Billions of 1958 Dollars										
Base	788.1	834.0	869.6	859.8	868.5	922.4	977.7	1012.2	1059.6	1090.8
NASA	790.2	836.5	871.7	862.1	871.7	928.6	988.0	1035.0	1077.4	1114.1
Change	2.1	2.5	2.1	2.3	3.2	6.2	10.3	13.8	17.8	23.3
% Change	.3	.3	.2	.3	.4	.7	1.1	1.4	1.7	2.1
Consumer Price Index, 1967 = 100.0										
Base	161.1	173.9	188.4	204.9	219.4	232.0	244.2	257.0	270.9	286.5
NASA	161.0	173.8	188.4	204.7	219.0	231.0	242.2	254.0	266.9	280.7
Change	-0.1	-0.1	0.0	-0.2	-0.4	-1.0	-2.0	-3.0	-4.0	-5.8
% Change	0.0	0.0	0.0	-0.1	-0.2	-0.5	-0.8	-1.1	-1.5	-2.0
Rate of Inflation, %										
Base	9.1	7.9	8.3	8.7	7.1	5.8	5.2	5.2	5.4	5.8
NASA	9.1	7.9	8.3	8.6	7.0	5.5	4.9	4.9	5.0	5.3
Change	0.0	0.0	0.0	-0.1	-0.1	-0.3	-0.3	-0.3	-0.4	-0.5
Unemployment Rate, %										
Base	9.0	8.2	7.4	8.6	9.9	9.2	8.0	7.1	6.5	6.0
NASA	8.9	8.0	7.3	8.5	9.8	9.1	7.7	6.8	6.1	5.6
Change	-0.1	-0.2	-0.1	-0.1	-0.1	-0.1	-0.3	-0.3	-0.4	-0.4
Employees on Payrolls, Millions										
Base	76.9	79.9	82.8	83.3	83.2	85.3	88.1	90.5	92.5	94.3
NASA	77.0	80.0	82.9	83.4	83.3	85.5	88.4	90.9	93.1	95.1
Change	0.1	0.1	0.1	0.1	0.1	0.2	0.3	0.4	0.6	0.8
% Change	0.1	0.1	0.1	0.1	0.1	0.2	0.3	0.4	0.6	0.8
Index of Industrial Production, Manufacturing Sector, 1967 = 100.0										
Base	109.1	120.2	129.6	125.3	122.4	132.6	145.3	154.6	162.2	168.6
NASA	109.9	121.2	130.5	126.3	123.5	134.3	148.1	158.1	166.5	174.0
Change	0.8	1.0	0.9	1.0	1.1	1.7	2.8	3.5	4.3	5.4
% Change	0.7	0.8	0.7	0.8	0.9	1.3	1.9	2.3	2.7	3.2
Index of Labor Productivity, 1967 = 100.0										
Base	110.2	112.1	113.3	112.5	115.2	120.1	123.9	126.9	129.9	132.0
NASA	110.3	112.2	113.4	112.7	115.5	120.8	125.1	128.6	132.0	134.7
Change	0.1	0.1	0.1	0.2	0.3	0.7	1.2	1.7	2.1	2.7
% Change	0.1	0.1	0.1	0.2	0.3	0.6	1.0	1.3	1.6	2.0
Change in Labor Productivity, %										
Base	-0.4	1.7	1.1	-0.7	2.4	4.3	3.2	2.4	2.4	1.6
NASA	-0.3	1.7	1.1	-0.6	2.7	4.6	3.6	2.7	2.7	2.0
Change	0.1	0.0	0.0	0.1	0.1	0.3	0.4	0.3	0.3	0.4

Base = baseline projection with current estimates of NASA R&D spending for next decade.
NASA = an increase of $1 billion in 1958 dollars in NASA R&D spending.
Change = NASA - Base
% Change = NASA - Base / Base — Since the unemployment rate is already given in percentage terms, we do not calculate this item for unemployment.

sector grows at a rate of 2.75% during the 1980–1984 period, compared to an average annual rise of 2.40% with no increase in spending. By 1984 the level of labor productivity is 2.0% higher than the baseline projection.

Further details and comparisons are given in Table 6 for a $1 billion increase in NASA R&D spending. We also calculated alternative runs for $0.5 and $0.1 billion and found that the results were approximately linear for other levels of spending change of equal or smaller magnitude. Similarly a decrease in NASA R&D spending of $1 billion would have reverse effects of the same magnitude on economic activity.

Significance and Reliability of Findings

Significance of Findings

One does not need an econometric model to show that an increase in government spending will raise GNP and lower unemployment. We learned many years ago that it is easy to spend our way out of a recession if no other constraints are involved. Yet having just recently come from the realm of double-digit inflation and the first postwar decline in labor productivity, [17] it is clear that alternative policies must be examined not only from the point of view of their effect on demand and employment but on the real growth rate and the rate of inflation as well.

NASA R&D spending increases the rate of technological change and reduces the rate of inflation for two reasons. First, in the short run, it redistributes demand in the direction of the high-technology industries, thus improving aggregate productivity in the economy. As a result, NASA R&D spending tends to be more stabilizing in a recovery period than general government spending.

Second, in the long run, it expands the production possibility frontier of the economy by increasing the rate of technological progress. This improves labor productivity further, which results in lower unit labor costs and hence lower prices. A slower rate of inflation leads in turn to a more rapid rise in real disposable income permitting consumers to purchase the additional goods and services being produced and generating greater employment.

In assessing these results, we once again stress the importance of distinguishing between demand and supply effects. A $1 billion increase in NASA spending will have an immediate effect on real GNP, raising it approximately $2.1 billion the first year and $2.5 billion the second year. These demand multiplier effects are not markedly different than those which would have occurred for a similar increase in other purchases of goods and services by the government sector or for release of funds to the private sector for construction projects. They are, however, substantially higher than the effects which would be obtained from a $1 billion increase in transfer payments or low-productivity jobs programs. In particular we have found that the demand multiplier is smallest and the increase in inflation is largest for a [18] unit change in transfer payments. When we turn to the supply side, however, the multiplier effects of lowering prices and increasing real income are more than twice as large. Other government spending programs which do not expand the production possibility frontier and improve productivity have no additional effect on the economy after the initial increase in demand.

Reliability of Findings

The results found for the equation estimating γ are all in agreement with economic theory, as the signs and magnitude of the coefficients are within the range expected from *a priori* expectations. Similarly, the statistical results indicate a high degree of correlation and no bias in the regression coefficients, or the goodness-of-fit statistics or the standard errors of estimate. In addition, the results are in accord with the findings of other econometric studies. Nevertheless, a number of criticisms have been raised about the final equation for γ, suggesting that the results might be significantly different if relatively minor changes were made to the function. These suggested changes focus on three areas; the choice of γ_C (the CEA series) instead of γ_D (the Denison series), the inclusion of the Cp term by itself and in conjunction with ORD, and the exclusion of the indexes of labor quality, particularly the level of education. To test the validity of these suggestions, we calculated sixty regression equations, including a "least favorable" case which incorporated all of the above changes. The sample period fits are some-what worse, indicating that γ_D

contains a larger random component than γ_C, but the coefficient of the term for NASA R&D spending is similar for these regressions. Even the "least favorable" case does not change the general conclusions of the study concerning either the rate of return or the economic impact of changes in NASA R&D spending.

Document III-5

Document title: Robert D. Shriner, Director of Washington Operations, Chase Econometrics, to Henry Hertzfeld, NASA, April 15, 1980.

Source: NASA Historical Reference Collection, NASA History Office, NASA Headquarters, Washington, D.C.

In 1980, NASA commissioned Chase Econometrics to redo its 1976 study (Document III-4), this time taking into account changes in the overall business cycle and possible statistical errors that might have accounted for the results of the earlier study. In the 1980 analysis, Chase found that the percentage of overall R&D for which NASA accounted was so small that the quantitative results that showed the benefits from NASA R&D were statistically insignificant. The study did not conclude that benefits did not occur, only that this type of economic tool was not capable of precise measurement, given the quality of the data available. This is the letter that accompanied the 1980 report, "The Economic Impact of NASA R&D Spending, an Update," by Chase's David M. Cross under Contract No. NASW-3345, dated March 1980.

April 15, 1980

Dr. Henry Hertzfeld
National Aeronautics and Space Administration
Room 6133, FOB#6
400 Maryland Avenue, S.W.
Washington, D.C. 20546

Dear Henry:

Chase Econometrics is pleased to transmit herewith 10 copies of the final report of our efforts under Contract NASW-3345 to update the 1976 study of the economic impact of NASA R&D spending.

As you are aware, major technical problems were encountered in the course of the study which ultimately led to the decision to abandon an attempt to prepare a 10-year macro simulation because the estimates of several variables developed in preliminary work were judged unstable and unreliable. The problems encountered in trying to replicate the prior study with an expanded time series calls into serious question the soundness of results obtainable from this sort of "macro" level approach to the estimation of returns to NASA R&D expenditures. While it is possible that some of these difficulties could be overcome if more time and effort were devoted to the task, there are conceptual simplifications implicit in the aggregate approach that will not disappear with more work. We have noted these in our report.

Our experience and that of other investigators in this general area suggests that further attention should be focused in the future on the examination of effects at a more micro level. Industry case studies (for which much data already exists) and interindustry studies would be mutually complementary and should provide significant new insights.

David Cross and I will be glad to discuss details of our analysis and conclusions with you at any time.

Cordially,

[hand-signed "Bob"]
Robert D. Shriner, Ph.D.
Director of Washington Operations

Document III-6

Document title: "Economic Impact and Technological Progress of NASA Research and Development Expenditures," Executive Summary, Midwest Research Institute, Kansas City, Missouri, for the National Academy of Public Administration, September 20, 1988, pp. 1–4.

Source: NASA Historical Reference Collection, NASA History Office, NASA Headquarters, Washington, D.C.

This Midwest Research Institute study of NASA economic impacts replicated the methodology of the 1971 analysis and calculated a nine-to-one return. Modifications of the economic methodology that had been developed subsequent to 1971 were employed, and the results were subjected to sensitivity analysis. This study was mainly a postscript to earlier studies—it did not add anything significantly new to the previous results, and it met with similar technical criticisms as had the prior macroeconomic analyses performed by Chase in 1976 (Document III-4), and the Midwest Research Institute in 1971 (Document III-3). What follows is the Executive Summary from Volume I, Executive Report.

Economic Impact and Technological Progress of NASA Research and Development Expenditures

[1]   EXECUTIVE SUMMARY

"Thirty years ago, there was no satellite communications industry. Today that industry generates gross annual revenues from sales of services and equipment exceeding $6 billion, provides an indispensable service to people, businesses, and governments throughout the world—and is responsible for returning more each year in tax revenues than the entire 30-year NASA investment cost to U.S. taxpayers.

"Perhaps even more significant, although not as obvious, is NASA's role in driving technologies which benefit the U.S. economy and the nation's security across the board. Requirements posed by NASA programs like Apollo, planetary exploration, and the Shuttle have produced miniaturized electronics, power systems and components, automatic checkout equipment, computers and software, high-volume data processing and communication, guidance and control systems, high-strength materials—the list is virtually endless. These technologies have transformed American business, spawned hundreds of new products and services, and made innumerable contributions to national defense."[1]

---

1.   "The Civil Space Program: An Investment in America," Report, American Institute of Aeronautics and Astronautics Workshop, Airlie House, Virginia, November 17–18, 1987.

In 1971 the National Aeronautics and Space Administration commissioned Midwest Research Institute (MRI) to conduct a macroeconomic analysis to measure the extent of the benefits of NASA R&D expenditures on growth in the U.S. economy. This research was augmented with two case studies, synchronous communication satellites and space crew support systems. The case studies detailed the substantive contributions and benefits of these NASA-related technologies to NASA programs, the private sector, and the American public in general. The study was well received and was used extensively by NASA in the 1970s to depict its role in U.S. economic growth and technology transfer.

In the fall of 1987, NASA commissioned the National Academy of Public Administration (NAPA), with MRI as a subcontractor, to conduct further research and to evaluate NASA contributions in the 1948–1986 time frame. The objectives of this latest study are to:

- Measure the impact of technological change on the economic growth of the nation and characterize NASA's contribution to the growth process.
- Identify linkages between the technology generated by selected NASA missions and the broader economic benefits.
- Identify and characterize future benefits of selected NASA programs.
[2] • Identify and characterize the economic impact of continued investment in NASA R&D programs.

This Executive Report presents MRI's findings and conclusions.

A primary objective of the earlier study by MRI was to measure the impact of R&D expenditures on the national economy. While R&D expenditures do have a nearly immediate economic impact through employment and payroll, the primary economic effects of R&D are felt over time. The 1971 study findings indicated that the average dollar spent on R&D returns about $7 in technology-induced economic gain over an 18-year period following the expenditure.

The approach MRI took in the 1971 study was based on methodologies developed by Dr. Robert Solow.[2] Dr. Solow was honored in 1987 with a Nobel Prize for Economics for his pioneering work in measuring total factor productivity. The approach MRI used, though relatively new at the time, has stood up under critical review in the 17-year period following the release of the report.

MRI's current study has been designed to be similar in approach and content to the 1971 study. The 1988 study primarily uses Dr. Robert Solow's approach in measuring the impact of technology on economic growth, but it incorporates refinements developed by other economists in recent years.[3] The 1971 study estimated economic impact during the period of 1948 to 1968; the 1988 study covers not only the original period but extends the estimates through 1986. The findings of the 1988 study, which incorporate essentially the same qualifying assumptions as in 1971, are:

- R&D expenditures have been an excellent national investment.
- On the average, each dollar spent on R&D returns about $9 in technology-induced economic gain over an 18-year period following the expenditure.

2.   Robert M. Solow, "Technical Change and the Aggregate Production Function," *The Review of Economics and Statistics,* August 1957.

3.   Edward F. Denison, *Why Growth Rates Differ: Postwar Experience in Nine Western Countries* (Washington, D.C.: Brookings Institution, 1967); *Accounting for U.S. Economic Growth, 1929–1969* (Washington, D.C.: Brookings Institution, 1974); *Accounting for Slower Economic Growth: The United States in the 1970s* (Washington, D.C.: Brookings Institution, 1979); and *Trends in American Economic Growth, 1929–1982* (Washington, D.C.: Brookings Institution, 1985). Also, reviews and comments by Drs. Z. Griliches, J. Kendrick, E. Denison, and N. Terleckyj.

- The discounted rate of return ranges from 19 to 35 percent annually (depending on the assumptions made regarding the time-lag relationships between an R&D expenditure and its contribution to productivity growth).
[3] • The $148 billion, in 1982 dollars, spent on NASA R&D during the 1960 to 1986 period has returned to the U.S. economy at least $950 billion through 1986 and will continue to produce payoff through 2004, at which time the total payoff will be an estimated $1,338 billion.

MRI used the payback coefficient of $9 to $1 to measure the economic impact of NASA R&D. The $9 to $1 payback is slightly higher than the estimate of $7 found in the earlier study. Differences in these results are attributable, for the most part, to methodological refinements developed in the 1971 to 1988 time period.

Any economic estimation requires an approach based on certain underlying assumptions. Critics of the production function approach as it is applied to NASA R&D in this study may have concerns that (1) the gains ascribed to the stock of technical knowledge measured by R&D expenditures are overstated and (2) there is no empirical evidence for the assumption that NASA R&D is representative of the average of all R&D.

To address the first issue, MRI conducted a sensitivity analysis which showed that a 10 to 30 percent overestimation of the economic gains attributable to R&D would reduce the R&D payback from $9 to a range of $8.50 to $6.50. If, in fact, the MRI payback figure is overestimated even by the worst case of 30 percent and the coefficient is more in the range of 6.5 to 1, the $148 billion spent on NASA R&D during the 1960 to 1986 period would return an estimated $966 billion rather than the $1,338 billion estimated return from a 9-to-1 payback. In either case, the payback from NASA R&D would be substantial.

The second principal assumption in the MRI study is that NASA R&D expenditures have the same economic payoff as the average of all R&D. In other words, NASA R&D is assumed to contribute as much to productivity growth as the average of all R&D. To illustrate and support this assumption, MRI selected two case studies to trace how technology developed by specific NASA missions has been applied commercially:

- Digital communications—including the use of error-correcting codes and data compression in processing digital signals for modern-day digital communication and data storage.
- Civil aeronautics performance and efficiency—centering on a series of advances in aerodynamic drag reduction, advances in propulsion, and advances in night control technology.

MRI chose these two from a list of over 250 major NASA technologies. Our results from these two case studies, as well as the knowledge we have gained in reviewing the 250 principal NASA technologies, indicate a very high payback from the NASA R&D investment. In narrowing the possibilities to two, MRI visited all of the major NASA R&D centers and reviewed NASA's research and technology operating plans (RTOPS), NASA's Tech Briefs and Spinoffs, and key NASA patents. During the course of the project, MRI staff interviewed over 200 NASA personnel familiar with past and current technological achievements. [4] The MRI study team chose the case studies to be representative of the breadth and diversity of NASA programs. The team sought to include both sides of NASA—aeronautics and space—and to select two vastly different technologies—incremental vs. leapfrogging advances—yet with common characteristics from the point of view of their beneficial offspring.

It is clear that without NASA's mission objective of communicating in deep space, digital communications would not be as advanced as it is today. "NASA pushed this technology further than any other entity."[4] The work of the MRI research team documents that NASA's support and extensive R&D funding made possible many comprehensive and ground-breaking advancements in coding theory. For many years coding was considered to be an esoteric and impractical approach to communications, yet it provided NASA an excellent alternative to adding weight, power, and complexity to spacecraft. The case study of digital communication/error-correcting codes illustrates how a technology advanced by NASA to meet the mission requirements for deep space communications has spawned a family of high performance and productivity-enhancing electronic devices with annual sales expected to reach over $17 billion by 1990.

Likewise, NASA's role in civil aeronautics is a good example of why the United States has a decided edge in the world's commercial aircraft market. Improvements in civil aeronautics performance and efficiency have spanned some 70 years since the early days of the National Advisory Committee [for] Aeronautics (NACA). This report summarizes a series of advances aimed at enhancing the performance and efficiency of civil aircraft. The cases cited are intended to illustrate the complex paths by which new knowledge applicable to the design, construction, and operation of modern aircraft comes into being; the interactions between the aerospace industry and government centers of research and technology; the numerous evolutionary changes and improvements that are contributed from many sources; and the often prolonged period of time required to validate, demonstrate, and refine technological advances before they become accepted commercially and widely used.

As a result of NASA's continuing R&D in aeronautics, man can fly farther, faster, higher, and more efficiently and safely than thought possible 20 years ago.

This Executive Report summarizes the economic impact of NASA's research. Part I explains the methodology, findings, and projections of economic benefits resulting from NASA R&D. Part II presents the technology advances and resulting benefits from digital communications, civil aeronautics performance and efficiency, and seven future technology areas. . . .

Document III-7

Document title: "NASA Report May Overstate the Economic Benefits of Research and Development Spending," Report of the Comptroller General of the United States, PAD-78-18, October 18, 1977, pp. i–iii.

Source: General Accounting Office Library, General Accounting Office, Washington, D.C.

Democratic Senator William Proxmire of Wisconsin commissioned the Government Accounting Office (GAO) to review the 1976 Chase study (Document III-4). Because of the study's somewhat unusual approach to macroeconomic forecasting, the GAO was critical of its findings and concluded that case studies of innovation and benefits were more telling of NASA impacts than overall budget/economy calculations. During this era, there was a general belief that R&D was an economic investment with a good payoff to society, but economic models, and particularly macroeconomic models, were not considered definitive evidence of long-term projected benefits. The following are the opening pages of the GAO report.

---

4.   Interview with Irving Reed at his office at the University of Southern California, April 28, 1988.

[i]

NASA Report May Overstate the Economic Benefits of Research and Development Spending . . .

The National Aeronautics and Space Administration (NASA) contracted with Chase Econometrics Associates, Inc., to evaluate how Government research and development spending, particularly NASA's, affects the U.S. economy.

Chase's report "The Economic Impact of NASA R&D Spending" concluded that this spending produced many benefits between 1960 and 1974. The study did not try to evaluate how effectively NASA carried out its primary objectives, such as space exploration and satellite communication. NASA cited the Chase study in its 1976 appropriations hearings as evidence of certain beneficial effects of research and development.

The Chase study does not prove convincingly that the benefits are as large as stated. The study is useful as exploratory research, but other types of studies are necessary to provide a complete evaluation of NASA research and development.

The most significant conclusion of the Chase study is that ". . . a $1 billion sustained increase in NASA R&D spending will raise real GNP $23 billion by 1984. . . ." Of this estimated increase, $21 billion would result from improved technology and productivity, and the rest would result from increased Government spending, which stimulates spending in different parts of the economy.

Since similar increases would result from Government spending on other projects, such as welfare programs, the "multiplier effects" alone do not justify more NASA research and development spending. For this reason, GAO focused on projected technological improvement.

[ii] The Chase conclusions are questionable from three points of view:

— The results, even if accepted as accurate, do not provide the type of information needed to determine whether NASA's spending should increase or decrease. The estimates are of average, rather than incremental, effects. Therefore they do not show whether more spending for research and development would result in as many benefits as before. Within NASA's budget, the most productive or unproductive projects or types of spending have not been spelled out.

— Even if the Chase approach is accepted, the techniques used had certain shortcomings. Plausible and seemingly minor changes in the study's assumptions lead to major changes in the results. Under some of these alternatives, NASA research and development seemed to have no great effect on productivity. Such sensitivity to small changes in methodology indicates considerable uncertainty in Chase's results.

— Basically, the results depend upon statistical correlation between NASA research and development spending and changes in a measure of gross productivity in the U.S. economy. No information on specific NASA projects or on the adoption of new techniques by private business is used in the study. Because of problems in measuring total productivity in the economy and because other possible causes of technological progress were ignored, the correlations may not indicate a true cause-and-effect relationship.

Although the methodology of a particular study may be questioned, the importance of evaluating NASA's programs is undiminished. NASA has clearly been an important source of technical progress in recent years.

[iii] <u>RECOMMENDATION TO THE ADMINISTRATOR OF NASA</u>

Future evaluation studies should look at specific innovations, their effect on specific industries, and the process by which NASA expenditures for research and development improve productivity in the economy. In other words, some of the more important innovations, rather than the total budget, could be examined, and individual technological improvements in individual industries, rather than gross national product figures, could be studied. GAO believes that such studies would give the Congress a more accurate picture of what taxpayers are getting for their money.

<u>MATTER FOR CONSIDERATION BY THE CONGRESS</u>

Technical studies are frequently presented to the Congress in support of agency budgets and as evidence for or against proposed legislation. When important questions are at stake, such studies should be subjected to independent examination and appraisal.

<u>AGENCY COMMENTS</u>

In general, NASA as well as Chase Econometrics, Inc., agreed with our assessment of the Chase report. . . .

Document III-8

Document title: Martin D. Robbins, John A. Kelley, and Linda Elliott, "Mission-Oriented R&D and the Advancement of Technology: The Impact of NASA Contributions," Final Report, Industrial Economics Division, Denver Research Institute, University of Denver, Contract NSR 06-004-063, May 1972, pp. iii–iv, 25–39, 59.

Source: NASA Historical Reference Collection, NASA History Office, NASA Headquarters, Washington, D.C.

During the late 1960s and early 1970s, the Denver Research Institute supported NASA's technology transfer program. This study reported the results of interviews with NASA personnel on the type and field of technology on which their work had an impact. The conclusions that impacts were indirect and were often due to the speed-up of bringing a new technology to the market presented a different perspective to the more aggregated national benefits measured by the 1971 Midwest Research Institute study. The following are excerpts of the report.

Mission-Oriented R&D and the Advancement of Technology: The Impact of NASA Contributions

[iii] MAJOR FINDINGS

Primary objective of this study was to identify and characterize the nature of NASA contributions to the advancement of major developments in several selected fields of technology. Major developments were identified through interviews with recognized leaders in each of the selected fields while NASA contributions were identified through interviews with NASA scientists, engineers and administrators and an extensive search of the NASA and non-NASA literature.

The major findings of the study follow:

1.   NASA contributions to the advancement of major developments in the selected fields of technology appear to be broader, more complex and more indirect than has been realized to date. The number of NASA contributions that find direct nonaerospace applications represent only a small fraction of the large number of contributions that advance the state of technology in a field.

2.   Ten dominant types of NASA contributions to the advancement of major developments were identified. Individual contributions identified in this study embodied from one to all ten types. The types include: developing new knowledge; developing new technology; demonstrating the application of new technology for the first time; augmenting existing technology; applying existing technology in a new context; stimulating industry to acquire or develop new technology; identifying problem areas requiring further research; and creating new markets. Certain types of NASA contributions appeared to be more dominant in some fields of technology than in others.

3.   The "significance" of most of the NASA contributions was to have caused the technological advancement to occur at an earlier time than it would have occurred otherwise. Significance of NASA contributions varied between fields of technology, with those fields most closely identified with NASA missions, such as cryogenics and telemetry, having the largest proportion of contributions leading to advancements that probably would not have occurred without the NASA contribution.

4.   The NASA contributions represented all levels of technology, including major step-changes in technology, incremental advances in technology, and consolidations of technology. Contributions that represented incremental or systematic advances in technology were the most frequent type of contribution, followed by contributions that represented a consolidation of knowledge. Contributions that represented major step-changes were relatively infrequent. Wide differences in these different classes of contribution to different fields of technology were found, depending upon the existing state of technology in the field and NASA's mission requirements.

[iv] 5.   NASA contributions were found in all stages of developmental activity studied, with more than one-half finding military or aerospace applications and almost one-quarter finding commercial applications.

6.   When impact was assessed on a linear scale of high, moderate or low, the technological impact of the NASA contributions was estimated to be moderate-to-high, the economic impact to be moderate, the scientific impact to be moderate-to-low, and the direct social impact to be low, with impact varying widely between different fields. . . .

[25] III. MEASURING THE IMPACT OF NASA CONTRIBUTIONS

Measuring innovative activity is itself a formidable problem which has not been solved to anyone's satisfaction. Moving from a measure of this activity to a measure of economic impact greatly increases the problem. Focusing the task on measurement of the economic impact of NASA's technological efforts reduces few problems and adds many. One recent attempt to measure the overall economic impact of technological progress appears to have met with some success.[1]

---

1.   *Economic Impact of Stimulated Technological Activity. Part I: Overall Economic Impact of Technological Progress—Its Measurement.* Kansas City, Missouri: Midwest Research Institute, Final Report 7, April 1970–15 April 1971 (NASA Contract NASW-2030).

A. Quantitative Assessment

The measurement process associated with quantifying the economic benefits of the NASA contributions in this study would require the three following steps: (1) identification of technological advances and NASA contributions; (2) estimation of economic benefits attributable to the advances; and (3) isolation of the links between the NASA contributions and the advances from among diverse other contributions. The first step has already been accomplished for a sizeable [sic] sample of technologies. Measuring the benefits of new technological advances—the second step—requires benefit/cost analysis. It is on the third step that the most difficult challenges arise. The knowledge and techniques needed to assess the relative economic contribution of one agency in a multi-agency development and diffusion process appears to be beyond the present state of the art.

1. Measuring the Benefits

During the past two decades considerable progress has been made on the theory and practice of benefit/cost analysis.[2] One pioneering example of the research and development area was Griliches' study of the economic benefits from hybrid corn research, in which he found the return on research investment to be approximately 700 percent.[3] Most of the NASA contributions that were identified in this study, however, present more difficult estimation problems. Perhaps most important, corn is a simple, nearly homogeneous product, while many of the NASA contributions have multiple nonaerospace applications. [26] In this respect a better prototype is A. W. Brown's study of economic benefits attributable to the development of atomic absorption spectroscopy by the Australian Commonwealth Scientific and Industrial Research Organization (CSIRO).[4] Brown identified more than a dozen applications for a new and more efficient spectrometer and, to estimate the benefits from these diverse applications, he interviewed some 37 different user organizations. An equally extensive interview schedule would undoubtedly be necessary to assess with tolerable precision the benefits from a typical NASA contribution. Even then, many NASA cases would pose special difficulties because they involve know-how and techniques not directly embodied in hardware, whereas Brown was able to devote his analysis to a single well-defined instrument.

2. Assessing NASA's Share of the Benefits

More fundamental conceptual problems must be solved in estimating how much of the benefits of an advance one can attribute to a NASA contribution. The crux of the matter is that few technological advances of any importance originate through the efforts of only a single person, group, or organization. Rather, numerous groups are likely to be at work on various aspects of the technology in a complex interacting way. This process can be illustrated with a well-known example in the field of nuclear physics. The basic experiment through which Otto Hahn and Fritz Strassmann discovered nuclear fission in late 1938 had been performed previously by Enrico Fermi in Rome during 1934, by Irene Joliot-Curie and associates in Paris during 1937 and 1938, and was being prepared by Philip Abelson at Berkeley when the nuclear age dawned. Knowledge flowed freely among the several research teams, influencing hypotheses and experimental design. In this multiple-paths environment, it seems fair to say that if Hahn and Strassmann had not achieved

    2.   For a general survey, see Roland N. McKean, *Efficiency in Government Through Systems Analysis* (New York: Wiley, 1958), Chaps. 8–12.
    3.   Zvi Griliches, "Research Costs and Social Return: Hybrid Corn and Related Innovations," *Journal of Political Economy*, October 1958, pp. 419–431.
    4.   A. W. Brown, "The Economic Benefits to Australia from Atomic Absorption Spectroscopy," *The Economic Record*, June 1969, pp. 158–180.

their momentous insight when they did, someone else surely would have done so later, and in all probability not much later.[5] Similar parallelism occurred in the experimental proof that chain reactions were possible and in the conception of isotopic separation methods. For instance, the gaseous diffusion process using uranium hexafluoride was outlined almost simultaneously and independently by George Kistiakowski in the United States and Franz Simon in England.

[27] Examination of the NASA contributions suggests that, like Hahn and Strassmann, NASA has typically not been working alone in the new fields of technology where it has made contributions. Instead, there was usually parallel and prior work sponsored by other federal government agencies, private firms, universities, and/or research institutes. For purposes of the present study, the key methodological question was, how should the credit be divided up among the several contributors?

This problem would require the use of marginal analysis, supplemented by network theory. The simplest theoretical illustration is the case of multiple research groups working in parallel, but independently, on a problem, with each group having, say, one chance in twenty of solving the problem during any given year's activity. Given these assumptions, the contribution of an additional equally competent and lucky research group can be evaluated through straightforward probability analysis. Thus, it can be shown that if four hypothetically identical groups are working in parallel and if a fifth path is added to the network, the average expected time from start to successful solution is reduced from 5.39 years to 4.42 years. In this case, the economic benefits from the 0.97 year speedup are correctly attributable to that fifth group, even though some other group turns out after the fact to have "won" the race, since ex ante a resource-allocating decision-maker could not have foretold which group would have been lucky enough to find the solution first.

To be sure, this illustration grossly oversimplifies reality. The typical real-world case characteristically involves an extremely complex network. If the network for some particular contribution process could be specified at least approximately, it would be possible to simulate the network on a computer, adding and withdrawing paths, modifying demand levels and hence resource allocations, etc., to determine how the presence or absence of a contributor like NASA affects the rate or character of technological progress. There is no reason why such an undertaking would not be feasible. Indeed, a group of British economists did some modest exploratory research in this direction.[6] Still the voids in our knowledge of the nodal functions and the interaction effects are so great that a major research effort would be required merely to estimate the network relationships for a single, relatively simple, technological advance. Even then, a high degree of quantitative precision could not be expected from such a pioneering venture. Such a venture was obviously beyond the scope of the study.

[28] B. Subjective Assessment

Given these difficulties in quantitatively assessing the impacts of various contributions on the advancement of a development, two subjective approaches were developed. The first was a simple subjective assessment of the actual and potential impact of each contribution on a linear scale of high, moderate and low. The second was a case study approach, describing the economic impact of several major technological advances on a particular development and tracing NASA's role in helping to bring about those advances. The scalar approach is described below, while the case studies are contained in section IV of this report.

---

5.   See F. M. Scherer, "Was the Nuclear Arms Race Inevitable?" *Co-Existence*, January 1966, pp. 59–69.
6.   I. C. R. Byatt and A. V. Cohen, *An Attempt to Quantify the Economic Benefits of Scientific Research*. London: United Kingdom Department of Education and Sciences, 1969.

The impact of a NASA contribution on the advancement of a major development was defined as the effect the contribution had on the development itself or on the environment in which the development exists. That is, did the contribution bring about a positive change? Four types of impacts were assessed:

- Technological impact. The [e]ffect a contribution has in changing the matrix of products, processes, techniques, methods and materials that make up the development.
- Scientific impact. The [e]ffect a contribution has in changing what we know or how well we understand the basic phenomena related to a technological development.
- Economic impact. The [e]ffect a contribution has in changing the economics of the development or the economics of the system in which the development is applied, including the availability and cost of the technology that makes up the development.
- Social impact. The [e]ffect a contribution has in changing the immediate social environment in which the development exists. (For the purposes of this study only first-order or primary social impacts were included.)

In interviews, NASA scientists, engineers and administrators were asked to subjectively assess the impact of their own contributions and the NASA contributions with which they were most familiar. Using a linear scale of 1 for high, 2 for moderate and 3 for low, an assessment was made for each contribution in each of the above four categories (technological, scientific, economic and social) for both actual (present) and potential impact. Each NASA assessment was reviewed by two members of the project team. In those cases where a NASA assessment of the impact of a contribution [29] could not be obtained, the assessment was made by the analyst on the project team responsible for that particular field of technology. These judgements [sic] were, in turn, reviewed by another member of the team.

Weighted averages for actual and potential impact were calculated for each field, in each of the four categories. Actual impact represents impact that has already been realized to date from the NASA contributions while potential impact represents the total impact of the contributions to be realized. Using potential impact as a measure of total impact to be realized does not mean to imply that the estimated total impact will definitely occur, since these are estimates of potential that might never be realized. Nevertheless, the assessment of impact already realized (actual) versus total impact (potential) is an important concept since it reveals much about the time lags experienced in applying NASA technology to nonaerospace needs.

Impacts of the NASA contributions, averaged for all twelve fields by category of impact, are shown in Figure 1. The technological impact of NASA contributions is greatest, followed by economic impact, scientific impact and social impact.

Technological impact of NASA contributions, by fields of technology, is shown in Figure 2. The technological impact does not appear to vary too greatly between fields, with assessments ranging from moderate to moderate to high. NASA contributions had the highest technological impact upon the field of telemetry and the lowest on integrated circuits.

The scientific impact of NASA contributions, shown in Figure 3, appears to vary much more by field of technology, with NASA having a very low impact on some fields such a machinery and joining, and a moderate impact on others such as microwave systems and internal gas dynamics.

Economic impact assessments, shown in Figure 4, ranged from telemetry with the highest potential economic impact of those studied, to energy conversion with the lowest potential impact.

The direct social impact of NASA contributions varies greatly between fields studied, as shown in Figure 5. For most fields, NASA's contributions have low social impact. For some other fields, however, such as gas dynamics, simulation, and telemetry, the impact of NASA's contributions approach moderate.

[30]

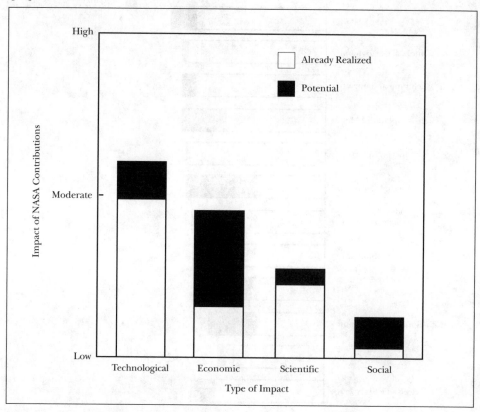

Figure 1. Impact of NASA Contributions on Major Developments in 12 Selected Fields of Technology

[31]

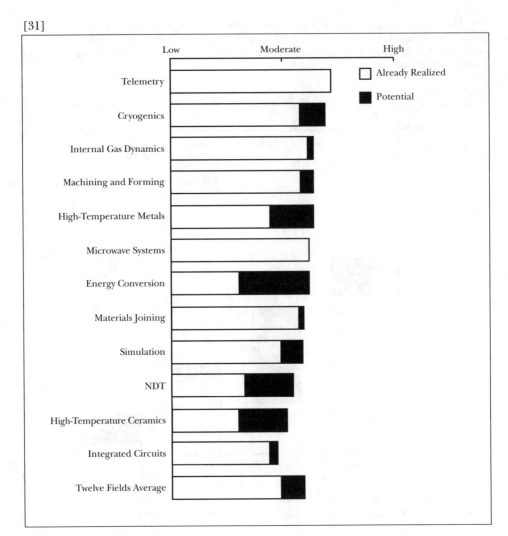

Figure 2. *Technological Impact of NASA Contributions on Major Developments in Selected Fields of Technology*

[32]

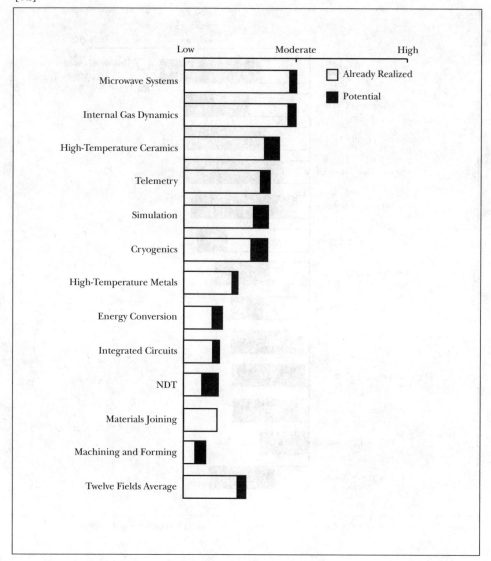

Figure 3. Scientific Impact of NASA Contributions on Major Developments in Selected Fields of Technology

[33]

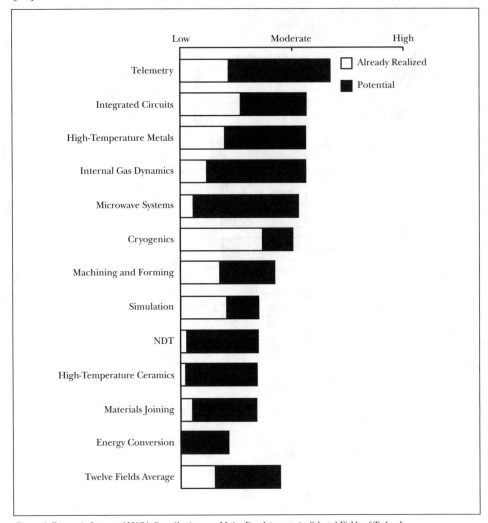

Figure 4. Economic Impact of NASA Contributions on Major Developments in Selected Fields of Technology

[34]

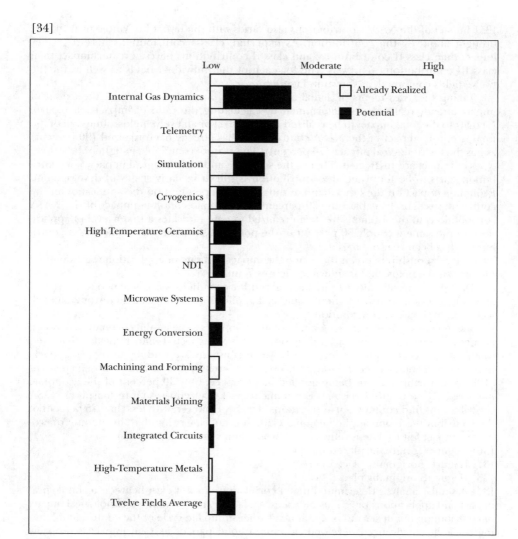

Figure 5. Social Impact of NASA Contributions on Major Developments in Selected Fields of Technology

[35] Impact of the NASA contributions also varies with the level of technological change brought about by the contribution. As expected, class I contributions have a greater impact than class II contributions, and class II contributions have a greater impact than class III contributions. This was true for each of the individual fields as well as for the twelve fields combined, as shown in Figure 6.

Taking the assessments of actual and potential impact for each field, the extent of impact already realized was determined by calculating the ratio of impact felt to date (actual) to the total impact to be realized (potential). For all twelve fields, almost all of the total scientific impact of the NASA contributions has already been realized (90 percent) as has the technological impact (70 percent). This is understandable in light of how these types of impact are [a]ffected. That is, the scientific and technological impact of a contribution starts to be felt when the contribution is still in its early stages of development. Economic impact on the other hand, is much more dependent on the application of the contributions. The fact that only 30 percent of the total economic impact of the NASA contributions to the 12 fields has been realized to date indicates a lower level of application. In the same way only 30 percent of the potential social impact of the NASA contributions has been realized to date.

The relationship between the economic impact and rate of application was examined for each of the fields, and is shown in Figures 7 and 8.

Data on economic impact already realized for each field was calculated as described above. Figures on aerospace and commercial applications were calculated from data collected on each NASA contribution.

Using least squares regression analysis, there appears to be a valid correlation between economic impact and rate of application. This is as expected, and indicates that even when a sizeable [sic] proportion of NASA contributions are finding aerospace applications, the percentage of economic impact realized is relatively low. With almost 55 percent of the NASA contributions being applied in aerospace, only 30 percent of the economic impact has been realized (Fig. 7). Economic impact realized rises more sharply as NASA contributions find commercial applications (Fig. 8). However, with less than 25 percent of the contributions being applied commercially, it is not surprising that the amount of economic impact felt to date is only a small proportion of its total potential.

[36] [Figure 6 originally placed here.]
[37] [Figure 7 originally placed here.]
[38] [Figure 8 originally placed here.]
[39] A similar analysis determined that a correlation did not exist between technological impact and applications rate. This indicates, as stated earlier, that the technological impact of contribution might start to be felt much earlier in the life cycle of the contribution, during research or advanced development, and a good part of its total impact is probably already realized by the time the contribution starts to find widespread application. . . .

[59] V. CONCLUSIONS

The general conclusions of this study concerning NASA contributions to the advancement of major developments in selected fields of technology include the following: (1) the NASA contributions are indirect and varied; (2) the NASA contributions become "embodied" in the advanced technology of a field; (3) the major effect of the NASA contributions was to cause technological advancement to occur earlier than it would have otherwise; (4) the NASA contributions represented all levels of technology, including step-changes, incremental advances and consolidations; (5) the NASA contributions were in all stages of developmental activity; and (6) the impact of the NASA contributions ranged from low to moderate-to-high.

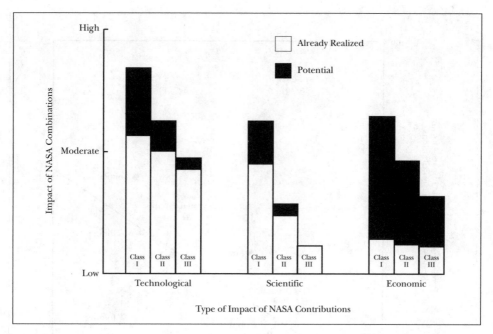

Figure 6. Impact of NASA Contributions on Major Developments in Twelve Fields of Technology for Different Classes of Technology

From these conclusions it becomes apparent that, in effect, NASA's role in advancing technology has been to create a demand for the technology to fill. Further, by creating this demand, NASA apparently aided industry, in several cases, in carrying out their own development efforts to further advance the technology in a field, resulting in new products and processes.

The impact of NASA contributions appears to be related to those factors which are inherent in the contribution as well as those factors which deal with the uses to which the contributions are put. If the NASA contribution can be thought of as a stimulus and the impact as a response, there does not appear to be too great a time lag between the stimulus provided by NASA's technological efforts which resulted in the contributions and the response in the form of technological and scientific impact brought about by these contributions. That is, most of the technological and scientific impact a contribution is going to have is felt within a reasonable time after the contribution occurs. For the identified contributions, 70 percent of the technological impact and 90 percent of the scientific impact has already been realized.

On the other hand, the rate at which the economic impact of NASA's contributions is felt, appears to be related to the rate at which the contributions find nonaerospace application. In many of the areas of NASA contributions, industry is ready to take immediate advantage of the technological and scientific stimulus provided, with a resulting economic impact. In many other areas, however, a gap apparently exists between the NASA stimulus and the ability of industry to respond. In these cases, only a small proportion of the potential economic impact inherent in the NASA contribution can occur.

The net result is that with only one-quarter of the identified NASA contributions being applied commercially, the amount of economic impact felt to date is less than one-third of its potential total impact. With a greater rate of commercialization, the economic impact of NASA contributions could be expected to rise sharply.

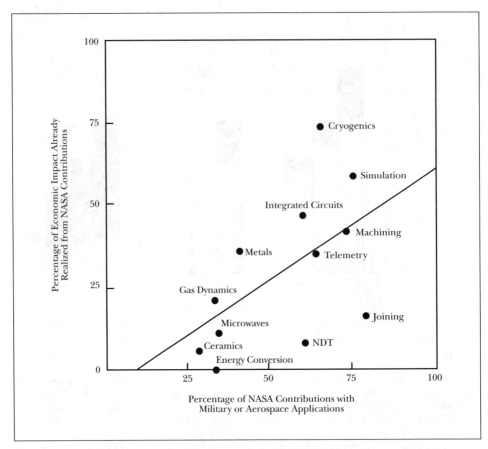

Figure 7. Relationship Between Economic Impact Already Realized from NASA Contributions and Proportion of Contributions Finding Military or Aerospace Applications

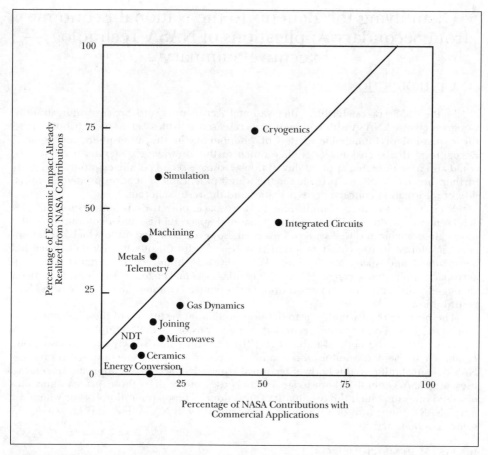

Figure 8. Relationship Between Economic Impact Already Realized From NASA Contributions and Proportion of Contributions Finding Commercial Applications

Document III-9

Document title: "Quantifying the Benefits to the National Economy from Secondary Applications of NASA Technology—Executive Summary," NASA CR-2674, Mathematica, Inc., March 1976.

Source: NASA Historical Reference Collection, NASA History Office, NASA Headquarters, Washington, D.C.

This study examined four cases of major technologies (cryogenics, gas turbines, integrated circuits, and the NASTRAN computer program) that NASA had advanced during the 1960s and showed, based on economic studies and interviews, that their benefits could be as high as $6.9 billion over ten years. Mathematica, Inc., relied on microeconomic consumer surplus theory, which is a more traditional approach to measuring benefits and more easily understood than production function approaches. Most of the benefits resulted from the earlier introduction of the specific technology rather than the development of the technology itself.

Quantifying the Benefits to the National Economy from Secondary Applications of NASA Technology— Executive Summary

[1] A. INTRODUCTION

In the 1958 law establishing the National Aeronautics and Space Administration, Congress charged NASA with conducting its research activities "so as to contribute . . . to the expansion of human knowledge of phenomena in the atmosphere and space." Recognizing that such knowledge, like much of the knowledge generated by research, could also have potential applicability in non-aerospace sectors of the economy, Congress further directed NASA to "provide for the widest practicable and appropriate dissemination of information concerning its activities and the results thereof."

NASA's success in accomplishing its aerospace objectives is unquestioned. The achievements of the satellite programs, manned space flights, and exploration of the moon are dramatic and well-known. Less clear, however, is the extent to which the knowledge developed in the NASA programs has been useful outside its originally intended aeronautical and space applications. While literally hundreds of instances of non-aerospace applications, ranging from the cardiac pacemaker to gas turbines, have been cataloged, hardly anything is known of the quantitative economic significance of NASA's contributions.

The purpose of this study was to develop preliminary estimates of the economic benefits to the U.S. economy from secondary applications of "NASA technology." If technology is defined as the body of knowledge [2] concerning how society's resources can be combined to yield economic goods and services, then NASA technology represents NASA's contribution to this body of technical knowledge. Secondary applications refer to uses of NASA generated knowledge for purposes other than those primary mission-oriented ones for which the original R&D was done. These applications occur whenever a non-NASA entity, with or without encouragement from NASA, uses this knowledge in some economic activity.

B. MEASUREMENT APPROACH

The development of procedures to quantify the economic benefits of secondary applications involved the adoption of concepts and tools that are theoretically sound and yet practically useful in dealing with a wide variety of applications of NASA technology. There are two key foundational elements of the approach: one, an understanding of how technological change generates economic benefits; and two, a determination of the role that NASA can play in the process of technological change.

1.   The Economic Benefits of Technological Change

In broad terms, the economic process involves the conversion of society's stock of resources into goods and services and the sale or exchange of these goods and services in the marketplace. This activity generates economic benefits by allowing people to consume desired combinations of goods and services. Advancing technology increases these benefits by allowing society to get more from the same stock of resources.

[3] The specific methods for quantifying economic benefits have been the subject of much discussion in the economics literature. The most widely accepted principle for evaluating economic benefits is founded on individuals' "willingness to pay" to move from a

"less" to a "more" preferred state. What this principle translates into for the purposes here is that the benefits of technological change can be measured as the cost savings generated by new or improved production processes plus the extra value that consumers attach to new or improved final products. Therefore, by determining how cost and demand for various products are affected by specific technological advances, one can estimate the benefits of these advances.

2. The Benefits Due to NASA

The research process by which technological advances are generated typically involves a complex interaction of various groups and individuals. In solving the particular problems associated with an advance, the individual research "actors" build on or combine their results with those generated by others. As a result, any "credit" for the benefits created by a particular advance should, in a real sense, be shared by the various contributors.

The goal here, of course, is to assign a particular share to NASA in some specific cases. The method for assigning this share is based on the premise that NASA R&D led to an earlier realization [4] of the particular technological advances being considered. In other words, had NASA not done its R&D—and had its failure to do so not led to changes in R&D by others—these technological advances would indeed have occurred, but at a later date.

If one accepts this view of technological advance—and it has been proposed and defended by a number of authors—then the measurement of benefits attributable to NASA becomes, at least theoretically, a rather straightforward task. These benefits can be measured as the difference between the present value of two benefits streams: one, the stream resulting from the advance as it has occurred, and two, the stream that would have resulted had NASA not been involved.

In each of the specific cases technical experts outside the NASA establishment were questioned about the speed-up resulting from NASA's role. Of course, there was some variation in their judgment. In order to allow for the inherent uncertainty in this aspect of the study, calculations were made based on alternative assumptions concerning the extent to which NASA accelerated the time stream of benefits. More specifically, benefits were calculated based on a minimum, a maximum, and a conservative "probable" speed-up due to NASA.

C. RESULTS OF CASE STUDIES

Before describing the results of the analysis, three points are worthy of mention. First, the case studies are by no means [5] a random selection from the possible cases which might have been examined. Cases were deliberately selected where data were available, where NASA's role was widely acknowledged, and where benefits were anticipated to be relatively large. Consequently, until additional experience with more case studies is acquired, it is not possible to draw inferences regarding the total secondary benefits of NASA's R&D. Second, because the case selection was not random and because of the innovative nature of the work, an effort was made to be conservative in the calculations. Third, the brief discussion of the cases which follows is not sufficient for a full understanding of the qualifications and limitations of the analyses. To acquire a more complete comprehension of these, a careful reading of the technical report is required.

1. Cryogenic Multilayer Insulation Materials

NASA's role in cryogenic technology is an outgrowth of the effort to minimize the weight, volume, and evaporative loss of gases used in launch and flight propulsion systems, life support systems, and power generation on board spacecraft. An integral part of this general concern was the design of improved insulation systems. The development of cryogenic insulation technology has contributed substantially to the rapid growth of the cryogenics industry.

In this case study, benefits are calculated as the cost savings generated by the use of multilayer insulation instead of the next best [6] insulation material (perlite) in the transport of liquid hydrogen, liquid helium, and liquid nitrogen. It should be emphasized that other savings arise from the use of multilayer insulation with liquid hydrogen, helium, and nitrogen, insofar as it is used to insulate storage tanks, piping and other equipment used in the production of these liquid gases.

The two principal sources of measured cost savings due to multilayer insulation are: reduced boil-off loss during the time the cryogen is transported, and reduced transportation costs due to the lighter weight of multilayer insulated tanks. Benefits were estimated using an engineering approach to specify the relevant technical relationships between evaporation loss, weight, and insulation characteristics. The "best guess" or "probable" estimate of benefits is $1,054 million.

2. Integrated Circuits

Prior to 1960, conventional electronic circuitry was based on the assembly of individually encapsulated circuit components such as transistors, resistors, capacitors, and diodes. Integrated circuit technology—the combination of these circuit functions on an inseparable, continuous base—provides significant advantages over conventional circuit technology, particularly in smaller size, lower power consumption, increased speed of operation, improved reliability, and reduced cost per electronic function. These features make integrated circuits especially attractive for space applications.

The introduction of integrated circuit technology produced significant changes in all electronic products, including consumer electronic products, and the estimates of the total benefits to advancing integrated circuit technology reflect its very widespread applicability. [7] Based on a simultaneous equation estimation of the demand for integrated circuits, estimates of benefits were derived. The "probable" estimate is $5,080 million.

3. Gas Turbines in Electric Power Generation

Since the early 1940's NASA (then NACA) has been intimately involved in gas turbine technology, primarily as it relates to improvement of jet engines for military and commercial aircraft. This basic research has also produced benefits in the production of electric power, as gas turbines have become more widely used as sources of peaking power and standby capacity.

The use of gas turbines in electric power generation undoubtedly confers may social benefits; e.g., gas turbines are relatively "clean" from an environmental point of view and also enhance the reliability of power production. Nevertheless, the estimates of benefits are based only on the fuel cost savings in power production produced by improvements in gas turbine performance. Improvements in gas turbine performance are assumed to result from advances in turbine technology; gas turbine vintage was used as a proxy for technology level in these calculations. Using standard regression analysis, the relationship between gas turbine vintage and the average cost of fuel consumed in the production of power was estimated; based on this relationship, a "probable" estimate of the total fuel cost savings of $111 million was determined.

[8] 4. NASTRAN

NASTRAN (NASA Structural Analysis) is a general purpose finite element computer software package for static and dynamic analysis of the behavior of elastic structures. Industrial users are generally product engineers in mechanical or civil engineering applications such as aircraft and automobile production, bridge construction, or power plant modeling.

NASA's Goddard Space Flight Center developed NASTRAN, through a combination of in-house and contracted research, over the period 1965 to 1970. It represented substantial improvement over similar extant programs, and was released to public users in November of 1970.

Because few published data exist regarding the extent of use of NASTRAN, estimates of cost savings from the use of NASTRAN were obtained from telephone interviews with a sample of users. From the sample responses the "probable" benefits accruing to the population of NASTRAN users were estimated at $701 million.

D. SUMMARY AND CONCLUSIONS

A summary of the "probable" estimates for each case study is presented in Table 1. It indicates that total benefits due to NASA for the four cases studied are probably on the order of $7,000 million.

[9]
Table 1
Results of Benefits Estimation

Technology	Interval of Benefits Estimation	Estimated Probable NASA Acceleration (Years)	Probable Benefits Attributable to NASA (Millions)
Gas Turbines	1969–1972	1.0	$ 111
Cryogenics	1960–1963	5.0	$1,054
Integrated Circuits	1963–1982	2.0	$5,080
NASTRAN	1971–1984	4.0	$ 701
Total	—	—	$6,946

[10] In interpreting the results of this study one should recognize that it is one of the first of its kind ever attempted. The results, while arrived at through careful and rigorous techniques, are sensitive to data uncertainties and analytical simplifications. Though one must necessarily view such results with caution, it seems that the following general observations could be safely made:

- Operational methods can be developed for estimating the secondary benefits of mission oriented R&D.
- Secondary benefits attributable to NASA's R&D programs may be impressively large. For example, the $7,000 million total for the four cases studied is more than twice NASA's present yearly budget.
- Because secondary benefits may indeed be significant, public decisions concerning the allocation of resources to research and development programs should, where possible, consider such benefits.

Document III-10

Document title: "Economic Effects of a Space Station: Preliminary Results," NASA, June 16, 1983, pp. 1–2, 20–21.

Source: NASA Historical Reference Collection, NASA History Office, NASA Headquarters, Washington, D.C.

This study, sponsored by the NASA Alumni League, reports on only the direct and indirect employment and income that were generated by NASA spending. It came as NASA was seeking White House approval to initiate a space station program. Although the impacts are presented by industry and state, the benefits only represent the multiplier effects of the spending patterns of NASA. These findings and similar studies were important in convincing legislators how important NASA's budget had become to their regions. The following text is the introduction and conclusions from the first draft of the report.

The Economic Effects of a Space Station: Preliminary Results

[1] 1.0 INTRODUCTION

This report discusses the economic effects of a manned Space Station. Since the major reasons for building a Space Station are not economic, this report will also discuss some of the non-economic benefits to be gained. Some of these benefits include:

- The Space Station will enhance the defense posture of the United States by demonstrating an ability to control the high ground of space.
- The Space Station will build respect for the industrial strength of the nation by proving the ability to initiate and maintain large-scale, high technology programs.
- The Space Station program will increase the nation's pride and confidence just as the Shuttle and other space programs have done.

Since few of these benefits can be quantitatively assessed, it is not feasible at this time to justify a Space Station on economic grounds alone.

To look at the Space Station program from an economic perspective, the program is divided into two major phases: a development phase and an operational phase. During the development phase, cash flows will consist primarily of government expenditures (to plan, design, construct, test, and deploy the Space Station hardware) and the resulting direct and multiplier effects of those expenditures on the economy. During the operational phase, the direct economic effects will diversify as a variety of government and private customers begin to use the Space Station.

An important point that needs to be stated is the level at which a Space Station will influence the economy. During the development phase, the expenditures will be too small to notice at a macroeconomic level (e.g., about .25% of the Gross National Product). However, the development expenditures will have significant effects for specific industries and locations. The operational phase should have a major influence on the overall economy as [2] the technology spreads throughout all sectors. A Space Station could open the door to the creation of a whole new industry based on space operations and may provide the stimulus for new and improved consumer products. The projected revenues for commercial space activities are highly uncertain due to the length of time (10–20 years) being addressed. The most likely near-term effect an operational Space Station will have is the reduced cost of performing certain missions in space.

The following sections will discuss the cost of building and operating a Space Station, the effect of Space Station expenditures on the economy, the potential savings on planned missions, the new or improved capabilities that will be provided by a Space Station, the influence of new technology on the economy, and the opportunities for new space industries. . . .

[20] 8.0 CONCLUSIONS

The acquisition of a Space Station will substantially enhance the ability of NASA to more effectively perform the planned space missions of the 1990's and will establish an exciting capability for the post 2000's that could foster the growth of a broad range of financially attractive space endeavors. The Space Station is a project that is consistent with existing NASA budget levels and does not conflict with the nation's economic policies of a balanced budget and reduced Federal deficits. In fact, the capabilities offered by the Space Station result in cost savings through the 1990's that are approximately equal to the cost of the project.[4] Larger economic benefits are expected beyond the year 2000.

The existence of a Space Station will also stimulate evolving commercial opportunities in space. It will provide a permanently manned and easily accessible research and development facility for new space ventures and ensure the nation's leadership in the industrial frontiers of space.

The real value of a Space Station, however, extends beyond economic benefits. The Space Station will also continue the unbroken chain of United States advances in space technology that has existed for 25 years. It will once again demonstrate the industrial prowess and military strength that has made the United States a great nation. It will also enhance the confidence and pride the people of this country have for their nation.

[21] REFERENCES . . .

4. "Space Station Needs, Attributes and Architectural Options Study: Final Report," Contract NASW-3680 (eight reports by various authors), 1983.

Document III-11

Document title: "The Economic Impact of the Space Program: A Macro and Industrial Perspective," prepared for Rockwell International by The WEFA Group, Bala Cynwyd, Pennsylvania, May 1994, pp. 1–4 (reprinted with permission).

Source: NASA Historical Reference Collection, NASA History Office, NASA Headquarters, Washington, D.C.

To support the space station program, Rockwell International commissioned WEFA (which had merged with Chase) to conduct an analysis of the industrial and economic impacts of the space station program expenditures. It did not calculate impacts of new technology, but instead focused on the total multiplier impact on jobs and income from NASA. This study was used to develop political support for the space station by illustrating that investments in high-tech industries were beneficial. It did not analyze impacts if the money had been spent elsewhere or if there had been a tax cut of equal proportions. The following is just the report's executive summary.

The Economic Impact of the Space Program:
A Macro and Industrial Perspective

[1] **Executive Summary**

In this study, the macroeconomic and industry impacts of the National Aeronautics and Space Administration Program on the U.S. economy are evaluated. In performing this study, The WEFA Group's macroeconomic model of the U.S. economy and its Industrial Analysis Service model were utilized. These simulations were generated by removing NASA expenditures from the current baseline projections of economic activity.

The NASA program provides the following economic impacts:

- By 1997, an estimated 380,000 jobs in the U.S. economy are generated by NASA related activity. **Summary Figure 1** depicts how many jobs would be lost in the U.S. economy if the NASA budget were eliminated beginning in fiscal year 1995. Aerospace, communications equipment, transportation equipment, industrial machinery, metals, research and consulting services, and construction jobs are highly dependent upon NASA. These employment estimates flow from direct sources (government contractors), indirect sources (the contractors' supplier network), and expenditure related (feedthrough of employment and income) sources. The direct and indirect employment totals **217,000,** with another **163,000** due to the multiplier impact on the rest of the economy. One important facet of this analysis is that many of the industries identified as likely beneficiaries of NASA programs are the industries that are important to the country's achievement of critical technologies, as identified by the Department of Commerce, the Office of the Vice President, and the Council on Competitiveness. These technologies include Applied Molecular Biology, Distributed Computing and Telecommunications, Flexible Manufacturing Electrical Supply and Distribution, Materials Synthesis and Processing, Microelectronics and Optoelectronics, and Software. While no one would make an argument that certain industries are sacred, it is true that the types of activities that NASA funds correlate highly with many technologies that have been identified as key to the U.S. economy's future competitiveness.

[Figure 1 originally placed here.]

[2] • An estimated **$23.49** billion of real GDP [Gross Domestic Product] (economic output) in the U.S. economy is tied to activities at NASA by 1997. Between 1995 and 2000, the space program will contribute a cumulative addition of **$130** billion to real GDP. This estimate does not include the increase to GDP that might result from technological innovation. **Summary Figure 2** displays the loss of real economic output that would occur if the NASA budget were eliminated beginning in fiscal year 1995. The positive impact on the economy is far greater than just federal outlays on NASA. Due to NASA's purchases of goods and services, firms increase their output and employment, promoting personal income growth. Higher personal income growth causes consumers to raise their purchases. Capital investment is stimulated due to its high sensitivity to output growth. Higher investment leads to a greater accumulation in the capital stock, promoting productivity growth and U.S. international competitiveness.

[Figure 2 originally placed here.]

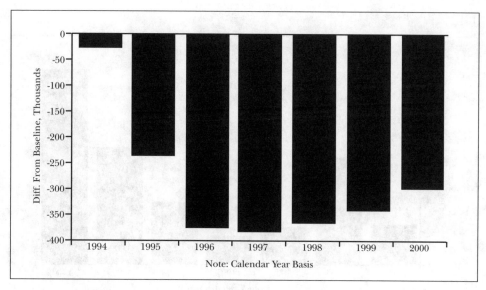

Figure 1. Establishment Employment NASA Simulation vs. Baseline

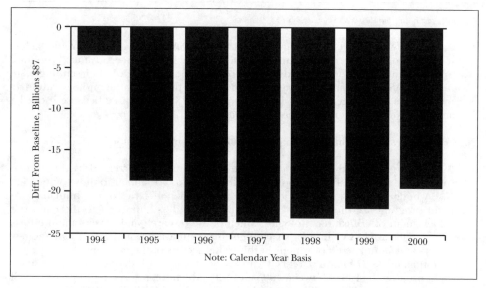

Figure 2. Real GDP: NASA Simulation I vs. Baseline

- If the NASA budget were eliminated, there is a loss of economic output, lower employment, and decreased corporate profits which reduce federal tax receipts. Payments for unemployment insurance and other related transfer payments rise, increasing government expenditures on these programs. The combination of lower tax receipts and higher transfer payments offsets much of the impact of the elimination of the NASA budget on the federal surplus. By 1997, the federal budget improves by only $1.60 billion relative to what it would be if NASA spending of $16.14 billion were not cut. **Summary Figure 3** highlights the extent of the improvement in the federal budget if the NASA budget were eliminated beginning in fiscal year 1995.

[3]

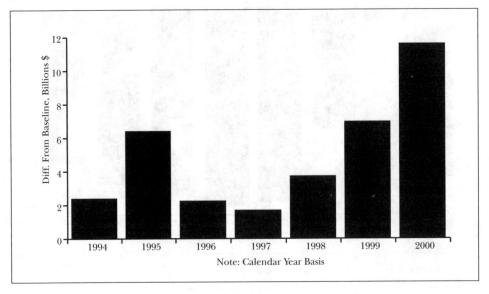

Figure 3: Federal Gov. Surplus: NASA Simulation I vs. Baseline

- Additionally, potential technology spinoffs from NASA-derived research would be lost if NASA programs were eliminated (such as artificial intelligence, advanced robotics, optical communication, and advanced computers). Because of their cutting-edge nature, NASA programs are highly conducive to promoting technology advancements. This store of technology is an important national resource because it can be adapted to develop new products and processes.

NASA's human space flight program provides the following economic impacts:

- By 1997, an estimated **179,000** jobs in the U.S. economy are created by the human space flight program. The composition of the job gains are similar to those of the total NASA budget. Many industries identified as likely beneficiaries of the human space flight program are those that have been determined as important to the country's achievement of critical technologies. Much of the research and development efforts taking place at NASA's human space flight program are devoted to high-technology sectors and stimulate employment in these vital sectors of the economy.
- An estimated **$8.37** billion of real GDP in the U.S. economy is tied to the human space flight program by 1997. Between 1995 and 2000, the human space flight program will contribute a cumulative addition of **$44.4** billion to real GDP. This estimate does not include the increase to GDP that might result from technological innovation. Though the transmission mechanism of federal expenditures on the human space flight program throughout the economy is similar to the first simulation, the impact is lower due to the smaller expenditures. One of the most critical benefits for the U.S. economy from NASA's human space flight program is the stimulus it provides to capital investment. We estimate that by 1997, capital investment is aided by $1.12 billion. Investment in equipment benefits the most and tends to foster productivity growth.
- [4] • If the human space flight program were eliminated, the federal deficit would not improve as much as the expenditure cuts. The combination of lower tax receipts and higher transfer payments offset much of the impact of the elimination of federal

expenditures on the human space flight program. By 1997, the federal budget improves by only $0.80 billion relative to what it would be if NASA spending on the human space flight program were not cut.

- Potential technology spinoffs would be lost if the human space flight program were eliminated. Much of the research in the human space flight program is taking place in artificial intelligence, advanced robotics, optical communication, and advanced computers. These are all areas that would have clear commercial applications.

NASA's space station program provides the following economic impacts:

- By 1997, an estimated **55,000** jobs in the U.S. economy are created by the space station program. Many are in high-technology sectors (such as high-tech capital goods, electronics, telecommunications, and software development). Employment at aerospace, communications equipment, transportation equipment, and industrial machinery manufacturers are dependent upon the continuation of space station program. Over the next couple of years, spending on the space station program will center on the final R&D and on manufacturing. While the estimated employment impacts are smaller than the other two simulations, the industries most impacted again show a high correlation with those sectors targeted as critical for national competitiveness.
- An estimated **$2.60** billion of real GDP in the U.S. economy is tied to the space station program by 1997. Between 1995 and 2000, the space station program will contribute a cumulative addition of **$13.8** billion to real GDP. This estimate does not include the increase to GDP that might result from technological innovation. The transmission mechanism throughout the economy of federal expenditures on the space station program are similar to the first two simulations, but the estimated impact is lower due to smaller expenditures. An important implication of the economic activity associated with the space station program is that it promotes proportionally more towards investment in equipment than either the total NASA budget or the human space flight program. Therefore, the productivity enhancing properties of these federal expenditures are high.
- If the space station program were eliminated, the federal deficit would not improve as much as the expenditure cuts. The combination of lower tax receipts and higher transfer payments offset much of the impact of the elimination of these federal expenditures. By 1997, the federal budget improves by only $0.26 billion relative to what it would be if spending on the space station program were not cut.
- Potential technology spinoffs would be lost if the space station program were eliminated. Similar to the human space flight program, much of the research is taking place in artificial intelligence, advanced robotics, optical communication, and advanced computers. All of these areas promise a high degree of commercial applications. . . .

Document III-12

Document title: Office of the Press Secretary, The White House, "The President's Space Policy and Commercial Space Initiative to Begin the Next Century," Fact Sheet, February 11, 1988.

Source: Documentary History Collection, Space Policy Institute, George Washington University, Washington, D.C.

By 1988, private sector activities in space were beginning to grow rapidly, particularly in the communications satellite and launch vehicle sectors. The Reagan administration issued this presidential directive, which, for the first time, made the creation of commercial opportunities in space a major component of national space policy. It prohibited NASA from operating an expendable launch vehicle program and encouraged the government to purchase commercial launch services. Also, it called for open private opportunities in space in microgravity, remote sensing, and other space ventures where there was the potential for commercial operations.

[no page number]

For Immediate Release   February 11, 1988

The President's Space Policy and Commercial Space Initiative to Begin the Next Century

FACT SHEET

The President today announced a comprehensive "Space Policy and Commercial Space Initiative to Begin the Next Century" intended to assure United States space leadership.

The President's program has three major components:
- Establishing a long-range goal to expand human presence and activity beyond Earth orbit into the Solar System;
- Creating opportunities for U.S. commerce in space; and
- Continuing our national commitment to a permanently manned Space Station.

The new policy and programs are contained in a National Security Decision Directive (NSDD) signed by the President on January 5, 1988, the FY 1989 Budget the President will submit shortly to Congress, and a fifteen point Commercial Space Initiative.

I. EXPANDING HUMAN PRESENCE BEYOND EARTH ORBIT

In the recent NSDD, the President committed to a goal of expanding human presence and activity in the Solar System. To lay the foundation for this goal, the President will be requesting $100 million in his FY 1989 Budget for a major new technology development program "Project Pathfinder" that will enable a broad range of manned or unmanned missions beyond the Earth's orbit.

Project Pathfinder will be organized around four major focuses:
— Exploration technology;
— Operations technology;
— Humans-in-space technology; and
— Transfer vehicle technology.

This research effort will give the United States know-how in critical areas, such as humans in the space environment, closed loop life support, aero braking, orbital transfer and maneuvering, cryogenic storage and handling, and large scale space operations, and provide a base for wise decisions on long term goals and missions.

Additional highlights of the NSDD are outlined in Section IV of this fact sheet.

[2] II. CREATING OPPORTUNITIES FOR U.S. COMMERCE IN SPACE

The President is announcing a fifteen point commercial space initiative to seize the opportunities for a vigorous U.S. commercial presence in Earth orbit and beyond—in research and manufacturing. This initiative has three goals:
- Promoting a strong U.S. commercial presence in space;
- Assuring a highway to space; and
- Building a solid technology and talent base.

Promoting a Strong U.S. Commercial Presence in Space

1. Private Sector Space Facility: The President is announcing an intent for the Federal Government to lease space as an "anchor tenant" in an orbiting space facility suitable for research and commercial manufacturing that is financed, constructed, and operated by the private sector. The Administration will solicit proposals from the U.S. private sector for such a facility. Space in this facility will be used and/or subleased by various Federal agencies with interest in microgravity research.

    The Administration's intent is to award a contract during mid-summer of this year for such space and related services to be available to the Government no later than the end of FY 1993.

2. Spacehab: The Administration is committing to make best efforts to launch within the Shuttle payload bay, in the early 1990s, the commercially developed, owned, and managed Shuttle middeck module: Spacehab. Manifesting requirements will depend on customer demand.

    Spacehab is a pressurized metal cylinder that fits in the Shuttle payload bay and connects to the crew compartment through the orbiter airlock. Spacehab takes up approximately one-quarter of the payload bay and increases the pressurized living and working space of an orbiter by approximately 1,000 cubic feet or 400 percent in useable research volume. The facility is intended to be ready for commercial use in mid-1991.

3. Microgravity Research Board: The President will establish, through Executive Order, a National Microgravity Research Board to assure and coordinate a broader range of opportunities for research in microgravity conditions.

    NASA will chair this board, which will include senior-level representatives from the Departments of Commerce, Transportation, Energy, and Defense, NIH, and NSF; and will consult with the university and commercial sectors. The board will have the following responsibilities:
    - To stimulate research in microgravity environments and its applications to commercial uses by advising Federal agencies, including NASA, on microgravity priorities, and consulting with private industry and academia on microgravity research opportunities;
    - To develop policy recommendations to the Federal Government on matters relating to microgravity research, including types of research, government/industry/academic cooperation, and access to space, including a potential launch voucher programs;
    - [3] To coordinate the microgravity programs of Federal agencies by:
        — reviewing agency plans for microgravity research and recommending priorities for the use of Federally-owned or leased space on microgravity facilities; and

— ensuring that agencies establish merit review processes for evaluating micro-gravity research proposals; and
- To promote transfer of federally funded microgravity research to the commercial sector in furtherance of Executive Order 12591.

NASA will continue to be responsible for making judgments on the safety of experiments and for making manifesting decisions for manned space flight systems.

4. External Tanks: The Administration is making available for five years the expended external tanks of the Shuttle fleet at no cost to all feasible U.S. commercial and non-profit endeavors, for uses such as research, storage, or manufacturing in space.

    NASA will provide any necessary technical or other assistance to these endeavors on a direct cost basis. If private sector demand exceeds supply, NASA may auction the external tanks.

5. Privatizing Space Station: NASA, in coordination with the Office of Management and Budget, will revise its guidelines on commercialization of the U.S. Space Station to clarify and strengthen the Federal commitment to private sector investment in this program.

6. Future Privatization: NASA will seek to rely to the great-est extent feasible on private sector design, financing, construction, and operation of future Space Station require-ments, including those currently under study.

7. Remote Sensing: The Administration is encouraging the development of commercial remote sensing systems. As part of this effort, the Department of Commerce, in con-sultation with other agencies, is examining potential opportunities for future Federal procurement of remote sensing data from the U.S. commercial sector.

Assuring a Highway to Space

8. Reliance on Private Launch Services: Federal agencies will procure existing and future required expendable launch services directly from the private sector to the fullest extent feasible.

9. Insurance Relief for Launch Providers: The Administration will take administrative steps to address the insurance concerns of the U.S. commercial launch industry, which currently uses Federal launch ranges. These steps include:

    - Limits on Third Party Liability: Consistent with the Administration's tort policy, the Administration will propose to Congress a $200,000 cap on noneconomic damage awards to individual third parties resulting from commercial launch acci-dents;
    [4] - Limits on Property Damage Liability: The liability of commercial launch opera-tors for damage to Government property resulting from a commercial launch accident will be administratively limited to the level of insurance required by the Department of Transportation. If losses to the Government exceed this level, the Government will waive its right to recover for damages. If losses are less than this level, the Government will waive its right to recover for those damages caused by Government willful misconduct or reckless disregard.

10. <u>Private Launch Ranges:</u> The Administration will consult with the private sector on the potential construction of commercial launch range facilities separate from Federal facilities and the use of such facilities by the Federal Government.

11. <u>Vouchers for Research Payloads:</u> NASA and the Department of Transportation will explore providing to research payload owners manifested on the Shuttle a one time launch voucher that can be used to purchase an alternative U.S. commercial launch service.

<u>Building a Solid Technology and Talent Base</u>

12. <u>Space Technology Spin-Offs:</u> The President is directing that the new Pathfinder program, the Civil Space Technology Initiative, and other technology programs be conducted in accordance with the following policies:
   • Federally funded contractors, universities, and Federal laboratories will retain the rights to any patents and technical data, including copyrights, that result from these programs. The Federal Government will have the authority to use this intellectual property royalty free;
   • Proposed technologies and patents available for licensing will be housed in a Pathfinder/CSTI library within NASA; and
   • When contracting for commercial development of Pathfinder, CSTI and other technology work products, NASA will specify its requirements in a manner that provides contractors with maximum flexibility to pursue innovative and creative approaches.

13. <u>Federal Expertise on Loan to American Schools:</u> The President is encouraging Federal scientists, engineers, and technicians in aerospace and space related careers to take a sabbatical year to teach in any level of education in the United States.

14. <u>Education Opportunities:</u> The President is requesting in his FY 1989 Budget expanding five-fold opportunities for U.S. teachers to visit NASA field centers and related aerospace and university facilities.

   In addition, NASA, NSF, and DOD will contribute materials and classroom experiments through the Department of Education to U.S. schools developing "tech shop" programs. NASA will encourage corporate participation in this program.

15. <u>Protecting U.S. Critical Technologies:</u> The Administration is requesting that Congress extend to NASA the authority it has given the Department of Defense to protect from wholesale release under the Freedom of Information Act those critical national technologies and systems that are prohibited from export.

[5] III. CONTINUING THE NATIONAL COMMITMENT TO THE SPACE STATION

In 1984, the President directed NASA to develop a permanently manned Space Station. The President remains committed to achieving this end and is requesting $1 billion in his FY 1989 Budget for continued development and a three year appropriation commitment from Congress for $6.1 billion. The Space Station, planned for development in cooperation with U.S. friends and allies, is intended to be a multi-purpose facility for the Nation's science and applications programs. It will permit such things in space as: research, observation of the solar system, assembly of vehicles or facilities, storage, servicing of satellites, and basing for future space missions and commercial and entrepreneur-ial endeavors in space.

To help ensure a Space Station that is cost effective, the President is proposing as part of his Commercial Space Initiative actions to encourage private sector investment in the Space Station, including directing NASA to rely to the greatest extent feasible on private sector design, financing, construction, and operation of future Space Station require-ments.

IV. ADDITIONAL HIGHLIGHTS OF THE JANUARY 5, 1988 NSDD

- Space Leadership: Leadership is reiterated as a fundamental national objective in areas of space activity critical to achieving U.S. national security, scientific, economic and foreign policy goals.
- Defining Federal Roles and Responsibilities: Government activities are specified in three separate and distinct sectors: civil, national security, and nongovernmental. Agency roles and responsibilities are codified and specific goals are established for the civil space sector; those for other sectors are updated.
- Encouraging a Commercial Sector: A separate, nongovernmental or commercial space sector is recognized and encouraged by the Federal Government actions shall not preclude or deter the continuing development of this sector. New guidelines are established to limit unnecessary Government competition with the private sector and ensure that Federal agencies are reliable customers for commercial space goods and services.
- The President's launch policy prohibiting NASA from maintaining an expendable launch vehicle adjunct to the Shuttle, as well as limiting commercial and foreign pay-loads on the Shuttle to those that are Shuttle-unique or serve national security or for-eign policy purposes, is reaffirmed. In addition, policies endorsing the purchase of commercial launch services by Federal agencies are further strengthened.
- National Security Space Sector: An assured capability for national security missions is clearly enunciated, and the survivability and endurance of critical national security space functions is stressed.
- Assuring Access to Space: Assured access to space is recognized as a key element of national space policy. U.S. space transportation systems that provide sufficient resiliency to allow continued operation, despite failures in any single system, are emphasized. The mix of space transportation vehicles will be defined to support mis-sion needs in the most cost effective manner.
- Remote Sensing: Policies for Federal "remote sensing" or observation of the Earth are established to encourage the development of U.S. commercial systems competitive with or superior to foreign-operated civil or commercial systems.

Document III-13

Document title: National Space Policy Directive 3, "U.S. Commercial Space Policy Guidelines," The White House, February 12, 1991.

Source: Documentary History Collection, Space Policy Institute, George Washington University, Washington, D.C.

These Bush administration guidelines expanded on the earlier Reagan administration presidential directive (Document III-12). They explicitly recognized the use of space for commercial purposes as an important element in developing the international competitiveness of the United States. They detail many government initiatives that range from a requirement to purchase commercially available space products and services to the avoidance of government regulations that could preclude or deter commercial space activities. In addition, they direct the government to enter into trade negotiations that encourage market-oriented competition on an international basis.

[no page number]

National Space Policy Directive 3
February 12, 1991

U.S. Commercial Space Policy Guidelines

A fundamental objective guiding United States space activities has been space leadership, which requires preeminence in key areas of space activity. In an increasingly competitive international environment, the U.S. Government encourages the commercial use and exploitation of space technologies and systems for national economic benefit. These efforts to encourage commercial activities must be consistent with national security and foreign policy interests; international and domestic legal obligations, including U.S. commitments to stem missile proliferation; and agency mission requirements.

United States space activities are conducted by three separate and distinct sectors: two U.S. Government sectors—the civil and national security—and a nongovernmental commercial space sector. The commercial space sector includes a broad cross section of potential providers and users, including both established and new market participants. There also has been a recent emergence of State government initiatives related to encouraging commercial space activities. The commercial space sector is comprised of at least five market areas, each encompassing both Earth- and space-based activities, with varying degrees of market maturity or potential:

Satellite Communications—the private development, manufacture, and operation of communications satellites and marketing of satellite telecommunications services, including position location and navigation;

Launch and Vehicle Services—the private development, manufacture, and operation of launch and reentry vehicles, and the marketing of space transportation services;

Remote Sensing—the private development, manufacture, and operation of remote sensing satellites and the processing and marketing of remote sensing data;

Materials Processing—the experimentation with, and production of, organic and inorganic materials and products utilizing the space environment; and

Commercial Infrastructure—the private development and provision of space-related support facilities, capabilities, and services.

In addition, other market-driven commercial space sector opportunities are emerging.

The U.S. Government encourages private investment in, and broader responsibility for, space-related activities that can result in products and services that meet the needs of Government and other customers in a competitive market. As a matter of policy, the U.S. Government pursues its commercial space objectives without the use of direct Federal subsidies. A robust commercial space sector has the potential to generate new technologies, products, markets, jobs, and other economic benefits for the Nation, as well as indirect benefits for national security.

Commercial space sector activities are characterized by the provision of products and services such that:

[III-20]—   private capital is at risk;
   — there are existing, or potential, nongovernmental customers for the activity;
   — the commercial market ultimately determines the viability of the activity; and
   — primary responsibility and management initiative for the activity resides with the private sector.

Implementing Guidelines

The following implementing guidelines shall serve to provide the U.S. private sector with a level of stability and predictability in its dealings with agencies of the U.S. Government. The agencies will work separately but cooperatively, as appropriate, to develop specific measures to implement this strategy. U.S. Government agencies shall, consistent with national security and foreign policy interests, international and domestic legal obligations, and agency mission requirements, encourage the growth of the U.S. commercial space sector in accordance with the following guidelines:

• U.S. Government agencies shall utilize commercially available space products and services to the fullest extent feasible. This policy of encouraging U.S. Government agencies to purchase, and the private sector to sell, commercial space products and services has potentially large economic benefits.
   — A space product or service is "commercially available" if it is currently offered commercially, or if it could be supplied commercially in response to a Government procurement request.
   — "Feasible" means that products and services meet mission requirements in a cost-effective manner.
   — "Cost-effective" generally means that the commercial product or service costs no more than governmental development or directed procurement where such Government costs include applicable Government labor and overhead costs, as well as contractor charges and operations costs.
   — However, the acquisition of commercial space products and services shall generally be considered cost effective if they are procured competitively using performance-based contracting techniques. Such contracting techniques give contractors the freedom and financial incentive to achieve economies of scale by combining their Government and commercial work, as well as increased productivity through innovation.

[III-21]
   — U.S. Government agencies shall actively consider, at the earliest appropriate time, the feasibility of their using commercially available products and services in agency programs and activities.
   — U.S. Government agencies shall continue to take appropriate measures to protect from disclosure any proprietary data which is shared with the U.S. Government in the acquisition of commercial space products and services.
• Government agencies shall promote the transfer of U.S. Government-developed technology to the private sector.
   — U.S. Government-developed unclassified space technology will be transferred to the U.S. commercial space sector in as timely a manner as possible and in ways that protect its commercial value.
   — U.S. Government agencies may undertake cooperative research and development activities with the private sector, as well as State and local governments, consistent with policies and funding, in order to fulfill mission requirements in a manner which encourages the creation of commercial opportunities.

— With respect to technologies generated in the performance of Government contracts, U.S. Government agencies shall obtain only those rights necessary to meet Government needs and mission requirements, as directed by Executive Order 12591.

• U.S. Government agencies may make unused capacity of space assets, services, and infrastructure available for commercial space sector use.

— Private sector use of U.S. Government agency space assets, services, and infrastructure shall be made available on a reimbursable basis consistent with OMB Circular A-25 or appropriate legislation.

• Government agencies may make available to the private sector those assets which have been determined to be excess to the requirements of the U.S. Government in accordance with U.S. law and applicable international treaty obligations. Due regard shall be given to the economic impact such transfer may have on the commercial space sector, promoting competition, and the long-term public interest.

• The U.S. Government shall avoid regulating domestic space activities in a manner that precludes or deters commercial space sector activities, except to the extent necessary to meet international and domestic legal obligations, including those of the Missile Technology Control Regime.

Document III-14

Document title: "Fact Sheet, National Space Policy," The White House, National Science and Technology Council, September 19, 1996.

Source: Documentary History Collection, Space Policy Institute, George Washington University, Washington, D.C.

While this addresses a full range of space policy issues, it is also the first presidential space policy directive that directly and specifically details the process by which the government can stimulate economic and business activity from space programs. It reflects the end of the Cold War, the shrinking federal discretionary budget, the maturity of some parts of the space program, and international competitive pressures.

---

[1] For Immediate Release   September 19, 1996

Fact Sheet
National Space Policy

<u>Introduction</u>

(1) For over three decades, the United States has led the world in the exploration and use of outer space. Our achievements in space have inspired a generation of Americans and people throughout the world. We will maintain this leadership role by supporting a strong, stable and balanced national space program that serves our goals in national security, foreign policy, economic growth, environmental stewardship and scientific and technical excellence. Access to and use of space is central for preserving peace and protecting U.S. national security as well as civil and commercial interests. The United States will pursue greater levels of partnership and cooperation in national and international space activities and work with other nations to ensure the continued exploration and use of outer space for peaceful purposes.

(2) The goals of the U.S. space program are to:
  (a) Enhance knowledge of the Earth, the solar system and the universe through human and robotic exploration;
  (b) Strengthen and maintain the national security of the United States;
  (c) Enhance the economic competitiveness, and scientific and technical capabilities of the United States;
  (d) Encourage State, local and private sector investment in and use of space technologies;
  (e) Promote international cooperation to further U.S. domestic, national security, and foreign policies.

(3) The United States is committed to the exploration and use of outer space by all nations for peaceful purposes and for the benefit of all humanity. "Peaceful purposes" allow defense and intelligence-related activities in pursuit of national security and other goals. The United States rejects any claims to sovereignty by any nation over outer space or celestial [2] bodies, or any portion thereof and rejects any limitations on the fundamental right of sovereign nations to acquire data from space. The United States considers the space systems of any nation to be national property with the right of passage through and operations in space without interference. Purposeful interference with space systems shall be viewed as an infringement on sovereign rights.

(4) The U.S. Government will maintain and coordinate separate national security and civil space systems where differing needs dictate. All actions undertaken by agencies and departments in implementing the national space policy shall be consistent with U.S. law, regulations, national security requirements, foreign policy, international obligations and nonproliferation policy.

(5) The National Science and Technology Council (NSTC) is the principal forum for resolving issues related to national space policy. As appropriate, the NSTC and NSC will co-chair policy processes.

This policy will be implemented within the overall resource and policy guidance provided by the President.

Civil Space Guidelines

(1) The National Aeronautics and Space Administration is the lead agency for research and development in civil space activities.

(2) NASA, in coordination with other departments and agencies as appropriate, will focus its research and development efforts in: space science to enhance knowledge of the solar system, the universe, and fundamental natural and physical sciences; Earth observation to better understand global change and the effect of natural and human influences on the environment; human spaceflight to conduct scientific, commercial, exploration activities; and space technologies and applications to develop new technologies in support of U.S. Government needs and our economic competitiveness.

(3) To enable these activities, NASA will:
  (a) Develop and operate the International Space Station to support activities requiring the unique attributes of humans in space and establish a permanent human

presence in Earth orbit. The International Space Station will support future decisions on the feasibility and desirability of conducting further human exploration activities.

(b) Work with the private sector to develop flight demonstrators that will support a decision by the end of the decade on development of a next-generation reusable launch system.

[3] (c) Continue a strong commitment to space science and Earth science programs. NASA will undertake:

    (i) a sustained program to support a robotic presence on the surface of Mars by year 2000 for the purposes of scientific research, exploration and technology development;

    (ii) a long-term program, using innovative new technologies, to obtain in-situ measurements and sample returns from the celestial bodies in the solar system;

    (iii) a long-term program to identify and characterize planetary bodies in orbit around other stars;

    (iv) a program of long-term observation, research, and analysis of the Earth's land, oceans, atmosphere and their interactions, including continual measurements from the Earth Observing System by 1998.

(d) In carrying out these activities, NASA will develop new and innovative space technologies and smaller more capable spacecraft to improve the performance and lower the cost of future space missions.

(4) In the conduct of these research and development programs, NASA will:

(a) Ensure safety on all space flight missions involving the Space Shuttle and the International Space Station.

(b) Emphasize flight programs that reduce mission costs and development times by implementing innovative procurement practices, validating new technologies and promoting partnerships between government, industry, and academia.

(c) Acquire spacecraft from the private sector unless, as determined by the NASA Administrator, development requires the unique technical capabilities of a NASA center.

(d) Make use of relevant private sector remote sensing capabilities, data, and information products and establish a demonstration program to purchase data products from the U.S. private sector.

(e) Use competition and peer review to select scientific investigators.

(f) Seek to privatize or commercialize its space communications operations no later than 2005.

[4] (g) Examine with DoD, NOAA and other appropriate federal agencies, the feasibility of consolidating ground facilities and data communications systems that cannot otherwise be provided by the private sector.

(5) The Department of Commerce (DoC), through the National Oceanic and Atmospheric Administration (NOAA), has the lead responsibility for managing Federal space-based civil operational Earth observations necessary to meet civil requirements. In this role, the DoC, in coordination with other appropriate agencies, will:

(a) acquire data, conduct research and analyses, and make required predictions about the Earth's environment;

(b) consolidate operational U.S. Government civil requirements for data products, and define and operate Earth observation systems in support of operational monitoring needs; and

(c) in accordance with current policy and Public Law 102–555 provide for the regulation and licensing of the operation of private sector remote sensing systems.

(6) The Department of the Interior, through the U.S. Geological Survey (USGS), will maintain a national archive of land remote sensing data and other surface data as appropriate, making such data available to U.S. Government and other users.

(7) The Department of Energy will maintain the necessary capability to support civil space missions, including research on space energy technologies and space radiation effects and safety.

National Security Space Guidelines

(1) The United States will conduct those space activities necessary for national security. These activities will be overseen by the Secretary of Defense and the Director of Central Intelligence (DCI) consistent with their respective responsibilities as set forth in the National Security Act of 1947, as amended, other applicable law, and Executive Order 12333. Other departments and agencies will assist as appropriate.

(2) Improving our ability to support military operations worldwide, monitor and respond to strategic military threats, and monitor arms control and non-proliferation agreements and activities are key priorities for national security space activities. The Secretary of Defense and DCI shall ensure that defense and intelligence space activities are closely coordinated; that space architectures are integrated to the maximum extent feasible; and will continue to modernize and improve their respective activities to collect against, and respond to, changing threats, environments and adversaries.

[5] (3) National security space activities shall contribute to U.S. national security by:
    (a) providing support for the United States' inherent right of self-defense and our defense commitments to allies and friends;
    (b) deterring, warning, and if necessary, defending against enemy attack;
    (c) assuring that hostile forces cannot prevent our own use of space;
    (d) countering, if necessary, space systems and services used for hostile purposes;
    (e) enhancing operations of U.S. and allied forces;
    (f) ensuring our ability to conduct military and intelligence space-related activities;
    (g) satisfying military and intelligence requirements during peace and crisis as well as through all levels of conflict;
    (h) supporting the activities of national policy makers, the intelligence community, the National Command Authorities, combatant commanders and the military services, other federal officials, and continuity of government operations.

(4) Critical capabilities necessary for executing space missions must be assured. This requirement will be considered and implemented at all stages of architecture and system planning, development, acquisition, operation, and support.

(5) The Department of Energy, in coordination with DoD, ACDA [the Arms Control and Disarmament Agency] and the DCI will carry out research on and development of technologies needed to effectively verify international agreements to control special nuclear materials and nuclear weapons.

(6) Defense Space Sector Guidelines:
    (a) DoD shall maintain the capability to execute the mission areas of space support, force enhancement, space control and force application.
    (b) In accordance with Executive Orders and applicable directives, DoD shall protect critical space-related technologies and mission aspects.

(c) DoD, as launch agent for both the defense and intelligence sectors, will maintain the capability to evolve and support those space transportation systems, infrastructure, and support activities necessary to meet national security requirements. DoD will be the lead agency for improvement and evolution of the current expendable launch vehicle fleet, including appropriate technology development.

[6] (d) DoD will pursue integrated satellite control and continue to enhance the robustness of its satellite control capability. DoD will coordinate with other departments and agencies, as appropriate, to foster the integration and interoperability of satellite control for all governmental space activities.

(e) The Secretary of Defense will establish DoD's specific requirements for military and national-level intelligence information.

(f) The Secretary of Defense, in concert with the DCI, and for the purpose of supporting operational military forces, may propose modifications or augmentations to intelligence space systems as necessary. The DoD may develop and operate space systems to support military operations in the event that intelligence space systems cannot provide the necessary intelligence support to the DoD.

(g) Consistent with treaty obligations, the United States will develop, operate and maintain space control capabilities to ensure freedom of action in space and, if directed, deny such freedom of action to adversaries. These capabilities may also be enhanced by diplomatic, legal or military measures to preclude an adversary's hostile use of space systems and services. The U.S. will maintain and modernize space surveillance and associated battle management command, control, communications, computers, and intelligence to effectively detect, track, categorize, monitor, and characterize threats to U.S. and friendly space systems and contribute to the protection of U.S. military activities.

(h) The United States will pursue a ballistic missile defense program to provide for: enhanced theater missile defense capability later this decade; a national missile defense deployment readiness program as a hedge against the emergence of a long-range ballistic missile threat to the United States; and an advanced technology program to provide options for improvements to planned and deployed defenses.

(7) Intelligence Space Sector Guidelines:

(a) The DCI shall ensure that the intelligence space sector provides timely information and data to support foreign, defense and economic policies; military operations; diplomatic activities; indications and warning; crisis management; and treaty verification, and that the sector performs research and development related to these functions.

(b) The DCI shall continue to develop and apply advanced technologies that respond to changes in the threat environment and support national intelligence priorities.

(c) The DCI shall work closely with the Secretary of Defense to improve the intelligence space sector's ability to support military operations worldwide.

[7] (d) The nature, the attributable collected information and the operational details of intelligence space activities will be classified. The DCI shall establish and implement policies to provide appropriate protection for such data, including provisions for the declassification and release of such information when the DCI deems that protection is no longer required.

(e) Collected information that cannot be attributed to space systems will be classified according to its content.

(f) These guidelines do not apply to imagery product, the protection of which is governed by Executive Order 12951.

(g) Strict security procedures will be maintained to ensure that public discussion of satellite reconnaissance by Executive Branch personnel and contractors is consistent with DCI guidance. Executive Branch personnel and contractors should refrain from acknowledging or releasing information regarding satellite reconnaissance until a security review has been made.

(h) The following facts are UNCLASSIFIED:

    (i) That the United States conducts satellite photoreconnaissance for peaceful purposes, including intelligence collection and monitoring arms control agreements.

    (ii) That satellite photoreconnaissance includes a near real-time capability and is used to provide defense-related information for indications and warning, and the planning and conduct of military operations.

    (iii) That satellite photoreconnaissance is used in the collection of mapping, charting, and geodetic data and such data is provided to authorized federal agencies.

    (iv) That satellite photoreconnaissance is used to collect mapping, charting and geodetic data to develop global geodetic and cartographic materials to support defense and other mapping-related activities.

    (v) That satellite photoreconnaissance can be used to collect scientific and environmental data and data on natural or man-made disasters, and such data can be disseminated to authorized federal agencies.

    (vi) That photoreconnaissance assets can be used to image the United States and its territories and possessions.

    (vii) That the U.S. conducts overhead signals intelligence collection.

[8]   (viii) That the U.S. conducts overhead measurement and signature intelligence collection.

    (ix) The existence of the National Reconnaissance Office (NRO) and the identification and official titles of its senior officials.

    All other details, facts and products of intelligence space activities are subject to appropriate classification and security controls as determined by the DCI.

(i) Changes to the space intelligence security policy set forth in the national space policy can be authorized only by the President.

Commercial Space Guidelines

(1) The fundamental goal of U.S. commercial space policy is to support and enhance U.S. economic competitiveness in space activities while protecting U.S. national security and foreign policy interests. Expanding U.S. commercial space activities will generate economic benefits for the Nation and provide the U.S. Government with an increasing range of space goods and services.

(2) U.S. Government agencies shall purchase commercially available space goods and services to the fullest extent feasible and shall not conduct activities with commercial applications that preclude or deter commercial space activities except for reasons of national security or public safety. A space good or service is "commercially available" if it is currently offered commercially, or if it could be supplied commercially in response to a government service procurement request. "Feasible" means that such goods or services meet mission requirements in a cost-effective manner.

(3) The United States will pursue its commercial space objectives without the use of direct Federal subsidies. Commercial Sector space activities shall be supervised or regulated only to the extent required by law, national security, international obligations and public safety.

(4) To stimulate private sector investment, ownership, and operation of space assets, the U.S. Government will facilitate stable and predictable U.S. commercial sector access to appropriate U.S. Government space-related hardware, facilities and data. The U.S. Government reserves the right to use such hardware, facilities and data on a priority basis to meet national security and critical civil sector requirements. Government Space Sectors shall:

    (a) Enter into appropriate cooperative agreements to encourage and advance private sector basic research, development, and operations while protecting the commercial value of the intellectual property developed.

[9] (b) Identify and propose appropriate amendments to or the elimination of applicable portions of United States laws and regulations that unnecessarily impede commercial space sector activities.

    (c) Consistent with national security, provide for the timely transfer of government-developed space technology to the private sector in such a manner as to protect its commercial value, including retention of technical data rights by the private sector.

    (d) To the extent feasible, pursue innovative methods for procurement of space products and services.

(5) Free and fair trade in commercial space launch services is a goal of the United States. In support of this goal, the United States will implement, at the expiration of current space launch agreements, a strategy for transitioning from negotiated trade in launch services towards a trade environment characterized by the free and open interaction of market economies. The U.S. Trade Representative, in coordination with the Office of Science and Technology Policy and the National Economic Council, will develop a strategy to guide this implementation.

(6) Consistent with Executive Order 12046 and applicable statutes, U.S. Government agencies and departments will ensure that U.S. Government telecommunications policies support a competitive international environment for space-based telecommunications.

Intersector Guidelines

The following paragraphs identify priority intersector guidance to support major United States space policy objectives.

(1) International Cooperation

The United States will pursue and conduct international cooperative space-related activities that achieve scientific, foreign policy, economic, or national security benefits for the nation. International agreements related to space activities shall be subject to normal interagency coordination procedures, consistent with applicable laws and regulations. United States cooperation in international civil space activities will:

    (a) Promote equitable cost-sharing and yield benefits to the United States by increasing access to foreign scientific and technological data and expertise and foreign research and development facilities;

    (b) Enhance relations with U.S. allies and Russia while supporting initiatives with other states of the former Soviet Union and emerging spacefaring nations;

    (c) Support U.S. technology transfer and nonproliferation objectives;

[10] (d) Create new opportunities for U.S. commercial space activities; and

  (e) Protect the commercial value of intellectual property developed with Federal support and ensure that technology transfers resulting from cooperation do not undermine U.S. competitiveness and national security.

  (f) In support of these objectives:

     (i) NASA and the Department of State will negotiate changes in the existing legal framework for International Space Station cooperation to include Russia in the program along with the United States, Europe, Japan, and Canada; and

     (ii) NASA, in coordination with concerned U.S. Government agencies, will explore with foreign space agencies and international organizations the possible adoption of international standards for the interoperability of civil research spacecraft communication and control facilities.

(2) Space Transportation

  (a) Assuring reliable and affordable access to space through U.S. space transportation capabilities is fundamental to achieving national space policy goals. Therefore, the United States will:

     (i) Balance efforts to modernize existing space transportation capabilities with the need to invest in the development of improved future capabilities;

     (ii) Maintain a strong transportation capability and technology base to meet national needs for space transport of personnel and payloads;

     (iii) Promote reduction in the cost of current space transportation systems while improving their reliability, operability, responsiveness, and safety;

     (iv) Foster technology development and demonstration to support a future decision on the development of next generation reusable space transportation systems that greatly reduce the cost of access to space;

     (v) Encourage, to the fullest extent feasible, the cost-effective use of commercially provided U.S. products and services that meet mission requirements; and

     (vi) Foster the international competitiveness of the U.S. commercial space transportation industry, actively considering commercial needs and [11] factoring them into decisions on improvements to launch facilities and vehicles.

  (b) The Department of Transportation (DoT) is the lead agency within the Federal government for regulatory guidance pertaining to commercial space transportation activities, as set forth in 49 U.S.C. § 701, et seq., and Executive Order 12465. The U.S. Government encourages and will facilitate U.S. private sector and state and local government space launch and recovery activities.

  (c) All activities related to space transportation undertaken by U.S. agencies and departments will be consistent with PDD/NSTC-4

(3) Space-based Earth Observation

  (a) The United States requires a continuing capability for space-based Earth observation to provide information useful for protecting public health, safety, and national security. Such a capability contributes to economic growth and stimulates educational, scientific and technological advancement. The U.S. Government will:

     (i) Continue to develop and operate space-based Earth observing systems, including satellites, instruments, data management and dissemination activities;

     (ii) Continue research and development of advanced space-based Earth observation technologies to improve the quality and reduce the costs of Earth observations;

      (iii) Support the development of U.S. commercial Earth observation capabilities by:

- pursuing technology development programs, including partnerships with industry;
- licensing the operation and, as appropriate, the export of private Earth observation systems and technologies, consistent with existing policy;
- providing U.S. Government civil data to commercial firms on a non-discriminatory basis to foster the growth of the "value-added" data enhancement industry; and
- making use, as appropriate, of relevant private sector capabilities, data, and information products in implementing this policy.

[12]   (iv) Produce and archive long-term environmental data sets.

  (b) The U.S. Government will continue to use Earth observation systems to collect environmental data and provide all U.S. Government civil environmental data and data products consistent with OMB Circular A-130, applicable statute and guidelines contained in this directive.

  (c) The U. S. Government will seek mutually beneficial cooperation with U.S. commercial and other national and international Earth observation system developers and operators, to:

    (i) define an integrated global observing strategy for civil applications;

    (ii) develop U.S. Government civil Earth observing systems in coordination with other national and international systems to ensure the efficient collection and dissemination of the widest possible set of environmental measurements;

    (iii) obtain Earth observation data from non-U.S. sources, and seek to make such data available to users consistent with OMB Circular A-130, national security requirements, and commercial sector guidance contained in the national space policy; and

    (iv) support, as appropriate, the public, non-discriminatory direct read-out of data from Federal civil systems.

  (d) The U.S. Government space sectors will coordinate, and where feasible, seek to consolidate Earth observation activities to reduce overlaps in development, measurements, information processing, and archiving where cost-effective and consistent with U.S. space goals.

    (i) In accordance with PDD/NSTC-2, DoC/NOAA, DoD, and NASA shall establish a single, converged, National Polar-Orbiting Environmental Satellite System (NPOESS) to satisfy civil and national security requirements.

    (ii) NASA, DoC/NOAA, DoD, the Intelligence Community, and DoE shall work together to identify, develop, demonstrate, and transition advanced technologies to U.S. Earth observation satellite systems.

    (iii) In accordance with PDD/NSTC-3, NASA, DoC/NOAA, and DoI/USGS shall develop and operate an ongoing program to measure the Earth's land surface from space and ensure the continuity of the Landsat-type data set.

[13]   (iv) Consistent with national security, the U.S. Government space sectors shall continue to identify national security products and services that can contribute to global change research and civil environmental monitoring, and seek to make technology, products and services available to civil agencies for such uses. Both unclassified and, as appropriate, classified data from national security programs will be provided through established mechanisms.

(4) Nonproliferation, Export Controls, and Technology Transfer

  (a) The MTCR [Missile Technology Control Regime] Guidelines are not designed to impede national space programs or international cooperation in such programs as long as such programs could not contribute to delivery systems for weapons of mass destruction. Consistent with U.S. nonproliferation policy, the United States will continue to oppose missile programs of proliferation concern, and will exercise particular restraint in missile-related cooperation. The United States will continue to retain a strong presumption of denial against exports of complete space launch vehicles or other MTCR Category I components.

(b) The United States will maintain its general policy of not supporting the development or acquisition of space launch vehicle systems in non-MTCR states.

(c) For MTCR countries we will not encourage new space launch vehicle programs which raise questions from a proliferation and economic standpoint. The United States will, however, consider exports of MTCR-controlled items to MTCR countries. Additional safeguard measures could also be considered for such exports, where appropriate. Any exports would remain subject to the non-transfer provisions of the INF [International Nuclear Forces] and START treaties.

(d) The United States will work to stem the flow of advanced space technology to unauthorized destinations. Executive departments and agencies will be fully responsible for protecting against adverse technology transfer in the conduct of their programs.

(e) In entering into space-related technology development and transfer agreements with other countries, Executive Departments and Agencies will take into consideration whether such countries practice and encourage free and fair trade in commercial space activities.

(5) Arms Control

The United States will consider and, as appropriate, formulate policy positions on arms control and related measures governing activities in space, and will conclude agreements on such measures only if they are equitable, effectively verifiable, and enhance the security [14] of the United States and our allies. The Arms Control and Disarmament Agency (ACDA) is the principal agency within the Federal government for arms control matters. ACDA, in coordination with the DoD, DCI, State, DoE, and other appropriate Federal agencies, will identify arms control issues and opportunities related to space activities and examine concepts for measures that support national security objectives.

(6) Space Nuclear Power

The Department of Energy will maintain the necessary capability to support space missions which may require the use of space nuclear power systems. U.S. Government agency proposals for international cooperation involving space nuclear power systems are subject to normal interagency review procedures. Space nuclear reactors will not be used in Earth orbit without specific approval by the President or his designee. Such requests for approval will take into account public safety, economic considerations, international treaty obligations, and U.S. national security and foreign policy interests. The Office of Science and Technology Policy, in coordination with the NSC staff, will examine the existing approval process, including measures to address possible commercial use of space nuclear systems

(7) Space Debris

(a) The United States will seek to minimize the creation of space debris. NASA, the Intelligence Community, and the DoD, in cooperation with the private sector, will develop design guidelines for future government procurement of spacecraft, launch vehicles, and services. The design and operation of space tests, experiments and systems, will minimize or reduce accumulation of space debris consistent with mission requirements and cost effectiveness.

(b) It is in the interest of the U.S. Government to ensure that space debris minimization practices are applied by other spacefaring nations and international organizations. The U.S. Government will take a leadership role in international fora to adopt policies and practices aimed at debris minimization and will cooperate internationally in the exchange of information on debris research and the identification of debris mitigation options.

(8) Government Pricing

The price charged for the use of U.S. Government facilities, equipment, and service, will be based on the following principles:

(a) Prices charged to U.S. private sector, state and local government space activities for the use of U.S. Government facilities, equipment, and services will be based on costs consistent with Federal guidelines, applicable statutes and the commercial guidelines contained within the policy. The U.S. Government will not seek to [15] recover design and development costs or investments associated with any existing facilities or new facilities required to meet U.S. Government needs and to which the U.S. Government retains title.

(b) Consistent with mission requirements, NASA and DoD will seek to use consistent pricing practices for facilities, equipment, and services.

(c) Tooling, equipment, and residual hardware on hand at the completion of U.S. Government programs will be priced and disposed of on a basis that is in the best overall interest of the United States while not precluding or deterring the continuing development of the U.S. commercial space sector.

Document III-15

Document title: "Commercial Space Industry in the Year 2000: A Market Forecast," The Center for Space Policy (CSP), Inc., Cambridge, Massachusetts, June 1985 (reprinted with permission).

Source: CSP Associates, Inc., Cambridge, Massachusetts.

In 1984, the Center for Space Policy, a small analytic group, projected a private space market of $60 billion by 2000. This projection was influential in promoting the potential of space as a commercial enterprise and in developing support for the space station program. This document is a revision of the 1984 projection, widely criticized for its optimism. The 1985 revisions are more detailed and reflect the broader range of space commercial markets rather than point estimates.

Commercial Space Industry in the Year 2000
A Market Forecast

[1]   **I. EXECUTIVE SUMMARY**

A. INTRODUCTION

One of the most remarkable facts of our industrial society is the accelerating rate of progress. Our scientific knowledge and understanding of the universe doubles now in less than a generation, and we have had more progress in the past fifty years than in the preceding millennium in terms of practical impact. It is interesting to note that economists predict that a substantial percentage of the jobs that will exist in the year 2010 (i.e. 25 years from now) do not exist now; rather, they will be created by the dynamism of new industries made possible by technologies which are only beginning to emerge from laboratories today.

Our nation's brief history in space is an unparalleled example of this trend. While the United States had been investigating space and began to develop rudimentary space technologies after the close of World War II, our civilian space programs did not begin in earnest until the creation of NASA in 1958. In a little more than a quarter of a century, we have come to understand the requirements of living and working in space.

Importantly, the way we look at the heavens is changing. For the first two decades, NASA's space programs had three primary objectives: the advancement of scientific knowledge about our universe; the development of an engineering capability which enabled us to conduct manned and unmanned operations in space; and the pursuit of heroic feats of exploration reminiscent of those of the maritime explorers of centuries ago. Now, we are beginning to look at space as a place of enterprise. The Administration and NASA have both come out strongly in support of commercial investment in space, and have expended considerable amounts of talent in order to understand what the Government can do to encourage the private sector to invest in a new industrial frontier.

This report focuses on the year 2000, and projects the commercial revenues for six space industries: satellite communications, materials processing in space (MPS), remote sensing, on-orbit services, space transportation, and ground–based support. Revenues are for U.S. industry only, and are stated in 1985 dollars. The projections do not take into account revenues accruing to the private sector through the R&D expenditures of the government (e.g. the NASA space station program is not included). However, government purchases of commercially developed space hardware and products are included (for example, government purchases of commercial upper stages). High and low scenarios have been generated for each industry. Under the low scenario, extremely conservative assumptions have been used in defining [2] the product or service and its market. Under the high scenarios, more optimistic assumptions have been used.

The satellite communications, MPS and remote sensing markets can be considered "Applications Markets" in that some aspect of space is critical to the provision of service. Satellite communications and remote sensing profit from the vantage point afforded by space; MPS utilizes other physical attributes of space (most notably microgravity) to produce materials which cannot be made on Earth. Space transportation, on-orbit services, and ground support, on the other hand, can be considered "Infrastructure Markets." They are not end products in themselves, but are necessary for the provision of other products.

B. SATELLITE COMMUNICATIONS

Satellite communications is the oldest and most mature commercial space industry. The three basic types of satellite services that have commercial applications are known as "fixed satellite service" (FSS), "broadcast satellite service" (BSS) and "mobile satellite service" (MSS). Total revenues projected for this industry are $8.8 billion under CSP's low market scenario, and $15.3 billion under the high scenario.

FSS provides a common carrier link for point-to-point and point-to-multipoint transmission. The primary FSS markets are the transmission of voice, video and data. The last two markets, addressing corporate needs for the development of private networks, are especially expected to grow to a $5.0–6.8 billion level by the turn of the century.

Direct broadcast satellite (DBS) service has had a spotty record in the U.S. Nevertheless, long term market prospects look tremendous for the industry (large up-front capital requirements, and the inability of DBS firms to secure affordable programming have been the major obstacles to date). By the 1990s, DBS services should be available; by the year 2000, annual revenues could be $2.6–6.6 billion.

Mobile satellite communications services will take advantage of the large footprint of a satellite to make thin-route mobile communications economically feasible. The market consists of two major segments: limited alphanumeric message services and full voice and data transmission. Annual revenues are projected to rise to $.8–1.5 billion by the year 2000.

CSP anticipates a domestic demand for approximately five new spacecraft per year by 2000. FSS satellites will comprise the largest component of market demand for spacecraft.

[3] C. MATERIALS PROCESSING IN SPACE

Commercial materials processing in space (MPS) will allow U.S. companies to take advantage of the properties of the space environment. There will be two general types of MPS activity: basic materials research, and the development and production of new products or processes that are possible only in space. This study includes only those revenues accruing from the second activity.

Basic economics limits severely the number of candidates for space processing to those which can justify high production costs, which include an estimated transportation cost of $10,000 per pound. In the next fifteen years, only three kinds of materials are likely to meet this threshold value while also generating sufficient demand to justify production. These materials include pharmaceuticals, semiconductor crystals, and halide optical fibers.

There will be significant research in numerous other materials fields, notably organic crystals, ceramics, and alloys. The knowledge gained from space research will be applied to terrestrial production techniques, or demand for products in these fields will be so limited that individual markets will be small (under $50 million). It must be noted that the development of a strong knowledge base is essential for the continued growth of commercial MPS revenues in the long term; current projections based on what is now known will likely prove conservative as the rate and quality of space-based materials research improves.

A key driver in the development of MPS markets will be the availability of appropriate research facilities on orbit. Present Shuttle facilities in the mid-deck and the payload bay are inadequate, especially for tasks requiring a high level of human interaction. The space station will alleviate this situation, but will be unavailable to researchers before 1995.

MPS is expected to generate revenues of $2.6 billion to $17.9 billion in the year 2000, based on 6 to 30 products. Most of these revenues will come from pharmaceutical products ($2.0 billion to $14. 9 billion); much of the rest will come from gallium arsenide crystal production ($500 million to $1 billion). These crystals will be employed primarily in defense applications. The balance of MPS products will come from other semiconductors ($1 billion in the high scenario) and halide optical fiber ($100 million to $1 billion).

[4] **D. REMOTE SENSING**

Remote sensing satellites provide data from space concerning the earth's surface and atmosphere. To date, satellites have been operated by the government as public goods. The government is still the largest user and supplier of Landsat data and services. The imminent launch of the first SPOT satellite, a foreign private system, and the current attempt to shift the American Landsat to the private sector will alter this situation dramatically.

The success or failure of the Landsat transfer will be the greatest single factor in the development of the American remote-sensing industry for the rest of this century. Delays in the process have already ensured that an interruption in service lasting at least two years will follow the expected 1987 failure of the current operating satellite, Landsat 5. This data gap could allow the SPOT system to preempt opportunities for a future American system.

The development of a "value-added" industry to process satellite data depends heavily on the existence of an American system on orbit. Without access to affordable data during these formative years, the value-added industry cannot get on its feet. The opportunities for private satellite raw data companies would then be limited severely: the demand for raw data will not support a private remote sensing satellite venture.

Total revenues for satellite remote sensing in the United States are expected to be between $500 million and $2.5 billion in the year 2000. In our low scenario, the only applications likely to be continued are those which require global coverage (such as the global crop assessment service). All U.S. revenues would be obtained from value-added services: $380 million is forecast for expenditures by the petroleum industry, while the remainder will derive from a number of smaller users, including paper and lumber companies, grain traders, commodity brokers, and some very large agricultural associations. Federal, state and local governments will continue to use Landsat data, processed by private contractors, for those surveying purposes where cost-effective alternatives are not available.

In our high scenario, a private remote sensing satellite is in operation, as well as private sensors on the space station's polar platform. The American value-added market also finds sufficient support for continued development. Computer and information technology improvements, increasing prices for non-renewable resources, and increasing competition in agricultural and forest-related industries support this development. Of the $2.5 billion generated in this scenario, $500 million accrues from raw data sales, while the remaining $2.0 billion are derived from value-added [5] services. Of the total revenues, $1 billion derives from non-renewable resource industries, $1.35 billion from renewable resource industries, and $150 million from other sources.

E. ON-ORBIT SERVICES

A large market opportunity will materialize in the coming years for companies providing on-orbit workspace and servicing. Several companies currently offer forerunners of the on-orbit hardware that can be expected to serve the MPS industry at the turn of the century, and NASA is developing servicing concepts and hardware that should be applicable to commercial use. Total yearly industry revenues are projected at $.6–2.8 billion.

Commercial workspace will be the location for all commercial production activities (R&D is assumed to take place aboard the NASA space station). Free flyers and Man-

Tended Facilities (MTFs) each offer specific technical advantages, depending upon the complexity of the process. Using the MPS market scenarios, requirements for on-orbit workspace were derived. Under the low scenario, five free flying platforms and two MTFs satisfy commercial demand; under the high scenario, these numbers climb to nineteen free flyers and twelve MTFs. The annual revenues earned by these facilities should be $.6–2.5 billion per year.

Satellite servicing revenues are comprised of commercial OMV [Orbital Maneuvering Vehicle] servicings of free flyers (MTFs must be serviced by the Shuttle; these revenues are included under the section on space transportation). Under the low scenario, no commercial OMV operations are predicted; hence there are no commercial revenues. Under the high scenario, seventy annual servicing missions generate revenues of almost $.3 billion.

F. SPACE TRANSPORTATION

Space transportation systems (STS) are required to support all space activities: launching payloads into their operational orbits, retrieving payloads and returning them to Earth, and performance of experiments and manufacturing activities on orbit. For the remainder of the century, the majority of space transportation requirements will focus on three orbital destinations: Low Earth Orbit (LEO—Shuttle "parking orbit"), geostationary orbit (GEO), and LEO polar orbits.

The space transportation systems available in the Western world to reach these orbits are the U.S. Space Shuttle, Expendable Launch Vehicles (ELVs—which include Europe's Ariane launch vehicle) and upper stages for use in conjunction with the Shuttle and ELVs. [6] In the low scenario, CSP assumes that the STS remains a government-owned and managed system, and hence no revenues accrue to commercial entities. Total demand of 12.8 STS equivalent launches in the year 2000 provides no "overflow" demand for the ELV industry and thus there is no inclusion of commercial ELV revenues. Upper stages are the only source of commercial revenues in the low scenario. Total revenues are expected to top $.2 billion.

The high scenario assumes a fleet of five Shuttle orbiters (each operating at six flights annually), with two operated by commercial entities to service man-tended facilities in LEO. The cost of an orbiter to the operator is a straight–line depreciation of the $1.7B purchase cost (in $1982) over 100 flights. An average of capital costs over remaining flights plus a gross margin of 30 percent brings the launch price to $76.2M. The strong demand for LEO transportation services generated by the MPS industry (especially for MTFs) is expected to use all of the available STS capacity; a strong market exists for ELVs to take up the overload. Under the high scenario, commercial orbiter revenues and ELV revenues reach over $1 billion each, and commercial upper stage revenues are projected at almost $300 million.

G. GROUND BASED SUPPORT SERVICES

Space operations require support services that are located on earth. The major services include: payload processing, earth station equipment manufacture, and space insurance. Each of these could become a significant market in its own right by the year 2000.

Payload processing must be performed on all commercial payloads prior to launch. Payloads include communications satellites, MPS free-flyers and resupply modules, and commercial OMVs and MTFs. By the year 2000, it is assumed that all commercial payloads will be processed by a commercial firm(s). The market for these services was calculated by evaluating the processing requirements for payloads generated in Chapters III–VI of the report. Under the low scenario, these revenues total $.03 billion; in our high scenario, the revenues rise to $.11 billion.

The earth station market includes transmitting and receiving antennas for satellite services, tracking and control facilities, and other ground equipment used with satellites. CSP expects that the principal sources of revenues will be sales of fixed and broadcast service earth station equipment. Revenues for this market are dominated by FSS services, although DBS and MSS equipment are expected to generate large revenues as well. Revenue totals are $3.7 billion in the low scenario, and $8.6 billion in the high scenario. [7] The insurance industry is a key element in the development of commercial space activity. Major forms of insurance to be offered include launch, operational lifetime, and liability. New types of insurance will need to be developed to support LEO operations for MPS and servicing. Industry revenues will depend directly on the level of development of other commercial space activities. Total insurance revenues are forecast at $.33 billion in the low scenario, and $1.64 billion in the high scenario.

Total revenues for the ground based support services are $4.1 billion in the low scenario and $10.4 billion in the high scenario.

H. SUMMARY

Table I-1 summarizes the revenue projections forecast in this report. Under the conservative assumptions used to generate the low scenarios, revenues in the year 2000 should be $16.8 billion; using the more optimistic conditions of the high scenario, revenues rise to $51.3 billion. As noted elsewhere in this report, the key determinants affecting the actual outcome are the continued commitment to developing the commercial space infrastructure, and the development of an MPS R&D base which will allow broad-scale industrial activity.

Table I-1
Commercial Space Revenues in the Year 2000

Activity	Low Scenario	High Scenario
Communications	$8.8B	$15.3B
MPS	2.6B	17.9B
Remote Sensing	5B	2.5B
On-Orbit Services	6B	2.8B
Space Transportation	2B	2.4B
Ground Based Services	4.1B	10.4B
Total Revenues	$1.6.8B	$51.3B . . .

[9] B. COMPARISON WITH EARLIER STUDIES

The market projections presented in this study are a revision of an earlier internal study conducted by CSP in January, 1984. The current figures (see below) are approximately [10] twenty percent lower than those presented a year ago. Two factors are responsible for the change. First, the earlier study assumed that the space station would be on orbit and functional in 1992, as predicated in NASA's original timeline. However, due to a slowdown in funding in the FY 1985 and FY 1986 budgets, the target date for space station has slipped until 1993. Reviewing historical evidence with the Shuttle program, it is not unreasonable to expect that the station will be delayed a year longer (i.e. 1994). This delay has a negative impact because of the critical role the space station will have on the amount of R&D activity which can be performed, and the economics of working on orbit. Much of the growth which had been projected for the late 1990s will still occur, but not until after the turn of the century.

The second factor has been a stronger methodological approach. As any statistician will note, it is difficult, if not impossible, to predict markets in the year 2000 for products and services which do not currently exist. Without an historical database, trend analysis is meaningless, and concepts such as focus groups are of limited value when talking of entirely new concepts that may not be on the market for a decade.

As a result, projection of commercial space markets must rely more upon developing a cohesive set of common sense assumptions about the basic economics of supply and demand. Earlier projections have been predicated on assumptions which tend to focus on the "supply side" of the market equation. hat is, many have made the assumption that once a product or service becomes possible, it will become profitable. This is a "technology push" approach to market projection which has often led to error.

[11] In addition to the supply side, the current analysis also addresses the issue of demand. It looks not only at what can be produced in space, but what advantage the space product has to the buyer as compared with other alternatives. In addition to looking at the relative advantages of space products, we have examined analogous situations (the introduction of technologically sophisticated, relatively high cost products into new markets) to determine what has happened in the past, and often have used these as models.

It should be noted that these projections are only as valid as their underlying assumptions. CSP has endeavored to use conservative assumptions in defining its scenarios, and has sought corroboration of these assumptions through lengthy interviews of experts in the aerospace industry. In short, CSP has used what might be termed a "modified-Delphi" approach.[1]

Actual revenues might be substantially lower than expected, due to an unforeseeable and fundamental change. For example, if new medical evidence proved that men cannot work in space without serious health drawbacks, the scope of feasible endeavors would be severely circumscribed. On the other hand, it should be noted that the revenue projections could be substantially lower than reality. Many of the activities contemplated in this study involve new materials that would be used in "high-end" applications. Thus, the situation might be analogous to projecting the future for the manufacture of silicon crystals in the early 1950s (i.e. [12] before the development of the transistor and the takeoff of the computer industry). Revolutionary products may create a substantial demand for space products that are not included in this study. Therefore, as with all long term projections, this study should be used as an indicator of potential, and not a prognosticator of fact.

C. GENERAL APPROACH

This report focuses on the year 2000, and projects the commercial revenues for six space industries: satellite communications, materials processing in space (MPS), remote sensing, on-orbit services, space transportation, and ground–based support. Revenues are for U.S. industry only, and are stated in 1985 dollars. The projections do not take into account revenues accruing to the private sector through the R&D expenditures of the government (e.g. the NASA space station program is not included). However, government pur-

---

1.   The Delphi method uses a group of experts in a field, who then develop their individual scenarios of the future. The experts then critique each other in order to develop a set of agreed-upon assumptions that serve as the basis for analysis of future trends. Historically, this method has proven to be the best means of predicting the change in situations where there are numerous variables and where data are scarce.

chases of commercially developed space hardware and products are included (for example, government purchases of commercial upper stages such as the PAM-D and the TOS). High and low scenarios have been generated for each industry. Under the low scenario, extremely conservative assumptions have been used in defining the product or service and its market. Under the high scenarios, more optimistic assumptions have been used.

The satellite communications, MPS and remote sensing markets can be considered "Applications Markets," in that some aspect of space is critical to the provision of service. Satellite communications and remote sensing profit from the vantage point afforded by space; MPS utilizes other physical attributes of space (most notably microgravity) to produce materials which cannot be made on Earth. Space transportation, on-orbit services, and ground support, on the [13] other hand, can be considered "Infrastructure Markets." They are not end products in themselves, but are necessary for the provision of other products.

It immediately becomes clear that the markets are inextricably linked. The demand for the applications markets is generated "externally" by individuals, companies, governments, and other organizations here on Earth. This demand is clearly influenced by the functional alternatives that can be produced on Earth. For example, the demand for satellite communications is dependent upon the economics and capabilities of terrestrial communications media. The demand for space infrastructure is a function of this applications demand.

Working in the opposite direction, the supply (and cost) of space infrastructure products and services is the dominant factor in determining which applications products and services can be produced economically (i.e. at a cost that the market will support, and which will earn the firm a minimum level of profit). In order to develop overall consistency within this report, the high and low scenarios for the applications markets are used in the baselines for determining the infrastructure markets.

Document III-16

Document title: William M. Brown and Herman Kahn, "Long-Term Prospects for Developments in Space (A Scenario Approach)," Hudson Institute, Inc., Croton-on-Hudson, New York, Contract NASW-2924, October 30, 1977, pp. 257–274.

Source: NASA Historical Reference Collection, NASA History Office, NASA Headquarters, Washington, D.C.

This study was very unusual for NASA. It developed several scenarios detailing possible long-term trends in space R&D, technology, defense, and environmental and economic development. These scenarios included pessimistic and optimistic growth and opportunities in space. Space is seen as a resource and a place for economic enterprise. The study did not detail economic benefits, but did make economic issues central to long-term space financing, exploration, and development. Few other studies of the mid-1970s were as far reaching as this one in vision.

———————

Long-Term Prospects for Developments in Space
(A Scenario Approach)

By

William M. Brown and Herman Kahn . . .

[257]   Chapter VII
                  REVIEW AND ASSESSMENT

A. Images of the Future

Let us first review briefly what has been attempted in the first six chapters. The first chapter offers a typology of various space scenarios and of various themes for constructing scenarios. The point is made that such scenarios have many uses, and that from NASA's viewpoint an important, if somewhat neglected, one is the systematic formulation and dissemination of appropriate images of the future. Hopefully these images would be realistically developed and become valuable to policymakers.

We believe that NASA should try, in a low-keyed manner, to formulate and promulgate a concept of future space development as part of the manifest destiny of humanity, and as an obvious next phase in an historical process which started in the 15th century with the age of exploration and which has led to today's modern world.

In the 19th century many Americans overtly believed in manifest destiny, a concept which encouraged the opening up of the West and extended this country to the Pacific. Through our scenarios we did find that space, to a rather remarkable degree, was likely to play roles similar to those which the frontier played in America's past. According to some historians, such as Frederick J. Turner, much of the character of American life—the egalitarianism, the feelings of independence and competence, the sense of openness and unlimited vistas, the upward mobility, and a deep belief in democracy[—]seem to have been dependent on, or strongly influenced by, the existence of a frontier. We believe that this characteristic of our past [258] may well be continued—possibly in a modified or weaker form—through the exploration, development and exploitation of space.

Whether the analogy is valid or not, an accepted *image* of the future can give rise to expectations that could materialize into real space projects. We also argued that for such images to have the greatest near-term impact in the U.S. they should emphasize the practical uses of space—i.e., its scientific and economic values—and should treat its important psychological, political, social, and cultural consequences as by-products.

We believe that basic images of the future such as we have presented in this report are generally unavailable in America, or elsewhere. It is clear there is a large and active group of science fiction fans and it is clear that publicists have been very instrumental in spreading particular concepts (such as Professor O'Neill, for space colonies). But much more can be done. Many potential space activities, even if they are unduly optimistic or exaggerated (for example, as some critics believe Professor O'Neill's estimates to be), are still useful as part of a social process. If supported by NASA they should be properly formulated and labeled. NASA should also furnish long-range estimates and images of the future which are more or less consonant with its official positions; these can be quite exciting and still be plausible, or even conservative, within NASA.

The authors believe that a basis exists for a popular but serious book that will reflect much of the material in this report. We believe our activity has been a very useful one, even though that is clearly a self-serving remark. However, we would not have entered this project unless we felt that it was useful from a broad national perspective. Enough is now happening in space to guarantee a moderate level of future activity. [259] This means that whatever unexpected treasures are yet to be found have some reasonable probability of being discovered even in a pessimistic context. We should add that the spirit and need of scientific inquiry, and the spirit and need for exploration, will remain as permanent forces which create varying pressures for some kind of activity almost everywhere—and in most of our scenarios these needs and pressures increase—if not in the United States then elsewhere.

B. Earth-Centered Perspective

In Chapter II we try to visualize the coming economic role of space developments in an earth-centered perspective. Our view lies between that of the more extreme space enthusiasts who feel that society's problems on earth are basically intractable and should not be allowed to hinder the future of space, and those who conceive of the space potential as very limited, and often as an activity to enthrall the young or the technostructure—and thus often a place for expensive, sterile, dangerous or foolish exploits.

We first offer evidence that the basic physical problems relating to the world's future needs can, in principle, be solved without recourse to outer space, that the earth has more than enough resources to supply an adequate standard of living for all. On the other hand it seems quite clear that cis-lunar space and possibly the rest of the solar system could turn out to be extraordinarily important in an economic and technological sense. This outcome appears to follow from just the current reasonably projected potential in space—that is, without having to conjure up unforeseeable great breakthroughs. Of course it also seems to be reasonable [260] to expect that future space development will yield some equivalents of Middle East oil or Klondike gold—that is, vast treasures which have not yet been dreamed of. Thus, it is almost certain that space exploration will lead to great benefits, and possibly to an extraordinary economic and technological impact.

In our basic Surprise-Free Earth-Centered Scenario we concluded that good long-term prospects existed for technological solutions to current concerns about the adequacy of the world's physical resources—although social and political problems could—and probably will—create many difficulties in applying such solutions. When potential space developments are added to the above scenario the outlook for the required solutions is further brightened. That is, over time space technology and spinoffs from it will certainly contribute to these solutions—perhaps enormously. With our necessarily poor vision into the future we can still list a few general ways in which space activities are likely to contribute. In each category below we include spinoffs and serendipities, since they are often the most productive, although intrinsically obscure, avenues:

Energy:
    1) Space-based electric power stations
    2) LANDSAT information for oil and gas exploration
    3) Spinoffs and serendipities

Materials:
    1) Superior materials from unique processing capabilities in space industries
    2) Lunar and asteroidal sources of minerals
    3) LANDSAT assistance in mineral exploration on earth
    4) Spinoffs and serendipities

Food & Water:
    1) Improving weather forecasts for days, weeks, months, and possibly over longer intervals
    2) LANDSAT information on crops, disease, insects, water, etc.
    3) Spinoffs and serendipities

[261] Environment:
1) Monitoring pollutants by satellites
2) LANDSAT information on environment and land use
3) Processing in space (e.g., nuclear power or other radioactive processes)
4) Spinoffs and serendipities

The economic and technological potential of space leads us "paradoxically" to conclude that the near-term "soft," or socio-psychological, effects may equal or outweigh the "hard" returns. As an analogous example, it might seem wise for the United States to devote, say, 1 percent of its GNP to building "pyramids" and "cathedrals" in order to improve public morale and national unity. However, a direct attempt toward this end would almost certainly be doomed to failure. In our culture it is also vital for most "grandeur-creating" projects to be economically sensible—otherwise the average citizen will reject them. No great national interest exists in "climbing mountains because they are there." Most Americans have to feel that a practical, scientific, military, economic or other purpose is served in "climbing mountains" before they will support and take pride in such activities. Isolated events may be exciting—and create temporary heroes—but a deep lasting pride and a solid sense of achievement usually require practical projects.

So many tangible economic and scientific opportunities do exist in outer space that the country can afford to pursue them with intensity—and also reap the important psychological, political, social and cultural benefits as "by-products" of the main effort. In a cost-benefit analysis the space projects cannot be given explicit credit for such "by-products," but it can note the potential which exists. It should be made clear that these "by-products" could be as valuable as the more tangible objectives, including any windfalls. That would be our judgment, at least for the balance of this century.

[262] C. <u>Space Technology</u>

Chapter III explores some of the technology expected to be associated with space in the near, medium and long term. Some readers may find this chapter exciting because the technological possibilities portrayed are greater than many relatively knowledgeable groups currently seem to understand. This chapter indicates that a basic change in the character of our practical activities in space is likely to occur by the late '80s or early '90s. As knowledgeable NASA personnel and other space-oriented professionals know, up until now space systems have attempted to keep the large, complex, expensive equipment on the ground, where possible, and place the smaller or cheaper equipment into space. (The present situation may be compared to the use of river ferries which are severely limited in the loads they can transport.) In ten or fifteen years, we expect to find many new systems which deploy the large complex equipment in space and keep the small inexpensive, but numerous, parts of the system on the ground. (To continue the analogy this change would be similar to the replacement of most river ferries with modern bridges. That change was basic and effective.)

Although Chapter III focuses strongly on technology, it also attempts to indicate that future developments in space, especially in terms of *manned* activities, are probably dependent more on a number of imponderables other than successful technology. For example, the personal health and safety of a space traveler—or tourist—is potentially crucial to many activities. Although safety is more or less a direct consequence of technology, future health in space is still a mystery which may involve risks that are subject to straightforward technological solution—or at least not for quite a long [263] time. At the moment there appears to be some useful information from prior life science studies, but these are reliable mainly for relatively short-term exposures to the space environment.

Space *tourism,* which appears in every scenario as a major new growth industry, should not be taken as quite that certain. That is, even assuming that problems of health, safety and cost do not become intrinsic deterrents, it is possible at least to conceive of other developments that might hamper that potential industry. For example, the *excitement* might vanish after the first few years. After all, will there be enough important experiences which an expensive tour can bring that the advanced electronic systems anticipated for the 21st century could not? Many of the visual experiences in space might be *better* perceived through electronics, and probably a lot more comfortably. Still the experience of weightlessness, or the knowledge of the new realities, such as being suspended in space, perhaps 25,000 miles above the earth, or a chance to walk on the moon, might prove to be priceless. Certainly electronics has not been a sufficient substitute to date. These uncertainties might possibly be resolved fully during this century.

Another consideration might be that of the possible fragility of the upper atmosphere. If it is found that this protective envelope would be seriously degraded beyond some calculated number of annual launches, then the future tourist industry could be greatly hampered since it would probably have a relatively low priority. This might not rule tourism out but could limit it severely or restrict it to only a very few high priority needs. On the other hand, the upper atmosphere might not prove to be fragile at all for properly designed propulsive systems.

[264] These potential issues are raised for balance in this discussion. Space tourism has an exciting long-term potential but it first needs to be developed and shown to yield sufficient benefits. During the next few decades unforeseen problems will undoubtedly arise and will need to be solved satisfactorily. Until that time space tours are likely to remain in the limbo of hopes or dreams.

If the health problems associated with *protracted* journeys into space should become relatively severe, various solutions may emerge over time that will permit space development to continue. Indeed, the major competitors to the human presence in space have been and undoubtedly will be the various automated devices—robots, in one form or another. Currently, according to Carl Sagan, "As a rule of thumb, a manned mission costs 50 to 100 times more than a comparable unmanned mission."[1] Over time, automation has become increasingly compact and effective. That is, the robots tend to shrink in size and increase the range of their activities. Humans can learn to do the latter, but they will have difficulty in shrinking.

Nevertheless, a human presence in space will be needed and is likely to grow. The economics of the competition with the robots generally will determine the relative balance only when the same tasks can be done by both. As space development proceeds, the balance may shift either way. However, as space industrialization grows in complexity the need for human specialists may grow in proportion. The first Space Shuttle decade, the '80s, should give us some early clues about the outcome of this long-term competition.

[265] D. <u>The Scenarios</u>

<u>Optimistic Scenario:</u> Some of the potential technology discussed in Chapter III is used in the Optimistic Scenario of Chapter IV. This scenario simply exercises the possible technological and economic muscles to show what could reasonably occur in an environment of sustained funding, high morale, dedication, cooperation, good management, and reasonable luck. It is intended to open up some vistas, to make it clear that extraordinary possibilities exist that are not necessarily Utopian. Although the events portrayed are generally not expected to happen as soon as indicated, we believe that the sequence is not intrinsically forbidden. It may only require a change in public attitudes, which is certain-

---

1.   *The New York Times Magazine,* July 10, 1977.

ly possible. The scenario not only is intended to give a sense of some ultimate possibilities but also of an *eventual* outcome. Almost all of the technological developments that "happen" in the first 100 years are likely to occur sooner or later—even in a pessimistic scenario, although probably much later and on a reduced scale. As is discussed below, the economic and technological progress associated with the Optimistic Scenario is amazing in its long-term outcome, but is relatively modest in any particular year or decade.

Pessimistic Scenario: Chapter V portrays the two New International Order Scenarios. Part I carries out a theme which today is widely accepted in the world, although not by the authors. This is a perspective which suggests that the developed nations generally are relatively "decadent" and that the developing nations have the energy and dynamism to push them aside and become the focus of the future. The second (Part II) version of [266] the scenario assumes that richer countries will, in part through the transfer of resources, greatly accelerate the development of the poorer countries and that these, as they attain comparable wealth and technological advancement, will adopt the social attitudes and ideology of their former benefactors. Both of these scenarios strike us as being relatively improbable—but they do raise important issues.

On the other hand, we do find that many formerly poor nations are now middle-income ones and progressing very rapidly. Scenario I gives special roles to China and Brazil in determining the world's future, each for different reasons. Scenario II gives a much weakened and modified form of this New International Order in which both the middle-income nations and the poor nations eventually become post-industrial, after which further economic progress is very slow.

In both scenarios, it is the middle-income nations that appear likely to "challenge" the lead of the U.S. and the Soviet Union in space activities by the end of the century or soon afterwards. In fact, in these scenarios the U.S. lead—as measured by budgets—is lost to both the Russians and the Chinese before the year 2000, and to the Brazilians soon afterwards. We believe that in this respect the two scenarios depicted are not implausible; it may well turn out that these newcomers in advanced technology and growing affluence will become technologically dominant, including space development. Perhaps nothing succeeds like success in a space "race." Moreover, the successes of the former middle-income countries in space could greatly increase their ability to become wealthier than the present developed nations. In these scenarios it occurs, not only because of [267] direct economic and technological achievements, but because their success creates a high morale and a sense of competence from the attainment of communal goals—attitudes that growing space activities might also engender in the [Organization for Economic Cooperation and Development] nations under appropriate circumstances.

Because visible signs of success are so important, many of the Third World countries often attempt to "fake" them. That is, they are attracted to four-lane highways and jet airliners in order to achieve the appearance of success before they have properly attended to their problems in rural roads, agriculture, employment, and education. The diversion of resources to showy projects can be tragic even if the showy projects are successful; usually they are not. Thus it may be undesirable for a Third World country to jump into space development rapidly. Unless both their economies and technological resources are substantial it can represent a serious and impractical diversion of scarce resources. However, the appropriate economic conditions and technological development can appear with astonishing rapidity. In fact, even S. Korea and Taiwan may be ready for certain specialized space ventures in the foreseeable future because of the rapidity with which they have been progressing. Potential economic and technological giants such as China and Brazil may also develop a space capability much more quickly than generally expected, if their recent progress continues.

In our most Pessimistic Scenario progress in economics and technology becomes very low after a country becomes post-industrial. Still, a surprising outcome (one which struck both authors forcibly) was that eventually, despite a general pessimism about space projects and other technological developments, space activities still become surprisingly extensive. [268] (Consider, as an analogy, that even if Queen Isabella had not financed Columbus, or if he had failed on his voyage, the Western Hemisphere would still be there, waiting to be found; since the time was ripe for worldwide exploration and exploitation other Europeans would have reached it before many more years had passed.)

The pressures for exploring and exploiting outer space basically derive from increasing wealth and advancing technologies. Over time the projects become easier to fund and, with advances in technology, less difficult to do. At some appropriate time, barring an almost religious aversion to new technology, a sufficient desire for space development will arise—even if long intervals occur when support is hard to find.

Another analogy might be made with the development of the U.S. railroads and the West. The railroads were stimulated by gifts of free land by the government and the belief that, as the railroads were built, the traffic would follow—that a great deal of industrial mining and agricultural development, including forestry, would occur quite rapidly and justify their investment. The time was ripe and it did.

A similar experience could occur with the Space Shuttle system. In space there may be no equivalent to the free 160-acre homesteads which were once available to the average American, but great opportunities are likely to exist for various "railroad" companies who "stake out claims" in outer space. At the moment the space frontier and its available resources seem relatively unlimited. Relatively few critical regions appear to exist that might eventually become the cause of major conflicts and hinder commercial development.

[269] Moderate Scenario: Chapter VI develops our Moderate Scenario which is intended to be more plausible than the others. It has an implied conviction that the progress represented is worth striving for—and can be achieved without undue reaching. It represents our "median" image of future developments in space. It contains some fictional elements which are intended to be prototypes of actual historical events, including individuals who are "movers" and "shakers" and who play central roles in forcing a more rapid space development.

The Moderate Scenario emphasizes two new commercial opportunities: *space industrialization and tourism,* both of which appear to have extremely large potential in all the scenarios. The term, "space tourism," alone may not convey the intended meaning. If we assume that space travel is the moving experience that it has been to many astronauts, at some future time various organizations or societies may wish to provide this experience to selected people. It may be a reward for dedicated service or special contributions, or a ritual associated with special religious groups. Or it could be arranged through an open or limited access lottery.

Shortly after the beginning of space tourism, assuming it is successful, we find the possibility of great interest in establishing permanent orbiting colonies—possibly with many of the same motivations. We have not dwelt upon the desire for utopias or choices for one's preferred life style. However, it may well turn out that one of the major motivations for colonies in space is similar to that which drove the Pilgrims to New England: the desire to choose their way of life with minimal interference from the home country. This outcome, of course, would depend very much on [270] the cost and viability of such colonies. But even relatively small sects, given a growing future affluence, could eventually finance a colony if they were sufficiently motivated—particularly if they had ability to tithe, like the Mormons or Black Muslims.[2]

---

2.   For example, if 1,000,000 families each contributed $2,500 a year for 40 years, without interest this [would] become $100 billion, enough to establish a substantial colony in the early to mid-21st century, according to our projections. The contribution could appear to a *believer* as a tax-exempt investment in the future rather than as a gift.

In many religions the wealth of the church is thought of as a common resource for its members to help those in difficulties and sometimes, even when some members are successful, to be used to facilitate additional success. To the extent that space colonies, in addition to their religious connotation, might look like profitable ventures then the investments would eventually increase the wealth of the church. Furthermore, through merit criteria or lotteries, every member would have a chance to become a colonist, or a tourist. The church might deem the high costs well worth the increased faith and activity of its members—particularly if these "tourists" returned with an inspired and zealous commitment to the church.

E.   Potential for Growth

Our very rough estimate of current world space project expenditures is about $9 billion per year on non-military developments and perhaps half that much on military space programs. That is, total expenditures on space projects are about .2 percent of the GWP (about .13 percent non-military, and .07 percent military).

In a pessimistic space scenario that small (.13 percent) fraction generally remains stable or decreases, despite the fact that the cost of transportation to space must almost certainly fall by more than a factor [271] of 10 during the next century. Still the GWP during the next 100 years is expected—in our earth-centered surprise-free projections—to rise by about a factor of 20. . . . If space expenditures merely keep pace with GWP they would be approximately $200–$300 billion by then and would almost double again in the following 100-year period. The "truly" pessimistic space scenario would, on average, maintain a slower growth in space investments than the GWP. *On average* means that "temporary" fluctuations may vary that projection somewhat over periods of a few decades or less.

A moderate scenario, in our view, would, on average, but probably with erratic fluctuations, show a growth rate in space which exceeds that on earth—perhaps by about a factor of 2 or 3. Thus, the world's space budget would have a mean growth rate of 3-1/2 percent to 5 percent, which would lead to space budgets between $200 and $1,200 billion at the Tricentennial. We arrived at $700 billion for our particular Moderate Scenario (Chapter VI).

However, we notice that the 2076 budget (or annual investment) estimate is extremely sensitive to the assumed average growth rate. Over a hundred years a *2-percent* change in this average affects the budget by about a factor of 7—and, over 200 years, by a factor of 50. Yet it appears to us that the inherent uncertainty in the average future growth rate is intrinsically greater than 2 percent and may be as high as 5 percent—the latter leads to a factor of more than 100 over a century and more than an astonishing *10,000* over two centuries!

[272] On the positive side—that is, from a space enthusiast's point of view—the high growth rates lead to an optimistic scenario. In the Optimistic Scenario of Chapter IV the average growth rate in worldwide space investments rose during the last quarter of the 20th century at about 10 percent per annum (about the average for a "glamour" or high technology industry currently). During the first 100 years after a brief rise to a 12-percent growth rate, the Gross Space Product (GSP) then declined steadily to about an 8-percent growth rate, and during the second 100 years to 5 percent. . . . We note that the growth rate in space, *per capita*, ends at about 2-1/2 percent—actually, less than the productivity increase of the average American worker in recent decades. Thus, the Optimistic Scenario portrayed would *not* be perceived as having an astonishing growth during any year of its history—except perhaps during 1980–2000, when a general change in societal attitudes toward space is assumed to occur. After the turn of the century it is just the assumed long-term steadiness of the slowly declining growth rate that brings about a "miraculous" transformation—first to a $30 trillion GSP and then, during the much slower growth of the

second 100 years, to an awesome $12,000 trillion. . . . Stated only in this last way the numbers tend to be hard to accept as having any reality; the *average* annual GWP/capita, including the GSPs after 200 years, reaches almost $1 million—about 600 times greater than the average today!

The 200-year progression of moderate successes in the Optimistic Scenario not only leads to fantastic developments in space, it also demands an extremely rich society on earth—one in which essentially everyone (except those who voluntarily opt out) is an active participant. To the [273] average citizen, today, this must represent an unbelievable outcome—even though the path to it is relatively straightforward. For example, to go from the current GWP of roughly $6.6 trillion to one of $6,700 trillion over 200 years (on earth . . .), requires "only" an average growth rate of 3.5 percent—*less* than that which the world as a whole has experienced during the last few decades! What is so astonishing about that? The answer appears to be *nothing* to an optimist, *everything* to a pessimist.

In the above sense of economic and technological progress, we have portrayed pessimistic, moderate, and optimistic scenarios. In the pessimistic one the intrinsically high-technology enterprises are treated as fascinating but dangerous tools to be kept under very tight control. Society responds to science and technology as if it were a "foreign" culture beyond its real understanding, and potentially fraught with great new risks.

The Moderate Scenario is more of a long-term business-as-usual perspective. Where space projects are profitable the economic rewards tend, over time, to dominate the socio-political restraints and, accompanied by numerous problems, difficulties and interruptions, space development erratically but slowly climbs the ladder of progress. But even such erratic slow progress *over a "mere" 100 years* brings about changes which in today's world could only seem amazing—for example, a $600-billion annual investment in space when transport costs are about 1/25 of the present ones, coupled with tremendous advances (to take a few examples) in automation, instruments, materials, and new designs (for vehicles, industries, communication systems, and processes in space). Thus, even [274] the slow, erratic *Moderate* Scenario reveals a potentially astounding transformation, one that may be almost impossible to comprehend fully or foresee accurately. That is, the projections we have made are likely to appear primitive 100 years from now—just as do the U.S. projections from 100 years ago that could not seriously imagine the general use of automobiles, let alone airliners, space flight, electronic computers, television, or nuclear energy—to name just a few of the evolved "miracles"—and all this with an average growth in per capita GNP of less than 2 percent.

How then, when the business-as-usual projection becomes shocking or incomprehensible, can we expect anything but incredulous reactions for any optimistic scenario? Our Optimistic Scenario requires, for the world as a whole, merely that human beings opt for growth and set about to obtain it with roughly the same, but *sustained,* vigor that on average we find exists today. That is all.

Document III-17

Document title: Robert Dunn, "NASA Policy to Enhance Commercial Investment in Space," internal NASA document, September 13, 1983.

Source: NASA Historical Reference Collection, NASA History Office, NASA Headquarters, Washington, D.C.

As part of a larger review of commercial space activities, this report reviews the various policies NASA had developed over the years concerning industrial and commercial involvement in space. Such documentation reveals a new way of thinking at NASA; the agency was beginning to consider the future users of space rather than just demonstrating national technological competence and exploration. This was partly in preparation for the space station, partly for continued pressure to find commercial R&D ventures for the Shuttle, and partly to bolster NASA support throughout the economy.

[no pagination]

NASA Policy to Enhance Commercial Investment in Space

It is national policy to effectively apply the resources of the nation to preserve the role of the United States as a leader in space science and technology and their applications. With the maturing of the Space Transportation System (STS) to a reliable and operational status and in light of recent initiatives in space industrialization, it is evident that the dawn of the era of wide-spread commercial activities in space is at hand. Maintenance of national leadership in space requires the support and expansion of commercial space activities.

The President's National Space Policy of July 4, 1982, directs NASA to expand United States private sector investment and involvement in civil space and space-related activities. In light of this directive and since substantial portions of the United States technological base and motivation reside in the United States private sector, NASA will invigorate its efforts to take necessary and proper actions to promote a climate conducive to expanded private sector investment and involvement in space by United States domestic concerns.

NASA views its role in the commercialization of space in light of the National Aeronautics and Space Act of 1958, as amended (Space Act), which establishes NASA as the agency responsible for the direction of civil "space . . . activities" of the United States. The legislative history of the Act states that "the term 'activities' should be construed broadly enough to enable the Administration . . . to carry on a wide spectrum of activities which relate to the successful use of outer space. These activities would include scientific discovery and research not directly related to travel in outer space but utilizing outer space, and the development of resources which may be discovered in outer space."

The Space Act also establishes that space activities will be conducted to make the "most effective utilization of the scientific and engineering resources of the United States . . . in order to avoid unnecessary duplication of effort, facilities, and equipment." NASA has made large scale use of private industry as contractors in carrying out its activities. It has provided space launch services for commercial purposes since 1962. Beginning in 1979 it has entered into "partnership" arrangements with private sector firms to enhance the commercial utilization of space resources. These and other activities carry on and expand the tradition of NASA's cooperation with industry and other private sector institutions which dates back to NASA's predecessor agency, the National Advisory Committee [for] Aeronautics.

In light of the Presidential policy of July 4, 1982, NASA will continue and expand its effort to facilitate private sector investment in outer space and will encourage commercial space activities consistent with that policy.

In order to more effectively encourage and facilitate private sector involvement and investment in civil space and space-related activities, *NASA will redirect a portion of its space* research and development activities to assure that its R&D program supports the research, development and demonstration of space technologies with commercial application.

To further support this objective, NASA will directly involve the private sector in initiatives which are consistent with NASA program objectives and which support commercial space activity.

These initiatives may include, but are not limited to: (1) engaging in joint arrangements with United States domestic concerns to operate on a commercial basis facilities or services which relieve NASA of an operational responsibility; (2) engaging in joint arrangements with U.S. domestic concerns to develop facilities or hardware to be used in conjunction with the STS or other aspects of the U.S. space program; and (3) by entering into transactions with United States concerns designed to encourage the commercial exploitation of space.

Principle NASA incentives available in joint arrangements may include in addition to making available the results of NASA research: (1) providing flight time on the space transportation system on appropriate terms and conditions as determined by the Administrator; (2) providing technical advice, consultation, data, equipment and facilities to participating organizations; and (3) entering into joint research and demonstration programs where each party funds its own participation. In making the necessary determination to proceed under this policy, the Administrator will consider the need for NASA funded support or other NASA action to commercial endeavors and the relative benefits to be obtained from such endeavors. The primary emphasis of these joint arrangements will be to provide support to ventures which result in or facilitate industrial activity in space when such activity would otherwise be unlikely to occur due to high technological or financial risk. Other ventures involving new commercial activities in space will also be supported. In either case, private capital must be at risk.

As major areas for NASA enhancement of total United States capability, including the private sector, may become apparent from time to time, the factors to be considered by NASA prior to providing incentives may include, but not be limited to, some or all of the following considerations: (1) the effect of the private sector activity on NASA programs; (2) the enhanced exploitation of NASA capabilities such as the Space Transportation System; (3) the contribution to the maintenance of United States technological superiority; (4) the amount of proprietary data or background information to be furnished by the concern; (5) the rights in date to be granted the concern in consideration of its contribution; (6) the impact of NASA sponsorship on a given industry; (7) provision for a form of exclusivity in special cases when needed to promote innovation; (8) recoupment of the contribution under appropriate circumstances; (9) support of socio-economic objectives of the government; and (10) the willingness and ability of the proposer to market any resulting products and services.

This policy supersedes the *NASA Guidelines Regarding Early Usage of Space for Industrial Purposes.* It does not affect existing programs (such as Materials Processing in Space) and relationships which are consistent with or outside the scope of the policy. This policy is not to be construed as authorizing or requiring NASA to perform a regulatory review of a proposed commercial use of space where no cooperative agreement or other appropriate arrangement between NASA and the commercial entity is contemplated. . . .

II.B.   Description and Discussion of Actions and Transactions

In order to understand NASA's view of its role in the commercialization of space, it should be noted that the National Aeronautics end Space Act of 1958, as amended (Space Act), establishes NASA as the agency responsible for the direction of civil "space . . . activities" of the United States. The legislative history states that "the term 'activities' should be construed broadly enough to enable the Administration to carry on a wide spectrum of activities which relate to the successful use of outer space. These activities would include scientific discovery and research not directly related to travel in outer space but utilizing outer space, and the development of resources which may be discovered in outer space."

The Space Act also establishes that space activities will be conducted to make the "most effective utilization of the scientific and engineering resources of the United States . . . in order to avoid unnecessary duplication of effort, facilities, and equipment." NASA has made large scale use of private industry as contractors in carrying out its activities. It has provided space launch services for commercial purposes since 1962. Beginning in 1979 it has entered into "partnership" arrangements with private sector firms to enhanced the commercial utilization of space resources.

In light of the Presidential policy of July 4, 1982, NASA will continue and expand its effort to facilitate private sector investment in space and will encourage commercial space activities consistent with that policy.

The following is a description of the mechanisms NASA uses to enter into cooperation with the private sector. These various actions and transactions are or may be used by NASA in varying degrees to assure the application of non-public resources to the exploitation of space for commercial purposes.

1. <u>Procurements.</u> NASA carries out most of its research and development (R&D) activities as well as space launch operations through the use of procurement contracts rather than through use of its own personnel and facilities. The United States Code prescribes that a procurement contract shall be used when:

"(1) the principle purpose of the instrument is to acquire (by purchase, lease, or barter) property or services for the direct benefit or use of the United States government; or (2) the agency decides in a specific instance that the use of a procurement contract is appropriate." 31 U.S.C. § 6303.

NASA's Procurement Regulations are found in Chapter 18, Title 41, Code of Federal Regulations. Unlike most civilian agencies NASA's procurement authority is based on the Armed Services Procurement Act, 10 U.S.C. § 2301, *et seq.* The Federal Property and Administrative Services Act, 40 U.S.C. § 471 *et seq.*, as it relates to disposal of property, is also applicable.

NASA implements the principles of OMB Circular A-76 "Performance of Commercial Activities" as it relates to commercial activities in support of R&D even though in most instances A-76 is not directly applicable to NASA functions. NASA's current approach to space commercialization is much broader than the A-76 concept.

2. <u>Cooperative Agreement (Chiles Act).</u> NASA seldom enters into cooperative agreements as defined in the Chiles Act (as distinguished from various Space Act cooperative arrangements). Such a cooperative agreement is used when:

"(1) the principal purpose of the relationship is to transfer a thing of value to the State, local governments or other recipient to carry out a public purpose of support or stimulation authorized by a law of the United States instead of acquiring (by purchase lease, or barter) property or services for the direct benefit or use of the United States Government; and

(2) substantial involvement is expected between the executive agency and the State, local government or other recipient when carrying out the activity contemplated in the agreement." 31 U.S.C. § 6305.

NASA is prepared to explore the possibility of expanding its use of such agreements to support commercialization.

3. <u>Grant Agreement.</u> NASA has made use of grants principally to fund university-sponsored research and development. Grants are used when:

"(1) the principal purpose of the relationship is to transfer a thing of value to the State or local government or other recipient to carry out a public purpose of support or stimulation authorized by a law of the United States instead of acquiring (by purchase, lease, or barter) property or services for the direct benefit or use of the United States Government; and

(2) substantial involvement is not expected between the executive agency and the State, local government, or other recipient when carrying out the activity contemplated in the agreement." 31 U.S.C. § 6304.

NASA intends to direct a portion of its grant-funded research toward areas with potential viability for commercial space ventures. NASA regulations covering research Grants and Cooperative Agreements are found in Part 1260, Title 14, Code of Federal Regulations.

4. <u>Space Act Arrangements.</u> Independent of its authority to enter into contracts for procurement, cooperative agreements or grants as defined above, the National Aeronautics and Space Act of 1958, as amended, authorizes NASA to enter into a variety

of flexible arrangements with public and private institutions. Sections 203(c)(5) and (6) of the Act provide:

"(5) without regard to section 3648 of the Revised Statutes, as amended (31 U.S.C. 529), to enter into and perform such contracts, leases, cooperative agreements or other transactions as may be necessary in the conduct of its work and on such terms as it may deem appropriate, with any agency or instrumentality of the United States, or with any State, Territory, or possession, or with any political subdivision thereof, or with any person, firm, association, corporation, or educational institution. To the maximum extent practicable and consistent with the accomplishment of the purposes of this Act, such contracts, leases, agreements and other transactions shall be allocated by the Administrator in a manner which will enable small-business concerns to participate equitably and proportionately in the conduct of the work of the Administration;

(6) to use, with their consent, the services, equipment, personnel and facilities of Federal and other agencies with or without reimbursement, and on a similar basis to cooperate with other public and private agencies and instrumentalities in the use of services, equipment, and facilities. Each department and agency of the Federal Government shall cooperate fully with the Administration in making its services, equipment, personnel, and facilities available to the Administration, and any such department or agency is authorized, notwithstanding any other provision of law, to transfer to or to receive from the Administration, without reimbursement, aeronautical and space vehicles, and supplies and equipment other than administrative supplies or equipment."

The following are examples of arrangements and transactions supporting space commercialization which NASA has entered into under its Space Act authority.

a.   <u>Early Initiatives in Commercial Space Activity.</u>

(1) <u>Communications Satellites.</u> On July 10, 1962, a NASA Delta launch vehicle launched the first privately-owned satellite which was also the world's first active communications satellite. Telstar I was a product of private industry, American Telephone & Telegraph Company, launched for AT&T by NASA on a reimbursable basis. This satellite enabled a whole continent to "see" across oceans. Television programs from and to Europe, for instance, brought new, real-time sights and sounds into the homes of millions. Even though Telstar's "mutual visibility"—the time during which signals could be sent and received—was relatively short (approximately 15 to 20 minutes), the portents of this new communications medium was immediate. With an elliptical orbit that crossed the Van Allen belts, Telstar I taught engineers a great deal about radiation damage to communications equipment. Telstar I's technology did not prove commercially feasible, however.

A legislative debate soon ensued on Capitol Hill as to how this new communications system was to be used operationally—by private industry, by a public utility, or by a Governmental agency. On August 31, 1962, President John F. Kennedy signed the "Communications Satellite Act of 1962." This law created a "communications satellite corporation for profit which will not be an agency or establishment of the United States Government, but which would have government representation on its Board of Directors and have many of its activities regulated by Government." A space-age development became a new business enterprise and marked a new form of Government-business collaboration.

Rapid strides were made by NASA in the area of improved communication techniques. Technological advances produced by AT&T's Telstar, NASA's Echo, Relay, and Syncom systems soon found further applications. In 1965 the control of Syncom II and Syncom III was transferred to the Department of Defense for operational communications and for study in design of military communications systems. Early Bird, the world's commercial communications satellite, was built by the Syncom contractor, Hughes Aircraft Co., for the Communications Satellite Corporation and was closely patterned on the earlier Syncom.

The communications satellite in geo-synchronous orbit has become the basis for subsequent government and commercial communications satellites. In less than twenty years a multi-billion dollar industry has grown up in international and domestic space telecommunications. Applications have resulted in the rapid expansion of the cable television industry and new applications of direct broadcast technology are in the offing.

One of NASA's most recent advances in communications satellite, the Tracking and Data Relay Satellite System (TDRSS), was originally proposed as a commercial venture but later reorganized as a more traditional procurement.

(2) Launch Vehicle Upgrade and Upper State Developments. In 1972 the United States committed itself to the development of the reusable Space Shuttle and a decision was made not to spend additional public funds to improve expendable launch vehicles such as the Delta. McDonnell Douglas, the Delta's manufacturer, undertook in 1973, a performance upgrade of the Delta called the 3914, which increased the Delta payload capability to 2000 lbs. into the geosynchronous transfer orbit. The emerging domestic communications satellite industry was very interested in the cost effectiveness of the additional Delta launch capability and this set the stage for McDonnell Douglas' first commercial space venture. RCA became the first customer and agreements were structured between RCA, McDonnell Douglas and NASA to operate the projects as follows:

- McDonnell Douglas agreed to design and develop the uprated vehicle at their own risk on commercial funds but with profit limitations.
- McDonnell Douglas agreed to recover its investment through a specified "not to exceed" customer charge for each commercial launch; however, there would be no "investment charge" for U.S. Government use of the vehicle.
- NASA agreed to contract for production and launch services of the improved vehicle as an integral part of the on-going Delta Program and would provide technical monitoring.
- RCA agreed to contract with NASA for three vehicles and launch services and with McDonnell Douglas for three user development amortization payments.

The first launch was successful in 1975 and as of mid-1983 there have been 23 launches of which seven have been U.S. government missions with the remainder being commercial and foreign customers.

McDonnell Douglas' second commercial venture in this field was the Payload Assist Module (PAM-D) undertaken in 1976. PAM-D is an upper stage vehicle for Delta class satellites designed to assist launch vehicle customers in planning launches in the era of transition between expendable launch vehicles and the Space Shuttle. The PAM is basically a part of the payload cargo element and as such it may be used either as a third stage on the Delta or as an upper stage on Shuttle to carry the satellite into the geosynchronous transfer orbit. This provides the customer with the flexibility to plan a launch on Delta or Shuttle and a relatively easy means to shift between the two if conditions should so warrant. Again, McDonnell Douglas began this development after reaching an agreement with NASA on how the program would be performed. The primary features of this agreement were as follows:

- McDonnell Douglas agreed to develop the system completely on commercial funds on a schedule compatible with the Shuttle operational requirements.
- McDonnell Douglas agreed to sell PAM commercially at a specified "not-to-exceed" ceiling price along with a fixed escalation for inflationary factors in addition to profit limitations.

- NASA agreed not to fund or formally solicit the development of competitive or alternate system.
- NASA agreed to provide suitable building facilities at [Kennedy Space Center] for PAM processing activities with reimbursement as a part of the Shuttle launch service contract with each customer using PAM.
- NASA agreed to provide interface data and technically monitor the project.

The first commercial PAM contract was signed between Hughes and McDonnell Douglas in 1978. The first flight was successful for Satellite Business Systems on Delta in November 1980. The first two flights on Shuttle were successfully completed in November 1982. To mid-1983, McDonnell Douglas had contracted 29 PAM-D missions and successfully flown 10 out of 10 scheduled.

In 1977 McDonnell Douglas signed a similar agreement to develop a larger PAM for Shuttle launch only called "PAM-A." This version has comparable payload capability to the Atlas Centaur ELV. After a competition with another aerospace firm, NASA awarded McDonnell Douglas a Firm Fixed Price Contract for 6 PAM-A launches which has subsequently been increased to 8. There are, however, no currently assigned missions for PAM-A and no commercial sales have been achieved.

In 1982 customers began requesting additional performance from the PAM-D system, but short of the more expensive PAM-A capability. As a result McDonnell Douglas decided to commercially undertake a growth version called "PAM-DII," which includes a new motor and raises the payload capability to approximately 4100 lbs. into the geosynchronous transfer orbit. An initial commercial contract resulted from a 1983 order by GTE Satellite Corporation. NASA has agreed to provide technical monitoring.

In early 1983 NASA entered into a cooperative agreement with Orbital Systems Corporation, an entirely new company, to provide for the production of another shuttle-compatible upper stage, the Transfer Orbit Stage (TOS). NASA has no current need for such an upper stage for its in-house programs but is cooperating with Orbital Systems to optimize Shuttle capabilities for commercial customers who may wish to purchase such an upper stage. The TOS is privately funded but has NASA technical monitoring and NASA's agreement not to build a competing upper stage.

b. Joint Endeavor Agreements. The Joint Endeavor Agreement (JEA) was originally developed to facilitate NASA's interest in involving the private sector in materials processing in space (MPS) but its basic provisions are applicable to other areas of space industrialization.

The JEA is a cooperative arrangement in which private participants and NASA share common program objectives, program responsibilities, and financial risk. The objective of a JEA is to encourage early space ventures and demonstrate the usefulness of space technology to meet marketplace needs. A JEA is a legal agreement between equal partners, and is not a procurement action; no funds are exchanged between NASA and the industrial partner. A private participant selects an experiment and/or technology demonstration for a joint endeavor which complies with MPS program objectives, conducts the necessary ground investigation, and develops flight hardware at company expense. As incentive for this investment, NASA agrees to provide free Shuttle flights for projects which meet certain basic criteria, such as technical merit, contribution to innovation, and acceptable business arrangements. As further incentive, the participant is allowed to retain certain proprietary rights to the results, particularly the nonpatentable information that yields a competitive edge in marketing products based on MPS results. However, NASA receives sufficient data to evaluate the significance of the results and requires that any promising technologies be applied commercially on a timely basis, or published.

The first JEA signed in 1980 involved McDonnell Douglas and OrthoPharmaceuticals in a project titled continuous flow electrophoresis which promises higher quantities and quality of certain pharmaceuticals produced in space. Successful experiments have been conducted on several Shuttle flights. A JEA for the purpose of producing Galium [sic] Arsenide crystals in space was signed between NASA and Micro Gravity Research Associates in 1983. Other MPS JEA's are under consideration.

An example of JEA extending outside the MPS area is the recent NASA-Fairchild JEA concerning the development of a space platform for lease. Other proposed JEA's deal with development of Shuttle payload carriers.

c. Technical Exchange Agreements. For companies interested in applying microgravity technology, but not ready to commit to a specific space flight experiment or venture, NASA has developed the Technical Exchange Agreement (TEA). Under a TEA, NASA and a company agree to exchange technical information and cooperate in the conduct and analysis of ground-based research programs. In this agreement, a firm can become familiar with microgravity technology and its applicability to the company product line at minimal expense. Under TEA, the private company funds its own participation, and derives direct access to and results from NASA facilities and research, with NASA gaining the support and expertise of the private company's industrial research capability.

Several MPS TEAs have been signed and others are proposed. NASA has provided microgravity drop tube use and aircraft flights, among other facilities, to support these efforts.

d. Industrial Guest Investigators. In an Industrial Guest Investigator (IGI) Agreement NASA and industry share sufficient mutual scientific interest that a company arranges for one of its scientists to collaborate (at company expense) with a NASA-sponsored principal investigator on a space flight MPS experiment. Once the parties agree to the contribution to be made to the objectives of the experiment, the IGI becomes a member of the investigation team, thus adding industrial expertise and insight to the experiment. A number of IGI agreements have been undertaken.

e. Commercial Launch Vehicles.

(1) Commercialization of NASA ELV Systems. Because of the nation's commitment to the reusable Space Shuttle both NASA and the Department of Defense plan to terminate use of expendable launch vehicles. In May 1983 the President decided that the private sector should be given the opportunity to operate these systems on a commercial basis. NASA is in the process of exploring ways to transfer the production and operation of its ELV systems to the private sector. By transferring these systems it is hoped that the existing production and launch facilities will remain a valuable national resource rather than be reduced to scrap but that government will be relieved of the cost of maintaining a redundant launch system. Transfer will require termination of various procurement contracts and execution of multi-faceted facility use agreements.

(2) Privately Developed Launch Vehicles. NASA personnel provided advice and guidance to Space Services, Inc. (SSI) in its attempts to develop a space rocket. In 1982 SSI successfully launched a sub-orbital rocket to demonstrate the ability of a private company in this field. Pursuant to a cooperative agreement, NASA provided the rocket motor for SSI's demonstration flight. It is believed that this demonstration generated great interest in the private operations of space launch vehicles and support system. SSI is continuing to develop a low cost orbital launch vehicle which will have slightly greater capability than NASA's Scout vehicle. SSI has requested NASA to agree not to restart Scout production or otherwise compete in the low weight, low earth-orbit commercial market.

NASA has also provided limited advice and technical assistance to ARC Technologies in its attempt to develop a launch vehicle. The proposed ARC vehicle is based on propulsion technology originally developed by the government but later abandoned. If successful, ARC may add to America's technological resources. In addition to NASA's advice ARC has requested tracking and data services from another government agency.

      f.   Other Transactions. NASA's current and projected programs and policies can facilitate commercial space ventures in a variety of ways.

      (1) Patents. NASA views its patent program as an integral portion of its mission responsibility to encourage new technology and foster the utilization and commercialization of NASA supported technologies. The statutory basis for the agency's patent policy is Section 305 of the Aeronautics and Space Act of 1958. Accordingly, NASA acquires title to all inventions under contract unless the Administrator decides that waiver of title to the contractor would be in the public interest (sec. 305(f)). Thus, the agency was granted broad waiver authority, but is required to retain a broad royalty-free license to all inventions under contract so that the waiver of title in actuality amounts to a waiver of commercial rights only. NASA patent regulations are found at Title 14 Code of Federal Regulations, Part 1245.

(i) Patent Waivers.

There are two types of domestic waivers granted by NASA: (1) advance waivers which are applicable to inventions made under a contract; and (2) waivers for inventions subsequently reported under a contract. The granting of waivers is authorized by the Administrator upon the recommendation of the Inventions and Contributions Board set up under section 305(f) of the 1958 Act. However, all waivers are subject to the retention of NASA of a broad, irrevocable royalty-free license and of "march-in rights." March-in rights permit the agency to intervene if it believes that inventions are being suppressed, that there is a danger to the public health and safety, or that a company is not meeting Government regulations. The agency also retains the authority to void waivers if a firm fails to report on its commercialization activities.

Seventy-five percent of the requests for waiver have been granted. However, to date this represents only a small number of waivers because there have been few requests.

(ii) Patent Licenses.

NASA patent licensing regulations were promulgated to use the patent system to promote the utilization of inventions arising from NASA supported research and development. An applicant is required to supply NASA with a satisfactory plan for development or marketing of the licensed inventions, or both, and with information about the applicant's capability to fulfill the plan.

(iii) Protecting Intellectual Property Rights in Commercial Space Activities.

In recognition of the substantial investment necessary to develop the electrophoresis experiment and other activities conducted as joint endeavors, NASA negotiated special clauses dealing with inventions and technical data. Typical clauses provide that as long as the party engaged in the joint endeavor with NASA continued to pursue the experiment, that party would retain all rights to inventions and proprietary technical data. NASA would not take a government license or any "march-in" rights to require licensing of others. The only exception is if the NASA Administrator, in response to a national emergency, determines that an invention made in the performance of the joint endeavor is urgently needed for public health reasons. NASA intends to continue to use its flexibility to accord full rights to inventions and proprietary technical information to private parties willing to invest substantial sums in joint endeavor agreements.

      (2) Privatization and Commercialization of Space Infra-structure. The transfer of Government ELV systems based on a Presidential decision was described above and men-

tion was made of privately financed upper-stages. These are prototypes for transferring existing infra-structure and creating new infra-structure through private investment.

NASA believes its policy of acquiring and operating its facilities, equipment, and technical services through industrial contractors has built a competence for supporting new initiatives and exploiting space technology in the private sector. For example, several proposals to NASA involve private financing of some of the shuttle infrastructure. These are opportunities to facilitate the commercialization process and reduce NASA funding requirements without posing a threat to NASA's principal mission—research and development. Therefore, in certain instances, after identifying a specific requirement for a product or service and determining that there is no compelling need to meet the requirement through a traditional NASA-controlled development program, NASA will advertise the need within the private sector as a commercial opportunity concurrently announcing that it will not initiate a competitive development. Difficulties encountered in the tracking and data relay satellite program in accommodating governmental and commercial function in a single spacecraft have made NASA keenly aware of the importance of a thorough examination of the government's interest before making a private sector commitment.

(3) The Aeronautics Model. Events over the past twenty years have established a two-fold R&D role for NASA in space applications—namely, to explore new opportunities for the application of space technology and to improve demonstrated technologies to achieve the extensive operational capabilities available today. NASA believes these activities support critical national needs; therefore, this NASA role should be continued and expanded. An element of this role is perhaps best illustrated by the need to advance communications technology including work in the 30–20 GHz frequency range. Since its demonstration in the early 1960's, the private sector has translated synchronous communications satellite technology into a highly successful growth industry. The communications satellite industry is the principal current example of commercialization of space technology, yet new technological breakthroughs are now required to maintain U.S. leadership and to realize continued economic benefits. The estimated cost of this advanced technology development exceeds the financial capability of any single firm in the industry. NASA intends to pursue this R&D requirement and any similar cases in space applications through demonstration of the applicable technology. In so doing, NASA will explore the feasibility of adapting the mutually beneficial experience accruing from the government/industry working relationships in conducting its aeronautical research programs, wherein NASA conducts R&D and in certain cases industry performs hardware fabrication and flight testing of new technologies under cost-sharing arrangements.

(4) Small Business Innovation Research. NASA has implemented the first stage of its SBIR program pursuant to the Small Business Innovation Development Act. The purpose of the statute is to set aside a portion of each agency's extramural R&D budget to assist in a substantial way small business in bringing to market advanced R&D products and services. The SBIR program though not specifically related to "in space" commercialization does attempt to develop space technology and its applications to the point of commercial viability.

The Small Business Innovation Development Act requires an escalated expenditure of an agency's R&D funds according to a set formula where the extramural R&D budget is over $100,000,000: in the first year 0.2 percent; 0.6 in the second; 1.0 in the third fiscal year; and, not less than 1.25 percent in all subsequent fiscal years. For NASA, this means $5 million in 1983 and $12 million in 1984, with a subsequent reduction in dollar amounts in later years as all Shuttle activities are treated as operational and not R&D for budget purposes.

The SBIR program is required to be conducted in three phases, only two of which will be funded by the Government. In the first phase, contractors are to prove the feasibility of their proposed scientific and technical ideas. Up to $50,000 will be awarded for selected Phase I proposals which will be conducted, normally, within six months. Price competition is not a factor in selection, but value to the Government is. In Phase II, embodying the principal research effort, awards will be made to Phase I contractors whose work shows promise of producing something of value for the agency in terms of technical merit and feasibility. Special consideration will be given to proposals which demonstrate funding commitments to development of commercial application of the idea. Phase II awards are expected to be in amounts of up to $500,000 for a period of performance, generally, not to exceed 24 months. In both Phase I and II contracts, a profit or fee may be included. Phase III, hopefully, will involve private market funding support of contractor efforts with ultimate commercialization of the product or service.

In order to be eligible to propose, a contractor must be a small business. And, the proposer must be the primary source of employment of the principal investigator. However, some subcontracting is permitted in both phases and joint ventures are permitted and even encouraged, providing small business eligibility standards for the proposers are maintained.

One of the most significant aspects of the SBIR contracts yet to be resolved is the exact nature of the data rights issue; neither the statute nor the SBA [Small Business Administration] have given clear guidance on this issue which must be resolved in a uniform manner throughout the SBIR program.

An SBIR program was established at NASA Headquarters in the Office of Aeronautics and Space Technology to coordinate and handle the development of topics, and to conduct the selection process. Topics were proposed by the NASA field centers, evaluated and culled at Headquarters for inclusion in a topic list sent out with the solicitation. Evaluation and selection of winning proposals are accomplished through panels of NASA scientific and technical experts. Awards of contracts will be made by the SBIR office with the administration at the field center from which the topic originated. Initial first phase contract awards have been made. If this program proves successful, NASA may extend the concept beyond the original statutory requirements as an adjunct of its commercialization program.

Document III-18

Document title: "Space Commercialization Meeting," memo with agenda, participants, and outline of policy issues, The White House, August 3, 1983.

Source: NASA Historical Reference Collection, NASA History Office, NASA Headquarters, Washington, D.C.

In 1983, the Reagan administration decided to encourage the private sector to invest in space research and commercialization and wanted to understand what government policies would provide the best climate for private investment in space. A meeting was held at the White House with business and government leaders to discuss measures that would stimulate private-sector industrial activity in space. This meeting was one of the first very high-level and visible signals from the government to business to plan for a new era in space of profit-making opportunities, manufacturing, and other activities. It was also a direct signal to business that the planned space station would be available for commercial opportunities and that to advance the plans for the station, the lobbying and support of the business sector would be important.

August 3, 1983

Space Commercialization Meeting

Agenda

1.	Introduction	Craig L. Fuller
2.	Welcome	Edwin Meese III
		James M. Beggs
3.	Review of Issue Outline	Craig L. Fuller
4.	Discussion of Commercial Space Issues	All Participants
5.	Lunch with the President	All Participants
6.	Summary	All Participants

Luncheon With the President

Old Family Dining Room
August 3, 1983
12 noon

Participants

Mr. John F. Yardley
President, McDonnell Douglas Astronautics
St. Louis, Missouri

Mr. Maxime A. Faget
President, Space Industries, Inc.
Houston, Texas

Mr. Robert A. Hanson
Chairman and Chief Executive Officer
Deere & Company
Moline, Illinois

Mr. Frederick W. Smith
Chairman and Chief Executive Officer
Federal Express Corporation
Memphis, Tennessee

Mr. George Jeffs
President of North American Space Operations
Rockwell International
El Segundo, California

Mr. George Skurla
Chairman and President
Grumman Aerospace Corporation
Bethpage, New York

Mr. David Thompson
President, Orbital Systems Corporation
Vienna, Virginia

Mr. David Hannah
President, Space Services Incorporated
Houston, Texas

Mr. Oliver C. Boileau
President, General Dynamics Corporation
St. Louis, Missouri

Dr. Klaus P. Heiss
New York, New York

Mr. John Latshaw
Executive Vice President and Managing Director
E.F. Hutton & Company, Inc.
Kansas City, Missouri

Dr. John W. Townsend, Jr.
President
Fairchild Space Company
Germantown, Maryland

Departments

The Honorable James M. Beggs
Administrator
National Aeronautics and Space Administration
Washington, D.C.

The Honorable Clarence J. Brown
Deputy Secretary-designate
Department of Commerce
Washington, D.C.

Mr. Llewellyn Evans
Assistant to the Associate Deputy Administrator
National Aeronautics and Space Administration
Washington, D.C.

White House Staff

Edwin Meese III
Richard G. Darman
Craig L. Fuller
George A. Keyworth, II
Gilbert D. Rye

Outline of Commercial Space Policy Issues

1. How does the Administration insure a consistent Federal space policy?
2. What constitutes a fair and favorable pricing policy?
3. What economic incentives should be considered to promote commercial space activity?
   - profit
   - tax credits
   - depreciation
   - low cost capital
   - free R&D flight time
   - risk sharing/risk reduction
   - insurance pools
4. What Federal funding/program commitments are important?
   - shuttle availability
   - space station: manned and unmanned
   - basic research
5. How will property rights be protected?
   - patent law
   - proprietary protections
6. What techniques should be used to expand the market for commercial space activities?
   - special government procurement policies
   - period of exclusivity for high risk, high-cost, high-benefit ventures
   - facilitate private sector access to government data
   - provide market guarantees for space products
   - heighten awareness about commercial space ventures
7. What is the appropriate role for NASA?
   - STS operations
   - research
   - regulator
8. What regulatory barriers exist that could retard commercial space development?
   - overlapping jurisdictions
   - unfavorable regulations
9. What national security issues affect commercial space ventures?
   - need to reexamine classification policies and procedures
   - reassess international space issues (technology transfer, foreign cooperative projects, etc.) given foreign competition

Document III-19

Document title: Craig L. Fuller, The White House, Memorandum for the Cabinet Council on Commerce and Trade, "Commercial Space Initiatives," April 10, 1984, with attached: "Private Enterprise in Space—An Industry View," pp. iv–v.

Source: Documentary History Collection, Space Policy Institute, George Washington University, Washington, D.C.

This memo and the attached document (below is only the introduction) summarized a series of issue papers prepared by business and government interests. The White House was a strong supporter of space commercialization, and this memo detailed the initiatives that industry felt would be necessary for the government to begin opening space to business opportunities. Craig Fuller was a White House staff member for the Cabinet Council on Commerce and Trade who had a particular interest in space commercialization.

April 10, 1984

Memorandum for the Cabinet Council on Commerce and Trade

FROM: Craig L. Fuller [hand-initialed: "CLF"]

SUBJECT: Commercial Space Initiatives

The President and Congress have expressed strong support of expanding private sector involvement in space. The success of recent shuttle flights has stimulated interest in possibilities of profitable free-enterprise businesses in space. The attached set of issue papers were developed by a diverse group of business leaders who met with the President last summer. The issue papers deal with initiatives that the Nation might take to help stimulate commercial space endeavors. With the Government as a partner, private sector enterprise can help turn space into an arena of immense benefits for our Nation.

In light of the President's desire to encourage such private investment in space, I would appreciate your having your staff review the attached issue papers. Please appoint a representative to serve on a Cabinet Council for Commerce and Trade Working Group that will be responsible for assuring appropriate coverage of critical agency concerns. NASA will chair the working group to discuss agency comments.

Please provide initial comments by c.o.b., April 16th.

cc: Members of SIG/Space

****

[this page and the following pages are rubber stamped "DRAFT"]

Private Enterprise in Space—An Industry View

The following analyses of potential commercial initiatives were drawn by a 15-member *Commercial Space Group* made up of representatives from diverse private sector firms. They examined opportunities in and impediments to the commercial use of space. . . .

[iv] <u>INTRODUCTION</u>

Historians may look at the 1980's as the beginning of an industrial revolution in space. They may pinpoint these years as the period in which U.S. business and Government joined in partnership to set up shop in orbit. The industrial move spaceward may presage a new economic and social expansion as well as a reemphasis of the United States' technological leadership.

Private undertakings in space promise the same rewards for our national welfare which free enterprise has historically bestowed on our people—jobs, higher living standards, new outlets for innovation and imagination, additional stimulation of technical education and new possibilities for investments and profits. It also can be expected to enhance our balance of payments and our national security and prestige.

Nine spectacularly successful flights by NASA's two Shuttles have shown that our nation is on the verge of a space transportation system sufficiently dependable to support space industries. Facilities for permanent manned operations in orbit are becoming feasible. For the first time it could become possible to assure industry of routine access to orbit and a suitable place to work once there.

Though the technology is ripe, manmade barriers block or slow private sector entrance to space. Laws and regulations enacted long before private investments in space were envisioned still govern commercial space operations. Onerous tax and tariff laws and regulations and outdated or inappropriate administrative mechanisms are discouraging even some of the staunchest advocates of investment in commercial space endeavors.

To these artificial handicaps must be added the natural high risks and costs inherent in space operations. Expenditures for building and launching research and manufacturing equipment for use in space can require investments from 10 to more than 100 times as large as for comparable facilities on the Earth. The danger of loss is many times greater in space. The possibilities of quick profit are low.

Yet, the ultimate social, economic and technological benefits for our nation and individual citizens justify the risks. Privately owned and operated, highly profitable communications satellites are already demonstrating that free enterprise in space works well.

Prolonged near-weightlessness and other unique attributes of space may make possible the manufacture of unprecedented products: medical preparations for fighting some of our most widespread diseases; alloys stronger yet lighter than any presently known; electronic components for faster and smarter computers and better electronic machines than are now available, and systems for almost universal information availability to increase the diffusion of knowledge.

[v] Domestic and international markets for space-based products and services are estimated to be immense and they may grow geometrically. Understandably, competitors abroad are experimenting in all of these fields. Foreign subsidies are often large and extend beyond research and development into production and marketing.

The accompanying 20 issue papers, in six categories, discuss each of the major problems connected with the commercialization of space:

National Commercial Space Policy. Because commercial developments in space often require many years to reach the production phase, entrepreneurs need assurances of consistent Government actions and policies over long periods.

Economic Incentives. Laws and regulations which discriminate against commercial space ventures need to be changed or eliminated.

Expanded Government Research. In partnership with industry and academia, NASA needs to expand basic and applied research and improve dissemination of research results which may have implications for investors aiming to develop marketable products and services.

Role of Federal Agencies. Responsibilities of U.S. Government Agencies relating to commercial space activities need to be firmly assigned and clearly defined.

Legal and Regulatory Barriers. Laws and regulations predating space operations need to be updated to accommodate space commercialization.

National Security Issues. Gaining access to Government-owned technical information and assuring fair international competition are among major concerns of prospective investors in space endeavors.

The entrance of free enterprise into space for commercial activities conforms with national traditions. Private initiative has been the foundation of our nation's development and progress from its beginning. Even during the earliest explorations of the North American continent, explorers and pioneers were followed by traders and craftsmen who came to serve new settlements. Now, industrial entrepreneurs are following our astronauts into the new realms.

Commercial expertise will perhaps do for space what the earliest American settlers did for our continent. They turned forbidding regions into prosperous and hospitable inhabited areas.

Commercialization will also perhaps do for space what Charles Lindbergh did for aviation. It will show that space is a vital arena for commercial and industrial activities. Outer space is perhaps the 21st-century equivalent of a new continent waiting to share its wealth. The partnership required for these undertakings by the Government, industry, academia and other sectors in our society can only strengthen our nation. Space commercialization is perhaps as much our nation's manifest destiny as was the taming of lands earlier in our history. . . .

Document III-20

Document title: "Feasibility Study of Commercial Space Manufacturing, Phase II Final Report," Volume I: Executive Summary, MDC E1625, McDonnell Douglas Astronautics Company, East St. Louis, Missouri, January 15, 1977, pp. 1–2, 8–20.

Source: NASA Historical Reference Collection, NASA History Office, NASA Headquarters, Washington, D.C.

This is one of a series of studies that were done for the Marshall Space Flight Center in the late 1970s. Initial NASA funding of industry to look at manufacturing goods in space led to further R&D by companies and to experiments on the Space Shuttle.

15 January 1977

Feasibility Study of Commercial Space Manufacturing

**Phase II
Final Report**

Volume I
Executive Summary

[1]   SPACE MANUFACTURING

1.0 INTRODUCTION

[originally set in two newspaper-style columns] Space processing experiments conducted during the Skylab and ASTP [Apollo-Soyuz Test Project] missions have shown that the space environment has some unique effects on materials processing. It is potentially possible to translate these effects into tangible benefits such as commercial products produced in space. To fully develop this potential, however, requires industry participation to guide, direct and implement space manufacturing operations. The purpose of this study was to examine the feasibility of establishing commercial space manufacturing operations, and thus assess the potential participation of commercial industry. This study analysis was divided into two phases. Phase I assessed the technical feasibility along with a preliminary economic evaluation; Phase II assessed the commercial and business aspects of implementing commercial space manufacturing.

The approach taken was to use a model product to assess the technical and economic feasibility of commercial space manufacturing. The principal data to aid the selection of a model product were the experimental results of space processing experiments on Skylab. Some of the most promising experimental results were in the area of crystal growth, such as uniform distribution of dopant elements(1) and crystal facets which were flat within a few hundred Angstroms(2). Because of these potential improvements, crystal processing was selected for further investigation. Since the most widely used crystal material is semiconductor silicon, and technical and economic data were available as a basis for study, it was selected as our candidate material. One economic consideration that led to silicon selection as a model product was that new products have a better chance of success if they are entering existing, growing markets(3) and silicon was found to be in a growing market.

With the selection of a model product for analysis, the following steps were found to be necessary in formulating an implementation plan for a space-manufactured product:

- Market analysis for system sizing   } Phase I
- Technical evaluation and plant design   } Phase I
- Product value assessment relative to earth product   } Phase I
- Financial analysis for commercial operations   } Phase II
- Risk assessment   } Phase II
- Adjust financial returns for risk   } Phase II
- Organizational evaluation   } Phase II

A nominal market analysis was completed during the Phase I study and is updated for this Phase II analysis. Evaluation of the first three steps were reported in detail in the Phase I final report and will only be summarized here for background information.

[2]   2.0 MARKET ANALYSIS

In order to determine how the value and market for space produced silicon could best be developed, an analysis of the semiconductor market was undertaken to both understand and identify market characteristics. These characteristics included type and extent of market segmentation, trends, product life cycle, and competition.

The market analysis results are illustrated by the projected world market for semiconductor devices as shown in Figure 1. This market is composed of discrete devices and integrated circuits, is currently about $5B, and is growing at an annual rate of 11%. Single crystal silicon for use in integrated circuits requires extremely high quality material in terms of purity and structure. The damaging effect of defects increases with circuit size and is greatest for large scale integrated circuits (LSI). These LSI would be a candidate application for space processed silicon if the promising experimental results could be realized in a developed process.

Crystal manufacturers estimate silicon material sales to be about 8% of device sales; therefore, materials sales [original placement of Figure 1] for LSI, the fastest growing market segment, is projected to be about $0.64B annually by 1985. Ten percent of the LSI segment of the 1985 market was set as a goal. Since there are approximately 10 companies currently producing single crystal silicon, and the market is growing at an 11% rate, this 10% market capture was felt to be a reasonable, conservative objective. This 10% market set our initial production rate which was used to size the overall production system. . . .

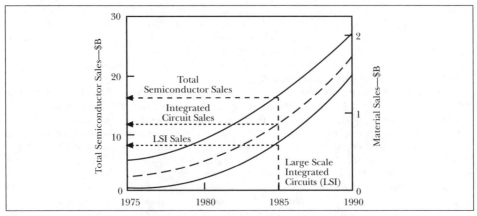

Figure 1. Project world semiconductor market

[8]   5.0 FINANCIAL ANALYSIS

    The ten plant economic analysis baseline was selected on the basis of satisfying 10% of the 1985 LSI silicon substrate market. The estimated 1985 Earth manufactured silicon sales for LSI of $0.64B shown in Figure 1 would correspond to an annual production of 0.47M kg of wafers at $1360 per kg, equivalent in kg to the output of 200 space manufacturing plants. Because of increased processing yield, however, only 110 plants would be required for the same number of integrated circuits with each plant satisfying about 1% of the 1985 LSI material market.

    To illustrate the potential economic feasibility of the space manufacture of silicon ribbon, the allowable cost of $7560/kg is compared with the manufacturing cost of $2930/kg. This comparison results in a positive economic benefit of $4630/kg. In Figure 11 all the costs and sales for the three years of design, development, test, engineering, and fabrication, the five years of production, and one year of run out are included over the five year production period for a ten plant operation.

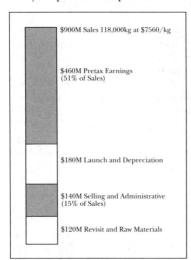

$900M Sales 118,000kg at $7560/kg

$460M Pretax Earnings
(51% of Sales)

$180M Launch and Depreciation

$140M Selling and Administrative
(15% of Sales)

$120M Revisit and Raw Materials

Figure 11. Pretax earnings (10 plants, 5 years)

    Figure 11 illustrates the distribution of all the calculated cost elements for space manufacturing and the resultant earnings. The pretax earnings are 51% of sales. The costs include launch (Shuttle user charge $21M/launch), depreciation, selling and administrative cost at 15% of sales, the shared-launch revisit charge of $840/kg, and raw material cost of $70/kg. If production continues after the fifth year when all the launch and depreciation costs are expensed, the pre-tax earnings for subsequent years would be 71% of sales.

    While it is important that the potential value of silicon ribbon manufactured in space is about two and one half (2-1/2) times the space manufacturing cost for a ten plant operation, space [original placement of Figure 11] manufacturing must also be evaluated in terms of the capital investment and risks involved. Capital investment decisions must take into account the time value of money; that is, a dollar spent on plant today is worth more than the promise of a dollar of profit from future operations. This is accomplished by discounting the future expenditures and receipts (cash flows) at a constant annual rate. If the sum of all the discounted cash flows for a project equals zero, then this rate is defined as the rate of Return on Investment (ROI) or Internal Rate of Return (IRR). This rate is analogous to an annual after-tax interest rate on the dollars at risk. The cash flows for a 10 plant operation are shown in Figure 12. The shaded areas represent the present values of the cash flows discounted at [9] [original placement of Figure 12] the rate of return on investment. The negative cash flows during the first three years represent the investment for plant construction. As shown, the ROI is 29.5%, which for an effective corporate tax rate of 48%, is analogous to a pretax return of 57%.

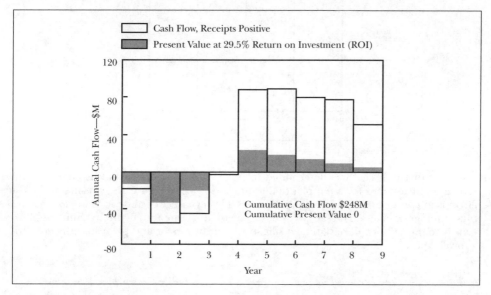

Figure 12. 10 plant cash flow

In a sensitivity analysis of rate of return on investment the number of plants, plant first unit cost, and integrated circuit processing yield (ribbon value) were found to be the major cost drivers, while Shuttle transportation cost was found to have a more moderate effect. The variation of return on investment with the total number of plants deployed is shown in Figure 13. Return on investment for one plant is relatively low, because the design, development, test and engineering (DDT&E) costs are approximately equal to the plant cost or $16M. As the number of plants increases, the DDT&E costs are spaced over more plants and the plant unit cost is decreased by a 91% learning curve, thus increasing return on investment. The return on investment for the selected 10 plant baseline (5% of 1985 market) is calculated as 29.5%. The number of plants also has an effect on the payback period or time required for the manufacturer to get his investment [original placement of Figure 13] back as shown in Figure 14. This figure shows cumulative cash flows, with the low points reflecting the maximum cash investment, the crossover points, the payback periods, and the end points the total after tax earnings. For the baselined 10 plants, the maximum investment is $120M, the payback period is 5.3 years, and the total net earnings are $248M.

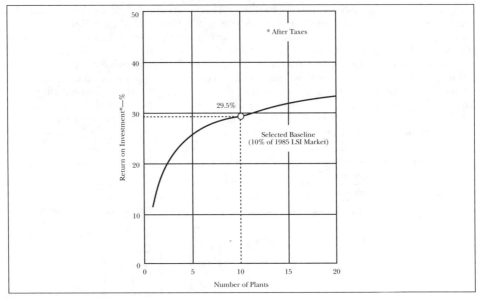

Figure 13. Return on investment

Plant first unit cost was determined to be a major cost driver and its effect on the return on investment for a ten plant operation is shown in Figure 15. This first unit cost effect is in contrast to DDT&E cost, which has a lesser effect because it is spread over the total number of plants. The sensitivity of return on investment to plant first unit cost illustrates the importance of designing a silicon ribbon processor and plant for efficient serial production.

[10]

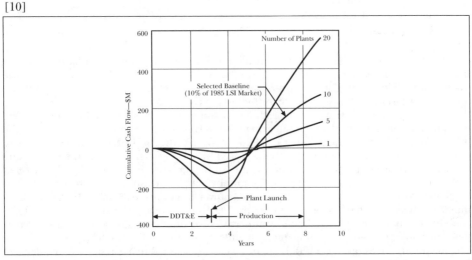

Figure 14. Cumulative cash flow

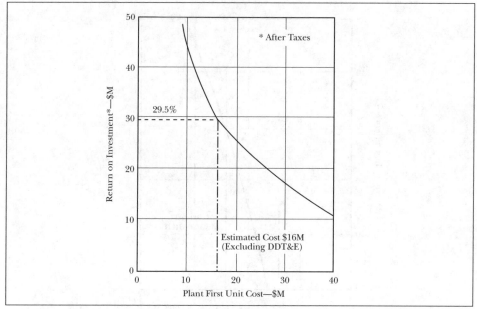

Figure 15. Plant cost sensitivity (10 plants)

The sensitivity of rate of return on investment to integrated circuit processing yield improvement is shown in Figure 16. As noted earlier, the integrated circuit processing cost savings determine its allowable cost premium over ground processed material. The importance of this parameter prompted an independent assessment of yield improvement from the Integrated Circuit Engineering Corporation (ICE), a consulting firm to the industry. For a composite market, with a ratio of MOS device sales to bipolar device sales of 3 to 1, the processing yield for .38 cm x .38 cm baseline chips using space processed material is expected to increase from 14% for ground material to 25% for space manufactured ribbon. It should be noted that a yield increase of only 5% would result in a rate of return on investment comparable to current industry operations.

[Figure 16 originally placed here]

[11] The sensitivity of return on investment to Shuttle user charge is shown in Figure 17. The estimated Shuttle user charge of $21M per launch amounts to a cost of about $840 per kg. This cost is a significant portion of the silicon ribbon manufacturing cost per kg, so return on investment decreases rapidly with increased Shuttle user charge.

[Figure 17 originally placed here]

[12]   6.0 RISK ASSESSMENT

To compare the 29.5% rate of return on investment for ten plants with a typical industry return on investment of 10% requires that it be adjusted for risk. In industry this is usually done by requiring a higher expected rate of return on investment for new projects than that resulting from current operations. In our survey of four crystal manufacturers

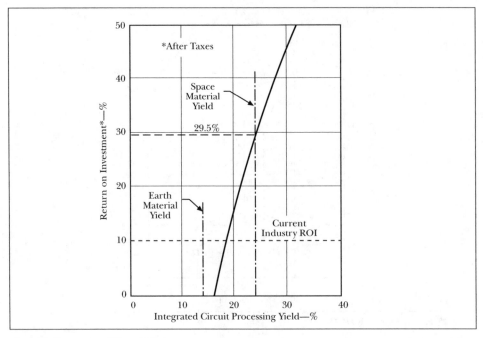

Figure 16. IC processing yield sensitivity

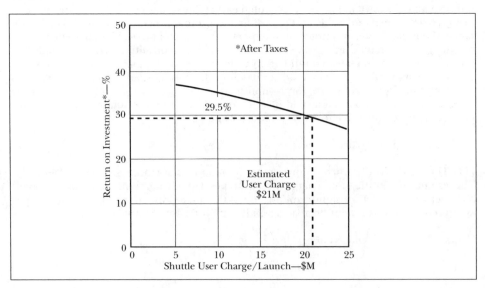

Figure 17. Shuttle user cost sensitivity

the desired average expected return on investment was 20.8%, or a risk premium of about 11% over current operations. The 29.5% projected return on investment for silicon ribbon manufacturing exceeds this criterion by a substantial margin.

Another more comprehensive method of risk analysis is to evaluate risk as a function of time and then risk-adjust the cash flows for use in calculating a risk adjusted rate of return. To do this, first the technical, legal, and market risks associated with the commercial space manufacturing of silicon ribbon are evaluated.

Evaluation of technical risk required definition of the research and development activities for the silicon ribbon process and process apparatus. The research and development required was divided into three phases, ground and sounding rocket research and development, Shuttle sortie development, and pilot plant demonstration. A proposed timetable for these activities is shown in Figure 18. The approximate cost of these activities is $36M, not including launch costs. For the purpose of risk analysis, three objectives were identified as being necessary to the technical implementation of space manufacturing; basic process development, manufacturing plant development, and mission operations. These objectives were divided into thirty-six technical risk elements in three levels of detail similar to a work breakdown structure. These elements were then classified as (1) proven space technology, (2) existing knowledge requiring development for space application, and (3) new technology requiring research. The applicable range of probability of success for these categories is: (1) greater than 0.99, (2) more than 0.95 but less than 0.99, and (3) less than 0.95, respectively. A subjective evaluation was made of where the risk for each element fell within the applicable range. The results of this analysis are shown in Figure 19. The probability of technical success increases with time from 0.22 today, to 0.48 at the completion of ground and sounding rocket research and development in 1981, to 0.69 at the completion of Shuttle sortie process demonstration, and finally to 0.95 at the completion of a pilot plant demonstration in 1985.

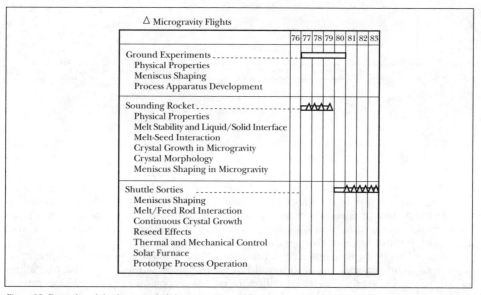

Figure 18. Research and development schedule

Three areas of concern were addressed in evaluating legal risk: patents, trade secrets and liabilities. These concerns were evaluated with respect to U.S. and international law as illustrated by Figure 20. The recommendations indicate [13] [original placement of Figure 19] [original placement of Figure 20] [14] that there are no insurmountable legal problems. Based on this evaluation, a subjective estimate was made of the probability of completion of the manufacturing program through 1990 without losses caused by legal problems (today). This probability of success was estimated as three chances in four. The amount of risk is a function of time[;] thus it decreases to zero as the 1990 completion date is approached. Early concentration on problem areas is expected to substantially reduce the risk to about 10% by the start of sortie missions.

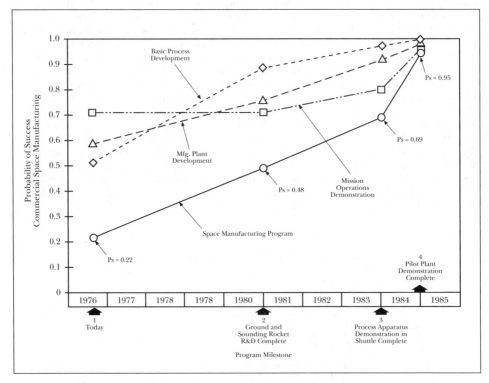

Figure 19. Technical risk assessment

Three areas of concern were evaluated to determine market risk. The first was that silicon might be replaced by another material, the second was that even if the projected market for high quality silicon existed, a competitive ground based process might be developed, and the third was that the space material might not command a premium price and thus not be price competitive.

Analysis of competitive materials failed to identify a material which would supplant silicon as the base substrate material for all LSI technologies. However, the risk was adjusted to account for potential competitive impact of silicon-on-sapphire (SOS) on the metal-oxide-semiconductor (MOS) LSI market segment. In terms of competitive processes, some ribbon processes are being developed as part of the government sponsored solar cell research program. These processes are directed towards the development of low-cost

	International		United States		Recommended Action
	Outer Space Treaty	1958 Space Act		Patent Law	
Patents	Article VIII • States have control • Ownership not affected by location	NASA owns funded development rights • Will grant conditional waivers		Owners have exclusive rights • U.S. laws apply only to U.S. territory	Congress extend U.S. territory to spacecraft (as for ships at sea)
Trade Secrets	Same as Patents	Possible disclosure through safety-of-flight documentation		Common law applies	NASA permit safety-of-flight certification by commercial operator
Liabilities	Article VI • States have responsibility and control Article VII • Owners are liable for damages	No provision		Not applicable	Government formulate licensing and liability regulations

Figure 20. Legal risk assessment

terrestrial solar cells for which the quality requirements are not as high as for integrated circuits. Although the projected material characteristics of solar cell silicon are not acceptable for integrated circuits, process improvements and new developments may occur at any time. Thus, a significant adjustment in market risk was made to account for the impact of ground processes competing for the same market segment.

Essential to the financial success of the space manufacturing venture is the ability to charge a price which provides sufficient revenue for an adequate return on investment. For space-produced silicon this involves the establishment of a price higher than its material cost and at a level equivalent to the value of processing yield improvement benefits.

An assessment of the cost trends of LSI devices identified that as device size increases, costs increase substantially. This is evidenced by the sharply higher cost for recently marketed devices such as a microprocessor as compared to the cost of a relatively simple baseline .38 cm x .38 cm calculator chip. Of greater significance is the increasing dependence upon processing yields to make ever larger LSI devices economically competitive. For future products such as computers-on-a-single chip and 64K Random Access Memories (RAMs), increased yields are critically important.

It is highly probable that, as compared to value of the yield improvement benefits of the relatively small baseline calculator chip, a much higher value will be associated with the use of space-produced silicon for these larger sized future products. Thus, a premium price can probably be charged to differentiate the space-produced silicon from ground-made material which cannot produce similar yield improvement values.

Based on these considerations, today's probability of market success was subjectively determined to be four chances in ten. This market risk is expressed as today's probability that space-produced silicon ribbon will capture 10% of the projected 1985 LSI material market. It is determined by multiplying the probabilities for each market risk factor together at each of the four major milestones prior to project completion. This is depicted in Figure 21 where the early [15] 1977 risk evaluation is established at 0.38, and increases to 1.0 in 1990 at program completion when all market functions are complete and known.

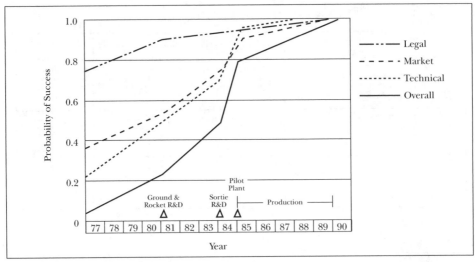

Figure 21. Space manufacturing implementation risk

[16]   7.0 RISK ADJUSTED FINANCIAL RETURN

The combined probability of successful implementation of the space manufacture of silicon ribbon is the product of the technical, legal, and market success probabilities given in Figure 21. The cash flows were adjusted for risk by considering each year as a fork of a decision tree. The value associated with continuing the project for a year is the cash flow for that year and the value associated with failure is zero. Risk adjusted cash flows are the product of the year's cash flow and the series probability of continuing the project through that year. Using these cash flows, risk adjusted returns on investment were calculated for the years of the commercial design, plant fabrication, and production, 1982–1990. The resulting risk adjusted rate of return on investment versus time is shown in Figure 22. While the expected rate of return after the five years of production in 1990 is 29.5%, the risk adjusted rate of return when the private sector would first commit to development in 1982 is 14%. This is greater than the rate of return for continuing ground based operations and could justify initiation of commercial development at that time. It should be realized that this favorable risk adjusted return on investment assumed government sponsored research and development through the pilot plant demonstration in 1984. If industry assumed these risks and cost, the risk adjusted rate of return today would be 4%, and because the 4% is less than the 10% typical return on investment for earth based operations, industry would probably not invest. This means that the government will most likely have to [original placement of Figure 22] sponsor the process research and development for space manufacturing to become a reality.

For the government to participate in any activity it is necessary that it be in the public interest. In general, advancing technology is considered to fall in this category because of the impetus it provides for continued economic growth. An example of government sponsorship of commercially applicable technology is NASA's work in satellite communication. Although Telstar in 1962 was advertised as a $50M investment by a commercial firm, AT&T, the government had already spent approximately ten times that amount to develop and demonstrate the potential of space communications. This technology was later applied by the Congressionally chartered Communications Satellite Corporation (COMSAT) in implementing improved international communications.

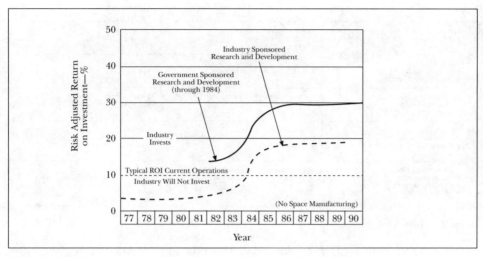

Figure 22. Risk adjusted return on investment

[17]   8.0 ORGANIZATIONAL EVALUATION

A COMSAT type congressionally chartered corporation was considered as a possible organizational arrangement for implementing commercial space manufacturing, and compared to a joint venture and existing separate company(s). While the financing of a Congressionally chartered corporation is aided by its special charter, and it or a joint venture would tend to spread the risk, existing separate companies were chosen, because of the tax advantage of being able to expense the cost of the required research and development against profits from operations as it is incurred. In the chosen arrangement, an aerospace company would design and fabricate plants for sale to electronics companies for the manufacture of silicon ribbon. Implementing space manufacturing through existing separate companies has the additional advantage that Congressional interest and oversight is not required as in the case of COMSAT, and possible antitrust action is avoided that could result from the formation of a joint venture including major industry producers.

[18]   9.0 CONCLUSIONS AND RECOMMENDATIONS

The following conclusions can be drawn about the implementation of space manufacturing:
• A method of assessing the feasibility of space manufacturing has been formulated to evaluate the associated technical, economic and risk factors.
• Product manufacturing in space appears to be feasible, but other product analyses and supporting research are necessary to verify assumptions.
• The cost of space material in comparison with earth material is not necessarily a valid comparison. The value-added from material improvements must be computed and assessed to evaluate economic feasibility.
• Ribbon manufacturing in space appears to be technically and economically feasible, but private initiative may be blocked by the long term, high risk development program required.
• The government should sponsor space processing research and [original placement of Figure 23] development in the interest of promoting future economic benefits to the U.S.

Activity	NASA	Industry

1. Identify Products
  • Space Benefits
  • Economic Potential
	Sponsor Systems Studies	Consultants

2. Proof of Concept
  • Verify Benefits
  • Design Data Base
	Sponsor Ground, Sounding Rocket
	Research and Development
		Consultants
		Experimenters

3. Apparatus Development
  • Processor Design
  • Processor Demonstration
	Sponsor Shuttle Demonstrations	Design Review
		Product Experimentation

4. Pilot Plant Development
  • Systems Design
  • Operational Demonstration
	Sponsor Demonstration
	(Aerospace Plant Design, Fabrication)
		Design Review
		Operations Control
		Characterize Product
		Test Market

5. Commercial Operations
  • Plant Fabrication
  • Orbital Operations
  • Marketing
	Operate Space Transportation System	Venture Capital & Control
		(Aerospace Plant Fabrication)
		Manufacturing Operations
		Marketing
		Commercial Carriage

Figure 23. Recommended roles

Recommendations—The recommendations, in the form of an implementation plan for the commercial space manufacturing of silicon ribbon, recognize the need for NASA sponsorship along with the need to get commercial industry involved. Without early involvement, industry will not have a data base for the subsequent commercial operations investment decision. Specific recommendations, supported by the results of a crystal manufacturer's survey conducted as part of the NASA study, are given in Figure 23. The message is that the government should sponsor space processing activities through pilot plant demonstrations, to decrease the implementation risk to a level commensurate with private venture capital commitment. The government in selecting processes for development can both stimulate economic growth and find solutions to problems of national concern such as the possible application of silicon ribbon in manufacturing solar cells for a solar power station. Without [19] this sponsorship it is possible that the potential economic benefits of space manufacturing may not be fully realized.

The milestone schedule shown in Figure 24 identifies the major activities necessary to the implementation of space manufacturing of silicon ribbon. Included is a recommendation for sponsorship of the activity and identification of the group or organization that should take the initiative for any action. Immediate action toward technical development is necessary if space manufacturing is to become a reality by the 1985 target.

[20]   REFERENCES

1. Witt, A. F., <u>M562 Indium Antimonide Crystals,</u> proc. Third Space Processing Symposium, Skylab Results, April 30–May 1, 1974, Huntsville, Ala.

2. Walter, H. U., <u>M560 Growth of Spherical Crystals,</u> proc. Third Space Processing Symposium, Skylab Results, April 30–May 1, 1974, Huntsville, Ala.

3. <u>Business Week,</u> February 16, 1976.

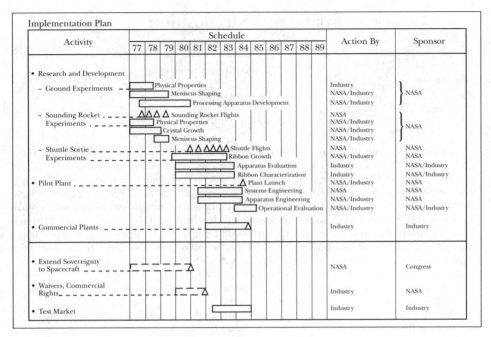

Figure 24. Milestone schedule

Document III-21

Document title: "Space Industrialization: Final Report," Volume 1. Executive Summary, SD 78-AP-0055-1, Rockwell International Space Division, Contract NAS8-32198, April 14, 1978, pp. 1–8.

Source: NASA Historical Reference Collection, NASA History Office, NASA Headquarters, Washington, D.C.

By the late 1970s, with the first launch of the Space Shuttle approaching, and with plans beginning to form for building a space station and/or larger platforms in space, NASA took particular interest in having industry study the possible uses of space for profit-making activities. This represented the first steps in thinking about the commercialization of space activities in areas other than satellite telecommunications. These "roadmaps" for the industrialization of space included economic analyses. Unlike the earlier benefit-cost projections, these futuristic looks emphasized more traditional business tools, including rates of return to investment and the relative demand for goods and services coupled with prices and costs. However, the true value of this and other studies of the era was the identification of space technologies with nonspace market demands that could be met through the use of space.

Space Industrialization
Final Report

Volume 1. Executive Summary

April 14,1978

[1]   INTRODUCTION

Space Industrialization can be defined as a new technology in which the special environmental properties of outer space are used for the social and economic benefit of the people on earth. These special properties include zero-g, hard vacuum, low vibration, wide-angle view, and a complete isolation from earth's biosphere. Design engineers have always been willing to go to great lengths to obtain those specific environmental conditions that fulfill their particular needs. For example, the Hale Observatory was constructed at the top of Mount Palomar so that it would be above a small portion of the earth's atmosphere. Eight million pounds of steel and cement were hauled up the side of a rugged mountain to achieve air density reductions of less than 20 percent. When industrial processes are transferred into space, the environmental conditions are typically modified to a far greater degree. In fact, in-space pressure levels of one-trillionth of an atmosphere are relatively easy to obtain.

Because so few experiments have been conducted in space, it is extremely difficult to envision all the benefits that might result from extremely low pressure levels there. But if the past is a reliable guide, pressures 12 orders of magnitude lower than those encountered at sea level should lead to previously unsuspected benefits. As Figure 1 shows, vacuum levels ranging from 10^{-2} to 10^{-10} atmospheres have already been used in a number of practical ways. These include food processing and preservation (including freeze-drying and refrigeration), metal distillation, x-ray devices, TV picture tubes, thin film deposition, and the manufacture of vacuum diodes and solid state electronic devices. Moreover, many orbiting satellites have already capitalized on the natural vacuum of outer space. For example, when the semi-rigidized Echo balloon was tested on the ground, more than 80,000 pounds of inflating gases were required to inflate it. When it was lofted into the vacuum of outer space, only 30 pounds of gases were needed.

The g-levels and the viewing areas achievable in space are also shown in Figure 1. In comparison with terrestrial conditions, these parameters are improved approximately six orders of magnitude. Precise g-level control is important in medical and chemical centrifuges, in crystal growth, electrophoretic separation, solidification and purification processes, and in the construction of extremely lightweight orbiting structures such as large-scale solar arrays and multibeam antennas. A wide-angle earth-oriented view can be extremely beneficial to meteorology, cartography, reconnaissance, communications, earth sensing, and wide-area navigation—all of these have already brought important benefits to the people on earth.

Thus, it is not hard to see why the aerospace engineer is so keenly interested in the beneficial environmental properties of outer space. But these properties can form the basis of a meaningful Space Industrialization program only if they can be exploited in practical ways. Businessmen are keenly interested in environmental properties, but they are much more concerned with profit-making opportunities to fill real human needs. The first task in our 18-month Space Industrialization study was thus quite clear. To match the needs of humanity with the opportunities for filling these needs through modern space technology.

As they look to the turn of the century and beyond, many people see increasingly bleak prospects for the future. The pressures of population growth continue, particularly in the less-developed countries. The people in these underdeveloped regions are surprisingly young with an average age of about 15, and (hopefully) they will live to see several generations of offspring. By contrast, in the developed countries like our own, the average age is about 29 and steadily increasing. The developed world has nearly reached that magic time when each couple replaces itself [2] [original placement of Figure 1] [3] with only two offspring. Worldwide, however, that is not the case, and it is not likely to be in the near future. As Figure 2 shows, the overwhelming majority of the world's population is in the underdeveloped countries; and within 100 years, their fractional share of world population will nearly double. Of course, their populations will also increase in absolute terms. The birth rates in many areas have recently undergone encouraging declines, but so many of the people in the world are below the critical childbearing ages that the earth is committed to supporting at least double, more likely triple, its present population. Most experts are convinced that the best way to cut population growth rates is to develop a healthy worldwide economy. Emerging affluence has always been accompanied by reductions in population growth rates.

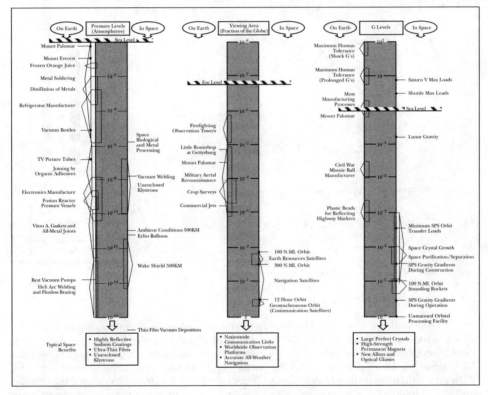

Figure 1. Environmental Properties in Space

One key to a healthy worldwide economy is expanded trade—especially trade that results in a reasonable balance between imports and exports. The United States and other industrialized countries sell large quantities of goods to the developing countries, but today only a discouraging trickle of trade flows in the opposite direction. This negative balance of trade endangers the economies of the underdeveloped countries; it also deprives us of a market that could be provided by the 2 billion people living in the underdeveloped regions. If United States investments (public and private) in the productive use of space could contribute to the economic growth and purchasing power of these poverty-stricken areas, this could have an important positive effect on the economy of our own country and on the rest of the world. Specifically, if we could develop two-way markets of only $17 for each world citizen, 2.7 million new American jobs would be created—enough to reduce our unemployment level to 4 percent of our work force. This would not be a difficult level of trade to attain if underdeveloped regions could be edged toward slightly higher socioeconomic conditions. In fact, as Figure 3 shows, it is only 20 percent of the per capita trade we are already achieving with West Germany and the United Kingdom.

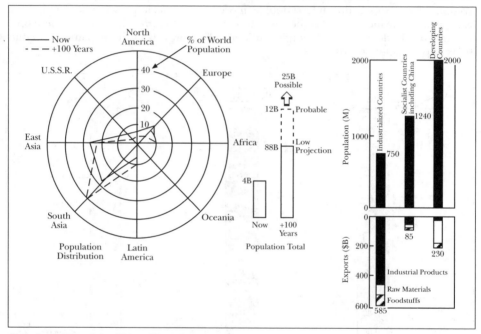

Figure 2. World Population Shifts and Trade Levels

[4] [original placement of Figure 3]

Healthy worldwide trade would also help assure us of uninterrupted supplies of needed raw materials. As indicated in the bar charts of Figure 4, we are spending more than $46 billion each year for imported petroleum products, and we currently import more than 50 percent of 14 important minerals, including platinum, tungsten, and magnesium. Without adequate exports to pay for these crucial substances, our country would quickly slip into a declining economic position. Over the long run, our trade balance has been quite favorable; however, in three of the last five years, our balance of payments has been negative, and in 1977 alone, we almost exceeded the deficits of all previous years combined.

Because of our high labor rates, high-technology items are essentially the only thing we, as a country, can export at competitive prices. Manufactured goods constitute about

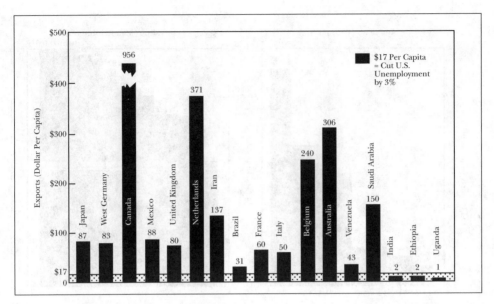

Figure 3. U.S. Exports per Capita to Various Countries

61 percent of our exports, and agricultural products make up another 19 percent.* Unfortunately, other countries of the world have recently been making unusually heavy investments in research so that many high-technology items that were once solid American exports are now becoming common imports. Television sets, steel ingots, precision optics, and automobiles are a few obvious examples. Thus, the only way we can maintain a positive balance of trade is to increase worker productivity or to stay in the forefront of advanced technology. As will be shown in the remainder of this report, space industrialization offers us many possibilities for exercising both of these important options.

In 1973, when the OPEC oil cartel successfully raised the price of petroleum, it was widely feared that other fuels and minerals might also experience substantial price hikes. So far, however, this has not occurred. Prices have remained reasonably stable because of the widespread distribution of most minerals, the tacit threat of substitutions, and the economies of large-scale mining operations. In many cases, however, the quality of ore has significantly declined. In particular, the copper ores now being mined are not nearly as rich as they once were. As Phillip Morrison pointed [5] [original placement of Figure 4] out in a recent *Science American* article, "The ancient miners looked for showy minerals with a copper content of 15 percent. The grade has steadily declined; it was 8 percent in Europe by the time of the Renaissance, and today most copper is won from low-grade ores, the U.S. average grade being about 0.65 percent."**

* Agricultural products are, in effect, high technology items: our farmers use the highest technology farming methods in the world.
** *Scientific American*, March 1978, p. 41.

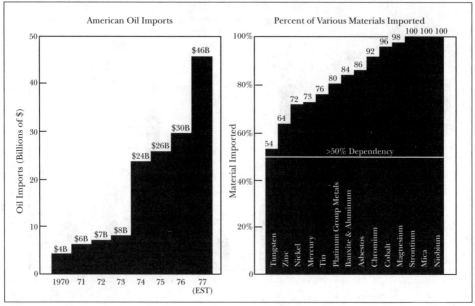

Figure 4. America's Dependency on Oil and Minerals

The recent book *The Limits to Growth* provides an alternate evaluation of the status of the world's future mineral supplies. Figure 5, taken from this popularized book, indicates that at least eight crucial minerals will be exhausted within the next fifty years at the present rates of increased usage. The assumptions made in this study on available reserves and recovery technology have been seriously challenged, and the projections are now widely regarded by most experts as needlessly pessimistic. For this reason and others, the Rockwell analysis team does not believe that these minerals will actually be exhausted in the indicated time frames. When supplies begin to run short, mankind will expend whatever energy and exploration efforts are required to locate and obtain needed supplies. Substitutions will also occur. Nevertheless, Figure 5 highlights a crucial problem: Our known mineral reserves are not infinite; large new supplies will be needed by future generations.

Fortunately, as is shown on the right-hand column of the chart, many techniques are available for expanding our recoverable supplies. These include intensified exploitation, the extraction of minerals from sea water, and the exploitation of ocean-floor reserves. These techniques could expand our mineral supplies to an essentially unlimited extent; however, a careful study of the list will reveal that each available technique requires larger inputs of energy than we are now expending. Thus, adequate energy supplies are again a key to a prosperous future for the United States and, indeed, for all mankind. As we shall see, space technology can help ensure that the needed energy will be available through conservation and through the production of abundant new supplies.

[6] [original placement of Figure 5]

In addition to their physical needs, human beings also have psychological needs. These are basically similar for people everywhere. We need to be productive and feel useful (i.e., to have a job). We need an acceptable standard of living that improves each year and a quality of life compatible with our individual heritage. We think the United States must play the role of leader in this worldwide enterprise. The key direction of this leadership should not merely be good stewardship of what we have, but the continued creation of wealth for ourselves and for the people throughout the rest of the world. In the

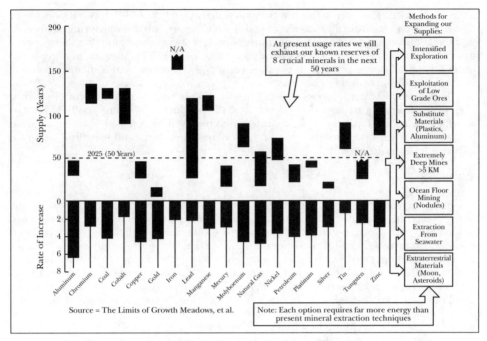

Figure 5. Potential Exhaustion of Selected Minerals

face of population growth, a more just and equitable distribution of scarcity is not enough. For prolonged scarcity makes the future look bleak and disappointing for the average world citizen. What is necessary, therefore, is new ways to create wealth—wealth that will make the world a healthier, more stable place to live.

[7]   THE OPPORTUNITIES FOR SPACE INDUSTRIALIZATION

During the course of this study, we attempted to look 50 years into the future and correlate real human needs with space opportunities. Our work proceeded down two parallel paths. Along one path, we looked into the future for meaningful trends in human needs; and along the other, we searched for practical and economically viable space opportunities. In general, these opportunities can be broken down into the following categories:

1. Services
    • Information transmission
    • Data acquisition

2. Products
    • Organic
    • Inorganic

3. Energy
    • Conservation
    • New energy sources

4. Human activities
    • Space careers
    • Frontier for mankind

As we proceeded into the study, we found that is [sic] was easier (and more fun) to think up new space projects than it was to reduce the list down to a more manageable number. We used the ideas that have been advanced by NASA (Outlook for Space), Ivan Bekey at the Aerospace Corporation, and many many others. We also added many new ideas of our own. The overall lists are presented in Tables 1, 2, and 3. It is not possible to discuss each one individually in any reasonably sized report, but it is reassuring to observe that the known opportunities are quite numerous. It is also encouraging that these opportunities respond to the needs of mankind to a major degree. This suggests that the space program should be considered as a mainstream activity rather than a matter of minor interest, benefiting only a few people.

In the next few paragraphs, we shall briefly discuss one or two opportunities from each of the four major categories: (1) services, (2) products, (3) energy, and (4) human activities. Once this has been done, an integrated plan will be revealed, showing what we believe to be the proper evolution for a relatively ambitious but entirely realistic program of Space Industrialization.

SERVICES

In a very real sense, the services area of space industrialization is already a reality. For several years, space platforms have been providing valuable communication, navigation, observation, and weather services for people worldwide. Some of these services have been earning comfortable profits for corporate shareholders. Today, communication satellites are owned and operated by more than a dozen countries, and more than 100 have their own Intelsat ground terminals. These terminals transmit messages and data to and from such unlikely places as Niger, Bangladesh, Cameroon, and French Guiana [sic].

The utilization of satellite technology in these primitive locations is based on simple economics. The cost of the hardware necessary to handle a satellite voice circuit (see Figure 6) has been declining by a factor of 100 every 12 years In 1966, an Early Bird voice circuit cost more than $20,000 per year. Today Westar provides equivalent voice circuits for a little over $200. Moreover, large-scale Antenna Farms launched by the Space Shuttle may soon bring about further important cost reductions. Hardware investment costs as low as $14 per circuit-year seem entirely within the realm of possibility. With costs at this level, we will begin to see numerous new applications of advanced communications.

[8] Complexity inversion . . . is quietly fostering another revolution in our approach to space communications. Complexity inversion refers to the concept of putting large and complicated hardware in space so that the units on the ground can be small and simple. This philosophy contrasts sharply with the approach that was adopted in the early days of the space program when every effort was made to keep the space segment of the system small and light. In order to do this, the corresponding ground segments had to be massive and complex. For example, the Telstar communication satellite weighed only 150 pounds. This compact design held down launch costs and simplified the satellite, but as a result, it could relay high-quality signals only between such massive ground installations as the 85-foot Goldstone antenna, which weighs 600,000 pounds. Numerous other examples can be found in both civilian and military programs in which huge ground antennas and major computer installations painstakingly processed raw data to extract useful information from weak and diffuse signals radiated by small, compact satellites.

Because of recent advances in space technology and improved transportation capabilities, it is now possible to enlarge the orbiting satellites and, in turn, shrink the ground user sets. In particular, modern electronic devices and multibeam antennas allow the space segment to be vastly more capable and complex, but still stay within reasonable launch cost limitations—especially considering the launch economics of the Space Shuttle.

Table 1. Attractive Opportunities in the Services Area

Communications

Information Relay
- Direct TV broadcast
- Electronic mail
- Education broadcast
- Rural TV
- Meteorological information dissemination
- Interagency data exchange
- Electronic cottage industries
- World medical advice center
- Centralized "distributed" printing systems
- Environmental information distribution
- Time and frequency distribution

Personal Communications
- National information services
- Personal communications wrist radio
- Voting/polling wrist set
- Diplomatic U.N. hot lines
- 3-D holographic teleconferencing
- Mobile communications relay
- Amateur radio relay
- "Telegraphing" personal communications systems
- Worldwide electronic ping pong tournaments
- Central computer service (for transmitting hand-held calculators)
- Urban/police wrist radio

Disaster Warning
- Disaster warning relay
- Pre-disaster data base (earthquake)
- Earthquake fault measurements
- Disaster communication set

Navigation, Tracking, and Control

Navigation
- Public navigation system
- Global position determination
- Coastal navigation control
- Global search and rescue locator

Tracking and Location
- Implanted sensor data collection
- Wild animal/waterfowl surveillance
- Marine animal migrations
- Vehicular speed limit control
- Rail anti-collision system
- Nuclear fuel locator
- Vehicle/package locator

Traffic Control
- Multinational air traffic control radar
- Surface ship tracking

Border Surveillance
- U.N. truce observation satellite
- Border surveillance
- Coastal anti-collision passive radar

Land Data

Agricultural Measurements
- Soil type classification
- Crop measurement
- Crop damage assessment
- Global wheat survey
- Crop identification/survey
- Agricultural land use patterns
- Crop harvest monitor
- Range land evaluation
- Crop stress detection
- Soil erosion measurement
- Agricultural acreage survey
- Soil moisture measurement
- Soil temperature monitor

Forest Management
- Timber site monitoring
- Logging residue inventory
- Forest stress detection
- Forest fire detection
- Rural/forest environment hazards
- Lightning contact prediction/detection

Hydrological Information System
- Snow moisture data collector
- Wet lands monitor
- Tidal patterns/flushing
- Water management surveillance
- Irrigation flow return
- Run-off forecasting
- Inland water/ice cover
- Subsurface water monitor
- Water resource mapping
- Soil moisture data collector
- Irrigation acreage measurement
- Aquatic vegetation monitoring

- Underwater vegetation survey
- Lake/river suspended solids
- Sediment measurements (rivers)
- Flooded area monitoring

Land Management
- Land capability inventory
- Land use mapping
- Wild land classification
- Range vegetation mapping
- Rangeland utilization/population
- Flood damage assessment
- Beach erosion

Pollution Data
- Advanced resources/pollution observatory
- Salt accumulations (irrigation)
- Agricultural pollutant monitoring
- Lake eutrophication monitor
- Great Lakes thermal mapping
- Effluent discharge patterns
- Toxic spill detector
- Air quality profilometer
- Air pollutant chemistry (Freon)
- Pollution detection and distribution
- Mosquito control (wetlands flooding)

Resource Measurements
- Oil/mineral location
- Drilling/mining operations monitor

Geographic Mapping
- Urban/suburban density
- Recreation site planning
- High-resolution earth mapping radar
- Wildland vegetation mapping
- Offshore structure mapping

Weather Data
- Atmospheric temperature profile sounder
- Rain monitor

Ocean Data
- Ocean resources and dynamics system
- Marine environment monitor
- Oil spill
- Shoreline ocean current monitor
- Algae bloom measurement
- Saline intrusion

Global Environment
- Glacier movement
- Ozone layer replenishment/protection
- Highway/roadway environment impact
- Radiation budget observations
- Atmospheric composition
- Energy monitor, solar terrestrial observatory
- Tectonic plate observation

Table 2. Attractive Opportunities in the Products Area

Organic
- Isozymes
- Genetic engineering of hybrid plants
- Urokinase
- Insulin
- New antibiotics via rapid mutation

Inorganic
- Large crystals
- Super-large-scale integrated circuits
- Transparent oxide materials
- Surface acoustic wave devices
- New glasses (including fiber optics)
- Tungsten X-ray target material
- Hollow ball bearings
- High-temperature turbine blades
- Separation of radioisotopes
- High strength permanent magnets
- Magnetic bubble memory crystal film
- Thin film electronic devices
- Filaments for high-intensity lamps
- Aluminum-lead lubricated alloys
- Continuous ribbon crystal growth
- Cutting tools
- Fusion targets
- Microspheres

Table 3. Attractive Opportunities in the Energy Area

Lunetta
- Night illumination for urban areas
- Night illumination for agriculture and industrial operations
- Night illumination for disaster relief operations

Soletta
- Night frost damage protection
- Local climate manipulation
- Reflected light for ground electricity conversion
- Ocean cell warning for climate control
- Controlled snow-pack melting
- Stimulation of photosynthesis process

Other
- Satellite power system (solar)
- Fusion in space
- Nuclear waste disposal

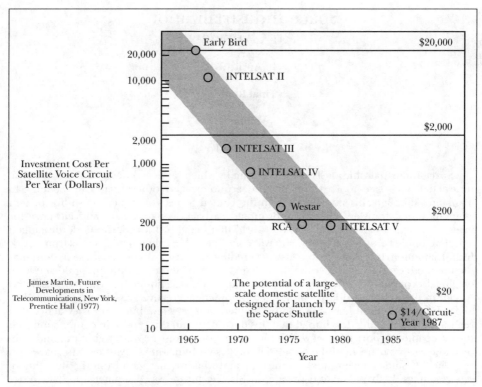

Figure 6. Investment in Satellite Voice Circuits

Document III-22

Document title: "Space Industrialization: An Overview," Final Report, Volume 1, SAI-79-662-HU, Science Applications, Inc., April 15, 1978, pp. 1–5, 10–12, 15–17.

Source: NASA Historical Reference Collection, NASA History Office, NASA Headquarters, Washington, D.C.

This study was similar to the Rockwell study of the same era (Document III-21). It examined detailed industrial sectors and the potential demand and revenues from sample space ventures. It also urged the aerospace industry to look at the possibilities of space industrialization. What follows are excerpts from sections 1, 3, and 4 and a figure from section 5.

Space Industrialization
An Overview

15 April 1978

Final Report

Volume 1

[1]   1. SUMMARY

Space Industrialization (SI) is the medium by which services, energy and products are returned from space to Earth to provide economic and other pragmatic benefits to mankind. Although this study focuses on the United States as the mechanism for benefit generation and transfer (with an appropriate payback to its industry and citizenry for investing resources and labor), it is the world that benefits. Indeed, the underdeveloped and developing countries are now, and will continue to be, prime beneficiaries from Space Industrialization. It is possible to construct credible scenarios which step these nations into the twentieth century equivalent of the U.S. in less than 100 years, without significant local or global economic or environmental damage. The great power for what is considered "good" in the western world (health, safety, knowledge, creative growth, etc.) afforded by Space Industrialization has been comprehended by a very few, but there is evidence that realization is spreading. It is hoped that this document and this report, in conjunction with the companion report by Rockwell International, will assist in this realization, and help promote early expansion of the beneficial returns on humanity's investment in space.

The SAI study concentrated on the U.S. and what we may gain from the investing of our resources, both public and private, in SI. The future was examined to characterize resource pressures, requirements and supply (population, energy, materials, food); also, the backdrop of probable events, attitudes and trends against which SI will evolve were postulated. The opportunities for space industry that would bring benefits to Earth were compiled and screened against terrestrial alternatives. Most survived, and a population of the survivors were [sic] examined to determine if SI would ever be "worth the investment." A cursory market survey was conducted for the selected services and products provided by these initiatives and the results were astounding. Space Industrialization is a billion dollar a year business now; in thirty years it could grow by 100 times that amount—or more!

[2] But, space is expensive. Might not the investment outweigh the gain? Programs of SI evolution corresponding to the postulated future scenarios were developed, and the investments compared to the revenues and their associated benefits. The program analysis results brought two observations: SI investments will be good investments and the sooner the investment, the better for all concerned (in terms of the pure mathematics). It was recognized, however, that certain other factors may control the practical rate of progress.

These "other factors" were examined to the extent practical in this study; a great deal remains to be done. The following observations are in order, however, based on this assessment.

(1) Foreign competition is becoming very strong in SI. It is no longer "our" domain and these pressures will increase. This may limit or spur U.S. increased involvement.

(2) The developing and underdeveloped nations of the world may consider the U.S. and SI a threat or a powerful tool for progress depending on how we promote it.

(3) Prospects for economic return to the government (public sector) are excellent, so long term investments should be justifiable A few billion dollars invested in the eighties will result in hundreds of billions in tax revenues, millions of jobs created, strong economic growth and good balance of trade impacts in twenty years or less.

(4) Although some U.S. industry will resist SI, a strong support base can be built among U.S. private enterprise.

(5) In both domestic and international law there are no legal entanglements which will seriously inhibit SI development, if we develop proper policies and stick to them!

(6) Although many social and political institutions will be affected by SI, the most significant are those institutions governing industry and government relations and those relating the U.S. to the rest of the world. Nothing precludes mutually beneficial arrangements in both of these arenas. Historically, such arrangements have taken several years to evolve.

[3] (7) The most important SI initiatives would appear to have rather high initial investments, and payback periods longer than normal for private investment. A mechanism for reducing initial risk and shortening these payback times is possible and will attract substantial industry support upon initiation.

Thus, in sum, this study has concluded that Space Industrialization exists and is substantial and sustained growth is highly desirable. From examination of the SI programs and their characteristics the following recommendations were drawn.

(1) Strong industry involvement in all areas of SI from planning to ultimate operations is necessary to return maximum benefits.

(2) A central group, perhaps under the Administrator of NASA, especially tasked to plan, integrate and advocate SI activities is needed badly. Such a group, located within the government, may indeed be essential if private enterprise can not meet the challenge on its own.

(3) Space Industries will need 25 to 75 KW of raw power in the early to mid-eighties, 100 to 500 KW in the latter eighties and 1–10 MW in the early to mid-nineties. A Solar Power Satellite prototype development program to prove technical/economic feasibility and environmental acceptability would have similar milestones and characteristics. Space power needs for products have a similar progression, with the possibility of a three to five year lag in demand relative to other requirements. A space power program designed to integrate and synergize these requirements should be initiated, beginning with development of the 25 KW Power Module currently proposed. The requirements for a concurrent large structures program is implicit to the power program.

[4] (4) The cost of space transportation to low Earth orbit must come down below shuttle projections by a factor of 10 to 100 to really open the products market in the nineties. The Shuttle is the key, but the longer term SI requirements are already apparent. Increases in flexibility and decreases in cost are needed by high orbit operations in the latter eighties for both services and energy initiatives. Propulsion and vehicle programs to meet these needs should be integrated into future transportation planning.

(5) The U.S. (probably through the NASA) should embark on an intensive data gathering and planning effort during FY 79, 80 and 81 in parallel to initiation of early projects such as 25 KW Power Module. This effort would culminate in a carefully coordinated, evolutionary Space Industrialization Plan with domestic and international as well as government and industry segments.

The above recommendations imply only modest budget commitments over the next three years (less than five million per year in studies and planning and less than fifty million per year in hardware commitments). The budget requirements for development and implementation of initiatives with early direct returns (mid to late eighties) plus long lead technology development for the nineties has a funding peak of less than four billion dollars annual. That cost could be shared in various ways between NASA, other government agencies, private industry and international (or foreign) organizations. The space technology peculiar funding requirements are less than two billion of the four billion total.

A great deal of work remains before Space Industrialization enters the main stream of government and industry planning, and a proper public understanding is achieved. A solid information base, a dedicated advocacy group and very hard work are the essential ingredients to accomplishing these objectives. The rewards will be worth the effort, and attaining these goals will turn Space Industrialization into the mechanism for achieving the next plateau of human development.

[5] The remainder of this document provides discussion in greater depth in the tasks of the study as outlined in the Summary. Volumes 2, 3 and 4 of this report contain the in-depth discussion and data. . . .

[10]   3. INDUSTRIAL OPPORTUNITIES IN SPACE

The establishment of future markets and a space industrialization program for each future scenario required a compilation of potential opportunities. These were established to a level of detail and breadth of application sufficient to allow gross market survey and preliminary program formulation.

The purpose of this compilation was not to create an exhaustive shopping list of opportunities but rather to key in certain indicative possibilities within each industrial activity identified (Information Services, Energy, Products, People). The goal was of sufficient breadth to insure represensative [sic] program formulation and appropriate market survey. The result of this is a compilation of over 200 potential applications for space related goods and services.

As previously noted, the opportunities and their identified representative usage were compiled under four industry activity categories: Information Services, Energy, Products and People (in space). Each of these categories was further subdivided into subcategories as follows.

Information Services	Energy
Communications	Solar Power Satellite
Observations	Redirected Isolation
Navigation	Nuclear Waste Disposal
Location	Nuclear Power/Breeder Satellite
Sensor Polling	Power Relay

Products	People
Biologicals	Tourism
Electronics	Medical
Electrical	Entertainment/Art
Structural	Recreation
Process	Education
Opticals	Support

[11]   4. THE TERRESTRIAL ALTERNATIVES

Thirty-two candidates for space utilization were compared to potential Earth based alternatives. Comparisons were based on examining the initial cost of installation on a first order basis and a cursory review of qualitative factors such as ease of use, reliability, technology requirements, etc. If costs and capability obtained appeared comparable between the alternatives, they were retained for further study. In certain instances the identified space uses exhibited much lower cost for similar capability or the reverse. These were identified as clearly viable candidates. Where cost and/or capability were clearly superior for the Earth alternative, the candidate was dropped from further consideration.

For five of the thirty-two the terrestrial alternative was deemed clearly superior, seven appeared more favorable accomplished from space and twenty depended too much on specific details (too close to call).

The generic lessons culminate with the conclusion that alternatives do exist, or can be visualized for most space initiatives. "Uniqueness" of the space candidates detailed was not deemed strong enough to warrant special consideration in a competitive environment. Significant technological "lead" for space options was found only in the area of earth resources. And, in the case of communications, implementation may be tipped already toward terrestrial options. In concert with these arguments it is concluded that market softness, in terms of systems requirements, remove the constraint that terrestrial alternative systems must duplicate exactly space products and services.

The implications of the above statements gives rise to the following observations on the viability of terrestrial alternatives.

(1) Complexity from detailed assessment of non-cost issues substantially reduces the opportunity to develop a "winning" mix of space efforts based on generalized benefits.

(2) In lieu of a mandate, space viability must be aggressively advocated/studied against competitors in the mid 1980's.

[12] (3) The current involvement of an existing industry will typically indicate which alternative would be favored by it unless forced by competition to change directions. New entries in an industry will select a path based on investment and risk considerations. Most space initiatives considered in this study will appear highly favorable over terrestrial alternatives only after steps toward risk reduction are implemented. . . .

[15]

**Projected Annual and Cumulative Revenue Potential for
Selected Information Services Initiatives
(1977 Dollars)**

	Potential Revenues (in Millions of Dollars)	
	Annual (Peak)	Cumulative (1985–2010)
Information Services		
Pocket Telephone	20,000	100,000
Teleconferencing	9,000	90,000
National Information Services	6,000	40,000
Electronic Mail	9,000	90,000
Disaster Communications Set	30	500
Advanced TV Broadcast	2,000	8,000
Vehicle Inspection	300	4,000
Global Search and Rescue	50	300
Nuclear Fuel Locators	3	40
Ocean Resources	2	50
Transportation Services (Equipment Sales)	70	400
Rail Anti-Collision System	40	600
Personal Navigation Sets (Equipment Sales)	100	400
Vehicle/Package Locator	300	5,000
Voting/Polling Wrist Set	40	200
	~$47B/Year	~$340 Billion

**Projected Annual and Cumulative Revenue Potential for
Selected Energy Initiatives
(1977 Dollars)**

	Potential Revenues (in Millions of Dollars)	
	Annual (Peak)	Cumulative (1985–2010)
Energy		
Solar Power Satellite (First SAT in 1996)		
49 5GW at 27 MILS/KWH	50,000	300,000
60 10GW at 11.5 MILS/KWH→7.1 MILS/KWH	30,000	200,000
60 10GW at 27 MILS/KWH	100,000	600,000
Urban Night Illuminator	200	2,000
Nuclear Waste Disposal	1,000	3,000
	~30–$100B	**~$200–$600B**

[16]

**Projected Annual and Cumulative Revenue Potential for
Selected Products
(1977 Dollars)**

	Potential Revenues (in Millions of Dollars)	
	Annual (Peak)	Cumulative (1985–2010)
Products		
Drugs and Pharmaceuticals	600	7,000
Electronics		
Semiconductors	2,000	20,000
Electrical		
Magnets	300	4,000
Superconductor (generating only)	2,000	20,000
Optical		
Fiber Optics	80	800
Special Metals		
Perishable Cutting Tools	800	8,000
Bearings and Bushings	200	2,000
Jewelry	100	2,000
	~$6B/Year	**~$64 Billion**

**Projected Annual and Cumulative Revenue Potential for
Selected People Initiatives
(1977 Dollars)**

	Potential Revenues (in Millions of Dollars)	
	Annual (Peak)	Cumulative (1985–2010)
People		
Space Tourism	50	900
Space Hotel	50	600
	~$100M/Year	~$1.5 Billion

[17]

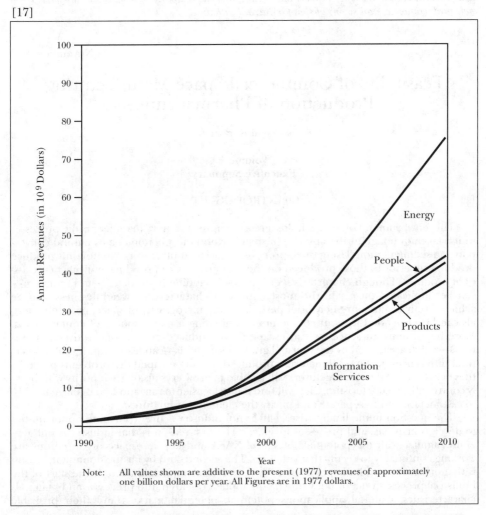

Figure 5-1. Projected Revenues for Space Industry Activities Assuming the Baseline Scenario for Terrestrial Background

Document III-23

Document title: "Feasibility of Commercial Space Manufacturing: Production of Pharmaceuticals," Final Report, Volume 1, Executive Summary, MDC E2104, McDonnell Douglas Astronautics Company, St. Louis Division, Contract NAS8-31353, November 9, 1978, pp. 1–3, 26–30.

Source: NASA Historical Reference Collection, NASA History Office, NASA Headquarters, Washington, D.C.

This is another one of the series of studies conducted in the late 1970s for the Marshall Space Flight Center. In the 1980s, McDonnell-Douglas teamed with Johnson and Johnson to experiment with an improved electrophoresis instrument that could make new drugs in microgravity aboard the Space Shuttle. Although the program was publicized as a privately funded effort, the genesis of the project can be traced back to the early support that NASA gave the company in looking at all aspects (technological, economic, and so on) of space industrialization.

9 November 1978

Feasibility of Commercial Space Manufacturing Production of Pharmaceuticals

Final Report

Volume 1
Executive Summary

[1]   1.0 INTRODUCTION

The environment of space holds great promise for new manufacturing processes which take advantage of the absence of such earthbound phenomena as natural convection and sedimentation. Using these processes, space manufacturers can not only produce products superior to those produced on the ground, they can produce entirely new classes of products. Though characteristics of space—including high vacuum and radiation—can be duplicated on earth, the most important characteristic, weightlessness, can be achieved only for an extremely brief period. In the microgravity of space, molten materials can be suspended without containers—eliminating a major source of contaminants. More importantly, in space we can escape gravity-induced convection. Convection currents—which are caused by the thermal gradients in fluids—can lead to undesirable structural differences in the solid materials produced. Having escaped the problems posed by these currents, space manufacturers will be able to grow crystals of great purity with highly controllable characteristics; they will find it much easier to mix and homogenize liquids, to cast metals, and to separate and purify the elements of mixtures.

The question immediately arises, why is not industry actively pursuing opportunities to develop materials and processes in space? The first reason is that industry is not generally familiar with the potentials of space. NASA and key aerospace organizations are working continually to rectify that situation. The second, and by far the dominant, reason is that observation of basic phenomena with potential application is only the start of the industrial process. A major body of data on applied research into processes and materials characteristics, material applications potential, potential markets and their probable growth, and the characteristics of production systems and logistics must be developed as a vital decision base. Before private industry will invest the money required to begin such untried processes, it must be reasonably confident that the product will have a high value, that the benefits of processing in space will be substantially greater than processing on the ground (i.e., capable of producing less expensive, more useful products, or producing products that cannot be made on earth). The investor must also be reasonably confident: that the space process can be developed in a given time at an affordable cost; that a market exists at a price which assures a reasonable return on investment; and that this market will [2] not disappear because a new product appears and captures the market, or because a breakthrough in the technology occurs that permits competitive ground production.

Because these risks are so difficult to assess, and because the required initial investment is so large, most industries adopt a "wait and see" attitude. Until more data are available, industries find it extremely difficult to assess the potential of new processes and products.

To address this problem, we approached NASA with a proposal to study the feasibility of commercial manufacturing of pharmaceuticals. The goal of this undertaking was to induce pharmaceutical firms to participate actively, on a continuing basis, in exploring the possibilities of using the unique environment of space to produce new products. The MDAC-St. Louis' approach was, first, to secure the initial commitment of these firms by providing key management and technical executives with preliminary data and forecasts of the business and technical potential of space processing. The second aspect of the approach was to foster the initial commitments by establishing continuing technical and management exchanges with the interested pharmaceutical companies to our mutual benefit.

Our enthusiasm for space processing focused on the promise shown by our company funded efforts with electrophoresis. In order to accomplish the facets of this goal we had to expand the data base we had developed—including significant laboratory work and an awareness of the state-of-the-art—and we had to target companies potentially interested in the benefits of the process.

In our early company funded work with electrophoresis, we learned how to separate relatively large quantities of test materials. We also experienced the adverse effects of gravity on the process—causing the vertically flowing stream to collapse on itself (if the sample were denser than the carrier fluid) or to ball up and float to the top of the chamber (if the sample is less dense than the carrier fluid). On the basis of these experiences we began developing, with MDAC-St. Louis funds, our own mathematical models of these effects so that we could predict effects of design changes and operating conditions, and ultimately forecast the benefits of operating in space. We also ran company funded preliminary mass balance calculations; these activities assured us that we could define [3] and demonstrate the types of requirements needed to characterize conceptual space and ground production systems, with their requisite logistics capabilities, in presentations to NASA and industry.

Under the contract, we addressed the problem of targeting pharmaceutical companies. Our first step was to engage the services of Price Waterhouse and Company to provide important drug industry data. The overall drug industry analysis provided by Price Waterhouse included: detailed assessments of the top twenty companies in the industry, focusing on their apparent commitment to innovation, their research and production emphasis on products having high potential for space production, and the prominence of their executives. Price Waterhouse also helped us prepare the presentation to be made to these companies, recommending a "businessman to businessman" approach.

Letters were written to the selected companies. These letters gave an overview of the feasibility study, listed some of the potential benefits to pharmaceutical manufacturing by processing in space, and requested an opportunity to make a presentation. Ten of fourteen companies requested the presentation.

Although the pharmaceutical company personnel were generally skeptical at first, once they understood the benefits of microgravity, the implications of the preliminary results of continuous flow electrophoretic separation, and the potential of an integrated space pharmaceutical production system manufacturing products of great value, they became increasingly intrigued. As a result of these initial contacts, six companies responded positively to our invitation for assistance and cooperation in this study. Two companies agreed to participate actively in the form of laboratory testing a product of specific interest to themselves. Four additional companies agreed to participate in a more passive mode by suggesting products, providing marketing information and reviewing the analysis of results. One of the conditions for their participation, however, was that the companies not be linked with any potential product or process data because of the highly competitive nature of the industry. With NASA concurrence, therefore, and participating company approval, we have deleted the names of any company associated with this study and, instead, emphasized the important product and process information obtained from them.

This report describes our method of obtaining pharmaceutical company involvement, the development of protocols with two of these companies, laboratory results of the separation of serum proteins by the continuous flow electrophoresis process, the selection and study of candidate products, and their production requirements. From the twelve candidate products discussed with, or suggested by, the visited pharmaceutical companies, six were selected for further evaluation: antihemophilic factor, beta cells, erythropoietin, epidermal growth factor, alpha-1-antitrypsin and interferon. Production mass balances for antihemophilic factor, beta cells, and erythropoietin were compared for space versus ground operation. Selection of the best mode of operation for these three representative products permitted a conceptual description of a multiproduct processing system for space operation. Production requirements for epidermal growth factor, alpha-1-antitrypsin and interferon were found to be satisfied by the system concept.

In the technical interchanges that occurred with these pharmaceutical companies, significant data were generated and many valuable lessons were learned. These data and lessons, detailed in this report, are intended to serve others interested in exploring the possibilities of space processing. . . .

[26]   6.0 LESSONS LEARNED

During this study a number of lessons have been learned about obtaining and fostering commercial producer participation in studies of space processing. These should be given due consideration in formulating plans for future studies of this nature. They are presented here in brief, and MDAC-St. Louis recommends that they be adopted as elements in the NASA model for exploring other market sectors considered for the commercialization of space.

The key to involving industry in space processing is to establish a fully business-like footing for their participation. In most cases, the producer industry is relatively unfamiliar with the space environment, operations in space and the requirements and techniques of designing and integrating systems hardware to be flown in space missions. Dealing directly with NASA would involve them in a new form of governmental interface to which they are not accustomed.

By establishing a buffer team between itself and the industry with which it desires to build participatory agreements, NASA can establish the businessman-to-businessman relationship so essential to nurturing commercial enterprise in space. The aerospace company chosen for the buffer team should have established a competence in dealing with the particular process NASA wishes to advance as a candidate for production operations in space. Moreover, the company should have made a significant commitment on its own, in terms of funds and manpower, to the development of that process before NASA chooses that firm to serve on the buffer team. The buffer team should also include an independent business analysis firm specializing in the particular industry to be approached. The right business analysis firm not only knows the industries of interest, but is familiar with the particular environment in which the producers operate. It has access to business documentation resources beyond the aerospace horizon; and, most importantly, such a firm will know key management and technical personnel of these companies plus the correct business basis on which to approach them.

The candidate producer firms identified by the buffer team should be subjected to a penetrating business analysis by the business consultant member of the team. This analysis should include such factors as: the firm's annual sales and growth; [27] the size of the company; the new products it has marketed; the tenure of the firm's senior officers; the surplus funds available for investment; the size of the firm's R&D budget; and the identifiable constraints on the firm's growth. In addition, there will be factors requiring evaluation which are peculiar to the specific class of industry being approached.

Having made contact with the companies by introductory letter, follow-up telephone calls should arrange for a formal presentation at the producer's own facility. After establishing a degree of rapport with their business or technical management, the presenting company should tailor each presentation to the interests of the key people in each producer company, i.e., the corporate decision makers and senior technical personnel.

Personnel making the presentation should be thoroughly familiarized with the segment of industry they will be visiting; at least one member should have credible experience in that producer industry. All members should be prepared to speak the vocabulary distinct to that industrial field of endeavor. The presentors [sic] should be a systems team that is capable of addressing all aspects of space processing to the audience's satisfaction. Not only must they be knowledgeable in the area of products and processes, but also familiar with space flight systems and how day-to-day activities in space are carried out. They must be thoroughly prepared, as well, to discuss resource requirements, costs and manpower, and schedules. Inclusion of a life sciences specialist in the team is highly desirable so that questions on man's contributions and requirements in space can be answered.

The presentation approach should reflect the businessman-to-businessman relationship desired between the buffer team and the manufacturing firm. It must reflect that industry is profit oriented rather than knowledge oriented; research must ultimately lead to increased corporate profit. By selecting products of particular interest to that company and presenting relevant market and business forecasts, a profit potential can be demonstrated in a way to engage both the technical and management attention of the audience. Using a conservative approach in the presentation, especially with technical and business values familiar to the audience, will give individuals an excellent chance to contribute to the discussions and to realize that their experience and participation would greatly enhance the program.

[28] In this situation, the presentation approach should reflect that a joint working arrangement between the visited company and the aerospace industry would be mutually beneficial, using the strengths of each partner to achieve new goals. It must convey the attitude: "We are deeply involved and would like you to join us" rather than "You tell us what we can do for you in space." If the presentation features working hardware, mathematical analyses and models, as well as preliminary product data with their related market and risk analyses, the audience will feel that the presenters have a strong corporate investment in the concept, both financially and in terms of manhours of effort.

During the presentation, the question is invariably asked of the presentor [sic], "Why should we manufacturers be interested in space processing?" Placing a good reason for their interest early in the presentation can forestall the inquiry. The reason can easily be developed by identifying the visited company with a candidate product that has the potential of being produced in space and that also complements their already existing product line. We have found it is also essential to indicate very early in our presentations that the products discussed are of the very low volume-very high value type. Many of the proprietary pharmaceutical companies think in terms of large volume-low cost products which are not applicable in the space operations we visualize.

Results of the presentations will develop relatively slowly. The companies visited will take time to digest what is presented and investigate the claims made. Experience in this study has shown that this phase will take about three months. If results of the initial investigation are favorable, the company will present the concept to their corporate management. This second phase usually will take two more months. The formal development of participating documentation (e.g., protocols, agreements, etc.) and approval of budgets will ordinarily consume an additional six to twelve months. All during this time, routine contact between the two companies should be maintained with pertinent management and technological data exchange by both sides as required.

During this period, it is extremely important not to exploit the manufacturing company names or the products they have under consideration. If such information [29] became generally known without the consent of the candidate participant, cooperation would probably be terminated. The privacy of a manufacturing company considering participation must be respected until the firm decides to announce publicly, for itself, its intent to participate in the exploration of space applications.

A data base concerning a potentially profitable candidate product must be developed upon which a technology interchange can be established between the interfacing company and the candidate producing company. Cooperative laboratory activities are essential tools in building the required data base. In this way MDAC-St. Louis established a technology interchange with the participating pharmaceutical companies. This, in turn, we believe, has enhanced the level of interest of these companies in the potential offered by processing pharmaceuticals in space.

Many potential products can be proposed for space processing by reviewing the literature and discussing the subject with professionals in the field of interest. Many of the suggestions may be of interest scientifically for their own sake but will have little chance of being rapidly adopted by the workers in that field if they offer no substantial improvement over existing materials. If the companies do not see a significant return on their investment in research and development of a product they will ignore that product.

Development of market data on products important to candidate participating firms is a key to securing their interest. A search of the literature, supplemented by consultations with clinical authorities will provide the information necessary to develop a picture of the current market open for the model products. Reasonable assumptions, based on the guidance of clinical research teams, will yield use market projections for advanced clinical uses. Finally, appropriate business risk analysis should be employed to assess the market risks for processes and products as they move from initial R&D commitment through ground and flight experimentation to achieve successful flight production demonstrations. By offering these preliminary analyses to prospective participators, much useful data can be obtained during the presentation itself. Most important, the potential participating firms will be assured that the presentor [sic] has a business understanding and an investment attitude appropriate for cooperative endeavors for their mutual benefit. [30] A hardware system that is capable of producing a variety of products with only minor operational changes, e.g., instrument settings and chemical substrates, would offer significant operational logistics and cost advantages to a producer. Using conservative assumptions based on data from the literature, tempered by laboratory experience, the requirements for such a system can be developed and its significant operating characteristics can be defined. This was the approach used by MDAC-St. Louis to define a system for processing pharmaceuticals in space. All of the twelve biological products reviewed in this study could exercise the system in part, or in its entirety, thus supporting the concept of a true multiproduct system.

While this study was done to assess the commercial feasibility of manufacturing pharmaceutical products in space, it serves as a model for those who wish to consider other processes or products in the same environment. The use of the mass balance analytical concept forces a delineation of what must be accomplished in the process for each product in a stepwise fashion. The calculated quantities of materials at each step will quickly determine if the process is feasible with current technology, where the areas of information must be obtained to fill in the gaps, and anticipated recurring transportation costs to haul the material to and from space. While it does not define the total cost of the system, it does give the prospective manufacturer and NASA a general idea of the size, power and weight of the processing equipment as well as the extent and type of storage requirements. The length of the missions will be defined to determine economic feasibility. This will have to be interwoven with the NASA program and schedules to determine if, and when, a vehicle capability will be available to support such a manufacturing facility. Legal and regulatory considerations will also have to be defined.

Recommendations for future work are presented. . . . It is recommended that drug firm involvement be continued and encouraged both in ground research and product evaluation. Because of companies' sensitivity about government interference and disclosure of trade secrets, each company should be dealt with on an individual basis with some other private firm serving as a buffer or interface between the individual companies and the government. Heavier involvement through evolutionary processes will probably lead to direct participation in space activities. Such participation will logically require a user development laboratory for these companies' product developments.

Document III-24

Document title: James Beggs, Administrator, NASA, to William Clark, Assistant to the President for National Security Affairs, August 26, 1983, with attached: John F. Yardley, President, McDonnell Douglas Astronautics Company, to James Beggs, Administrator, NASA, August 23, 1983.

Source: Ronald W. Reagan Library, Sima, California.

After an August 3, 1983, meeting with President Ronald Reagan, John Yardley, the president of McDonnell Douglas Astronautics Company, offered to be the first commercial user of a NASA space station. In the interim, before the station became operational, Yardley proposed manufacturing by using free-flying spacecraft. The business plan revolved around the successful results of experiments in drug production aboard the Space Shuttle.

August 26, 1983

Honorable William Clark
Assistant to the President for
   National Security Affairs
The White House
Washington, D.C. 20506

Dear Bill:

I just received the [handwritten underlining] <u>enclosed letter from John Yardley of McDonnell Douglas.</u> ["This is our" crossed out by hand and replaced by handwritten "re:

the"] [handwritten underlining] <u>first formal industrial commitment to use Space Station commercially.</u> I am confident there will be many more commitments of this kind as we move into planning and implementing a Space Station.

     With best personal regards.

                                   Sincerely,

                                   [hand-signed: "Jim"]
                                   James M. Beggs
                                   Administrator

[handwritten note: "McDonnell will go further than this if asked. They also would make this public if we desire it." hand-initialed: "B"]

cc:
Commerce - Mr. Baldridge
CIA - Mr. Casey

                         ****

23 August 1983

Mr. James M. Beggs
Administrator
NASA
4th and Maryland Avenues, S.W.
Washington, D.C. 20546

Dear Jim:

     After participating in the recent White House meeting on commercial space activity, I thought it appropriate to review the McDonnell Douglas Electrophoresis Operations in Space (EOS) Program relative to the potential development of a man-habited [sic] space station by NASA.

     As you know, McDonnell Douglas and Johnson & Johnson are actively pursuing the development of an electrophoresis process that will use the gravity-free environment of space to produce pharmaceutical products that cannot be economically produced on Earth. We are now developing our first protein product, a natural hormone currently unavailable. Since 1976, we have spent many millions of dollars on this effort. We have been successful in proving the validity of our concept in our first three Shuttle flights with our continuous flow electrophoresis research equipment. We are now designing a production version of our system which will fly in the payload bay of the Shuttle in 1985 and 1986 and therefore, our expenditure rate has greatly accelerated.

     We believe that the potential for manufacturing new and improved pharmaceuticals in space is real and attainable. While shuttle-based research has been successful, this method is slow and laborious. [handwritten underlining] <u>With the opportunity for research and development of new products that a space station would provide, we could during the 1990s bring five times the number of new breakthrough pharmaceuticals to market.</u> Also, the costs of development and production of these new products can be greatly reduced.

Our recent work with live cell material such as the islets (beta cells) being studied in cooperation with Washington University School of Medicine as a potential cure for diabetes, leads us to believe that it will be impossible to automate a facility that could successfully separate live cells by electrophoresis. Unlike protein materials, the sensitivity of [2] living organisms, i.e., beta cells, to operating conditions within the systems dictates a man interface during processing to ensure their survival. If the current treatment under investigation proves successful, it follows that without a space station the probability of achieving a population-wide cure for diabetes is low.

As you know, we are striving to begin commercial production of our first protein product in early 1987 or before. Because a space station would not be available until the early 1990s, we are planning to use dedicated unmanned free flying spacecraft for increased production. Serious negotiations are presently under way with three companies willing to invest private funds to build this spacecraft. We look at this initial commercial step as being only interim.

We have been encouraged by the progress NASA has been making in defining such a space station program. [handwritten underlining] Consider this a formal request for the McDonnell Douglas EOS Program to be included as the first commercial user of a NASA space station.

Assuming our continued success in this activity, you may consider this a formal commitment to use the space station as the major base of operations for carrying out and expanding this new industry.

Sincerely yours,

John F. Yardley
President
MCDONNELL DOUGLAS ASTRONAUTICS COMPANY

Document III-25

Document title: L. Smith, McDonnell Douglas Corporation, "Electrophoresis Operations in Space," briefing charts, September 1983, pp. 6–7, 30.

Source: Documentary History Collection, Space Policy Institute, George Washington University, Washington, D.C.

This McDonnell Douglas viewgraph summary of the EOS program illustrates that the company had prepared detailed business plans and saw potential profits in drug production in space using its electrophoresis instrument. It emphasizes the potential market growth and demand, as well as the relative efficiency of microgravity production compared to similar terrestrial production of these drugs.

_____

[6]   WHY SPACE?

- Gravity Limits the Full Potential of the Continuous Flow System as a Commercial Process
- Gravity Limits Sample Concentration, Flow Volume, and Purity in Continuous Flow Electrophoresis
  - 100 to 800 Times More Throughput in Space for Same Degree of Purity (Varies for Different Product Materials)
  - Five Times Improvement in Purification Potential in Space

- Electrophoresis Operations in Space Necessary to Satisfy Patient Population
  — Ground Scale-up Costs Prohibitive to Achieve Meaningful Plant Output
  — Price of Ground Derived Product Would Be Substantially Higher
  — Purity Levels That Are Achievable Only in Space May Be Required

[7]   INITIAL BUSINESS STRATEGY

Objectives	Accomplishments
Find Pharmaceutical Company as Active Partner	Ortho Pharmaceutical Became Active Partner in July 1978
Obtain Free Flights from NASA to Verify Concepts and Equipment	MDAC/NASA Joint Endeavor Agreement Signed January 1980
Optimize the Electrophoresis Process for Ground Research	Five Years Development of Electrophoresis Technology 1977 Thru 1981
Ensure Proprietary Nature of Process and Hardware	Five Invention Disclosures Written, Two Patents Issued, One Pending
Identify at Least One Product for Development and Marketing	Product Identified and Detail Market Research of Ortho Confirmed Large Market
Assess Feasibility of EOS Program Including Economic Viability of Commercial Phase	Economic Feasibility Established and Presented in Business Plan in June 1981

[no page number]   PARAMETERS OF THE STUDY

Number of Products
- 1 Product Industry
- 6 Product Industry
- 12 Product Industry

Market Capture
- 25% Domestic
- Pharmaceutical Products Only

Modes of Operation
- Shuttle Sortie
- Shuttle + Unmanned Platform
- Shuttle + Space Station

Time Frame
- Until the Year 2000

[no page number]

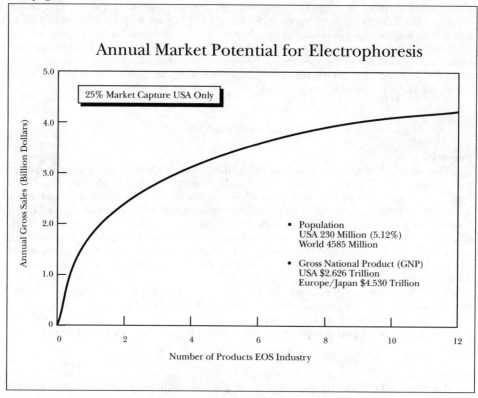

Annual Market Potential for Electrophoresis

25% Market Capture USA Only

Annual Gross Sales (Billion Dollars)

- Population
  USA 230 Million (5.12%)
  World 4585 Million

- Gross National Product (GNP)
  USA $2.626 Trillion
  Europe/Japan $4.530 Trillion

Number of Products EOS Industry

[30]   CONCLUSIONS

- Potential for Manufacturing New and Improved Products in Space Is <u>Real</u>
- Without Long Duration Capability Market Penetration for Any One Product Is Limited
- Unmanned Free Flight Support Will Allow Market Development for One or More Products Within the Limitations of the Space Transportation System
- Manned Long Duration Facility Can Provide the Basis for Industry Growth With Improved Economics

Document III-26

Document title: U.S. General Accounting Office, "Commercial Use of Space: Many Grantees Making Progress, but NASA Oversight Could be Improved," Executive Summary, GAO/NSIAD-91-142, May 1991, pp. 2–5.

Source: Documentary History Collection, Space Policy Institute, George Washington University, Washington, D.C.

In 1986, NASA initiated a program establishing Centers for the Commercial Development of Space at a number of universities. It was aimed at encouraging universities and business to form partnerships to perform R&D on space-related topics. NASA would initially fund the program with the expectation that the business sector would eventually take over the funding responsibilities. NASA would benefit from having customers who would want to perform experiments in space, and the universities and private sector would benefit from the knowledge—and profit potential—of space activities. As this General Accounting Office (GAO) report suggests, this technological development program had some success, but it did not fulfill all of its initial goals.

---

[2]   **Executive Summary**

Purpose

The National Aeronautics and Space Administration (NASA) has long recognized that to help the United States maintain a technological edge throughout the world, it must find ways to encourage and support the development of a domestic commercial space industry. In 1985, NASA began to provide grants and other types of support to the Centers for the Commercial Development of Space to encourage the melding of the resources and talents of government, industry, and academic institutions for researching and developing space-related technologies that have potential commercial applications. After a limited period of grant support of 5 to 7 years, NASA expected the centers to become self-sufficient.

The Chair, Subcommittee on VA [Veterans Affairs], HUD [Housing and Urban Development], and Independent Agencies, Senate Committee on Appropriations, asked GAO to review the extent of private sector involvement in the centers' programs, the centers' progress toward and prospects for self-sufficiency, and NASA's management of the program.

Background

Through 1990, NASA has provided about $81 million in grants to 16 centers, most of which are located at state and private universities. The centers work in one of seven areas of specialization: materials processing, life sciences, remote sensing, automation and robotics, space structures and materials, space propulsion, and space power. The centers, which have operated from 3 to 5 years, reported that by the beginning of 1990 they had established about 300 affiliations with other organizations and companies, and they had completed over 750 flight tests and other experiments, including 18 conducted in space. At that time, they were also planning over 300 more flight tests.

Results in Brief

Since the inception of the program, NASA has had some success in establishing centers capable of attracting and sustaining industry interest and support. It is too soon to gauge the extent to which the program may ultimately achieve its goals, although it is clear that the centers will not become self-sufficient in 5 to 7 years. However, such a fixed period of support applicable to all centers fails to recognize differences among the centers. Recognizing such differences would require NASA to establish grant support goals for the individual centers based primarily on the 3 to 5 years' operating experience each center has had.

NASA also has opportunities to make improvements elsewhere in the program. With the expectation for significant growth in the number of [3] future experiments requiring access to space, the process for evaluating the centers' payload requests should be examined to ensure that it efficiently provides the desired mix of expertise to adequately review requests. Also, NASA needs to examine the adequacy of the internal controls it employs to ensure that its accounting system contains timely, complete, and accurate information reported by grantees on their uses of federal funds.

Principal Findings

Growth of Industry Involvement and Support

Since the inception of the program, the centers have been increasing the number of organizations and companies with which they have become affiliated. More importantly, the number of such affiliates that represent industry has also been increasing, from 63 reported by 6 centers for 1986 to an estimated 199 reported by 16 centers for 1990.

The level of cash support the centers have received from their industry affiliates has also been increasing. In 1986, industry affiliates provided less than $1 million. By 1990, the amount of cash support from industry was estimated at $4.1 million for the 13 centers that received such support. The industry affiliates that were working with a center in 1989 had been doing so for an average of 2.3 years, and almost all of them had provided cash or other types of support to their center.

Centers Will Not Be Self-Sufficient Soon

The proportion of centers' support provided by NASA grants has been increasing, not decreasing. For example, NASA provided 28 percent of the centers' total support in 1986, but by 1990 NASA's share was estimated at 47 percent. The centers' heavy reliance on NASA grants will continue for the foreseeable future. The main reason for this pattern of increasing support is that NASA's overall grant support to help the centers fund the cost of access to space and the cost of unique hardware and facilities has increased.

None of the center directors believe that their centers will be able to continue at their present levels of activity if grant support is withdrawn before 1995. The most optimistic directors believe their centers can achieve self-sufficiency somewhere between 1995 and 2000. At the opposite end of the scale, five directors do not ever foresee a time when their centers will be able to do without NASA grant assistance.

[4] Structure of the Payload Review Process Should Be Reviewed

For about 2 years NASA has used a Payload Selection Board to assist in reviewing the centers' requests for flying their payloads on the Space Shuttle. However, little specific guidance has been provided to Board members about the review process and what they were expected to contribute to it. Some Board members expressed uncertainty and concern about the process and their role in it. In addition, the Board's membership, which was initially planned to include three members representing industry, has not had more than one.

Availability of Good Fiscal Information Should Be Ensured

Timely, complete, and accurate fiscal information on grantees is not routinely available from NASA's accounting system because reporting requirements on the use of federal funds are not effectively enforced. Even after the reports are received, the information is not routinely entered into the agencywide accounting records in a timely fashion.

NASA accounting personnel have been able to get the centers to voluntarily correct various reporting problems, including a number of instances of erroneous and incomplete reporting. However, late reporting has proven to be much more difficult to deal with. NASA accounting personnel estimate that a majority of the required quarterly reports are late.

Information from grantees' financial reports is used to update the agencywide accounting records. However, sometimes such updates are not done until two or more quarterly reports are on file. NASA accounting personnel frequently receive complaints about the lack of current financial information on grantees in the agencywide data base.

Recommendations

GAO recommends that the Administrator, NASA,

- establish, in consultation with each center, a grant support goal with interim targets for tracking progress toward self-sufficiency and for determining the need for, and to help measure the results of, corrective actions;
- review the flight request and approval process to ensure that the expertise needed for such reviews is available in the most efficient manner possible and that those who are asked to assess flight requests fully understand the intended scope of their participation; and
- assess and, as necessary, strengthen the internal controls for ensuring that timely, complete, and accurate fiscal information on grantees is available in NASA's accounting system.

Agency Comments and GAO's Evaluation

In commenting on a draft of GAO's report, NASA said that it provides a useful commentary on one of NASA's newest and fastest growing commercial space programs. NASA noted that GAO's recommendations were reasonable and could be implemented. However, while recognizing the slower-than-anticipated pace of the program, NASA said that determining how and when to establish grant support time limits would be considered in the future. GAO believes that a grant program that is essentially intended to be self-liquidating must include a constantly visible grant support goal to focus and encourage each grantee's efforts to develop alternative revenue sources. GAO recognizes that support goals may change as circumstances warrant, but each such change should be a highly visible management action subject to review and to a determination that the change in the goal, rather than grant termination, is justified.

NASA also offered other specific suggestions, which GAO incorporated into the report where appropriate. . . .

Document III-27

Document title: Leo S. Packer, Special Assistant to Associate Administrator, Office of Advanced Research and Technology, NASA, "Proposal for Enhancing NASA Technology Transfer to Civil Systems," September 26, 1969, pp. 1–9.

Source: NASA Historical Reference Collection, NASA History Office, NASA Headquarters, Washington, D.C.

This Apollo-era document specifically relates to technology transfer. Economic benefits are deemphasized in favor of public and social benefits. The objective of the technology transfer program was primarily to use NASA technology to help solve problems in "public fields." Although technological help for business is discussed, it is generally dismissed as not being appropriate or useful.

Proposal for Enhancing NASA Technology Transfer to Civil Systems

September 26, 1969 . . .

[1]   CHAPTER III

Objectives

Since the assignment for this study was phrased in very general terms, it was helpful initially to break down the subject matter into manageable categories. The following list indicates the scope of the study, with the understanding that it represents a selection of the most important elements, without which one cannot obtain a valid picture. The objectives flow quite naturally from the list.

1) Examine the current status of Federal public policy for R&D, with emphasis on congressional and public attitudes and the guidance of the White House advisory organizations.

2) Understand the history and nature of NASA as a national R&D resource.

3) Obtain some feeling for the diversified and scattered NASA activities in non-aerospace and non-aeronautical R&D. How did these projects arise, how are they justified and funded, and why are they relatively inconspicuous within NASA's program structure?

4) What is the current best understanding of the nature of technology transfer among government, industry, universities, and institutes? What is NASA's perceived role and effectiveness?

5) Assemble a list of national, civil, social problem areas that have some obvious technology components.

6) Develop generalized criteria for evaluation of the above in order to reduce the list to a fewer number suitable for NASA interest and investigation.

7) Recommend a small number of problems for serious consideration as NASA challenges, with appropriate suggestive pros and cons for each.

8) Propose policy, organizational adaptation, and actions that NASA might consider to enhance its responsiveness and contributions to the public welfare, in addition to its major responsibility for assigned missions.

[2]   CHAPTER IV

Observations and Recommendations

I - Observations

1. NASA possesses certain unique talents and experience that are relevant to technology needs of public problems, either directly or with minor modification. These are generally in the categories of specific technology, systems engineering, and organization and management.

2. NASA is faced with an opportunity to take new organizational and program initiatives to apply a small portion of its resources, say two percent, on a continuing basis to technology applications in public fields.

3. There are many indications that the public and government environment is now generally favorable to such initiatives if they are convincingly explained and vigorously advocated. Although some political and jurisdictional problems will be encountered,

an appropriate and timely action by NASA is expected to be welcomed and supported. I believe it would also strengthen NASA's "mainstream" plans and programs.

4. There is a reservoir of potential support among NASA people, based on response to challenge and social sensitivity and altruism. An overt organizational step toward social application of technology, no matter how restricted and cautious, would generate considerable enthusiasm. There would also be opposition based on resistance to change, administrative obstacles, and dislike of unfamiliar, difficult and frustrating problems.

5. I find no difficulty in extending NASA's technology charter to include direct participation in civil problems of national scope. NASA does study, penetrate and exploit unconventional and hostile environments for man, such as zero gravity, underseas, radiation, aeronautical flight conditions, closed life cycle, extreme temperatures, vacuum, etc. Since hostile environments can be natural or social, we can also include crowding, malnourishment, air and water pollution, noise pollution, violence and insecurity, fear, economic dislocation, resource depletion, earthquakes, destructive storms, and one can go on as far as one wishes to include most of the social ills of our time.

6. There should be no doubt that NASA's primary job is space exploration and space operations as well as aeronautical R&D. I believe that the Apollo achievement provides an appropriate time to propose that technology transfer to serve public and social purposes is now a major concern of the Agency, that it will be pursued with the same mission-oriented concentration that characterized the space program. This new policy does not preclude a careful approach to unfamiliar application environments. We have to try to structure as favorable an environment as possible in order that our technology contributions have a maximum impact.

[3] 7. NASA needs a strong and visible focus for people and activities involved in new arrangements, new technology and new applications to civil and social problems.

Recommendations

1. I recommend the formation of a new Program Office (Office of Civil Systems Technology) to assume responsibility for all of NASA's technology activities for public programs, with the exception of those directly tied to aeronautics, space exploration and space operations. The basic components of the Program Office are:

    a) An Advisory Council for policy guidance, composed of people from industry, from other NASA Program Offices, the Administrator's office, Office of DOD and Interagency Affairs, Office of Policy, [National Science Foundation, President's Science Advisory Committee, Office of Science and Technology], Bureau of the Budget, and possibly other agencies.

    b) Technology Utilization. I believe that this activity would be more effective in a line technical organization directly related to its function, rather than in an administrative staff organization.

    c) Market Research and Requirements Analysis Division to provide initial technical and feasibility analysis of proposed problems, evaluation of NASA capabilities, projection of evolving technical needs in public problem areas, interaction with NAE, [National Academy of Sciences] and other agencies and industry, state-of-the-art studies, and exploratory technology studies to provide a basis for NASA decisions.

    d) A program management organization that will coordinate projects in being, initiate new efforts, channel information from projects to management and other agencies, and act as "customer representatives" or "account executives" for outside agencies concerned with NASA work under the Program Office.

e) A Special Mission Development Division that will concern itself with implementing standard and innovative institutional arrangements with other agencies and interests, such as special-purpose institutes, seminars, training agreements, development of R&D cadres for other agencies, NASA-industry cooperation for specific purposes, assistance to public affairs objectives, and other administrative and management support for the Program Office.

I emphasize that no scientific and technological work, other than the Market Research and Mission Requirements Analysis, is to be performed in the Program Office. The reservoir of technology resources will be in the Centers, industry, universities and institutes. The Program Office acts as the integrative mechanism for policy, decision making, planning, allocation of resources, program control, communication and progress reporting.

[4] 2. After deciding on its desired course, NASA management should conduct a discreet program of persuasion among leaders of Congress, Bureau of the Budget, Office of Science and Technology, other government agencies, private industry, and others to gain prior acceptance of the policy.

3. NASA management should identify at Headquarters and in the Centers existing and potential capabilities applicable to those programs and opportunities deemed worthy of NASA participation. Of particular interest are people who would wish to apply their skills and experience in new need environments.

4. NASA should, after application of proper criteria and adequate problem definition studies, select a limited number of promising challenges, negotiate the required agreements, develop objectives, assign resources and move ahead under a new major Program Office as it normally does when assuming new missions. NASA identification, in the scientific and technical communities and in the public eye, with a limited number of major programs of perceived urgency will follow naturally.

5. The new activity should receive separate funding as a line item titled, "non-aerospace technology transfer and applications" to maintain its identity and permit adequate congressional exposure.

II - Observation

Although a broad management consensus and policy are lacking, NASA currently has an impressive number of projects, some of them quite promising, relating directly to public problem areas. Many of them do not show up in the formal management control system. Some of them are disguised, some are bootlegged, whereas some are shown explicitly in the formal system. It would be a formidable task to assemble them for consideration as a single group, but it would be extremely useful to do so, if only to provide integrative management and source information for a splendidly cogent answer to the question of what NASA is now doing for the common man.

Recommendation

NASA should organize a team effort to visit all the Centers, dig into and underneath the formal documentation, interview key people, and assemble a current catalog of efforts applicable to public problem fields, in accordance with clearly understood criteria. This information should be kept current and be made available to NASA public affairs, congressional relations and top level NASA staff, as well as to other government agencies, [Office of Science and Technology, Bureau of the Budget], etc. It is important to make this largely invisible activity respectable and subject to evaluation, planning, and management awareness.

[5] III - Observation

It is a fact that NASA is held up as a model of spectacular achievement in difficult problem areas. It does not matter to the public that there is a wide difference between NASA's technology accomplishments and desired accomplishments in social areas and that comparison of the two is illogical and uninformed. The reality of the situation is that NASA is perceived by the public (and the Congress) as a ray of hope and a source of potential leadership and help with problems that are deeply and emotionally felt. Arguing that space problems and social problems are vastly different and that the latter are much more difficult will have little effect on the public other than causing bitterness toward NASA and the space program.

Recommendation

NASA should make a clear (but not defensive) statement to include:
a) An unequivocal determination to continue primary work in space exploration and aeronautics.
b) A persuasive summary of NASA's impact on science, technology, the nation's economy, and the quality of life in this country.
c) What NASA is doing today to help solve social and civil problems of national importance. This is actually quite impressive when properly presented.
d) An intention to develop new areas for NASA participation in solving civil systems problems, with an outline of organizational and policy steps taken or to be taken.

IV - Observation

NASA does a poor job of bringing its scientific and general technology activities to public attention.

Recommendation

A special public relations effort should be mounted in connection with NASA's work for public welfare and social progress. Enough material for successful exploitation exists at present and more should be available later. Recent speeches by NASA officials have been less than inspirational on the subject of NASA and its relationship to the needs of the nation. We must appreciate and counter the fact that the space program, although exciting and challenging to the imagination, is unfortunately remote from the daily concerns of the common man.

[6] V - Observation

As far as the public is concerned, NASA has expertly demonstrated the "what" and the "how" of space exploration, but has not been as articulate or successfully communicative with the "why" of space exploration. Similarly, there is inadequate understanding of the impact that NASA has had and is causing in technology, the economy, and the quality of life, although some perceptive observers have recently begun to understand this question in its truly dramatic sense.

Recommendation

A comprehensive study of the national impact of space exploration and technology, far broader than anything yet attempted, would be extremely valuable. In Chapter IX,

"NASA Social Impact," I have suggested some of the unique and impressive contributions attributable to NASA. This list could serve as a tentative outline for such a study.

VI - Observation

One frequently hears and sees in print the statement, "If we can put a man on the moon, we should be able to do so-and-so." "So-and-so" usually is a complex social problem.

Recommendation

The proper response to this hostile or uninformed statement should be to point out that:

a) The Apollo program had a clear and unambiguous objective, a realistic time period, an unrestricted long-range allocation of resources, consistent support and a continuing commitment, an available source of people, technology and building blocks of organization, high morale and committed people, and a central authority to run the program. It also had no opposing vested interests.

b) The typical social problem has no clear and unambiguous objectives, no long-range allocation of resources, spotty and controversial support, inadequate numbers of skilled people, inadequate technology and lack of applications experience, no measures of progress, inadequate or non-existent organizations to lean on, and generally no central authority to organize and run the program. A powerful space technology cannot solve all these problems. It can only help when the social problem environment is ready to accept and use the technology.

[7] VII - Observation

In social problems, there are many hazards and obstacles to the successful application of technology.

Recommendations

1. In selecting technology areas to work on, primary emphasis should be placed on those that derive from space capabilities in a rather direct manner. Then we should consider those that require talents and technology unique to NASA. Then we should consider minor modifications and conversions of NASA technology. Following that, we would consider major modifications of NASA technology and significant investments in applying NASA technology to new needs. Finally, we might consider the generation of new technology that does not exist, that no one is working on, and where we have reason to expect a high probability of success. I use the term "technology" to include both hardware and software as well as organizational, management, procurement, legal, personnel, and leadership skills residing within NASA. (Specific criteria for evaluation of proposed opportunities are discussed in Chapter XI, and some of the pitfalls in social fields are mentioned in Chapter VII.)

2. Certain safeguards and cautions must be applied to prevent premature, inadequate or technically unsound approaches to problems. Certain kinds of problems, particularly where technology is not the dominant deficiency, should not be touched until the environment is more favorable for achievement. Certain problems are and will be intractable or unattractive for some years. These factors are discussed further in Chapter XI of this report.

3. In general, I feel that NASA should avoid basic research, hardware development that can be done by industry, social, psychological, behavioral or sociological studies,

operational functions, minor or trivial projects, anything in which industry or government already has a heavy investment and on-going work, anything lacking a direct link to NASA skills and experience, undertakings lacking definable goals and with a very long-range payoff, and projects without a reassuring prospect of success for the overall (not only the technological) objectives.

VIII - Observation

NASA has a very creditable record of interagency cooperation and coordination. Some innovative and imaginative initiatives have been taken by Headquarters and by the Centers but have not exhausted the possibilities for further development of technology transfer arrangements.

[8] Recommendations

1. NASA should critically examine its current work for other government agencies with a view to trimming it down to significant, challenging, and promising efforts for which NASA has a unique capability. We should not be a generalized R&D job shop for other agencies, nor should we do in-house work that can be purchased on contract from industry or universities.
2. NASA should broaden its policy of interagency cooperation to accommodate a spectrum of modes to satisfy different needs and conditions. I would include:
    a) On-the-job training of personnel from other agencies on NASA activities—an excellent start has been made with the Army.
    b) Dedication of specialized personnel to specific tasks for other agencies.
    c) Transfer of facilities and operating staff under certain conditions to other agencies.
    d) Exchange personnel with other agencies by sabbaticals and training assignments.
    e) Joint creation with other agencies of special-purpose research institutes. (For example, possibilities might be aircraft structures, urban systems engineering, crime technology, building systems research, highway safety, synthetic food research, and air traffic control.)
    f) Creation within NASA of cadre R&D organizations to work on technology problems of other agencies with a commitment to transfer a productive, mature and viable activity to the other agency after a number of years. Although many administrative problems will be encountered, I believe they can be solved if the basic policy enjoys strong management support.

IX - Observations

1. One hears a frequent criticism that spinoff from the space program has done little or nothing for business. I take a dim view of prospects of dramatic success in this area. In general, small business is not interested in knowledge per se, it merely wants a specific product or production technique problem to be solved. Our [Technology Utilization] program, which dispenses knowledge, information and reports finds an unresponsive recipient in small business, in spite of its many innovative attempts to identify, package and push its product. Since we cannot send [9] government or contract engineers to small business to solve their specific problems, we will continue to hear their complaints for many years. Until small business realizes that it must develop the recipient capability either by individual or group initiatives (industry or trade associations), it will look longingly and grudgingly at government R&D expenditures. Nevertheless, government must continue to try to develop, by every feasible means, its flow of tech-

nology information to small business. Big business can take care of itself since it is frequently the depository for technology or it knows how to obtain and use it.

2. NASA's Technology Utilization program has been the subject of contentious discussion over the years. What many people fail to realize is that technology transfer is a
social communication process that is just now beginning to be understood. The capability of the source to direct and push application of technology is severely limited.
The entrepreneurial element is often lacking and the receptor environment is often
unresponsive. Neither can be controlled by the source of the technology.
Documentation, screening, identification and dissemination are absolutely necessary
but are not sufficient to [e]nsure the use of the technology. These activities are
among the least potent factors in stimulating the movement and use of technology.

Recommendation

NASA should re-affirm its commitment to technology transfer in its broadest sense, to
the private sector of the economy and to other government entities. There should be less
emphasis on devices, techniques, materials and components since we know that repeated
enumeration of these items, while impressive to engineers, is less than persuasive to the
public. The major emphasis should be on direct technology support of a small number of
major programs and missions of government and industry, especially in innovative
arrangements that help other organizations to apply existing technology more effectively.
The cooperative efforts of people should be stressed rather than the dissemination of
technical documentation and the natural, long-term, diffusion of space technology. NASA
needs to accept the principle that there is no easy shortcut method of technology transfer
(Chapter VI). The most effective methods involve the generous giving of our resources
with no other consideration than being of service. This idea runs counter to conventional indoctrination, but it produces new challenge and the kind of dynamism in an R&D
organization that NASA needs to foster at this point in history. . . .

Document III-28

**Document title: F. Douglas Johnson, Panayes Gastseos, and Emily Miller, with assistance
from Charles F. Mourning, Thomas Basinger, Nancy Gundersen, and Martin Kokus,
"NASA Tech Brief Program: A Cost Benefit Evaluation," Executive Summary, University
of Denver Research Institute, Contract NASW-2892, May 1977, pp. i–iii.**

**Source: NASA Historical Reference Collection, NASA History Office, NASA
Headquarters, Washington, D.C.**

*In the mid-1970s, while NASA's budget was declining from its peak spending period, a series of eco-
nomic evaluation studies was commissioned. The main purpose of these studies was to bolster the
arguments for increased funding based on the premise that the cumulative benefits from NASA R&D
were large enough for the nation to continue to invest in space. Because NASA had an active technology transfer program in place, the monitoring of this program for economic impacts and case
analyses was important information to document the spinoff benefits. The University of Denver
Research Institute had the ongoing contract for collecting this information. This study was performed
to estimate aggregate benefits from the information-dissemination-based technology transfer program.*

NASA Tech Brief Program:
A Cost Benefit Evaluation

May 1977

[i]   EXECUTIVE SUMMARY

A cost benefit study of the NASA Tech Brief Program was conducted by the Denver Research Institute under contract to the Technology Utilization Office. Net benefits to public and private sector organizations due to Technical Support Package (TSP) requests between 1971 and mid-1976 were statistically estimated from random sample data. Program operating costs for the same time period were based on a unit cost analysis conducted by the [Technology Utilization Office] Program Evaluation and Control Division. The study objectives, methodology and results are summarized below.

Objectives

The Tech Brief/TSP Program is one of several operational mechanisms in the NASA Technology Utilization [TU] Program designed to transfer aerospace technology to both public and private sectors of the economy. It is, however, the oldest of these mechanisms, dating back to 1963, and has been one of the principal mainstays of NASA's technology transfer efforts over the years. Tech Briefs and other new technology announcements published by the TU Program have generated an annual average of over 26,000 inquiries since 1964. In addition, NASA has maintained, under contract, a data bank on requests and applications for new technology announced by Tech Briefs since 1968. This data bank contains over 120,000 entries and provides one of the most complete records of any technology transfer program operated by the Federal Government. Based on the availability of data and the request by Congress in the FY 1977 NASA Authorization Bill to conduct "a cost benefit follow-up analysis," the Agency elected to study its Tech Brief/TSP Program. The second objective for this study was to develop an evaluation method which satisfies the Office of Management and Budget guidelines for evaluation managements.

Methodology

Between 1971 and mid-1976, 72,500 TSP requests due to Tech Briefs were recorded in the data bank and 15,500 questionnaires had been returned from the ongoing six month mail questionnaire survey. A two-tiered random sample of questionnaires was selected to assure a 95 percent confidence level for extrapolating the sample data to the entire population of TSP requests. Structured telephone interviews were conducted for the second tier random sample cells defined by request year and questionnaire responses.
[ii] The interview data included responses to the following questions:
   a) What specific use was made of the TSP (e.g., information source on solar energy or developed new computer control software for chemical processing)?
   b) What costs and gross benefits are directly attributed to the particular TSP, how were these quantities estimated, and when did they occur (e.g., number of hours saved in 1972 times the hourly rate including overhead)?
Only data which satisfied Federal guidelines on costs and benefits were accepted for analysis. Standard statistical methods were used to estimate three probability distributions for the sample data, and an expected net benefit value per TSP request was calculated from these distributions. The expected net benefit per request was multiplied by the total requests to obtain the estimated total benefits from requests made between 1971 and mid-

1976. This figure includes net benefits which are expected to occur after 1976, with some net benefit streams continuing into the 1980's.

NASA costs were calculated for each operating year by multiplying the total units (e.g., Tech Briefs published and mailed, TSP's reproduced) times the cost per unit. Unit costs were estimated by experienced TU personnel for all direct and indirect cost factors.

Total net benefits to users were divided by NASA costs to calculate a benefit-to-cost ratio for the Program.

Results

The benefit-to-cost ratio for the Tech Brief/TSP Program is between 10:1 and about 11:1. The total NASA costs, discounted to 1976, were $6.4 million for the five and one-half year period. Total net benefits, discounted to 1976, were between $63.8 million and $72.5 million for requests made in the same time period. Federal tax revenues due to corporate taxes only for these net benefits were estimated to be from one and one-half to three times the Program costs, which indicates that these costs are more than recovered without charging for the documents.

Applications for TSP's were characterized in four application modes, each having an expected net benefit and probability of occurrence:

Mode 0 - no application, $0 net benefit; 34% chance.
[iii] Mode 1 - information acquisition, $100 net benefit, 54% chance.
Mode 2 - improved process, product or service, $5,000 net benefit, 11% chance.
Mode 3 - new process, product or service, $22,600 to $31,100 net benefit, 1% chance.

The expected net benefit per TSP request is about $875, but three out of five requests produce net benefits less than $100.

The benefit-to-cost ratio is quite good for any type of government program, and it compares very favorably with the results from other technical information dissemination programs. The overall assessment for the Tech Brief/TSP Program, based on qualitative data from the interview sample, is also good. A high potential for improving the Program was indicated by further statistical analyses of the data and opportunities for doing so are recommended.

Document III-29

Document title: Robert J. Anderson, Jr., William N. Lanen, and Carson E. Agnew, with Faye Duchin and E. Patrick Marfisi, "A Cost-Benefit Analysis of Selected Technology Utilization Office Programs," Executive Summary, MathTech, Contract NASW-2731, November 7, 1977, pp. 1–6.

Source: NASA Historical Reference Collection, NASA History Office, NASA Headquarters, Washington, D.C.

As a followup to the March 1976 Mathematica study, MathTech, the successor company of Mathematica, analyzed the benefits and costs of several successful technology transfer office projects. The study added to the succession of very positive benefit-cost ratios that NASA was generating during this era in its budget support activities and, in particular, for the support of the technology transfer budget. The economic methodology was straightforward and well documented. However, the narrow focus, which included only the costs associated with technology transfer activities, tended to overstate the ratios and results.

A Cost-Benefit Analysis of Selected Technology Utilization Office Programs

November 7, 1977

[1]   EXECUTIVE SUMMARY

Since its establishment in 1958, the National Aeronautics and Space Administration (NASA) has played a major role in technology transfer through activities which encourage the adoption, by other sectors of the economy, of technologies or techniques developed for the space program. To provide a formal program to support and monitor technology transfer, NASA, in 1962, established its Industrial Applications Office, the predecessor of today's Technology Utilization Office (TUO).

This summary briefly reports the results of a study applying standard methods of cost-benefit analysis to selected program activities managed by TUO. Our primary objective in the study is to analyze the costs and benefits of selected TUO activities, based upon available data.

In order to meet this objective, we have selected a subset of TUO's projects or activities for analysis. The main criterion for selection was the availability of data. To ease the burden of data gathering, we have further limited the activities examined to those conducted during the period 1970 to 1976. Nevertheless, the available data for the projects we have selected are sometimes incomplete, or are subject to considerable uncertainty.

The individual activities that we have analyzed are grouped into two general categories: information activities and applications projects. Information activities are directed toward the production and dissemination of documents describing NASA technology as well as computer programs and documentation. Application projects are designed to support the transfer of a specific technology or technique by participating with others in the [2] development of a new product or process.

We have estimated two different indicators of value for each activity/project we examine. For *both* information activities and applications projects, we estimate the benefits which are directly associated with TUO's costs. For convenience we will call these "TUO Benefits." The "cost-benefit test" which we make in each of our analyses of TUO activities consists of comparing TUO benefits with the TUO costs of the activity generating them. If TUO benefits exceed TUO costs, the activity passes the cost-benefit test. This is an indication that society gained more from TUO's activities (in the form of new information, new processes, or new products) than it lost in the taxes, user costs, and user charges which were incurred in the provision of use of these activities.

While the primary results of this study estimate the costs and benefits of TUO's technology transfer activities, we also present two other special indicators of the economic impact of these activities to provide some perspective. These indicators differ for information activities and applications projects.

For information activities, we estimate the sum of TUO benefits, user charges, and user costs. This number provides an indication of the value society places on the information contained in the transfer media. For convenience, we refer to this measure as "activity scale." The measure of activity scale for information activities is intended to indicate the resources others are willing to spend to extract the information contained in the various media. As such, it provides one (albeit imperfect) estimate [3] of the value of the technology contained in the information. For applications projects, we estimate the sum of TUO benefits and those benefits that are attributable to other participants. We refer to this measure as "applications benefits." Applications benefits are intended to estimate the value to society of the new project or process to which TUO is a contributor.

It is important to recognize that only the TUO benefits measure can be used to pass on the merits of TUO's activities. In addition, activity scale and applications benefits do not measure comparable values. In our estimate of applications benefits, we measure the value of the product or process which includes both the technology transferred through TUO and the technology contributed by other participants. The estimate of activity scale is an estimate of the value of technology contained in the transfer mechanism alone.

A summary of our findings is presented in Table 1. Employing standard methods of cost-benefit analysis, we find that the TUO benefits of those activities we have examined are greater than the TUO costs incurred.

In Table 1, estimates of both types of all measures are expressed in present values in 1976 of the stream of benefits over the relevant period, measured in 1976 constant dollars. For each of the individual activities, TUO benefit results are presented in both dollar terms and as ratios to the corresponding activity cost borne directly by TUO.

Each of the estimates reported in columns 2–4 of Table 1 has been adjusted by a realization probability factor, which is reported in column 1. This realization factor reflects our estimate of the likelihood that positive [4] [original placement of Table 1] [5] benefits have or will derive from the activities we have examined. For the information activities, the realization probabilities are 1.0, because expected benefits of the activities may be inferred directly from actual market data on user demand for the activities. Our estimates of realization probabilities for applications projects are generally less than one. This is because most of the project technologies have yet to reach the marketplace, and accordingly there is uncertainty about whether they will. The method by which these realization probabilities were estimated for the applications projects is discussed in Chapter V. The ratio shown in column 4 is the ratio of TUO benefits, i. e., benefits attributable to TUO, to TUO activity costs. This ratio shows whether or not the TUO benefits of a given TUO activity are greater than its cost. If this ratio is greater than one, then the cost-benefit test is passed.

Table 1

Summary of Estimated Benefits and Benefit-Cost Ratios
of Selected Activities Initiated 1970–1976[1]
(All Benefits in Millions of 1976 Dollars Discounted to 1976)

a. Information Activities

	1 Realization Probability	2 TUO Benefits Plus User Costs and Charges	3 TUO Benefits	4 TUO Benefits- TUO Cost Ratio
Technical Support Packages	1.0	$ 83.0	$ 2.0	1.2
COSMIC	1.0	307.0	6.1	4.1

1.   For Information Programs, the estimates are for the period 1970–1976. For the Applications Projects, the estimates are for ten years after expected (or actual) commercialization.

b. Applications Projects

	1 Realization Probability	2 TUO Benefits Plus Other	3 TUO Benefits	4 TUO Benefit- TUO Cost Ratio
Biomedical:				
Cataract Tool	0.5	31.0	6.4	41.0
Burns Diagnosis	0.5	2.7	1.8	8.2
Meal Systems	0.10	10.5	.8	5.8
Pacemaker	1.00	72.0	.7	4.1
Human Tissue				
Stimulator	0.30	516.0	2.6	9.6
Engineering:				
Nickel-Zinc Battery	0.50	328.0	15.0	68.0
Zinc-Rich Coatings	0.80	68.0	14.6	340.0
Track-Train Dynamics2	0.20	98.0	.02	2.6
Firefighter's Breathing				
System[2]	1.00	6.1	3.8	3.6

The total TUO benefits from the applications projects analyzed are estimated to be $44.9 million with a benefit-cost ratio of 22. However, because the applications projects evaluated do not constitute a random sample, the results reported in Table 1 cannot be used to impute benefits to the overall applications program. The corresponding total and ratio for the information activities we examined are respectively $8.1 million and 2.5. Remembering that the information activity estimates are based upon data on transactions that actually took place during the period 1970–1976, while the applications project estimates are generally for projects that will be completed after 1976, an approximate overall estimate of the TUO benefits of the activities analyzed can be derived from the sum of the benefits shown in the table. When this is done, we obtain estimated TUO benefits for these [6] selected activities of $53 million and a benefit-cost ratio of 10 for TUO's activity.

It is important to understand that our estimates of benefits and benefit-cost ratios should be taken as averages about which some uncertainties exist. There are three main sources of this uncertainty: possible measurement error in the data; possible errors in our modeling of relationships using the data; and errors in forecasting the future. Because the last of these types of uncertainty does not pose a problem for our estimates of the benefits associated with the information activities, those benefits probably have less uncertainty associated with them. However, the nature of the data sources and the compounding of many random events prevents any quantitative estimates of these error bounds from being made.

Based on the information in Table 1, it can be concluded that *for those programs and projects that we have analyzed,* the contribution of TUO in the form of benefits gained through its technology transfer programs is greater than the costs it incurred in the process. . . .

---

2.   Estimated parametrically. . . .

Document III-30

Document title: Richard L. Chapman, Loretta C. Lohman, and Marilyn J. Chapman, "An Exploration of Benefits from NASA 'Spinoff,' " Chapman Research Group, Contract 88-01 with NERAC, Inc., June 1989, pp. 1–5, 23–28.

Source: Chapman Research Group, Inc., Littleton, Colorado.

Since 1976, NASA has annually published the book Spinoff, which reports on successful cases of technology transfer. This study examined the various technologies that have been featured in Spinoff. The Chapman Research Group concluded that the benefits from more than 400 cases may have been as great as $21 billion in sales. These figures do not include any costs, nor any unsuccessful technologies. The primary purpose of this study was to update and expand on the earlier studies of the technology transfer program that were used to support both the NASA budget and the technology transfer budget, which historically has always been under scrutiny.

An Exploration of Benefits
From NASA "Spinoff"

June 1989

[1] The focus of this study has been to explore those applications of NASA technology (or NASA-assisted technology transfer) that have been reported in the annual report, *Spinoff*. The primary purpose has been to identify what benefits resulted from those applications, and, further, to quantify benefits (where possible) toward which the applications made a contribution.

Part I of this report, "Study Approach and Conduct," summarizes the methodology used and the challenges faced by the study team. . . . However, the reader should be aware of several important, general conditions which affect the scope and inclusiveness of this study in terms of *how fully* it captures the benefits of NASA-furnished technology.

First, the *Spinoff* magazine does not include even all of the "good" examples known. Some examples have not been published simply because they are difficult to illustrate in a meaningful way to the general public. Such is the case with the many uses of NASTRAN—a computer program initially developed by NASA for structural analysis of large rockets, and considerably modified for literally thousands of non-NASA applications.

Second, in working "backwards" from known applications, one misses those applications where NASA technology is "embedded" into whatever was applied. That is, the original NASA-furnished technology may have been the basis for a series of modifications during which the original technology, now embedded in the changes, has been "lost" as to its origins.

[2] Third, these benefits resulted from the contributions of only 259 applications of NASA-sponsored or furnished technology. It *excludes* a number of important benefits which should be obvious to even the casual observer: (1) *direct NASA or Department of Defense use:* such as NASA commercialization programs, mission-directed applications (such as weather satellites, communication satellites and the like); and, (2) *social benefits:* such as lives saved, lengthened or improved; labor days saved from illness, accident or death; improvements in the environment or the quality of life; productivity improvements and the like.

As revealed in this study, the technology transfer process includes not only the hardware end of technology, but managerial and economic aspects as well. It includes suppliers and users, inputs and outputs, products and processes. Working back, in an historical sense, also provides challenges of information gaps where people move or forget, organizations which have changed or disappeared, and where there may be a reluctance to fully acknowledge a particular benefit or its origins. In many respects this study represents a serious probe into the complexity (and difficulty) of capturing "spinoff" applications. It certainly demonstrates the need for *early* and *systematic* attention to means for identifying and tracking potential spinoff applications—if only to more fully understand this phenomenon and its contributions to the Nation.

[3]   **PART I. STUDY APPROACH AND CONDUCT**

An examination of the benefits of spinoffs of NASA technology presents a particular challenge in the delineation of study parameters because the scope is vast and the documentation is sparse. Since *Spinoff* articles provide the only continuous source of technology transfer information, this research has as its principal source the articles which appeared in the annual *Spinoff* report between 1978 and 1986.

Defining and Locating Information Sources

The basic information for the study was composed of persons, companies and institutions or agencies that had been mentioned in any *Spinoff* article or on an accompanying list from the *Spinoff* files. Information available from these files was cross-checked with former Denver Research Institute (DRI) case files and with available directories such as *Moody's, Thomas Register, Dun & Bradstreet,* and the *Corporate Technology Directory 1987,* to obtain the most recent corporate or business address, CEO, telephone number, and any other relevant information.

Development of an Interview Guide

An interview guide was developed concurrently with the basic study contact list. A study of the *Spinoff* articles and old case [4] files in conjunction with the study's statement of work and discussions with various NASA Technology Utilization Office and Industrial Application Center personnel contributed to defining what data was needed and what data might be possible to obtain. Earlier studies about the benefits of NASA (and other) research and development were reviewed for content and completeness. From this process guideline questions were drafted, discussed and revised. The interview guide was accompanied by a one page instruction sheet for interviewers.

Data Collection Through Telephone Interviews

Eight months of telephone interviews yielded some 600 useful interviews involving 400 companies. Approximately 2500 outgoing and 500 return calls were made during the course of the study. It took an average of five contacts to obtain a completed interview and the average interview length (including all contacts) was approximately 15 minutes.

There were almost no refusals to cooperate, but it was a challenge to persist until someone with an appropriate corporate memory could be contacted. Contacts were almost universally responsive to a request for help in a study for NASA.

Standardization of Interview Data

A spreadsheet system of record keeping, which placed the technology application into categories determined by "end use," [5] was developed by extrapolating from an array of earlier benefits studies. Monetary data was standardized by using current estimates and converting labor-saving information into dollars whenever possible. Relevant information was highlighted using key words or phrases. . . .

[23]   **PART III. BENEFITS: CONTRIBUTION TO SALES OR SAVINGS**

The primary focus of the study has been the nature and extent of benefits from the application of NASA developed or NASA provided technology. This has been expressed, where it was possible to make estimates of quantification, in terms of either sales or in savings—stated in dollars or as a percentage of business. Where dollar savings could not be elicited from the respondents, emphasis was placed upon man months or man years, and also savings that might be estimated resulting from materials, utilities, equipment, maintenance and even avoided research and development costs. As noted in the section on the study approach, [Chapman Research Group] researchers have attempted to "standardize" these sales and savings benefits (where they were provided) so that the resulting figures presented in this study represent total dollars in sales or savings, even though the initial answers may have been given on a yearly basis, on a percentage of sales, or in man years of effort.

The term sales includes such items as new products, additional sales because of an improved product, or increased sales because of NASA use. No attempt was made to isolate the specific economic contributions of the particular technology or assistance to the full range of sales. However, this report excludes gross sales/savings figures that probably included other products or processes. . . . The complete assurance of [24] excluding all but directly "provable" benefits can only be done through detailed case studies and examination of accounting information from the particular firms involved. Since all of these interviews involved telephone interviews, the researchers relied upon the estimates of the respondents and often accepted total sales figures of a particular product where the technology was used. This means that the NASA technology *contributed* to the sales, but that contribution can vary substantially from a relatively small percentage of the total sales or saving figure to one where a new product or process was completely dependent upon the NASA technology.

Savings include such concerns as increased efficiency, labor saved, reduction in materials, maintenance, utilities and processing costs and research and development avoided.

The various applications were categorized according to end use as described in the *Spinoff* article. This resulted in nine categories: communication/data processing, energy, industrial (manufacturing and processes), medical, consumer products, public safety, transportation, environmental, and other. Leaving aside the "other" category, the largest contributions were made to industrial use, followed by transportation, medical, and consumer products. See Table I, "Benefits Realized from NASA-furnished Technology, Case Applications from *Spinoff* Reports, By Categories of End Use, Sales or Savings," for a breakdown by end use description, showing number of cases, sales and savings.

[25]   **Table I**

**Benefits Realized from NASA-Furnished Technology
Case Applications from *Spinoff* Reports
By Categories of End Use, Sales or Savings, $ (000)***

End Use Description	Number of Cases	Number of cases with sales or savings	Benefits Realized $(000)		
			Sales	Savings	Total
Communi-cation/Data Processing	51	32	171,007	51,964	222,971
Energy	30	13	203,500	15,613	219,113
Industrial (mfg & process)	170	107	5,767,649	67,837	5,835,486
Medical	61	31	2,003,036	30,613	2,033,649
Consumer Products	24	18	1,278,294	524	1,278,818
Public Safety	27	16	347,888	555	348,443
Transportation	40	18	9,887,865	116,623	10,004,488
Environmental	16	11	16,962	21,788	38,750
Other	22	13	1,654,989	10,232	1,665,221
Total	441	259	$21,331,190	$315,749	$21,646,939

*   Estimates were obtained from company officials, or derived from company estimates of manpower or other types of savings. . . . The 441 cases were reported in Spinoff magazine, 1978–86; of these 368 had acknowledged sales or savings, but 109 cases could not be estimated as to extent.

[26] A few comments are appropriate regarding the distribution by end use. It is not unusual, given the nature of NASA's high technology, that the largest share would be directed toward manufacturing and processing where the principal user (in a direct sense) of the NASA technology is a supplier or manufacturer. The ultimate user may be a consumer at the end of a particular chain. However, the end use description here is determining the nature of the use prior to that process being completed elsewhere or the product moving on for a further refinement or other use. For example, the "consumer product" end use description was used only in those instances where the company making the application actually produced consumer goods.

One might also anticipate that transportation would rank high (second) in the use of technology since NASA is one of the primary if not the principal producer of technology for aeronautics and aerospace. Aviation uses of technology clearly were the most predominant within this category of transportation.

Finally, the medical end use category also rates high (here, third) and not unexpectedly so, because of the virtual explosion in medical use of such computer technology as digital-imaging techniques and the like. The microminiaturization of electronic circuits as well as mechanical features, are especially adaptable to medical needs. The development of the Programmable Implantable Medication System (PIMS) and its substantial potential, along with digital imaging used for both brain and whole body scans and subsequent diagnostic procedures are only partial evidence of the [27] explosive growth of this type of technology in the future.

Of 441 separate instances of the application of NASA-sponsored or provided technology, the study team was able to identify 368 cases where the respondents acknowledged that there were contributions toward savings or sales—this amounted to 83 percent of the total cases identified. Of the cases in which sales or savings were acknowledged, 109 (25 percent) involved circumstances in which the respondent either could not estimate sales or savings or was unwilling to because of the proprietary nature of the information.

Of the 259 cases in which the respondents were able to identify sales or savings, it was possible to identify contributions toward sales of $21.3 billion ($21,331,190,000). Contributions toward savings were $315.7 million ($315,749,000). Total contributions toward sales and savings were $22 billion. This figure *excludes* nearly $12 billion in sales that included NASA-furnished technology, but which were given as total sales figures for a company, including all products. . . .

Discussions with corporate officials revealed 67 instances in which a product, process, or even an entire company would *not* have come into existence had it not been for the NASA-furnished technology. These represented 18 percent of all cases involving sales/savings and amounted to $5.1 billion in sales/savings. . . .

[28] *Other Benefits of Economic Value*

Once one has an estimate of additional revenues, it is possible to postulate the revenues or jobs (created or saved) associated with that revenue. Using standard economic projection procedures, it is estimated the Federal Government received corporate income tax receipts of nearly $356 million as a result of these spinoffs and that over 352,000 jobs were created or saved. And these jobs were in relatively high skilled categories. . . .

Document III-31

Document title: H. R. Hertzfeld, "Technology Transfer White Paper," internal NASA document, June 23, 1978.

Source: Documentary History Collection, Space Policy Institute, George Washington University, Washington, D.C.

This internal paper was intended to assess a program called FEDD (For Early Domestic Dissemination) that NASA initiated in the 1970s during the early years of the Nixon administration. The intent was to make U.S. space technology available to American firms first. The U.S. balance of trade was turning negative during the period, and such programs represented an effort to stem the flow of American technology abroad. The FEDD program was ineffective, and this memo addressed the issues involved.

[1]   [handwritten note: "6/23/78"]

Technology Transfer White Paper

INTRODUCTION AND DEFINITIONS

* Technology

Technology is a term loosely applied to know-how, end products and even the benefits of technology. Dictionaries and the aerospace industry define it effectively as know-how applied to practical purposes.

Technology is generally distinguished from basic (scientific) knowledge and from end products and their uses, but there are important exceptions.

The Bucy report[1] defined technology as know-how for design, development, manufacturing, quality control and testing, performance analysis, maintenance and repairs, etc. Closely associated with such know-how may be instrumentation and basic knowledge necessary to its use. However, basic knowledge is generally widely available and not restricted. It therefore becomes of concern for the control or dissemination of technology only in rare cases where fundamental knowledge is critical to its application and is new in character or not yet generally known.

Thus, technology as used here refers primarily to know-how, supplemented as necessary by the equipment and scientific knowledge required for its implementation—and all at relatively sophisticated levels.

[2] • Technology transfer

Technology transfer is a term referring to all types of exchanges involving technology, including both those that the nation takes pains to control or limit, and those which the nation wishes to make (domestically and internationally). Thus, we have export licensing regulations to prevent or condition the commercial transfer of certain non-military, unclassified technologies and products abroad. We also have policies designed to encourage transfers in the interest of US industry and government entities or in the interests and support of developing countries and military allies. For the sake of clarity, the control of transfer will be discussed separately from the encouragement of transfer.

* NASA's charter

NASA's charter (under Sec. 102 (c) of the Space Act) directs the agency "to contribute materially to . . . the following objectives: . . .

(2) The improvement of the usefulness, performance, speed, safety, and efficiency of aeronautical and space vehicles; . . .

(4) The establishment of long-range studies of the potential benefits to be gained from, the opportunities for, and the problems involved in the utilization of aeronautical and space activities for peaceful and scientific purposes

(5) The preservation of the role of the United States as a leader in aeronautical and space science and technology and in the application thereof to the conduct of peaceful activities within and outside the atmosphere."

Thus, NASA is in effect directed to generate technology, so that the US will be a leader in world aerospace technology, and to contribute to the utilization and application of that technology.

[3] Section 203 (a) states that "The Administration, in order to carry out the purpose of this Act, shall— . . .

---

1.   *An Analysis of Export Control of U.S. Technology—A DOD Perspective*, A Report of the Defense Science Board Task Force on Export of U.S. Technology, J. Fred Bucy, Jr., Chairman, February 4, 1976.

(3) provide for the widest practicable and appropriate dissemination of information concerning its activities and the results thereof."

Note that the provision does *not* state, as sometimes suggested, that NASA must provide the widest dissemination of information but only that which is practicable and appropriate. "Practicable" presumably means within the limits of budgets, personnel, and communications and dissemination systems. "Appropriate" would mean within the limits of security classification, audience and user characteristics, commercial considerations and national self-interest.

NASA's statutory obligations to develop and lead in aeronautics and space technology and to disseminate results widely may sometimes conflict. NASA may properly develop more technology than it need report, reporting only that which is practicable and appropriate. (Obviously, NASA cannot "preserve" US technological leadership if it publishes the full details of its technological activities and expertise in all circumstances.)

GENERATION AND TRANSFER OF TECHNOLOGY

• Generation of technology

The generation of technology (in NASA) derives from in-house and contracted work in space programs and projects, aeronautical programs and projects, energy programs and projects and the developmental aspects of the technology utilization program. Supporting research and technology and the NASA capabilities in quality assurance and management also significantly contribute to the technology base.

[4] • Transfer of technology

The transfer of technology occurs both deliberately and inadvertently, and transfers may be prime objectives or entirely incidental to other activities. Examples of the variety of transfer mechanisms which operate are the following:

Transfer Mechanisms Involving Personal Contact
  Advisory committees
  NASA seminars and workshops
  Professional activity
  Personnel mobility
  Ongoing technical exchanges
  Personal discussions
Transfer Mechanisms Involving Agreements, Contracts and Patents
  Cooperative NASA programs:
    with governmental agencies
    with US industry
    with foreign nations
  Industry use of NASA facilities
  NASA RFP's and work statements (US and foreign)
  Contract and Subcontract implementation
  Patents, licenses, waivers, etc.
Transfer Mechanisms Involving Technical Literature
  Publications:
    Tech Briefs, STAR, [Industrial Applications Center] searches,
    Technology for Aviation and Space
  Documentation provided with licenses
  Literature informally provided upon request
  Test reports and analyses
  COSMIC

[5] Other Transfer Mechanisms
    Exhibits
    Technology Utilization program projects
    Theft, leaks, espionage
- Effectiveness of different transfer mechanisms

The effectiveness of different transfer mechanisms was rated according to the following table (see page 6) reproduced from the Bucy Report (which focused primarily on the problem of the export control of DOD technology, but has broader implications).

CURRENT NASA POLICIES/PRACTICES
PROMOTING TECHNOLOGY TRANSFER

- Domestic policies

In aeronautics, NASA carries on the tradition of NACA in working closely with the Defense agencies and with US industry to contribute an advancing technology base upon which both can draw. Recent examples relating to Defense are the dual inter-active flight simulation at LaRC [Langley Research Center], the helicopter program at ARC [Ames Research Center], the Hi-Mat project at DFRC [Dryden Flight Research Center], etc. More broadly relating to the aeronautical industry are the energy-efficient engine and composite primary aircraft structures programs. These programs are most often shaped, organized and funded in NASA, then carried out through contractor work or a combination of in-house and contractor work. Where the defense agencies are involved, there is generally joint funding; even in non-defense work, industry may contribute in some part to the funding requirements.

Also in aeronautics, NASA actively supports developmental or operational research of interest to the Department of Transportation, generally managing the projects and providing or sharing the funding.

[6] [Effectiveness of Technology Transfer According to Industry and Transfer Mechanism originally placed here]

[7] *In applications,* NASA carries out a variety of programs designed to transfer technology to the private and public sectors. Congress, as well as states and local governments, exert pressures on NASA to develop affirmative and aggressive programs of technology transfer. NASA's efforts to transfer technology parallel those of other government agencies such as the Department of Agriculture and the Department of Commerce.

Space applications in the field of communications and meteorology, for example, were developed in NASA, demonstrated successfully and then taken up in the first case by both public agencies and private corporations and in the second case by the Department of Commerce for operational use. Currently, NASA is developing satellite remote sensing and assisting particularly the public sector to assimilate and apply the analytical techniques required to use the satellite data product.

The NASA applications program stretches over a broad range of activities. New technology is developed, particularly in the civil systems area. There are R&D projects which apply NASA know-how to non-aerospace civil sector problems. Projects such as the activated carbon water treatment system represent potential contributions to the solution of significant scientific, social and economic problems. Others such as the system for underwater survey and exploration represent the unique contribution of NASA technology and expertise to the solution of non-aerospace problems.

Application System Verification and Test (ASVT) programs are cooperative projects with Federal and non-Federal public sector agencies. Potential ASVT projects are identified at the field centers, with final selection at the Headquarters office of NASA. The field

Effectiveness of Technology Transfer According to Industry and Transfer Mechanism*

Transfer Effectiveness	Instrumentation	Semiconductor	Jet Engine	Airframe	Transfer Mechanism	
Highly Effective (Tight Control)	H	H	H	H	Turnkey Factories	Active
	H	H	H	H	Licenses with Extensive Teaching Effort	
	H	H	H	H	Joint Ventures	
	H	H	H	H	Technical Exchange with Ongoing Contact	
	H	H	H	H	Training in High-Technology Areas	
	MH	H	M	M	Processing Equipment (with Know-how)	
Effective	M	H	MH	MH	Engineering Documents and Technical Data	Donor Activity
	M	H	MH	MH	Consulting	
	M	MH	M	M	Licenses (with Know-how)	
Moderately Effective	L	L	M	M	Proposals (Documented)	
	L	MH	L	L	Processing Equipment (without Know-how)	
	L	LM	L	L	Commercial Visits	
Low Effectiveness (Decontrol)	L	L	L	L	Licenses (without Know-how)	
	L	L	L	L	Sale of Products (without Maintenance & Operations Data)	
	L	L	L	L	Proposals (Undocumented)	Passive
	L	L	L	L	Commercial Literature	
	L	L	L	L	Trade Exhibits	

L   = Low Effectiveness
LM = Low to Medium Effectiveness
M   = Medium Effectiveness
MH = Medium to High Effectiveness
H   = High Effectiveness

* Taken from 1976 Bucy Report

centers manage the projects, working closely with the future user of the system. New R&D is often necessary. The immediate problem as well as the future use of the technology is worked out with the cooperating agency, and therefore the commitment to transfer the technology developed is strong. There is also the necessary personnel interaction to make the transfer fast and effective.

[8] The university applications program is managed from NASA Headquarters. Universities identify state and local problems that can be aided by the application of remote sensing techniques. The transfer process is directly from NASA to the universities (not the ultimate users) and much of the program involves the support of undergraduate and graduate courses in remote sensing techniques. Therefore, this program is aimed at educating a potential user community rather than directly transferring NASA technology.

The regional applications program, like the university program, primarily involves the use of Landsat data. NASA personnel directly market the data and remote sensing data processing techniques to the various states, but it is the user (state) that identifies problems and proposes the cooperative projects. NASA directly trains the users of the data, but there is no new NASA R&D involved.

Beyond this, the Office of Space and Terrestrial Applications conducts an active and aggressive program for the identification of promising technologies resulting from both in-house and contracted work with a view to stimulating the application of those technologies, where appropriate, to non-space uses in American industry or the public sector. This Technology Utilization program utilizes a number of university-based and other centers around the country as technical data centers servicing industrial subscribers and others interested in exploring and possibly acquiring the technologies identified for potential non-space uses.

The Technology Utilization program also includes "applications teams" of experts who conduct new R&D for projects which have commercial potential. These projects are managed by NASA field centers and Headquarters. They are cooperative projects, with joint funding with other Federal agencies where appropriate. If a commercial vendor is involved in the development of the new technology, there is joint participation with the vendor in the development stage.

[9] In a more general sense, the advent of the *Space Transportation System* has been shaped by NASA so as to encourage the use of launch services for public and private R&D, thus stimulating the development of new technologies by users but not necessarily transferring actual technology (know-how) to them.

- International policies

NASA support for other government agency programs includes support for [Department of] State/AID [Agency for International Development] objectives in extending the benefits of advanced technology to the developing countries. Thus, NASA in cooperation with [the Department of the] Interior makes Landsat data available to all users and provides technical guidance and training for the data processing and analysis techniques required to make use of it. The same thing is done in the communications field where, in addition, launch services are provided to permit foreign acquisition of domestic Comsat systems.

CURRENT NASA POLICIES/PRACTICES
LIMITING TECHNOLOGY TRANSFER

Limitations on NASA technology transfer apply to both in-house and contractor developed technology and are all essentially predicated on considerations of security or foreign competition.

- In-house Technology

In-house technology is rarely classified but this control is available where (defense-related) security is applicable. To control the transfer of unclassified in-house technology, the agency may also obtain patents which combine disclosure with controls in the US and abroad. Patent policy, though, has the stated objective of encouraging commercialization of an invention.

[10] A broader attempt to limit the wholesale dissemination of unclassified NASA technology was initiated in 1973 in response to two stimuli: (1) the suggestion in some congressional forums that NASA was a major conduit of technology to competitors of the United States, and (2) the possibility that the supercritical wing, as a valuable US technology, may have been compromised by early open publication.

The FEDD program was the result.[2] It provides for identifying certain technology as having possible early and significant commercial potential, then marking any documentation on that technology For Early Domestic Dissemination. Efforts to benefit US industry and defer general availability to foreign industry then could be implemented for the technology so identified. The Departments of State and Commerce were consulted and endorsed the program as a desirable experiment.

Similar controls on the COSMIC computer software distribution program were adopted recently.[3] These regulations call for the identification of computer programs by their potential commercial usefulness. Restrictions on distribution to foreign nationals are placed on the most critical software programs for a designated period of years. Less critical programs may be exchanged or sold.

The transfer of NASA-generated technology is further controlled in the context of international cooperative space programs. It is a general precondition of such programs that the foreign partner have the essential technological capability required to discharge the responsibilities which he undertakes in the cooperative agreement. Then, where appropriate, it is specified that in the event that the foreign partner discovers a need for technical assistance, NASA may refer that need to commercial sources in the US (where it becomes subject to export controls). NASA also reserves the right to require that any technical assistance given be provided in "black box" form (as end product rather than [11] technology). In general, however, there are relatively clean interfaces in international space projects and the transfer of technological know-how is not involved or required between the cooperating agencies. There is, however, substantial commercial involvement directly with US aerospace companies (see section below).

A certain amount of discretion is exercised at both NASA Headquarters and field centers with regard to the subject matter and treatment of technical papers to be presented by NASA personnel abroad, thereby controlling in some degree transfer of know-how by this means.

Also, NASA requires the field centers to report and clear invitations to foreign nationals attending symposia. This requirement is informal and probably warrants formalization and centralization.

There has been a change in administration of the NASA foreign resident research associate program. The policy now in practice discourages associates in technical disciplines in favor of those in scientific areas.

Finally, in connection with export controls discussed below, NASA, although *not* subject to the Munitions Controls of the Department of State, nevertheless takes steps on those rare occasions when technology transfer is contemplated, to assure through informal coordination with the Office of Munitions Control that such transfer would be consistent with government practice.

• Contracted Technology

Limitations on the export of contracted technology are essentially outside NASA's responsibilities, since the export of space technology by a private contractor falls under the International Traffic in Arms Regulations Act administered by the Office of Munitions

---

2. NASA NMI 2210.1 dated December 13, 1973.
3. NASA NMI 2210.2 dated April 24, 1978.

Control of the State Department. The same applies to aeronautical technology of primarily military character. Aeronautical technology of dual use [12] (military and civilian), plus a very extensive list of other technologies covering sensors, data processing, communications equipment, etc., is covered by Department of Commerce export controls. NASA provides technical advice to both State and Commerce when requested and participates in the development of lists of controlled items, but the responsibility for granting or denying export licenses rests with those agencies.

If technical material is published (made generally available), it is considered to have a general license for export and is not then controlled. Therefore, NASA and its contractors have some responsibility to consider whether the publication of particular technical data could compromise the intent of State and Commerce control activities. In this connection, both the NASA patent program and the FEDD program, described above, are relevant and apply to contracted technology as well as to in-house technology.

Since unilateral US export controls would obviously have little effect if a foreign purchaser could simply turn to other nations for his needs, the United States has been active in organizing COCOM, an effort of the NATO nations and Japan to concert their export controls on critical items vis-à-vis the Communist world. This system is in some sense an international projection of the US Munitions Control procedure, but suffers from considerable differences of view and competitive pressures among the participating nations.

Finally, of course, NASA-contracted technology may be classified on defense grounds, but this is relatively rare.

To generate advice for State or Commerce on specific export license requests, the central coordinating point in NASA is the International Affairs Division. That Division draws on the technical expertise of the entire agency for this purpose. The final recommendation is treated according to general guidelines, contained in a policy paradigm approved by the NASA Deputy Administrator several years ago. The guideline distinguishes technical know-how from end product. Know-how which is uniquely available from the US would presumptively [13] be denied export. But if the know-how is readily available from other foreign sources, the presumption would be that US industry should be allowed to compete. End products, whether uniquely available from the US or not, are presumptively exportable unless unique know-how could be extracted from them. Of course, there may be over-riding considerations where, e.g., an end product might be critical to an objectionable end-use, as in the case of missile components destined for a country thought to be developing a nuclear weapons delivery system, or in the case of a computer whose capacity could be diverted to weapons system design. The policy paradigm, which appears on the following page, has been incorporated in a classified [National Security Decision Memorandum] as an available guideline for the government as a whole.

DISCUSSION

- Issues
  The principal questions which have been raised with respect to NASA policy and practice in technology transfer, whether positive or restrictive, appear to be these:
    – Should NASA be developing aeronautical technology for defense and for private industry? To what extent and according to what criteria?
    – What role should NASA play in stimulating the industrial R&D needed to produce innovations consistent with national goals, and are significant changes needed in current policies to maximize industry's investment in this R&D?
    – Similarly, in fields such as space communications and remote sensing, should NASA be developing and promoting an advancing technology base for other agencies and for private industry and under what criteria?

[14]   POLICY PARADIGM* FOR EXPORT AND MUNITIONS CONTROL

	KNOW-HOW	END PRODUCT
1. Items unique to U.S. or *critical* to U.S. commercial lead	1A Presumptively, no release	1B Consider export if: – end-use conditions met – cannot be reverse-engineered If reverse-engineering feasible; evaluate risk and decide accordingly.
2. Items available elsewhere	2A Consider export if end-use conditions met	2B Export should be approved if end-use conditions met
3. Items involved in U.S. coop. programs	3A Consider export (even if unique) in light of relevant provisions of cooperative agreement, *quid pro quo*, and/or U.S. interest	3B Should be approved unless critical elements of technology unique to U.S. are subject to reverse-engineering, in which case we consider as in 3A and reserve right to deliver in black box state.

*   Intended to serve as fundamental guidance subject to exception in special cases in the national interest.

[15] –   Should NASA maintain a technology utilization program directed to non-space uses and under what criteria?

–   To what extent should NASA continue and develop the multiplicity of public and private sector experiment-and-demonstration projects now begun in specific technology applications? How controlled or flexible should these activities be? What mechanism should coordinate them, if any?

–   With respect to overseas participation and benefit, as in the Landsat program, what policies should be followed?

–   Should NASA continue to try to control unclassified technical data (as in the FEDD and COSMIC programs) and how?

–   Is NASA's foreign patent filing program cost-effective? Should it be continued?

–   To what extent should NASA "market" its technology and in what style? What role should cost-benefit studies have in this connection?

• Options

The foregoing issues are briefly discussed in the paragraphs below:

–   Aeronautical technology transfer. The Space Act obligates NASA to foster US leadership in aeronautics. Leadership in aeronautics equates with broad national security objectives which encompass both defense and commercial interests. Since NASA's fundamental mission is conceived in R&D terms, clearly it has no choice under current legislation but to develop technology for the defense and commercial aspects of the aeronautics industry. It follows that NASA must find suitable mechanisms for transferring its technology output to industry so that it can best serve the spirit and the objectives of the Act.

As the Bucy report makes quite clear, the most effective means of transfer entail integral relationships in the [16] various phases of the R&D process. NASA certainly practices most of the mechanisms identified for such transfer in the aeronautical field under a fairly well-understood set of criteria. These generally make the greatest government support available for public interest purposes (environmental and safety objectives, e.g.) and require more industrial (or defense) support to the degree that end product development is approached. Where government regulation or private profit motivation can elicit technical advances from the industry, NASA should presumably leave those developments to the industry. Where national considerations of safety, environment, or strategic competition in the military or commercial fields becomes critical, or where the cost, risk and facilities for technical advance are beyond industry's capacity in a competitive system, NASA as the government's agent must consider filling the gap.

In sum, it is difficult to conceive of a very different role for NASA than its traditional one in the development of aeronautical technology; however, its transfer to industry and defense must be responsive to national needs and international pressures on US industry. Fine tuning of the operating policies, of course, is always appropriate. Major innovations may be required for one-of-a-kind situations.

   –   Space applications. The principal options are to leave the development of supporting and advancing technologies to operational interests once the operational stage is reached or, in the alternative, to continue technology development in some degree. What technology should NASA generate for transfer to the commercial sector? If a commercial industry exists, should we consider any further development?

The communications field provides a useful example. It is clear that the operational interests are indeed carrying forward R&D for much of their needs, thereby removing justification for government R&D in those particular areas. But there are public interest areas where the private sector may have no motivation to do R&D. In communications these areas include technological developments for conserving the spectrum and also for conserving the use of the geosynchronous orbit.

[17] It would therefore seem that NASA should consider continuing its R&D program (including test and demonstration projects) in such areas. These programs in the communications area should be carefully and narrowly defined. A positive answer should be contingent upon thorough advance discussion with those government agencies and potential user communities which may have real or fancied concerns in a given area to develop the necessary degree of support. Cost-benefit considerations should be included but (a) will often be very difficult to measure for not-yet-existing applications, and (b) should not necessarily be determinative in public interest areas in any case, since intangible values may be of overriding importance.

Where NASA itself may be given operational responsibilities for space applications, as could occur in the Landsat program, NASA would, of course, determine the R&D requirements to meet the needs of the user agencies, working in conjunction with them.

   –   Technology utilization, ASVT programs. The major options are to run a push program or a pull program. In either case, NASA must, under the terms of the Space Act, be prepared to transfer technology to public and private interests of this country in the most effective and appropriate way, whether for space or non-space applications. NASA's experience to date seems abundantly clear that technology transfer is a sometimes slow and difficult-to-measure process. Energetic measures can be taken to speed-up the process by identifying possibilities, bringing them to the attention of possible users, organizing a body of data, establishing a retrieval system and, in many cases, testing and demonstrating the applications. This is particularly true for the public sector (cities and states, e.g., and other government agencies) where non-space needs may be strongly felt but the expertise required for adaption [sic], test, demonstration and implementation may be entirely lacking.

At the same time, experience also suggests that the dangers of internal enthusiasms for applications in fields outside the agency's experience and expertise dictate some measure [18] of conservatism. It would seem, therefore, that the agency should continue substantially along its present course but with greater visibility for management for the many non-space applications explored and tested.

This means that both headquarters and centers should continue their initiatives and responses but that a system should be established to permit parallel awareness of specific activities so there may be independent consideration of sensitivities, implications and additional coordination requirements of a public, international, governmental, congressional or industry affairs character. The system should not be designed to delay or screen activities but rather to permit "flagging" problems on a timely basis if they should be perceived.

Cost-benefit analysis may be the appropriate tool for evaluating some applications. Before resources are committed to doing such studies, the criteria noted under the paragraph above on Space Applications should be considered. But there is a continuing need for such cost-benefit studies for many potential projects. Traditionally these studies have been done by the various program offices. As more are done, there is a need for uniform techniques to be applied to cost-benefit analysis so that the results of the various studies can be compared. These criteria should be determined by a central office and then applied with the assistance of the various program offices or field centers.

–   International. The Space Act mandates a program of international cooperation, and the Outer Space Treaty calls for the sharing of the *benefits* of space activity. There is no requirement to transfer technology itself abroad.

The options are to conduct "give-away" or support and aid-type programs or to seek cost-sharing or other economic return. Until the Landsat program, the precedents established for cooperative programs were generous, but most often were subjected to requirements of scientific validity and mutual interest.

[19] Thus, in the first experimentation with communications satellites, foreign states were permitted to participate in the *testing* phase (not the R&D) on the basis that they funded the necessary overseas ground stations. NASA thereby saved the expense of funding facilities at both ends of the experiments. Even in the weather satellite field, the initial cooperation entailed comparison flights by foreign aircraft coordinated with spacecraft passes and other activities designed to help calibrate and validate the first satellite data analyses.

Landsat. In the Landsat case, it was felt that a peculiarly economic application was involved and that cost-sharing for the space segment should be established in principle. (The foreign users, of course, fund all their ground-based facilities and activities.) Before cost-sharing was even broached, however, NASA's agreement to program Landsats to read out for foreign stations was offset by several quid-pro-quos: the foreign stations undertook to supply data free to [principal investigators] selected by NASA; they were obligated to provide data of interest to NASA on request; and they represented valuable insurance to NASA for desired foreign coverage when the spacecraft tape recorders should fail.

Technology transfer, in the sense of industrial know-how, is not involved in the above Landsat type foreign involvement. Here we are speaking more of the transfer of the benefits of a technology, but some transfers of data processing and analysis know-how are required by training personnel. NASA does not operate training programs as such, although limited on-the-job training opportunities are made available under certain conditions. Industry and [the Agency for International Development], along with other specialized government agencies, often provide training services.

FEDD. The options are whether to continue to try to control unclassified technical outputs by NASA and its contractors or not, and if so, whether through the FEDD program or another.

[20] The arguments for control are these: NASA has been criticized for its large output of technical information available to foreign competitors. The circumstances of release of supercritical wing information at least suggested that major innovations with significant military and/or commercial potential might be compromised and that professional motivation and tradition may operate to obscure the national interest. The fact that public funds are used to underwrite such R&D would, with other factors, seem to require some regard for the national interest.

The arguments against FEDD are that it carries no sanctions, that it "flags" the most important items so that foreign interests can focus on obtaining them, that the "troops" at the centers do not like it, that it creates a useless work load, etc.

In a series of NASA Middle Management Seminars, FEDD was discussed in depth. The participants recognized that "something ought to be done" to assure preferential use by US interests of unclassified technology which might have significant early commercial potential. When asked if they could think of something better than FEDD, the answers were negative with a single exception. Langley has felt, given the inability to control FEDD publications and the danger that the FEDD label might simply target them, that colloquiums bringing together the interested US firms are to be preferred. Significant items should be discussed in depth anyhow. This approach appears to be meritorious and has been commended to the other field centers.

An important question raised by efforts to implement the FEDD program is this: Since only a handful of publications has been "FEDD'ed" by NASA, are we doing the job of review and identification very badly, is NASA admitting that only a minuscule percentage of its technical reports possess any significant early commercial potential or is the present program simply unworkable? It would seem essential that a thorough review of the FEDD program as it now stands in the Office of Aeronautics and Space Technology be undertaken, along with a comparison and evaluation of any similar programs [21] in other government agencies. Suggestions for revising the program, coordinating it with other government programs and policies, or abolishing it could then be more rigorously evaluated.

Foreign patent program. Since the current cost of the foreign patent program greatly exceeds the very small revenues obtained, it might be argued that the program should be dropped. On the other hand, it would seem worthwhile to evaluate experience with the supercritical wing patents over the next few years to better judge the future of this program.

– Marketing. As noted above, the dangers of internal enthusiasms for applications in fields beyond the agency's own experience and competence dictate some measure of conservatism in pushing such applications. Therefore, "markets" must be carefully explored in advance with the best-informed user groups; cost-benefit considerations should be included but are especially difficult and may often have limited validity in connection with innovations in new fields or where public interests override.

It remains true, nevertheless, that NASA's statutory obligation to contribute to technological advance will not implement itself. Therefore, the agency must undertake well-considered programs to inform possible users, to experiment with, test, and on occasion, demonstrate space and non-space applications in the national interest.

Because of the implications for the agency's image, its congressional and other government agency relationships, its industry, international and university relationships, it is important that headquarters and center undertakings looking to new user groups and markets be given timely visibility for management. A "flagging" system that does not inhibit activities or establish new clearance requirements is therefore important, to permit control by exception.

Document III-32

**Document title: "NASA Technology Transfer: Report of the Technology Transfer Team,"
December 21, 1992.**

Source: NASA Historical Reference Collection, NASA History Office, NASA Headquarters, Washington, D.C.

NASA's technology transfer program has been lauded as one of the more successful of such government programs, and at the same time it has been severely criticized as not being very effective. Traditionally, NASA has focused on technology development for space and has downplayed employee rewards for the transfer of technology outside NASA. This report, however, which was the result of an intensive in-house review of the program, recommended improving internal incentives for managers to stimulate the transfer of technology.

[no pagination]

NASA Technology Transfer
Report of the Technology Transfer Team

December 21, 1992

BASIS FOR RECOMMENDATIONS

1. **NASA is accountable to transfer its special capabilities and technology.** This is an important mission of the agency.
2. **Success in technology transfer requires deliberate dedicated effort.** Thus NASA must initiate technology transfer activities.
3. **Technology transfer occurs mainly in the context of an appropriate person-to-person relationship between the providers and recipients.**
4. **Experience suggests that technology transfer is most successful when recipients *want* technology for their needs.** *Effective, proactive outreach* creates this *desire*. **Thus a marketing model for technology transfer has greater potential for success.** A passive diffusion model leaves much to chance.
5. **Technology transfer is inseparable from the technology development process.**
6. **The influence of customer interests on NASA R&D goals is a vital indicator of potential success.** This influence shows early recipient involvement, and shows that a technology transfer relationship exists that is more likely to succeed.
7. **For technology transfer process management and improvement, *effectiveness* metrics are better than *activity* metrics. Activity metrics having strong, causative influence on effectiveness are useful.**
8. **The technology transfer process should be conducted such that employees' interests are benefited (ideally) and protected (at minimum).**

TECHNOLOGY TRANSFER IS BEST ACHIEVED AS A MARKET-ORIENTED, TECHNICALLY CONDUCTED, LEGALLY SUPPORTED ACTIVITY

[Why What We Found Was There to Be Found/What We Suggest Doing About It originally placed here]

RECOMMENDATIONS

All NASA elements must implement and be evaluated on their technology transfer program
1. Each center must manage to the recommended metrics . . . or define and manage to more effective set
2. Headquarters must implement a unified plan to support technology transfer
   Specified roles and missions of each office

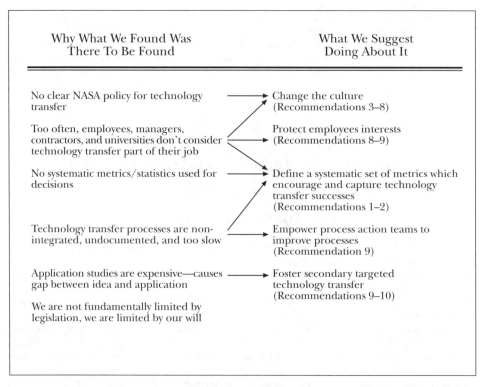

Why What We Found Was There To Be Found	**What We Suggest Doing About It**

No clear NASA policy for technology transfer → Change the culture (Recommendations 3–8)

Too often, employees, managers, contractors, and universities don't consider technology transfer part of their job → Protect employees interests (Recommendations 8–9)

No systematic metrics/statistics used for decisions → Define a systematic set of metrics which encourage and capture technology transfer successes (Recommendations 1–2)

Technology transfer processes are non-integrated, undocumented, and too slow → Empower process action teams to improve processes (Recommendation 9)

Application studies are expensive—causes gap between idea and application → Foster secondary targeted technology transfer (Recommendations 9–10)

We are not fundamentally limited by legislation, we are limited by our will

Provide infrastructure activities supporting all centers (SBIR, Tech Brief, COS-MIC, . . .)

Institute a proactive effort to change the agency's technology transfer culture and ensure broader participation by all employees.

3. NASA should specifically mention technology transfer in V-M-V statement . . .
4. Administrator should send a directive to [Associate Administrators] and [Center Directors] stating that technology transfer is a mission of NASA and specifically that secondary targeted and non-targeted are fully valued, important NASA missions which should be managed accordingly . . .
5. Administrator should continue strong technology transfer support and measure overall agency performance . . .
6. Each center should include technology transfer in their mission statement
7. Each center should provide technology transfer training for all employees . . .
8. Assess, promote, and reward employees according to metrics/contributions
9. Form and empower at least the following process action and process development teams: . . .

Tech Briefs—information acquisition to publication
Patent applications and licensing
Software distribution and transfer
Conversion of non-targeted to secondary targeted } Including use of
Conversion/integration of primary targeted to } jointly sponsored
secondary targeted } research activities
Execution of secondary targeted programs }
Define relationship of centers to [Centers for the
Commercial Development of Space]
Employee motivation and incentive for technology transfer activities

10. Secondary technology transfer activities should be proactively sought. The budget allocated to each center for its use in secondary targeted transfer programs should grow and be taken "off the top" as is SBIR.

Biographical Appendix

William A. Anders (1933–) was a career U.S. Air Force officer, although a graduate of the U.S. Naval Academy. Chosen with the third group of astronauts in 1963, he was the backup pilot for Gemini 11 and the lunar module pilot for Apollo 8. He resigned from NASA and the Air Force (active duty) in September 1969, when he became executive secretary of the National Aeronautics and Space Council. Joined the Atomic Energy Commission in 1973 and became chair of the Nuclear Regulatory Commission in 1974. He was named U.S. ambassador to Norway in 1976. Later, he worked as a vice president of General Electric and then as senior executive vice president of operations for Textron, Inc. Anders retired as chief executive officer of General Dynamics in 1993, but he remained chair of the board. See "Anders, W.A.," biographical file, NASA Historical Reference Collection, NASA History Office, NASA Headquarters, Washington, D.C.

Clinton P. Anderson (1895–1975) (D–NM) was elected to the House of Representatives in 1940 and served through 1945, when he was appointed secretary of agriculture. He resigned from that position in 1948 and was elected to the Senate, where he served until 1973. See *Biographical Directory of the United States Congress, 1774–1989* (Washington, DC: U.S. Government Printing Office, 1989).

Peter Badgley (1925–) received a Ph.D. from Princeton University in 1951 and was a specialist in geology and tectonics. He served for a time in the 1960s as chief of the Earth Resources Survey Program within NASA's Space Applications Programs Office and as chief of advanced missions for the Manned Space Science Program within NASA's Office of Space Science and Applications. See "Badgley, Peter," biographical file, NASA Historical Reference Collection.

D. James Baker (1937–) has served as the administrator of the National Oceanographic and Atmospheric Administration in the Department of Commerce since 1993. See *Who's Who in America 1996* (New Providence, NJ: Marquis Who's Who, 1996).

Malcolm Baldridge (1922–1987) served as secretary of commerce from 1981 until his death. See *Who's Who in America, 44th edition, 1986–1987* (Wilmette, IL: Marquis Who's Who, 1987).

James E. Beggs (1926–) served as NASA's administrator from July 10, 1981, to December 4, 1985, when he took an indefinite leave of absence pending disposition of an indictment from the Justice Department for activities taking place prior to his tenure at NASA. This indictment was later dismissed, and the U.S. attorney general publicly apologized to Beggs for any embarrassment. His resignation from NASA was effective on February 25, 1986. Prior to NASA, Beggs had been executive vice president and a director of General Dynamics in St. Louis. Previously, he had served with NASA in 1968–1969 as associate administrator for advanced research and technology. From 1969 to 1973, he was under secretary of transportation. He went to Summa Corporation in Los Angeles as managing director of operations and joined General Dynamics in January 1974. Before joining NASA the first time, he had been with Westinghouse Electric in Sharon, Pennsylvania, and Baltimore, Maryland, for thirteen years. A 1947 graduate of the U.S. Naval Academy, he served with the Navy until 1954. In 1955, he received a master's degree from the Harvard Graduate School of Business Administration. See "Beggs, James E.," biographical file, NASA Historical Reference Collection.

Anatoli A. Blagonravov (1895–1975) was head of an engineering research institute in the Soviet Union. As Soviet representative to the United Nations Committee on the Peaceful Uses of Outer Space (COPUOS) in the early 1960s, he was also senior negotiator, with NASA's Hugh L. Dryden, for cooperative space projects at the height of the Cold War in the early 1960s. He worked on developing infantry and artillery weapons in World War II and on rockets afterward. See "Blagonravov, A.A.," biographical file, NASA Historical Reference Collection.

Ralph Braibanti joined the Department of State in 1972 and held a number of assignments related to Latin American and East Asian affairs before joining the State Department's Bureau of Oceans and International Environmental and Scientific Affairs in 1985. He currently serves as director of that bureau's space and advanced technology staff. See biographical sketch provided by Ralph Braibanti, NASA Historical Reference Collection.

Zbigniew Brzezinski (1928–) served as the President Carter's national security advisor from 1977 to 1981. See *Who's Who in America 1996*.

George E. Brown, Jr. (1920–) (D–CA), served in the House of Representatives from January 3, 1963, to January 3, 1971, and then again from January 3, 1973, to the present. He chaired the House Committee on Science, Space, and Technology for a number of years and currently is its ranking minority member. See *Biographical Directory of the American Congress, 1774–1996* (Alexandria, VA: CQ Staff Directories, Inc., 1997).

Ronald Brown (1941–1996) served as secretary of commerce from 1993 until his death in a plane crash. See *Who's Who in America 1996*.

David K. Bruce (1898–1977) was one of the most notable diplomats of the twentieth century. He served in World War I, and he was admitted to the Maryland bar in 1921 before turning his attention to farming in 1928. In 1941, he helped organize the Office of Strategic Services and later became the director of the economic cooperation mission, charged with the task of administering the Marshall Plan. He served in several coveted ambassadorial posts, most notably France (1948–1952), West Germany (1957–1959), Great Britain (1961–1969), and NATO (1974–1976). He was also a representative to the Vietnam Peace Conference in Paris (1970–1971) and was liaison officer to Communist China from 1973 to 1974. See "Bruce, David K.," in John S. Bowman, ed., *The Cambridge Dictionary of American Biography* (Cambridge, Eng.: The Cambridge University Press, 1995).

C

William Casey (1913–1987) served as chief of secret intelligence in Europe for the Office of Strategic Services during World War II. After the war, he became a wealthy businessman. From 1971 to 1975, he served successively as the chair of the Security and Exchange Commission, as under secretary of state for economic affairs, and as head of the Export-Import Bank. He also served on the Foreign Intelligence Advisory Board under President Ford. He was President Reagan's first presidential campaign manager and then served as Reagan's director of central intelligence until his death at the height of the Iran-Contra scandal. See "Casey, William," biographical file, NASA Historical Reference Collection.

Emanuel Celler (1888–1981) (D–NY) graduated from Columbia Law School in 1912 and immediately began practicing law in New York City. During World War I, he served as an appeal agent on the draft board. Following the war, he made a successful run for the House of Representatives in 1923 and served as a Democratic until 1973, for a time as chair of the powerful Judiciary Committee. Following his defeat in 1973, he joined a commission to revise the federal appellate court system and, in 1975, returned to his law practice. See "Celler, Emanuel," in Bowman, *The Cambridge Dictionary of American Biography*.

Arthur C. Clarke (1917–), one of the most well-known science fiction authors, has also been an eloquent writer on behalf of the exploration of space. In 1945, before the invention of the transistor, Clarke wrote an article in *Wireless World* describing the possibility of a geosynchronous orbit and the development of communications relays by satellite. He also wrote several novels, the most known being *2001: A Space Odyssey*, based on a screenplay of the same name that he prepared for director Stanley Kubrick. The movie is still one of the most realistic depictions of the rigors of spaceflight ever filmed. See "Clarke, A.C.," biographical file, NASA Historical Reference Collection.

Edgar M. Cortright (1923–   ) earned a master of science degree in aeronautical engineering from Rensselaer Polytechnic Institute in 1949, the year after he joined the staff of Lewis Laboratory. He conducted research at Lewis on the aerodynamics of high-speed air induction systems and jet exit nozzles. In 1958, he joined a small task group to lay the foundation for a national space agency. When NASA came into being, he became chief of advanced technology at NASA Headquarters, directing the initial formulation of the agency's meteorological satellite program, including the TIROS and Nimbus projects. Becoming assistant director for lunar and planetary programs in 1960, Cortright directed the planning and implementation of such projects as Mariner, Ranger, and Surveyor. He became deputy director and then deputy associate administrator for space science and applications in the next few years and was deputy associate administrator for manned spaceflight in 1967. In 1968, he became director of the Langley Research Center, a position he held until 1975, when he went to work for private industry, becoming president of the Lockheed-California in 1979. See "Cortright, Edgar M.," biographical file, NASA Historical Reference Collection.

D

John M. Deutch (1938–   ) served as deputy secretary of defense from 1994 to 1995 and then as director of central intelligence from 1995 to 1997. He received a Ph.D. from the Massachusetts Institute of Technology and also served as that school's dean of science and provost. See "Deutch, John M.," biographical file, NASA Historical Reference Collection.

Hugh L. Dryden (1898–1965) was a career civil servant and an aerodynamicist by discipline who had also begun life as something of a child prodigy. He graduated at age 14 from high school and went on to earn an A.B. in three years from Johns Hopkins (1916). Three years later (1919), he earned his Ph.D. in physics and mathematics from the same institution, even though he had full-time employment at the National Bureau of Standards since June 1918. His career there, which lasted until 1947, was devoted to studying airflow, turbulence, and particularly the problems of the boundary layer—the thin layer of air next to an airfoil that causes drag. In 1920, he became chief of the bureau's aerodynamics section. His work in the 1920s on measuring turbulence in wind tunnels facilitated research in the National Advisory Committee for Aeronautics (NACA) that produced the laminar flow wings used in the P-51 Mustang and other World War II aircraft. From the mid-1920s to 1947, his publications became essential reading for aerodynamicists around the world. During World War II, his work on a glide bomb named the Bat won him a Presidential Certificate of Merit. He capped his career at the Bureau of Standards by becoming its assistant director and then associate director during his final two years there. He then served as director of the NACA from 1947 to 1958, after which he became deputy administrator of NASA under T. Keith Glennan and James E. Webb. See Richard K. Smith, *The Hugh L. Dryden Papers, 1898–1965* (Baltimore, MD: The Johns Hopkins University Library, 1974).

E

Dwight D. Eisenhower (1890–1969) was president of the United States between 1953 and 1961. Previously, he had been a career U.S. Army officer and, during World War II, was supreme allied commander in Europe. As president, he was deeply interested in the use of space technology for national security purposes and directed that ballistic missiles and reconnaissance satellites be developed on a crash basis. For more information on Eisenhower's space efforts, see Rip Bulkeley, *The Sputniks Crisis and Early United States Space Policy* (Bloomington, IN: Indiana University Press, 1991); R. Cargill Hall, "The Eisenhower Administration and the Cold War: Framing American Astronautics to Serve National Security," *Prologue: Quarterly of the National Archives* 27 (Spring 1995): 59–72; Robert A. Divine, *The Sputnik Challenge: Eisenhower's Response to the Soviet Satellite* (New York, NY: Oxford University Press, 1993).

John D. Erlichman was a senior assistant to the president during the Nixon administration. See John Erlichman, *Witness to Power: The Nixon Years* (New York, NY: Simon and Schuster, 1982).

James Exon (1921–   ) served as the governor of Nebraska from 1971 to 1979. Since 1979, he has served as a Democratic Senator from Nebraska. See *Who's Who in America 1996*.

F

Philip J. Farley (1916–) earned a Ph.D. from the University of California at Berkeley in 1941 and was on the faculty at Corpus Christi Junior College from 1941 to 1942 before entering government work for the Atomic Energy Commission (1947–1954) and for the State Department (1954–1969). From 1957 until 1961, he was a special assistant to the secretary of state for disarmament and atomic energy, and from 1961 to 1962, his responsibilities shifted to atomic energy and outer space. After several years of assignment to the North Atlantic Treaty Organization (NATO), he returned to Washington and became deputy secretary of state for political-military affairs (1967–1969). Then from 1969 to 1973, he became deputy director of the U.S. Arms Control and Disarmament Agency. See "Farley, P.J.," biographical file, NASA Historical Reference Collection.

James Brown Fisk (1910–1981) received his Ph.D. in physical science from the Massachusetts Institute of Technology in 1935 and served in a variety of educational and industry positions. He was heavily involved in work at Bell Telephone Laboratories, ultimately becoming its president. See "Fisk, J.B.," biographical file, NASA Historical Reference Collection.

Peter M. Flanigan (1923–) was an assistant to the president on the White House staff from 1969 to 1974. Previously, he had been involved in investment banking with Dillon, Read, and Co. He returned to business when he left government service. His position in the White House involved him in efforts to gain approval to build the Space Shuttle during the 1969–1972 period. See "Miscellaneous Other Agencies," biographical file, NASA Historical Reference Collection.

James C. Fletcher (1919–1991) received an undergraduate degree in physics from Columbia University and a doctorate in physics from the California Institute of Technology. After holding research and teaching positions at Harvard and Princeton Universities, he joined Hughes Aircraft in 1948 and later worked at the Guided Missile Division of the Ramo-Wooldridge Corporation. In 1958, Fletcher co-founded the Space Electronics Corporation in Glendale, California, which after a merger became the Space General Corporation. He was later named systems vice president of the Aerojet General Corporation in Sacramento. In 1964, he became president of the University of Utah, a position he held until he was named NASA's administrator in 1971. He served until 1977. He also served as NASA administrator a second time, for nearly three years following the loss of the Space Shuttle *Challenger,* from 1986 until 1989. During his first administration at NASA, Dr. Fletcher was responsible for beginning the Shuttle effort. During his second tenure, he presided over the effort to recover from the *Challenger* accident. See Roger D. Launius, "A Western Mormon in Washington, D.C.: James C. Fletcher, NASA, and the Final Frontier," *Pacific Historical Review* 64 (May 1995): 217–41.

Arnold W. Frutkin (1918–) was deputy director of the U.S. National Committee for the International Geophysical Year in the National Academy of Sciences when NASA hired him in 1959 as director of international programs, a title that changed in 1963 to assistant administrator for international affairs. In 1978 ,he became associate administrator for external relations, a post he relinquished in 1979 when he retired from federal service. During his career, he had been NASA's senior negotiator for almost all of the important international space agreements. See "Frutkin, Arnold," biographical file, NASA Historical Reference Collection.

G

John H. Gibbons (1929–) headed the Office of Technology Assessment under Congress for fourteen years before becoming President Clinton's science advisor and head of the White House Office of Science and Technology Policy in 1993. He received a Ph.D. in physics from Duke University in 1954. See "Gibbons, John," biographical file, NASA Historical Reference Collection.

T. Keith Glennan (1905–1995) was NASA's first administrator. He was educated at Yale University and worked in the sound motion picture industry with the Electrical Research Products Company. He was also studio manager of Paramount Pictures and Samuel Goldwyn Studios in the 1930s. Glennan joined Columbia University's Division of War Research in 1942, serving through the war, first as administrator and then as director of the U.S. Navy's Underwater Sound Laboratories at New London, Connecticut. In 1947 he became president of the Case Institute of Technology in Cleveland. During his administration, Case rose from a primarily local institution to rank with the top engineering schools in the nation. From October 1950 to November 1952, Glennan served as a member of the Atomic Energy Commission. He also served as administrator of NASA while on leave from Case, between August 7, 1958 and January 20, 1961. After leaving NASA, he returned to the Case, where he was continued to serve as president until 1966. See J.D. Hunley, ed., *The Birth of NASA: The Diary of T. Keith Glennan* (Washington, DC: NASA SP-4105, 1993).

H

James C. Hagerty (1909–1981) had been on the staff of the *New York Times* from 1934 to 1942, the last four years as legislative correspondent at the paper's Albany bureau. He served as executive assistant to New York Governor Thomas Dewey from 1943 to 1950 and then as Dewey's press secretary for the next two years before becoming press secretary for President Eisenhower from 1953 to 1961. See "Miscellaneous Other Agencies," biographical file, NASA Historical Reference Collection.

Irwin P. Halpern served as the director of NASA's policy analysis staff in the mid-1960s. He previously worked at the Central Intelligence Agency on Soviet and Chinese political-military affairs and doctrine. He received a Ph.D. in soviet history from Columbia University. See "Miscellaneous NASA," biographical file, NASA Historical Reference Collection.

Henry R. Hertzfeld (1943–) is a senior research scientist at the George Washington University's Space Policy Institute. Previously, he served as the senior economist at NASA and as a policy analyst at the National Science Foundation. He received a Ph.D. in economics from Temple University and a J.D. degree from George Washington University. See "Miscellaneous NASA," biographical file, NASA Historical Reference Collection.

Walter J. Hickel (1919–) was governor of Alaska and then secretary of the interior from 1969 to 1970. See *Who's Who in America 1996*.

J

Leonard Jaffe joined the National Advisory Committee for Aeronautics in 1948 and worked for it and its successor organization, NASA, for thirty-three years before moving to the private sector in 1981. He primarily worked in the field of space applications, overseeing many of NASA's early efforts in remote sensing and satellite communications. See "Jaffe, Leonard," biographical file, NASA Historical Reference Collection.

Karl G. Jansky (1905–1950) was a researcher for Bell Laboratories in New Jersey who, while studying the static that often disrupted radio communications, discovered interstellar radio waves. Thus the field of radio astronomy was born. See "Karl G. Jansky," biographical file, NASA Historical Reference Collection.

David S. Johnson was the Environmental Satellite Center director at the Environmental Science Service Administration in the mid-1960s. See "Miscellaneous Other Agency," biographical file, NASA Historical Reference Collection.

John A. Johnson (1915–), after completing law school at the University of Chicago in 1940, practiced in Chicago until 1943, when he entered military service with the Navy. From 1946 to 1948, he was an assistant for international security affairs in the Department of State. He joined the office of the general counsel of the Department of the Air Force in 1949 and served until October 7, 1958 (for the last six years as the general counsel), when he accepted the general counsel position at NASA. In 1963, he left the space agency to become director of international arrangements at the Communications Satellite Corporation (Comsat). The next year, he became a vice president of Comsat and then, in 1973, senior vice president and later chief executive officer. He retired in 1980. See "Johnson, J.A.," biographical file, NASA Historical Reference Collection.

Lyndon B. Johnson (1908–1973) (D–TX) was elected to the House of Representatives in 1937 and served until 1949. He was a senator from 1949 to 1961, U.S. vice president from 1960 to 1963, and president from then until 1969. Best known for the social legislation he passed during his presidency and for his escalation of the war in Vietnam, he was also highly instrumental in revising and passing the legislation that created NASA and in supporting the U.S. space program as chair of the Committee on Aeronautical and Space Sciences and of the preparedness subcommittee of the Senate Armed Services Committee. He later as chaired the National Aeronautics and Space Council when he was vice president. On his role in support of the space program, see Robert A. Divine, "Lyndon B. Johnson and the Politics of Space," in Robert A. Divine, ed., *The Johnson Years: Vietnam, the Environment, and Science* (Lawrence, KS: University of Kansas Press, 1987), pp. 217–53; Robert Dallek, "Johnson, Project Apollo, and the Politics of Space Program Planning," unpublished paper delivered at a symposium on "Presidential Leadership, Congress, and the U.S. Space Program," sponsored by NASA and American University, March 25, 1993.

Nicholas L. Johnson is NASA's chief scientist for orbital debris at the Johnson Space Center. Previously, he worked in private industry and was considered an expert on the Soviet space program. See "Johnson, Nicholas L.," biographical file, NASA Historical Reference Collection.

Roy W. Johnson (1906–1965) was named the first director of the Department of Defense's Advanced Research Projects Agency and served from 1958 to 1959. As such, he was head of Defense Department's initial space efforts. Prior to joining the government, he worked for General Electric and retired as an executive vice president. See "Johnson, Roy W.," biographical file, NASA Historical Reference Collection.

K

Frederick R. Kappel was the chair of the board of directors of the American Telephone and Telegraph Company in 1963. See "Miscellaneous Industry," biographical file, NASA Historical Reference Collection.

Nicholas Katzenbach (1922–   ) was twenty-one when he was captured by the Germans during World War II, and he was a prisoner of war for two years until the war ended. He returned to the United States and became a Rhodes Scholar in 1947. When he returned from England in 1950, he was admitted to the New Jersey bar. He became a law professor at Yale University in 1952 and then at the University of Chicago from 1956 until 1960. He joined the Justice Department in 1961 as assistant attorney general, and he was promoted to deputy attorney general in 1962. He remained in that position until the end of 1964 and was instrumental in drafting the Civil Rights Act of that same year. He became attorney general in 1965 and under secretary of state in 1966. He left government service to work for IBM in 1969, where he stayed until 1986. He returned to private practice and was named chair of the failing Bank Credit and Commerce International in 1991. See "Katzenbach, Nicholas (de Belleville)," in Bowman, *The Cambridge Dictionary of American Biography*.

Estes Kefauver (1903–1963) (D–TN) served in the House of Representatives from 1939 to 1949 and in the Senate from 1949 to 1963. He ran unsuccessfully as Adlai Stevenson's vice presidential choice in 1956. See *Biographical Directory of the United States Congress, 1774–1989*.

John F. Kennedy (1916–1963) (D–MA) was U.S. president from 1961 to 1963. A senator from Massachusetts between 1953 and 1960, he successfully ran for president as the Democratic candidate, with party wheelhorse Lyndon B. Johnson as his running mate. Using the slogan "Let's get this country moving again," Kennedy charged the Republican Eisenhower administration with doing nothing about the myriad social, economic, and international problems that festered in the 1950s. He was especially hard on Eisenhower's record in international relations, taking a "cold warrior" position on a supposed "missile gap" (which turned out not to be the case) wherein the United States lagged far behind the Soviet Union in intercontinental ballistic missile technology. On May 25, 1961, President Kennedy announced to the nation the goal of sending an American to the Moon before the end of the decade. The human spaceflight imperative was a direct outgrowth of it; Projects Mercury (at least in its latter stages), Gemini, and Apollo were each designed to execute it. On this subject, see Walter A. McDougall, . . . *The Heavens and the Earth: A Political History of the Space Age* (New York, NY: Basic Books, 1985); John M. Logsdon, *The Decision to Go to the Moon: Project Apollo and the National Interest* (Cambridge, MA: MIT Press, 1970).

Robert S. Kerr (1896–1963) (D–OK) had been governor of Oklahoma from 1943 to 1947 and was elected to the Senate the following year. From 1961 until 1963, he chaired the Aeronautical and Space Sciences Committee. See Anne Hodges Morgan, *Robert S. Kerr: The Senate Years* (Norman, OK: University of Oklahoma Press, 1977).

Nikita S. Khrushchev (1894–1971) was premier of the Soviet Union from 1958 to 1964 and first secretary of the Communist party from 1953 to 1964. He was noted for an astonishing speech in 1956 denouncing the crimes and blunders of Joseph Stalin and for gestures of reconciliation with the West in 1959–1960, ending with the breakdown of a Paris summit with President Eisenhower and the leaders of France and Great Britain in the wake of Khrushchev's announcement that the Soviets had shot down an American U-2 reconnaissance aircraft over the Urals on May 1, 1960. Then in 1962, Khrushchev attempted to place Soviet medium-range missiles in Cuba. This led to an intense crisis in October, following which Khrushchev agreed to remove the missiles if the United States promised to make no more attempts to overthrow Cuba's Communist government. Although he could be charming at times, Khrushchev was also given to bluster (extending even to shoe-pounding at the United Nations). He was also a tough negotiator, although he believed, unlike his predecessors, in the possibility of Communist victory over the West without war. For further information about him, see his own *Khrushchev Remembers: The Last Testament* (Boston, MA: Little, Brown, 1974), as well as Edward Crankshaw, *Khrushchev: A Career* (New York, NY: Viking, 1966); Michael R. Beschloss, *Mayday: Eisenhower, Khrushchev and The U-2 Affair* (New York, NY: Harper and Row, 1986); Robert A. Divine, *Eisenhower and the Cold War* (New York, NY: Oxford University Press, 1981).

Henry A. Kissinger (1923–) was assistant to the president for national security affairs from 1969 to 1973 and secretary of state thereafter until 1977. In these positions, he was especially involved in international aspects of spaceflight, particularly the joint Soviet-American flight, the Apollo-Soyuz Test Project, in 1975. See "Kissinger, Henry," biographical file, NASA Historical Reference Collection.

Christopher C. Kraft, Jr. (1924–), was a long-standing official with NASA throughout the Apollo program. He received a bachelor of science degree in aeronautical engineering from Virginia Polytechnic University in 1944 and joined the Langley Aeronautical Laboratory of the NACA the next year. In 1958, still at Langley, he became a member of the Space Task Group developing Project Mercury and moved with the group to Houston in 1962. He was flight director for all of the Mercury and many of the Gemini missions and directed the design of Mission Control at the Manned Spacecraft Center, which was renamed the Johnson Space Center in 1973. He was named the Manned Spacecraft Center's deputy director in 1970 and its director two years later, a position he held until his retirement in 1982. Since then, he has remained active as an aerospace consultant. See "Kraft, Christopher C., Jr.," biographical file, NASA Historical Reference Collection.

L

William E. Lilly (1921–) entered federal civilian service in 1950 as a budget and program analyst with the Navy Ordnance Test Station in California and held a variety of positions with the Navy and the Bureau of Standards until 1960, when he joined NASA as chief, plans and analysis, for the Office of Launch Vehicles. He served NASA for twenty-one years, becoming its first comptroller—a position with associate administrator status—in 1973. He retired in 1981 with thirty-seven years of federal service, including service in the Navy from 1940 to 1946. See "Lilly, W.E.," biographical file, NASA Historical Reference Collection.

George M. Low (1926–1984), a native of Vienna, Austria, came to the United States in 1940 and received an aeronautical engineering degree from Rensselaer Polytechnic Institute in 1948 and a master of science in the same field from that school in 1950. He joined the NACA in 1949, and at Lewis Flight Propulsion Laboratory, he specialized in experimental and theoretical research in several fields. He became chief of manned spaceflight at NASA Headquarters in 1958. In 1960, Low chaired a special committee that formulated the original plans for the Apollo lunar landings. In 1964, he became deputy director of the Manned Spacecraft Center in Houston, the forerunner of the Johnson Space Center. He became deputy administrator of NASA in 1969 and served as acting administrator from 1970 to 1971. He retired from NASA in 1976 to become president of Rensselaer, a position he still held until his death. In 1990, NASA renamed its quality and excellence award after him. See "Low, G.M.," Deputy Administrator files, NASA Historical Reference Collection.

M

Leonard H. Marks was one of the original Comsat incorporators appointed by President Kennedy. He resigned from Comsat's board of directors in 1965 to become director of the U.S. Information Agency. See "Miscellaneous Industry," biographical file, NASA Historical Reference Collection.

Robert P. Mayo (1916–) was an economist and President Nixon's first director of the Bureau of the Budget. On July 1, 1970, when the Bureau of the Budget was replaced with the Office of Management and Budget, Mayo was shifted to the White House as a presidential assistant. Shortly thereafter, he left Washington to assume the presidency of the Federal Reserve Bank of Chicago. See "Mayo, Robert P(orter)," *Current Biography 1970*, pp. 282–84.

Richard C. McCurdy (1909–), an engineer specializing in petroleum, was associate administrator for organization and management at NASA Headquarters from 1970 to 1973 and a consultant to the agency from 1973 until 1982. See "McCurdy, R.C.," biographical file, NASA Historical Reference Collection.

Newton Minow (1926–) was a lawyer in Chicago before being appointed as chair of the Federal Communications Commission by President Kennedy. He gained a reputation by attacking the quality of television programming and threatening to revoke broadcast licenses based on programming. He returned to his law practice following Kennedy's assassination and joined the Public Broadcasting System board in 1973. He became chair of that organization in 1978, and then he became director of the Annenberg Communications Program in Washington in 1987. See "Minow, Newton (Norman)," in Bowman, *The Cambridge Dictionary of American Biography*.

George E. Mueller (1918–) was NASA's associate administrator for manned spaceflight from 1963 to 1969; as such, he responsible for overseeing the completion of Project Apollo and beginning the development of the Space Shuttle. He moved to General Dynamics as senior vice president in 1969, where he remained until 1971. He then became president of the Systems Development Corporation (1971–1980) and then its chair and corporate executive officer (1981–1983). See "Mueller, George E.," biographical file, NASA Historical Reference Collection.

Karl Mundt (1900–1974) (R–SD) served in the House of Representatives from January 3, 1939, until December 30, 1948. He then served as a senator from December 31, 1948, until January 3, 1973. See *Biographical Directory of the United States Congress, 1774–1996*.

N

Homer E. Newell (1915–1983) earned his Ph.D. in mathematics from the University of Wisconsin in 1940 and served as a theoretical physicist and mathematician at the Naval Research Laboratory from 1944 to 1958. During part of that period, he was science program coordinator for Project Vanguard and was acting superintendent of the atmosphere and astrophysics division. In 1958, he transferred to NASA to assume responsibility for planning and developing the new agency's space science program. He soon became deputy director of spaceflight programs. In 1961, he became director of the Office of Space Sciences, and in 1963, he became associate administrator for space science and applications. Over the course of his career, he became an internationally known authority in the field of atmospheric and space sciences as well as the author of numerous scientific articles and seven books, including *Beyond the Atmosphere: Early Years of Space Science* (Washington, DC: NASA SP-4211, 1980). He retired from NASA at the end of 1973. See "Newell, Homer," biographical file, NASA Historical Reference Collection.

Richard M. Nixon (1913–1994) was U.S. president between January 1969 and August 1974. Early in his presidency, Nixon appointed a Space Task Group under the direction of Vice President Spiro T. Agnew to assess the future of spaceflight in the nation. Its report recommended a vigorous post-Apollo exploration program culminating in a human expedition to Mars. Nixon did not approve this plan, but he did decide in favor of building one element of it, the Space Shuttle, which was approved on January 5, 1972. See Roger D. Launius, "NASA and the Decision to Build the Space Shuttle, 1969–72," *The Historian* 57 (Autumn 1994): 17–34.

Herman Noordung (1892–1929) was a pseudonym for Herman Potôcnik. He was a relatively obscure officer in the Austrian army who became an engineer. Encouraged by Hermann Oberth, he wrote an early seminal book titled *The Problem of Space Travel: The Rocket Motor,* which mostly focused on the engineering aspects of space stations. See Hermann Noordung, edited by Ernst Stuhlinger and J.D. Hunley with Jennifer Garland, *The Problem of Space Travel: The Rocket Motor* (Washington, DC: NASA SP-4026, 1995).

Robert G. Nunn, Jr. (1917–1975), earned a law degree from the University of Chicago in 1942. After four years in the Army during World War II, then private practice of law for eight years in Washington, D.C., and in his hometown of Terre Haute, Indiana, he joined the office of general counsel of the Air Force in 1954. He became NASA's assistant general counsel in November 1958 and then special assistant to T. Keith Glennan in September 1960. He helped draft many legal and administrative regulations for NASA, and then he went to work for the Washington law firm of Sharp and Bogan. Later, he formed the firm of Batzell and Nunn, specializing in energy legislation and administrative law. See "Nunn, R.G., Jr.," biographical file, NASA Historical Reference Collection.

P

John R. Pierce (1910–   ) is commonly referred to as the inventor of the communications satellite in 1954. He worked for thirty-five years as an engineer at Bell Laboratories and then worked at the California Institute of Technology and its Jet Propulsion Laboratory. See "Pierce, J. R.," biographical file, NASA Historical Reference Collection.

Frank Press (1924–   ) served as President Carter's science advisor. From 1981 to 1993, he served as president of the National Academy of Sciences. He received a Ph.D. in geophysics from Columbia University in 1949. See "Press, Frank," biographical file, NASA Historical Reference Collection.

William Proxmire (1915–   ) (D–WI) served as a senator from August 29, 1957, to January 3, 1989. He chaired the Senate Committee on Banking, Housing, and Urban Affairs for several sessions of Congress. He was also an outspoken critic of wasteful government spending, often awarding his infamous "Golden Fleece" award to those people or government organizations he felt wasted taxpayer money. See *Biographical Directory of the American Congress 1774–1996;* "Proxmire, William," biographical file, NASA Historical Reference Collection.

R

Harold A. Rosen (1926–   ) was one of the key scientists at the Hughes Aircraft who developed Syncom, the first geosynchronous communications satellite, for NASA. He received the National Medal of Technology in 1985. "Rosen, Harold," biographical file, NASA Historical Reference Collection.

S

William C. Schneider joined NASA in June 1963 and was the Gemini mission director for seven of the ten manned Gemini missions. From 1967 to 1968, he served as Apollo mission director and the Apollo program's deputy director for missions. He then served from 1968 to 1974 as the Skylab program's director. After that, he worked as the deputy associate administrator for space transportation systems for almost four years. From 1978 to 1980, he served as the associate administrator for space tracking and data systems. He received a Ph.D. in engineering from Catholic University. See "Schneider, William C.," biographical file, NASA Historical Reference Collection.

Robert C. Seamans, Jr. (1918–), had been involved in aerospace issues since he completed his Sc.D. degree at the Massachusetts Institute of Technology in 1951. He was on the faculty at MIT's department of aeronautical engineering from 1949 to 1955, when he joined the Radio Corporation of America (RCA) as manager of the Airborne Systems Laboratory. In 1958, he became the chief engineer of the Missile Electronics and Control Division and joined NASA in 1960 as associate administrator. In December 1965, he became NASA's deputy administrator. He left NASA in 1968, and in 1969, he became secretary of the Air Force, serving until 1973. Seamans was president of the National Academy of Engineering from May 1973 to December 1974, when he became the first administrator of the new Energy Research and Development Administration. He returned to MIT in 1977, becoming dean of its School of Engineering in 1978. In 1981, he was elected chair of the board of trustees of Aerospace Corporation. See "Seamans, Robert C., Jr.," biographical file, NASA Historical Reference Collection; Robert C. Seamans, Jr., *Aiming at Targets: The Autobiography of Robert C. Seamans, Jr.* (Washington, DC: NASA SP-4106, 1996).

Willis H. Shapley (1917–), the son of famous Harvard astronomer Harlow Shapley, earned a bachelor of arts degree from the University of Chicago in 1938. From that point until 1942, he did graduate work and performed research in political science and related fields at the University of Chicago. He joined the Bureau of the Budget in 1942 and became a principal examiner in 1948. From 1956 to 1961, he was assistant chief (Air Force) in the bureau's military division, becoming progressively deputy chief for programming (1961–1965) and deputy chief (1965) in that division. He also served as special assistant to the director for space program coordination. In 1965, he moved to NASA as associate deputy administrator, with his duties including supervision of the public affairs, congressional affairs, Department of Defense and interagency affairs, and international affairs offices. He retired in 1975 but rejoined NASA in 1987 to help it recover from the *Challenger* disaster. He served as associate deputy administrator (policy) until 1988, when he again retired but continued to serve as a consultant to the administrator. See "Shapley, W.H.," biographical file, NASA Historical Reference Collection.

George P. Shultz (1920–) served as director of the Office of Management and Budget after 1970, during the Nixon administration. Before that time, he had been Nixon's secretary of labor. During the Reagan administration (1981–1989), Shultz served as secretary of State. See "Shultz, George P.," 1988 *Current Biography Yearbook*, pp. 525–30.

Eugene Skolnikoff served on the staff of the White House science advisor from 1958 to 1963. Afterward, he went to the Massachusetts Institute of Technology, where he served as a political science professor specializing in science, technology, and foreign policy issues. See "Skolnikoff, Eugene," biographical file, NASA Historical Reference Collection.

Jacob E. Smart rose to the rank of general in the U.S. Army, serving as deputy commander of the U.S. European Command. He joined NASA in 1966 as special assistant to the administrator. He then became the acting assistant administrator for administration, the assistant administrator for policy analysis, and the assistant administrator for the Department of Defense and interagency affairs. He graduated from the U.S. Military Academy at West Point in 1931. See "Smart, Jacob," biographical file, NASA Historical Reference Collection.

Cyrus R. Smith (1899–1990) worked in banking until he became manager of Texas Air Transport, a subsidiary of the Texas-Louisiana Power Company. In 1934, this company reorganized, becoming American Airlines. He was chief executive until he became secretary of commerce in 1968, a position he held for one year. He retired in 1969. See "Smith, C.R. (Cyrus Rowlett)," in Bowman, *The Cambridge Dictionary of American Biography*.

T

Robert A. Taft (1889–1953) (R–OH), the son of President William Taft, served as a senator from January 3, 1939, until July 31, 1953. See *Biographical Directory of the United States Congress, 1774–1996*.

Konstantin E. Tsiolkovskiy (1857–1935) became enthralled with the possibilities of interplanetary travel as a boy and, at age fourteen, started independent study using books from his father's library on natural science and mathematics. He also developed a passion for invention, and he constructed balloons, propelled carriages, and other instruments. To further his education, his parents sent him to Moscow to pursue technical studies. In 1878, he became a teacher of mathematics in a school north of Moscow. Tsiolkovskiy first started writing on space in 1898, when he submitted for publication to the Russian journal, *Nauchnoye Obozreniye* (Science Review), a work based on years of calculations that laid out many of the principles of modern spaceflight. The article, "Investigating Space with Rocket Devices," presented years of calculations that laid out many of the principles of modern spaceflight and opened the door to future writings on the subject. In it, Tsiolkovskiy described in depth the use of rockets for launching orbital spaceships. There followed a series of increasingly sophisticated studies on the technical aspects of spaceflight. In the 1920s and 1930s, he proved especially productive, publishing ten major works, elucidating the nature of bodies in orbit, developing scientific principles behind reaction vehicles, designing orbital space stations, and promoting interplanetary travel. He also furthered studies on many principles commonly used in rockets today: specific impulse to gauge engine performance, multistage boosters, fuel mixtures such as liquid hydrogen and liquid oxygen, the problems and possibilities inherent in microgravity, the promise of solar power, and spacesuits for extravehicular activity. Significantly, he never had the resources—nor perhaps the inclination—to experiment with rockets himself. After the Bolshevik revolution of 1917 and the creation of the Soviet Union, Tsiolkovskiy was formally recognized for his accomplishments in the theory of spaceflight. Among other honors, in 1921 he received a lifetime pension from the state that allowed him to retire from teaching at the age of sixty-four. Thereafter, he devoted full time to developing his spaceflight theories studies. His theoretical work greatly influenced later rocketeers both in his native land and throughout Europe. While less well known during his lifetime in the United States, Tsiolkovskiy's work enjoyed broad study in the 1950s and 1960s, when Americans sought to understand how the Soviet Union had accomplished such unexpected success in its early spaceflight efforts. See "Tsiolkovskiy, K.E.," biographical file, NASA Historical Reference Collection.

W

James E. Webb (1906–1992) was NASA's administrator between 1961 and 1968. Previously, he had been an aide to a congressman in New Deal Washington, an aide to Washington lawyer Max O. Gardner, and a business executive with the Sperry Corporation and the Kerr-McGee Oil Company. He had also been director of the Bureau of the Budget between 1946 and 1950 and under secretary of state from 1950 to 1952. See W. Henry Lambright, *Powering Apollo: James E. Webb of NASA* (Baltimore, MD: Johns Hopkins University Press, 1995).

Caspar W. Weinberger (1917–   ), longtime Republican government official, was a senior member of the Nixon, Ford, and Reagan administrations. For Nixon and Ford, he was deputy director (1970–1972) and director (1972–1976) of the Office of Management and Budget. In this capacity, he had a leading role in shaping the direction of NASA's major effort of the 1970s, the development of a reusable Space Shuttle. For Reagan, he served as secretary of defense, in which he oversaw the use of the Shuttle in the early 1980s for the launching of classified Department of Defense payloads into orbit. See "Weinberger, Caspar W(illard)," *Current Biography 1973*, pp. 428–30.

Edward C. Welsh (1909–1990) had a long career in various private and public enterprises. He had served as legislative assistant to Democratic Senator Stuart Symington of Missouri from 1953 to 1961, and he was the executive secretary of the National Aeronautics and Space Council through the 1960s. See "Welsh, E.C.," biographical file, NASA Historical Reference Collection.

Harry Wexler (1911–1962) worked for the U.S. Weather Bureau from 1934 until his death. He was one of the first scientists to envision using satellites for meteorological purposes and was known as the father of the TIROS satellite. From 1955 to 1958, he was also the chief scientist for the U.S. expedition to Antarctica for the International Geophysical Year. In 1961, he was a lead negotiator for the United States in drafting plans for joint U.S.-Soviet Union use of meteorological satellites. He received a Ph.D. in meteorology from the Massachusetts Institute of Technology in 1939. See "Wexler, Harry," biographical file, NASA Historical Reference Collection.

Robert M. White (1923–) served as head of the U.S. Weather Bureau and the Environmental Science Services Administration in the 1960s, as administrator of the National Oceanic and Atmospheric Administration in the 1970s, and as head of the National Academy of Engineering in the late 1980s. See "White, Robert M.," biographical file, NASA Historical Reference Collection.

Clay T. Whitehead was a White House staff assistant during the Nixon administration from 1969 to 1972 who was heavily involved in space policy associated with the decision to build the Space Shuttle and post-Apollo planning for NASA. See Launius, "NASA and the Decision to Build the Space Shuttle, 1969–72"; Launius, "A Western Mormon in Washington, D.C."

Donald D. Williams (1931–1966) was instrumental in the development of the Early Bird and Syncom communications satellites. He was employed by Hughes Aircraft and was named one of America's ten outstanding young men of 1965 by the U.S. Junior Chamber of Commerce. On February 21, 1966, Williams committed suicide. See "Academic and Scientific Miscellaneous," biographical file, NASA Historical Reference Collection.

Y

John Yardley began his career in aerospace in 1946, when he joined McDonnell Aircraft. While with that company, he assumed a major role in the Mercury and Gemini programs. In 1974, he came to NASA as the associate administrator for manned spaceflight. His title then changed to associate administrator for spaceflight, and in 1978, he became associate administrator for space transportation systems. In 1981, he returned to the private sector as president of the McDonnell-Douglas Astronautics Company. See "Yardley, John," biographical file, NASA Historical reference Collection.

John D. Young (1919–) earned a master of science degree from Syracuse in 1943 and served as an officer in the Marine Corps from 1942 to 1945. He worked for various government agencies in the next few years and then became a management consultant with McKinsey & Co. from 1954 to 1960. He served as NASA's director of management analysis from 1960 to 1961 and then became, successively, deputy director for administration and deputy associate administrator at NASA Headquarters. He left NASA in 1966 for a series of management positions in the Bureau of the Budget and the Department of Health, Education, and Welfare. Thereafter, he became a professor of public management at American University. See "Young, J.D.," biographical file, NASA Historical Reference Collection.

Index

A

Advanced Communications Technology Satellite (ACTS), 9, 145-47; and "Federal Research and Development for Satellite Communications," 135-45
Advanced Research Projects Agency (ARPA), 158, 203-04
Agriculture, Department of, 167, 168-69
Air Force, United States, 157
American Broadcasting Company (ABC), 106
American Rocket Society, 62
American Securities Corp., 62
American Telephone & Telegraph (AT&T), 2, 3, 5, 39-41; and "Administrative and Regulatory Problems Relating to the Authorization of Commercially Operable Space Communications Systems," 61-64; and "Establishment of Domestic Communications-Satellite Facilities by Non-Governmental Entities," 120-32; and "Exotic Radio Communications," 22-30; and "FCC Relation to Space Communication," 42-45; and "Federal Research and Development for Satellite Communications," 135-45; and F.R. Kappel, 45-60; and legislation about, 67-72; and Newton Minnow, 60-61, 76; and State Department, 85-89, 91-95; and Telstar, 89-90
Anders, William A., 135
Anderson, Clinton P., 269-71
Andrews, Duane P., 375, 377-78
Anik, 10, 138
Applications Technology Satellite (ATS), 9, 161
APSTAR, 10
ARABSAT, 10
Army Ballistic Missile Agency (ABMA), 157
ASIASAT, 10
ASTRA, 10
Atlas launch vehicle, 2
Atomic Energy Commission (AEC), 385
ATS-6, 139
Augenstein, Bruno, 282-93

B

Badgley, Peter C., 237-40; and "Current Status of NASA's Natural Resources Program" (1966), 226-37
Baker, D. James, 213-14, 224-25
Baldridge, Malcolm, 151-53
Ball Aerospace Corp., 7
Beggs, James M., 396, 539-41; and "NASA Policy to Enhance Commercial Investment in Space," 488-98; and "Space Commercialization Meeting," 498-501
Bell Laboratories, 31, 32; and "Exotic Radio Communications," 22-30; and F.R. Kappel, 45-60; and Telstar, 89-90
Berg, Otto, 157
Best, G.L., 46, 48-49, 51
Boileau, Oliver C., 498-501
Bolster, Edward A., 85-89
Braibanti, Ralph, 375, 378
British Interplanetary Society, 11
Brown, Clarence J., 498-501
Brown, George E., 164, 213-14, 368-72
Brown, Ron, 164, 213, 215-16
Brown, William M., and "Long-Term Prospects for Developments in Space," 480-88
Bruce, David, 96-99
Brzezinski, Zbigniew, 294-95

Bullington, Kenneth, 25

Bush, George W., 173, 175-76; and "Land Remote Sensing Policy Act of 1992," 173-75, 352-68; and "Landsat Remote Sensing Policy," 345-47; and "U.S. Commercial Space Policy Guidelines," 460-63

Byerly, Radford, 145-47

C

California Institute of Technology (CIT), 29

Cape Canaveral, FL, 35

Carter, James E. (Jimmy), 171; and "Civil Operational Remote Sensing," 294-95; and "Planning for a Civil Operational Land Remote Sensing Satellite System," 296-306

Cedar Rapids, IA, 24

Celler, Emanuel, 67-71

Chase Econometric Associates, Inc., 389, 390, 391, 426, 427, 431, 451; and "The Economic Impact of NASA R&D Spending," 414-26

Chance-Vought Corp., and Hughes "Commercial Satellite Communication Project," 31-35; and Rosen's *Commercial Communications Satellite*, 35-39

Chapman Research Group, and "An Exploration of Benefits from NASA 'Spinoff,'" 559-63

Clark, Tim, 145-47

Clarke, Arthur C., 1, 4; and "Extra-Terrestrial Relays," 16-22; and "The Space Station," 11-15

Clinton, William J., and "Convergence of U.S.-Polar-orbiting Operational Environmental Satellite Systems," 165, 221-23; and "Land Remote Sensing Policy Act of 1992," 173-75, 352-68; and "Landsat Remote Sensing Strategy," 372-75; and "National Space Policy" (1996), 463-73; and "U.S. Policy on Licensing and Operation of Private Remote Sensing Systems," 379-81

Commerce, Department of, 171-72; and "Basic Agreement Between U.S. Department of Commerce and the National Aeronautics and Space Administration Concerning Operational Meteorological Satellite Systems," 206-11; and "Civil Operational Remote Sensing," 294-95; and "Commercial Space Industry in the Year 2000," 473-80; and "Convergence of U.S.-Polar-orbiting Operational Environmental Satellite Systems," 165, 221-23; and "Earth Information from Space by Remote Sensing," 282-93; and Earth resource satellite commercialization, 175-76; and "Earth Resources Survey Program," 248-50, 250-52, 253-56; and "Land Remote-Sensing Commercialization Act of 1984," 329-44; and "Land Remote Sensing Policy Act of 1992," 173-75, 352-68; and "Landsat Remote Sensing Policy," 345-47; and "Landsat Remote Sensing Strategy," 372-75; and National Performance Review, 164-65, 216-21; and "National Plan for a Common System of Meteorological Observation Satellites," 204-206; and "Observing the Weather from a Satellite Vehicle," 177-83; and "Planning for a Civil Operational Land Remote Sensing Satellite System," 296-306; and "The President's Space Policy and Commercial Space Initiatives to Begin the Next Century," 455-60; and "Report of the Government Technical Review Panel on Industry Responses on Commercialization of the Civil Remote Sensing Systems," 309-21; and "Resolution of Issues Related to Private Sector Transfer of Civil Land Observing Satellite Activities," 306-08; and "Transfer of Civil Meteorological Satellites," 321-29; and "U.S. Policy on Licensing and Operation of Private Remote Sensing Systems," 379-81

Communications, satellite, 1-153; and "Administrative and Regulatory Problems Relating to the Authorization of Commercially Operable Space Communications Systems," 61-64; and Advanced Communications Technology Satellite (ACTS), 9; and Anik, 10; and Applications Technology Satellite (ATS) program, 9; and APSTAR, 10; and ARABSAT, 10; and ASIASAT, 10; and ASTRA, 10; and AT&T proposals, 45-60; and brief history of, 6-10; and Clarke's "Extra-Terrestrial Relays," 16-22; and Clarke's "The Space Station," 11-15; Communications Satellite Act of 1962, 5, 61, 63, 76-85, 85-89; and Communications Satellite Corp. (Comsat), 4-5, 72-76, 85-89; and Communications Technology Satellite program, 9; and Defense Satellite Communications System (DSCS), 10; and early concepts of, 2-4; and "Establishment of Domestic Communications-Satellite Facilities by Non-Governmental Entities," 120-32; and European Telecommunications Satellite Organization (EUTELSAT), 10; and "FCC Relation to Space Communication," 42-45; and "Federal Research and Development for Satellite Communications," 135-45; and F.R. Kappel, 45-60; and "A Global System of Satellite Communications," 99-108; and Hughes "Commercial Satellite Communication Project," 31-35; and INMARSAT, 11; and INTELSAT, 4-5, 6-10, 11, 91-95; and international relations, 4-10; and International Telecommunications Union (ITU), 10-11, 65, 89-90; and legislation about, 67-72; and National Aeronautics and Space Council, 65-67, 76-77; and "National Space Policy" (1996), 463-73;

and Newton Minnow, 60-61, 76; and Orion Satellite System, 9; and PanAmSat, 10; and Pierce's "Exotic Radio Communications," 22-30; and "Policy Concerning U.S. Assistance in the Development of Foreign Communications Satellite Capabilities," 91-95; and "Report of the United States Delegation . . . on Definitive Arrangements for the International Telecommunications Satellite Consortium," 108-20; and Rosen's *Commercial Communications Satellite,* 35-39; and Soviet Union, 85-89; and State Department, 85-89, 91-95; and Tracking and Data Relay Satellite System (TDRSS), 9; and Telstar, 89-90; and T. Keith Glennan, 29, 39-41; and "Transfer of U.S. Communications Satellite Technology," 96-99; and "U.S. Assistance in the Early Establishment of Communications Satellite Service," 95-96; and "White Paper on New International Satellite Systems," 147-53; and "U.S. Commercial Space Policy Guidelines," 460-63

Communications Satellite Act of 1962, 5, 61, 63, 76-85; and "Establishment of Domestic Communications-Satellite Facilities by Non-Governmental Entities," 120-32; and State Department, 85-89, 91-95

Communications Satellite Corp. (Comsat), 4-5, 72-76, 155; and ACTS, 145-47; and "Establishment of Domestic Communications-Satellite Facilities by Non-Governmental Entities," 120-32; and "Federal Research and Development for Satellite Communications," 135-45; and "A Global System of Satellite Communications," 99-108; and "Policy Concerning U.S. Assistance in the Development of Foreign Communications Satellite Capabilities," 91-95; and "Report of the United States Delegation…on Definitive Arrangements for the International Telecommunications Satellite Consortium," 108-20; and State Department, 85-89, 91-95; and Telstar, 89-90; and "Transfer of U.S. Communications Satellite Technology," 96-99; and "U.S. Assistance in the Early Establishment of Communications Satellite Service," 95-96; and "White Paper on New International Satellite Systems," 147-53

Communications Technology Satellite, 9

Cortright, Edgar M., and "Earth Resources Survey Program," 253-56

Courier 1B, 3, 90

Craven, T.A.M., 43-44

Crawford, A.B., 25

Cross, David M., 426

Cygnus Corp., and "White Paper on New International Satellite Systems," 147-53

D

Darman, Richard G., 498-501

Deere & Company, 498-501

Defense Early Warning (DEW) line, 25

Defense, Department of (DOD), 2, 5, 7, 9, 31; and "Administrative and Regulatory Problems Relating to the Authorization of Commercially Operable Space Communications Systems," 61-64; and "Basic Agreement Between U.S. Department of Commerce and the National Aeronautics and Space Administration Concerning Operational Meteorological Satellite Systems," 206-11; and "Civil Operational Remote Sensing," 294-95; and "Convergence of U.S.-Polar-orbiting Operational Environmental Satellite Systems," 165, 221-23; and "Earth Information from Space by Remote Sensing," 282-93; and Earth resource satellites, 167-76; and "Establishment of Domestic Communications-Satellite Facilities by Non-Governmental Entities," 120-32; and "FCC Relation to Space Communication," 42-45; and "Federal Research and Development for Satellite Communications," 135-45; and F.R. Kappel, 45-60; and "A Global System of Satellite Communications," 99-108; and "Earth Resources Survey Program," 248-50, 250-52, 253-56; and "Land Remote-Sensing Commercialization Act of 1984," 329-44; and "Land Remote Sensing Policy Act of 1992," 173-75, 352-68; and "Landsat Remote Sensing Strategy," 372-75; and "Management Plan for the Landsat Program," 347-52; meteorological satellites, 156-63; and National Performance Review, 164-65, 216-21; and "National Plan for a Common System of Meteorological Observation Satellites, 204-206; and "National Space Policy" (1996), 463-73; and "Planning for a Civil Operational Land Remote Sensing Satellite System," 296-306; and "Policy Concerning U.S. Assistance in the Development of Foreign Communications Satellite Capabilities," 91-95; and "Report of the United States Delegation . . . on Definitive Arrangements for the International Telecommunications Satellite Consortium," 108-20; and "Resolution of Issues Related to Private Sector Transfer of Civil Land Observing Satellite Activities," 306-08; and State Department, 85-89, 91-95; and Telstar, 89-90; and TIROS, 157, 158, 161-62, 203-06; and "Transfer of U.S. Communications Satellite Technology," 96-99; and "U.S. Assistance in the Early Establishment of Communications Satellite Service," 95-96; and "White Paper on New International Satellite Systems," 147-53

Defense Satellite Communications System (DSCS), 10
Defense Meteorological Satellite Program (DMSP), 158, 215-16
Delta launch vehicle, 35
Denver Research Institute, 390; and "Mission Oriented R&D and the Advancement of Technology," 432-45
Deutch, John , 370
Dingman, J.E., 56-57, 58-59
Dryden, Hugh L., 47, 58-60, 88, 203-04; and "Basic Agreement Between U.S. Department of Commerce and the National Aeronautics and Space Administration Concerning Operational Meteorological Satellite Systems," 206-11
Dunn, Robert, and "NASA Policy to Enhance Commercial Investment in Space," 488-98
Dutton, Frederick G., 71-72

E

Early Bird, 5
Earth Observation Satellite Company (EOSAT), 172, 344-45; and "Land Remote Sensing Policy Act of 1992," 173-75, 352-68; and "Landsat Remote Sensing Policy," 345-47
Earth Resources Observation Satellite (EROS) Program, 244-48, 256-57, 275-76
Earth Observing Satellite (EOS), 163, 396
Earth Resources Technology Satellite (ERTS), see Landsat
Echo 1, 2-3, 22, 30, 50
Echo 2, 3
Eisenhower, Dwight D., 2, 4, 39, 41-42, 86, 158
Elliott, Linda, and "Mission Oriented R&D and the Advancement of Technology," 432-45
Energy, Department of (DOE), 385
European Organisation for the Exploitation of Meteorological Satellites (EUMETSAT), 165-67; and Earth resource satellites, 167-76, 224-25
European Posts and Telecommunications, Committee on, 6
European Space Agency (ESA), 165-66; and Earth resource satellites, 167-76, 224-25
European Telecommunications Satellite Organization (EUTELSAT), 10
Evans, Llewellyn, 498-501
Evans, Michael K., and "The Economic Impact of NASA R&D Spending," 414-26
Exon, James, 164, 213, 215-16

F

Faget, Maxime A., 498-501
Fairchild Aviation Corp., 7, 498-501; and "Establishment of Domestic Communications-Satellite Facilities by Non-Governmental Entities," 120-32
Farley, Philip, 85-89
Faucett, Jack G., 388, 401-402
Federal Communications Commission (FCC), 9, 10; and "Administrative and Regulatory Problems Relating to the Authorization of Commercially Operable Space Communications Systems," 61-64; and Communications Satellite Act of 1962, 5, 61, 63, 76-85; and "Establishment of Domestic Communications-Satellite Facilities by Non-Governmental Entities," 120-32; and "FCC Relation to Space Communication," 42-45; and "Federal Research and Development for Satellite Communications," 135-45; and "A Global System of Satellite Communications," 99-108; and legislation about, 67-72; and National Aeronautics and Space Act, 72-85, 392; and National Aeronautics and Space Council, 65-67, 76-85; and "National Space Policy" (1996), 463-73; and Newton Minnow, 60-61, 76; and "Policy Concerning U.S. Assistance in the Development of Foreign Communications Satellite Capabilities," 91-95; and "Report of the United States Delegation . . . on Definitive Arrangements for the International Telecommunications Satellite Consortium," 108-20; and Soviet Union, 85-89; and State Department, 85-89, 91-95; and "Transfer of U.S. Communications Satellite Technology," 96-99; and "U.S. Assistance in the Early Establishment of Communications Satellite Service," 95-96; and "White Paper on New International Satellite Systems," 147-53
Federal Express Corp., 498-501
Felkel, Ed, 33

Fletcher, James C., 132-35, 259-62, 269-71, 277-81
Ford Aerospace Corp., 7
Ford Foundation, 106
Friis, H.T., 24, 25
Frutkin, Arnold W., 259-62, and "Foreign Policy Issues Regarding Earth Resource Surveying by Satellite," 262-69
Fuller, Craig L., 394, 498-501, 501-04
Fuqua, Don, 321-29

G

General Electric Corp., 7, 62, 168-69
General Accounting Office (GAO), 269-71; and "Commercial Use of Space," 543-46; and "NASA Report May Overstate the Economic Benefits of Research and Development Spending," 430-32
General Dynamics Corp., 498-501
Geological Survey, U.S., 168, 240; and EROS, 275-76; and Landsat, 257-59; and "Meeting at the U.S. Geological Survey . . . Regarding Remote Sensing and South America," 240-44
Gibbons, John H., 368-72
Glennan, T. Keith, 29, 39-41; and AT&T, 45-60
Goldstone, CA, 29
Goodall, W.M., 24-25
Goddard Space Flight Center, MD, 51, 54-56, 449
Gore, Al, 164-65
Government Performance and Results Act, 400
Granger, John V.N., 262-69
Green, E.I., 46, 49-50
Green Bank, WV, 24, 45
Grumman Aerospace Corp., 498-501

H

Haeff, A.V., and "Commercial Satellite Communication Project," 31-35
Hannah, David, 498-501
Halpern, Irwin P., and "Earth Resources Survey Program," 248-50
Hanson, Robert A., 498-501
Harper, Ed, 306-308
Hayden Planetarium, 177
Heiss, Klaus P., 498-501
Hertzfeld, Henry, 426-27; 563-74
Hodges, Luther H., and "Basic Agreement Between U.S. Department of Commerce and the National Aeronautics and Space Administration Concerning Operational Meteorological Satellite Systems," 206-11
Hogg, D.C., 25
Holman, Mary, 388
Holmdel Laboratory, 25, 50-50
Hough, Roger W., 402-07
Houston, TX, 133
Hudson Institute, 394; and "Long-Term Prospects for Developments in Space," 480-88
Hudspeth, Thomas, and "Commercial Satellite Communication Project," 31-35
Hughes Aircraft Corp., 3-4, 5, 7; and ACTS, 145-47; and "Commercial Satellite Communication Project," 31-35; and EOSAT, 172; and "Establishment of Domestic Communications-Satellite Facilities by Non-Governmental Entities," 120-32; and "Land Remote-Sensing Commercialization Act of 1984," 329-44; and Rosen's *Commercial Communications Satellite*, 35-39
Hutton, E.F., and Co., 498-501

I

Indiana University, 388
Initial Defense Satellite Communication System (IDSCS), 5
INMARSAT, 11
"Inquiry into the Feasibility of Weather Reconnaissance from a Satellite Vehicle," 157, 185-202
INTELSAT I, 5
Interior, Department of the, and "Civil Operational Remote Sensing," 294-95; and "Convergence of U.S.-Polar-orbiting Operational Environmental Satellite Systems," 165, 221-23; and "Earth's Resources to be Studied from Space," 244-46; and Landsat, 168-73; and "Landsat Remote Sensing Policy," 345-47; and "Land Remote Sensing Policy Act of 1992," 173-75, 352-68; and "National Space Policy" (1996), 463-73; and "Operational Requirements for Global Resource Surveys by Earth-Orbital Satellites," 246-48; and "Planning for a Civil Operational Land Remote Sensing Satellite System," 296-306
International Telecommunications Satellites Consortium (INTELSAT), 4-5, 6-10, 11, 91-95; and "FCC Relation to Space Communication," 42-45; and "Federal Research and Development for Satellite Communications," 135-45; and "A Global System of Satellite Communications," 99-108; and "Land Remote-Sensing Commercialization Act of 1984," 329-44; and "Policy Concerning U.S. Assistance in the Development of Foreign Communications Satellite Capabilities," 91-95; and "Report of the United States Delegation . . . on Definitive Arrangements for the International Telecommunications Satellite Consortium," 108-20; and "Transfer of U.S. Communications Satellite Technology," 96-99; and "U.S. Assistance in the Early Establishment of Communications Satellite Service," 95-96; and "White Paper on New International Satellite Systems," 147-53
International Telecommunications Union (ITU), 10-11, 65, 89-90
International Telephone and Telegraph (ITT), 5; and "Administrative and Regulatory Problems Relating to the Authorization of Commercially Operable Space Communications Systems," 61-64; and State Department, 85-89, 91-95
I.R.E, Proceedings of the, 25

J

Jaffe, Leonard, 237-40, and "Earth Resources Survey Program," 248-50, 250-52, 253-56; and "Meeting at the U.S. Geological Survey . . . Regarding Remote Sensing and South America," 240-44
Jansky, Karl G., 24
Janus, Project, 157-58
Jeffs, George, 498-501
Jet Propulsion, 22; and "Exotic Radio Communications," 22-30
Jet Propulsion Laboratory (JPL), 7, 29
Johns Hopkins Applied Physics Laboratories, 7
Johnson, David S., 296-306
Johnson, E. Douglas, 553-55
Johnson, John, 72
Johnson, Lyndon B., and "Establishment of Domestic Communications-Satellite Facilities by Non-Governmental Entities," 120-32; and "FCC Relation to Space Communication," 42-45; and "A Global System of Satellite Communications," 99-108; and "Policy Concerning U.S. Assistance in the Development of Foreign Communications Satellite Capabilities," 91-95; and "Report of the United States Delegation . . . on Definitive Arrangements for the International Telecommunications Satellite Consortium," 108-20; and "Transfer of U.S. Communications Satellite Technology," 96-99; and "U.S. Assistance in the Early Establishment of Communications Satellite Service," 95-96
Johnson, Ray W., 203-04

K

Kahn, Herman, and "Long-Term Prospects for Developments in Space," 480-88
Kappel, F.R., 45-60
Kefauver, Estes, 5, 72
Kelley, John A., and "Mission Oriented R&D and the Advancement of Technology," 432-45

Kennedy, John F., 2, 4-5, 46, 60-61, 65, 67-72, 87, 89, 159
Kerr, Robert S., 5, 72-76, 87
Keyworth, George A., III, 498-501
Khrushchev, Nikita A., 85-89
Kissinger, Henry A., 135
Kraft, Christopher C., Jr., 281-82

L

Large Area Crop Inventory Experiment (LACIE), 170
Landsat, 133, 168-73; and "Civil Operational Remote Sensing," 294-95; and "Convergence of U.S.-Polar-orbiting Operational Environmental Satellite Systems," 165, 221-23; and "Crop Forecasting by Satellite," 272-75; and "Earth Information from Space by Remote Sensing," 282-93; and "Foreign Policy Issues Regarding Earth Resource Surveying by Satellite," 262-69; and "Land Remote-Sensing Commercialization Act of 1984," 329-44; and "Landsat Remote Sensing Policy," 345-47; and "Landsat Remote Sensing Strategy," 372-75; and "Management Plan for the Landsat Program," 347-52; and "Planning for a Civil Operational Land Remote Sensing Satellite System," 296-306; and "Private Sector Operation of Landsat Satellites," 281-82; and "Report of the Government Technical Review Panel on Industry Responses on Commercialization of the Civil Remote Sensing Systems," 309-21; and "Resolution of Issues Related to Private Sector Transfer of Civil Land Observing Satellite Activities," 306-08; and "Some Recent International Reactions to ERTS-1," 259-62
Latshaw, John, 498-501
Lawrence Livermore National Laboratory, 175
Lincoln Experimental Satellite (LES), 4, 9-10, 142
Lincoln Laboratory, 4, 7, 9-10, 141
Lockheed-Martin Corp., 7, 62, 381-83; and commercialization of Earth resource satellites, 176; and "Establishment of Domestic Communications-Satellite Facilities by Non-Governmental Entities," 120-32
Lovell, Robert, 145-47
Low, George M., 132-35, 259-62
Luce, Charles F., and "Operational Requirements for Global Resource Surveys by Earth-Orbital Satellites," 246-48
Luton, Jean-Marie, and Earth resource satellites, 224-25
Lutz, S.G., and "Commercial Satellite Communication Project," 31-35

M

McDonnell Douglas Corp., 396, 498-501, 539-41; and "Electrophoresis Operations in Space," 541-43; and "Feasibility Study of Commercial Space Manufacturing," 504-17; and "Feasibility Study of Commercial Space Manufacturing: Production of Pharmaceuticals," 534-39
McGhee, George, 85-89
Madison, John J., 145-47
Mandel, J.T., 33
Markey, David J., 147-53
Marks, Leonard H., 108-20
Marshall Space Flight Center, 395
Massachusetts Institute of Technology (MIT), 141, 389
Mathematica, Inc., 390; and "Quantifying the Benefits to the National Economy from Secondary Applications of NASA Technology," 445-49
MathTech, Inc., 399; and "Cost-Benefit Analysis of Selected Technology Utilization Office Programs," 555-58;
Mayo, Robert P., 257-59
Meese, Edwin III, 498-501
Meteorological satellites, 156-63; and "Basic Agreement Between U.S. Department of Commerce and the National Aeronautics and Space Administration Concerning Operational Meteorological Satellite Systems," 206-11; and "Inquiry into the Feasibility of Weather Reconnaissance from a Satellite Vehicle," 157, 185-202; and METOP, 166; and "National Plan for a Common System of Meteorological Observation Satellites," 204-06; and Nimbus Weather Satellite, 159-60, 168; and TIROS, 157, 158, 161-62, 203-06; and "Transfer of Civil Meteorological Satellites," 321-29

METOP satellite series, 166

Midwest Research Institute, 389, 390, 391, 415; and "Economic Impact of Stimulated Technological Activity," 408-14; and "Economic Impact and Technological Progress of NASA Research and Development Expenditures," 427-30

Miernyk, William, 388

Mignogno, Michael, 375, 378

Minnow, Newton, 60-61, 76

Molniya ("Lightning"), 5

Morgan, John, and Earth resource satellites, 224-25

Morrill, William, 133

Motorola Corp., and ACTS, 145-47

Mueller, George E., 388; and Earth resources survey program, 248-50, 250-52, 253-56

Mundt, Karl, 169

N

NASA Alumni League, 391; and "The Economic Effects of a Space Station," 450-51

National Academy of Public Administration (NAPA), 389, 391

National Academy of Sciences, and "Federal Research and Development for Satellite Communications," 135-45

National Advisory Committee for Aeronautics (NACA), 2, 397

National Aeronautics and Space Act of 1958, 72-76, 392, 397

National Aeronautics and Space Administration (NASA), 2, 4, 7; and ACTS, 9, 145-47; and "Administrative and Regulatory Problems Relating to the Authorization of Commercially Operable Space Communications Systems," 61-64; and AT&T, 45-60; and "Basic Agreement Between U.S. Department of Commerce and the National Aeronautics and Space Administration Concerning Operational Meteorological Satellite Systems," 206-11; and "Civil Operational Remote Sensing," 294-95; and "Commercial Space Industry in the Year 2000," 473-80; and "Commercial Use of Space," 543-46; and "Convergence of U.S.-Polar-orbiting Operational Environmental Satellite Systems," 165, 221-23; and "Cost-Benefit Analysis of Selected Technology Utilization Office Programs," 555-58; and "Crop Forecasting by Satellite," 272-75; and "Earth Information from Space by Remote Sensing," 282-93; and Earth resource satellites, 167-76; and "Earth Resources Survey Program," 248-50, 250-52, 253-56; and "Earth's Resources to be Studied from Space," 244-46; and "The Economic Effects of a Space Station," 450-51; and "The Economic Impact of the Space Program," 451-55; and "Economic Impact of Stimulated Technological Activity," 408-14; and Eisenhower, 39-41; and "Economic Impact and Technological Progress of NASA Research and Development Expenditures," 427-30; and "Electrophoresis Operations in Space," 541-43; and "Establishment of Domestic Communications-Satellite Facilities by Non-Governmental Entities," 120-32; and EOS, 163, 396; and "An Exploration of Benefits from NASA 'Spinoff,'" 559-63; and "FCC Relation to Space Communication," 42-45; and "Feasibility Study of Commercial Space Manufacturing," 504-17; and "Feasibility Study of Commercial Space Manufacturing: Production of Pharmaceuticals," 534-39; and "Federal Research and Development for Satellite Communications," 135-45; and "Foreign Policy Issues Regarding Earth Resource Surveying by Satellite," 262-69; and F.R. Kappel, 45-60; and "A Global System of Satellite Communications," 99-108; and Hughes "Commercial Satellite Communication Project," 31-35; and "Land Remote-Sensing Commercialization Act of 1984," 329-44; and "Landsat Remote Sensing Policy," 345-47; and "Land Remote Sensing Policy Act of 1992," 173-75, 352-68; and Landsat, 133, 168-73; and "Landsat Remote Sensing Strategy," 372-75; and Large Area Crop Inventory Experiment (LACIE), 170; and "Long-Term Prospects for Developments in Space," 480-88; and "Management Plan for the Landsat Program," 347-52; and "Meeting at the U.S. Geological Survey . . . Regarding Remote Sensing and South America," 240-44; and Meteorological Satellites, 156-63; and "Mission Oriented R&D and the Advancement of Technology," 432-45; and "NASA Policy to Enhance Commercial Investment in Space," 488-98; and "NASA Report May Overstate the Economic Benefits of Research and Development Spending," 430-32; and "NASA Tech Brief Program," 553-55; and "NASA Technology Transfer: Report of the Technology Transfer Team," 574-77; and National Aeronautics and Space Act, 72-76, 392; and National Aeronautics and Space Council, 65-67, 76-77; and National Performance Review, 164-65, 216-21; and "National Plan for a Common System of Meteorological Observation Satellites," 204-06; and "National Space Policy" (1996), 463-73; and Newton Minnow, 60-61, 76; and Nimbus Weather Satellite, 159-60, 168; and NOAA, 163-67; and "Operational Requirements for Global Resource Surveys by Earth-Orbital Satellites," 246-48; and "Planning for a Civil Operational Land Remote Sensing Satellite System," 296-306; and Polar-orbiting Operational Environmental Satellites (POES), 163; and "Policy Concerning U.S. Assistance in the Development of Foreign Communications Satellite Capabilities," 91-95; and "The President's Space Policy and Commercial Space Initiatives to Begin the Next Century," 455-60; and "Private Sector Operation of Landsat Satellites," 281-82; and "Proposal for Enhancing NASA Technology Transfer to Civil Systems," 546-53; and remote sensing, 155-383; and "Quantifying the Benefits to the National Economy from Secondary Applications of NASA Technology," 445-49; and "Report of the United States Delegation . . . on Definitive Arrangements for the International Telecommunications Satellite Consortium," 108-20; and "Report of the Government Technical Review Panel on Industry Responses on Commercialization of the Civil Remote Sensing Systems," 309-21; and "Resolution of Issues Related to Private Sector Transfer of Civil Land Observing Satellite Activities," 306-308; and Rosen's *Commercial Communications Satellite*, 35-39; and "Some Recent International Reactions to ERTS-1," 259-62; and "Space Commercialization Meeting," 498-501; and space economics, 385-577; and "Space Industrialization: An Overview," 527-33; and "Space Industrialization Final Report," 517-27; and State Department, 85-89, 91-95; and "Technology Transfer White Paper," 563-74; and TIROS, 158-60, 161-62, 203-06; and T. Keith Glennan, 29, 39-41, 203-04; and "Transfer of Civil Meteorological Satellites," 321-29; and "Transfer of U.S. Communications Satellite Technology," 96-99; and "U.S. Assistance in the Early Establishment of Communications Satellite Service," 95-96; and "U.S. Commercial Space Policy Guidelines," 460-63; and "U.S. Policy on Licensing and Operation of Private Remote Sensing Systems," 379-81; and "White Paper on New International Satellite Systems," 147-53

National Aeronautics and Space Council, 65-67, 76-77; and "Policy Concerning U.S. Assistance in the Development of Foreign Communications Satellite Capabilities," 91-95; and State Department, 85-89

National Oceanographic and Atmospheric Administration (NOAA), 215-16; and "Basic Agreement Between U.S. Department of Commerce and the National Aeronautics and Space Administration Concerning Operational Meteorological Satellite Systems," 206-11; and "Civil Operational Remote Sensing," 294-95; and "Commercial Space Industry in the Year 2000," 473-80; and "Earth Information from Space by Remote Sensing," 282-93; and Earth resource satellites, 167-76, 224-25; and "Earth Resources Survey Program," 248-50, 250-52, 253-56; and "Earth's Resources to be Studied from Space," 244-46; and NASA, 163-67; and "Convergence of U.S.-Polar-orbiting Operational Environmental Satellite Systems," 165, 221-23; and "Foreign Policy Issues Regarding Earth Resource Surveying by Satellite," 262-69; and "Land Remote-Sensing Commercialization Act of 1984," 329-44; and "Land Remote Sensing Policy Act of 1992," 173-75, 352-68; and Landsat, 133, 168-73; and "Landsat Remote Sensing Policy," 345-47; and "Landsat Remote Sensing Strategy," 372-75; and Large Area Crop Inventory Experiment (LACIE), 170; and "Management Plan for the Landsat Program," 347-52; and National Performance Review, 164-65, 216-21; and "National Plan for a Common System of Meteorological Observation Satellites," 204-06; and "National Space Policy" (1996), 463-73; and "Operational Requirements for Global Resource Surveys by Earth-Orbital Satellites," 246-48; and "Planning for a Civil Operational Land Remote Sensing Satellite System," 296-306; and Polar-orbiting Operational Environmental Satellites (POES), 163; and "Report of the Government Technical Review Panel on Industry Responses on Commercialization of the Civil Remote Sensing Systems," 309-21; and "Resolution of Issues Related to Private Sector Transfer of Civil Land Observing Satellite Activities," 306-308; and "Some Recent International Reactions to ERTS-1," 259-62; and "Transfer of Civil Meteorological Satellites," 321-29; and "U.S. Policy on Licensing and Operation of Private Remote Sensing Systems," 379-81

National Performance Review (NPR), 164-65, 216-21

National Radio Observatory, 24, 45

National Research Council, and "Federal Research and Development for Satellite Communications," 135-45

National Science Foundation (NSF), 7, 240, 385

National Weather Service, 173

Naval Research Laboratory, 157

Newell, Homer E., 211-13; and "Earth Resources Survey Program," 248-50, 250-52, 253-56; and "Meeting at the U.S. Geological Survey . . . Regarding Remote Sensing and South America," 240-44; and "Some Recent International Reactions to ERTS-1," 259-62

Nimbus Weather Satellite, 159-60, 168

Nixon, Richard M., 133-35

Noordung, Hermann, 11

Nunn, Robert, 39

O

O'Connell, James D., 91-95, 99-108
Orbital Solar Observatory (OSO)-1, 133
Orbital Systems Corp., 498-501
Orion Satellite System, 9; and "White Paper on New International Satellite Systems," 147-53

P

Packer, Leo S., 546-53
PanAmSat, 10
Pasadena, CA, 29
Pennsylvania, University of, 388
Pharsalia, NY, 25
Pierce, John R., 2-3, 48, 90; and "Exotic Radio Communications," 22-30
Potôcnik, Herman, 11
"Preliminary Design of an Experimental World-Circling Spaceship," 156, 183
Proxmire, William, and "NASA Report May Overstate the Economic Benefits of Research and Development Spending," 430-32

R

Radio Corporation of America (RCA), 3, 5, 7; and ACTS, 145-47; and "Administrative and Regulatory Problems Relating to the Authorization of Commercially Operable Space Communications Systems," 61-64; and EOSAT, 172; and "Establishment of Domestic Communications-Satellite Facilities by Non-Governmental Entities," 120-32; and Hughes "Commercial Satellite Communication Project," 31-35; and "Land Remote-Sensing Commercialization Act of 1984," 329-44; and meteorological satellites, 157; and Rosen's *Commercial Communications Satellite*, 35-39; and "White Paper on New International Satellite Systems," 147-53
RAND Corp., 156, 157; and "Earth Information from Space by Remote Sensing," 282-93; and "Inquiry into the Feasibility of Weather Reconnaissance from a Satellite Vehicle," 157, 185-202
Reagan, Ronald, and ACTS, 145-47; and "Basic Agreement Between U.S. Department of Commerce and the National Aeronautics and Space Administration Concerning Operational Meteorological Satellite Systems," 206-11; and "Convergence of U.S.-Polar-orbiting Operational Environmental Satellite Systems," 165, 221-23; and "Land Remote-Sensing Commercialization Act of 1984," 329-44; and "Land Remote Sensing Policy Act of 1992," 173-75, 352-68; and "The President's Space Policy and Commercial Space Initiatives to Begin the Next Century," 455-60; and "Report of the Government Technical Review Panel on Industry Responses on Commercialization of the Civil Remote Sensing Systems," 309-21; and "Resolution of Issues Related to Private Sector Transfer of Civil Land Observing Satellite Activities," 306-08; and space commercialization, 498-501, 539-41; and space economics, 392-400; and "Transfer of Civil Meteorological Satellites," 321-29; and "White Paper on New International Satellite Systems," 147-53
Reeves, Robert G., and "Meeting at the U.S. Geological Survey . . . Regarding Remote Sensing and South America," 240-44
Reichdelerfer, F.W., and "National Plan for a Common System of Meteorological Observation Satellites," 204-06
Relay 1, 3, 35
Remote sensing, 155-383; and "Basic Agreement Between U.S. Department of Commerce and the National Aeronautics and Space Administration Concerning Operational Meteorological Satellite Systems," 206-11; and "Civil Operational Remote Sensing," 294-95; and commercialization of, 175-76; and "Commercial Space Industry in the Year 2000," 473-80; and "Convergence of U.S.-Polar-orbiting Operational Environmental Satellite Systems," 165, 221-23; and "Crop Forecasting by Satellite," 272-75; and "Current Status of NASA's Natural Resources Program" (1966), 226-37; and "Earth Information from Space by Remote Sensing," 282-93; and Earth resource satellites, 167-76; and "Earth Resources Survey Program," 248-50, 250-52, 253-56; and "Earth's Resources to be Studied from Space," 244-46; and "Foreign Policy Issues Regarding Earth Resource Surveying by Satellite," 262-69; and "Inquiry into the Feasibility of Weather Reconnaissance from a Satellite Vehicle," 157, 185-202; and "Land Remote Sensing Policy Act of 1992," 173-75, 352-68; and "Land Remote-Sensing Commercialization Act of 1984," 329-44; and Landsat, 133, 168-73; and "Landsat Remote Sensing Policy," 345-47; and "Landsat Remote Sensing Strategy," 372-75; and Large Area Crop Inventory Experiment (LACIE), 170; and "Management Plan for the Landsat Program," 347-52; and meteorological satellites,

156-63; and "National Space Policy" (1996), 463-73; and Nimbus Weather Satellite, 159-60, 168; and "Observing the Weather from a Satellite Vehicle," 177-83; and "Planning for a Civil Operational Land Remote Sensing Satellite System," 296-306; and Polar-orbiting Operational Environmental Satellites (POES), 163; and "Private Sector Operation of Landsat Satellites," 281-82; and "Report of the Government Technical Review Panel on Industry Responses on Commercialization of the Civil Remote Sensing Systems," 309-21; and "Resolution of Issues Related to Private Sector Transfer of Civil Land Observing Satellite Activities," 306-08; and Satellite Pour l'Observation de la Terre (SPOT), 171, 173, 476; and TIROS, 157, 158, 161-62, 203-06; and "Some Recent International Reactions to ERTS-1," 259-62; and "U.S. Policy on Licensing and Operation of Private Remote Sensing Systems," 379-81

Richardson, Elliot L., 111

Robbins, Martin D., and "Mission Oriented R&D and the Advancement of Technology," 432-45

Robinove, Charles J., 275-76

Rockwell International, Inc., 391; and "The Economic Impact of the Space Program," 451-55; and "Space Commercialization Meeting," 498-501; and "Space Industrialization Final Report," 517-27

Rosen, Harold A., and *Commercial Communications Satellite*, 35-39; and Hughes "Commercial Satellite Communication Project," 31-35

Rye, Gilbert D., 498-501

S

Satellite Pour l'Observation de la Terre (SPOT), 171, 173, 476

Satellite Military Observation System (SAMOS), 157

Saturn launch vehicle, 32

Sawhill, John C., 277-81

Schmookler, Jacob, 408

Schneider, William, 147-53

Science Applications, Inc., and "Space Industrialization: An Overview," 527-33

SCORE (Signal Communication by Orbital Relay Equipment), Project, 2, 89-90

Scott, Walter S., 375-77

Scout Launch Vehicle, 32, 33, 35

Seamans, Robert C., Jr., 237-40; and "Earth Resources Survey Program," 248-50, 250-52, 253-56; and "Operational Requirements for Global Resource Surveys by Earth-Orbital Satellites," 246-48

Shapley, Willis, 133, 259-62, 282-93, 401-02

Shriner, Robert D., 426-27

Shultz, George P., 151-53, 277

Shupe, Walter C., 269-71

Signal Corps, U.S., 2; and Hughes "Commercial Satellite Communication Project," 31-35

Sioux Falls, SD, 169, 259-62

Skellett, A.M., 24-25

Skolnikoff, Eugene B., 282-93

Skurla, George, 498-501

Skylab, 134

Smart, Jacob E., 250-52

Smith, Albert E., 381-83

Smith, Bromley, 95-96

Smith, C.R., 54-56, 59

Smith Frederick W., 498-501

Solow, Robert, 389

Southworth G.C., 24

Space economics, 385-577, and "Commercial Space Industry in the Year 2000," 473-80; and "Commercial Use of Space," 543-46; and "Cost-Benefit Analysis of Selected Technology Utilization Office Programs," 555-58; and "The Economic Effects of a Space Station," 450-51; and "The Economic Impact of NASA R&D Spending," 414-26; and "The Economic Impact of the Space Program," 451-55; and "Economic Impact of Stimulated Technological Activity," 408-14; and "Economic Impact and Technological Progress of NASA Research and Development Expenditures," 427-30; and "Electrophoresis Operations in Space," 541-43; and "An Exploration of Benefits from NASA 'Spinoff,'" 559-63; and "Feasibility Study of Commercial Space Manufacturing," 504-17; and "Feasibility Study of Commercial Space Manufacturing: Production of Pharmaceuticals," 534-39;

and "Long-Term Prospects for Developments in Space," 480-88; and measuring impact of, 387-392; and "Mission Oriented R&D and the Advancement of Technology," 432-45; and NASA activities, 386-87; and "NASA Report May Overstate the Economic Benefits of Research and Development Spending," 430-32; and "NASA Tech Brief Program," 553-55; and "NASA Technology Transfer: Report of the Technology Transfer Team," 574-77; and national policy, 392-96; and "National Space Policy" (1996), 463-73; and "The President's Space Policy and Commercial Space Initiatives to Begin the Next Century," 455-60; and "Quantifying the Benefits to the National Economy from Secondary Applications of NASA Technology," 445-49; and "Proposal for Enhancing NASA Technology Transfer to Civil Systems," 546-53; and "Some Major Impacts of the National Space Program," 402-407; and "Space Industrialization: An Overview," 527-33; and "Space Industrialization Final Report," 517-27; and stimulation of technology transfer, 396-400; and "Technology Transfer White Paper," 563-74; and "U.S. Commercial Space Policy Guidelines," 460-63

Space Electronics Corp., and Hughes "Commercial Satellite Communication Project," 31-35

Space Industries, Inc., 498-501

Spaceflight, 11; and "The Space Station," 11-15

"Space Station, The," 11-15

Space Services, Inc., 498-501

Space Systems/Loral, 7

Sputnik 1, 2, 26

Stanford Research Institute, 389

State, Department of, 85-89; and "Civil Operational Remote Sensing," 294-95; and "Convergence of U.S.-Polar-orbiting Operational Environmental Satellite Systems," 165, 221-23; and "Earth Resources Survey Program," 248-50, 250-52, 253-56; and "Foreign Policy Issues Regarding Earth Resource Surveying by Satellite," 262-69; and "A Global System of Satellite Communications," 99-108; and "National Space Policy" (1996), 463-73; and "Planning for a Civil Operational Land Remote Sensing Satellite System," 296-306; and "Policy Concerning U.S. Assistance in the Development of Foreign Communications Satellite Capabilities," 91-95; and "Report of the United States Delegation . . . on Definitive Arrangements for the International Telecommunications Satellite Consortium," 108-20; and "Resolution of Issues Related to Private Sector Transfer of Civil Land Observing Satellite Activities," 306-308; and "Transfer of U.S. Communications Satellite Technology," 96-99; and "U.S. Assistance in the Early Establishment of Communications Satellite Service," 95-96; and "White Paper on New International Satellite Systems," 147-53

Sterling, VA, 24

Stewart, Irvin, 85

Sugar Grove, WV, 45

Syncom 1, 4, 5, 35

Syncom 2, 4, 5

Syncom 3, 4, 5

T

Television Infrared Operational Satellite (TIROS), 155, 157, 158, 203-06; and NASA, 158-62; and "National Plan for a Common System of Meteorological Observation Satellites," 204-06

Telstar, 3, 22, 72, 89-90

Thompson, David, 498-501

Thor launch vehicle, 2

Titan launch vehicle, 2

Townsend, John W., Jr., 498-501

Tracking and Data Relay Satellite System (TDRSS), 9

TRW Aerospace, 7; and ACTS, 145-47

Tsiolkovskiy, Konstantin, 11

U

United Nations, and APSTAR, 10; and ARABSAT, 10; and ASIASAT, 10; and ASTRA, 10; and Communications Satellite Act of 1962, 5, 61, 63, 76-85, 85-89; and Communications Satellite Corp. (Comsat), 4-5, 72-76, 85-89; and European Telecommunications Satellite Organization (EUTELSAT), 10; and "A Global System of Satellite Communications," 99-108; and INMARSAT, 11; and INTELSAT, 4-5, 6-10, 11, 91-95; and international relations, 4-10; and International Telecommunications Union (ITU), 10-11, 65, 89-90; and PanAmSat, 10;

and "Policy Concerning U.S. Assistance in the Development of Foreign Communications Satellite Capabilities," 91-95; and "Report of the United States Delegation...on Definitive Arrangements for the International Telecommunications Satellite Consortium," 108-20; and "Some Recent International Reactions to ERTS-1," 259-62; and State Department, 85-89, 91-95; and "Transfer of U.S. Communications Satellite Technology," 96-99; and "U.S. Assistance in the Early Establishment of Communications Satellite Service," 95-96

V

V-2 rocket, 157
Vanguard, Project, 2
Viking, Project, 134

W

Waple, Ben F., 56-57, 61-64; and "Establishment of Domestic Communications-Satellite Facilities by Non-Governmental Entities," 120-32
Washington University, George, 388
Weather Bureau, U.S., 158-60, 168; and "Basic Agreement Between U.S. Department of Commerce and the National Aeronautics and Space Administration Concerning Operational Meteorological Satellite Systems," 206-11; and "National Plan for a Common System of Meteorological Observation Satellites," 204-06; and "Observing the Weather from a Satellite Vehicle," 177-83
Webb, James E., and AT&T, 45-60; and "Basic Agreement Between U.S. Department of Commerce and the National Aeronautics and Space Administration Concerning Operational Meteorological Satellite Systems," 206-11; and "Earth Resources Survey Program," 248-50, 250-52, 253-56
WEFA Group, Inc., 391; and "The Economic Impact of the Space Program," 451-55
Weinberger, Caspar, 135
Welsh, Edward C., 71, 76-77
Wenk, Edward, Jr., 86
Western Electric Co., 54-56, 69
Western Union International Corp., 5; and "Administrative and Regulatory Problems Relating to the Authorization of Commercially Operable Space Communications Systems," 61-64; and "Establishment of Domestic Communications-Satellite Facilities by Non-Governmental Entities," 120-32
Wexler, Harry, 156; and "Observing the Weather from a Satellite Vehicle," 177-83
White, Robert M., 211-13
White Sands, NM, 157
Williams, Donald D., and *Commercial Communications Satellite*, 35-39; and "Commercial Satellite Communication Project," 31-35
Winokur, Robert S., 381-83
Wireless World, 1; and "Extra-Terrestrial Relays," 16-22
Withee, Gregory W., 375-78
Wolf, Francis Colt de, 89
Wright, George W., 401-02

Y

Yardley, John F., 396, 498-501, 539-41

Z

Zimmerman, James V., 262-69

The NASA History Series

Reference Works, NASA SP-4000

Grimwood, James M. *Project Mercury: A Chronology* (NASA SP-4001, 1963).

Grimwood, James M., and Hacker, Barton C., with Vorzimmer, Peter J. *Project Gemini Technology and Operations: A Chronology* (NASA SP-4002, 1969).

Link, Mae Mills. *Space Medicine in Project Mercury* (NASA SP-4003, 1965).

Astronautics and Aeronautics, 1963: Chronology of Science, Technology, and Policy (NASA SP-4004, 1964).

Astronautics and Aeronautics, 1964: Chronology of Science, Technology, and Policy (NASA SP-4005, 1965).

Astronautics and Aeronautics, 1965: Chronology of Science, Technology, and Policy (NASA SP-4006, 1966).

Astronautics and Aeronautics, 1966: Chronology of Science, Technology, and Policy (NASA SP-4007, 1967).

Astronautics and Aeronautics, 1967: Chronology of Science, Technology, and Policy (NASA SP-4008, 1968).

Ertel, Ivan D., and Morse, Mary Louise. *The Apollo Spacecraft: A Chronology, Volume I, Through November 7, 1962* (NASA SP-4009, 1969).

Morse, Mary Louise, and Bays, Jean Kernahan. *The Apollo Spacecraft: A Chronology, Volume II, November 8, 1962–September 30, 1964* (NASA SP-4009, 1973).

Brooks, Courtney G., and Ertel, Ivan D. *The Apollo Spacecraft: A Chronology, Volume III, October 1, 1964–January 20, 1966* (NASA SP-4009, 1973).

Ertel, Ivan D., and Newkirk, Roland W., with Brooks, Courtney G. *The Apollo Spacecraft: A Chronology, Volume IV, January 21, 1966–July 13, 1974* (NASA SP-4009, 1978).

Astronautics and Aeronautics, 1968: Chronology of Science, Technology, and Policy (NASA SP-4010, 1969).

Newkirk, Roland W., and Ertel, Ivan D., with Brooks, Courtney G. *Skylab: A Chronology* (NASA SP-4011, 1977).

Van Nimmen, Jane, and Bruno, Leonard C., with Rosholt, Robert L. *NASA Historical Data Book, Volume I: NASA Resources, 1958–1968* (NASA SP-4012, 1976; rep. ed. 1988).

Ezell, Linda Neuman. *NASA Historical Data Book, Volume II: Programs and Projects, 1958–1968* (NASA SP-4012, 1988).

Ezell, Linda Neuman. *NASA Historical Data Book, Volume III: Programs and Projects, 1969–1978* (NASA SP-4012, 1988).

Gawdiak, Ihor Y., with Fedor, Helen, compilers. *NASA Historical Data Book, Volume IV: NASA Resources, 1969–1978* (NASA SP-4012, 1994).

Astronautics and Aeronautics, 1969: Chronology of Science, Technology, and Policy (NASA SP-4014, 1970).

Astronautics and Aeronautics, 1970: Chronology of Science, Technology, and Policy (NASA SP-4015, 1972).

Astronautics and Aeronautics, 1971: Chronology of Science, Technology, and Policy (NASA SP-4016, 1972).

Astronautics and Aeronautics, 1972: Chronology of Science, Technology, and Policy (NASA SP-4017, 1974).

Astronautics and Aeronautics, 1973: Chronology of Science, Technology, and Policy (NASA SP-4018, 1975).

Astronautics and Aeronautics, 1974: Chronology of Science, Technology, and Policy (NASA SP-4019, 1977).

Astronautics and Aeronautics, 1975: Chronology of Science, Technology, and Policy (NASA SP-4020, 1979).

Astronautics and Aeronautics, 1976: Chronology of Science, Technology, and Policy (NASA SP-4021, 1984).

Astronautics and Aeronautics, 1977: Chronology of Science, Technology, and Policy (NASA SP-4022, 1986).

Astronautics and Aeronautics, 1978: Chronology of Science, Technology, and Policy (NASA SP-4023, 1986).

Astronautics and Aeronautics, 1979–1984: Chronology of Science, Technology, and Policy (NASA SP-4024, 1988).

Astronautics and Aeronautics, 1985: Chronology of Science, Technology, and Policy (NASA SP-4025, 1990).

Noordung, Herman. *The Problem of Space Travel: The Rocket Motor.* Stuhlinger, Ernst, and Hunley, J.D., with Garland, Jennifer, editors (NASA SP-4026, 1995).

Astronautics and Aeronautics, 1986–1990: A Chronology (NASA SP-4027, 1997).

Management Histories, NASA SP-4100

Rosholt, Robert L. *An Administrative History of NASA, 1958–1963* (NASA SP-4101, 1966).

Levine, Arnold S. *Managing NASA in the Apollo Era* (NASA SP-4102, 1982).

Roland, Alex. *Model Research: The National Advisory Committee for Aeronautics, 1915–1958* (NASA SP-4103, 1985).

Fries, Sylvia D. NASA *Engineers and the Age of Apollo* (NASA SP-4104, 1992).

Glennan, T. Keith. *The Birth of NASA: The Diary of T. Keith Glennan.* Hunley, J.D., editor (NASA SP-4105, 1993).

Seamans, Robert C., Jr. *Aiming at Targets: The Autobiography of Robert C. Seamans, Jr.* (NASA SP-4106, 1996)

Project Histories, NASA SP-4200

Swenson, Loyd S., Jr., Grimwood, James M., and Alexander, Charles C. *This New Ocean: A History of Project Mercury* (NASA SP-4201, 1966).

Green, Constance McL., and Lomask, Milton. *Vanguard: A History* (NASA SP-4202, 1970; rep. ed. Smithsonian Institution Press, 1971).

Hacker, Barton C., and Grimwood, James M. *On Shoulders of Titans: A History of Project Gemini* (NASA SP-4203, 1977).

Benson, Charles D. and Faherty, William Barnaby. *Moonport: A History of Apollo Launch Facilities and Operations* (NASA SP-4204, 1978).

Brooks, Courtney G., Grimwood, James M., and Swenson, Loyd S., Jr. *Chariots for Apollo: A History of Manned Lunar Spacecraft* (NASA SP-4205, 1979).

Bilstein, Roger E. *Stages to Saturn: A Technological History of the Apollo/Saturn Launch Vehicles* (NASA SP-4206, 1980).

SP-4207 not published.

Compton, W. David, and Benson, Charles D. *Living and Working in Space: A History of Skylab* (NASA SP-4208, 1983).

Ezell, Edward Clinton, and Ezell, Linda Neuman. *The Partnership: A History of the Apollo- Soyuz Test Project* (NASA SP-4209, 1978).

Hall, R. Cargill. *Lunar Impact: A History of Project Ranger* (NASA SP-4210, 1977).

Newell, Homer E. *Beyond the Atmosphere: Early Years of Space Science* (NASA SP-4211, 1980).

Ezell, Edward Clinton, and Ezell, Linda Neuman. *On Mars: Exploration of the Red Planet, 1958–1978* (NASA SP-4212, 1984).

Pitts, John A. *The Human Factor: Biomedicine in the Manned Space Program to 1980* (NASA SP-4213, 1985).

Compton, W. David. *Where No Man Has Gone Before: A History of Apollo Lunar Exploration Missions* (NASA SP-4214, 1989).

Naugle, John E. *First Among Equals: The Selection of NASA Space Science Experiments* (NASA SP-4215, 1991).

Wallace, Lane E. *Airborne Trailblazer: Two Decades with NASA Langley's Boeing 737 Flying Laboratory* (NASA SP-4216, 1994).

Butrica, Andrew J., editor. *Beyond the Ionosphere: Fifty Years of Satellite Communication* (NASA SP-4217, 1997).

Butrica, Andrews J. *To See the Unseen: A History of Planetary Radar Astronomy* (NASA SP-4218, 1996).

Reed, R. Dale, with Lister, Darlene. *Wingless Flight: The Lifting Body Story* (NASA SP-4220, 1997).

Center Histories, NASA SP-4300

Rosenthal, Alfred. *Venture into Space: Early Years of Goddard Space Flight Center* (NASA SP-4301, 1985).

Hartman, Edwin, P. *Adventures in Research: A History of Ames Research Center, 1940–1965* (NASA SP-4302, 1970).

Hallion, Richard P. *On the Frontier: Flight Research at Dryden, 1946–1981* (NASA SP- 4303, 1984).

Muenger, Elizabeth A. *Searching the Horizon: A History of Ames Research Center, 1940–1976* (NASA SP-4304, 1985).

Hansen, James R. *Engineer in Charge: A History of the Langley Aeronautical Laboratory, 1917–1958* (NASA SP-4305, 1987).

Dawson, Virginia P. *Engines and Innovation: Lewis Laboratory and American Propulsion Technology* (NASA SP-4306, 1991).

Dethloff, Henry C. *"Suddenly Tomorrow Came . . .": A History of the Johnson Space Center* (NASA SP-4307, 1993).

Hansen, James R. *Spaceflight Revolution: NASA Langley Research Center from Sputnik to Apollo* (NASA SP-4308, 1995).

Wallace, Lane E. *Flights of Discovery: 50 Years at the NASA Dryden Flight Research Center* (NASA SP-4309, 1996).

Herring, Mack R. *Way Station to Space: A History of the John C. Stennis Space Center* (NASA SP-4310, 1997).

Wallace, Harold D., Jr. *Wallops Station and the Creation of the American Space Program* (NASA SP-4311, 1997).

General Histories, NASA SP-4400

Corliss, William R. *NASA Sounding Rockets, 1958–1968: A Historical Summary* (NASA SP-4401, 1971).

Wells, Helen T., Whiteley, Susan H., and Karegeannes, Carrie. *Origins of NASA Names* (NASA SP-4402, 1976).

Anderson, Frank W., Jr. *Orders of Magnitude: A History of NACA and NASA, 1915–1980* (NASA SP-4403, 1981).

Sloop, John L. *Liquid Hydrogen as a Propulsion Fuel, 1945–1959* (NASA SP-4404, 1978).

Roland, Alex. A *Spacefaring People: Perspectives on Early Spaceflight* (NASA SP-4405, 1985).

Bilstein, Roger E. *Orders of Magnitude: A History of the NACA and NASA, 1915–1990* (NASA SP-4406, 1989).

Logsdon, John M., editor, with Lear, Linda J., Warren-Findley, Jannelle, Williamson, Ray A., and Day, Dwayne A. *Exploring the Unknown: Selected Documents in the History of the U.S. Civil Space Program, Volume I: Organizing for Exploration* (NASA SP-4407, 1995).

Logsdon, John M., editor, with Day, Dwayne A., and Launius, Roger D. *Exploring the Unknown: Selected Documents in the History of the U.S. Civil Space Program, Volume II: External Relationships* (NASA SP-4407, 1996).

☆ U.S. GOVERNMENT PRINTING OFFICE: 1998 440–153